Bioactive Glasses
Fundamentals, Technology and Applications

RSC Smart Materials

Series Editors:
Professor Hans-Jörg Schneider, *Saarland University, Germany*
Professor Mohsen Shahinpoor, *University of Maine, USA*

Titles in this Series:
1: Janus Particle Synthesis, Self-Assembly and Applications
2: Smart Materials for Drug Delivery: Volume 1
3: Smart Materials for Drug Delivery: Volume 2
4: Materials Design Inspired by Nature
5: Responsive Photonic Nanostructures: Smart Nanoscale Optical Materials
6: Magnetorheology: Advances and Applications
7: Functional Nanometer-Sized Clusters of Transition Metals: Synthesis, Properties and Applications
8: Mechanochromic Fluorescent Materials: Phenomena, Materials and Applications
9: Cell Surface Engineering: Fabrication of Functional Nanoshells
10: Biointerfaces: Where Material Meets Biology
11: Semiconductor Nanowires: From Next-Generation Electronics to Sustainable Energy
12: Supramolecular Materials for Opto-Electronics
13: Photocured Materials
14: Chemoresponsive Materials: Stimulation by Chemical and Biological Signals
15: Functional Metallosupramolecular Materials
16: Bio-Synthetic Hybrid Materials and Bionanoparticles: A Biological Chemical Approach Towards Material Science
17: Ionic Polymer Metal Composites (IPMCs): Smart Multi-Functional Materials and Artificial Muscles Volume 1
18: Ionic Polymer Metal Composites (IPMCs): Smart Multi-Functional Materials and Artificial Muscles Volume 2
19: Conducting Polymers: Bioinspired Intelligent Materials and Devices
20: Smart Materials for Advanced Environmental Applications
21: Self-cleaning Coatings: Structure, Fabrication and Application
22: Functional Polymer Composites with Nanoclays
23: Bioactive Glasses: Fundamentals, Technology and Applications

How to obtain future titles on publication:
A standing order plan is available for this series. A standing order will bring delivery of each new volume immediately on publication.

For further information please contact:
Book Sales Department, Royal Society of Chemistry, Thomas Graham House, Science Park, Milton Road, Cambridge, CB4 0WF, UK
Telephone: +44 (0)1223 420066, Fax: +44 (0)1223 420247
Email: booksales@rsc.org
Visit our website at www.rsc.org/books

Bioactive Glasses
Fundamentals, Technology and Applications

Edited by

Aldo R. Boccaccini
University of Erlangen-Nuremberg, Erlangen, Germany
Email: aldo.boccaccini@ww.uni-erlangen.de

Delia S. Brauer
Friedrich Schiller University, Jena, Germany
Email: delia.brauer@uni-jena.de

Leena Hupa
Åbo Akademi University, Turku, Finland
Email: leena.hupa@abo.fi

ROYAL SOCIETY OF CHEMISTRY

THE QUEEN'S AWARDS FOR ENTERPRISE: INTERNATIONAL TRADE 2013

RSC Smart Materials No. 23

Print ISBN: 978-1-78262-976-4
PDF eISBN: 978-1-78262-201-7
EPUB eISBN: 978-1-78262-169-0
ISSN: 2046-0066

A catalogue record for this book is available from the British Library

Published by the Royal Society of Chemistry,
Thomas Graham House, Science Park, Milton Road,
Cambridge CB4 0WF, UK

Registered Charity Number 207890

For further information see our web site at www.rsc.org

Printed in the United Kingdom by CPI Group (UK) Ltd, Croydon, CR0 4YY, UK

Preface

Since their development by the late Prof. Larry L. Hench in the early 1970's and their first clinical use in the mid-1980's, bioactive glasses have constantly attracted the research interest of materials scientists, biologists, tissue engineers and clinicians. This is not surprising, as bioactive glasses have a range of fascinating properties: They were the first man-made material to form an integrated bond with bone. Owing to their inherent release of biologically active ions, such as silicon ions, when in contact with body fluids and tissue, they stimulate cell behaviour and can obviate or reduce the need for supplementation with growth factors. Bioactive glass implants degrade over time and allow for bone to be regenerated. Owing to their amorphous structure, therapeutic ions can be incorporated into, and released from, bioactive glasses easily, widening the therapeutic spectrum considerably.

Lately, bioactive glasses have started to be increasingly successful, both clinically and commercially, with Larry Hench's first bioactive glass, *Bioglass 45S5*, still being the most used composition to date. It is now a widely used synthetic bone graft in the United States (*e.g.* under the names of *NovaBone* or *PerioGlas*), while another bioactive glass, *S53P4* (or *BonAlive*), which was developed in Finland, has been used with great clinical success to combat bone infections. Other uses for bioactive glasses have emerged, the major commercial success being the use of 45S5 (*NovaMin*) as a remineralising additive in toothpastes for the treatment of dentine hypersensitivity. 45S5 powder (*Sylc*) is also sold as a product for air abrasion and air polishing in oral healthcare.

Besides their traditional applications in the orthopaedic and dental fields, a new field of uses for bioactive glasses, supported by their angiogenic effects, is emerging in soft-tissue repair and wound healing. A borate bioactive glass should be mentioned here, which is used for wound healing

RSC Smart Materials No. 23
Bioactive Glasses: Fundamentals, Technology and Applications
Edited by Aldo R. Boccaccini, Delia S. Brauer and Leena Hupa

Published by the Royal Society of Chemistry, www.rsc.org

applications in animals (*RediHeal*) while also awaiting FDA approval for use in humans for treating slow-healing wounds, particularly in patients suffering from diabetes.

Owing to these clinical successes, new biomaterials based on bioactive glasses are constantly being investigated. Orthopaedic applications currently focus on non-load bearing implants in granulate form, but porous bioactive glass scaffolds are being increasingly considered in bone regeneration strategies. One important point here is vascularisation, and several new bioactive glass compositions which release ions promoting this are being investigated. Another current research area is bioactive glasses as drug delivery vehicles, *e.g.* in mesoporous form derived from sol–gel methods. The recent use of the bottom-up sol–gel approach for the synthesis of organic/inorganic hybrids allowed for the combination of an organic and inorganic phase at an atomic scale, adding to the field of traditional glass/polymer composite materials. In order to widen the field of bioactive glass applications, their compositional range has to be extended, and current research not only includes the traditional Hench-type phosphosilicate bioactive glasses but also borate glasses, phosphate glasses or sol–gel derived compositions and nanomaterials.

This book aims to bridge the different scientific communities associated with the field of bioactive glasses with a focus on the materials science point of view, covering both traditional and emerging bioactive glass applications. For the reader familiar with medicine or dentistry, the book will describe the possibilities and challenges of glasses and bioactive glass-based materials to be developed and manufactured for healthcare applications. For tissue engineers, the book will reveal the emerging field of applications of bioactive glass scaffolds for the regeneration of hard and soft tissue. For the materials scientist, the most relevant requirements which the dynamic human body environment imposes onto bioactive glasses will be discussed, and concepts for tailoring or predicting bioactive glass properties will be addressed. In addition, combinations of bioactive glasses with polymers will be covered, mainly for soft tissue applications, drug delivery and other advanced uses in regenerative medicine.

The book is divided into three main parts: Fundamentals, Technology and Applications, each including a series of chapters written by international authorities in the respective fields. We hope that this book will contribute to extending the field of bioactive glasses, and that it will be useful not only to established researchers but particularly to those students and young researchers who are starting their careers in the biomaterials field.

Sadly, Prof. Larry L. Hench, the inventor of bioactive glasses, passed away in December 2015, at the time the chapters for this book were being written. As Larry had originally agreed to write an introduction for our book, it is our great honour to dedicate this book to his memory.

Aldo R. Boccaccini, Erlangen, Germany
Delia S. Brauer, Jena, Germany
Leena Hupa, Turku, Finland

Acknowledgements

This book would not exist without the effort and dedication of the individual authors, not only in writing their respective chapters but also in their continuing research. In addition, we acknowledge the work of the referees, who contributed with their timely and detailed reviews to improve the quality of the chapters. The editors would like to extend their gratitude to the Royal Society of Chemistry team, in particular Lindsay McGregor, for her patience and enormous help, making the timely publication of this book possible.

RSC Smart Materials No. 23
Bioactive Glasses: Fundamentals, Technology and Applications
Edited by Aldo R. Boccaccini, Delia S. Brauer and Leena Hupa

Published by the Royal Society of Chemistry, www.rsc.org

Contents

RSC Smart Materials No. 23
Bioactive Glasses: Fundamentals, Technology and Applications
Edited by Aldo R. Boccaccini, Delia S. Brauer and Leena Hupa

Published by the Royal Society of Chemistry, www.rsc.org

CHAPTER 1

Melt-derived Bioactive Silicate Glasses

SUSANNE FAGERLUND[a] AND LEENA HUPA*[b]

[a] Paroc Group Oy, 21600 Parainen, Finland; [b] Johan Gadolin Process Chemistry Centre, Åbo Akademi University, 20500 Turku, Finland
*Email: leena.hupa@abo.fi

1.1 Introduction

1.1.1 Glass – A Versatile Biomaterial

This chapter introduces some fundamental chemical and physical properties of glasses to be taken into account when designing and fabricating products based on bioactive glasses to be implanted inside the human body. The main emphasis is to explain the constraints to be taken into account from the materials science and chemical engineering points of view. The ultimate goal is to deliver the basic principles of glass science to serve as a platform for the various disciplines ranging from materials science to molecular biology, biochemistry, medicine, *etc.* The vision is that the steadily increasing multi-disciplinary experience provides the crucial knowledge needed for developing implants and scaffolds for controlled predetermined performance in the target application.

At first sight, the inherent brittle nature of glass does not make it a feasible material for implantable medical devices. At the same time, glass has several useful properties which support its utilization as a biomaterial. What is a biomaterial? Most biomaterials based on glasses or ceramics are

RSC Smart Materials No. 23
Bioactive Glasses: Fundamentals, Technology and Applications
Edited by Aldo R. Boccaccini, Delia S. Brauer and Leena Hupa

Published by the Royal Society of Chemistry, www.rsc.org

designed to improve human health and the quality of life by restoring the function of living tissue and organs in the body. The single most important factor for a biomaterial is that it is able to be in contact with tissues of the human body without causing an unacceptable degree of harm to that body, *i.e.* the material is biocompatible.[1] Recently, considerable research efforts have been directed to tailor highly porous tissue engineering scaffolds not only for bone tissue but also for emerging soft tissue applications. Detailed understanding of the nature and properties of glass provides a thorough basis for assessing its potential in prospective biomedical applications. In general, choice of a material for a particular application is based on its performance, properties, fabricability, and manufacturing costs.

1.1.2 Glass and Properties

Due to its amorphous structure glass possesses several features which make it an optimal material for manifold applications. Simplified, conventional inorganic glasses are homogeneous mixtures of oxides of alkalis, alkaline earths, aluminium, boron, silicon, *etc.* Most modern uses of these so-called soda lime glasses, including flat glass, hollowware glass and fibre glass, rely on the transparency to visible light combined with some other property, such as good mechanical strength, adequate chemical durability or electrical resistivity. Essentially, these commercial glass types are fabricated *via* the inexpensive melting route. They are easy to shape into various product forms at high temperatures when present as viscous liquids. In addition to the conventional glasses, specialty glasses possessing certain functional properties are essential in the modern everyday environment.

The functionality of the speciality glasses is often connected with the optical properties. Bioactive glasses used in contact with the living body as implants or as tissue engineering scaffolds belong to the family of speciality glasses. In contrast to most other applications, the transparency is not an essential property for the current applications of bioactive glass. Bioactivity of glasses may be defined in different ways, but common to all of these definitions is the requirement that the surface composition and morphology of the glasses change upon implantation. Simultaneously, the concentrations of the inorganic ions in the surrounding extracellular fluid change. Only glasses within certain limited composition range fulfil the requirements of bioactivity, *i.e.* show the desired interaction with the living tissue. Interestingly, whether it be transparency or the controlled surface reaction, the origin of the functionality of the glass is the same—the amorphous glass structure. As explained in Chapter 6 the amorphous glass structure enables adjusting the physical and chemical properties of glasses within certain limits merely by changing the constituent oxides or their ratios. This gives interesting possibilities to tailor the glass composition for various clinical applications.

1.1.3 Bioactivity of Glass

Bioactive glasses consist of the same oxides as conventional soda lime glasses but in proportions which give rise to large differences between several properties of these types of glass. The most marked difference dictating the bioactivity is the chemical durability; the bioactive glasses dissolve in aqueous environments at markedly higher rates than the soda lime glasses used, for example, in containers or windows. For the glass to be bioactive the dissolution rate must be compatible with the cellular processes so that the dissolving glass supports and enhances tissue regeneration and growth, while we expect the traditional glasses to be inert in their normal applications. Bioactive glasses thus provide temporary support to tissue healing and regeneration. Brittleness is one of the biggest challenges for utilization of bioactive glasses, especially in load-bearing applications. In future the problems with the brittleness may be solved by using the bioactive glass in composites together with polymers, or developing tissue engineering scaffolds with special architectures adapted to the requirements of loaded bone.

The first bioactive glasses studied as prosthetic materials were melt-derived compositions within the system Na_2O–CaO–P_2O_5–SiO_2.[2] Glass was an interesting and smart choice for a material to be in contact with the human skeleton. The pioneering idea by Professor Hench, the inventor of bioactive glasses, was to develop a material that consists of elements abundant in the human body. In addition, the ratio between the oxides in the compositions studied was selected to favour rapid initial dissolution of alkalis from the glass surface in aqueous solutions followed by precipitation of an outer layer rich in calcium and phosphorus at the inner alkali-depleted silica layer. Virtually all glasses gradually dissolve in aqueous solutions, but the ion leaching rates and the tendency to form surface layers varies markedly depending on the total composition.[3] It was hypothesized that if the composition of the calcium phosphate surface layer is similar to the hydrated calcium phosphate component in bone tissue, hydroxyapatite (HCA), the glass would not be rejected by the body.[4] The compositions developing a HCA layer *in vivo* are today known as bioactive glasses.

Accordingly, the bioactive glasses were proven as advantageous materials in skeletal repair. Today, extensive research efforts are made to develop new compositions of bioactive glasses for bone and soft tissue engineering.[5] Increasing knowledge of the influence of various ions released from the glasses on the tissue regenerative capability has inspired researchers to adjust and tailor bioactive glass compositions far beyond the original bone tissue related applications toward emerging areas in soft tissue regeneration. Though the history of bioactive glasses is longer than forty years, most commercial products today are used for treatment of trauma, disease or injury of bone tissue.[6]

1.2 Properties Essential for Fabrication

1.2.1 Definition of Glass

For any utilization of a material we must understand its features and properties. All glasses, whether manufactured *via* melt quenching, sol–gel processing or some other suitable method, possess two common characteristics: the glass structure and the gradual change of several properties when heated or cooled between the solid form, glass, and the liquid form, melt. Glass is often defined as "*an amorphous solid completely lacking long range, periodic arrangement of the atomic structure, and exhibiting a region of glass transformation behaviour*".[7] In this transformation the properties of the liquid gradually change into solid state properties. Some general rules of glass formation and the high temperature properties to be considered in glass forming processes are explained below. A detailed discussion of glass structure and its influence on glass properties is given in Chapter 3.

1.2.2 Glass Transformation and Liquidus Temperatures

Melt capable of forming a glass maintains its liquid-like structure as a supercooled liquid below the melting point of the crystal, to transform into a brittle, elastic glass on further cooling. The transformation behaviour of a glass forming melt is depictured in Figure 1.1. The figure gives changes in the volume or the enthalpy at characteristic temperatures when the melt converts into either a crystalline state or forms a supercooled melt which transforms into a non-crystalline solid, a glass, on continued cooling. Glass transformation takes place at a certain temperature called the glass transition temperature or glass transformation temperature T_g. This temperature depends on the composition of the glass but also on the cooling rate; the higher the rate the higher the transformation temperature. Glass transformation is thus a time-dependent behaviour and the value of T_g is always dependent on the thermal history of the glass.

Glass transformation temperature is measured normally from graphs recorded in thermal analysis or dilatometry of glasses, *i.e.* in heating glasses from the solid state into the molten state. The T_g value obtained depends on the heating rate, the instrumental method used in the measurement, and the thermal history of the glass. Below T_g the glass behaves as an elastic solid, around T_g the glass shows viscoelastic behaviour, and above T_g the glass softens and starts to behave as a viscous liquid. The glass transformation temperature is thus an important parameter for selecting the experimental conditions in thermal treatment of a particular composition into various products above T_g. In addition, the annealing curve, *i.e.* controlled time–temperature history for relaxing any stresses present in the glass after the forming operations, is based on the T_g value. For bioactive glasses, the T_g value is of interest when sintering porous implants from

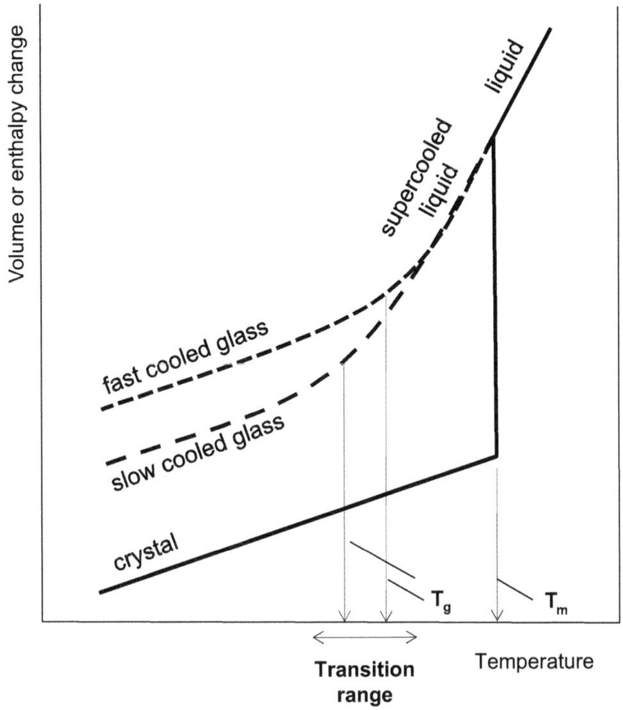

Figure 1.1 Typical volume or enthalpy changes in cooling of a glass forming melt.

powdered fractions or when estimating a suitable annealing curve for a monolith. However, T_g has no practical significance, *e.g.* in the quenching of glass melts into water to give glass particles. In contrast, if internal stresses in the particles caused by the quenching are to be relaxed, then an additional annealing step at around T_g is required.

If the melt were to crystallize during the cooling, abrupt change in the properties takes place when the crystal, with long range, periodic arrangements of the atoms, forms at the melting point (T_m in Figure 1.1), independently of the cooling rate. Since glass forming liquids have typically high viscosity at the melting point they easily form a supercooled liquid on cooling. For glass forming liquids, the melting point is often referred to as the liquidus temperature T_{liq}. The liquidus temperature is the highest temperature at which crystals can be in thermodynamic equilibrium with the melt. Thus, the value of T_{liq} (or T_m) does not depend on the melting history of the glass. From the manufacturing point of view, there is always a risk of crystallization if the melt is processed below T_{liq}. In commercial soda lime–silica melts, crystals form within a few hours if held below the liquidus temperature.[8]

The liquidus temperature is traditionally measured from glass samples treated in a furnace with controlled temperature profile for periods long enough to precipitate crystals which can be identified optically in quenched samples. The high tendency of bioactive glasses to crystallize in thermal treatments combined with the slow melting or dissolution of the crystals eventually formed during the heating questions the accurate determination of liquidus temperature in bioactive glasses. Thermal analysis is frequently used to estimate the temperature range in which the crystals melt.

1.2.3 Crystallization

A defect-free amorphous glass contains no crystals. Nevertheless, all glasses crystallize but at different rates in the temperature window between glass transition T_g and liquidus T_{liq}. Detailed understanding of crystallization is vital in fabrication processes requiring any longer thermal treatment in this critical temperature window. Although bioactive silicate glasses and soda lime–silica glasses are composed to a large extent from the same oxides, the markedly lower content of the main glass network forming oxide, SiO_2, in bioactive silicate glasses gives easy crystallization.

Crystallization of a liquid happens *via* two processes: nucleation and crystal growth. In nucleation, a sufficient quantity of atoms form an ordered first structure (the nucleus) after which crystal growth is facilitated by new layers of atoms forming around the nucleus. For crystalline solids, these processes take place at the melting point, while for typical glass forming melts both nucleation and crystal growth show maximum values within certain temperature ranges below the melting point (Figure 1.2). All melt-derived glasses pass this critical crystallization range during the fabrication. The overlapping of the rate curves depends on glass composition and partly controls the suitability of a particular composition to thermal treatments in the temperature range from glass transition to melting point.

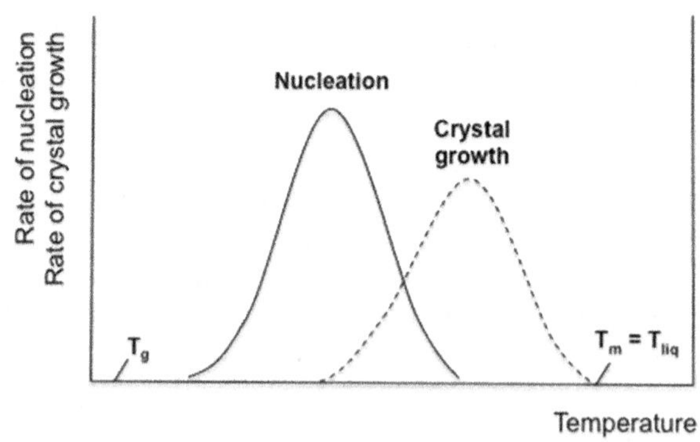

Figure 1.2 Schematic showing the effect of temperature on nucleation and crystal growth rates.

1.2.4 Crystallization of Bioactive Glasses 45S5, S53P4 and 13-93

Detailed information on crystallization mechanism, *i.e.* nucleation and crystal growth parameters, is available only for a few melt-derived bioactive glass compositions. In this chapter, the results from measurements of the characteristic values for bioactive glasses 45S5, S53P4 and 13-93 are discussed. Table 1.1 gives the oxide compositions of these three glasses. The original bioactive glass 45S5 Bioglass® developed by Professor Hench *et al.*[2] and S53P4 BonAlive® developed by Andersson *et al.*[9] are commercially available for certain clinical applications. The clinical applications of bioactive glasses are discussed in Chapters 14 and 19. Glass 13-93 was originally developed by Brink *et al.*[10,11] to enable easier fabrication into shapes that are challenging from the glass manufacturing point of view. Since then, glass 13-93 has been used to manufacture continuous fibres and porous sintered implants or scaffolds by several research groups.[12–21]

Crystallization of S53P4 and 13-93 takes place *via* surface nucleation and crystal growth mechanisms while for 45S5 the crystallization is more complex.[22–24] For 45S5, smaller and larger particles are reported to show different nucleation mechanisms,[23–25] and the crystallization proceeds rapidly from surface to the bulk.[23,26] Phase separation above T_g into immiscible liquids precedes the nucleation in 45S5.[25–27]

The crystallization mechanism is most often based on activation energy values, Johnson–Mehl–Avrami values, and nucleation-like curves determined from thermal spectra which are measured using certain temperature–time treatment in thermal analysis.[28–30] The maximum nucleation rate for the primary crystals which form above T_g was around 560–580 °C for 45S5, while a slightly higher temperature of around 610 °C was suggested for S53P4.[23] Correspondingly, the highest nucleation rate took place at 700 °C for 13-93.[22] Crystal growth rate is markedly lower in 13-93 than in the two other compositions. In thermal treatment of 45S5 and S53P4 more than one crystalline phase may form depending on the temperature.[23–27] Since this chapter focuses on properties and characterization of bioactive glasses, the crystallization is given only as a limiting factor for the fabrication window, while the phase separation and formation on secondary crystals are not considered.

Table 1.1 Nominal oxide compositions of glasses 45S5, S53P4, and 13-93.

	45S5		S53P4		13-93	
Oxide	wt%	mol%	wt%	mol%	wt%	mol%
Na_2O	24.5	24.4	23	22.7	6	6
K_2O					12	7.9
MgO					5	7.7
CaO	24.5	26.9	20	21.8	20	22.1
P_2O_5	6	2.6	4	1.7	4	1.7
SiO_2	45	46.1	53	53.8	53	54.6

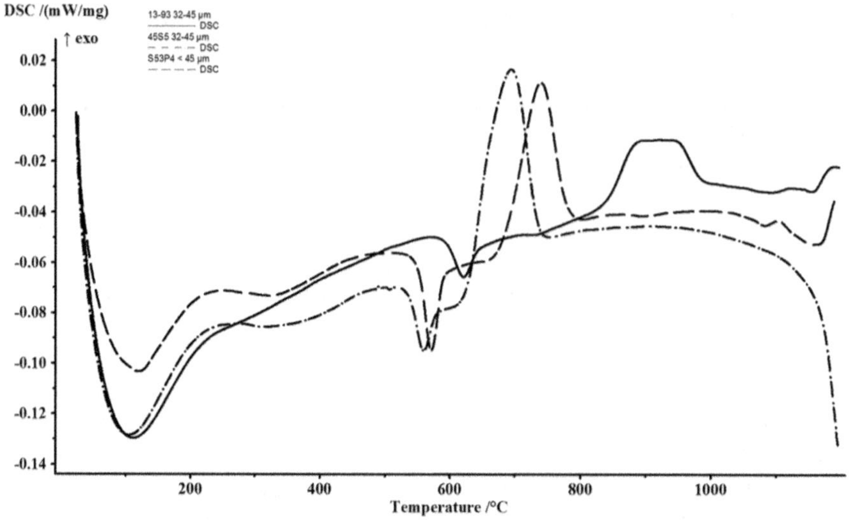

Figure 1.3 Typical DTA traces showing the thermal effects associated with glass transition, crystallization and liquidus temperatures of glasses 45S5, S53P4 and 13-93.
(Åbo Akademi University, 2015).

The overall crystallization procedure is usually determined using differential thermal analysis or differential scanning calorimetry. A certain size fraction of glass particles is heated using a controlled rate (5–20 °C min^{-1}) and thermal effects compared to an inert reference material are recorded as a function of temperature. Figure 1.3 shows typical thermal spectra for the three bioactive glasses 45S5, S53P4 and 13-93. The endothermic peak in the thermal spectra at around 500–600 °C gives the glass transformation while the exothermic peaks at around 600–800 °C (45S5 and S53P4) and 800–1000 °C (13-93) give the temperature spans at which crystals form upon heating.

The last thermal effects in Figure 1.3 are associated with the melting of the crystal, *i.e.* T_{liq} of the glasses. The thermal effect related to the liquidus is typically low and thus not always clearly shown in the thermal spectrum. The values for the thermal effects in Figure 1.3 and values reported in the literature for the three glasses are summarised in Table 1.2. The minor differences in the measured values can be explained by differences in the experimental conditions (thermal history of the glass, heating rate, particle size) and using the inflection point, the first deviation from baseline or the maximum value of the thermal effect to describe the property value.

1.2.5 Glass Stability

Glass stability is an often-used indicator describing the resistance to crystallization during heating. Glass stability is typically given as the difference between the crystallization onset value (T_x) and the glass transition

Table 1.2 Measured high temperature properties for 45S5, S53P4, and 13-93 using DTA. A temperature range given for a specific temperature value indicates values measured using different heating rates for the same sample.[37]

Property	45S5	S53P4	13-93	Ref.
Glass transition			612	22
temperature T_g (°C)	552	561		23
	550			26, 27
	530	541	600	31, 35
		550		34
	530			36
	505–551			37
		547		38
	538		591	39
	538	552	595	Figure 1.3
Crystallization			800	22
temperature T_x (°C)	610			26
		630		34
	647	693	853	35
	635			36
	720			39
	626	682	841	Figure 1.3
Crystallization peak			1038	22
temperature T_p (°C)	715	748		23
	740			36
	650–690			37
		796		38
	694	739	920	Figure 1.3
Temperature at			700	22
maximum nucleation rate (°C)	566–575	608		23
Crystal melting			1150–1250	22
range (°C)	1180–1248			23
	1070–1278			26
	1160–1260	1170–1230	870–1020	32
		1067–1225		38
		1051–1196	1144–1195	Figure 1.3

Table 1.3 Typical ranges of mechanical properties of bioactive glass 45S5 and human bone.[42,46,51,53–55]

Material	Young's modulus (GPa)	Compression strength (MPa)	Bending strength (MPa)	Fracture toughness, K_{1C} (MPa $m^{-1/2}$)
Glass 45S5	30–50	500	40–60	0.5–1
Cancellous bone	0.1–0.5	2–12	10–20	
Cortical bone	6–20	100–180	50–193	2–12

temperature (T_g). Figure 1.3 and Table 1.3 give the stability ranges of 88 °C for 45S5, 130 °C for S53P4, and 246 °C for 13-93. Accordingly, 13-93 can be thermally treated within a wide temperature range without crystallization while 45S5, showing a temperature difference less than 100 °C, crystallises easily. Correspondingly, S53P4 allows limited sintering above T_g.[34]

1.2.6 Time–Temperature–Transformation

Ultimately, crystallization during heating is a kinetic phenomenon and suitable processing parameters can be given as time–temperature–transformation (*TTT*) diagrams giving the maximum allowed duration at various temperatures to avoid crystallization. The *TTT* curves of 45S5 indicate an increase of transformation from 10% crystals after 200 s to 90% after 2000 s at 622 °C.[26] The transformation rate increases with temperature: 90% of crystals were measured already after 250 s just at around 680 °C.[26] No detailed *TTT* curves have been published for S53P4. The crystallization studies indicate that S53P4 has a markedly lower crystallization rate than 45S5. No crystals were detected in samples sintered of S53P4 at 650 °C for one hour while at 700 °C a distinct, crystallized layer had formed at the particle surfaces.[34] The crystallization parameters for 13-93 have been characterized in more detail.[22] The crystal growth was very slow still at 800 °C but rapidly increased above 850 °C.

1.2.7 Crystals in Thermal Treatment

The primary crystals, which are formed during heating of bioactive glasses belong mainly to two different composition areas: crystals consisting of Na_2O, CaO and SiO_2 (NCS) in various ratios, and crystals consisting of CaO and SiO_2 (CS).[33] In addition, secondary crystals may form at higher temperatures.[23–27,34,37] Though the exact crystal composition is of relevance in the manufacture of glass-ceramics, the primary crystal type, NCS or CS, correlates with the crystallization tendency of the glass.[33] The glasses forming NCS type crystals are much more sensitive to thermal treatments than the glasses forming CS type crystals. X-ray measurements on thermally treated glasses are used to characterize the exact phase composition of the crystals.

1.2.8 Viscosity

The basis of glass forming processes is the flow of the melt and heat transfer to complete the forming without deformation under gravity after the required shape is reached. The viscosity of the melt and its change with temperature are used to estimate the suitability of a particular melt composition to various forming operations. The viscosity of glass melts varies strongly with temperature but is also influenced by the glass composition. Since bioactive glass melts achieve optimal viscosity values for different forming operations at temperatures close to the crystallization range, time dependence must also be considered in the manufacture of crystal-free products. The standard viscosity points for soda lime–silica glasses are often used to describe the viscosity–temperature behaviour. For the manufacture of bioactive glasses, the following viscosity standard points are of interest: glass melting at $10^{2.0}$ dPa s, glass working temperature at 10^4 dPa s, glass transformation temperature around $10^{11.3}$ dPa s, and annealing around 10^{13} dPa s.[8] Several additional standard points are defined for soda lime–silica glasses between the glass transformation and liquidus temperatures.[8] However, these viscosity standard points are irrelevant for bioactive glasses with a strong crystallization tendency.

For bioactive glass products based on melt-derived compositions, two forming routes are obvious: (i) direct forming from a viscous melt into the desired product either *via* casting into moulds, quenching into water, or drawing into continuous fibres, and (ii) reheating glass above T_g to allow sintering of particles together into desired porous architectures, drawing of fibres from a pre-form or marbles, or sealing particles to give a layer on a substrate. In general, crystallization below the liquidus is slow for melts with high viscosity at T_{liq}. Thus, high viscosity at T_{liq} favours forming processes.

1.2.8.1 Particles and Monoliths

Today, most commercial melt-derived bioactive glasses are manufactured as particles of different size fractions, or as rather small monoliths. Powdered fractions are often obtained *via* quenching the melt into water to give small granules, which are then dried and sieved to desired size fractions. If manufactured *via* crushing of monoliths, rapid transfer into the annealing furnace is required after casting to avoid crystallization. However, there are no strict viscosity requirements for the glass melts used to produce particles. In contrast, in casting monolithic implants, a rapid increase of the viscosity guarantees geometric stability at the high temperatures immediately following the forming.

1.2.8.2 Continuous Fibres

Drawing of continuous fibres from bioactive glass melts is challenging. Fibres would be of interest as components in composites or other special structures utilising their very high surface area to induce bioactivity.

Drawing of continuous fibres from commercial fibreglass is characterized by a rather narrow viscosity range of $10^{3.5}$–$10^{4.0}$ dPa s.[8] The difference between the fibre-forming and liquidus temperatures should be at least 50 °C to avoid crystallization during the drawing. The forming can be performed below the liquidus if the nucleation and crystal growth temperatures are not overlapping to a great extent. If the nucleation temperature range is well below the crystal growth temperature range, fibre drawing can be realized below the liquidus.

Fibres are also drawn from a pre-form or from marbles, *i.e.* glass reheated to the fibre drawing temperature corresponding to the optimal viscosity range of $10^{3.5}$–10^4 dPa s.[8] During the heating, the glass passes through the nucleation temperature range. If nuclei are easily formed, fibres can be drawn only over a limited period of time. The nuclei and growing crystals may also affect the composition of the fibres manufactured from a pre-form of a given composition. The strong crystallization tendency between T_x and T_{liq} limits the suitability of bioactive glasses to the fibre-drawing processes.

1.2.8.3 Sintered Scaffolds

Viscosity also controls the suitability of a certain glass composition to sintering of amorphous porous scaffolds. A porous tissue engineering scaffold has an open continuous porosity which supports ingrowth of fully vascularised new tissue. A suitable viscosity range in sintering of glass particles into porous bodies *via* viscous flow is 10^8–$10^{8.8}$ dPa s.[8] When manufacturing the scaffolds *via* the sintering route, the particle size is one critical factor for the commencement of the densification process. Careful optimisation of the time–temperature history is required for achieving defect-free scaffolds with adequate strength.

1.2.8.4 Viscosity of 45S5, S53P4 and 13-93

It is well known that 45S5 and S53P4 are prone to crystallization during any prolonged thermal processing in the temperature window between the glass transition and the liquidus temperatures. Accordingly, the commercial products of these glasses are typically melt-quenched products or small monoliths which do not require any specific processing within the critical temperature window.[6] In contrast, glass 13-93 is one composition in a series of 30 glasses that were designed to enable hot-working.[10] *In vivo* studies confirmed that this composition is not only suitable for versatile thermal treatments but also shows similar reactivity with bone as glass S53P4.[9,11,39] Later, this composition was frequently studied in various tissue engineering applications as a composition allowing for manufacture of amorphous products without crystallization.[12–21] Viscosity measurements explain the differences in the suitability of the three glasses 45S5, S53P4 and 13-93 to various hot-working methods. Since the viscosities vary by several orders of magnitude between the practical melting temperatures ($10^{2.0}$ dPa s at 1300–1400 °C) and T_g ($10^{11.3}$ dPa s at 500–600 °C), several methods are needed to describe the viscosity at different temperature ranges.

The low temperature viscosities of bioactive glasses have been measured for increasing temperatures above T_g using beam bending viscometry[40] and hot-stage microscopy.[10,41] The measured viscosity–temperature values for 45S5, S53P4 and 13-93 are shown in Figure 1.4. Only a few values could be measured for 45S5 and S53P4 using hot-stage microscopy while for 13-93 several values were obtained.[41] For each composition, crystallization interrupted the measurements at increasing temperature. The high temperature viscosities were measured using a rotation viscometer for decreasing temperatures (Figure 1.4).[40,41] Also at the high temperature range, the viscosity of 13-93 could be measured for a clearly higher range than for 45S5 and

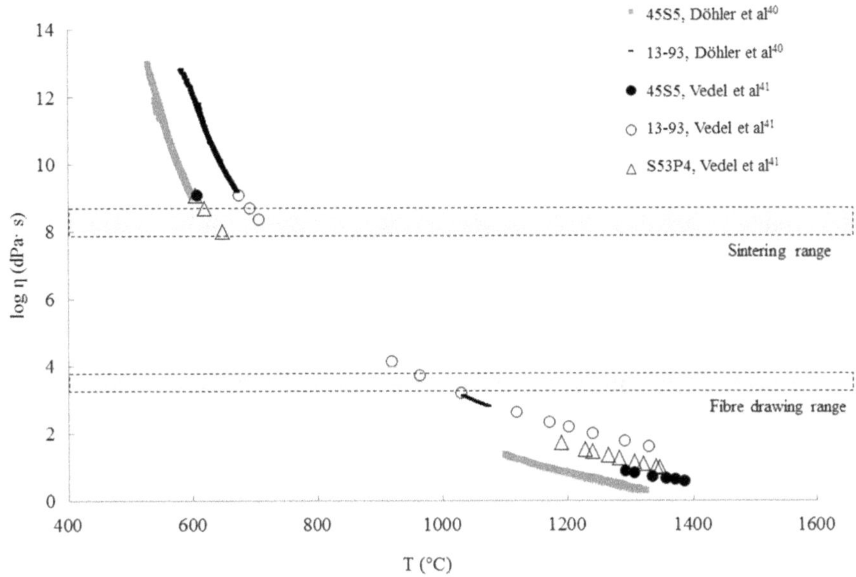

Figure 1.4 Measured viscosity values for 45S5, S53P4 and 13-93 at low and high temperature ranges.[40,41] The grey areas give typical viscosity ranges for drawing of continuous fibres or sintering of porous bodies.

S53P4. Between the low and high temperature ranges, the glasses exist as mixtures of varying amounts of melt and crystals and no true viscosity for the melt can be specified. The ranges for sintering and fibre drawing are also marked in Figure 1.4. 45S5 melt crystallizes at very low viscosity during cooling. Accordingly, normal fibre drawing processes cannot be used to achieve continuous fibres of 45S5. In contrast, the 13-93 melt is resistant to crystallization below the liquidus down to around typical fibre drawing viscosities. Similarly, 13-93 can be sintered into amorphous porous bodies, while 45S5 and S53P4 easily crystallize.

1.3 Properties Essential in the Final Application

Today, bioactive glasses are mainly used to fill defects and to promote and support bone tissue regeneration. The current research trend is to fabricate glass and glass-ceramic scaffolds with architectures mimicking the three-dimensional interconnected porosity of natural bone.[42–49] The porous scaffold not only actively induces bone regeneration but, optimally, also degrades at a rate matching the tissue ingrowth rate. The requirements of the porous tissue engineering scaffolds are discussed in detail in Chapter 21. In general, adequate mechanical strength and controlled, predictable reaction and dissolution rates are the most important properties for the bioactive glass product in its final application.

1.3.1 Mechanical Strength

The theoretical strength of flawless solid silicate glass is ≈35 GPa.[50,51] However, the ability of a glass to resist fracture when a crack is present, *i.e.* the fracture toughness, is low. Thus, already small flaws decrease the strength noticeably and typical strengths of common glass products are only around 14–70 MPa.[51] Glasses are brittle and fail without yielding as indicated by the high Young's modulus of silicate glasses, 45–100 GPa.[52] Table 1.3 summarizes typical mechanical property values reported for glass 45S5 and bone.[42,46,51,53–55]

Mechanical durability is a major drawback for porous bioactive glass scaffolds both because of the intrinsic brittleness of glasses and the high interconnected porosity required for cell penetration, vascularization, and nutrient flow.[43–49] Thin struts, large pore size, and high total porosity may give scaffolds which have too low mechanical strength even in low-load applications. The strength must also be high enough to sustain the surgery without mechanical failure. After implantation, the scaffold starts to dissolve which may lead to crack formation in the struts especially when subjected to a load or mechanical stresses.

Since most melt-derived glasses partly crystallize in the sintering processes applied in scaffold fabrication, most data available on mechanical properties are for glass-ceramics. Mechanical properties of scaffolds prepared *via* different approaches, such as the foam replication technique and various 3D printing techniques from the parent glass 45S5, have been reported in several studies.[42–44,55–60] In contrast, only limited information is available on porous scaffolds sintered of glass S53P4.[34] The mechanical properties of amorphous scaffolds based on glass 13-93 have been tested frequently *in vitro* and also *in vivo*.[16,21,44–46,61–66] The porous scaffolds are often coated with a biodegradable polymer for increased mechanical strength. The degradation of mechanical strength *in vivo* depends on several variables: the composition of the glass, the pore architecture and total porosity of the scaffold, the degradation rate of the polymer coating, and the effects of the interaction of the polymer and glass on the degradation of both materials. One additional paramount factor determining the suitability of the scaffold to the regeneration of loaded bone is the influence of the rate of new bone formation on the overall mechanical properties of the implant. The strength properties of the porous scaffolds are discussed in Chapter 21.

1.3.2 Dissolution Reactions

1.3.2.1 Hydroxyapatite Formation and Ion Release

Bioactive glasses are designed to dissolve and react in a controlled manner in the body environment. Traditionally, bioactive glasses have been described as surface-active materials capable of forming a mechanically strong chemical bond with living tissue, mainly bone.[4,6,45,67] The basis for the glass to form chemical bonding with tissue is the time-dependent dissolution and precipitation reactions of the glass with its surrounding solution. The changes in the ion concentrations of the solution around the reacting glass lead to the formation of the carbonated hydroxyapatite (HCA) interfacial layer. The HCA layer is biomimetic with the inorganic mineral apatite in the bone and gives thus a chemical fixation of the implant with the surrounding tissue instead of a mechanical fixation provided by the fibrous capsules around biostable implants. Biomimetic crystals not only have a similar composition to the apatite in living bone but also their crystallite size is compatible with the nano-sized bone or dentine apatite crystals.[68,69] The principles of the bone bonding mechanisms have been described in several papers.[2,45,67,70]

The HCA layer surface also makes the glass osteoconductive. Since the bioactive glasses dissolve with time while supporting apatite precipitation at the dissolving surface and nearby tissue, the bioactive glasses can also be classified as resorbable materials. The ion release products from the bioactive glasses are known to activate several families of genes, among others the genes that regulate osteogenesis and the production of growth factors, to affect adsorption of proteins and cell attachment at the surface of the reacting glass.[71–75] Thus, bioactive glasses, which stimulate regeneration of new bone *via* ionic dissolution products, are not only osteoconductive but also osteostimulative. The dissolution and precipitation reactions of bioactive glasses *in vitro* and *in vivo* lead to similar surface template structures, which favor precipitation of hydroxyapatite.[9,76] Accordingly, *in vitro* formation of HCA at the glass surface is usually taken as an indication of the bioactivity.[77] The glass surface first partly dissolves and then serves as a substrate for precipitation of HCA. Primarily, the precipitate is amorphous but converts into carbonated hydroxyapatite with time.

Interestingly, the utilization of bioactive glasses is not restricted only to applications where bone bonding provided by HCA crystals is desired, *i.e.* bone tissue engineering. The impact of the released ions on activation stimulation of genes and cells involved in hard and soft tissue regeneration, wound healing, and angiogenesis has been discussed in several papers.[4,6,45,78–80] Bioactive glasses have potential also in the regeneration of cardiac tissue, lung tissue, nerves, gastrointestinal tissue, *etc.*[5] The HCA layer which forms at the surface of the bioactive glass provides a bonding interface also to soft tissues. However, the mechanisms for interactions between the bioactive glass and the cells that compose the different soft tissues are still poorly understood.

The pH and the ion concentrations of the interfacial solution increase rapidly around the dissolving glass especially when using high concentrations and/or small particles. This pH effect and the released ions inhibit and prevent bacterial growth around the glass.[81–91] The antibacterial effect may be enhanced by doping the glass with Ag, Ce, Co, Zn, *etc.*[79] These findings indicate that a detailed knowledge of the dissolution mechanisms is also essential when assessing the antibacterial effects of various size fractions of bioactive glasses.

1.3.2.2 *Initial Dissolution Mechanisms*

The glass dissolution reaction mechanisms are well described in the literature.[3,92–101] Glass can either dissolve congruently, *i.e.* uniformly, which implies that the ratios of elements in the solution are the same as in the dissolving glass, or incongruently, *i.e.* preferential or selective leaching of some of the elements. The reaction mechanisms depend on the glass composition and environmental conditions, such as the surface area to volume ratio, and on the pH of the solution. In general, all silicate glasses dissolve in aqueous solutions but the extent of the reactions strongly depends on the composition of the glass and solution. In more general terms, these basic dissolution mechanisms can be divided into primary reactions, leading to the formation of a subsurface zone consisting of a silica-rich layer, and secondary precipitation reactions, giving an outer surface layer or an alternate surface layer structure. These reaction mechanisms also explain the formation of the typical dual layer at the bioactive glasses.

The reactions leading to the formation of the bone bonding are usually described as a sequence of eleven reaction stages at the surface of the glass.[97] The first five rapid reaction stages take place at the surface of a bioactive glass with the highest level of bioactivity within 24 h. These reactions explain the formation of a dual layer of hydrated silica (silica-rich gel) and polycrystalline HCA at the glass surface. At the subsequent reaction stages, bonding with bone *via* a series of biological reactions takes place and within 6 to 12 days the final product, a collagen-HCA matrix containing mature osteocytes, is formed.[97] Simplified, the HCA at the glass surface is similar to the apatite in natural bone and thus is able to form mechanically strong chemical bonding with the living hard tissue. The five first bioactivity reactions stages are summarized below and in Figure 1.5.

At $pH < 9$, a selective leaching characterized by an ion exchange reaction of alkali and alkaline-earth ions with protons from the surrounding solution dominates (Reaction I and Figure 1.5, insert 1). The field strength and the radius of the network-modifying mobile ions affect their leaching rate.[3] Accordingly, alkali ions and the more weakly bonded ions with larger radii are leached out more easily than the smaller ions. The silica network structure may also react with water *via* hydration (Reaction 2 and Figure 1.5, insert 2). The ion exchange and the hydration reactions result in a slightly porous hydrated silica-rich layer and an increase in the pH of the surrounding solution. The hydroxyl ions act as a catalyst for network dissolution and, at $pH > 9$, the hydrolysis reaction starts to dominate leading to depolymerisation of the glass network structure (Reaction 3, Figure 1.5, insert 3). The silanol groups (≡Si–OH) may gradually depolymerise further and

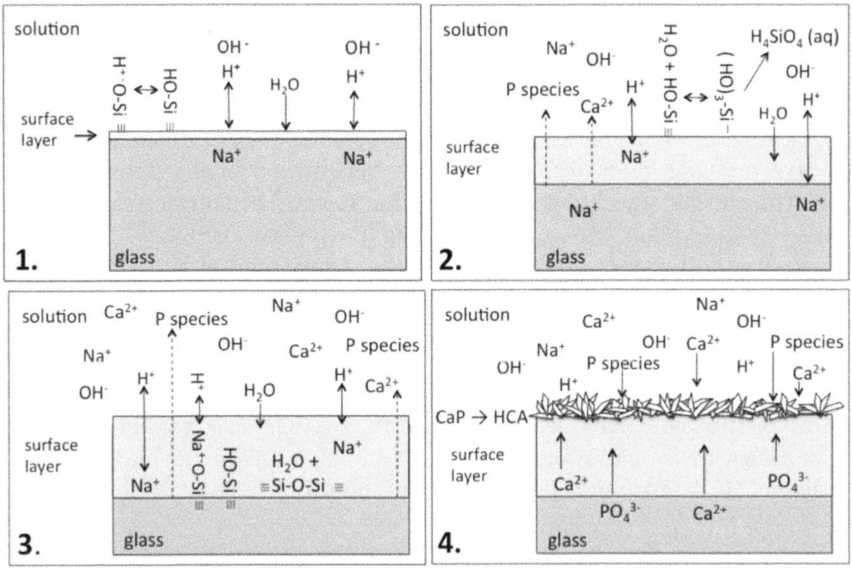

Figure 1.5 Dissolution and precipitation reactions at the surface of a bioactive silicate glass (modified from Conradt,[101] Hupa and Fagerlund[102]).

release soluble silica to the solution. The silanol groups can also condense to give a silica-rich gel at the glass surface (Reaction 4, Figure 1.5, insert 3).

Reaction 1, ion-exchange:

$$\equiv\text{Si–O}^{-}\text{M}^{+}(\text{glass}) + \text{H}^{+}(\text{aq}) \rightarrow \equiv\text{Si–OH}(\text{glass}) + \text{M}^{+}(\text{aq})$$

Reaction 2, hydration:

$$\equiv\text{Si–O}^{-}\text{M}^{+}(\text{glass}) + \text{H}_2\text{O} \rightarrow \equiv\text{Si–(OH)}(\text{glass}) + \text{M}^{+}(\text{aq}) + \text{OH}^{-}(\text{aq})$$

Reaction 3, hydrolysis:

$$\equiv\text{Si–O–Si}\equiv (\text{glass}) + \text{OH}^{-}(\text{aq}) \rightarrow \equiv\text{Si–(OH)}(\text{glass}) + \equiv\text{Si–O}^{-}(\text{glass})$$

Reaction 4, condensation:

$$\equiv\text{Si–OH}(\text{glass}) + \text{OH–Si}\equiv(\text{glass}) \rightarrow \equiv\text{Si–O–Si}\equiv(\text{glass}) + \text{H}_2\text{O}$$

The sub-surface structure that develops in the aforementioned reactions is often described as a silica-rich gel layer as it may be porous and contains water and silanol groups. The secondary reaction type, precipitation, starts when the solubility of the dissolved ions is exceeded. The silica-rich gel provides suitable nucleation sites for Ca^{2+} and PO_4^{3-} migrating from the bulk glass or precipitating from the solution to form the amorphous calcium phosphate film, which gradually crystallizes into HCA (Figure 1.5. insert 4). Also, other anions, such as F^-, OH^-, and CO_3^{2-}, may be incorporated into the crystalline structure.[97] This crystallized HCA is similar to that in natural bone and it is able to form chemical bonds with natural bone *via* a series of biological reactions. The formation of an HCA layer on the leached silica-rich layer may act as a diffusion barrier, depending on the density of the layer. This decreases the driving force for further dissolution and might partially help explain why, with some bioactive glass compositions, intact core glass is still found after several years of implantation.[103]

For melt-derived glasses, the layered surface structure of SiO_2 and HCA forms rapidly on glasses with low silica content. The network structure of these glasses consists mainly of Q^2 and Q^3 units, *i.e.* an open silicate network with poor chemical durability.[104,105] However, composition alone does not determine the bioactivity, the sample form, particle size, concentration of the particles in the solution, fluid flow, *etc.* also affect the bioactivity expressed as HCA formation *in vitro* or bone bonding *in vivo*.

1.3.3 Dissolution Measurements

Several analysis methods have been utilized to gain information on the dissolution behaviour and reactions of bioactive glasses *in vitro*. These are based on studying the changes occurring both in the glass and in the solution upon incubation of the glass samples in phosphate buffered aqueous solutions, such as a simulated body fluid.

Recently, the Technical Committee 4 (TC04) of the International Commission of Glass (ICG) suggested a static *in vitro* method for testing the apatite-forming ability of bioactive glasses.[106] In the test, glass powders are immersed in simulated body fluid (SBF) using a fixed mass to volume solution ratio in airtight polyethylene containers. The containers are held at 37 °C in an incubating orbital shaker for various time periods. At the end of each immersion time period, the particles and solution are separated and analysed. The ion concentrations in the solution are measured using an inductively coupled plasma optical emission spectrometer (ICP-OES). In addition, the pH of the solution is measured. The glass samples are characterized using Fourier transform infrared spectrometry (FTIR), X-ray diffraction (XRD), and/or scanning electron microscopy (SEM). The method was tested at several laboratories and was found to give appropriate values for apatite formation, especially on samples consisting of particles or glasses with high surface area. Time points for detection of HCA nucleation on the sample surface and changes in phosphate concentration were found to correlate. The test method is recommended when comparing new glasses with known compositions reported in the literature.[106]

1.3.4 Dissolution in Continuously Flowing Solution

Dissolution tests in static or agitated solutions give comparable values between different glass compositions but do not provide accurate information of the glass dissolution kinetics. However, the human body is not a static environment and the fluid flow conditions vary in different locations. Different continuous *in vitro* testing methods have been developed to better imitate the conditions in the dynamic body environment. These have utilized fluid flow rates ranging from 0.03 mL min^{-1} to 2 mL min^{-1} to study the *in vitro* reactions.[39,107–113] In most studies, the solution flow slowly changes the fluid in a relatively large container containing the sample. In several studies, the same solution has been circulated in the system causing changes to the original solution composition, but fresh solutions have also been used. The solution has been analysed for samples collected at predetermined time intervals. By connecting a dynamic measurement method directly into a fast analysis method, such as pH or ion measurement, on-line data can be collected. Inductively coupled plasma optical emission spectroscopy (ICP-OES) offers a sensitive on-line method for qualitative and quantitative determination of metals and certain non-metals in solution.[39,113]

1.3.5 *In vitro* Studies of 45S5, S53P4 and 13-93

Three phenomena are of interest when assessing the *in vitro* bioactivity of glasses: (i) the increase in the pH of the fluid around the glass, (ii) the ion concentrations in the fluid of the ions releasing from the dissolving glass, and (iii) the layers that form at the dissolving glass surface.

The layer formation on 45S5, S53P4 and 13-93 has been reported frequently but seldom in the same studies or using the same experimental conditions. Although typical HCA and Si-rich surface layers form on each composition, clearly thicker layers form rapidly on 45S5, while the layer thickness is least on 13-93.[39,114–116]

The ability of a glass to form reaction layers correlates with the release of ions from the glass and changes in the pH of the immersion solution.[39] Figure 1.6 shows the initial release of Ca and Si into a fresh TRIS buffer solution fed continuously through a bed of particles of 45S5, S53P4 and 13-93.[39] The solution flow rate of 0.2 mL min^{-1} is close to the value reported for fluid flow in human muscles.[108]

The amount of Na ions released from 45S5 was very high and the values exceeded the detection limit. For S53P4, the Na release values were also close to the limit but decreased after the initial peak value to around 110 mg L^{-1}. Lower amounts of Na dissolved from 13-93, around 10 mg L^{-1}, while the amount of K was 20 mg L^{-1} at 15 min. For all glasses, the concentration of Ca ions stabilized after a minor initial peak and was around 130 mg L^{-1} for 45S5, 100 mg L^{-1} for S53P4, and 50 mg L^{-1} for 13-93 at 15 min. At this time point, the concentration of Mg ions releasing from 13-93 was around 9 mg L^{-1}. The initial release of soluble Si ions was highest from 45S5 while the concentration increased steadily for S53P4 and 13-93. Finally, the initial phosphate release was around 10 mg L^{-1} for 45S5 and S53P4 while slightly lower concentrations were measured for 13-93.

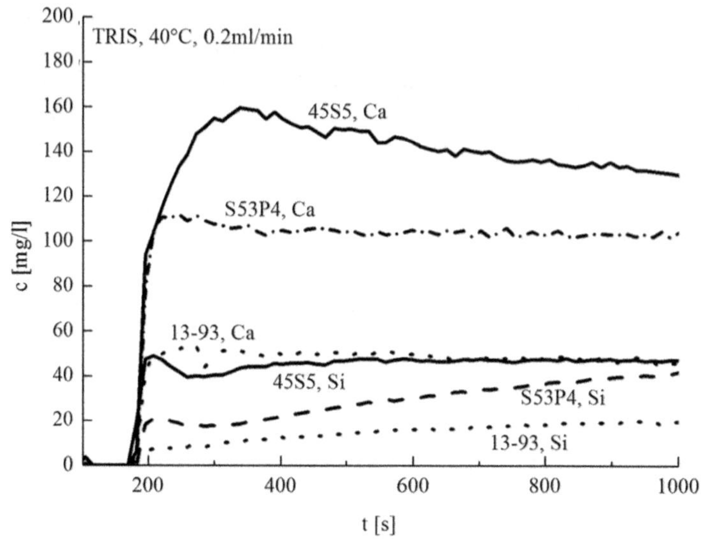

Figure 1.6 Average initial dissolution of Ca and Si from 300–500 μm particles of 45S5, S53P4 and 13-93 into continuously flowing TRIS. The variation was less than 5% for all ions.
Data from Fagerlund *et al.*[39]

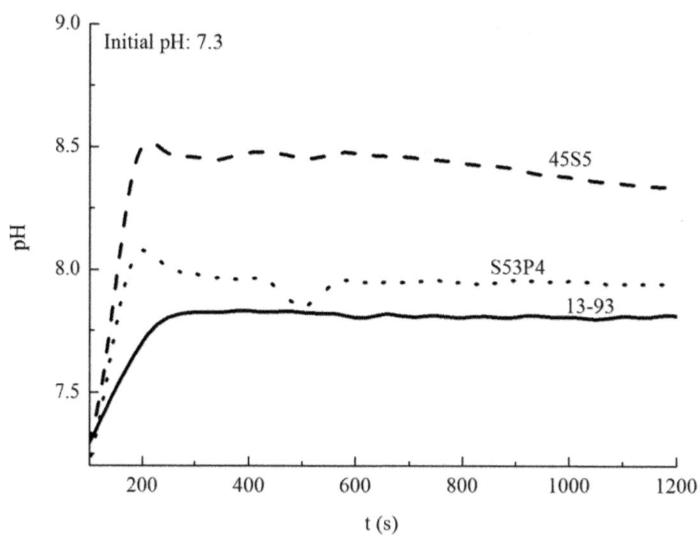

Figure 1.7 pH profiles in initial dissolution of 45S5, S53P4 and 13-93 in a continuous flow of TRIS.
Data from Fagerlund *et al.*[39]

The pH values were recorded in similar experimental conditions using a flow-through micro volume pH electrode. The pH curves for the glasses verify the highest overall ion dissolution from 45S5 and would also be typical for systems in which reaction layers have been observed (Figure 1.7).

The initial ion dissolution and pH profiles were in line with the observations of the *in vitro* and *in vivo* layer formation on these glasses. The high initial release of calcium ions from 45S5 suggests that the saturation limit in the solution is exceeded already after the first minutes of the contact with the 45S5 particles and leads to rapid calcium phosphate precipitation while slower precipitation is likely on the other glasses. The reaction rate also depends on the environmental conditions such as the surface area to volume ratio or flow rate of the solution, both of which affect the pH of the solution.

The initial ion release can also be used to estimate the capability of the glasses to affect cellular processes. For osteostimulation, the release rates of biologically active soluble Ca and Si ions are 60–90 ppm for Ca and 15–30 ppm for Si.[71,72,78] If present in a critical concentration, these ions also activate, or up-regulate, seven families of genes in osteogenic cells.[71–73,78] The initial concentrations released from 45S5 and S53P4 into the TRIS buffer solution were higher than these critical values while the calcium concentration was less than the critical concentration (Figure 1.5). Whether the ion concentrations stay at critical levels for long enough time periods should be verified with longer test times. Relatively high concentrations of potassium ions were released from 13-93 during the first minutes. Extensive potassium release may limit the utilization of 13-93 or other potassium oxide containing glasses in product forms giving high surface area to volume ratio to avoid any deleterious cellular effects.[117] The dynamic dissolution tests give additional interesting information for assessing the dissolution kinetics and measuring the ion release rates from bioactive glasses.

1.4 Conclusions and Outlook

Melt-derived bioactive glasses are commercially utilized mainly as granules or particles to repair and treat bone injuries and defects. In contrast, extensive research effort is devoted to developing porous scaffolds for soft and bone tissue regeneration. Thorough understanding of the material glass, its physical and chemical properties is essential when designing novel scaffolds and devices with desired porosity, shape, and controlled, pre-determined ion release. Fabrication of monoliths, porous scaffolds or fibres is limited by the viscosity–temperature behaviour and crystallization tendency of the glass in hot-working. Biological responses of the glass can be estimated by studying the ability of the glass to form layered structures of silica and hydroxyapatite at the surface *in vitro*. Ion release rates from the glass give additional information for estimating the dissolution rate of the glass and its capability to stimulate and support tissue regeneration.

References

1. D. F. Williams, *Biomaterials*, 2008, **29**, 2941.
2. L. Hench, R. J. Splinter, W. C. Allen and T. K. Grenlee Jr., *J. Biomed. Mater. Res.*, 1971, **2**, 117.

3. L. L. Hench, *J. Non-Cryst. Solids*, 1975, **19**, 27.
4. L. L. Hench, *J. Mater. Sci.: Mater. Med.*, 2006, **17**, 967.
5. V. Miguez-Pacheco, L. L. Hench and A. R. Boccaccini, *Acta Biomater.*, 2015, **13**, 1.
6. J. R. Jones, *Acta Biomater.*, 2013, **9**, 4457.
7. J. E. Shelby, *Introduction to Glass Science and Technology*, The Royal Society of Chemistry, Cambridge, UK, 1997, ch. 1, pp. 1–7.
8. F. T. Wallenberger and A. Smrck, *Int. J. Appl. Glass Sci.*, 2010, **1**, 151.
9. Ö. H. Andersson, G. Liu, K. H. Karlsson, L. Niemi, J. Miettinen and J. Juhanoja, *J. Mater. Sci.: Mater. Med.*, 1990, **1**, 219.
10. M. Brink, *J. Biomed. Mater. Res.*, 1997, **36**, 109.
11. M. Brink, T. Turunen, R. Happonen and A. Yli-Urpo, *J. Biomed. Mater. Res.*, 1997, **37**, 114.
12. A. Itälä, H. O. Ylänen, J. Yrjas, T. Heino, T. Hentunen, M. Hupa and H. T. Aro, *J. Biomed. Mater. Res., Part A*, 2002, **62**, 404.
13. A. Itälä, V. V. Välimäki, R. Kiviranta, H. O. Ylänen, M. Hupa, E. Vuorio and H. T. Aro, *J. Biomed. Mater. Res., Part B*, 2003, **65B**, 163.
14. R. F. Brown, D. E. Day, T. E. Day, S. Jung, M. N. Rahaman and Q. Fu, *Acta Biomater.*, 2008, **4**, 387.
15. A. J. Asikainen, J. Hagström, T. Sorsa, J. Noponen, M. Kellomäki, H. Juuti, C. Lindqvist, J. Hietanen and R. Suuronen, *J. Biomed. Mater. Res., Part A*, 2007, **83A**, 530.
16. Q. Fu, M. N. Rahaman, B. S. Bal, R. F. Brown and D. E. Day, *Acta Biomater.*, 2008, **4**, 1854.
17. Q. Fu, M. N. Rahaman, B. S. Bal, K. Kuroki and R. F. Brown, *J. Biomed. Mater. Res., Part A*, 2010, **95**, 235.
18. R. Meszaros, R. Zhao, N. Travitzky, T. Fey, P. Greil and L. Wondraczek, *Glass Technol.: Eur. J. Glass Sci. Technol., Part A*, 2011, **52**, 111.
19. N. D. Doiphode, T. Huang, M. C. Leu, M. N. Rahaman and D. E. Day, *J. Mater. Sci.: Mater. Med.*, 2011, **22**, 515.
20. X. Liu, M. N. Rahaman and Q. Fu, *Acta Biomater.*, 2013, **8**, 4889.
21. S. Eqtesadi, A. Motealleh, A. Pajares, F. Guilberteau and P. Miranda, *J. Non-Cryst. Solids*, 2016, **432**, 111.
22. S. Fagerlund, J. Massera, L. Hupa and M. Hupa, *J. Eur. Ceram. Soc.*, 2012, **32**, 2731.
23. J. Massera, S. Fagerlund, L. Hupa and M. Hupa, *J. Am. Ceram. Soc.*, 2012, **95**, 607.
24. S. Fagerlund and L. Hupa, *Ceram. Mat. J. Pol. Ceram. Soc.*, 2010, **62**, 349.
25. R. Golovchak, P. Thapar, A. Ingram, D. Savytskii and H. Jain, *Acta Biomater.*, 2014, **10**, 4878.
26. L. Lefebvre, J. Chevalier, L. Grémillard, R. Zenati, G. Thollet, D. Bernache-Assollant and A. Govin, *Acta Mater.*, 2007, **55**, 3305.
27. L. Lefebvre, L. Gremillard, J. Chevallier, R. Zenati and D. Bernache-Assolant, *Acta Biomater.*, 2008, **4**, 1894.
28. H. E. Kissinger, *Anal. Chem.*, 1957, **29**, 1702.

29. J. A. Augis and J. E. Bennett, *J. Therm. Anal. Calorim.*, 1978, **13**, 283.
30. A. Marotta, A. Buri and F. Branda, *J. Mater. Sci.*, 1981, **16**, 341.
31. H. Arstila, L. Hupa, E. Vedel, H. Ylänen and M. Hupa, *Glass Technol.*, 2005, **46**, 138.
32. H. Arstila, M. Tukiainen, S. Taipale, M. Kellomäki and L. Hupa, *Adv. Mater. Res.*, 2008, **39–40**, 287.
33. H. Arstila, L. Hupa and K. H. Karlsson, *J. Non-Cryst. Solids*, 2008, **354**, 722.
34. S. Fagerlund, J. Massera, N. Moritz, L. Hupa and M. Hupa, *Acta Biomater.*, 2012, **8**, 2331.
35. H. Arstila, E. Vedel, L. Hupa and M. Hupa, *Glass Technol.: Eur. J. Glass Sci. Technol., Part A*, 2008, **49**, 260.
36. M. Tylkowski and D. S. Brauer, *J. Non-Cryst. Solids*, 2013, **376**, 175.
37. O. Bretcanu, X. Chatzistavrou, K. Paraskevopoulos, R. Conradt, I. Thompson and A. R. Boccaccini, *J. Eur. Ceram. Soc.*, 2009, **29**, 3299.
38. J. Massera, L. Hupa and M. Hupa, *J. Non-Cryst. Solids*, 2012, **358**, 270.
39. S. Fagerlund, L. Hupa and M. Hupa, *Acta Biomater.*, 2013, **9**, 5400.
40. F. Döhler, D. Groh, S. Chiba, J. Bierlich, J. Kobelke and D. S. Brauer, *J. Non-Cryst. Solids*, 2016, **432**, 130.
41. E. Vedel, H. Arstila, H. Ylänen, L. Hupa and M. Hupa, *Glass Technol.: Eur. J. Glass Sci. Technol., Part A*, 2008, **49**, 251.
42. K. Rezwan, Q. Z. Chen, J. J. Blaker and A. R. Boccaccini, *Biomaterials*, 2006, **27**, 3413.
43. L. C. Gerhardt and A. R. Boccaccini, *Materials*, 2010, **3**, 3867.
44. Q. Fu, E. Saiz, M. N. Rahaman and A. P. Tomsia, *Mater. Sci. Eng., C*, 2011, **31**, 1245.
45. M. N. Rahaman, D. E. Day, B. S. Bal, Q. Fu, S. B. Jung, L. F. Bonewald and A. P. Tomsia, *Acta Biomater.*, 2011, **7**, 2355.
46. Q. Fu, E. Saiz, M. N. Rahaman and A. P. Tomsia, *Adv. Funct. Mater.*, 2013, **23**, 5461.
47. A. J. Salinas and M. Vallet-Regí, *RSC Adv.*, 2013, **3**, 11116.
48. G. Kaur, O. P. Pandey, K. Singh, D. Homa, B. Scott and G. Pickrell, *J. Biomed. Mater. Res., Part A*, 2014, **102A**, 254.
49. G. Kaur, G. Pickrell, N. Sriranganathan, V. Kumar and D. Homa, *J. Biomed. Mater. Res., Part B*, 2016, **104B**, 1248.
50. C. R. Kurkjian, P. K. Gupta and R. K. Brow, *Int. J. Appl. Glass Sci.*, 2010, **1**, 27.
51. A. K. Varshneya, Fundamentals of Inorganic Glasses, Sheffield, The Society of Glass Technology, 2006.
52. S. Inaba, S. Fujino and K. Moringa, *J. Am. Ceram. Soc.*, 1999, **82**, 3501.
53. P. K. Vallittu, T. O. Närhi and L. Hupa, *Dent. Mater.*, 2015, **31**, 371.
54. J. A. Juhasz and S. M. Best, *J. Mater. Sci.*, 2012, **47**, 610.
55. Q. Z. Chen, I. D. Thompson and A. R. Boccaccini, *Biomaterials*, 2006, **27**, 2414.

56. Q. Z. Chen and A. R. Boccaccini, *J. Biomed. Mater. Res., Part A*, 2006, **77A**, 445.
57. O. Bretcanu, S. Misra, I. Roy, C. Renghini, F. Fiori, A. R. Boccaccini and V. Salih, *J. Tissue Eng. Regen. Med*, 2009, **3**, 139.
58. Q. Z. Chen, A. Efthymiou, V. Salih and A. R. Boccaccini, *J. Biomed. Mater. Res., Part A*, 2008, **84A**, 1049.
59. P. Tesavibul, R. Felzmann, S. Gruber, R. Liska, I. Thompson, A. R. Boccaccini and J. Stampfl, *Mater. Lett.*, 2012, **74**, 81.
60. M. Montazeri, S. Karbasi, M. R. Foroughi, A. Monshi and R. Ebrahimi-Kahirzsangi, *J. Mater. Sci.: Mater. Med.*, 2015, **26**, 1.
61. Q. Fu, M. N. Rahaman, B. S. Bal and R. F. Brown, *J. Biomed. Mater. Res., Part A*, 2010, **93**, 1380.
62. X. Liu, M. N. Rahaman, Q. Fu and A. P. Tomsia, *Acta Biomater.*, 2012, **8**, 415.
63. K. C. R. Kolan, M. C. Leu, G. E. Hilmas and M. Velez, *J. Mech. Behav. Biomed. Mater.*, 2012, **13**, 14.
64. X. Liu, M. N. Rahaman and Q. Fu, *Acta Biomater.*, 2013, **9**, 4889.
65. X. Liu, M. N. Rahaman, G. E. Hilman and B. S. Bal, *Acta Biomater.*, 2013, **9**, 7025.
66. Y. Cao, B. Yang, C. Gao, P. Feng and C. Shuai, *Sci. Sintering*, 2015, **47**, 31.
67. L. L. Hench, *J. Am. Ceram. Soc.*, 1998, **8**, 1705.
68. D. Groh, F. Döhler and D. S. Brauer, *Acta Biomater.*, 2014, **10**, 4465.
69. M. D. O'Donnell, S. J. Watts, R. G. Hill and R. V. Law, *J. Mater. Sci.: Mater. Med.*, 2009, **20**, 1611.
70. L. L. Hench, J. W. Hench and D. C. Greenspan, *J. Aust. Ceram. Soc.*, 2004, **40**, 1.
71. L. L. Hench, J. M. Polak, I. D. Xynos and L. D. K. Buttery, *Mater. Res. Innovations*, 2000, **3**, 313.
72. I. D. Xynos, M. V. J. Hukkanen, J. J. Batten, L. D. Buttery, L. L. Hench and J. M. Polak, *Calcif. Tissue Int.*, 2000, **67**, 321.
73. I. D. Xynos, A. J. Edgar, L. D. Buttery, L. L. Hench and J. M. Polak, *Biochem. Biophys. Res. Commun.*, 2000, **276**, 461.
74. I. D. Xynos, A. J. Edgar, L. D. K. Buttery, L. L. Hench and J. M. Polak, *J. Biomed. Mater. Res.*, 2001, **55**, 151.
75. P. Valerio, M. M. Pereira, A. M. Goes and M. F. Leite, *Biomaterials*, 2004, **25**, 2941.
76. Ö. H. Andersson and I. Kangasniemi, *J. Biomed. Mater. Res.*, 1991, **25**, 1019.
77. T. Kokubo and H. Takadama, *Biomaterials*, 2005, **27**, 2907.
78. L. L. Hench, *J. Eur. Ceram. Soc.*, 2009, **29**, 1257.
79. A. Hoppe, N. S. Güldal and A. R. Boccaccini, *Biomaterials*, 2011, **32**, 2757.
80. M. Ojansivu, S. Vanhatupa, L. Björkvik, H. Häkkänen, M. Kellomäki, R. Autio, J. Ihalainen, L. Hupa and S. Miettinen, *Acta Biomater.*, 2015, **21**, 190.
81. P. Stoor, E. Söderling and R. Grénman, *J. Biomed. Mater. Res., Part B*, 1999, **48**, 869.

82. I. Allan, H. Newman and M. Wilson, *Biomaterials*, 2001, **22**, 1683.
83. M. Zehnder, T. Waltimo, B. Sener and E. Söderling, *Oral Surg. Oral Med. Oral Pathol. Oral Radiol. Endod.*, 2006, **101**, 530.
84. M. Zehnder, G. Baumgartner, K. Marquardt and F. Paqué, *Oral Surg. Oral Med. Oral Pathol. Oral Radiol. Endod.*, 2007, **103**, 423.
85. E. Munukka, O. Leppäranta, M. Vaahtio, T. Peltola, D. Zhang, L. Hupa, H. Ylänen, J. I. Salonen, M. K. Viljanen and E. Eerola, *J. Mater. Sci.: Mater. Med.*, 2008, **19**, 27.
86. O. Leppäranta, M. Vaahtio, T. Peltola, D. Zhang, L. Hupa, M. Hupa, H. Ylänen, J. Salonen, M. K. Viljanen and E. Eerola, *J. Mater. Sci.: Mater. Med.*, 2008, **19**, 547.
87. D. Zhang, O. Leppäranta, E. Munukka, H. Ylänen, M. K. Viljanen, E. Eerola, M. Hupa and L. Hupa, *J. Biomed. Mater. Res. A*, 2010, **93A**, 475.
88. N. Lindfors, *J. Biotech. Biomater*, 2011, **1**, 111.
89. M. M. Echezarreta-Lopez and M. Landin, *Int. J. Pharm.*, 2013, **453**, 641.
90. L. Drago, C. Vassena, S. Fenu, E. De Vecchi, V. Signori, R. De Francesco and C. L. R. Romanò, *Future Microbiol.*, 2014, **9**, 593.
91. L. Drago, E. De Vecchi, M. Bortolin, M. Toscano, R. Mattina and C. L. Romano, *Future Microbiol.*, 2015, **10**, 1293.
92. R. W. Douglas and T. M. M. El-Shamy, *J. Am. Ceram. Soc.*, 1967, **50**, 1.
93. L. L. Hench and D. E. Clark, *J. Non-Cryst. Solids*, 1975, **19**, 27.
94. M. F. Dilmore, D. E. Clark and L. L. Hench, *J. Am. Ceram. Soc.*, 1978, **61**, 439.
95. B. C. Bunker, G. W. Arnold, E. K. Beauchamp and D. E. Day, *J. Non-Cryst. Solids*, 1983, **58**, 295.
96. H. Sholze, *J. Non-Cryst. Solids*, 1988, **102**, 1.
97. L. L. Hench, *J. Am. Ceram. Soc.*, 1991, **74**, 1487.
98. B. C. Bunker, *J. Non-Cryst. Solids*, 1994, **179**, 300.
99. C. Cailleteau, C. Weigel, A. Ledieu, P. Barboux and F. Devreux, *J. Non-Cryst. Solids*, 2008, **354**, 117.
100. W. H. Casey, *Nat. Mater.*, 2008, **7**, 930.
101. R. Conradt, *J. Am. Ceram. Soc.*, 2008, **91**, 728.
102. L. Hupa, S. Fagerlund, *Handbook of Oral Biomaterials*, ed. J. P. Matinlinna, Pan Stanford Publishing, 2014, ch. 8, pp. 281–315.
103. N. C. Lindfors, I. Koski, J. T. Heikkilä, K. Mattila and A. J. Aho, *J. Biomed. Mater. Res., Part B*, 2010, **94**, 157.
104. M. Edén, *J. Non-Cryst. Solids*, 2011, **357**, 1595.
105. R. G. Hill and D. S. Brauer, *J. Non-Cryst. Solids*, 2011, **357**, 3884.
106. A. L. B. Maçon, T. B. Kim, E. M. Vallliant, K. Goetschius, R. K. Brow, D. E. Day, A. Hoppe, A. R. Boccaccihi, I. Y. Kim, C. Ohtsuki, T. Kokubo, A. Osaka, M. Vallet-Regí, D. Arcos, L. Fraile, A. J. Salinas, A. V. Teixeira, Y. Vueva, R. M. Almeida, M. Miola, C. Vitale-Brovarone, E. Verne, W. Höland and J. R. Jones, *J. Mater. Sci.: Mater. Med.*, 2015, **26**, 115.
107. A. Rámila and M. Vallet-Regí, *Biomaterials*, 2001, **22**, 2301.

108. P. Siriphannon, Y. Kameshima, A. Yasumori, K. Okada and S. Hayashi, *J. Biomed. Mater. Res.*, 2002, **60**, 175.
109. Y. R. Duan, Z. R. Zhang, C. Y. Wang, J. Y. Chen and X. D. Zhang, *J. Mater. Sci.: Mater. Med.*, 2005, **16**, 795.
110. A. H. De Aza, P. Velásquez, M. I. Alemany, P. Pena and P. N. De Aza, *J. Am. Ceram. Soc.*, 2007, **90**, 1200.
111. D. Rohanová, A. R. Boccaccini, D. M. Yunos, D. Horkavcová, L. Březovská and A. Helebrant, *Acta Biomater.*, 2011, **7**, 2623.
112. S. Yue, P. D. Lee, G. Poologasundarampillai and J. R. Jones, *Acta Biomater.*, 2011, **7**, 2637.
113. S. Fagerlund, P. Ek, L. Hupa and M. Hupa, *J. Am. Ceram. Soc.*, 2012, **95**, 3130.
114. L. Hupa, K. H. Karlsson, H. Aro and M. Hupa, *Glass Technol.: Eur. J. Glass Sci. Technol., Part A*, 2010, **51**, 89.
115. D. Zhang, M. Hupa, H. T. Aro and L. Hupa, *Mater. Phys. Chem*, 2008, **111**, 497.
116. L. Varila, S. Fagerlund, T. Lehtonen, J. Tuominen and L. Hupa, *J. Eur. Ceram. Soc.*, 2012, **32**, 2757.
117. L. Hupa, S. Fagerlund, J. Massera and L. Björkvik, *J. Non-Cryst. Solids*, 2016, **432**, 41.

CHAPTER 2

Bioactive Glass-ceramics: Processing, Properties and Applications

MAZIAR MONTAZERIAN* AND EDGAR DUTRA ZANOTTO

Department of Materials Engineering (DEMa), Center for Research, Technology and Education in Vitreous Materials (CeRTEV), Federal University of São Carlos (UFSCar), São Carlos, SP 13.565-905, Brazil
*Email: maziar_montaz@yahoo.com

2.1 Glass-ceramic Fabrication Methods

Glass-ceramics (GCs) are polycrystalline materials that contain one or more crystal phases embedded into a residual glass and are generally produced by two methods: melting and sol–gel followed by some heat treatment.[1–4] Figure 2.1 shows the main stages of glass-ceramic synthesis *via* these two methods. Through the melting route, oxides such as SiO_2, CaO, P_2O_5, Na_2O or their chemical compounds are weighed and mixed. Then, they are usually melted in a platinum crucible in the range of 1300–1450 °C, depending on the composition of bioactive GCs. Bulk pieces are formed by casting the melt in graphite or steel molds. After casting, annealing at approximately 450–550 °C is crucial because of the high coefficient of thermal expansion of the bioactive glass compositions. Finally, glass-ceramics are obtained after controlled heat treatment to induce crystallization. If a granulated or powdered material is required, the melt can be quenched into water or air before grinding and sieving into the desired particle sizes to obtain a frit. The glass frit should be rapidly dried to avoid corrosion while in contact with

RSC Smart Materials No. 23
Bioactive Glasses: Fundamentals, Technology and Applications
Edited by Aldo R. Boccaccini, Delia S. Brauer and Leena Hupa

Published by the Royal Society of Chemistry, www.rsc.org

water. An advantage of the glass fritting route is that mixtures of compositions can be made and then formed into shapes and densified by sintering. Sintering usually proceeds concurrently with crystallization, when the free surfaces of the glass frits encourage crystallization. Sinter-crystallization is the process followed in the manufacture of some glass-ceramics that show poor internal crystallization.[1–4]

The sol–gel method has been used to make a new generation of bioactive gel-derived glasses and glass-ceramics and offers promise for tailoring their composition to match the requirements of specific applications.[5] Brinker and Scherer[6] and Hench *et al.*[7,8] describe the details of sol–gel processing. Sol–gel is a chemically-based method for producing glass-ceramics at much lower temperatures than the traditional processing methods described above.[6–8] Often, some steps shown in Figure 2.1 are also involved in making glasses or glass-ceramics by the sol–gel method. The first step is mixing the reagents in a solvent (water or alcohol) which forms a low viscosity sol. In the case of silicate-based bioactive glasses, the silicate reagent could be an alkoxide, such as tetraethyl orthosilicate (TEOS), or similar. If other components apart from silica are required in the glass composition, they are added to the sol either as other alkoxides or as salts, *e.g.* phosphate is

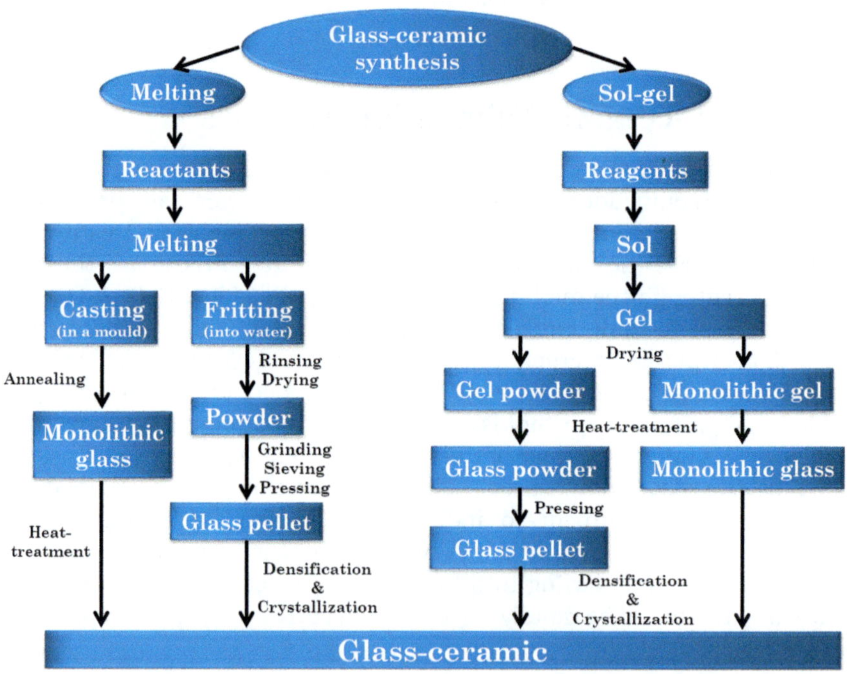

Figure 2.1 Schematic diagram of the main stages in the synthesis of glass-ceramics. Reproduced from ref. 4 with permission from John Wiley and Sons. © 2016 Wiley Periodicals.

incorporated by adding triethyl phosphate (TEP) and calcium by adding calcium nitrate tetrahydrate. Prior to completion of the network formation, the sol can be applied as a coating, be pulled into a fiber, impregnated into a composite, formed into powders or cast into a mold with a precise shape and surface features. Gelation occurs in the mold or on the surface of a substrate forming a solid object or a surface coating. The three-dimensional gel network is completely filled with liquid. Aging involves holding the gel in its liquid for several hours at 25–80 °C. This leads to homogenizing and re-precipitation of the solid network, which increases the thickness of the inter-particle necks and the density and strength of the gel. A gel is dried when the physically adsorbed water is completely eliminated from the pores. This requires heating at controlled rates at temperatures of 120–180 °C. Powder or monolithic gels can be obtained after aging and drying steps. Stabilization of a dried gel is necessary to control the stability of the material. Thermal treatment in the range of 500–900 °C removes silanols (Si-OH) and decomposes sub-products, mainly nitrate ions. Stabilization also increases the density, strength and hardness of the gels and converts the gel to a glass with a structure similar to melt-derived glasses. Finally, densification of gel-derived glasses, which is usually accompanied by some degree of crystallization, is completed in the range of 900–1300 °C, leading to glass-ceramic development. Glass-ceramics normally contain a residual glassy phase and one or more embedded crystalline phases. The crystallinity varies between 0.5 and 99.5 percent, most frequently between 30 and 70 percent.[6–8]

2.2 Bioactivity

Bioactive glass-ceramics undergo a particular biological reaction at the interface, which stimulates cell proliferation, gene response and the formation of a bond between living tissue and the material.[9] A common feature of bioactive GCs is that their surface develops a biologically active hydroxycarbonate apatite (HCA) layer that bonds to bone. The HCA phase that forms on bioactive glass-ceramics is chemically and structurally equivalent to the mineral phase of bone and teeth. This similarity is key to interfacial bonding.[9]

The basis of the bone-bonding ability of bioactive glass-ceramics is the chemical surface reactivity of the residual glass and crystalline phases in the body fluid. The surface chemical reactions result in the formation of an HCA layer to which bone can bond. On immersion of a bioactive glass-ceramic in an aqueous body fluid-like solution, three general processes happen: leaching, dissolution and precipitation. These processes are illustrated schematically in Figure 2.2. Leaching is characterized by release of alkali or alkaline earth elements such as Na^+ or Ca^{2+}, which is accelerated by cation exchange with H^+ or H_3O^+ ions (Figure 2.2a). The release of network-modifying ions is rapid for residual bioactive glasses.[9] This ion exchange process leads to an increase in pH to values greater than 7.4. Network

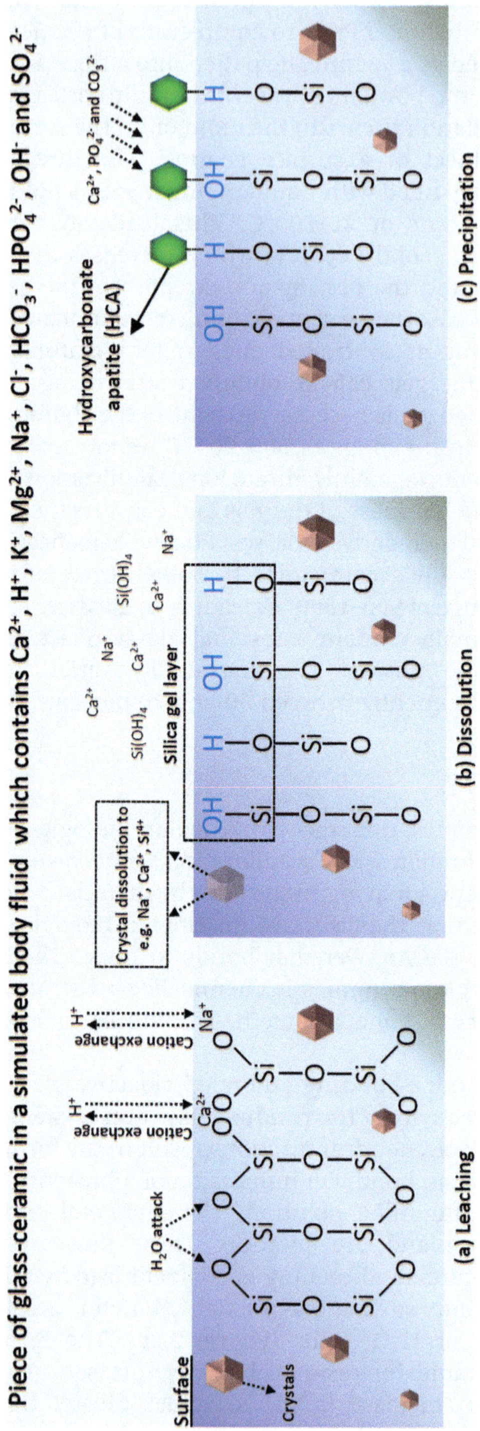

Figure 2.2 Schematic illustration showing the different stages and surface reactions of a bioactive glass-ceramic in a simulated body fluid.

dissolution occurs concurrently, by the breaking of –Si–O–Si–O–Si– bonds through the action of hydroxyl (OH) ions (Figure 2.2b). Breakdown of the network occurs locally and releases silica into solution in the form of silicic acid [$Si(OH)_4$]. The rate of dissolution of silica depends very much on glass-ceramic composition and structure. In some cases, the dissolution of crystals (*e.g.* Ca/Na-silicate crystals) embedded in residual glass gives rise to the release of alkali or alkaline earth ions (Figure 2.2b).[9] The hydrated silica (SiOH) formed on the surface undergoes polycondensation and forms a silica gel layer. In the precipitation stage, calcium and phosphate ions released from the glass-ceramic, together with those from the solution migrate to the surface and form a calcia-phosphate-rich (CaP) layer. The calcium phosphate phase that accumulates in the gel surface is initially amorphous (a-CaP) (Figure 2.2c). It finally crystallizes to an HCA structure by incorporating carbonate anions from solution within the a-CaP phase.[9]

The apatite-forming ability of bioactive glass-ceramics is typically tested *in vitro* before their translation into *in vivo* or clinical trials.[9] *In vitro* bioactivity tests have been widely performed according to the methods suggested by Kokubo and Takadama,[10] the International Standards Organization (ISO)[11] and, more recently, Maçon *et al.*[12] These methods evaluate the bone-bonding ability of a bioactive material by examining the ability of apatite to form on its surface in a simulated body fluid (SBF) with ion concentrations nearly equal to those of human blood plasma.[10–12]

The ability to activate the apatite formation from a supersaturated solution like SBF has been accepted to imply the bioactivity *in vivo*. However, the reader should be warned about the debate proposed by Bohner and Lemaitre[13] and Pan *et al.*[14] that questions the validity of these methods. They believe that the way these experiments are conducted requires major improvements. They have recommended several modifications through using arguments and the available facts.[13,14]

Figure 2.3 shows HCA growth on a glass-ceramic surface. Spherical aggregates of HCA developed on the sample surface after immersion in simulated body fluid. The details of HCA formation could be identified by scanning electron microscopy (SEM) when the rate of HCA growth was reduced for a gel-derived SiO_2–CaO–P_2O_5–ZrO_2 bioactive glass-ceramic.[15]

After positive *in vitro* results of biocompatibility assessments, including investigation of apatite-forming ability, solubility, ion release and cell culture, a bioactive glass-ceramic is considered for final long-term *in vivo* examinations (implanting in a animal's body) prior to approval for the clinical tests in humans.[9]

The level of *in vivo* bioactivity depends on the rate of development of the interfacial bond between an implant and bone. Hench[9] introduced an index of bioactivity as a measure of this. This bioactivity index is given by $I_B = (100/t_{0.5bb})$, where $t_{0.5bb}$ is the time for more than 50% of the surface to be bonded to bone. The I_B varies between 0 and 12. Bioactive materials with high I_B values ($10 < I_B < 12$) show soft tissue bonding.[9]

Figure 2.3 Hydroxycarbonate apatite (HCA) formation on the SiO_2–CaO–P_2O_5–ZrO_2 BGC surface (partially sintered powder) after 24 h exposure to simulated body fluid (SBF).[15]

2.3 Background

The first melt-derived bioactive glass, *Bioglass 45S5*, was discovered by Hench in 1969.[9] This was followed by bioactive gel-derived glasses, also proposed by Hench in 1991.[9] Most recently, bioactive glasses with ordered mesoporosity were designed by Vallet-Regi *et al.*[16] They show both osteoconduction and osteoinduction properties and can be used in a variety of applications, such as bone grafting, scaffolding, drug delivery, coatings and soft tissue engineering.[9,16–18] However, despite their excellent biochemical properties, the major drawbacks of bioactive glasses are their low mechanical strength and low fracture toughness (a bending strength of approximately 70 MPa and K_{Ic} of 0.5 MPa $m^{1/2}$).[19,20] These characteristics restrict their use to very low load-bearing conditions. To improve their mechanical strength, various types of glasses that undergo precipitation of different crystalline phases, known as bioactive glass-ceramics (GCs), have been developed. The most famous and commercially available bioactive GCs are Cerabone®, Biosilicate®, Ceravital® and Bioverit®.

Ceravital® has apatite (CaO–P_2O_5) precipitates in a glass matrix and was first developed by Brömer *et al.* in 1973.[9] Kokubo *et al.*[19,20] developed glass-ceramic A-W in the late 1980s, commercially known as Cerabone®, in which apatite and wollastonite are crystallized. Bioverit® containing apatite and mica crystals was developed by Höland *et al.* in 1985.[21] All these glass-ceramics are composed of an apatite-like crystalline phase and are much less soluble than Bioglass® 45S5. In 1996, Peitl *et al.*[22] developed an apatite-free glass-ceramic based on the crystallization of (a somewhat modified) Bioglass®. They called their product Biosilicate® and demonstrated that controlled crystallization that led to a well-designed microstructure of a base glass could increase its average 4-point bending strength from 75 MPa to 210 MPa. This value is similar to that found for the A-W glass-ceramic (215 MPa), which exhibits the best mechanical performance of all commercial bioactive glass-ceramics.[22,23] Figure 2.4 shows that through the advancement of crystallization in the base glass (Biosilicate®), mechanical strength increases.[24] Crystallization can be considered as a main reason behind the strengthening of bioactive glasses. Crystals embedded in glasses always produce a residual stress field. When they reach a certain critical size, cracks may be spontaneously generated. In the particular case of Biosilicate®, cracks propagate within the residual glass phase and are deflected by

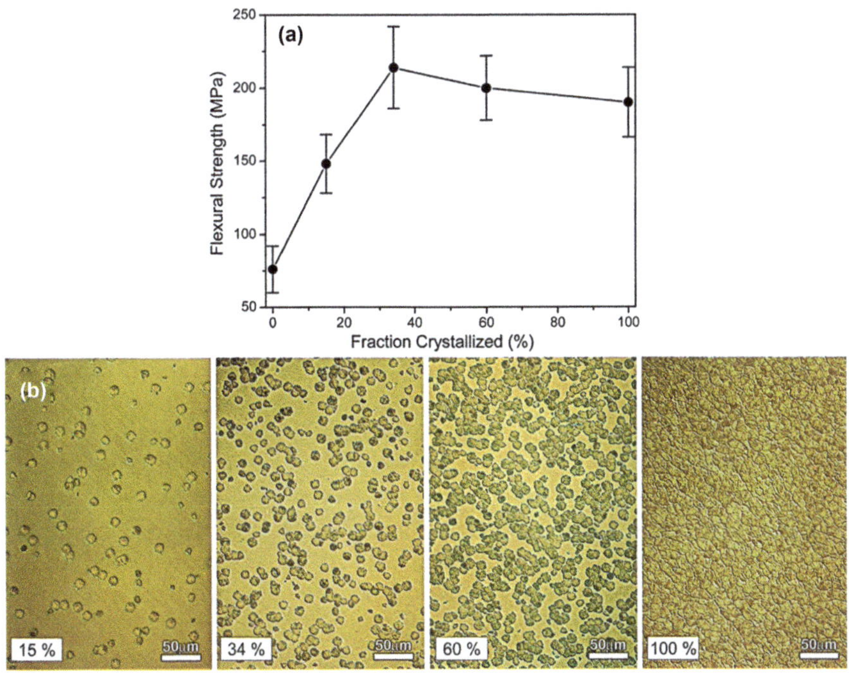

Figure 2.4 (a) 3-point or 4-point bending strength *versus* crystalline volume fraction for the Biosilicate® glass-ceramic with a constant crystal size of 13 μm; (b) Optical micrographs of the corresponding microstructures for 15%, 34%, 60% and fully crystallized samples.
Reprinted from M. C. Crovace *et al.*, Biosilicate® – A multipurpose, highly bioactive glass-ceramic. In vitro, in vivo and clinical trials, *J. Non-Cryst. Solids*, **432**, 90–110, Copyright 2016 with permission from Elsevier.[24]

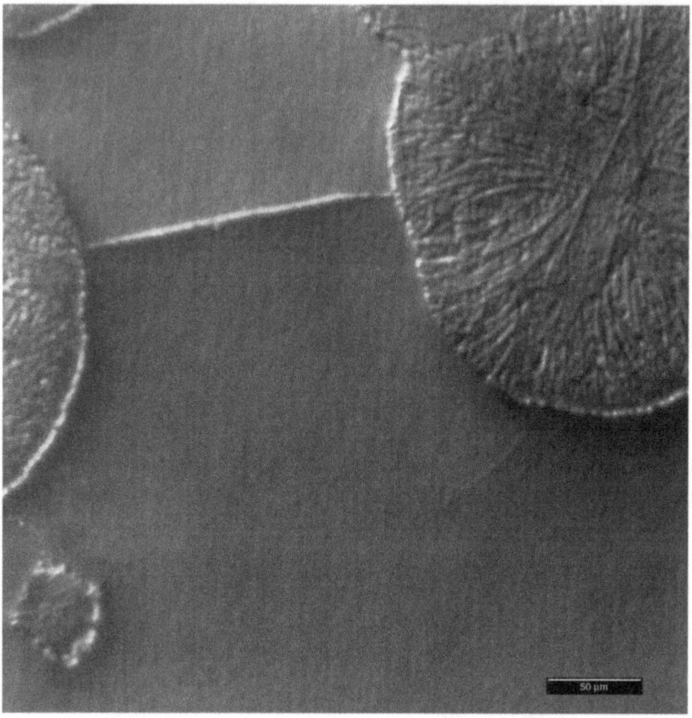

Figure 2.5 Crack deflection by the crystals within the glassy matrix of a bioactive glass-ceramic (Biosilicate®).
Reproduced from ref. 25 with permission from John Wiley and Sons. © 2013 John Wiley & Sons, Inc. and The American Ceramic Society.

the crystals (Figure 2.5), which enhances the toughness and strength of material.[23,24]

In this chapter, we will summarize the properties, applications and fabrication methods of the above-cited bioactive GCs in Section 2.4. In addition to these commercially available bioactive GCs, there are promising GCs based on canasite-apatite, K-fluorrichterite, apatite-mullite, oriented apatite, chlorapatite, calcium pyrophosphate, rhenanite, *etc.*, which will be briefly introduced in Section 2.5. The history and issues associated with these bioactive GCs are also addressed in this section. Magnetic bioactive GCs, radiopaque bioactive GCs, coatings, composites, scaffolds, gel-derived GCs and their relevant open issues will be briefly explained in Sections 2.6–2.11. Finally, possible developments and trends will be highlighted in Section 2.12.

2.4 Commercial Bioactive Glass-ceramics

Commercial bioactive glass-ceramics, such as Cerabone®, Biosilicate®, Ceravital® and Bioverit®, are usually produced *via* the melting method. The base glass compositions are designed in such a way that the desirable crystalline phases are evolved after the controlled heat treatments of nucleation and crystallization. We summarize the inventor, composition and crystalline phases contained in these glass-ceramics in Table 2.1.

A-W glass-ceramic, commercially available with the brand name Cerabone®, is the most extensively and successfully used bioactive glass-ceramic for bone replacement. Some years ago, Kokubo *et al.*[19] envisaged that they should develop a glass-ceramic that contains a reinforcing crystal phase. They selected β-wollastonite ($CaO \cdot SiO_2$) phase for this purpose.[19] Sintering with concurrent crystallization proceeded at 1050 °C to obtain A-W glass-ceramic. β-wollastonite was crystallized in the glass with a fine fibrous and needle-like morphology.[19] This morphology reinforced the glass-ceramic through activating various toughening mechanisms, such as crack deflection, bridging and pull-out.[19] Ceravital® was developed just after Bioglass® 45S5 to improve the mechanical properties.[9] Melting-casting followed by double stage heat treatments at 600–750 °C was employed for Ceravital® fabrication.[9] Biosilicate® was designed to improve mechanical properties of bioactive glasses like Bioglass® 45S5.[24] It was fabricated after controlled double stage heat treatments at 540–700 °C.[24] Bioverit® glass-ceramics were designed to benefit from the machinability of mica crystals and other phosphate phases in the case of Bioverit® III.[3] Nucleation at 650–750 °C and crystallization at 1050–1100 °C were triggered in order to fabricate these glass-ceramics.[3] Therefore, each commercial glass-ceramic has been designed to contain especial crystalline phases with a variety of chemical and mechanical characteristics. The reader is encouraged to refer to the references in Table 2.1 which comprehensively describe the history, processing and properties of these glass-ceramics.

Commercial glass-ceramics were originally developed to overcome one marked weakness of bioactive glasses—brittleness. Therefore, their mechanical properties compared to bone were of great interest for scientists in this community. The mechanical properties of commercial GCs are summarized in Table 2.2.

At a glance, comparison of the fracture toughness of commercial BGCs (0.6–2.1 MPa $m^{1/2}$) and cortical bone (2–12 MPa $m^{1/2}$) reveals that this property still needs to be significantly improved. The modulus of elasticity (45–150 GPa) is also higher than the desirable value of cortical bone (7–30 GPa). Therefore, it seems that current commercial BGCs are prone to fracture or causing stress shielding in high load-bearing conditions. However, at present, bioactive GCs are in several unique clinical applications which do not demand high fracture toughness.[9] Ceravital® was a short-lived bioactive glass-ceramic used for only a few years as middle ear prostheses.[9] Cerabone® has been used to fabricate various types of prostheses and spacers used in spinal and tumor surgeries.[19] In particular, lumbar intervertebral spacers and cervical laminoplastic spacers have been used in thousands of clinical cases with excellent clinical, radiological and biomechanical results.[19]

Positive *in vitro* results make Biosilicate® a potential candidate for applications in which bone bonding or bone regeneration is desired.[24]

Table 2.1 Glass composition, inventors, and crystalline phases of commercial bioactive glass-ceramics.

Bioactive glass-ceramic	Inventors	Glass composition (wt%)									Crystalline phases	Ref.
		SiO_2	P_2O_5	CaO	MgO	Na_2O	K_2O	Al_2O_3	F	TiO_2/ZrO_2 *etc.*		
Ceravital	Brömer *et al.*	40–50	10–15	30–35	2.5–5	5–10	0.5–3	—	—	—	Apatite ($Ca_{10}(PO_4)_6(OH)_2$)	9
Cerabone	Kokubo *et al.*	34.2	6.2	44.7	4.6	—	—	—	1	—	Apatite ($Ca_{10}(PO_4)_6(OH)_2$) and wollastonite ($CaSiO_3$)	19
Biosilicate	Zanotto *et al.*	48.5	4	23.75	—	23.75	—	—	—	—	$Na_2CaSi_2O_6$ and $NaCaPO_4$	24
Bioverit I	Höland *et al.*	29–50	8–18	13–28	6–28	Na_2O/K_2O 5.5–9.5		0–19.5	2–7	Minor	Mica ($KMg_3AlSi_3O_{10}F_2$), and fluorapatite ($Ca_{10}(PO_4)_6(F)_2$)	3
Bioverit II	Höland *et al.*	43–50	0.1–5	0.1–3	11–15	Na_2O/K_2O 7–10.5		26–30	3.3–4.8	—	Mica ($(K,Na)Mg_3AlSi_3O_{10}F_2$)	3
Bioverit III	Höland *et al.*	—	45–55	13–19	—	11–18	—	6–18	—	1.5–10	Apatite, $AlPO_4$, and complex phosphates like $Na_5Ca_2Al(PO_4)_4$ and $Na_{27}Ca_3Al_5(P_2O_7)_{12}$	3

Table 2.2 Mechanical properties and bioactivity index of commercial bioactive glass-ceramics and cortical bone.[3,9,19,24]

Bioactive glass-ceramics	Compressive strength (MPa)	Bending strength (MPa)	Fracture toughness (MPa $m^{1/2}$)	Young's modulus (GPa)	Vickers hardness (HV GPa)	Bioactivity index (I_B)
Ceravital	500	100–150	—	150	—	5.6
Cerabone	1080	215	2.0	118	680 (HV)	6
Biosilicate	—	210	1.0	60–80	—	>8
Bioverit I	500	140–180	1.2–2.1	70–88	5 (GPa)	<8
Bioverit II	450	90–140	1.2–1.8	70	8 (GPa)	—
Bioverit III	—	60–90	0.6	45	—	—
Cortical bone	100–200	50–150	2–12	5–20	—	—

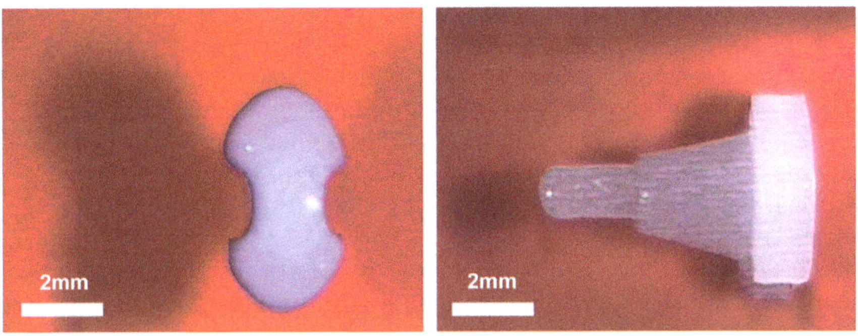

Figure 2.6 Biosilicate® prosthesis (Otosilicate).
Reprinted from M. C. Crovace *et al.*, Biosilicate® – A multipurpose, highly bioactive glass-ceramic. In vitro, in vivo and clinical trials, *J. Non-Cryst. Solids*, **432**, 90–110, Copyright 2016 with permission from Elsevier.[24]

After 20 years of research, Biosilicate® was evaluated in different situations and by various *in vitro*, *in vivo* and clinical tests.[24] In one of the first clinical tests, Biosilicate® was used to occlude dentin tubules, rapidly reacting with saliva and the surrounding tissue and successfully eliminating the cause of dentin hypersensitivity. In this clinical study, 142 patients received different desensitizing treatments, Sensodyne®, SensiKill® or Biosilicate® dispersed in a gel suspension or in a solution with distilled water. Over a period of 6 months, 232 teeth were evaluated, and the study was performed using pain assessments (patients used a visual analogue scale of pain, VAS, from 1 to 10). Regarding the global diminution of pain over the course of the study, Biosilicate® mixed with distilled water displayed the greatest effect, and could alleviate the pain in the very first periods of the experiment, followed by SensiKill®, Sensodyne® and Biosilicate® dispersed in gel.[26] After the *in vivo* tests in guinea pigs, Biosilicate® was compared with remodeled anvil ossiculoplasty (total and partial prostheses, Figure 2.6) using audiometric tests three months after surgery.[26]

14 partial ossicular replacement prostheses (PORP), 7 total ossicular replacement prostheses (TORP) and 8 bone PORP were performed using Biosilicate®. The results indicated that there was an improvement in the air-bone gap (≤20 dB) in all of the study groups, and all of the groups were statistically significant ($P \leq 0.001$) with 50% in PORP Biosilicate®, 29% in TORP Biosilicate® and 50% in bone PORP. It was concluded that an Otosilicate (the name given to these special shapes of monolithic Biosilicate® pieces) prosthesis is an effective substituent for ossicles, not only for its biological properties but also for its machinability.[27] Recently, after successful *in vivo* tests on rabbits, twelve patients who had undergone enucleation or evisceration procedures were implanted with Biosilicate® orbitals.[24] According to a preliminary report by the surgical team, none of the patients showed postoperative complications or signs of inflammation, dehiscence or extrusion. The laboratory tests of patients who have completed the study have revealed no changes in vital organs. They were also followed up by computed tomography examinations, which showed no migration, formation of abscesses or inflammation around the implants.[24]

The Bioverit® series was also successfully applied in different orthopedic surgeries (especially as different types of spacers), head and neck surgery (especially middle ear implants), stomatology (especially tooth root and veneer laminates).[28,29]

As overviewed above, commercial bioactive GCs still suffer from relatively poor mechanical properties, specifically fracture toughness. Therefore, numerous attempts have been made to overcome this drawback leading to the development of some promising bioactive GCs which will be addressed in the next section.

2.5 Miscellaneous Bioactive Glass-ceramics

In addition to those well known bioactive GCs, other bioactive GCs have been proposed for improving different properties, such as strength or bioactivity. Table 2.3 shows the bioactive GCs suggested as bone substitute. We will also explain the history and issues regarding these bioactive GCs in the next few paragraphs.

In the mid-1990s, R.G. Hill *et al.*[30] introduced apatite–mullite glass-ceramic as a potential bioactive glass-ceramic for bone replacement (Table 2.3). They have reported fracture toughness values of 1.0 to 3.3 MPa $m^{1/2}$, a bending strength of 90–330 MPa and a Young's modulus of 70–90 GPa.[30–32] Later, Goodridge *et al.*[33] reported that some apatite–mullite glass-ceramics did not form HCA *in vitro* after immersion in SBF. There were no pH changes following immersion and no significant ion release into the SBF solution. However, they show osseointegration *in vivo* with no sign of fibrous capsule formation. On the contrary, some glass-ceramics are degradable and form HCA in SBF but do not osseointegrate *in vivo*. The *in vitro* and *in vivo* responses of this glass-ceramic depend on the presence of the crystalline and residual glass phases. The glass-ceramic which crystallized to apatite

Table 2.3 Some bioactive glass-ceramics proposed by numerous researchers for future applications.

Bioactive glass-ceramic	Glass compositions (wt%)									Crystalline phases	Ref.
	SiO_2	P_2O_5	CaO	MgO	Na_2O	K_2O	CaF_2	Al_2O_3	Others		
Apatite–mullite glass-cermaic	25.2	19.8	15.7	—	—	—	10.9	28.5	—	Needle-like fluorapatite ($Ca_{10}(PO_4)_6(F)_2$) and mullite ($Si_2Al_6O_{11}$)	30–34
Canasite–apatite glass-ceramic	24–70	2–13	20–30	4.6	6–12	3–10	F-3–11	$Al_2O_3/B_2O_3/ZrO_2$ <6 wt.%		Fluorapatite and fluorocanasite ($Ca_5Na_4K_2Si_{12}O_{30}F_4$)	35–37
K-fluorrichterite glass-ceramic	55.5	—	0.0	19.8	3.6	5.4	15.7	—	—	K-fluorrichterite ($KNaCaMg_5Si_8O_{22}F_2$), fluorite (CaF_2) and diopside ($CaMgSi_2O_6$)	38–42
	56.6	—	4.9	20.2	3.6	5.5	9.1	—	—		
Rhenanite glass-ceramic	29–70	2–23	5.5–23	—	6–27	—	F 0–1.5	<0.01–0.1	—	Rhenanite (β-$NaCaPO_4$)	43, 44
Calcium pyrophosphate glass-ceramic	42.2	49.9	—	—	5.1	—	—	—	TiO_2 2.8	β-$Ca_3(PO_4)_2$ (β-tricalcium phosphate) and β-$Ca_2P_2O_7$ (β-calcium pyrophosphate)	45–49
Chlorapatite glass-ceramic	25–35	9–13	34–47	—	—	—	0.0–15		$CaCl_2$ 1.9–31.4	Mixed chlor/fluor-apatite ($Ca_{10}(PO_4)_6(F,Cl)_2$) and wollastonite ($CaSiO_3$)	50
Diopside–apatite glass-ceramic	33.3	17.95	36.81	11.17	—	—	0.77	—	—	Diopside ($CaMgSi_2O_6$) and fluorapatite ($Ca_{10}(PO_4)_6(F)_2$)	51–53

but had not crystallized to mullite osseointegrated poorly. The sample crystallized to both apatite and mullite exhibited better osseointegration and evidence of osteoconduction. Increasing the apatite volume fraction in the glass appears to increase the osseointegration ability. The residual glass phase may add to osseointegration by promoting apatite formation or may hinder osseointegration if the glass phase degrades and releases ions such as Al^{3+}, which are known to inhibit biological mineralization. In the case of aluminum-containing glass-ceramic compositions, it is important that all the aluminum ions be either locked away in a chemically inert glass phase or in a chemically inert crystalline phase (even trace amounts of the order of 1 ppm of aluminum are known to inhibit the mineralization of a newly forming osteoid). There is a belief that the SBF test proposed by Kokubo is applicable for bioactive glasses and A-W glass-ceramics, but its validity with regard to aluminum-containing glasses and glass-ceramics should be questioned.[30] It seems that extensive research is still being conducted in the UK and Ireland on these glass-ceramics as bioactive bone substitutes or implants.[30–34]

In 1994, Wolcott[35] managed to develop apatite–canasite glass-ceramic (Table 2.3). Apatite–canasite glass-ceramics have high fracture toughness, up to 3.9 MPa $m^{1/2}$, combined with *in vitro* bioactivity. Therefore, the material has been suggested for use as a biomaterial for replacing bone tissue.[35] In 2002, Miller *et al.*[36] and Barros *et al.*[37] investigated the *in vitro* and *in vivo* properties of canasite-based glass-ceramic. Miller *et al.*[36] showed that a crystalline apatite layer was formed on the surface of a P_2O_5-containing canasite after 5 days and after 3 days for calcium-enriched canasite.[36] However, *in vivo* bone tissue response to canasite glass-ceramic evaluated by Barros *et al.*[37] revealed that the canasite formulation is not osteoconductive and degrades in the biological environment. They therefore concluded that the canasite formulation was inappropriate for use as implants.[36] Further work is required to improve the biocompatibility of these materials with bone tissue. This could be achieved by reducing the solubility of the glass and glass-ceramic[36] because there should be an optimum level of solubility. Highly soluble materials will not remain in the body after implantation.

Later, Mirsaneh *et al.*[38] at Sheffield University, tried to develop new glass-ceramics based on another chain silicate mineral, K-fluorrichterite (Table 2.3). In view of their excellent mechanical properties, MOR = 217–227 MPa and $K_{Ic} = 2.1–2.7$ MPa $m^{1/2}$, these glass-ceramics were considered potential candidates for the fabrication of custom medical devices in restorative dentistry and moderate load-bearing reconstructive bone surgery.[38] Bhakta *et al.*[39–41] investigated some *in vitro* and *in vivo* characteristics of K-fluorrichterite glass-ceramics. They suggested that an apatite-forming ability of the materials (seen in SBF tests) is responsible for osteoconductivity.[39] Some promising compositions also showed direct bone tissue contact with evidence of new bone formation *in vivo* and cell proliferation along the implant surface into the bone space. There was no evidence of bone necrosis or fibrous tissue encapsulation around the test specimens. Therefore, this glass-ceramic has greatest promise as a bone substitute for load-bearing conditions due to its osteoconductive potential and good mechanical properties, but further investigations, *e.g.* clinical tests, are still required.[40–42]

Apel, Höland and Rheinberger[43] have registered a patent for a new bioactive glass-ceramic, rhenanite (Table 2.3). For this glass-ceramic, residual glass phase contributes to quick apatite formation due to its high solubility in SBF, whereas the osteoconductive rhenanite crystal phase leads to a slower apatite formation but guarantees the improved mechanical stability of an implant. The solubility of this particular glass-ceramic is lower than that of known bioactive glasses. Its reduced solubility prevents it from being transported away from the implantation site before new bone can form. As a result of the variation of the crystalline proportion of rhenanite in the glass-ceramic, control of the kinetics of apatite formation is possible, and glass-ceramics with 4 to 50 wt% rhenanite are preferred.[43,44]

In 1998, Kasuga *et al.*[45] developed bioactive calcium phosphate glasses and glass-ceramics in the pyrophosphate region (Table 2.3). First, bone-like apatite particles were detected on the phosphate invert $60CaO–30P_2O_5–10TiO_2$ glass.[45,46] Later, Kasuga *et al.* included Na_2O in the glass composition to enhance its bioactivitiy and sinterability. Apatite formed on the $60CaO–30P_2O_5–3TiO_2–7Na_2O$ sintered glass-ceramic after soaking in SBF. This glass-ceramic shows moderate mechanical properties (MOR = 160 MPa and $K_{Ic} = 1.8$ MPa $m^{1/2}$) and Kasuga believes that the main advantage of his material over HA or β-TCP bioceramics is that it can be easily machined using conventional tools rather than diamond tips. The machinablity is suggested to result from the interlocking plate-like microstructure of β-calcium pyrophosphate crystals dispersed in the glassy matrix phase.[45,46] It seems that experiments to prepare biomedical material devices for dental and plastic surgery, and coating using these calcium pyrophosphate glass-ceramics are still in progress.[46–48] For example, rod-shaped, glass-ceramic coated Ti–29Nb–13Ta–4.6Zr implants of 5 mm diameter and 10 mm length have been implanted into the femurs of Japanese rabbits. Contact microradiography (CMR) microscopic images showed new bone formation around the sample 4 weeks after implantation, and the bone tissue showed direct contact with the implants.[45]

More recently, Chen and Hill[50] reported on the development and characterization of new bioactive glass-ceramics based on chlorapatite (Table 2.3). They believed that chlorapatite glass-ceramic would be more interesting than fluorapatite for medical and dental applications because it is more soluble than fluorapatite and will convert completely to hydroxyapatite in the presence of water. They concluded that the chloride-containing glasses are very reactive and appear to react with atmospheric water. Therefore, further investigation is required to develop practical chlorapatite and mixed chlor/fluor-apatite glass-ceramics, and to study their *in vitro* and *in vivo* biocompatibility.

In recent years, Prof. J. M. F. Ferreira's group in the University of Aveiro has tried to design and develop an alkali metal-free series of bioactive glasses and glass-ceramics within the $CaO–MgO–SiO_2–P_2O_5–CaF_2$ system, alongside the diopside ($CaMgSi_2O_6$)–fluorapatite ($Ca_5(PO_4)_3F$)–tricalcium phosphate ($3CaO.P_2O_5$) composition (Table 2.3).[51–53] They were successful in sintering their glasses to an appropriate density with good bending strength. They believe that the unfavorable effects of alkali metal ions on the sintering and crystallization behaviors of glasses can be avoided by choosing alkali metal-free compositions. Additionally, crystallization of a chain silicate mineral like diopside with elongated and interlocking morphology would contribute to the enhancement of mechanical properties. It seems that their materials have the proper structure and crystalline phases (diopside and hydroxyapatite) to trigger bioactivity and improve mechanical properties. The preliminary *in vitro* tests and mechanical strength measurements are promising, but *in vivo* experiments need to be performed.[51–53]

2.6 Magnetic Bioactive Glass-ceramics

Hyperthermia is a method for destroying cancer cells by heating them, normally between 41–46 °C. Ferro- or ferrimagnetic materials locally generate heat by hysteresis loss under an alternating magnetic field, damaging only tumor cells and not healthy ones.[9] The power loss produced by these magnetic materials is dissipated in the form of heat. Ferrimagnetic and bioactive glass-ceramics are good candidates as thermoseeds for hyperthermia of cancer.[9] These glass-ceramic powders are usually included in polymer-based bone cements.[9] Luderer *et al.*[54] first reported a non-bioactive glass-ceramic system with lithium ferrite ($LiFe_5O_8$) and magnetite (Fe_3O_4) in an $Fe_2O_3–P_2O_5–Li_2O–SiO_2–Al_2O_3$ glassy matrix. Then, Ebisawa *et al.*[55] prepared another non-bioactive glass-ceramic with Fe_3O_4 in a matrix of $FeO–Fe_2O_3–CaO–SiO_2$ glass. In their next study, their new glass-ceramics with 3 wt% addition of B_2O_3 and P_2O_5 were found to show bioactivity.[56,57] More magnetic glass-ceramics with bioactivity have been reported since then. Bretcanu *et al.*[58] prepared a bioactive glass-ceramic in the system $SiO_2–Na_2O–CaO–P_2O_5–FeO–Fe_2O_3$ (wt%). Singh *et al.*[59,60] and Shah *et al.*[61] synthesized the system $ZnO–Fe_2O_3–CaO–SiO_2–P_2O_5–Na_2O$ containing zinc ferrite. Hsi *et al.*[62] reported the system $Li_2O–MnO_2–CaO–P_2O_5–SiO_2–Fe_2O_3$ containing a (Li, Mn) ferrite phase. Wang *et al.*[63] prepared a degradable glass-ceramic in the $Na_2O–CaO–P_2O_5–SiO_2–Fe_2O_3$ system. Li *et al.*[64] managed to develop magnetic bioactive glass-ceramics through precipitation of magnesium ferrite and Mn–Zn ferrite in A-W-based glass, and a good magnetic property and bioactivity were observed. In their later study, a magnetic bioactive glass-ceramic was synthesized by doping MnO_2 and Fe_2O_3 in the same glass.[65] Then, Leenakul *et al.*[66–68] suggested two new glass compositions of $SrFe_{12}O_{19}–SiO_2–CaO–Na_2O–P_2O_5$ and

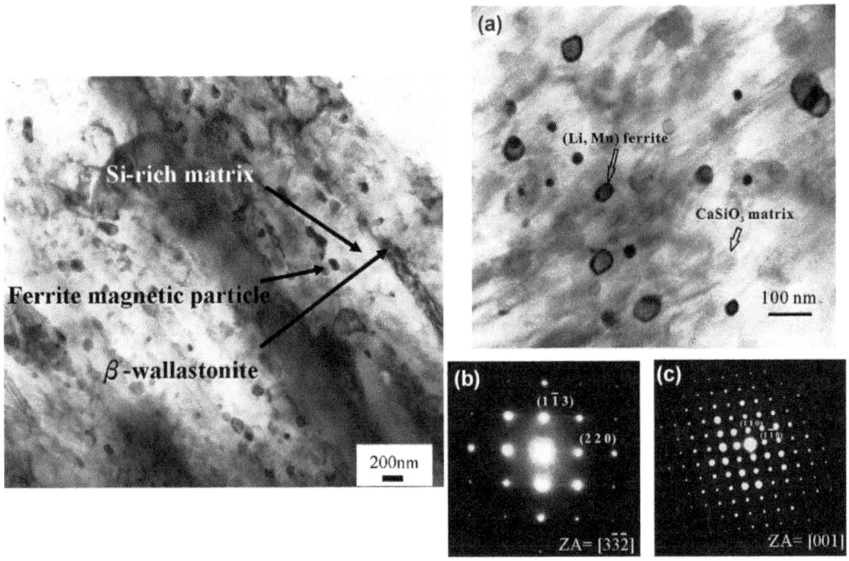

Figure 2.7 TEM micrograph of the $25Li_2O$–$8MnO_2$–$20CaO$–$2P_2O_5$–$45SiO_2$–$4Fe_2O_3$ (at %) glass crystallized at 850 °C for 4 h. (a) Bright field images and (b) SAED pattern of the (Li,Mn) ferrite with the $[3\bar{3}\bar{2}]$ zone axis and (c) β-wollastonite with the [001] zone axis.
Reprinted from C.-S. Hsi *et al.*, Crystallization kinetics and magnetic properties of iron oxide contained $25Li_2O$-$8MnO_2$-$20CaO$-$2P_2O_5$-$45SiO_2$ glasses, *J. Eur. Ceram. Soc.*, **27**, 3171–3176, Copyright 2007 with permission from Elsevier.[62]

$BaFe_{12}O_{19}$–SiO_2–CaO–Na_2O–P_2O_5 in which Fe_3O_4, $BaFe_{12}O_{19}$, $SrFe_{12}O_{19}$ and sodium–calcium silicate/phosphate are crystallized.

The above parent glasses are prepared by the melting route. After controlled heat treatment of the parent glasses, the magnetic phases mentioned above are crystallized in the bioactive residual glass.[54–68] For example, the transmission electron microscope (TEM) image in Figure 2.7 shows ferrite nano-crystals surrounded by the residual glass and a β-wollastonite phase.[62]

Generally, iron should remain in the crystal structure, because otherwise it would diminish the apatite-forming ability of the residual glass. In order to satisfy this criterion, biphasic materials containing a melt-derived magnetic glass-ceramic ($45SiO_2$–$45CaO$–$10Fe_2O_3$ in mol%) and a gel-derived bioactive glass ($58SiO_2$–$6P_2O_5$–$36CaO$ in mol%) were suggested by Vallet-Regí's team.[69–71] The gel-glass gives rise to porosity, which plays a fundamental role in the promotion of the apatite-forming ability on the surface.[69] On the other hand, as the gel-glass content increases, the magnetic properties change due to the diffusion of Fe ions to the glassy phases of the biphasic materials. Additionally, the biphasic nature of these materials allows proper changing of both properties, depending on the requirements of the patient.[69–71] It thus seems that further *in vivo* and clinical tests in this interesting field are required.

2.7 Radiopaque Bioactive Glass-ceramics

Radiopacity is an essential feature of bioactive bone cements that allows easy follow-up of the treated patient under radiographic control. Recently, Tallia *et al.*,[72] Vitale-Brovarone *et al.*[73] and Montazerian *et al.*[74] developed ZrO_2-containing bioactive mesoporous glasses and glass-ceramics *via* a simple sol–gel method. These researchers incorporated zirconium (a radiopaque element) into the structure of a mesoporous bioactive glass. They avoided melting at high temperature by synthesizing SiO_2–CaO–P_2O_5–ZrO_2 glass powders *via* the sol–gel method. These glass and derivative glass-ceramic powders are mesoporous, bioactive, biocompatible and encourage osteoblast cell proliferation.[72–74]

2.8 Bioactive Glass-ceramic Coatings

The majority of the commercial bioactive glass-ceramics and some promising ones already discussed in Sections 2.4 and 2.5 (Tables 2.1 and 2.3) form a mechanically strong bond with bone. The bond's strength is proportional to, or even greater than, that of the bone. In this manner, bioactive glass-ceramics usually have outstanding biochemical similarity. However, their fracture toughness is less than, and their elastic modulus is greater than, that of bone, showing that most bioactive glass-ceramics have a problematic biomechanical similarity when employed as a load-bearing part of the body. Two approaches examined to overcome this problem were the development of bioactive glass-ceramic composites and the development of coatings, which are briefly addressed in the next sections.[75]

Bioactive glass-ceramic coatings on a mechanically tough substrate can help to reduce the mechanical limitations in load-bearing situations. Good examples of research to develop bioactive glass-ceramic coatings on metallic alloy, alumina and zirconia implants were reported by Verné and her co-workers.[76–78] She has also recently reviewed properties and challenging issues related to bioactive glass-ceramic-coated implants.[79]

Verné *et al.*[76] prepared double-layer bioactive glass-ceramic coatings on Ti–6Al–4V substrates by dip-coating and firing. A glass based on $61.1SiO_2$–$12.6CaO$–$10.3Na_2O$–$7.2MgO$–$6.0P_2O_5$–$2.8K_2O$ (wt%) was used as the first layer in direct contact with the metallic substrate, and a glass-ceramic based on $26.2SiO_2$–$17.9Al_2O_3$–$17.5P_2O_5$–$10.5K_2O$–$19.6CaO$–$8.3F^-$ (wt%) was used as the outer bioactive layer. Their design allowed the low-melting intermediate glass composition to work below the $\alpha \rightarrow \beta$ transformation of the Ti alloy and ensured the adhesion of the outer bioactive layer. The intermediate layer was also chosen to match the thermal expansion coefficient of both the metal and the outer glass-ceramic. The process used for the coating did not affect the bioactivity of the glass-ceramic surface and did not modify either the nature of the crystalline phases or their shape (needle-like fluorapatite). Their method has been recommended to coat Ti–6Al–4V screws for dental applications.[76]

Verné *et al.*[77] also managed to coat pure and dense Al_2O_3 substrates with a similar glass-ceramic. Through a careful optimization of the coating process, they obtained coatings approximately 200 μm thick. The results revealed that to obtain adherent and low-porosity coatings on alumina substrates, the above glass-ceramic should be used, with very short thermal treatments at a temperature slightly above 1300 °C. Lower temperatures did not guarantee a satisfactory degree of adherence between the substrate and the coating and moreover gave the glass-ceramic coating an excessively high degree of porosity. By optimizing the time and temperature schedule, they could minimize the alumina diffusion through the glass-ceramic. A further thermal treatment at 1200 °C was required to promote the needle-like fluorapatite crystallization.[77]

Ferraris *et al.*[78] reported that glazing can be successfully used to coat zirconia by bioactive glass-ceramics. An intermediate layer often forms between the coating and the substrate, for example, between a glass coating and a zirconia substrate, there is a "composite" layer made of glassy phase and zirconia particles. During the thermal treatment above its liquidus, the glass diffuses within the zirconia substrate; hence, the zirconia granules are surrounded by a glassy matrix, leading to the formation of a "composite" layer, which assures the continuity of the thermal and mechanical properties from the zirconia substrate to the glass coating. The osseointegration of bioactive glass-coated zirconia cylinders has been evaluated in an animal model and compared to uncoated cylinders.[78] After 30 days, the bone bonding was better than that of the uncoated cylinders, but after 60 days, the difference was within statistical uncertainty.[78]

More recently, Marghussian[80] classified nano-structured bioactive glass-ceramic coatings for biomedical applications. He concluded that *"all the coatings that are produced as amorphous or glassy layers in the first stage of their fabrication process, and subsequently are subjected to a controlled heat treatment, in order to transform them into partially crystalline products, can structurally be classified as glass-ceramic coatings"*.[80] As an example of the development of this type of nano-structured coating, Xiao *et al.*[81] attempted to deposit bioactive glass-ceramic coatings on titanium substrates by the liquid precursor plasma spraying (LPPS) process. Tetraethyl orthosilicate, triethyl phosphate, calcium nitrate and sodium nitrate solutions were mixed together to form a sol after hydrolysis, and the liquid suspension was used as the feedstock for the plasma spraying of P_2O_5–Na_2O–CaO–SiO_2 bioactive glass-ceramic coatings. Bioactive glass-ceramic coatings with nano-structures were successfully synthesized, and the coatings showed rapid formation of a nano-structured HCA layer after being soaked in SBF. Overall, their results indicate that the LPPS process is an effective and simple method to synthesize nano-structured bioactive glass-ceramic coatings with good *in vitro* bioactivity.[81] Wang *et al.*[82] utilized the same technique to coat a Ti-alloy with nano-structured bioactive hardystonite ($Ca_2ZnSi_2O_7$) and sphene ($CaTiSiO_5$) glass-ceramic.

2.9 Bioactive Glass-ceramic Composites

Materials with high fracture toughness, such as metallic or ceramic phases, are utilized to reinforce bioactive glasses or glass-ceramics. The reinforced bioactive glass-ceramics are usually improved in both the bending strength and the fracture toughness. For example, Kasuga *et al.*[83,84] reinforced well known A-W glass-ceramic with zirconia particles, performed *in vivo* tests, and reported that by using a controlled crystallized glass powder and zirconia powder as raw materials, a sintered body densified to near its theoretical density by hot isostatic pressing (HIP). This bioceramic exhibited an extremely high bending strength of 400 to 1000 MPa and a fracture toughness of 3 to 5 MPa $m^{1/2}$ for 30 to 80 vol% of zirconia.[83,84] Bioverti® I and III have been also reinforced by Ti particles.[85–87] Mechanical property measurements revealed the toughening benefits of Ti addition and no obvious adverse effects on *in vitro* biological properties.[84–86] Montazerian *et al.*[88,89] also employed zirconia to reinforce mica and mica–apatite glass-ceramics, and both the bending strength and the fracture toughness were improved. The addition of 15 wt% ZrO_2 to the glass-ceramics increased the 3-point bending strength from 50 MPa to 130 MPa, and the fracture toughness was doubled from 0.7 MPa $m^{1/2}$ to 1.4 MPa $m^{1/2}$.[88,89] The SEM image of the mica glass-ceramic reinforced with 10 wt% ZrO_2 is illustrated in Figure 2.8(a). The microstructure of the glass-ceramic contains plate-like mica crystals with conventional interlocking structure. Both fine dispersed zirconia particles (<1 μm) and agglomerated ones are observed.[88,89] The fracture surface of the glass-ceramic is also shown in Figure 2.8(b). As can be clearly observed, the fractured surface is rough. Extensive cleavage planes of mica crystals around

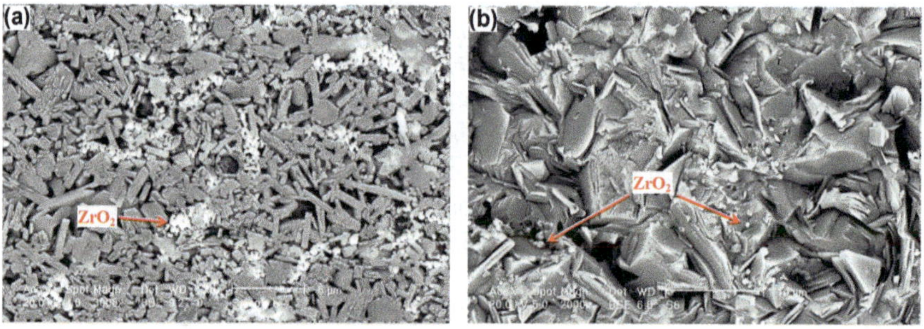

Figure 2.8 (a) Microstructure of the mica glass-ceramic reinforced with 10 wt% fine (<1 μm) zirconia particles (etched with 5 wt% HF solution for 45 s). (b) Fracture surface of the composite. (a) Reproduced from M. Montazerian *et al.*, Pressureless sintering and mechanical properties of mica glass-ceramic/Y-PSZ composite, *J. Eur. Ceram. Soc.*, **28**, 2687–2692, Copyright 2008 with permission from Elsevier[88] and (b) Reprinted from Journal of the European Ceramic Society, 28, Processing and properties of a mica-apatite glass-ceramic reinforced with Y-PSZ particles, 2693–2699, Copyright 2008 with permission from Elsevier.[89]

the fine zirconia particles shows that this fracture surface results from many deflections of the propagating cracks. It seems that crack deflection by zirconia crystals is the main toughening mechanism in the composite.[88,89]

Verné *et al.*[90] have also employed monoclinic and tetragonal ZrO_2 particles to reinforce mica-based glass-ceramic (Bioverit® I) through the viscous flow sintering route. They have reported dissolution and agglomeration of the zirconia particles within the matrix as the major drawbacks of this method.[90]

Rawlings[91] reviewed early researches related to metal-reinforced bioactive glasses and glass-ceramics in 1993, when there was a growing interest in this topic. He reported on the processing and properties of glass-ceramic matrix composites reinforced with silver, titanium, Co–Cr alloy, stainless steel, and aluminum. He showed that the strength of the composites was 26–47% greater than that of the monolithic matrix.[91] The *in vitro* and *in vivo* characteristics of such composites have also been investigated, *e.g.*, Ag- and Ti-reinforced bioactive glass-ceramics showed no adverse effects on the osteointegration or biocompatibility of the composite.[92–94] However, further *in vivo* tests are required to accept these composites for the future clinical applications. Additionally, one disadvantage of glass-ceramics reinforced with ceramic or metallic phases is that they are generally denser; they all have elastic moduli much greater than bone and thus may cause stress shielding of a bone under stress.

It has been demonstrated that bio-polymers have a closer modulus of elasticity to bone than ceramics and metals. Therefore, stress shielding effects diminish when these polymers are used in the vicinity of bone. However, the lack of mechanical strength and bioactivity are well known drawbacks of polymers that limit their applications for load-bearing conditions. Therefore, numerous attempts have been made to develop a bio-polymer matrix composite reinforced with bioactive glass-ceramics.[95]

It has been discussed that Bioglass® is highly bioactive, but its mechanical properties are poor. In contrast, *e.g.*, Cerabone® and Biosilicate® have much better mechanical properties while still possessing high bioactivity. Particulate Cerabone® can be used as a stiffer reinforcement in the composite while still providing the composite with a much higher bioactivity than hydroxyapatite (HA) particles.[95–97] Therefore, the processing technology established for HA/polyethylene composites (also known as HAPEX® since 1995, when Smith & Nephew Richards Inc. introduced their series of middle ear implants made of the composite) has also been used for producing the Cerabone®/polyethylene composite.[95–97] Young's modulus and the microhardness of the composite also increased with an increase in glass-ceramic volume fraction while the tensile strength and fracture strain decreased. Even with 40 vol% of glass-ceramic particles, the composite exhibited considerable ductility.[95–97]

Bioactive glass-ceramics have also been loaded in polymer-based bone fillers, glues or cements in order to stimulate bioactivity, improve mechanical properties or even induce especial features like ferro- or ferrimagnetic properties.[98–102] For example, injectable composites based on Biosilicate® and alginate (a biocompatible natural polysaccharide) have been developed by Gabbai-Armelin *et al.* as a bone cements.[98] They envisaged that Biosilicate® would induce bioactivity and mechanical strengthening, while alginate promotes flowability of the composite in bone defects with irregular and complex shapes.[98] Based on the results of the their study, injectable Biosilicate®/alginate formulations seem suitable for bone regenerative applications, for which future studies should aim at biological evaluation in animal experimental models.[98] Bioactive glass-ceramic powders have been added to PMMA-based (poly(methyl methacrylate)) bone cements.[99,100] The results are promising and bioactive glass-ceramics are good reinforcing agents and the composite can be considered for future *in vivo* tests.[99,100] Moreover, ferrimagnetic bioactive SiO_2–Na_2O–CaO–P_2O_5–FeO–Fe_2O_3 glass-ceramics, which contain magnetite (Fe_3O_4), have been loaded in commercial PMMA matrix (Palamed®) for hyperthermia of cancer.[101,102] The interesting research field of hyperthermia of bone cancer demonstrates promising results and is being pursued in Italy by Verné's team.[101,102]

Despite the enormous numbers of studies on bio glass/polymer composites, there are few studies related to bio glass-ceramic/polymer composites.[103] It seems that, in most cases, the conversion of glasses to glass-ceramics reduces the bioactivity as an essential role of these reinforcements. However, the highly bioactive glass-ceramics cited above can be successfully used to develop polymer-based composite or new hybrid materials.

2.10 Bioactive Glass-ceramic Scaffolds

Scaffolds should possess 3D interconnected porosity to support vascularization to encourage cells to grow into the required physical structure. An interconnected porosity along with a pore diameter of at least 100 μm is required for the proper vascularization of the tissue. For making scaffolds, bioactive glasses may be sintered to a specific density and shape, which leads to the nucleation and growth of crystalline phases embedded in a residual glass matrix. These crystallized phases must not induce any cytotoxic effect or hinder any bioactive process inside the cell/tissue. An advantage of crystallization is that it improves the mechanical properties.[104]

A comprehensive review of the state of the art in bioactive glasses and glass-ceramics scaffolds was authored by Gerhardt and Boccaccini[105] and Fu *et al.*[106] Moreover, there is now a stronger focus on engineered scaffolds from bioactive glass/glass-ceramic combined with bio-polymers. Fracture toughness improvement can be achieved by producing composites. Polymer-based scaffolds reinforced with bio glasses or glass-ceramics represent a suitable option due to the possibility to tailor their various properties (*e.g.*, mechanical and structural behavior, degradation kinetics and bioactivity). Polymers exhibit generally high ductility, toughness, favorable formability and processability and plasticity. The glass or glass-ceramic phase adds stiffness and adequate mechanical strength to the composite. For further reading, comprehensive review papers and book chapters in ref. 105–110 are recommended.

2.11 Gel-derived Bioactive Glass-ceramics

The main advantages of the gel-derived glasses over melt-derived glasses are:[111–113]

- lower temperatures are required for their synthesis,
- easy powder technology production,
- improved homogeneity and purity of the powder,
- wider range of bioactive compositions with increased bioactivity of compositions up to 90 mol% SiO_2,
- they exhibit meso-porosity and an ability to form macroporous structures (scaffolds),
- sol–gel processing is a versatile technique to develop thin layers of bioactive glasses/glass-ceramics on bulk materials with high homogeneity and mechanical and chemical stability.

Upon adequate heat treatment, gel-glasses can be converted to glass-ceramics (Figure 2.1). Proper heat treatments, which may induce crystallization, are required to stabilize the glass structure or develop mesoporous granules, coatings and scaffolds with interconnected pore structure. On the other hand, crystallization may reduce the surface area but it is helpful for modulating mechanical properties and bioactivity.

Gel-derived glass-ceramics have started to emerge in various applications in medicine and dentistry, including tissue engineering and bone regeneration, drug delivery and hybrid materials. Figure 2.9 shows typical gel-derived macroporous scaffolds with the material surface functionalized with protein or growth factors.[112]

The interested reader is encouraged to refer to comprehensive papers, reviews and book chapters authored by Vallet-Regí, Jones, Boccaccini, Hench, Vitale-Brovarone and their colleagues, which deal with gel-derived glasses and glass-ceramics and their derivative products. They believe that biomimetic regeneration of the complex structures of bone and teeth demand 3D porous structures and the incorporation of biofactors, drugs and seeded stem cells. As the example described above demonstrates, gel-derived bioactive glasses and glass-ceramics are promising to meet these requirements.[16,69,110–113]

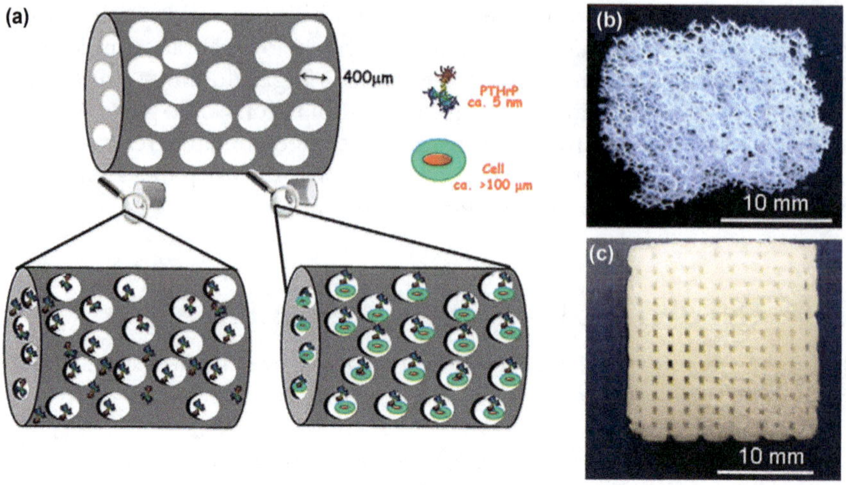

Figure 2.9 (a) Scheme of gel-derived macroporous scaffolds with functionalized material surface with parathyroid hormone-related protein (PTHrP). (b) Scaffold prepared by foaming processing. (c) Scaffold prepared by rapid prototyping process.
Reprinted from D. Arcos and M. Vallet-Regí, Sol-gel silica-based biomaterials and bone tissue regeneration, *Acta Biomater.*, **6**, 2874–2888, Copyright 2010 with permission from Elsevier.[112]

2.12 Conclusions and Outlook

Bioactive glass-ceramics were originally developed to overcome one marked weakness of bioactive glasses—brittleness. However, a comparison of the fracture toughness of commercial glass-ceramics (1–2 MPa $m^{1/2}$) and cortical bone (2–12 MPa $m^{1/2}$) reveals that this property still needs to be significantly improved. Furthermore, highly bioactive glasses and glass-ceramics with adequate biological and mechanical properties should be developed for soft-tissue engineering or drug delivery and for preventive treatments to slow down deterioration and maintain tissue health. These two general objectives can be achieved by the efficient engineering of new glass-ceramic compositions, processing methods and heat treatments. A few key areas for further exploration are summarized below:

- Chemistry-based processes can be employed to expand the bioactive glass-ceramic composition ranges. Recently, it has been demonstrated by the authors that new gel-derived glasses with a high ZrO_2 content can be converted to bioactive nano-apatite–zirconia glass-ceramics with improved mechanical properties.[74,114]
- Adequate matrices can be engineered for the development of hierarchical nanostructured bioactive glass-ceramics based on variations

in the size, shape, distribution, and composition of nano-sized crystals.

- New or improved crystallization processes, such as microwave heating, the biomimetic assemblage of crystals, textured crystallization, laser crystallization, and electron beam crystallization, should be developed.
- Coating properties should be improved. For example, degradation over time, which leads to detachment of coating, is a noticeable drawback.
- Further investigations are required to improve polymer/glass-ceramic composites (hybrids). There is a lack of *in vivo* testing in this interesting area. The use of bioactive glass-ceramic nano-particles and their combination with bioresorbable polymers is of great interest. However, the toxicity of nano-particles will have to be investigated.
- Doping bioactive glass-ceramics with various elements that provide specific biological responses or cell gene expression should be thoroughly investigated before being translated into clinical applications.
- Many issues, including sterilization, packaging and international standards for the production of bioactive glass-ceramics for clinical applications, need to be resolved to make new glass-ceramics commercially viable. For example, conventional sterilization is an issue regarding polymer/glass-based scaffold because it changes the molecular weight of resorbable polymers.
- Not all toughening mechanisms have yet been thoroughly activated in bioactive glass-ceramics. There are still several toughening mechanisms, such as transformation toughening, bridging, microcracking, and pull out, that can be stimulated by the controlled crystallization of different crystals with a variety of morphologies and structures.[115]
- Borate- and phosphate-based bioactive glasses are promising biodegradable materials for soft-tissue engineering, *e.g.*, wound healing.[116,117] However, the effects of crystallization on their biological and mechanical properties are unknown. Controlled crystallization may control their degradation rate and improve their mechanical properties.
- (Expensive, time-consuming) clinical tests have been very limited so far and should be encouraged.
- Furthermore, a graphical abstract in Figure 2.10 shows various bioactive glass-ceramics and respective research areas.

Finally, we fully agree with an inspiring sentence by Prof. Larry L. Hench: *"Previous revolutions in healthcare, prevention of death and replacement of tissues, need to be augmented with two new, innovative revolutions: tissue regeneration and prevention of tissue deterioration"*.[118,119]

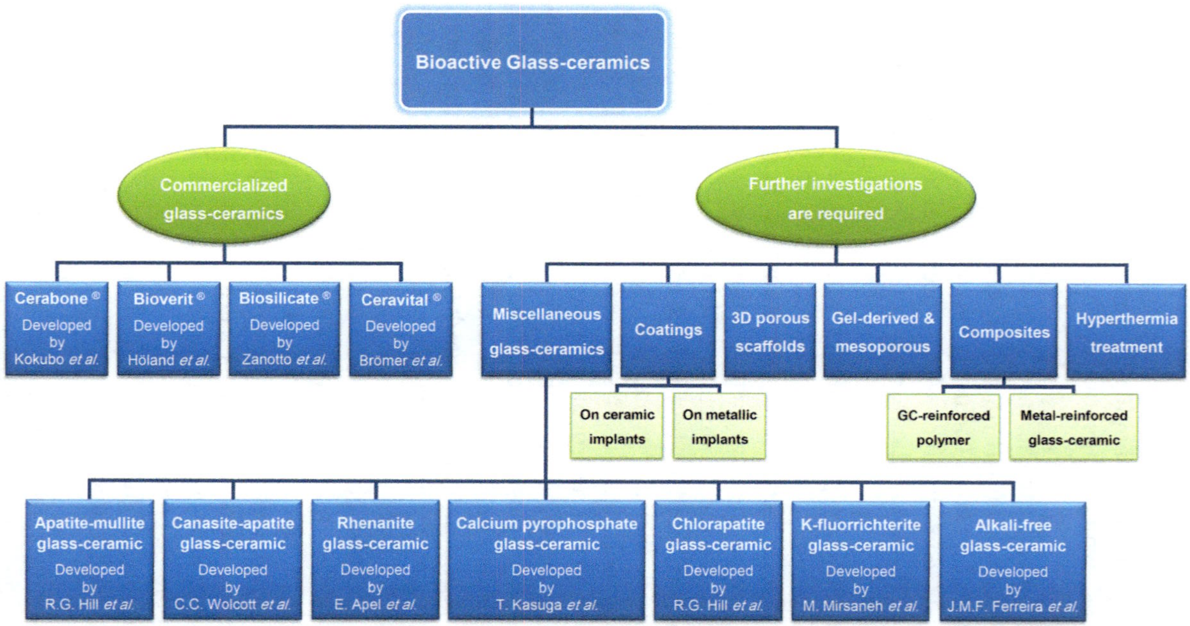

Figure 2.10 Various bioactive glass-ceramics and respective open research areas.

Acknowledgements

The authors are grateful to the São Paulo Research Foundation – FAPESP, # 2013/07793-6 – for the financial support of this work and funding a post-doctoral fellowship for Maziar Montazerian (# 2015/13314-9).

We dedicate this chapter to the late Prof. Larry L. Hench, who not only invented bioactive glass, but also introduced E.D. Zanotto to the field of bioactive glass-ceramics about 20 years ago.

References

1. E. D. Zanotto, A bright future for glass–ceramics, *Am. Ceram. Soc. Bull.*, 2010, **89**, 19–27.
2. M. Montazerian, S. P. Singh and E. D. Zanotto, An analysis of glass-ceramic research and commercialization, *Am. Ceram. Soc. Bull.*, 2015, **94**, 30–35.
3. W. Höland and G. Beall, *Glass-ceramic Technology*, The American Ceramic Society & John Wiley & Sons, Inc., USA, 2nd edn, 2012.
4. M. Montazerian and E. D. Zanotto, History and trends of bioactive glass-ceramics, *J. Biomed. Mater. Res., Part A*, 2016, **104**, 1231–1249.
5. L. L. Hench, The future of bioactive ceramics, *J. Mater. Sci.: Mater. Med.*, 2015, **26**, 1–4.
6. C. J. Brinker, G. W. Scherer, *Sol–Gel Science*, Academic Press, San Diego, CA, 1990.
7. L. L. Hench and J. K. West, The sol–gel process, *Chem. Rev.*, 1990, **90**, 33–72.
8. L. L. Hench, *Sol–Gel Silica: Properties, Processing and Technology Transfer*, Elsevier, New York, 1996.
9. *An Introduction to Bioceramics*, ed. L. L. Hench, Imperial College Press, London, 2nd edn, 2013.
10. T. Kokubo and H. Takadama, How useful is SBF in predicting in vivo bone bioactivity? *Biomaterials*, 2006, **27**, 2907–2915.
11. ISO 23317, Implants for surgery: *in vitro* evaluation for apatite-forming ability of implant materials, 2007.
12. A. L. B. Maçon, T. B. Kim, E. M. Valliant, K. Goetschius, R. K. Brow, D. E. Day, A. Hoppe, A. R. Boccaccini, I. Y. Kim, C. Ohtsuki, T. Kokubo, A. Osaka, M. Vallet-Regí, D. Arcos, L. Fraile, A. J. Salinas, A. V. Teixeira, Y. Vueva, R. M. Almeida, M. Miola, C. Vitale-Brovarone, E. Verné, W. Höland and J. R. Jones, A unified in vitro evaluation for apatite-forming ability of bioactive glasses and their variants, *J. Mater. Sci.: Mater. Med.*, 2015, **26**, 115–125.
13. M. Bohner and J. Lemaitre, Can bioactivity be tested in vitro with SBF solution? *Biomaterials*, 2009, **30**, 2175–2179.
14. H. Pan, X. Zhao, B. W. Darvell and W. W. Lu, Apatite-formation ability – Predictor of "bioactivity"?, *Acta Biomater.*, 2010, **6**, 4181–4188.
15. M. Montazerian, Effect of ZrO_2 addition on properties of gel-derived bioactive SiO_2-CaO-P_2O_5 glass-ceramics, PhD dissertation, Iran University of Science and Technology, 2015.

16. A. J. Salinas and M. Vallet-Regí, Bioactive ceramics: from bone grafts to tissue engineering, *RSC Adv.*, 2013, **3**, 11116–11131.
17. F. Baino, G. Novajra, V. Miguez-Pacheco, A. R. Boccaccini and C. Vitale-Brovarone, Bioactive glasses: Special applications outside the skeletal system, *J. Non-Cryst. Solids*, 2016, **432**, 15–30.
18. V. Miguez-Pacheco, L. L. Hench and A. R. Boccaccini, Bioactive glasses beyond bone and teeth: Emerging applications in contact with soft tissues, *Acta Biomater.*, 2015, **13**, 1–15.
19. T. Kokubo, *Bioceramics and Their Clinical Applications*, Woodhead Publishing Limited, England, 2008.
20. T. Kokubo, Bioactive glass-ceramics: properties and applications, *Biomaterials*, 1991, **12**, 155–163.
21. W. Höland, W. Vogel, K. Naumann and J. Gummel, Interface reactions between machinable bioactive glass-ceramics and bone, *J. Biomed. Mater. Res.*, 1985, **19**, 303–312.
22. O. Peitl, G. P. LaTorre and L. L. Hench, Effect of crystallization on apatite-layer formation of bioactive glass 45S5, *J. Biomed. Mater. Res.*, 1996, **30**, 509–514.
23. O. Peitl, E. D. Zanotto and L. L. Hench, Highly bioactive P_2O_5–Na_2O–CaO–SiO_2 glass-ceramics, *J. Non-Cryst. Solids*, 2001, **292**, 115–126.
24. M. C. Crovace, M. T. Souza, C. R. Chinaglia, O. Peitl and E. D. Zanotto, Biosilicate® - A multipurpose, highly bioactive glass-ceramic. In vitro, in vivo and clinical trials, *J. Non-Cryst. Solids*, 2016, **432**, 90–110.
25. E. D. Zanotto, *Crystals in Glass: A Hidden Beauty*, The American Ceramic Society & Wiley, USA, 2013.
26. C. Tirapelli, H. Panzeri, E. H. Lara, R. G. Soares, O. Peitl and E. D. Zanotto, The effect of a novel crystallised bioactive glass ceramic powder on dentine hypersensitivity: a long-term clinical study, *J. Oral Rehabil.*, 2011, **38**, 253–262.
27. E. T. Massuda, Evaluation of Otosilicate as Middle Ear Prothesis (Thesis) University of São Paulo (FMRP-USP), São Paulo, Brazil, 2007. (62 pp., Available at: http://www.dominiopublico.gov.br/pesquisa/DetalheObraForm.do?select_action=&co_obra=87478).
28. C. Turck, G. Brandes, I. Krueger, P. Behrens, H. Mojallal, T. Lenarz and M. Stieve, Histological evaluation of novel ossicular chain replacement prostheses: an animal study in rabbits, *Acta Otolaryngol.*, 2007, **127**, 801–808.
29. J. C. Vogt, G. Brandes, N. Ehlert, P. Behrens, I. Nolte, P. P. Mueller, Y. Lenarz and M. Stieve, Free Bioverit® II implants coated with a nanoporous silica layer in a mouse ear model - a histological study, *J. Biomater. Appl.*, 2009, **24**, 175–191.
30. R. G. Hill, Bioactive glass–ceramics, in *Comprehensive Biomaterials, Volume 1: Metallic, Ceramic and Polymeric Biomaterials*, ed. P. Ducheyne, 2011, pp. 181–186.
31. R. G. Hill and D. J. Wood, Apatite–mullite glass ceramics, *J. Mater. Sci.: Mater. Med.*, 1995, **6**, 311–318.

32. A. Clifford and R. G. Hill, Apatite–mullite glass-ceramics, *J. Non-Cryst. Solids*, 1996, **196**, 346–352.
33. R. D. Goodridge, D. J. Wood, C. Ohtsuki and K. W. Dalgarno, Biological evaluation of an apatite-mullite glass-ceramic produced via selective laser sintering, *Acta Biomater.*, 2007, **3**, 221–231.
34. K. T. Stanton, K. P. O'Flynn, S. Kiernan, J. Menuge and R. G. Hill, Spherulitic crystallization of apatite-mullite glass-ceramics: Mechanisms of formation and implications for fracture properties, *J. Non-Cryst. Solids*, 2010, **356**, 1802–1813.
35. C. C. Wolcott, Canasite-apatite glass-ceramic, Eur. Pat. No. EP 0 641 556, 1994.
36. C. A. Miller, T. Kokubo, I. M. Reaney, P. V. Hatton and P. F. James, Formation of apatite layers on modified canasite glass-ceramics in simulated body fluid, *J. Biomed. Mater. Res.*, 2002, **59**, 473–480.
37. V. M. D. R. Barros, L. A. Salata, C. E. Sverzut, S. P. Xavier, R. Van Noort, A. Johnson and P. V. Hatton, *In vivo* bone tissue response to a canasite glass-ceramic, *Biomaterials*, 2002, **23**, 2895–2900.
38. M. Mirsaneh, I. M. Reaney, P. F. James and P. V. Hatton, Effect of CaF_2 and CaO substituted for MgO on the phase evolution and mechanical properties of K-fluorrichterite glass ceramics, *J. Am. Ceram. Soc.*, 2006, **89**, 587–595.
39. S. Bhakta, D. K. Pattanayak, H. Takadama, T. Kokubo, C. A. Miller, M. Mirsaneh, I. M. Reaney, I. M. Brook, R. van Noort and P. V. Hatton, Prediction of osteoconductive activity of modified potassium fluorrichterite glass–ceramics by immersion in simulated body fluid, *J. Mater. Sci.: Mater. Med.*, 2010, **89**, 2979–2988.
40. S. Bhakta, K. Hurrell-Gillingham, M. Mirsaneh, C. A. Miller, I. M. Reaney, I. M. Brook, R. van Noort and P. V. Hatton, *In vitro* biocompatibility of modified potassium fluorrichterite and potassium fluorrichterite–fluorapatite glass–ceramics, *J. Mater. Sci.: Mater. Med.*, 2011, **22**, 2065–2070.
41. S. Bhakta, P. E. Faira, L. A. Salata, P. J. D. O. Neto, C. A. Miller, R. van Noort, I. M. Reaney, I. M. Brook and P. V. Hatton, Determination of relative *in vivo* osteoconductivity of modified potassium fluorrichterite glass–ceramics compared with 45S5 bioglass, *J. Mater. Sci.: Mater. Med.*, 2012, **23**, 2521–2529.
42. E. El-Meliegy, R. van Noort, *Glasses and glass ceramics for medical applications*, Springer, New York, 2011, ch. 14, pp. 229–238.
43. E. Apel, W. Höland and V. Rheinberger, Bioactive rhenanite glass ceramic, United States Patent 7074730, 2006.
44. M. Höland, A. Dommann, W. Höland, E. Apel and V. Rheinberger, Microstructure formation and surface properties of a rhenanite-type glass-ceramic containing 6.0 wt% P_2O_5, *Glass Sci. Technol.*, 2005, **78**, 153–158.
45. T. Kasuga, Bioactive calcium pyrophosphate glasses and glass-ceramics, *Acta Biomater.*, 2005, **1**, 55–64.
46. T. Kasuga, Development of phosphate glass-ceramics for biomedical applications, *J. Ceram. Soc. Jpn.*, 2007, **115**, 455–459.

47. T. Kasuga, T. Mizuno, M. Watanabe, M. Nogami and M. Niinomi, Calcium phosphate invert glass-ceramic coatings joined by self development of compositionally gradient layers on a titanium alloy, *Biomaterials*, 2001, **22**, 577–582.
48. A. Obata, M. Sasaki and T. Kasuga, Control of ß-tricalcium phosphate formation in macroporous phosphate glass-ceramic composites, *Mater. Trans.*, 2007, **48**, 313–316.
49. H. Maeda, T. Miyajima, S. Lee, A. Obata, K. Ueda, T. Narushima and T. Kasuga, Preparation of calcium pyrophosphate glass-ceramics containing Nb_2O_5, *J. Ceram. Soc. Jpn.*, 2014, **122**, 122–124.
50. X. Chen, R. G. Hill and N. Karpukhina, Chlorapatite glass-ceramics, *Int. J. Appl. Glass Sci.*, 2014, **5**, 207–216.
51. S. Kapoor, A. Goel, M. J. Pascual and J. M. F. Ferreira, Alkali-free bioactive diopside-tricalcium phosphate glass-ceramics for scaffold fabrication: Sintering and crystallization behaviours, *J. Non-Cryst. Solids*, 2016, **432**, 81–89.
52. A. Goel, S. Kapoor, R. R. Rajagopal, M. J. Pascual, H. Kim and J. M. F. Ferreira, Alkali-free bioactive glasses for bone tissue engineering: A preliminary investigation, *Acta Biomater.*, 2012, **8**, 361–372.
53. S. Kapoor, A. Goel, M. J. Pascual and J. M. F. Ferreira, Thermomechanical behaviour of alkali free bioactive glass-ceramics co-doped with strontium and zinc, *J. Non-Cryst. Solids*, 2013, **375**, 74–82.
54. A. A. Luderer, N. F. Borrelli, J. N. Panzarino, G. R. Mansfield, D. M. Hess, J. L. Brown, E. H. Barnett and E. W. Hahn, Glass-ceramic-mediated, magnetic-fieldinduced localized hyperthermia: response of a murine mammary carcinoma, *Radiat. Res.*, 1983, **94**, 190–198.
55. Y. Ebisawa, Y. Sugimoto, T. Hayashi, T. Kokubo, K. Ohura and T. Yamamuro, Crystallization of (FeO,Fe_2O_3)-CaO-SiO_2 glasses and magnetic properties of their crystallized products, *J. Ceram. Soc. Jpn.*, 1991, **99**, 7–13.
56. Y. Ebisawa, F. Miyaji, T. Kokubo, K. Ohura and T. Nakamura, Bioactivity of ferrimagnetic glass-ceramics in the system FeO-Fe_2O_3-CaO-SiO_2, *Biomaterials*, 1997, **18**, 1277–1284.
57. K. Ohura, M. Ikenaga, T. Nakamura, T. Yamamuro, Y. Ebisawa, T. Kokubo, Y. Kotoura and M. Oka, A heat-generating bioactive glass-ceramic for hyperthermia, *J. Appl. Biomater.*, 1991, **2**, 153–159.
58. O. Bretcanu, S. Spriano, E. Verné, M. Cöisson, P. Tiberto and P. Allia, The influence of crystallised Fe_3O_4 on the magnetic properties of coprecipitation-derived ferrimagnetic glass-ceramics, *Acta Biomater.*, 2005, **1**, 421–429.
59. R. K. Singh and A. Srinivasan, Apatite-forming ability and magnetic properties of glass-ceramics containing zinc ferrite and calcium sodium phosphate phases, *Mater. Sci. Eng., C*, 2010, **30**, 1100–1106.
60. R. K. Singh and A. Srinivasan, Magnetic properties of bioactive glass-ceramics containing nanocrystalline zinc ferrite, *J. Magn. Magn. Mater.*, 2011, **323**, 330–333.

61. S. A. Shah, M. U. Hashmi, S. Alam and A. Shamim, Magnetic and bioactivity evaluation of ferrimagnetic $ZnFe_2O_4$ containing glass ceramics for the hyperthermia treatment of cancer, *J. Magn. Magn. Mater.*, 2010, **322**, 375–381.
62. C. S. Hsi, H. Z. Cheng, H. J. Hsu, Y. S. Chen and M. C. Wang, Crystallization kinetics and magnetic properties of iron oxide contained $25Li_2O$–$8MnO_2$–$20CaO$-$2P_2O_5$–$45SiO_2$ glasses, *J. Eur. Ceram. Soc.*, 2007, **27**, 3171–3176.
63. T. W. Wang, H. C. Wu, W. R. Wang, F. H. Lin, P. J. Lou, M. J. Shieh and T. H. Young, The development of magnetic degradable DP-Bioglass for hyperthermia cancer therapy, *J. Biomed. Mater. Res.*, 2007, **83**, 828–837.
64. G. D. Li, D. L. Zhou, Y. Lin, T. H. Pan, G. S. Chen and Q. D. Yin, Synthesis and characterization of magnetic bioactive glass-ceramics containing Mg ferrite for hyperthermia, *Mater. Sci. Eng., C*, 2010, **30**, 148–153.
65. G. D. Li, S. Feng and D. Zhou, Magnetic bioactive glass ceramic in the system CaO–P_2O_5–SiO_2–MgO–CaF_2–MnO_2–Fe_2O_3 for hyperthermia treatment of bone tumor, *J. Mater. Sci.: Mater. Med.*, 2011, **22**, 2197–2206.
66. W. Leenakul, P. Kantha, N. Pisitpipathsin, G. Rujijanagul, S. Eitssayeam and K. Pengpat, Structural and magnetic properties of SiO_2–CaO–Na_2O–P_2O_5 containing BaO-Fe_2O_3 glass-ceramics, *J. Magn. Magn. Mater.*, 2013, **325**, 102–106.
67. W. Leenakul, J. Ruangsuriya, P. Jantaratana and K. Pengpat, Fabrication and characterization of ferrimagnetic bioactive glass-ceramic containing $BaFe_{12}O_{19}$, *Ceram. Int.*, 2013, **39**, 201–205.
68. W. Leenakul, P. Intawin, J. Ruangsuriya, P. Jantaratana and K. Pengpat, Magnetic bioactive $SrFe_{12}O_{19}$–SiO_2–CaO–Na_2O–P_2O_5 glass-ceramics for hyperthermia treatment of bone cancer, *Integr. Ferroelectr.*, 2013, **148**, 81–89.
69. D. Arcos, R. P. Del Real and M. Vallet-Regí, A novel bioactive and magnetic biphasic material, *Biomaterials*, 2002, **23**, 2151–2158.
70. D. Arcos, R. P. Del Real and M. Vallet-Regí, Biphasic materials for bone grafting and hyperthermia treatment of cancer, *J. Biomed. Mater. Res., Part A*, 2003, **65**, 71–78.
71. M. Serrano, M. T. Portolés, R. Pagani, J. S. De Guinoa, E. Ruiz-Hernández, D. Arcos and M. Vallet-Regí, In vitro positive biocompatibility evaluation of glass-glass ceramic thermoseeds for hyperthermic treatment of bone tumors, *Tissue Eng.*, 2008, **14**, 617–627.
72. F. Tallia, M. Gallo, L. Pontiroli, F. Baino, S. Fiorilli, B. Onida, G. C. Anselmetti, A. Manca and C. Vitale-Brovarone, Zirconia-containing radiopaque mesoporous bioactive glasses, *Mater. Lett.*, 2014, **130**, 281–284.
73. C. Vitale-Brovarone, E. Verné, M. Bergui, B. Onida, F. Baino, M. Miola, S. Ferraris and F. Tallia, Injectable osteoinductive bone cements, World Patent Application, WO 2011141889A1, 2011.
74. M. Montazerian, B. Eftekhari Yekta, V. K. Marghussian, C. F. Bellani, R. L. Siqueira and E. D. Zanotto, Bioactivity and cell proliferation in radiopaque gel-derived CaO–P_2O_5–SiO_2–ZrO_2 glass and glass-ceramic powders, *Mater. Sci. Eng., C*, 2015, **55**, 436–447.

75. W. Cao and L. L. Hench, Bioactive materials, *Ceram. Int.*, 1996, **22**, 493–507.
76. E. Verné, C. Fernández Vallés, C. Vitale Brovarone, S. Spriano and C. Moisescu, Double-layer glass-ceramic coatings on Ti6Al4V for dental implants, *J. Eur. Ceram. Soc.*, 2004, **24**, 2699–2705.
77. E. Verné and C. Vitale, brovarone, C. Moisescu, E. Ghisolfi, E. Marmo, Coatings on Al_2O_3 by bioactive glass-ceramics, *Acta Mater.*, 2000, **48**, 4667–4671.
78. M. Ferraris, E. Verné, P. Appendino, C. Moisescu, A. Krajewski, A. Ravaglioli and A. Piancastelli, Coatings on zirconia for medical applications, *Biomaterials*, 2000, **21**, 765–773.
79. E. Verné, Bioactive glass and glass-ceramic coatings, in *Bio-glasses: An introduction*, ed. J. R. Jones and A. G. Clare, John Wiley and Sons Ltd., UK, 1st edn, 2012, pp. 107–120.
80. V. Marghussian, *Nano-glass ceramics: Processing, properties and applications*, Elsevier, 1st edn, 2015.
81. W. Xiao, L. Song, X. Liu, Y. Huang, T. Huang, Y. Wu, Y. Chen and F. Wu, Nanostructured bioactive glass–ceramic coatings deposited by the liquid precursor plasma spraying process, *Appl. Surf. Sci.*, 2011, **257**, 1898–1905.
82. G. Wang, Z. Lu, X. Liu, X. Zhou, C. Ding and H. Zreiqat, Nanostructured glass-ceramic coatings for orthopaedic applications, *J. R. Soc., Interface*, 2011, **8**, 1192–1203.
83. T. Kasuga, M. Yoshida, A. J. Ikushima, M. Tuchiya and H. Kusakari, Bioactivity of zirconia-toughened glass-ceramics, *J. Am. Ceram. Soc.*, 1992, **75**, 1884–1888.
84. T. Kasuga, K. Nakajima, T. Uno and M. Yoshida, Preparation of zirconia-toughened bioactive glass-ceramic composite sinter-hot isostatic pressing, *J. Am. Ceram. Soc.*, 1992, **75**, 1103–1107.
85. E. Verné, M. Ferraris and C. Jana, Pressureless sintering of bioverit® III/Ti particle biocomposites, *J. Eur. Ceram. Soc.*, 1999, **19**, 2039–2047.
86. E. Verné, M. Ferraris, C. Jana and L. Paracchini, Bioverit® I base glass/Ti particulate biocomposite: "In situ" vacuum plasma spray deposition, *J. Eur. Ceram. Soc.*, 2000, **20**, 473–479.
87. M. Ferraris, P. Rabajoli, L. Paracchini and F. Brossa, Vacuum plasma spray deposition of titanium particle/glass-ceramic matrix biocomposites, *J. Am. Ceram. Soc.*, 1996, **79**, 1515–1520.
88. M. Montazerian and P. Alizadeh, B. Eftekhari Yekta, Pressureless sintering and mechanical properties of mica glass–ceramic/Y-PSZ composite, *J. Eur. Ceram. Soc.*, 2008, **28**, 2687–2692.
89. M. Montazerian, P. Alizadeh and B. Eftekhari Yekta, Processing and properties of a mica–apatite glass–ceramic reinforced with Y-PSZ particles, *J. Eur. Ceram. Soc.*, 2008, **28**, 2693–2699.
90. E. Verné, R. Defilippi, G. Carl, C. Vitale Brovarone and P. Appendino, Viscous flow sintering of bioactive glass-ceramic composites toughened by zirconia particles, *J. Eur. Ceram. Soc.*, 2003, **23**, 675–683.

91. R. D. Rawlings, Bioactive glasses and glass-ceramics, *Clin. Mater.*, 1993, **14**, 155–179.
92. C. Müller-Mai, H. J. Schmitz, V. Strunz, G. Fuhrmann, T. Fritz and U. M. Gross, Tissues at the surface of the new composite material titanium/glass-ceramic for replacement of bone and teeth, *J. Biomed. Mater. Res.*, 1989, **23**, 1149–1168.
93. E. Claxton, B. A. Taylor and R. D. Rawlings, Processing and properties of a bioactive glass-ceramic reinforced with ductile silver particles, *J. Mater. Sci.*, 2002, **37**, 3725–3732.
94. D. K. Pattanayak, R. C. Prasad, B. T. Rao and T. R. Rama Mohan, Apatite-wollastonite-titanium biocomposites: Synthesis and in vitro evaluation, *J. Am. Ceram. Soc.*, 2006, **89**, 2172–2176.
95. M. Wang, Developing bioactive composite materials for tissue replacement, *Biomaterials*, 2003, **24**, 2133–2151.
96. J. A. Juhasz, S. M. Best, M. Kawashita, N. Miyata, T. Kokubo, T. Nakamura and W. Bonfield, Bonding strength of the apatite layer formed on glass-ceramic apatite-wollastonite-polyethylene composites, *J. Biomed. Mater. Res., Part A*, 2003, **67**, 952–959.
97. J. A. Juhasz, S. M. Best, R. Brooks, M. Kawashita, N. Miyata, T. Kokubo, T. Nakamura and W. Bonfield, Mechanical properties of glass-ceramic A-W-polyethylene composites: Effect of filler content and particle size, *Biomaterials*, 2004, **25**, 949–955.
98. P. R. Gabbai-Armelin, D. A. Cardoso, E. D. Zanotto, O. Peitl, S. C. G. Leeuwenburgh, J. A. Jansen, A. C. M. Renno and J. J. J. P. Van Den Beucken, Injectable composites based on biosilicate® and alginate: Handling and in vitro characterization, *RSC Adv.*, 2014, **4**, 45778–45785.
99. S. Shinzato, T. Nakamura, K. Ando, T. Kokubo and Y. Kitamura, Mechanical properties and osteoconductivity of new bioactive composites consisting of partially crystallized glass beads and poly(methyl methacrylate), *J. Biomed. Mater. Res.*, 2002, **60**, 556–563.
100. H. A. Samad, M. Jaafar, R. Othman, M. Kawashita and N. H. A. Razak, New bioactive glass-ceramic: Synthesis and application in PMMA bone cement composites, *Bio-Med. Mater. Eng.*, 2011, **21**, 247–258.
101. M. Bruno, M. Miola, O. Bretcanu, C. Vitale-Brovarone, R. Gerbaldo, F. Laviano and E. Verné, Composite bone cements loaded with a bioactive and ferrimagnetic glass-ceramic. part I: Morphological, mechanical and calorimetric characterization, *J. Biomater. Appl.*, 2014, **29**, 254–267.
102. E. Verné, M. Bruno, M. Miola, G. Maina, C. Bianco, A. Cochis and L. Rimondini, Composite bone cements loaded with a bioactive and ferrimagnetic glass-ceramic: Leaching, bioactivity and cytocompatibility, *Mater. Sci. Eng., C*, 2015, **53**, 95–103.
103. E. Zeimaran, S. Pourshahrestani, I. Djordjevic, B. Pingguan-Murphy, N. A. Kadri and M. R. Towler, Bioactive glass reinforced elastomer composites for skeletal regeneration: A review, *Mater. Sci. Eng., C*, 2015, **53**, 175–188.

104. G. Kaur, O. P. Pandey, K. Singh, D. Homa, B. Scott and G. Pickrell, A review of bioactive glasses: Their structure, properties, fabrication, and apatite formation, *J. Biomed. Mater. Res., Part A*, 2014, **102**, 254–274.
105. L. Gerhardt and A. R. Boccaccini, Bioactive glass and glass ceramic scaffolds for bone tissue engineering, *Materials*, 2010, **3**, 3867–3910.
106. Q. Fu, E. Saiz, M. N. Rahaman and A. P. Tomsia, Toward strong and tough glass and ceramic scaffolds for bone repair, *Adv. Funct. Mater.*, 2013, **23**, 5461–5476.
107. F. Baino, E. Verné and C. Vitale-Brovarone, 3-D high-strength glass-ceramic scaffolds containing fluoroapatite for load-bearing bone portions replacement, *Mater. Sci. Eng., C*, 2009, **29**, 2055–2062.
108. X. Chatzistavrou, P. Newby and A. R. Boccaccini, Bioactive glass and glass-ceramic scaffolds for bone tissue engineering, in *Bioactive Glasses*, ed. H. O. Ylänen, Woodhead Publishing Limited, Cambridge, UK, 2011, pp. 107–128.
109. V. Miguez-Pacheco, S. K. Misra and A. R. Boccaccini, Biodegradable and bioactive polymer/inorganic phase nanocomposites for bone tissue engineering (BTE), in *Tissue Engineering Using Ceramics and Polymers*, ed. A. R. Boccaccini and P. X. Ma, Woodhead Publishing Limited, Cambridge, UK, 2nd edn, 2014, pp. 115–150.
110. J. R. Jones, Review of bioactive glass: from Hench to hybrids, *Acta Biomater.*, 2013, **9**, 4457–4486.
111. X. Chatzistavrou, E. Kontonasaki, K. M. Paraskevopoulos, P. Koidis and A. R. Boccaccini, Sol–gel derived bioactive glass ceramics for dental applications, in *Non-Metallic Biomaterials for Tooth Repair and Replacement*, ed. P. Vallittu, Woodhead Publishing Limited, Cambridge, UK, 2013, pp. 194–231.
112. D. Arcos and M. Vallet-Regí, Sol–gel silica-based biomaterials and bone tissue regeneration, *Acta Biomater.*, 2010, **6**, 2874–2888.
113. A. J. Salinas and M. Vallet-Regí, Glasses in bone regeneration: A multiscale issue, *J. Non-Cryst. Solids*, 2016, **432**, 9–14.
114. M. Montazerian, J. F. Schneider, B. Eftekhari Yekta, V. K. Marghussian, A. M. Rodrigues and E. D. Zanotto, Sol–gel synthesis, structure, sintering and properties of bioactive and inert nano apatite-zirconia glass-ceramics, *Ceram. Int.*, 2015, **41**, 11024–11045.
115. F. C. Serbena, I. Mathias, C. E. Foerster and E. D. Zanotto, Crystallization toughening of a model glass-ceramic, *Acta Mater.*, 2015, **86**, 216–228.
116. M. N. Rahaman, D. E. Day, B. S. Bal, Q. Fu, S. B. Jung, L. F. Bonewald and A. P. Tomsia, Bioactive glass in tissue engineering, *Acta Biomater.*, 2011, **7**, 2355–2373.
117. D. S. Brauer, Phosphate glasses, in *Bio-glasses: An Introduction*, ed. J. R. Jones and A. G. Clare, John Wiley and Sons, Ltd., UK, 2012, pp. 45–64.
118. L. L. Hench, Glass and glass-ceramic technologies to transform the world, *Int. J. Appl. Glass Sci.*, 2011, **2**, 162–176.
119. L. L. Hench, Opening paper 2015 – Some comments on Bioglass: Four eras of discovery and development, *Biomed. Glasses*, 2015, **1**, 1–11.

CHAPTER 3

Introduction to the Structure of Silicate, Phosphate and Borate Glasses

DELIA S. BRAUER* AND DORIS MÖNCKE

Otto Schott Institute of Materials Research, Friedrich Schiller University Jena, Fraunhoferstr. 6, 07743 Jena, Germany
*Email: delia.brauer@uni-jena.de

3.1 Introduction

Glasses are amorphous and their structure, unlike that of crystalline solids, does not show any long-range order or any significant symmetry in the atomic arrangement.[1] But this absence of periodic order does not mean that glasses have no "structure"; on the contrary, the atoms constituting the glass are arranged in a certain short-range order which depends on the glass composition. Variations in the ratio of network modifiers, *i.e.* metal oxides such as alkali or alkaline earth oxides, to network formers, *e.g.* SiO_2, P_2O_5 or B_2O_3, have a pronounced influence on the glass structure as they influence the polymerisation of the glass network. In addition, certain superstructural units such as rings or chains are commonly observed in glasses. This atomic arrangement directly affects many glass properties. A thorough understanding of these structures can therefore give us a tool to predict many glass properties, interpret experimental findings or design glasses which show specific properties such as solubility, ion release or bioactivity.

RSC Smart Materials No. 23
Bioactive Glasses: Fundamentals, Technology and Applications
Edited by Aldo R. Boccaccini, Delia S. Brauer and Leena Hupa

Published by the Royal Society of Chemistry, www.rsc.org

This chapter explains the basic structural arrangements in three groups of glasses used as biomaterials: silicate glasses, phosphate glasses and borate glasses, and how structural variations control the glass properties. In a final section, mixed glass systems, such as phosphosilicate or borosilicate glasses, are considered.

3.2 Basics of Silicate and Phosphate Glasses

Although glasses do not have a long-range order or a periodic arrangement of atoms over longer distances, they still show short-range order. In silicate and phosphate glasses, the basic structural unit is either a silicon or phosphorus atom surrounded tetrahedrally by four oxygen atoms (Figure 3.1). In silicate glasses, this SiO_4 tetrahedron can be connected to a maximum of four other tetrahedra *via* its four corners. In phosphate glasses, each phosphate tetrahedron is connected to a maximum of three other phosphate tetrahedra. Such covalent bonds between tetrahedra, ≡Si–O–Si≡ bonds in silicate and ≡P–O–P≡ bonds in phosphate glasses, are called *bridging oxygen atoms* (bO). The average number of bO per network-forming atom in a glass defines its *network connectivity*, a parameter describing the network polymerisation.

If *network modifiers*, such as alkali or alkaline earth oxides, are added to a phosphate or silicate glass, bridging oxygen atoms are turned into *non-bridging oxygen atoms* (nbO) as shown in the following pseudo-reactions:

$$\equiv\text{Si–O–Si}\equiv + M_2O \rightarrow 2\ \equiv\text{Si–O}^-\ M^+ \tag{3.1}$$

or

$$\equiv\text{P–O–P}\equiv + M_2O \rightarrow 2\ \equiv\text{P–O}^-\ M^+ \tag{3.2}$$

where M_2O is the oxide of a monovalent modifier cation. With increasing concentrations of modifiers in the glass the average number of nbO per network forming atom (Si or P) increases, resulting in a decrease in the number of bO and network connectivity and thus a less polymerised network. With increasing modifier concentrations, the dominating type of chemical bonds present in a silicate or phosphate glass changes as well, from predominantly covalent to predominantly ionic, with modifier cations acting as ionic bridges between nbO (Figure 3.2).[2]

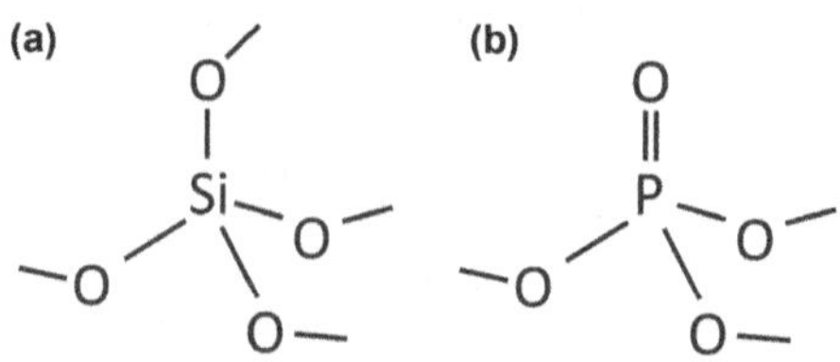

Figure 3.1 Basic building units of (a) silicate and (b) phosphate glasses.

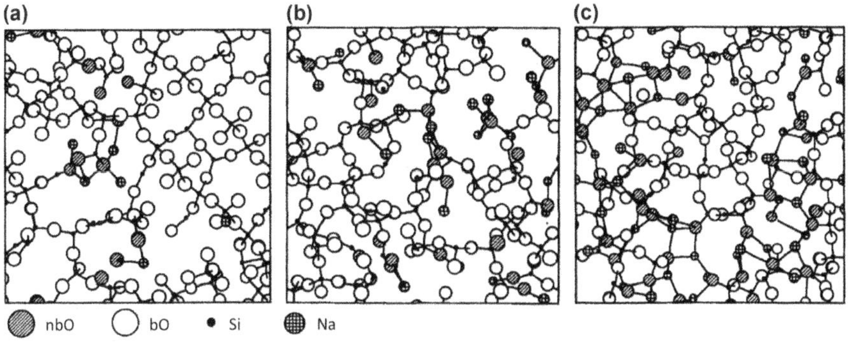

Figure 3.2 The atomic structure of $SiO_2 \cdot Na_2O$ glasses containing (a) 5, (b) 15 and (c) 25 mol% Na_2O as obtained from molecular dynamics calculations showing increasing concentrations of non-bridging oxygen atoms (nbO) and decreasing concentrations of bridging oxygen atoms (bO) as the Na_2O content increases.[2] The network connectivity of these glasses is (a) 3.89, (b) 3.65 and (c) 3.33.
(Reprinted with permission from C. D. Huang, A. N. Cormack, The structure of sodium silicate glass, *J. Chem. Phys.*, 1990, **93**, 8180–8186, Copyright 1990, AIP Publishing Ltd.)

3.3 Structure and Properties of Silicate Glasses

SiO_2 is a typical network former. In vitreous silica, each SiO_4 tetrahedron is linked to other SiO_4 tetrahedra at each of its four corners, which corresponds to four bridging oxygen atoms per tetrahedron. Tetrahedra with four bO we commonly refer to as Q^4 groups (Figure 3.3).

As soon as modifier oxides are introduced into the glass, bridging oxygen atoms are turned into non-bridging ones, and Q^3 groups (Figure 3.3) appear. Their relative amount increases and the relative amount of Q^4 decreases with increasing modifier oxide concentration until all Q^4 groups transform into Q^3 groups (Figure 3.4). If the amount of modifier oxides increases further, the relative amount of Q^3 decreases while, simultaneously, Q^2 groups start to form (Figure 3.4). Figure 3.2 illustrates this increasing disruption of the silicate network with decreasing bO and increasing nbO concentrations for silicate glasses containing 5, 15 or 25 mol% Na_2O. At 50 mol% SiO_2, *i.e.* a network connectivity of 2, theoretically the silicate network should consist of Q^2 groups only, corresponding to a structure consisting of chains or rings only, rather than a network. Owing to their pronounced tendency to crystallise, silicate compositions consisting of short silicate chains (with large numbers of Q^1 chain end groups or even showing the presence of some isolated Q^0 groups) cannot be obtained in a glassy state using conventional melting techniques. In crystalline silicates, by contrast, such groups do exist.[3]

^{29}Si magic-angle-spinning nuclear magnetic resonance (MAS NMR) spectroscopy has been used successfully to characterise the silicate structure of

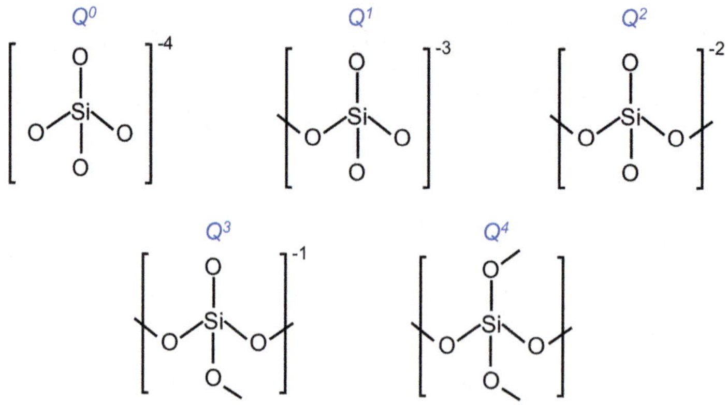

Figure 3.3 Silicate tetrahedral species, referred to as Q^n groups, where n denotes the number of bridging oxygen atoms.

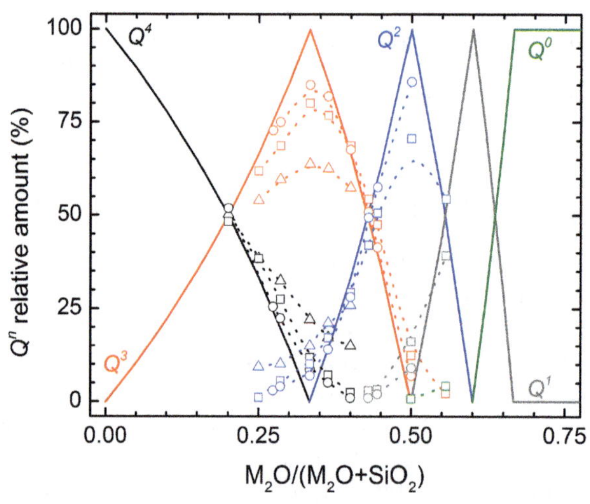

Figure 3.4 Changes in the relative amounts of Q^n groups in $SiO_2 \cdot M_2O$ silicate glasses with increasing modifier (M_2O) content (Q^4: black, Q^3: red, Q^2: blue, Q^1: grey and Q^0: green): solid lines represent theoretical percentages calculated assuming a binary distribution of Q^n groups; symbols represent experimental amounts obtained from solid-state ^{29}Si MAS NMR[4] on lithium (triangles), sodium (squares) and potassium (circles) silicate glasses (dotted lines are visual guides only).

bioactive glasses, as different Q species are characterised by different chemical shifts.[5] Owing to broad signals, however, deconvolution of the spectra is usually necessary to obtain information about the relative amounts of different Q species (Figure 3.5).

The polymerisation of the silicate network, *i.e.* the average number of bO per SiO_4 tetrahedron, is usually described as the *network connectivity* of the glass. The actual network connectivity can also be obtained from structural investigations giving the distribution of Q^n groups (*e.g.* by ^{29}Si MAS NMR,

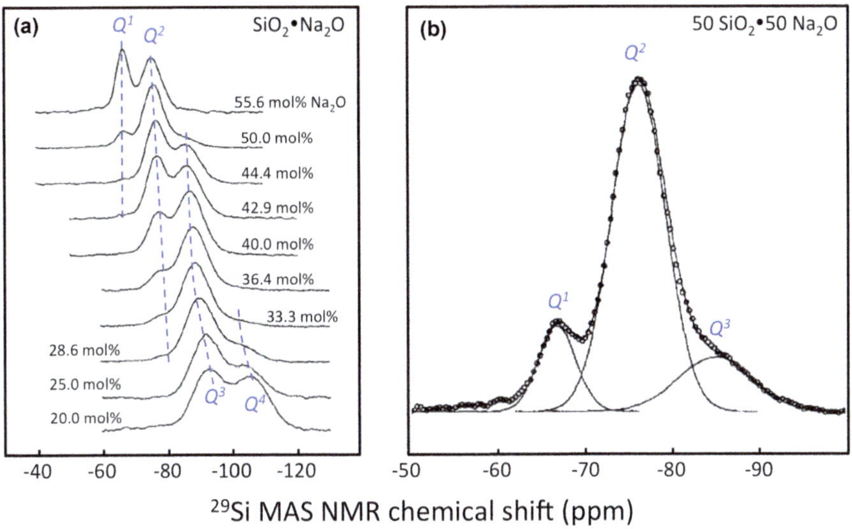

Figure 3.5 (a) ^{29}Si MAS NMR spectra of $SiO_2 \cdot Na_2O$ glasses: with increasing Na_2O content, peak positions and relative peak intensities change, owing to variations in the relative amounts of *Q* species. (b) ^{29}Si MAS NMR spectrum of a sodium metasilicate (50 $SiO_2 \cdot$ 50 Na_2O) glass (dotted line) and deconvolution into Gaussian peaks (solid lines) for Q^1, Q^2 and Q^3 groups.[4]
(Reprinted from H. Maekawa *et al.*, The structural groups of alkali silicate glasses determined from ^{29}Si MAS-NMR, *J. Non-Cryst. Solids*, **127**, 53–64, Copyright 1991 with permission from Elsevier.)

Figure 3.5), but the network connectivity of simple silicate glasses (NC_{Si}) can also be estimated from the nominal composition, as shown in eqn (3.3):

$$NC_{Si} = \frac{4[SiO_2] - 2([M_2O] + [M'O])}{[SiO_2]} \quad (3.3)$$

where $[SiO_2]$, $[M_2O]$ and $[M'O]$ represent the molar percentages of silica, alkali and alkaline earth oxides, respectively.

The network connectivity of bioactive glasses tends to be low, with typical values lying between 2.0 and 2.6. A network connectivity of 2.0 corresponds to a metasilicate glass consisting of a chain or ring structure of Q^2 groups exclusively. An increase in network connectivity describes how the silicate structure becomes more polymerised with increasing amounts of Q^3, and subsequently also Q^4, groups.

The network connectivity has a pronounced influence on many glass properties, including the thermal behaviour. As vitreous silica has a network connectivity of four, it has very high melting (1723 °C[3,6]) and glass transition temperatures (about 1170 °C[6]). As we incorporate modifier oxides into the glass, the resulting compositions can be melted or shaped at lower temperatures. Figure 3.6 illustrates this trend, showing that the softening

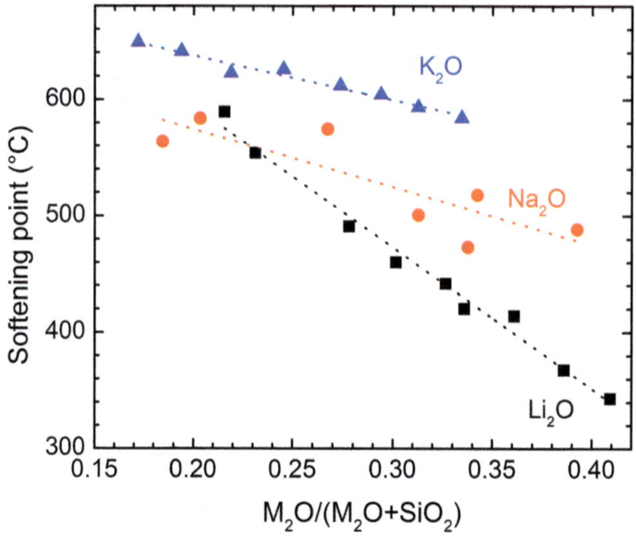

Figure 3.6 Littleton softening point (corresponding to a viscosity of $10^{6.6}$ Pa s) of binary alkali silicate ($SiO_2 \cdot Na_2O$) glasses showing a decrease in softening temperature with increasing modifier content (after Ray[7]).

temperature decreases as the modifier content increases from about 15 to 40 mol%,[7] corresponding to a decrease in network connectivity. However, as the network connectivity decreases, the glasses usually crystallise more easily either during cooling from the melt or upon heating the glass. This means that with decreasing network connectivity, the compositions are more difficult to obtain in a glassy (*i.e.* an amorphous) state, and they are also more difficult to process at elevated temperatures without crystallising.

3.3.1 Structure, Solubility and Bioactivity

It is not only the thermal properties that vary with network connectivity. Vitreous silica is chemically very stable at neutral or acidic pH[8] (with the exception of hydrofluoric acid) while the glasses become less chemically stable with increasing modifier oxide content. At acidic to slightly alkaline pH (below about pH 10[8]) modifier oxide-containing silicate glasses dissolve by an ion exchange mechanism: modifier cations, *e.g.* Ca^{2+} or Na^{+}, but also ions having therapeutic properties (*cf.* Chapter 16), are released and replaced by protons (H^{+}, originating from water dissociation) from the surrounding aqueous medium.[8] This results in a pH rise in the dissolution medium (owing to an excess of hydroxyl ions, OH^{-}). In solutions containing phosphate ions (including body fluids *in vivo* as well as cell culture medium or simulated body fluid *in vitro*[9]), even phosphate-free bioactive silicate glasses can form an apatite surface layer, a process which is usually considered a prerequisite for *in vivo* bioactivity and bone bonding.[10]

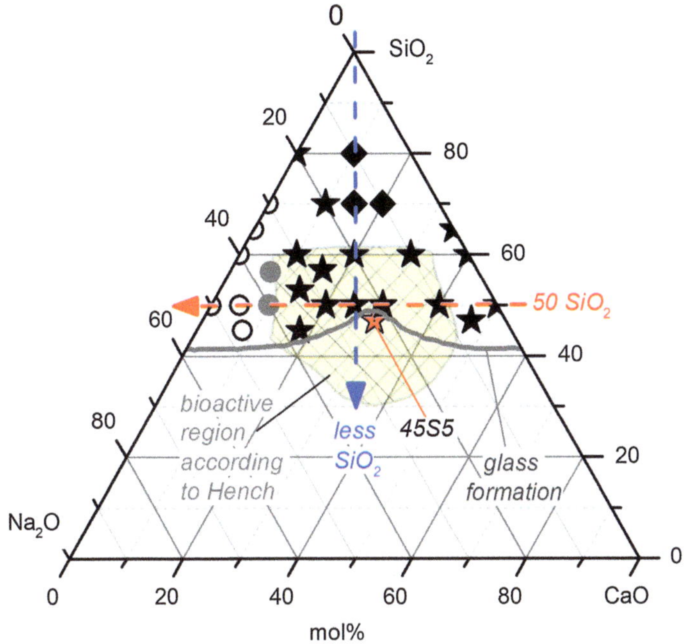

Figure 3.7 Compositional dependence of dissolution and apatite formation of $SiO_2 \cdot CaO \cdot Na_2O$ glasses immersed in simulated body fluid (SBF) for 30 days (diamonds: no apatite formation; stars: apatite formation; solid spheres: pronounced dissolution; circles: complete dissolution).[11] The composition of Bioglass 45S5 (containing additional 2.6 mol% of P_2O_5) is given for reference. The area in the centre is the bioactivity region according to Hench.[13] The area above the solid grey line is the glass-forming region. The blue dashed vertical arrow represents glasses with decreasing SiO_2 content, *i.e.* decreasing network connectivity, and constant $Na_2O:CaO$ ratio of 1 : 1. The red dotted horizontal arrow represents glasses with a constant network connectivity ($NC_{Si} = 2.0$; 50 mol% SiO_2) and increasing $Na_2O:CaO$ ratio.
(Plotted after data published in ref. 11.)

Figure 3.7 shows results from *in vitro* immersion of $SiO_2 \cdot CaO \cdot Na_2O$ glasses in simulated body fluid (SBF).[11] Glasses were classified depending on their degradation behaviour: showing complete dissolution, pronounced dissolution, formation of an apatite surface layer or no apatite formation when immersed in SBF for 30 days. For glasses of the composition $(80-x)SiO_2 \cdot (10+x/2)CaO \cdot (10+x/2)Na_2O$ (with x ranging from 0 to 30 mol%), *i.e.* glasses with decreasing network connectivity and a constant $Na_2O:CaO$ ratio (vertical dashed arrow in Figure 3.7) the solubility increases, with the first two compositions not forming apatite in SBF while the next two do. This is caused by their more disrupted silicate network, characterised by a network connectivity decreasing from 3.5 to 2.2 with decreasing SiO_2 content, allowing for easier water penetration owing to larger concentrations of nbO.[12] This means that for bioactive glasses, which are desired to release ions and degrade when in contact with body fluids, compositions with relatively high modifier contents are preferred.

Figure 3.8 shows a similar effect caused by changes in network connectivity: during immersion of $SiO_2 \cdot CaO \cdot Na_2O$ glasses in SBF, apatite formation was delayed as the network connectivity increased from 2.0 to 3.14. (For the glass with the highest network connectivity, the actual time-point of first apatite formation was not detected, but no apatite formation had occurred within 28 days.[10]) The glass network connectivity in this system also affected the *in vivo* bioactivity, shown as the relative amount of newly formed bone when the glasses were implanted into rabbit femurs: With increasing network connectivity, significantly less bone was formed[10] (Figure 3.8).

The network connectivity thus has a pronounced influence on silicate glass properties, but it is not the only contributing factor. The field strength (or, more simply, the charge-to-size ratio) of the modifier cation also directly affects glass properties. Calcium (Ca^{2+}) and sodium ions (Na^{+}) have comparable ionic radii[14] but they differ in their charge. Modifier ions of larger charge-to-size ratio form more effective ionic bridges between non-bridging oxygen ions. As a result, for glasses with constant network connectivity, the glass transition temperature decreases with increasing $Na_2O:CaO$ ratio, while their crystallisation tendency increases. Simultaneously, the glasses become less chemically stable. In Figure 3.7, glasses in the system $50SiO_2 \cdot (50+y)CaO \cdot yNa_2O$ (with y ranging from 0 to 50 mol%) with a constant network connectivity of 2 (horizontal dashed arrow) become more

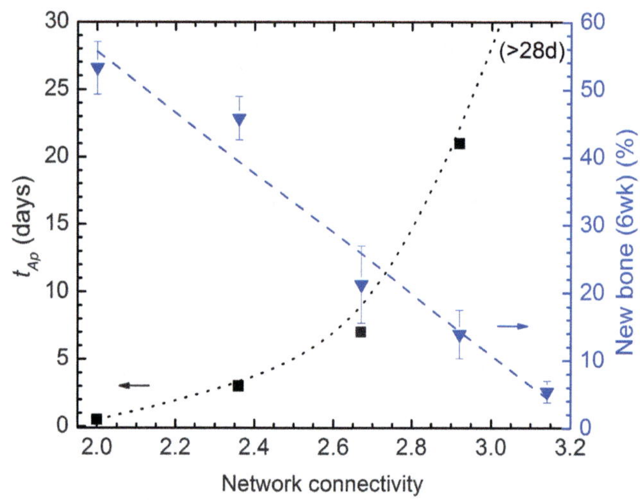

Figure 3.8 First time-point of apatite detection (t_{Ap}) during *in vitro* immersion studies of $SiO_2 \cdot CaO \cdot Na_2O$ glasses in SBF (black squares; left axis) and relative amount of newly formed bone at 6 weeks post implantation of the same glass compositions into the femur of rabbits (blue triangles; right axis).
(Calculated and plotted from data by Fujibayashi *et al.*[10])

degradable with increasing Na_2O content, *i.e.* with increasing $Na_2O:CaO$ ratio;[11] glasses containing between 0 and 30 mol% Na_2O formed apatite. The composition containing 40 mol% Na_2O showed pronounced dissolution, while the two compositions with higher Na_2O contents dissolved completely. Figure 3.7 also shows that the *in vitro* bioactivity of $SiO_2 \cdot CaO \cdot Na_2O$ glasses is not limited to the area originally indicated by Hench[13] but actually extends significantly beyond it.[11]

3.3.2 Sol–Gel Silicate Glasses

Silicate sol–gel glasses are obtained *via* polycondensation reactions from organic precursors (alkoxides such as tetraethyl orthosilicate) rather than by a melt–quench procedure. Owing to this preparation, by reaction in a water-based sol, the silicate network contains many Si–OH groups. Here the nominal glass composition does not represent the actual one, because water constitutes an additional glass component. ^{29}Si and ^{1}H MAS NMR investigations showed that, depending on the sintering temperature, the OH content varies but that up to two OH groups can be present per silicon atom in sol–gel bioactive glass 70S30C of the nominal composition $70SiO_2 \cdot 30CaO$ (mol%).[15] Sol–gel glasses tend to be more degradable and more bioactive than melt-derived glasses of the same nominal composition. The reason is that the large number of Si–OH groups present in sol–gel glasses not only lowers their network connectivity but also increases ion release and degradation. For example, a melt-derived glass of the composition $70SiO_2 \cdot 30CaO$ (mol%) is not bioactive while the sol–gel glass 70S30C has been shown to form apatite *in vitro*.[15]

3.4 Structure and Properties of Phosphate Glasses

Structurally, the effect of network modifier oxide incorporation into phosphate glasses is comparable to the effect in silicate glasses: P–O–P bonds are broken and non-bridging oxygen atoms are formed. The network connectivity of simple phosphate glasses (NC_P) can be calculated according to eqn (3.4):

$$NC_P = \frac{3[P_2O_5] - [M_2O] - [M'O]}{[P_2O_5]} \tag{3.4}$$

where $[P_2O_5]$, $[M_2O]$ and $[M'O]$ represent the molar percentages of phosphate, alkali and alkaline earth oxides, respectively. Vitreous P_2O_5 consists of $Q_P{}^3$ groups (Figure 3.9) exclusively. It therefore has a network connectivity of 3, meaning that it is less polymerised than vitreous silica. The reason is that the PO_4 tetrahedron has a terminal oxygen atom, which is connected to the phosphorus atom *via* a double bond and thus does not form a link to other phosphate tetrahedra. As modifier oxides are incorporated, $Q_P{}^2$ groups are formed and the relative amount of $Q_P{}^3$ species decreases (Figure 3.10).

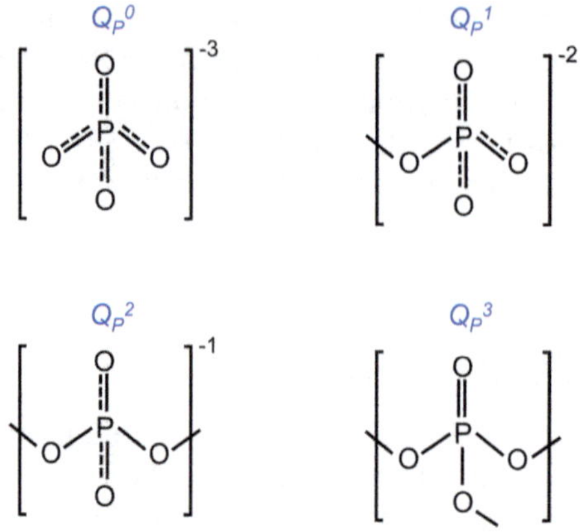

Figure 3.9 Phosphate tetrahedral species ($Q_P{}^n$) that can exist in phosphate glasses. Dashed lines denote delocalised electrons.

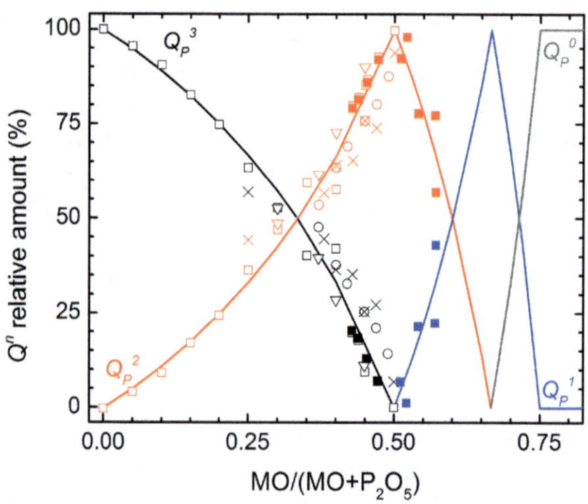

Figure 3.10 Changes in the relative amounts of $Q_P{}^n$ groups in $P_2O_5 \cdot MO$ phosphate glasses with increasing modifier (MO) content ($Q_P{}^3$: black, $Q_P{}^2$: red, $Q_P{}^1$: blue and $Q_P{}^0$: grey): solid lines represent theoretical percentages calculated assuming a binary distribution of $Q_P{}^n$ groups; symbols represent experimental amounts obtained from solid-state ^{31}P MAS NMR on calcium (triangles), strontium (circles), barium (crosses)[18] and sodium (squares, where solid squares represent glasses containing water as an additional modifier) phosphate glasses.[16,19,20]

In $Q_P{}^2$ groups the terminal oxygen and the non-bridging oxygen are indistinguishable, as the negative charge is delocalised over both oxygen atoms.[16] Similarly, in $Q_P{}^1$ and $Q_P{}^0$ groups no terminal (double-bonded) oxygen can be distinguished (Figure 3.9). Unlike for silicate glasses, highly disrupted phosphate glasses, where the ratio of modifier oxides : P_2O_5 is >0.5, can be obtained in an amorphous state and have been shown to contain $Q_P{}^1$ and $Q_P{}^0$ groups.[17]

Raman spectra allow for identification of different Q_P^n species present in a phosphate glass (Figure 3.11) while the relative amounts of Q_P^n groups can be quantified using ^{31}P MAS NMR (Figure 3.12). Similar to what is described above in Section 3.3, deconvolution of the NMR spectra allows for quantification of the relative amounts of Q_P^n, from which, subsequently, parameters such as experimental chain length, cross-link density or network connectivity can be calculated.

The phosphate tetrahedra can be arranged to form branching networks, chains or rings, and depending on their O : P ratio and network connectivity, phosphate glasses can be divided into different groups: ultraphosphate, metaphosphate and polyphosphate glasses.[23] Ultraphosphate glasses have an O : P ratio between 2.5 and 3.0, and their structure consists of a two-dimensional phosphate network having network connectivities between 3 and 2. With increasing amounts of modifier oxides the number of nbO, and thus the O : P ratio, increases. At an O : P ratio of 3.0, the glasses consist of Q_P^2 groups only,[21] *i.e.* chains of infinite length or rings, corresponding to a network connectivity of 2.0. Such glasses are usually referred to as metaphosphate glasses.

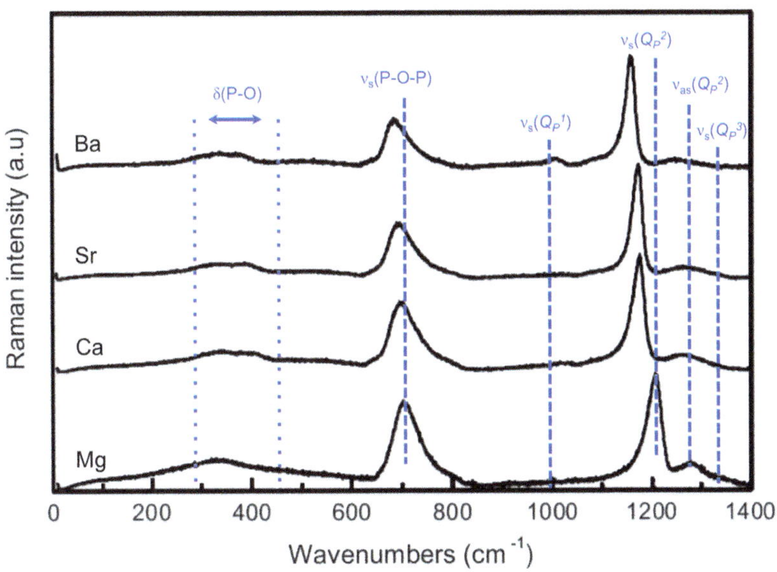

Figure 3.11 Raman spectra of alkaline earth metaphosphate glasses $M(PO_3)_2$ (or $P_2O_5 \cdot MO$), where M is either Mg, Ca, Sr or Ba. These spectra show the presence of Q_P^n related bands (symmetric and asymmetric stretch modes, ν_s or ν_{as}) corresponding to Q_P^2 mostly, with some low intensity bands of Q_P^1 (for $Ba(PO_3)_2$) or Q_P^3 (for $Mg(PO_3)_2$) present as well. A band corresponding to symmetric stretch modes of bridging oxygen atoms between phosphate tetrahedra, ν_s(P–O–P), can also be identified. The spectra also show that the band frequency depends on the field strength of the cation, as the bands move to lower wavenumbers in the order Mg > Ca > Sr > Ba.[21]

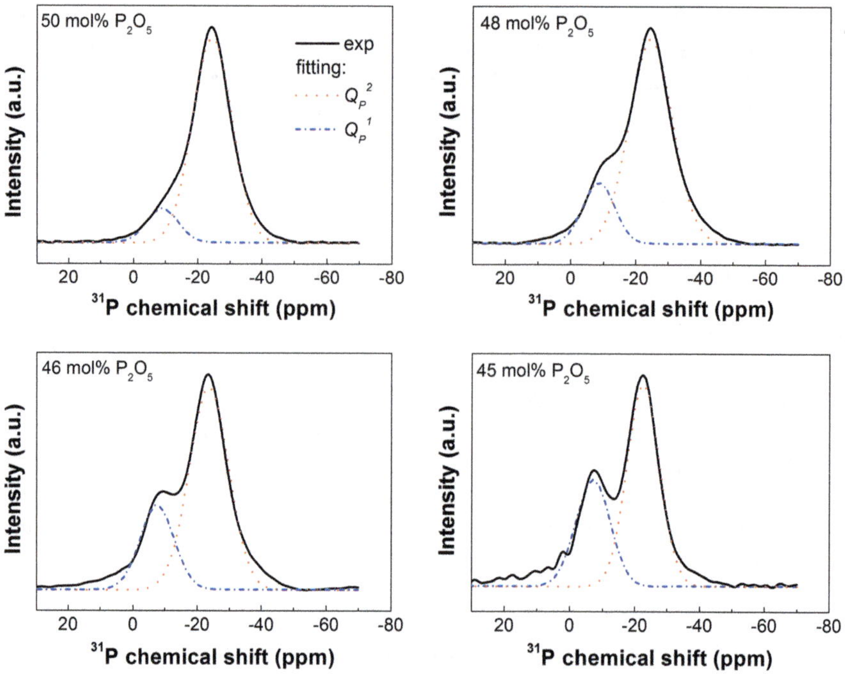

Figure 3.12 ^{31}P MAS NMR spectra of phosphate glasses in the system $P_2O_5 \cdot CaO \cdot MgO \cdot Na_2O \cdot TiO_2$ with varying phosphate content (50 to 45 mol%), decomposed into Gaussian curves for the positions of the Q_P^1 (about −4 to −6 ppm) and Q_P^2 peak (−17 to −22 ppm). Quantification of the relative amounts of Q_P^1 and Q_P^2 groups indicates a change from 12% Q_P^1 and 88% Q_P^2 (50 mol% P_2O_5) to 35% Q^1 and 65% Q^2 (45 mol% P_2O_5), corresponding to a decrease in chain length from 17 (50 mol% P_2O_5) to 6 phosphate tetrahedra (45 mol% P_2O_5).[22]

With increasing modifier content, *i.e.* increasing O : P ratio, the chains become shorter and the network connectivity decreases. Such glasses are referred to as polyphosphate glasses, and they contain more modifiers than phosphate. For this reason, they are often also called *invert* glasses. Here, the glass structure does not consist of an actual network any longer, and the ionic bonds between nbO and modifier cations, rather than the covalent P–O–P bonds, dominate the glass properties.[24] Owing to this increasing ionic character of the chemical bonds in the glass structure, the glass transition temperature (T_g) of $P_2O_5 \cdot MgO$ polyphosphate glasses[25] increases with increasing magnesium oxide content (Figure 3.13).

3.4.1 Structure and Dissolution

Vitreous P_2O_5 is very reactive and hygroscopic and adding network modifiers such as calcium oxide increases phosphate glass durability.[26] Still, ultraphosphate glasses (with >50 mol% P_2O_5) can contain large amounts of water

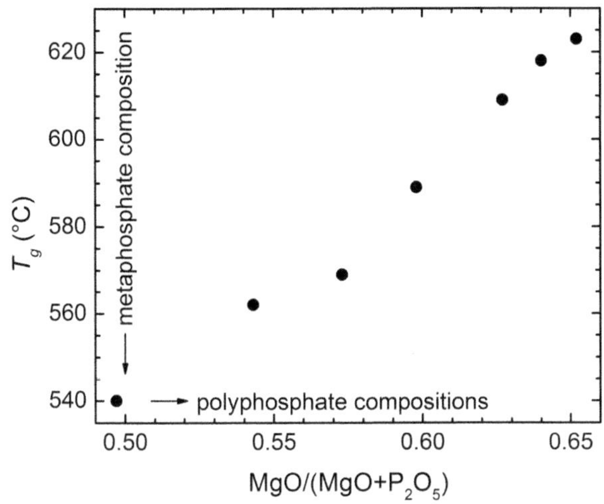

Figure 3.13 The glass transition temperature (T_g) of binary magnesium polyphosphate glasses ($P_2O_5 \cdot MgO$) increases with increasing MgO content owing to increasing ionic contributions to the chemical bonds in the glass (after Walter *et al.*[25]).

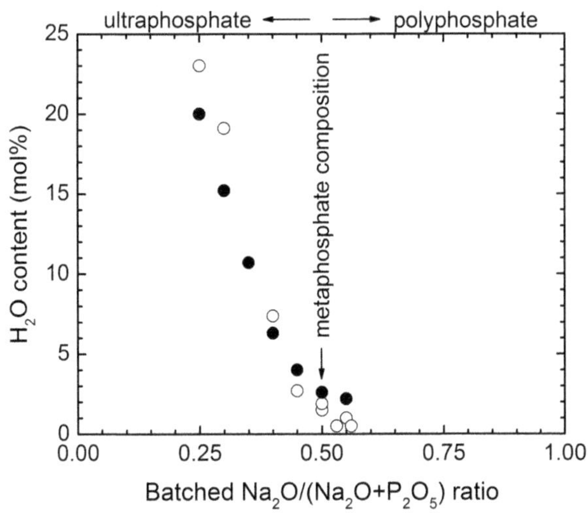

Figure 3.14 The water content of sodium phosphate glasses melted in open crucibles (after Brow *et al.*[20] and Gray & Klein[27]) varies with modifier content: ultraphosphate glasses contain large amounts of water when melted in open crucibles while meta- and polyphosphate glasses contain only small amounts.

(Figure 3.14), acting as an additional modifier forming P–OH groups, unless glasses are melted in a sealed ampoule.[20] Glasses with lower phosphate content (meta- or polyphosphate glasses) usually contain small amounts of water only, if any.

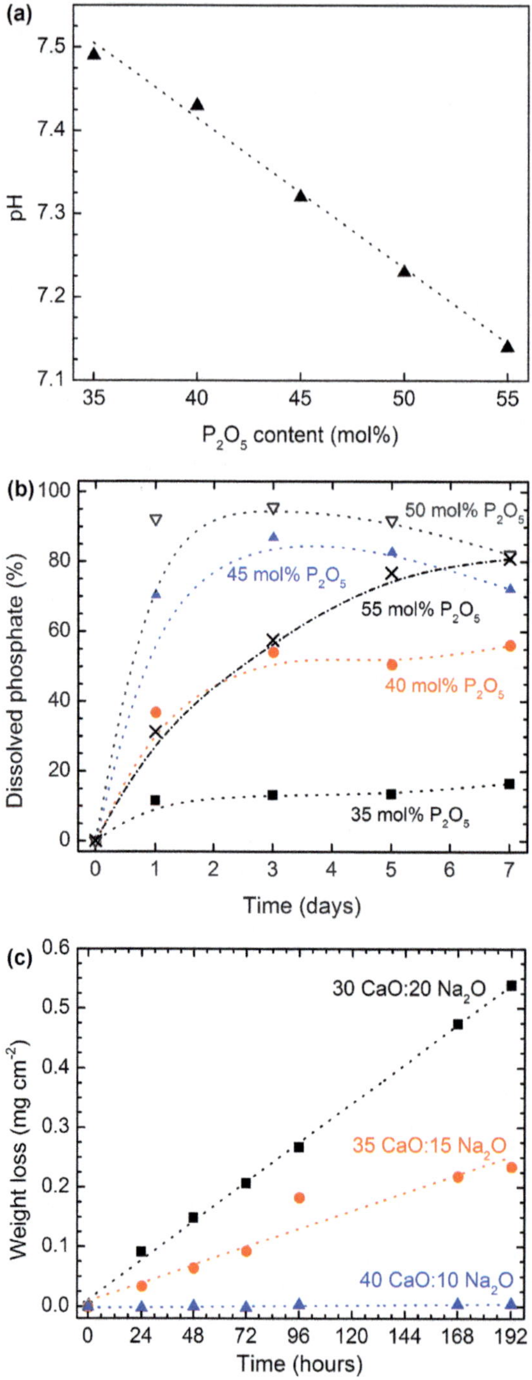
(a)
pH
7.5
7.4
7.3
7.2
7.1
35
40
45
50
55
P2O5 content (mol%)
(b)
Dissolved phosphate (%)
100
80
60
40
20
0
50 mol% P2O5
45 mol% P2O5
55 mol% P2O5
40 mol% P2O5
35 mol% P2O5
0
1
2
3
4
5
6
7
Time (days)
(c)
Weight loss (mg cm-2)
0.6
0.5
0.4
0.3
0.2
0.1
0.0
30 CaO:20 Na2O
35 CaO:15 Na2O
40 CaO:10 Na2O
0
24
48
72
96
120
144
168
192
Time (hours)

This is in contrast to silicate glasses, where addition of modifiers decreases the durability. The reason is that phosphate glasses and silicate glasses degrade by different mechanisms. When silicate glasses are in contact with water, modifier ions are released by an ion exchange process, leaving an ion-depleted silica gel behind.[8] Phosphate glasses, by contrast, dissolve by actual dissolution of the phosphate network, including both hydration of entire chains[28] and P–O–P hydrolysis.[29] Q_P^3 groups (present in vitreous P_2O_5 and ultraphosphate glasses) are more susceptible to water attack and hydrolysis[23] owing to the π electrons of the double-bonded terminal oxygen atom not being delocalised (Figure 3.9). P–O–P hydrolysis results in a pH decrease of the surrounding medium, and this pH decrease is more pronounced with increasing phosphate content in the glass (Figure 3.15a).

Variations in phosphate content have a pronounced influence on the dissolution behaviour (Figure 3.15b). For glasses with $\leq$50 mol% P_2O_5, solubility decreases with decreasing phosphate content. Glasses with >50 mol% P_2O_5 show a slower dissolution rate initially, but at later time-points they dissolve even faster than polyphosphate glasses. The reason is that for the dissolution of ultraphosphate glasses, P–O–P links need to be hydrolysed, while for polyphosphate glasses entire chains can be hydrated and go into solution directly (while some P–O–P hydrolysis may still occur).[28] The charge-to-size ratio of the modifier ion also affects dissolution, shown by a decrease in glass solubility with increasing CaO : Na_2O ratio (Figure 3.15c). This is caused by the higher field strength of the Ca^{2+} ion,[31] owing to its higher charge and similar ionic radius compared to the Na^+ ion. With its higher charge, Ca^{2+} binds together phosphate tetrahedra and phosphate network fragments more strongly than the Na^+ ion, and with its higher coordination number connects to more phosphate fragments.[24]

3.5 Borate Glasses

Vitreous B_2O_3 consists of planar $BØ_3$ trigonal groups. (In this context, "Ø" denotes bridging oxygen atoms.) The majority, *i.e.* up to 70–85%, of the $BØ_3$

Figure 3.15 (a), (b) Results of immersion experiments of $x\ P_2O_5\cdot(100-x)/2$ $CaO\cdot(100-x)/2\ Na_2O$ glasses in Tris buffer solution:[29] (a) pH of Tris buffer solution at 5 days (initial pH 7.4): with increasing phosphate content, the glasses cause a pH decrease owing to P–O–P hydrolysis. (b) Percentage of phosphate dissolved: The meta- and polyphosphate glasses (35 to 50 mol% P_2O_5) show rapid dissolution over the first day, with concentrations remaining constant afterwards; solubility decreases with decreasing phosphate content. The ultraphosphate glass (55 mol% P_2O_5) shows a different dissolution pattern from the other glasses, with slower dissolution at early time-points. (c) Weight loss of $50P_2O_5\cdot xCaO\cdot(50-x)Na_2O$ glasses immersed in deionised water (after Ahmed *et al.*[30]): with increasing CaO : Na_2O ratio, the dissolution rate decreases.

trigonals form six-membered boroxol rings (3*x* $BØ_3$; Figure 3.16a).[32–35] Trigonal $BØ_3$ units, which connect these rings, are often, if confusingly, named "isolated" $BØ_3$ units, as they are situated outside ring structures.

In contrast to phosphate or silicate glass, the initial addition of M_2O modifier oxides to B_2O_3 does not result in the formation of nbO on BO_3 groups. Instead, it increases the boron coordination from 3-fold to 4-fold, as neutral $BØ_3$ trigonals transform into negatively charged $[BØ_4]^-$ tetrahedra, which are charge balanced by the modifier cation M^+ (Figure 3.17). This means that the polymerisation degree, *i.e.* the network connectivity, increases upon initial incorporation of modifiers. These borate tetrahedra can be arranged in borate rings (Figure 3.16b) with one or two $[BØ_4]^-$ tetrahedra or in even larger superstructural units such as diborate (Figure 3.16c).

For even higher oxide additions, tetrahedral borate units convert back to trigonal borate groups and the network becomes more depolymerised as the number of nbO ions increases (Figure 3.17). Here, ring and chain isomers (Figure 3.16(d) and (e), respectively) of trigonal metaborate can be distinguished, while further depolymerisation results in the presence of pyroborate dimers (Figure 3.16f). For orthoborate, a second boron anomaly has been demonstrated as trigonal $BO_3{}^{3-}$ groups may convert into tetrahedral $[BØ_2O_2]^{3-}$ units (Figure 3.17), which fuse together into three membered orthoborate rings $[B_3O_9]^{9-}$.[39–41]

The equilibrium between trigonal and tetrahedral isomers depends on the type of modifier cation, which has therefore a significant impact on the

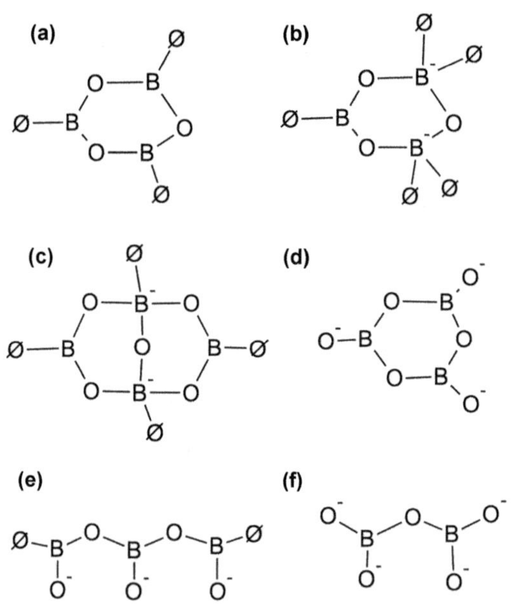

Figure 3.16 Superstructural units present in borate glasses: (a) boroxol ring, (b) borate ring, (c) diborate group, (d) ring metaborate, (e) chain metaborate and (f) pyroborate.[36–38] (Ø denotes bridging oxygen atoms.)

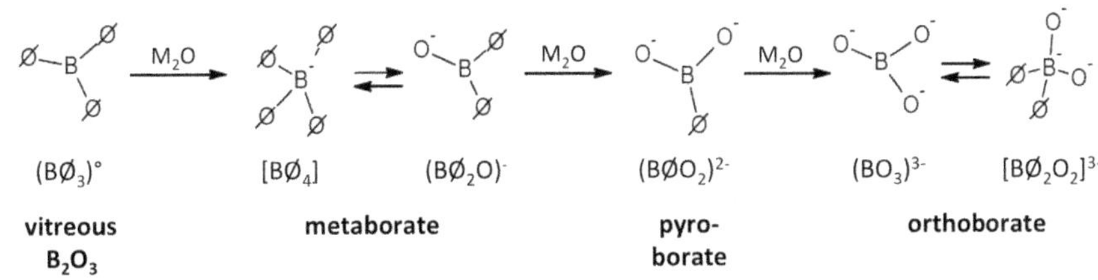

Figure 3.17 Changes in the short-range order of borate glasses with modifier (M_2O) addition: Initial addition of modifier oxides to vitreous B_2O_3 converts trigonal $BØ_3$ groups (Ø: bridging oxygen atom) to tetrahedral $[BØ_4]^-$ metaborate units. An equilibrium exists between tetrahedral ($[BØ_4]^-$) and trigonal ($BØ_2O^-$) metaborate entities. Further modifier addition converts metaborate groups to trigonal pyroborate and, subsequently, orthoborate groups. Trigonal $BO_3{}^{3-}$ groups may convert into tetrahedral $[BØ_2O_2]^{3-}$ units.

overall degree of polymerisation and, subsequently, on the glass properties.[3,36,39,40,42,43] For example, NMR and infrared spectroscopic studies[44] showed that the $BØ_2O^- \leftrightarrows [BO_4]^-$ isomerisation equilibrium shifts to the left as the cation radius increases from Li^+ to Cs^+ in $M_2O \cdot 2\ B_2O_3$ glasses.[45–47]

The polymerisation degree of the borate network can be described by the fraction of tetrahedrally coordinated boron, N_4:

$$N_4 = B_4/(B_3 + B_4) \tag{3.5}$$

where B_3 and B_4 denote three- and four-fold coordinated boron, respectively. For modifier contents below 33.3 mol%,

$$N_4 = R = M_2O/B_2O_3 \tag{3.6}$$

For binary $x\ M_2O \cdot (1-x)\ B_2O_3$ glasses with $x < 0.33\ M_2O$, the fraction of tetrahedral borate can therefore be estimated using eqn (3.7):[48]

$$N_4 = x/(1-x) \tag{3.7}$$

^{11}B MAS NMR can easily distinguish tetrahedral from trigonal borate units. Additionally, NMR can quantify the number of nbO connected to trigonal borate groups or determine if ring or out-of-ring configurations are present. For tetrahedral borate, the shift of the ^{11}B NMR signal is related to the number and type of next-nearest neighbours (see Section 3.6). Infrared spectroscopy also allows for differentiation of tetrahedral and trigonal borate entities, since the B–O stretching modes of trigonal borate groups are found at higher energies than those of $[BØ_4]^-$ (Figure 3.18).

3.5.1 Effect of Structure on Borate Glass Properties

The non-linear change in network connectivity with modifier content described above is the explanation for an interesting feature of borate glasses: if the modifier content in binary $M_2O \cdot B_2O_3$ borate glasses is varied, many

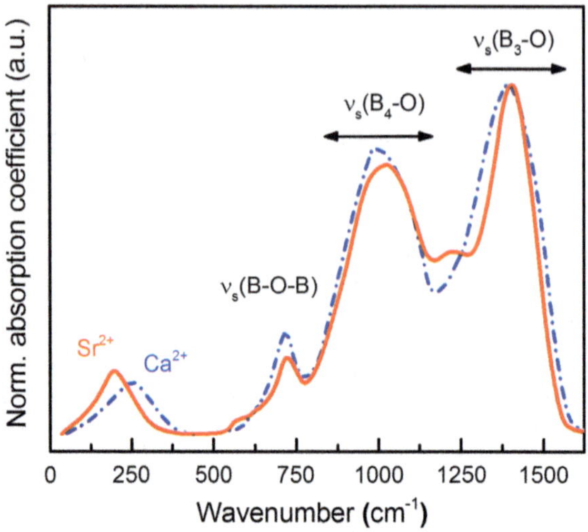

Figure 3.18 Normalised IR spectra of $55B_2O_3 \cdot 45MO$ glasses (M = Ca or Sr) showing the band envelopes of B–O stretching modes of trigonal (B_3; 800–1250 cm^{-1}) and tetrahedral (B_4; 1250–1600 cm^{-1}) borate units, the B–O–B bending mode near 715 cm^{-1}, as well as the cation motion bands in the far IR (data from ref. 49).

properties show a minimum or maximum for a certain $R = M_2O : B_2O_3$ ratio,[3] shown in Figure 3.19. As this behaviour is in contrast to that of silicate or phosphate glasses, which show continuous changes in properties, the phenomenon is commonly called the boron anomaly. As the polymerisation degree increases initially from 3 to 4, the glass transition temperature, T_g, increases and the coefficient of thermal expansion (CTE) decreases (Figure 3.19a and b). The subsequent depolymerisation for even higher modifier contents causes a decrease in T_g and an increase in CTE (Figure 3.19a and b). Tetrahedral metaborate groups also enhance glass strength, as $[BØ_4]^- Na^+$ plays a comparable role in this glass network to the fully bridged SiO_4 (Q^4) tetrahedra in silicate glasses. Trigonal BO_3 groups, by contrast, lower the glass strength.

Although the polymerisation degree of the borate network increases upon the conversion of $BØ_3$ into $[BØ_4]^-$ groups, the extrema observed in the properties of binary borate glasses do not necessarily coincide with the maximum fraction of tetrahedral boron, N_4.[38] The glass properties depend just as strongly on the cross-linking capability of the modifier cations.[49,50]

Borate glasses are soluble in water[52] and have recently attracted considerable interest for use as degradable wound dressings.[53] Borate glass solubility is affected by the borate network polymerisation in a similar way as the thermal properties: solubility decreases upon initial incorporation of modifier oxides (Figure 3.19c) caused by increasing amounts of $[BØ_4]^- M^+$ groups. Once the modifier content increases further, solubility increases again owing to a more depolymerised borate network containing trigonal

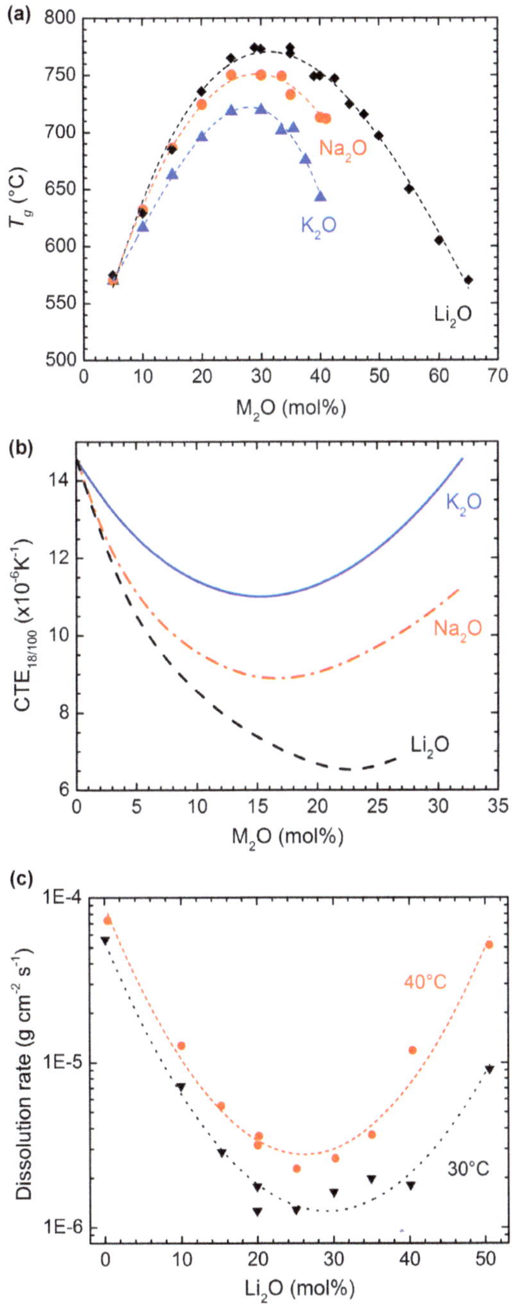

Figure 3.19 Variations of (a) transition temperature T_g (after Kamitsos *et al.*[50]) and (b) coefficients of thermal expansion CTE (after Vogel[3]) as a function of the modifier oxide concentration in binary $B_2O_3 \cdot M_2O$ glasses with M = Li, K, Na. (c) Dissolution rate of $B_2O_3 \cdot Li_2O$ glasses as a function of Li_2O content (after Velez *et al.*[51]).

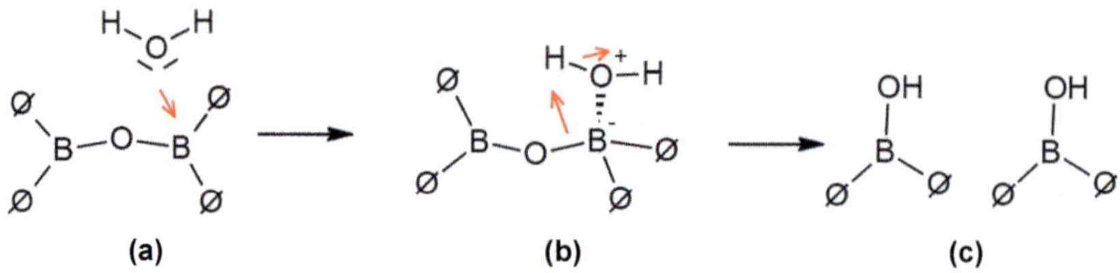

Figure 3.20 Trigonal BO_3 groups act as Lewis acids and are easily attacked by water molecules, acting as Lewis bases, resulting in hydrolysis of the borate network (Ø: bridging oxygen atom).

BO_3 groups. Figure 3.19c illustrates this for $B_2O_3 \cdot Li_2O$ glasses, with the dissolution rate going through a minimum at about 20 to 30 mol% Li_2O.

An explanation for this behaviour is that tetrahedral metaborate groups enhance glass chemical stability; trigonal BO_3 groups, on the other hand, enhance hydrolysis (Figure 3.20). Water, a Lewis base, donates an electron to [BØ$_3$], a Lewis acid, and a new bond between O and B is formed. This newly generated, larger compound, a Lewis acid–base complex, undergoes a re-arrangement of its bonds, resulting in the formation of trigonal borate groups and B–OH groups. This hydrolysis reaction is favoured by the effortless coordination change of boron, the Lewis acid character of trigonal BO_3, as well as easy access of water molecules to the planar BO_3 unit.

3.6 Mixed Glass Former Systems

Many bioactive glasses, among them Larry Hench's first bioactive glass, Bioglass 45S5, are soda lime–silicate glasses with small amounts (usually up to 3 mol%) of P_2O_5. As it was originally thought that these glasses consisted of a mixed silicate–phosphate network, they were called **phosphosilicate glasses**. From structural investigations with, for example, solid-state NMR spectroscopy,[54] we know that most (if not all) phosphate is actually present as orthophosphate (Q_P^0) in these glasses (Figure 3.21a), charge-balanced by modifier ions.[54] This structure is visualised in Figure 3.22. A more correct name would therefore be phosphate silicate glasses. The structural role of phosphate, with orthophosphate (PO_4^{3-}) scavenging modifier cations from the silicate network, needs to be taken into account when calculating the network connectivity of bioactive phosphosilicate glasses ($NC_{Si(P)}$), as shown in eqn (3.8):

$$NC_{Si(P)} = \frac{4[SiO_2] + 6[P_2O_5] - 2([M_2O] + [M'O])}{[SiO_2]} \tag{3.8}$$

Network connectivities of bioactive phosphosilicate glasses are similar to those for bioactive silicate glasses (Section 3.3), typically lying between 2.0 and 2.6. Therefore the silicate network of these glasses is highly disrupted, consisting mostly of Q^2 groups, with smaller amounts of Q^3 (Figure 3.21b) and possibly Q^1. Well-known examples include Bioglass 45S5 ($NC_{Si(P)} = 2.11$)

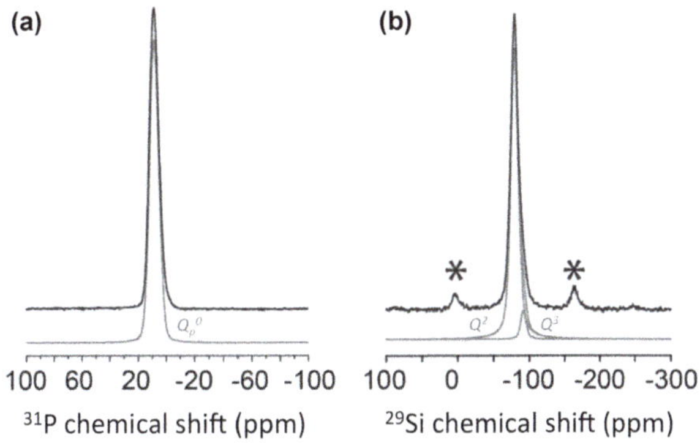

Figure 3.21 Results from (a) ^{31}P and (b) ^{29}Si MAS NMR on Bioglass 45S5 (ref. 58 – Published by the PCCP Owner Societies): (a) ^{31}P MAS NMR spectrum (black line) and its simulation (grey line) giving a single peak at 8.0 ppm corresponding to orthophosphate (Q_P^0); (b) ^{29}Si MAS NMR spectrum (black line; * spinning sidebands) and the deconvolution into Q^2 (−79.9 ppm; 90%) and Q^3 (−94.1 ppm; 10%) components (grey lines).

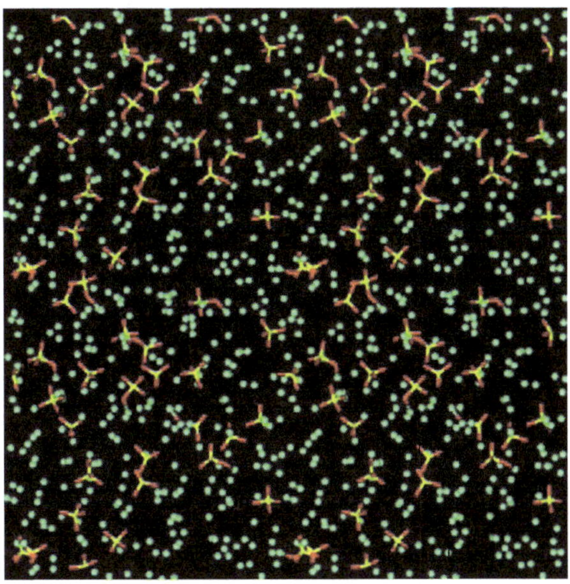

Figure 3.22 Phosphate in Bioglass 45S5 obtained from molecular dynamics simulations[59] showing phosphate present as orthophosphate groups. Orthophosphate groups are shown as ball-and-stick models with phosphorus atoms shown in yellow and oxygen in red; silicon atoms are presented as turquoise spheres and modifier cations and oxygen atoms of the silicate network are omitted.
(Reprinted with permission from A. Tilocca and A. N. Cormack, Structural effects of phosphorus inclusion in bioactive silicate glasses, *J. Phys. Chem. B*, 2007, **111**, 14256–14264. Copyright (2007) American Chemical Society.)

or BonAlive S53P4 ($NC_{Si(P)} = 2.54$).[55] Orthophosphate is readily released from such glasses[56] and has been shown to accelerate and improve apatite formation *in vitro*.[57]

Borosilicate glasses are among the most important technical glass systems, being used in a range of applications from laboratory glassware (Duran, Pyrex) to optical glasses (BK7). Small amounts of borate (typically up to 3 mol%) have been incorporated into a range of bioactive silicate and phosphosilicate glasses,[60] while in a series of bioactive glasses based on composition 13-93[60] silica was partially to fully replaced by B_2O_3.[61]

For low concentrations of alkali oxides, BO_3 entities are transformed into $[BØ_4]^-$ tetrahedra, which play a similar structural role as fully polymerised Q^4, *i.e.* SiO_4, tetrahedra. In these glasses, the silicate and borate groups are linked preferentially *via* trigonal borate groups, while tetrahedral borate groups are preferentially connected to BO_3 rather than to silicate groups (Figure 3.23). As a consequence low alkaline borosilicate glasses are prone to phase separation.[39,64–66] At higher modifier oxide concentrations non-bridging oxygens form, which are preferentially present at silicate rather than at borate groups.[67] Several models can be found in the literature for the calculation of N_4 in borosilicate glasses, either derived from a thermodynamic approach[68] or based on NMR studies.[69–71]

Another group of mixed glass former glasses, **borophosphate glasses**, have also been suggested for biomedical use.[72] Here, phosphate and borate

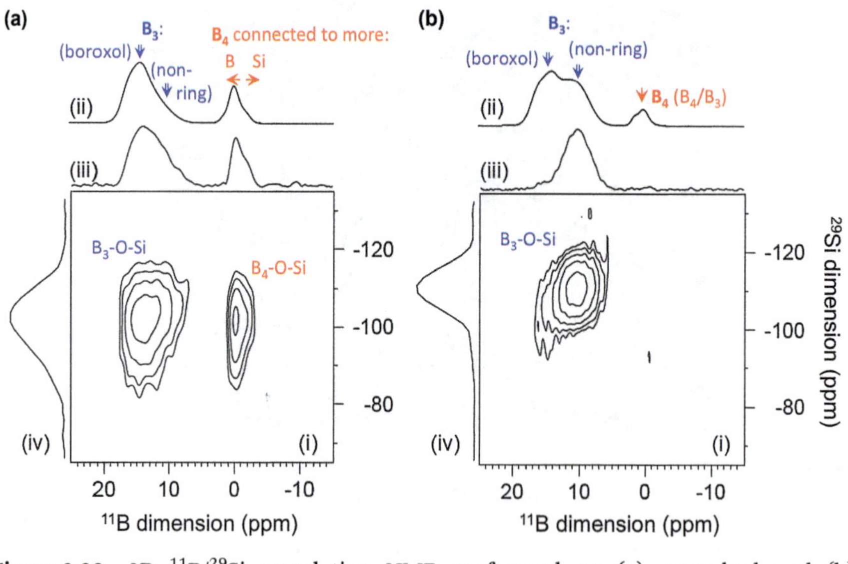

Figure 3.23 2D $^{11}B/^{29}Si$ correlation NMR performed on (a) quenched and (b) annealed sodium borosilicate glass ($15Na_2O \cdot 42.5B_2O_3 \cdot 42.5SiO_2$). The 2D maps (i) are accompanied by (ii) the ^{11}B MAS NMR spectrum and (iii) the ^{11}B and (iv) ^{29}Si 2D map projections.[62] B_4-O–Si bonds are present in the quenched glass but disappear after annealing, while the number of B_3-O–Si bonds is only slightly reduced.[63]

groups form a mixed network containing B–O–P bonds.[72,73] **Sulfophosphate glasses,**[74] which contain isolated sulphate groups (SO_4^{2-}) between the phosphate chains rather than a mixed phosphate–sulphate network,[75] have recently been introduced as a new class of bioactive glasses.[76]

3.7 Conclusions and Outlook

Taking a more structural approach when designing bioactive glasses will allow us to design custom-made bioactive glasses which exhibit properties tailored to their application. For simple glass compositions, the glass structure can be estimated from the composition, while for more complicated glasses, particularly mixed glass former systems, detailed structural investigations using vibrational spectroscopy or solid-state NMR (or a combination of both) is usually necessary. Understanding how the glass structure, *e.g.* the network polymerisation, influences and determines the glass properties may help to further improve bioactive glass properties and widen their range of applications.

References

1. J. E. Shelby, *Introduction to Glass Science and Technology*, The Royal Society of Chemistry, Cambridge, 2nd edn, 2005.
2. C. D. Huang and A. N. Cormack, The structure of sodium silicate glass, *J. Chem. Phys.*, 1990, **93**, 8180–8186.
3. W. Vogel, *Glass Chemistry*, Springer, Berlin, Heidelberg, New York, London, 2nd edn, 1994.
4. H. Maekawa, T. Maekawa, K. Kawamura and T. Yokokawa, The structural groups of alkali silicate glasses determined from ^{29}Si MAS-NMR, *J. Non-Cryst. Solids*, 1991, **127**, 53–64.
5. K. J. D. MacKenzie and M. E. Smith, *Multinuclear Solid-state NMR of Inorganic Materials, Pergamon*, Elsevier Science, Amsterdam, 2002.
6. R. K. Mishra and K. S. Dubey, Glass forming ability of materials: A thermodynamic approach, *J. Non-Cryst. Solids*, 2009, **355**, 2199–2204.
7. N. H. Ray, Composition-property relationships in inorganic oxide glasses, *J. Non-Cryst. Solids*, 1974, **15**, 423–434.
8. L. L. Hench and D. E. Clark, Physical chemistry of glass surfaces, *J. Non-Cryst. Solids*, 1978, **28**, 83–105.
9. T. Kokubo, H. Kushitani, S. Sakka, T. Kitsugi and T. Yamamuro, Solutions able to reproduce *in vivo* surface-structure changes in bioactive glass-ceramic A-W, *J. Biomed. Mater. Res.*, 1990, **24**, 721–734.
10. S. Fujibayashi, M. Neo, H. M. Kim, T. Kokubo and T. Nakamura, A comparative study between *in vivo* bone ingrowth and in vitro apatite formation on Na_2O-CaO-SiO_2 glasses, *Biomaterials*, 2003, **24**, 1349–1356.
11. H. M. Kim, F. Miyaji, T. Kokubo, C. Ohtsuki and T. Nakamura, Bioactivity of Na_2O-CaO-SiO_2 glasses, *J. Am. Ceram. Soc.*, 1995, **78**, 2405–2411.

12. A. Tilocca and A. N. Cormack, The initial stages of bioglass dissolution: a Car-Parrinello molecular-dynamics study of the glass-water interface, *Philos. Trans. R. Soc., A*, 2011, **467**, 2102–2111.
13. L. L. Hench, Bioceramics: From concept to clinic, *J. Am. Ceram. Soc.*, 1991, **74**, 1487–1510.
14. R. D. Shannon, Revised effective ionic radii and systematic studies of interatomic distances in halides and chalcogenides, *Acta Crystallogr.*, 1976, **A32**, 751–767.
15. J. R. Jones, T. F. Kemp and M. E. Smith, Effect of OH content on the bioactivity of sol-gel derived glass foam scaffolds, *Key Eng. Mater.*, 2006, **309–311**, 1031–1034.
16. R. K. Brow, Review: the structure of simple phosphate glasses, *J. Non-Cryst. Solids*, 2000, **263**, 1–28.
17. T. Kasuga and Y. Abe, Calcium phosphate invert glasses with soda and titania, *J. Non-Cryst. Solids*, 1999, **243**, 70–74.
18. P. Losso, B. Schnabel, C. Jager, U. Sternberg, D. Stachel and D. O. Smith, ^{31}P NMR investigations of binary alkaline earth phosphate glasses of ultra phosphate composition, *J. Non-Cryst. Solids*, 1992, **143**, 265–273.
19. R. K. Brow, D. R. Tallant, J. J. Hudgens, S. W. Martin and A. D. Irwin, The short-range structure of sodium ultraphosphate glasses, *J. Non-Cryst. Solids*, 1994, **177**, 221–228.
20. R. K. Brow, R. J. Kirkpatrick and G. L. Turner, The short-range structure of sodium-phosphate glasses 1. MAS NMR-studies, *J. Non-Cryst. Solids*, 1990, **116**, 39–45.
21. L. L. Velli, C. P. E. Varsamis, E. I. Kamitsos, D. Möncke and D. Ehrt, Structural investigation of metaphosphate glasses, *Phys. Chem. Glasses*, 2005, **46**, 178–181.
22. D. S. Brauer, C. Rüssel and J. Kraft, Solubility of glasses in the system P_2O_5-CaO-MgO-Na_2O-TiO_2: Experimental and modeling using artificial neural networks, *J. Non-Cryst. Solids*, 2007, **353**, 263–270.
23. J. R. van Wazer, *Phosphorus and Its Compounds*, Interscience, New York, 1958.
24. U. Hoppe, G. Walter, R. Kranold and D. Stachel, Structural specifics of phosphate glasses probed by diffraction methods: a review, *J. Non-Cryst. Solids*, 2000, **263**, 29–47.
25. G. Walter, J. Vogel, U. Hoppe and P. Hartmann, Structural study of magnesium polyphosphate glasses, *J. Non-Cryst. Solids*, 2003, **320**, 210–222.
26. D. S. Brauer, Phosphate glasses, in *Bio-glasses – An Introduction*, ed. J. R. Jones and A. G. Clare, Wiley, New York, Chichester, Brisbane, Singapore, Toronto, 2012, pp. 45–63.
27. P. E. Gray and L. C. Klein, The chemical durability of sodium ultraphosphate glasses, *Glass Technol.*, 1983, **24**, 202–206.
28. B. C. Bunker, G. W. Arnold and J. A. Wilder, Phosphate glass dissolution in aqueous solutions, *J. Non-Cryst. Solids*, 1984, **64**, 291–316.
29. F. Döhler, A. Mandlule, L. van Wüllen, M. Friedrich and D. S. Brauer, ^{31}P NMR characterisation of phosphate fragments during dissolution of

calcium sodium phosphate glasses, *J. Mater. Chem. B*, 2015, **3**, 1125–1134.
30. I. Ahmed, M. Lewis, I. Olsen and J. C. Knowles, Phosphate glasses for tissue engineering: Part 1. Processing and characterisation of a ternary-based P_2O_5-CaO-Na_2O glass system, *Biomaterials*, 2004, **25**, 491–499.
31. J. K. Christie, R. I. Ainsworth, D. Di Tommaso and N. H. de Leeuw, Nanoscale chains control the solubility of phosphate glasses for bio-medical applications, *J. Phys. Chem. B*, 2013, **117**, 10652–10657.
32. M. A. Bursukova, E. P. Kaschieva, Y. Y. Ivanova and Y. B. Dimitriev, Electron diffraction study of B_2O_3 glasses obtained at different cooling rates, *Phys. Chem. Glasses*, 2000, **41**, 337–338.
33. A. C. Hannon, D. I. Grimley, R. A. Hulmc, A. C. Wright and R. N. Sinclair, Boroxol groups in vitreous boron oxide: new evidence from neutron diffraction and inelastic neutron scattering studies, *J. Non-Cryst. Solids*, 1994, **177**, 299–316.
34. B. N. Meera, A. K. Sood, N. Chandrabhas and J. Ramakrishna, Raman study of lead borate glasses, *J. Non-Cryst. Solids*, 1990, **126**, 224–230.
35. F. L. Galeener and A. E. Geissberger, Raman studies of B_2O_3 glass structure: $^{10}B \rightarrow {}^{11}B$ isotopic substitution, *J. Phys.*, 1982, **43**, 343–346.
36. E. I. Kamitsos and G. D. Chryssikos, Borate glass structure by Raman and infrared spectroscopies, *J. Mol. Struct.*, 1991, **247**, 1–16.
37. B. N. Meera and J. Ramakrishna, Raman spectral studies of borate glasses, *J. Non-Cryst. Solids*, 1993, **159**, 1–21.
38. A. C. Wright, Borate structures: crystalline and vitreous, *Phys. Chem. Glasses: Eur. J. Glass Sci. Technol., Part B*, 2010, **51**, 1–39.
39. D. Möncke, E. I. Kamitsos, D. Palles, R. Limbach, A. Winterstein-Beckmann, T. Honma, Z. Yao, T. Rouxel and L. Wondraczek, Transition and post-transition metal ions in borate glasses: Borate ligand speciation, cluster formation, and their effect on glass transition and mechanical properties, *J. Chem. Phys.*, 2016, DOI: 10.1063/1.4962323, in press.
40. A. Winterstein-Beckmann, D. Möncke, D. Palles, E. I. Kamitsos and L. Wondraczek, Structure and properties of orthoborate glasses in the Eu_2O_3–(Sr,Eu)O–B_2O_3 quaternary, *J. Phys. Chem. B*, 2015, **119**, 3259–3272.
41. A. Winterstein-Beckmann, D. Möncke, D. Palles, E. I. Kamitsos and L. Wondraczek, Structure–property correlations in highly modified Sr, Mn-borate glasses, *J. Non-Cryst. Solids*, 2013, **376**, 165–174.
42. G. D. Chryssikos, E. I. Kamitsos, W. M. Risen and A. Raman, investigation of cadmium borate and borogermanate glasses, *J. Non-Cryst. Solids*, 1987, **93**, 155–168.
43. A. Vegiri, C. P. E. Varsamis and E. I. Kamitsos, Composition and temperature dependence of cesium-borate glasses by molecular dynamics, *J. Chem. Phys.*, 2005, **123**, 014508.
44. E. I. Kamitsos, Y. D. Yiannopoulos, C. P. Varsamis and H. Jain, Structure-property correlation in glasses by infrared reflectance spectroscopy, *J. Non-Cryst. Solids*, 1997, **222**, 59–68.

45. E. I. Kamitsos, M. A. Karakassides and G. D. Chryssikos, Structure of borate glasses. 1. Raman-study of cesium, rubidium, and potassium borate glasses, *Phys. Chem. Glasses*, 1989, **30**, 229–234.
46. E. I. Kamitsos, Modifying role of alkali-metal cations in borate glass networks, *J. Phys. Chem.*, 1989, **93**, 1604–1611.
47. E. I. Kamitsos, G. D. Chryssikos and M. A. Karakassides, Glass-transition phenomena and cation vibrations in alkali borate glasses, *Phys. Chem. Glasses*, 1988, **29**, 121–126.
48. P. J. Bray and M. L. Lui, NMR study of structure and bonding in glasses, in *Structure and Bonding in Noncrystalline Solids*, ed. G. E. Walrafen and A. G. Revesz, Springer, New York, 1986, pp. 285–301.
49. Y. D. Yiannopoulos, G. D. Chryssikos and E. I. Kamitsos, Structure and properties of alkaline earth borate glasses, *Phys. Chem. Glasses*, 2001, **42**, 164–172.
50. E. I. Kamitsos, C. P. E. Varsamis and A. Vegiri, *Spectroscopic Studies of Mobile Cations in Glass*, International Congress of Glass, Society of Glass Technology, Edinburgh, Scotland, 2001, pp. 234–246.
51. M. H. Velez, H. L. Tuller and D. R. Uhlmann, Chemical durability of lithium borate glasses, *J. Non-Cryst. Solids*, 1982, **49**, 351–362.
52. J. L. George and R. K. Brow, *In-situ* characterization of borate glass dissolution kinetics by μ-Raman spectroscopy, *J. Non-Cryst. Solids*, 2015, **426**, 116–124.
53. S. B. Jung, Bioactive borate glasses, in *Bio-glasses – An Introduction*, ed. J. R. Jones and A. G. Clare, Wiley, New York, Chichester, Brisbane, Singapore, Toronto, 2012, pp. 75–94.
54. M. W. G. Lockyer, D. Holland and R. Dupree, NMR investigation of the structure of some bioactive and related glasses, *J. Non-Cryst. Solids*, 1995, **188**, 207–219.
55. D. S. Brauer, Bioactive glasses—structure and properties, *Angew. Chem., Int. Ed.*, 2015, **54**, 4160–4181.
56. L. Bingel, D. Groh, N. Karpukhina and D. S. Brauer, Influence of dissolution medium pH on ion release and apatite formation of Bioglass® 45S5, *Mater. Lett.*, 2015, **143**, 279–282.
57. M. D. O'Donnell, S. J. Watts, R. G. Hill and R. V. Law, The effect of phosphate content on the bioactivity of soda-lime-phosphosilicate glasses, *J. Mater. Sci.: Mater. Med.*, 2009, **20**, 1611–1618.
58. R. A. Martin, H. L. Twyman, G. J. Rees, J. M. Smith, E. R. Barney, M. E. Smith, J. V. Hanna and R. J. Newport, A structural investigation of the alkali metal site distribution within bioactive glass using neutron diffraction and multinuclear solid state NMR, *Phys. Chem. Chem. Phys.*, 2012, **14**, 12105–12113.
59. A. Tilocca and A. N. Cormack, Structural effects of phosphorus inclusion in bioactive silicate glasses, *J. Phys. Chem. B*, 2007, **111**, 14256–14264.
60. M. Brink, T. Turunen, R. P. Happonen and A. Yli-Urpo, Compositional dependence of bioactivity of glasses in the system Na_2O-K_2O-MgO-CaO-B_2O_3-P_2O_5-SiO_2, *J. Biomed. Mater. Res.*, 1997, **37**, 114–121.

61. M. N. Rahaman, D. E. Day, B. Sonny Bal, Q. Fu, S. B. Jung, L. F. Bonewald and A. P. Tomsia, Bioactive glass in tissue engineering, *Acta Biomater.*, 2011, **7**, 2355–2373.
62. D. Möncke, G. Tricot, A. Winterstein-Beckmann, L. Wondraczek and E. I. Kamitsos, On the connectivity of borate tetrahedra in borate and borosilicate glasses, *Phys. Chem. Glasses: Eur. J. Glass Sci. Technol., Part B*, 2015, **56**, 203–211.
63. D. Möncke, D. Ehrt, H. Eckert and V. Mertens, Influence of melting and annealing conditions on the structure of borosilicate glasses, *Phys. Chem. Glasses*, 2003, **44**, 113–116.
64. M. P. Brungs and E. R. Cartney, Structure of sodium borosilciate glasses, *Phys. Chem. Glasses*, 1975, **16**, 48–52.
65. L. S. Du and J. F. Stebbins, Solid-state NMR study of metastable immiscibility in alkali borosilicate glasses, *J. Non-Cryst. Solids*, 2003, **315**, 239–255.
66. B. C. Bunker, D. R. Tallant, R. J. Kirkpatrick and G. L. Turner, Multinuclear nuclear magnetic resonance and Raman investigation of sodium borosilicate glass structures, *Phys. Chem. Glasses*, 1990, **31**, 30–41.
67. A. Winterstein-Beckmann, D. Möncke, D. Palles, E. I. Kamitsos, L. Wondraczek and A. Raman-spectroscopic, study of indentation-induced structural changes in technical alkali-borosilicate glasses with varying silicate network connectivity, *J. Non-Cryst. Solids*, 2014, **405**, 196–206.
68. N. M. Vedishcheva, I. G. Polyakova and A. C. Wright, Short and intermediate range order in sodium borosilicate glasses: a quantitative thermodynamic approach, *Phys. Chem. Glasses: Eur. J. Glass Sci. Technol., Part B*, 2014, **55**, 225–236.
69. S. Feller, T. Mullenbach, M. Franke, S. Bista, A. O'Donovan-Zavada, K. Hopkins, D. Starkenberg, J. McCoy, D. Leipply, J. Stansberry, E. Troendle, M. Affatigato, D. Holland, M. E. Smith, S. Kroeker, V. K. Michaelis and J. E. C. Wren, Structure and properties of barium and calcium borosilicate glasses, *Phys. Chem. Glasses: Eur. J. Glass Sci. Technol., Part B*, 2012, **53**, 210–218.
70. Y. H. Yun and P. J. Bray, Nuclear magnetic resonance studies of the glasses in the system Na_2O-B_2O_3-SiO_2, *J. Non-Cryst. Solids*, 1978, **27**, 363–380.
71. W. J. Dell and P. J. Bray, ^{11}B NMR studies and structural modeling of Na_2O-B_2O_3-SiO_2 glasses of high soda content, *J. Non-Cryst. Solids*, 1983, **58**, 1–16.
72. D. Carta, D. Qiu, P. Guerry, I. Ahmed, E. A. Abou Neel, J. C. Knowles, M. E. Smith and R. J. Newport, The effect of composition on the structure of sodium borophosphate glasses, *J. Non-Cryst. Solids*, 2008, **354**, 3671–3677.
73. N. Sharmin, M. S. Hasan, C. D. Rudd, D. Boyd, U. Werner-Zwanziger, I. Ahmed and A. J. Parsons, Effect of boron oxide addition on the viscosity-temperature behaviour and structure of phosphate-based

glasses, *J. Biomed. Mater. Res., Part B*, 2016, DOI: 10.1002/jbm.b.33610, in press.
74. K. Ghosh, G. K. DasMohapatra and N. SoodBiswas, Glass formation in K_2SO_4-CaO-P_2O_5 system, *Phys. Chem. Glasses*, 2003, **44**, 313–318.
75. A. Thieme, D. Möncke, R. Limbach, S. Fuhrmann, E. I. Kamitsos and L. Wondraczek, Structure and properties of alkali and silver sulfophosphate glasses, *J. Non-Cryst. Solids*, 2015, **410**, 142–150.
76. D. C. Bassett, R. Meszaros, D. Orzol, M. Woy, Y. Ling Zhang, K. Tiedemann, L. Wondraczek, S. Komarova and J. E. Barralet, A new class of bioactive glasses: Calcium–magnesium sulfophosphates, *J. Biomed. Mater. Res., Part A*, 2014, **102**, 2842–2848.

CHAPTER 4

Molecular Dynamics Simulations of Bioactive Glass Structure and In vitro *Reactivity*

ANTONIO TILOCCA

Department of Chemistry, University College London,
London WC1H 0AJ, UK
Email: a.tilocca@ucl.ac.uk

4.1 Introduction

Tailoring the bioreactivity of a biomaterial for different specific goals and applications is a key but complicated task, because of our limited understanding of the fundamental links between the chemical composition and the behaviour of a material in a biological environment. Whereas adjusting the composition of amorphous biomaterials such as bioactive glasses,[1] for instance, by introducing ionic dopants with favourable properties,[2–4] is relatively straightforward, the effect of specific compositional changes on the bioactivity is not as clear. Therefore, lengthy trial-and-error procedures are often required to explore a wide range of potentially interesting compositions and identify a few suitable candidates for a new application. The *a priori* knowledge of the effect of a particular substitution within a reference composition would save a large amount of time and resources, by allowing researchers to screen out unsuitable compositions and focus on

RSC Smart Materials No. 23
Bioactive Glasses: Fundamentals, Technology and Applications
Edited by Aldo R. Boccaccini, Delia S. Brauer and Leena Hupa

Published by the Royal Society of Chemistry, www.rsc.org

synthesizing, characterizing, and testing a narrower range of samples. How can this *a priori* knowledge be achieved, or how can we improve our ability to *predict* the biological performance of a glass? One possible route is through the glass structure. The way in which a glass will react to the stimuli exerted by the physiological host in which it is implanted is ultimately controlled by several structural features of the glass. Therefore, in principle, predicting the way in which these features are influenced by the composition could represent a powerful tool to optimise the composition itself. A major issue here is that the accurate determination is feasible only for some structural features of a glass. For instance, the network connectivity (defined as the average number of bridging oxygen atoms per network former) is known to play a central role in the dissolution rate of silicate and phosphate glasses,[5–7] and can be extracted from the Q_n distribution (where Q_n denotes a network former bonded to *n* bridging oxygens) obtained by NMR experiments. There are other structural properties with a potential impact on the dissolution behaviour of a glass, such as the extent of ion clustering[8] or that of non-covalent cross-linking between separate portions of the silicate/phosphate network.[9] However, unlike the network connectivity, the experimental determination of these properties is not a common or simple task. A direct consequence is that their impact on the glass durability and bioactivity is not always obvious, and for many systems it remains to be determined. This is precisely where computer-generated atomistic models of the glass can be crucial. Molecular dynamics (MD) simulations nowadays allow researchers to reproduce the melt-and-quench synthesis of melt-derived silicate- and phosphate-based glasses, thus producing three-dimensional models from which virtually *any* structural feature can be extracted with simple post-processing tools.[10] This not only enables the direct determination of composition–structure trends for the non-standard structural descriptors mentioned above,[11–13] but also opens the possibility to identify new structural features which steer the biological behaviour of a glass. This kind of information is rarely available from other experimental sources and represents a key contribution to develop improved predictions of composition–activity relationships in bioglasses. Robust predictions of how a new composition will perform in practical conditions can be made by extracting key structural features from MD models of glasses of known bioactivity/durability and identifying the corresponding composition–structure–bioactivity relationships for each family of glasses.

The static structure is one key aspect, and probably the first one that should be taken into account in any rational effort to develop new bioactive glasses. However, the bioactive process is a truly dynamic process wherein the glass structure rapidly transforms under the physical and chemical cues from the surrounding biological environment. This means that a better understanding of the behaviour of different bioactive glass compositions can only be achieved by also taking into account these dynamical transformations. Whereas an atomistic MD simulation of the whole sequence of physical and chemical processes behind the activity of these materials is out of reach even with modern high-performance computing facilities, the current simulations can address several components of the whole picture. For instance, the early transformations following contact of the glass surface with an aqueous environment,[14,15] the migration of network modifiers in the glass matrix,[16] the hydration and the effect of alkali depletion upon ion exchange occurring at the surface[17,18] have all been recently investigated through MD.

This chapter illustrates the success and capabilities of MD simulations in the areas discussed above, focusing on the rationalisation of the composition–properties relationship through both structural and dynamical studies.

4.2 Structure

4.2.1 Simulation Methods and Aims

The local (short-range) structure of a glass can be reproduced with high accuracy using parameter-free *ab initio* Molecular Dynamics (AIMD),[19,20] which has the advantage of being completely unbiased by any arbitrary choice of an empirical force field, whereas an inappropriate choice of the force field may affect models obtained using classical MD.[21]

Bulk bioactive glasses have been successfully modelled through either approach.[22–25] Unfortunately, due to their substantial computational requirements, it is not possible to apply AIMD to model systems of size large enough to provide a statistically accurate description of medium-range structural features such as Q_n distribution, ion clustering, *etc.*, which are most relevant to understand the bioreactivity of a glass; in fact, AIMD cannot satisfy the minimal requirement in terms of sample size and cooling rate necessary to produce medium-range properties not dominated by statistical fluctuations.[26] *Ab initio* calculations are suitable for the calculation of simulated vibrational and NMR spectra[27,28] or for characterizing the local environment surrounding certain species.[29] The AIMD-generated short-range structure can also be used to assess and if necessary refine the force field used in classical MD.[30] Upon confirming the good quality of the latter, classical MD can be performed on systems large enough to span the distances relevant to reproduce medium-range properties.[31] Following this approach, accurate force fields have been developed and applied to determine the medium-range structural features in bioactive glasses.[31] These applications have mainly concerned melt-derived Hench compositions.[24,32,33] The analysis generally focuses on determining the Q_n distribution, network connectivity, and modifier ion coordination of different compositions, and linking these atomic-scale properties to the expected bioactivity.[10] The relationship between these structural features and the bioactivity is now relatively well understood, thanks to the structural data extracted from the simulations for compositions of known bioactivity.[34,35]

4.2.2 Structural Descriptors of Bioactivity and Glass Durability

The network connectivity remains the most common structural descriptor linking the experimental bioactivity to the glass composition, based on the empirical observation that highly bioactive compositions have low network connectivity denoting a fragmented silicate matrix with most fragments being silicate chains (Figure 4.1).[7] This open matrix will be more prone to hydrolysis and thus degrade rapidly in a physiological environment. From a chemical/thermodynamical perspective, this observation simply reflects the basic fact that detaching and releasing linear $-O-[Si-O-]_n$ units requires breaking a lower number of covalent Si–O bonds compared to the release of more cross-linked Q_3 and Q_4 silicate units from a highly ramified structure (Figure 4.2).[36]

The MD models play an important role in this context, particularly where the determination of the Q_n speciation by standard experimental techniques is difficult. For instance, whereas conventional NMR probes frequently fail to detect the signatures of Si–O–P linkages in the 45S5 glass, MD simulations constantly reveal the presence of a small but not negligible fraction of Q_1 phosphate groups linked to an adjacent silicon tetrahedron by a single bridging oxygen.[26,27,33,37] This apparent controversy has been solved by advanced solid-state NMR techniques,[38] which confirmed the presence of PO_4 groups linked to a Si through an Si–O–P bond, as predicted by the MD simulations. This shows the important role of MD models to clarify the connectivity of glasses featuring low phosphate amounts such as the 45S5 composition, where the accurate determination of Q_1 signals for low-phosphorus compositions by standard ^{31}P NMR remains problematic.

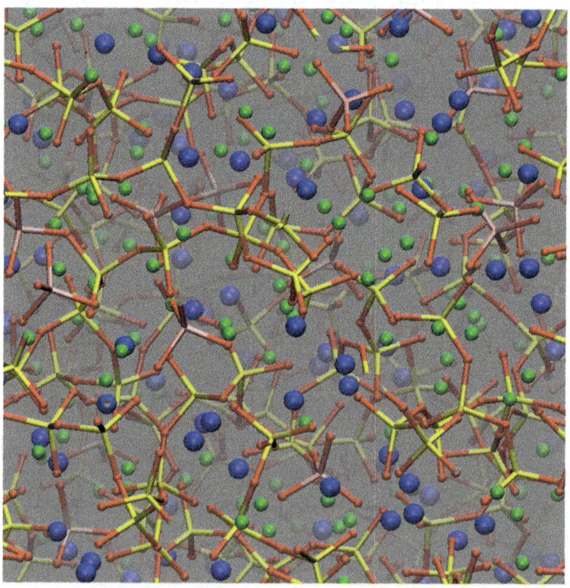

Figure 4.1 Large model of 45S5 glass structure generated by MD.[26] Silicon, phosphorus, sodium, calcium, and oxygen atoms are coloured yellow, brown, green, blue, and red, respectively.

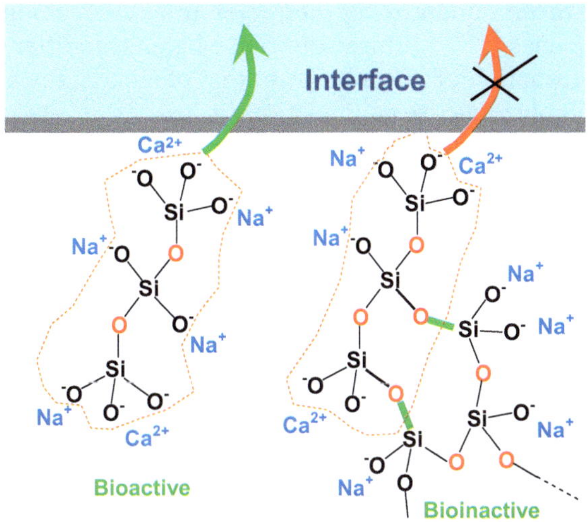

Figure 4.2 Schematic illustration of the easier release of chain-like fragments from a low-connectivity bioactive glass structure compared to a more cross-linked bioinactive composition.

The release of ions and ionic fragments from the glass into the surrounding environment is known to represent a key factor for triggering various processes leading to growth of new tissues.[39] This is also of crucial importance for the activity of bioactive glasses doped with ions such as cerium, gallium, fluorine, zinc, strontium, and cobalt in order to gradually deliver them to the target tissues.[2,32,35,40–43] A related, less common branch of applications which is also affected by the ion release rate is *in situ* radiotherapy. In these applications, silicate glasses are employed as vectors of radioisotopes, such as yttrium or holmium, to a cancerous site.[44] To minimise their adverse effects on the organism, the radioactive species should remain trapped within the glass carrier during the treatment. Bioactive glasses doped with yttrium may represent an ideal combination between treatment and biocompatibility/tissue regeneration, and some studies have been aimed at assessing the dissolution rate of these systems.[45,46] Whereas the network connectivity obviously plays a role also in the release of non-network forming cations, the latter process is also influenced by other structural parameters describing the coordination of the ions of interest. For instance, MD simulations revealed a correlation between the reduction in the experimental bioactivity of Hench-type bioglasses and the appearance (highlighted in the models) of nanosegregated calcium phosphate regions, which recalls the increased resistance to dissolution of phase-separated glasses.[24,47] A possible interpretation for this correlation is that segregation in clusters reduces the long-range mobility of ionic species by breaking down continuous ion migration channels (required for fast migration) into spatially separated and thus disconnected ionic clusters.[48] Whereas the experimental determination of the extent of association of ionic species in a cluster is also possible,[38] the MD models can be analyzed to explicitly extract properties, such as cluster size distributions,[26] as well as statistical parameters quantifying the preferential association or avoidance between two species.[23,24,32] It turns out that whereas modifier cations always favour the coordination of phosphorus to that of silicon, the actual extent of clustering in highly bioactive compositions is low.[26] Ca^{2+}–PO_4^{3-} association and corresponding Ca clustering become substantial in less bioactive compositions with a higher network connectivity.[10] The inverse relationship between clustering and bioactivity is effectively illustrated in the models of Figure 4.3. It is important to consider that, as the cluster size increases,

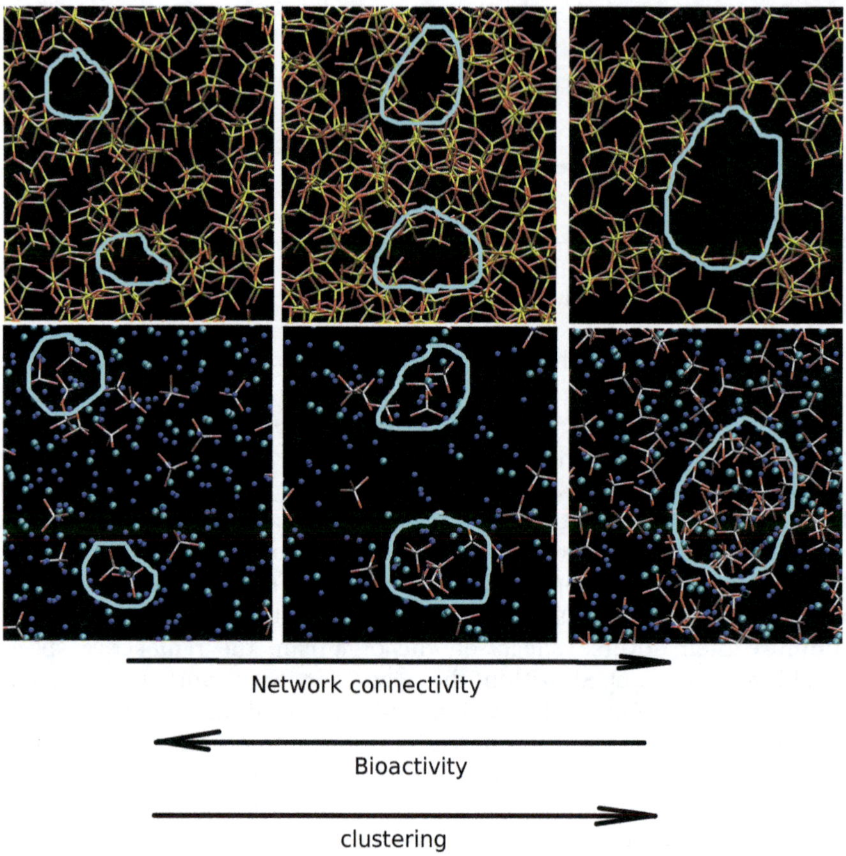

Figure 4.3 Inverse relationship between bioactivity and clustering illustrated by the MD-generated models of three compositions of different network connectivity. Compared to 45S5 (left panel, 46.1 SiO_2; 24.4 Na_2O; 26.9 CaO; 2.6 P_2O_5, mol%), the network connectivity of both middle panel (66.9 SiO_2; 14.5 Na_2O; 16 CaO; 2.6 P_2O_5, mol%) and right panel (36.3 SiO_2; 24.4 Na_2O; 27.1 CaO; 12.2 P_2O_5, mol%) compositions is higher; the latter also has a higher phosphorus content, which further enhances clustering.[24] The top and bottom panels separately show the silicate matrix and the Na/Ca/PO_4 species of each structure, respectively. The circled areas highlight regions of low silica density populated by cation-phosphate clusters.

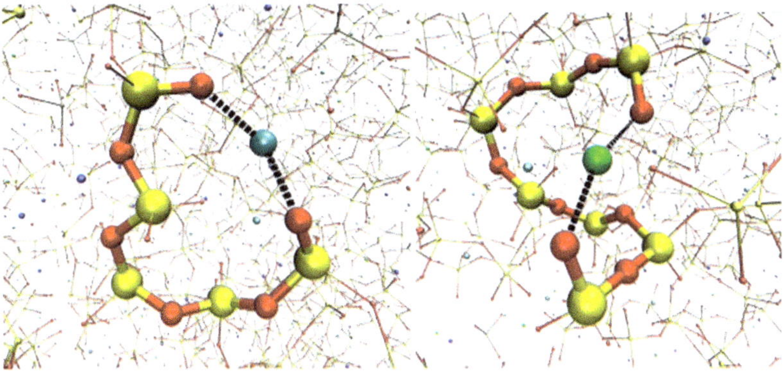

Figure 4.4 Examples of a non-network former cation (Ca in the left and Y in the right panel) joining together two tetrahedra not directly connected in an yttrium-doped bioactive glass.[11]
Adapted from ref. 11 with permission from the Royal Society of Chemistry.

adjacent clusters may eventually coalesce and re-establish the ion-conducting channels that had been broken by the nanosegregation.[49] Therefore, the cluster size is a key factor to consider in evaluating the effect of clustering on the bioactivity.

The complex network of interactions within a bioactive glass involves a balance of different bonding and non-bonding interatomic forces. Whereas theories based on the network connectivity focus on the strong covalent bonds involving network formers, noncovalent interactions are less frequently taken into account. However, the T–O···M···O–T cross-links between two tetrahedral network formers T bridged by a modifier M are known to play an important role in strengthening the network by improving the cohesion between fragments not directly connected through covalent bonds (Figure 4.4).[9] This concept has prompted the development of predictive models to estimate the durability of a glass from structural descriptors of the network strength that also incorporate the properties of intertetrahedral "bonds", such as the noncovalent cross-links mentioned above.[11,13] The latter can explicitly be obtained from the MD models, by studying the properties of different tetrahedra connected through a central modifier cation M. A suitable approach thus involves fitting available experimental datasets reporting the dissolution rate of different compositions with a linear combination of standard (network connectivity) and non-standard (clustering, noncovalent linkages) structural descriptors. The fitting curve can then be employed to interpolate the unknown durability of a new composition.[13]

4.2.3 Bioglass Nanoparticles

The biological properties of bioglasses are affected by the size of the glass particles; in particular, reducing the particle size below the conventional micrometre scale leads to an enhancement in various aspects related to the bioreactivity, such as protein adsorption, remineralisation rate, and antibacterial activity.[50–52] As atomistic simulations have been frequently employed to model nanosized materials including glasses,[53,54] their application to nanosized bioactive glasses is relatively straightforward, yet it can be of great help to rationalise the improved properties associated to the reduced size. Models of a 45S5 spherical nanoparticle revealed that some of the structural features present on the surface of compact 45S5 Bioglass samples most beneficial for their bioactive behaviour are further enhanced in the 45S5 nanoparticles.[55] For instance, the high fragmentation of the silicate network further increases on the nanoparticle surface, which exhibits a higher Q_1/Q_3 ratio that could promote a faster release of soluble silica species (a key effect for the tissue-regeneration properties of these materials[56]) from nanosized BG particles (Figure 4.5). Moreover, the 45S5 nanoparticle surface also hosts an increased fraction of three-membered silicate rings which have previously been indicated as nucleation sites for calcium phosphate.[57] The MD models also highlighted a marked enrichment in undercoordinated Na^+ on the surface of the nanoparticle compared to a compact reference. As also discussed below, these surface sites will accelerate the initial hydrolysis and ion exchange stages of the biodegradation process.

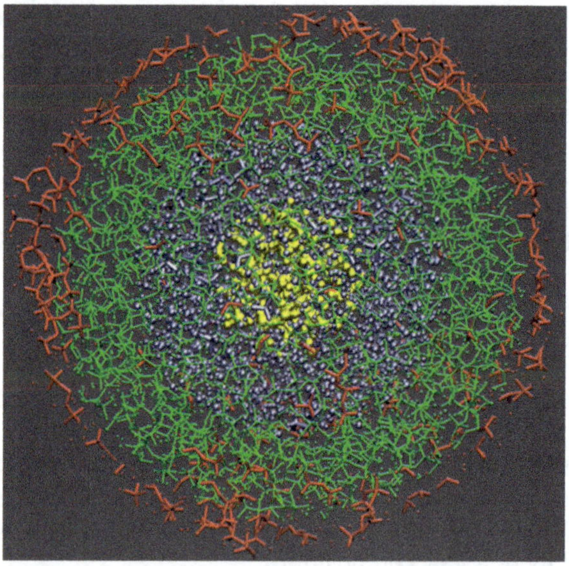

Figure 4.5 Model of a 6 nm 45S5 nanoparticle.[55] Spherical shells with increasing fragmentation moving from the core towards the surface of the particle are represented in different colours.

4.3 Dynamics and Reactivity

4.3.1 Surface and Ion-exchange Processes

Most computational studies of bioactive glasses focus on the bulk structure, driven by the hypothesis that the biodegradation and bioactive behaviour are controlled by the structural features of the bulk material (such as those described in the previous section). This simplifies, and at the same time makes possible, the computational investigations because it shifts the problem from that of simulating all the actual physical and chemical processes involved to that of just simulating the bulk structure and using it to interpret existing data and predict the behaviour of other compositions. In other words, common current approaches to model bioglasses are *indirect*, because they do not directly observe the dynamical transformation occurring at the interface. While this is certainly a suitable approach, there is no doubt that a more complete picture can be obtained by integrating the insight based on the glass structure with the explicit modelling of key dynamical processes. The timescales typical of the overall bioactive process remain prohibitive even for state-of-the-art computational resources. However, some important steps in this direction have been taken in the past years. *Ab initio* simulations have been used to examine the reactivity of typical bioactive glass surfaces.[58,59] Following a widely established approach to model crystalline and amorphous oxide surfaces and interfaces,[59–61] the dry surface is modelled as a two-dimensional slab cut from the (3D) bulk: this represents the most suitable procedure, as long as care is taken in selecting realistic cut directions (out of the virtually infinite possibilities for an amorphous system) and surface terminations, for instance preserving neutrality. The reactivity of the pristine surface is assessed by testing the strength of possible active sites based on their interaction with a water molecule used as a representative "probe" of the surface sites. The water physisorption or chemisorption energy can then be calculated and used to identify the most active sites present on the pristine surface of the glass, *i.e.*, the first sites that will be involved in the biodegradation process, following contact of the glass with the physiological environment. Models of the 45S5 surface have revealed that particularly stable surface sites are represented by undercoordinated silicon atoms and non-bridging oxygens (NBOs) connected to (undercoordinated and thus highly reactive) Na or Ca modifier cations.[14,58] The high hydrophilicity of these cations creates favourable paths where water can penetrate, gaining access to the inner regions of the glass structure where it can cause further corrosion (by breaking Si–O–Si bridges), thus accelerating the biodegradation process (Figure 4.6). This effect is achieved through the high fragmentation of the 45S5 network, which first creates Na/Ca-rich pockets exposed on the surface, and then permits water molecules initially adsorbed in these pockets to easily access the inner glass

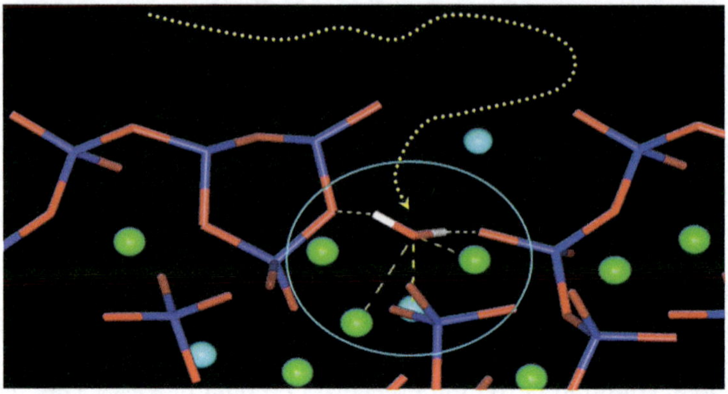

Figure 4.6 Possible mechanism of water penetration in a 45S5 surface mediated by strong interaction with exposed pockets populated by modifier cations/non-bridging oxygens.
Adapted from ref. 10 with permission from the Royal Society of Chemistry.

regions without having to break or significantly distort Si–O–Si bonds (as would happen in more cross-linked structures) in order to penetrate inside the bulk structure. The facile penetration of water contributes to the fast initial dissolution of BG45, which is a key feature of this biomaterial.

In this context, the fate of a water molecule once it reaches the inner (bulk) regions of the glass is of particular interest to understand the biodegradation mechanism. Recent AIMD simulations showed that a water molecule is rather mobile in bulk 45S5, so that it can move between a number of different local minima.[17] One of the most stable water configurations identified by the simulations involves a cage similar to the Na–Ca–NBO pockets found on the surface. Here the molecule can dissociate through proton transfer to an NBO and form a free hydroxyl (f-OH) not bonded to either Si or P and stabilized by surrounding modifier cations.[17] This water dissociation mechanism is fundamentally different from that effective for instance in amorphous SiO_2, because it does not involve Si–O bond breaking. The stability of the f-OH's revealed by the AIMD runs was also confirmed in further MD simulations of the Na^+/H^+-exchanged 45S5 Bioglass.[18] These simulations probed the structural changes induced by the first stages of the biodegradation process, which involve release of alkali cations replaced by protons from the aqueous contact medium.[62–64] The models reveal that a large fraction of hydroxyl groups enter the structure without initially forming covalent bonds with Si and P network formers, *i.e.*, as f-OH stabilised by modifier metal cations; however, substantial Si–OH and P–OH bonding is eventually observed as the Na^+/H^+ exchange proceeds (Figure 4.7).[18] This finding suggests an alternative (or parallel) direct route for the formation of a repolymerised silica-rich gel in the early stages of the bioactive mechanism, which does not require sequential repeated breaking of Si–O–Si bonds and silanol condensations, as in the typical proposed mechanism.[39] In the alternative mechanism, new Si–O–Si bridges do not

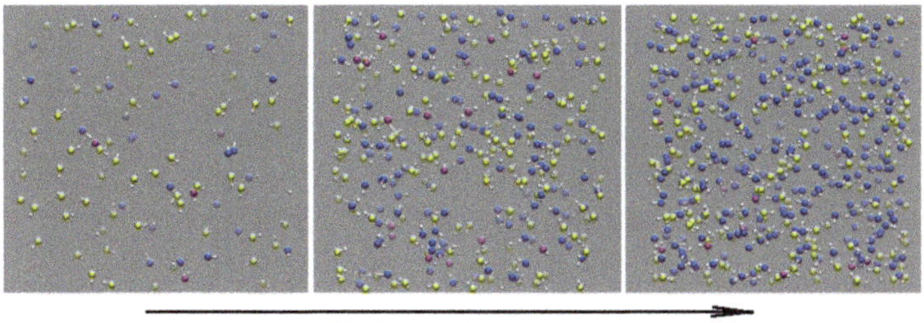

Figure 4.7 Hydroxyl configuration in Na^+/H^+-exchanged 45S5 glass structures with increasing exchange levels. Free (non-bonded), Si-bonded, and P-bonded OH groups are highlighted in yellow, blue, and purple, respectively, while hydrogen atoms are white. For clarity, Si, Na, Ca, P, and non-hydroxyl O atoms are not shown.
Adapted from ref. 18 with permission from the Royal Society of Chemistry.

result from condensation of Si–OH groups, but from f-OH-induced Si–O–Si cross-linking. In practice, it is possible that a transition between the two silica gel polymerisation mechanisms occurs as the Na leaching proceeds and the silica-rich gel layer is progressively formed.

4.3.2 Ion Migration

The peculiar structural features of highly bioactive compositions essentially create a favourable background for the rapid transformation of the glass in a physiological environment. Ions, chains, and fragments move within this "enabling framework" rapidly gaining access to the surface to be released in the surrounding medium where they can take part in the bioactive processes. Molecular Dynamics methods are naturally suited to follow the diffusion of these species (the main limitation being again related to the timescales involved) and have been widely applied to model ion migration in glasses.[33,65–69]

The application of MD to 45S5 revealed that modifier ion migration in this material is deeply affected by the relatively flexible nature of the fragmented phosphosilicate matrix.[16] This creates ion migration pathways that are generally not permitted in the more cross-linked network of common, higher-silica and bio-inactive glasses. In particular, the simulations showed how the migration mechanism of a sodium cation in the 45S5 matrix involves discrete hops along vacant transient sites created by the temporary lateral displacement of another Na or Ca cation from their metastable sites (Figure 4.8). The formation of these temporary sites, and thus this mechanism, would not be energetically favourable in the more rigid network of a higher-silica glass. As each individual hopping event involves a marked change to the coordination environment of an ion, the formation of intermediate configurations that are significantly distorted with respect to the stable initial and final states will control the frequency of the jumps and thus the overall diffusivity.[70] Figure 4.9 shows how the shape and population

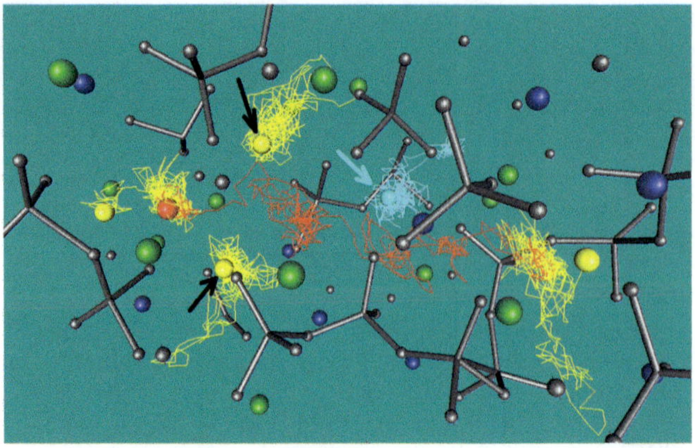

Figure 4.8 Traces of selected Na and Ca ions in 45S5 during an MD trajectory. The migration of an individual Na^+ coloured in red is facilitated by the correlated displacements of several other ions (Na and Ca, coloured yellow and cyan, respectively).
Reprinted with permission from A. Tilocca, *J. Chem. Phys.*, 2010, **133**, 014701–014710.[16] Copyright 2010, AIP Publishing LLC.

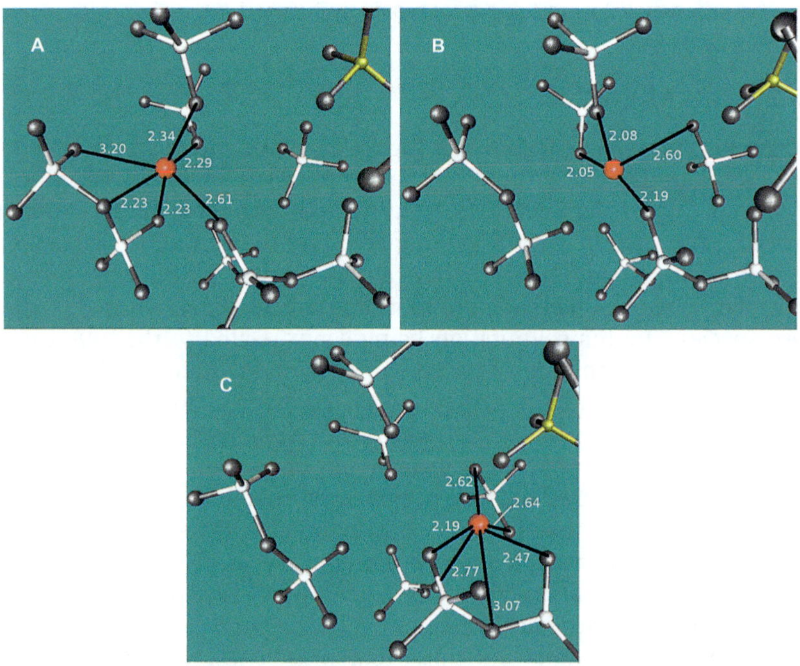

Figure 4.9 Evolution of the oxygen coordination shell of a Na ion (red sphere) during a hopping event between two stable sites (A) and (C) in 45S5 Bioglass, passing through an intermediate state (B). Dark lines are used to mark the Na–O interactions within the Na coordination shell with their corresponding distances.
Reprinted with permission from A. Tilocca, *J. Chem. Phys.*, 2010, **133**, 014701–014710.[16] Copyright 2010, AIP Publishing LLC.

of the oxygen coordination shell of a migrating Na ion gradually changes during a single hopping event; the energetics and rate of the process is thus controlled by the concerted breaking and formation of Na···O links.

4.4 Conclusions and Outlook

Molecular Dynamics simulations have played an increasingly important role in the field of bioactive glasses in recent years, evolving from simple tools to assess the structure of these materials at a higher (atomic-scale) resolution than that permitted by experimental techniques, to powerful methods to access unusual structural and dynamical features of known and new compositions. The present chapter illustrates some of these important achievements, which highlight how the simulations today represent a key component to support a more rational approach to the development of bioactive glass materials, based on a fundamental understanding of the interactions and mechanisms that control their behaviour.

In order to bring the simulations another step closer to the experimental work, future modelling studies of bioactive glasses should be focused on resolving the timescale limitations that currently do not allow direct MD observations of "slow" (relative to the longest duration of MD trajectories affordable at present) bioactive processes, such as biodegradation of the glass matrix and growth of bone–bonding surface layers from the contact medium. Enhanced-sampling computational methods,[71] developed to accelerate the exploration of the configurational space of activated, infrequent processes, could be of great help in this context. Other key advances may involve tackling different interactions, such as those with biomolecules, and processing approaches, such as the sol–gel synthesis.[72]

Acknowledgements

The author wishes to thank the Royal Society for financial support (University Research Fellowship).

References

1. L. L. Hench, *J. Am. Ceram. Soc.*, 1998, **81**, 1705–1728.
2. J. M. Smith, R. A. Martin, G. J. Cuello and R. J. Newport, *J. Mater. Chem. B*, 2013, **1**, 1296–1303.
3. J. Lao, E. Jallot and J.-M. Nedelec, *Chem. Mater.*, 2008, **20**, 4969–4973.
4. I. Kansal, A. Goel, D. U. Tulyaganov, L. F. Santos and J. M. F. Ferreira, *J. Mater. Chem.*, 2011, **21**, 8074–8084.
5. Z. Strnad, *Biomaterials*, 1992, **13**, 317–321.
6. R. Hill, *J. Mater. Sci. Lett.*, 1996, **15**, 1122–1125.
7. A. Tilocca, *Proc. R. Soc. A*, 2009, **465**, 1003–1027.
8. U. Voigt, H. Lammert, H. Eckert and A. Heuer, *Phys. Rev. B: Condens. Matter Mater. Phys.*, 2005, **72**, 064207.
9. Y. Vaills, Y. Luspin and G. Hauret, *Mater. Sci. Eng., B*, 1996, **40**, 199–202.

10. A. Tilocca, *J. Mater. Chem.*, 2010, **20**, 6848.
11. J. K. Christie and A. Tilocca, *J. Mater. Chem.*, 2012, **22**, 12023–12031.
12. A. Tilocca, *Phys. Chem. Chem. Phys.*, 2014, **16**, 3874–3880.
13. J. K. Christie and A. Tilocca, *J. Phys. Chem. B*, 2012, **116**, 12614–12620.
14. A. Tilocca and A. N. Cormack, *ACS Appl. Mater. Interfaces*, 2009, **1**, 1324–1333.
15. A. Tilocca and A. N. Cormack, *Proc. R. Soc. A*, 2011, **467**, 2102–2111.
16. A. Tilocca, *J. Chem. Phys.*, 2010, **133**, 014701–014710.
17. E. Berardo, M. Corno, A. N. Cormack, P. Ugliengo and A. Tilocca, *RSC Adv.*, 2014, **4**, 36425–36436.
18. A. Tilocca, *Phys. Chem. Chem. Phys.*, 2015, **17**, 2696–2702.
19. R. Car and M. Parrinello, *Phys. Rev. Lett.*, 1985, **55**, 2471–2474.
20. C. Massobrio and A. Pasquarello, *Phys. Rev. B: Condens. Matter Mater. Phys.*, 2008, **77**, 144207.
21. M. P. Allen and D. J. Tildesley, *Computer Simulation of Liquids*, Clarendon, Oxford, 1989.
22. A. Pedone, T. Charpentier and M. C. Menziani, *J. Mater. Chem.*, 2012, **22**, 12599–12608.
23. Y. Xiang and J. Du, *Chem. Mater.*, 2011, **23**, 2703–2717.
24. A. Tilocca and A. N. Cormack, *J. Phys. Chem. B*, 2007, **111**, 14256–14264.
25. A. Tilocca, *Phys. Rev. B: Condens. Matter Mater. Phys.*, 2007, **76**, 224202.
26. A. Tilocca, *J. Chem. Phys.*, 2013, **139**, 114501.
27. M. Corno, A. Pedone, R. Dovesi and P. Ugliengo, *Chem. Mater.*, 2008, **20**, 5610–5621.
28. A. Pedone, E. Gambuzzi and M. C. Menziani, *J. Phys. Chem. C*, 2012, **116**, 14599–14609.
29. J. K. Christie, A. Pedone, M. C. Menziani and A. Tilocca, *J. Phys. Chem. B*, 2011, **115**, 2038–2045.
30. A. Tilocca, N. H. de Leeuw and A. N. Cormack, *Phys. Rev. B: Condens. Matter Mater. Phys.*, 2006, **73**, 104209.
31. A. Tilocca, *J. Chem. Phys.*, 2008, **129**, 084504.
32. G. Malavasi, A. Pedone and M. C. Menziani, *J. Phys. Chem. B*, 2013, **117**, 4142–4150.
33. J. Du and Y. Xiang, *J. Non-Cryst. Solids*, 2012, **358**, 1059–1071.
34. G. Lusvardi, G. Malavasi, M. Cortada, L. Menabue, M. C. Menziani, A. Pedone and U. Segre, *J. Phys. Chem. B*, 2008, **112**, 12730–12739.
35. S. Kapoor, A. Goel, A. Tilocca, V. Dhuna, G. Bhatia, K. Dhuna and J. M. F. Ferreira, *Acta Biomater.*, 2014, **10**, 3264–3278.
36. D. Arcos, D. C. Greenspan and M. Vallet-Regí, *Chem. Mater.*, 2002, **14**, 1515–1522.
37. A. Pedone, G. Malavasi and M. C. Menziani, *J. Phys. Chem. C*, 2009, **113**, 15723–15730.
38. F. Fayon, C. Duée, T. Poumeyrol, M. Allix and D. Massiot, *J. Phys. Chem. C*, 2013, **117**, 2283–2288.
39. J. R. Jones, *Acta Biomater.*, 2013, **9**, 4457–4486.

40. K. Fujikura, N. Karpukhina, T. Kasuga, D. S. Brauer, R. G. Hill and R. V. Law, *J. Mater. Chem.*, 2012, **22**, 7395–7402.
41. A. J. Salinas, S. Shruti, G. Malavasi, L. Menabue and M. Vallet-Regi, *Acta Biomater.*, 2011, **7**, 3452–3458.
42. J. C. Du, L. Kokou, J. L. Rygel, Y. S. Chen, C. G. Pantano, R. Woodman and J. Belcher, *J. Am. Ceram. Soc.*, 2011, **94**, 2393–2401.
43. Y. Shaharyar, E. Wein, J.-J. Kim, R. E. Youngman, F. Munoz, H.-W. Kim, A. Tilocca and A. Goel, *J. Mater. Chem. B*, 2015, **3**, 9360–9373.
44. M. Kawashita, *Mater. Sci. Eng. C*, 2002, **22**, 3–8.
45. D. Cacaina, H. Ylänen, M. Hupa and S. Simon, *J. Mater. Sci.: Mater. Med.*, 2006, **17**, 709–716.
46. J. K. Christie, J. Malik and A. Tilocca, *Phys. Chem. Chem. Phys.*, 2011, **13**, 17749–17755.
47. H.-F. Wu, C.-C. Lin and P. Shen, *J. Non-Cryst. Solids*, 1997, **209**, 76–86.
48. P. Jund, W. Kob and R. Jullien, *Phys. Rev. B: Condens. Matter Mater. Phys.*, 2001, **64**, 134303.
49. A. Pradel, C. Rau, D. Bittencourt, P. Armand, E. Philippot and M. Ribes, *Chem. Mater.*, 1998, **10**, 2162–2166.
50. S. K. Misra, D. Mohn, T. J. Brunner, W. J. Stark, S. E. Philip, I. Roy, V. Salih, J. C. Knowles and A. R. Boccaccini, *Biomaterials*, 2008, **29**, 1750–1761.
51. M. Vollenweider, T. J. Brunner, S. Knecht, R. N. Grass, M. Zehnder, T. Imfeld and W. J. Stark, *Acta Biomater.*, 2007, **3**, 936–943.
52. V. Mortazavi, M. M. Nahrkhalaji, M. H. Fathi, S. B. Mousavi and B. N. Esfahani, *J. Biomed. Mater. Res., Part A*, 2010, , **94A**, 160–168.
53. D. Spagnoli, J. P. Allen and S. C. Parker, *Langmuir*, 2011, **27**, 1821–1829.
54. N. N. Linh and V. Van Hoang, *J. Phys.: Condens. Matter*, 2008, **20**, 265005.
55. A. Tilocca, *J. Mater. Chem.*, 2011, **21**, 12660–12667.
56. I. D. Xynos, A. J. Edgar, L. D. K. Buttery, L. L. Hench and J. M. Polak, *J. Biomed. Mater. Res.*, 2001, **55**, 151–157.
57. N. Sahai and M. Anseau, *Biomaterials*, 2005, **26**, 5763–5770.
58. A. Tilocca and A. N. Cormack, *J. Phys. Chem. C*, 2008, **112**, 11936–11945.
59. E. Berardo, A. Pedone, P. Ugliengo and M. Corno, *Langmuir*, 2013, **29**, 5749–5759.
60. M. P. Gaigeot, M. Sprik and M. Sulpizi, *J. Phys.: Condens. Matter*, 2012, **24**, 124106.
61. P. Masini and M. Bernasconi, *J. Phys.: Condens. Matter*, 2002, **14**, 4133.
62. C. Y. Kim, A. E. Clark and L. L. Hench, *J. Non-Cryst. Solids*, 1989, **113**, 195–202.
63. C. R. Mariappan, D. M. Yunos, A. R. Boccaccini and B. Roling, *Acta Biomater.*, 2009, **5**, 1274–1283.
64. J. Mei, R. M. Shelton and P. M. Marques, *J. Mater. Sci.: Mater. Med.*, 1995, **6**, 703–707.
65. W. Smith, T. R. Forester, G. N. Greaves, S. Hayter and M. J. Gillan, *J. Mater. Chem.*, 1997, **7**, 331–336.
66. H. Lammert, M. Kunow and A. Heuer, *Phys. Rev. Lett.*, 2003, **90**, 215901.

67. J. Habasaki and K. L. Ngai, *Phys. Chem. Chem. Phys.*, 2007, **9**, 4673–4689.
68. A. N. Cormack, J. Du and T. R. Zeitler, *Phys. Chem. Chem. Phys.*, 2002, **4**, 3193–3197.
69. A. Pedone, G. Malavasi, M. C. Menziani, U. Segre and A. N. Cormack, *J. Phys. Chem. C*, 2008, **112**, 11034–11041.
70. A. N. Cormack, J. Du and T. R. Zeitler, *J. Non-Cryst. Solids*, 2003, **323**, 147–154.
71. P. Shaffer, O. Valsson and M. Parrinello, *Proc. Natl. Acad. Sci. U. S. A.*, 2016, **113**, 1150–1155.
72. D. Arcos and M. Vallet-Regi, *Acta Biomater.*, 2010, **6**, 2874–2888.

CHAPTER 5

Sol–Gel Glass and Nano–Macro Porous Bioscaffolds

MATTHIAS M. FALK,[a] TIA J. KOWAL,[a] RUI M. ALMEIDA,[b] MANAL SAAD,[c] MONA K. MAREI,[c] UKRIT THAMMA[d] AND HIMANSHU JAIN*[d]

[a] Department of Biological Sciences, Lehigh University, Bethlehem, PA, USA; [b] Departamento de Engenharia Quımica/CQE, Instituto Superior Tecnico, University of Lisbon, Lisbon, Portugal; [c] Tissue Engineering Laboratories, Faculty of Dentistry, Alexandria University, Alexandria, Egypt; [d] Department of Materials Science and Engineering, Lehigh University, Bethlehem, PA, USA
*Email: H.Jain@Lehigh.Edu

5.1 Introduction

Tissue engineering and regenerative medicine are no longer futuristic dreams[1,2] but society's realistic hope for treating patients requiring tissue or organ replacements.[3–5] In this fundamentally superior approach, a defective tissue is replaced such that a 3D scaffold structure helps the patient regenerate new natural tissue (biologic regeneration) that will last for the whole lifetime,[6] as opposed to a permanent foreign implant. The scaffold is critical to the success of this approach as it helps regenerate natural tissue and then disappears.[7] Indeed, several organs have been regenerated, and regenerative medicine is expected to grow exponentially in the future.[8] These expectations have introduced unprecedented opportunities for patients as well as challenges that need the coordinated efforts of materials scientists,

RSC Smart Materials No. 23
Bioactive Glasses: Fundamentals, Technology and Applications
Edited by Aldo R. Boccaccini, Delia S. Brauer and Leena Hupa

Published by the Royal Society of Chemistry, www.rsc.org

biologists, chemists, clinicians and engineers.[9] So far, initial success has been realized with soft tissues using polymer scaffolds,[10,11] but challenges remain for the biologic regeneration of hard tissues (bone and dental). A variety of scaffold materials and fabrication techniques has been developed for bone regeneration: ceramics, synthetic and natural polymers, metals, *etc.* Depending on the biomaterial, often unsatisfactory effectiveness in clinical repair is found because of the limited porosity, inflammatory response, mechanical instability during loading, unacceptable rates of degradation, *etc.* Nonetheless, encouraging results have been obtained with bioactive glasses (BGs) and glass-ceramics,[12] including synthetic hydroxyapatite, 45S5 Bioglass®, 55S4.3 bioactive glass, Ceravital, A-W glass ceramic, *etc.*[13–15] Overall, with properly selected composition, the glass bonds with both soft and hard tissue without the formation of an intervening fibrous layer; and it shows only minimal systemic or local toxicity, inflammation, and immunogenic response. Among BGs, the CaO–SiO_2 system has become the base glass for such applications to which other components may be added to meet further requirements.[16,17] The solid form of original Bioglass® 45S5, however, has been found to be of little use for hard tissue regeneration.[18,19] Interestingly, only BG has shown osteo-stimulation for bone-cell differentiation by ions leaching from the resorbing scaffolds, thus offering advantages of cellular repair at the gene level over other candidate materials.[20,21]

5.1.1 Tailored Amorphous Multi-Porous (TAMP) Bioscaffolds

The design of a bioactive scaffold is challenging because it must possess a number of highly relevant characteristics. An ideal bioscaffold for hard tissue reconstruction, for example, must satisfy the following, very diverse requirements:[17,22,23] (a) biocompatibility; (b) biodegradation at a rate that matches the tissue growth rate; (c) the leachate or dissolution products may not be harmful to the body; and (d) high porosity with interconnected 'macro' pores (≥100 μm) to allow ingrowth of cells, vascularization and nutrient delivery to the new tissue. In addition, the scaffold designs should be tailored for the specific patient needs, which vary depending on their age, stature, and the type and location of the defect to be repaired. Chemical composition of the scaffold material would largely control the first three characteristics of the bioscaffold, whereas porosity is to be accomplished by using appropriate fabrication methods. The glass and glass-ceramic materials mentioned above satisfy the biocompatibility and dissolution of products requirements, and have been fabricated with suitable porosity. However, they are unacceptable in regard to degradation rate. After all, for many years glass researchers have tried improving the chemical durability of glass to last thousands of years, whereas we would like BG to degrade in weeks. An attractive solution to the challenge of increasing the degradation rate without changing the chemical composition is to increase the surface area of the bioscaffold that would be exposed to body fluids. In practice, this can be achieved by introducing interconnected nanoporosity (~10 nm) along with much larger scale macroporosity (~100 μm).[†] The desired bioscaffold material will then have bimodal nano–macro porosity. Furthermore, it will be desirable to design this porosity to match the degradation rate and other site-specific requirements for a particular type of patient. A bioscaffold meeting all these requirements for hard tissue regeneration has been termed generally: tailored amorphous multi-porous (TAMP) material.[24]

The coexistence of nano and macro pores next to each other is thermodynamically unstable, making the fabrication of nano–macro structures a non-trivial task. On one hand, there are several approaches to introduce interconnected macroporosity in a biocompatible material, which is necessary for it to function as a scaffold for tissue regeneration. These include replication of a sponge,[25–27] 3D printing,[28] dry pressing,[29,30] freeze casting,[31] glass fiber sintering,[32] *etc.* On the other hand, there are two main approaches to introduce interconnected nanoporosity in a glass. First is the classic 'Vycor' process applicable to certain glass compositions that undergo spinodal phase separation.[33] Here a homogeneous glass prepared by the melt–quench method is heat-treated to produce spinodal decomposition into two interconnected phases. Then one phase is preferentially etched in a suitable acid, leaving behind glass with 3D porosity. In the process, the composition of the porous glass is changed from the starting composition as a part of it is leached out. The other approach is based on the sol–gel (SG) method, where the gel is comprised of a polymerized network of glass intermixed with a solution of water, alcohol and an acid or base (see Section 5.2). To obtain glass, the gel is dried and heat-treated to form a glass network (without full densification), which inevitably results in a nanoporous glass. Insofar as the liquid phase of the gel does not include any volatile component of the intended glass, the final composition should be very close to the target value.

For the above-stated reasons an SG method is preferred for making nanoporous glass bioscaffolds. To fabricate TAMP bioscaffolds then, the challenge becomes how to incorporate macroporosity—it is feasible either during the SG process or as a separate fabrication step. In this regard, note that in the conventional SG method of glass making, the nanopores are eliminated by viscous flow during sintering, as the dried gel is heated at sufficiently high temperature to obtain monolithic glass; the high surface area of nanopores serves as a strong driving force for their removal. Therefore, so long as interconnected nanoporosity is preserved, TAMP bioscaffolds can be fabricated simply by a two-step process, where nanoporous powder of desired specific surface area is prepared first by drying the gel followed by one of the various methods of introducing macroporosity mentioned above,

---

[†]The terms macro, meso, micro and nano porosity have different meanings in different fields. In keeping with IUPAC's basic designation we have used nanoporosity for pore size <50 nm (includes micro + meso porosity) and macroporosity for size >50 nm, although our macropores will be usually larger than 10's of microns.

such as replication of a sponge,[24,34] 3D/robotic printing,[35] dry pressing, *etc.* However, from the viewpoint of processing, a preferred method will be the one in which both nano and macro porosity are introduced in one step, *viz.* the sol–gel process. This integrated approach for fabricating TAMP bioscaffolds was introduced successfully by foaming a sol with a surfactant.[15,36] When the pore volume fraction is sufficiently large, isolated bubbles join and interconnected macroporosity is introduced in the gel. Subsequent drying of this gel produces a nano–macro porous bioscaffold as seen in Figure 5.1. Note that here the 'doorway' size is much smaller than the size of pores, and the pore volume fraction must be high (>80%) to establish connectivity between adjacent pores. As a result the mechanical strength of foamed bioscaffolds is inherently poor and the microstructure is not optimized for tissue ingrowth.[12]

Finally, a new approach based on a modified SG method was developed at Lehigh University for fabricating TAMP bioscaffolds, which focused on tailorable biodegradability by controlling the nano and macro porosity independently.[37,38] In this process, an additional spinodal phase separation ensues on the macro scale during the gelation process which produces nanoscale phase separation. Extensive *in vitro* studies show that while nanoporosity allows for control of degradation rate, additionally the cells respond more favorably than to the glass of same composition without nanoporosity. *In vitro* tests with bone forming cells and *in vivo* tests in New Zealand rabbits and dogs have shown sufficiently promising results that

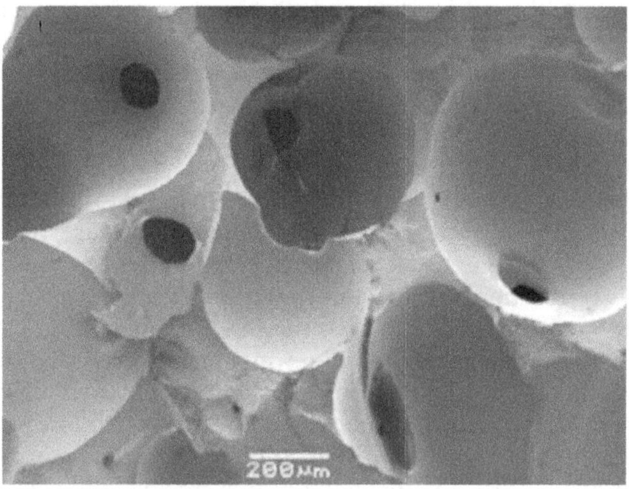

Figure 5.1 SEM micrograph showing interconnected macroporosity in 30CaO–70SiO_2 glass. Macroporosity is introduced *via* foaming of gel with a surfactant followed by sintering at 700 °C. The struts are nanoporous, so that overall the material is nano–macro porous.[15]
(Reprinted from Jones *et al.*, Optimising bioactive glass scaffolds for bone tissue engineering, *Biomaterials*, 27, 964–973, Copyright 2006, with permission from Elsevier, ref. 15.)

a clinical trial has been initiated at the Tissue Engineering Laboratories, Alexandria University (NIH clinical trial identifier: NCT01878084). For these reasons, we believe that these nano–macro porous structures provide superior TAMP bioscaffolds for bone regeneration, especially under non-load bearing conditions. Accordingly, in this chapter we present an overview of this new process for fabricating nano–macro porous scaffolds, and key observations of performance under both *in vitro* and *in vivo* conditions. The results are discussed using the example of 30CaO–70SiO_2 as the simple model system, although nano–macro porous glasses of more complex ternary compositions have also been prepared.[35,39]

5.2 Chemistry and Mechanisms of Fabrication of Sol–Gel Bioactive Glasses

5.2.1 Basic Concepts of Sol–Gel Processing

SG processing has emerged as a *bottom-up* technique for fabricating specialty materials.[40,41] It is used to produce a wide range of glass and ceramic materials (mainly oxides) across a large spectrum of dimensional architectures, including nanoparticles and powders, fibers, thin films, membranes, monoliths and composites. The SG process starts by the formation of a colloidal suspension (sol), its gelation and the removal of the liquid within the porous gel, which is then consolidated by heat treatment. The gelation, drying and densification steps are all critical for the control of the SG product. A basic SG process via the polymeric route may be described using the example of silica glass. Briefly, it starts with the hydrolysis of an alkoxide such as tetraethyl orthosilicate (TEOS) in the presence of a catalyst, most commonly an acid like HCl, HNO_3 or CH_3COOH but base catalysts are also employed:

$$Si(OR)_4 + nH_2O \rightarrow Si(OR)_{4-n}(OH)_n + nROH \tag{5.1}$$

with the formation of a sol. The rate of hydrolysis increases with the decrease in pH, starting from neutral solution of pH 7. Acid-catalyzed sols need longer gelling times than base-catalyzed sols, yielding gels with large volume shrinkage. By comparison, when a base such as ammonia is used as a catalyst, the gel generally shrinks less, yielding lower density products than those synthesized under acidic conditions. For silica, a pH ~2–3 is required, since its isoelectric point occurs at pH ~2, where the time to form a gel is the longest, rapidly decreasing for less acidic or basic conditions. Further polymerization occurs through condensation, with the formation of either water or alcohol as a by-product:

$$Si(OEt)_3OH + HOSi(OEt)_3 \rightarrow (EtO)_3SiOSi(OEt)_3 + H_2O \tag{5.2}$$

$$Si(OEt)_3OR + HOSi(OEt)_3 \rightarrow (EtO)_3SiOSi(OEt)_3 + ROH \tag{5.3}$$

The rate of condensation is also the slowest at the silica isoelectric point (pH∼2). Hydrolysis and condensation occur simultaneously, rather than sequentially. As condensation proceeds (*ageing* step), the molecules become so large that the material stiffens, with a marked increase in viscosity, and forms a gel, which may be dried subsequently and heat-treated into a dense silica glass. The overall process can be written as:

$$Si(C_2H_5O)_4 + 2H_2O = SiO_2 + 4C_2H_5OH \rightarrow \text{dry gel} \rightarrow \text{dense } SiO_2 \text{ glass} \quad (5.4)$$

Because of the hydrophobic nature of the ethoxy groups, TEOS and water are not miscible and the presence of a co-solvent (*e.g.* an alcohol like ethanol) is necessary to attain miscibility between the reactants. The water : alkoxide ratio (R) determines the amount of co-solvent required and it controls the hydrolysis rate, together with the pH value. The stoichiometric value of R for complete hydrolysis is 4, but less water can be used since the condensation reactions may lead to water formation (eqn (5.2)).

For the synthesis of multicomponent gels, the polymeric solution route offers great possibilities. The miscibility of different alkoxide compounds allows, in principle, complete polymerization of all metal species, yielding highly homogeneous products. However, distinct rates of metal alkoxide hydrolysis may cause inhomogeneities and phase separation in the final gel. Nevertheless, if two or more different metal alkoxide precursors are used, a sequential addition procedure in which the least reactive alkoxide is pre-hydrolyzed to some extent, before the more reactive one is added, can prevent inhomogeneities from appearing. The situation becomes complicated, however, for a system like $70SiO_2$–30CaO (in mol%), when a silica precursor alkoxide like TMOS (tetramethyl orthosilicate) is pre-hydrolyzed and then mixed with calcium nitrate,[35] whose dissolution and hydration shell formation occur at a rate much faster than the hydrolysis of TMOS, so it becomes more difficult to create a truly homogeneous mixture. In this case, in fact, Fourier transform infrared (FTIR) spectroscopy indicates that the structure of the final material closely resembles that of pure silica, suggesting phase separation of the gel into silica-rich and Ca-rich phases.[42] Thus the main challenge for multicomponent gel preparation is always to control the reactivity of the more reactive precursors in order to obtain a homogeneous sol. This goal may be addressed through different strategies, namely: (1) pre-hydrolysis of less reactive compounds;[43] (2) use of complexing agents like acetylacetone;[44] (3) use of less reactive organometallic compounds like calcium acetate and triethyl orthophosphate instead of $Ca(NO_3)_2$ and P_2O_5, respectively.[37]

5.2.2 Modified Sol–Gel Process: Introduction of Multimodal Porosity *via* Multiscale Spinodal Phase Separation

The SG method produces an inherently nanoporous material initially, which starts losing porosity upon heating. Indeed, nanoporosity would persist provided that full densification were not achieved by a prolonged heat treatment at sufficiently high temperature. Different approaches have been pursued in order to combine both nano and macro porosity in the same SG-derived glassy material: (1) Nakanishi[45] developed porous silica monoliths with a bimodal meso/macro (~0.1–40 μm) pore size distribution for high performance liquid chromatography (HPLC) column fillings. The method includes hydrolysis and polycondensation of alkoxysilanes in the presence of water-soluble polymers, which lead to polymer-induced phase separation. (2) Jones *et al.*[15,46] created interconnected macroporosity in an otherwise nanoporous phosphosilicate SG material through foaming with the help of a surfactant. The resulting material exhibits rather poor mechanical strength. (3) Marques *et al.*,[35,36,47] Vueva *et al.*,[39] Almeida *et al.*,[48] Wang *et al.*[24] and Jain *et al.*[49] developed a modified SG method to generate interconnected macroporosity combined with nanoporosity in calcium silicate and calcium phosphosilicate glasses with a coral-like morphology. Their method exploits phase separation induced by a water-soluble polymer, polyethylene oxide (PEO). (4) Maekawa *et al.*[50] produced nano/macroporous inorganic oxide monoliths with 5–10 nm nanopores and 0.1–5 μm macropores, using a polystyrene macroporous template and a self-assembling block-copolymer–SG mixture. Similarly, Wang and Jain[34] utilized the powder prepared by the SG method with polymer replication and subsequent sintering to obtain monolithic nano–macro porous samples for hard tissue regeneration, with nanopores smaller than 10 nm and macropores in the range of 300–500 μm.

Glass-in-glass phase separation is a well-known phenomenon in several modified silicate, borate and borosilicate glasses.[51] For immiscibility to occur, the separation of a liquid into two phases has to lead to a decrease in the Gibbs free energy. Depending on the initial composition, phase separation may occur by *nucleation* and *growth* of small *droplets* within a continuous glass matrix, or through a spontaneous separation into two interconnected phases by *spinodal decomposition*. The latter morphology is well suited for creating interconnected porosity when one of the phases is leached out selectively. Clearly, the choice of composition is critical for accomplishing the right kind of phase separation. In the method of Marques *et al.*,[37,38,47] spinodal phase separation is induced by a water-soluble polymer simultaneously with gelation that is catalyzed by acetic and hydrofluoric acids. The gel so produced has a polymerized skeleton separated by a liquid phase on the scale of a few nm, just as in a classical SG process. On this structure is superimposed spinodal phase separation that is created by the addition of water-soluble polymer like PEO. Removal of liquid phase from the gel made by this modified SG process leads to a coral-like interconnected morphology which includes both interconnected macropores (~10–300 μm) and nanopores (~5–50 nm),[36] as shown in Figure 5.2(a) and (b), respectively. This sample was prepared by adding appropriate quantities of TMOS and $Ca(NO_3)_2 \cdot 4H_2O$ into a solution made of PEO in acetic acid.[22] After vigorous stirring, HF was added to catalyze gelation. The sol at this stage was immediately cast into the wells of tissue culture plates.

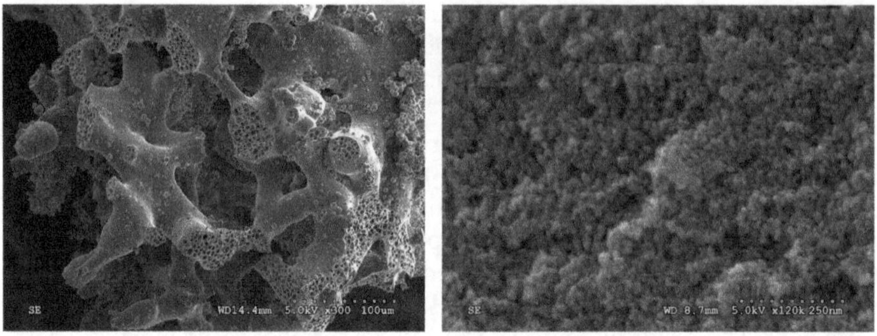

Figure 5.2 SEM micrographs of a heat-treated scaffold of composition 30CaO–70SiO_2 (mol%), prepared by the modified sol–gel process that includes PEO for creating macroscale phase separation: (a) magnification of 300× (scale bar = 100 μm); (b) magnification of 120 000× (scale bar = 250 nm). (Adapted from Marques *et al.*, *J. Mater. Res.*, 2009, ref. 38 with permission from Materials Research Society.)

Gelled samples were aged at 40 °C, and then soaked in distilled water or ammonia solution (if larger nanopores were desired) for solvent exchange before they were dried and stabilized by sintering at 700 °C. To preserve the concentration of water soluble calcium nitrate, it is important to not discard the solution but to dry it up gradually. The drying process can introduce a calcium concentration gradient locally, which may be reduced significantly during the sintering step.

Figure 5.2(a) also shows the presence of micron-sized small spheres and isolated pores which are typical of most materials prepared by this method. The composition of these microspheres is very similar to that of the matrix. Therefore, their biochemical properties should be similar to that of the surrounding scaffold. The isolated distribution of spheres and pores suggests their formation by a nucleation and growth mechanism. Apparently, they result from a droplet-like secondary phase separation together with the primary spinodal separation described above.[45] Upon drying, these droplets become micro-spheres which then dislodge from the matrix and leave behind spherical pores as seen in Figure 5.2(a).

The overall porosity exhibits a bimodal pore size distribution, with nanopores peaking at ~10 nm and macroporosity peaking around 100 μm, as shown in Figure 5.3 for a calcium phosphosilicate glass of 36CaO–4P_2O_5–60SiO_2 composition. The total volume fraction of porosity, determined by mercury intrusion porosimetry, is ~75%. In the modified SG method, the volume fraction of porosity, pore size and pore interconnectivity can be tailored to a large degree. On the other hand, a major challenge for the fabrication of bioactive scaffolds with interconnected macroporosity is due to the inherently poor mechanical properties when the material is used in monolithic form. This will not be a problem when the desired product is a powder or an injectable paste, for example.

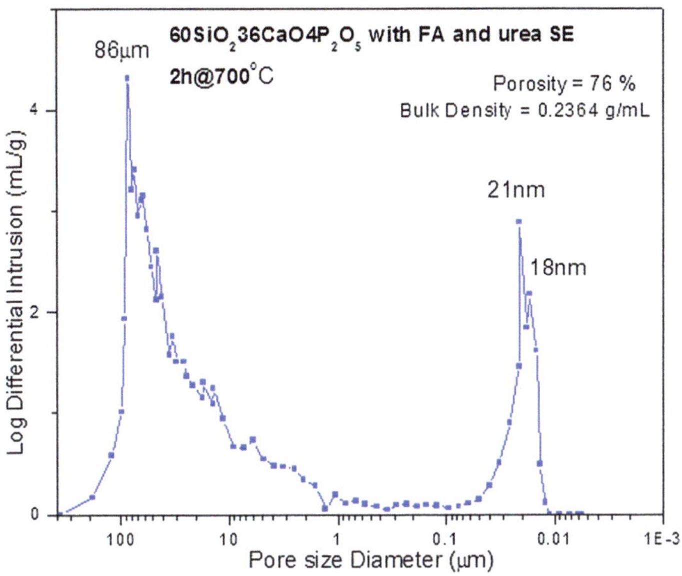

Figure 5.3 Pore size distribution in a typical nano-macro porous glass fabricated by modified sol-gel process. The data are for $36CaO–4P_2O_5–60SiO_2$ porous glass as determined using mercury porosimetry.
(Reprinted from Vueva *et al.*, *J. Am. Ceram. Soc.*, 2010, ref. 39 with permission from Wiley.)

Finally, we make note of an important chemical property of calcium silicate TAMP material. When exposed to simulated body fluid, it is readily covered with an interfacial layer of hydroxyapatite (HA, $Ca_{10}(PO_4)_6(OH)_2$)-like precipitate.[40] This mineral is known to promote the formation and bonding of bone to the substrate. Its presence is a preliminary indicator that the so-formed scaffold is biocompatible.[52,53] More detailed *in vitro* studies with bone forming cells and *in vivo* studies with animals described in Sections 5.5 and 5.6 confirm this prediction.

5.3 Biodegradation of TAMP Bioscaffolds Fabricated by Modified Sol–Gel Method

The drive behind the development of a high surface area TAMP bioscaffold was the ability to accelerate, in a controlled manner, its degradation in the body at a rate comparable to that of tissue growth. Naturally, one would like to establish how well this expectation is realized by the nano-macro porous bioactive glass, such as $30CaO–70SiO_2$ fabricated by the modified SG process described in Section 5.2.2. In this regard, we note that the basic steps of corrosion of silicate glasses in aqueous medium, such as the body fluid, which often involves both uniform dissolution and selective leaching of certain species, are fairly well established.[54] They include ion exchange between the mobile alkali or alkaline earth cations in the glass and hydrated protons in solution, formation of a silica gel layer at the surface, dissolution of the gel layer, followed by condensation and repolymerization of the silica-rich layer. For the case of dissolution of bioactive glass in body fluids, additional reactions follow: migration of Ca^{2+} and PO_4^{3-} ions (which may be supplied by the scaffold or present in body fluid) from glass to the surface through the silica gel layer to form a layer at the top rich in CaO and P_2O_5. The calcium and phosphates in body fluid then help constitute this layer into an amorphous CaO–P_2O_5 layer, which then crystallizes into hydroxyl carbonate apatite (HCA). In the applications of solid bioactive glass as an implant, the HCA layer becomes the foundation of subsequent biological activities that lead to the formation of new bone on top of the glass. A strong bond between the glass and bone forms because HCA mimics the inorganic part of the natural bone. Soft tissue also bonds to bioactive glass as collagen fibrils chemisorb on the porous silica-rich layer while HCA forms on them as well as on the glass.

Although there is a reasonable understanding of the corrosion of bioactive monolithic glass implants in simulated body fluid (SBF),[52] there is very little information available on how a nano–macro porous scaffold will behave *in vivo*. Of course, the much larger surface area will enhance the corrosion rate, but we may expect significant qualitative changes as well due to the dramatic difference in diffusion through nanopores *vs.* open surface. Nucleation and growth of the HCA layer and other precipitates are also likely to be modified significantly (as seen in Figure 5.4) creating a complex degradation process.[55]

Zhang *et al.*[55] performed the first investigation of the degradation of nano–macro porous 30CaO–70SiO_2 in SBF under quasi-dynamic conditions such that the solution was changed every day over a period of 60 days. The degradation was assessed from the dissolution of silica from the glass into

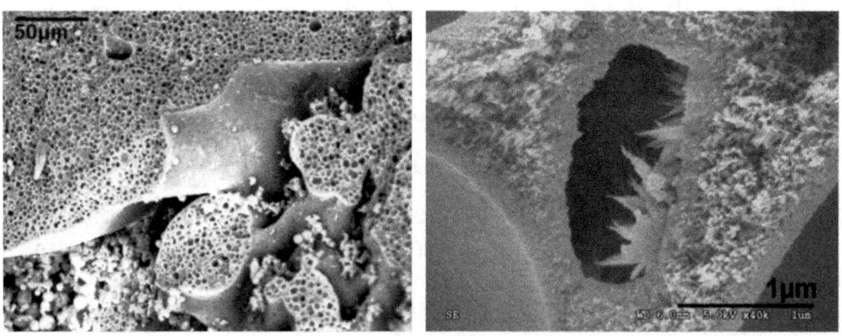

Figure 5.4 SEM micrographs of the cross-sectional surface of 30CaO–70SiO_2 (mol%) scaffolds at 28 days in SBF. The needle-like crystals are calcium phosphate precipitates.
(Reprinted from Zhang *et al.*, *J. Am. Ceram. Soc.*, 2012, ref. 55 with permission from Wiley.)

the SBF which did not contain any silicon at the beginning, while the *in vitro* bioactivity was characterized by precipitation of calcium phosphate (CaP) on and inside the scaffold. The amount of silicon in solution was determined every day, which led to the degradation degree (DD) of the scaffold defined as percent weight of silicon in SBF solution relative to that in the original solid scaffold. In parallel, the integrity of scaffolds was followed by measuring specific surface area and pore size distribution as a function of exposure time. This study yielded the following key conclusions about the degradation of TAMP under quasi-dynamic test conditions:

1. Overall DD follows an empirical exponential law with a half-life of 15.4 days, which is a reasonable degradation rate for tissue growth. Incidentally, this trend parallels that of porous scaffolds of biodegradable polymers such as poly(D, L-lactide-*co*-glycolide).[56]
2. A closer inspection of surface area (SA) over the period of experiments indicates three mechanistic stages of the degradation process.[53] In stage I, the fluid gradually enters the scaffold over a period of about a day, indicating slow ingression into the nano–macro porous structure. During this wetting stage the sample sinks gradually as the nanopores are filled with fluid. It is characterized by an increase in solution pH due to the fast release of calcium, but little change in SA. The changes in the characteristic properties are caused mainly by the reactions, such as the precipitation of calcium phosphate, on the external surfaces of the scaffold. In stage II, the SA increases, as degradation is dominated by CaP precipitation on both the external and internal surfaces that are exposed concurrently *via* interconnected porosity. In stage III, degradation intensifies with the gradual rupturing of the structure and elimination of the nanopores as the network dissolves and CaP precipitates throughout the material. During this stage, the SA reduces and pH of the solution decreases to an almost constant level. The glass network dissolves congruently instead of calcium ion leaching; the nanopores that were exposed to the SBF gradually rupture or consolidate thus decreasing the SA. The precipitation of the CaP layer continues on the exterior, but more inside the scaffold at a higher rate as indicated by the changes in the concentration of phosphorous.
3. Although SA decreases in stage III and pore structure is modified with prolonged immersion, the overall integrity and microporous structure of the remaining scaffold are maintained throughout the immersion period. This is an important observation for the continued functioning of TAMP material as a scaffold for tissue engineering.
4. A direct relationship is found between the degradation rate and specific surface area, as seen in Figure 5.5 for two samples. Here different surface areas were obtained by using different sintering temperatures. This result demonstrates that by using appropriate processing parameters, it should be possible to tailor the degradation rate of TAMP scaffolds to match a patient's anticipated tissue regeneration rate.

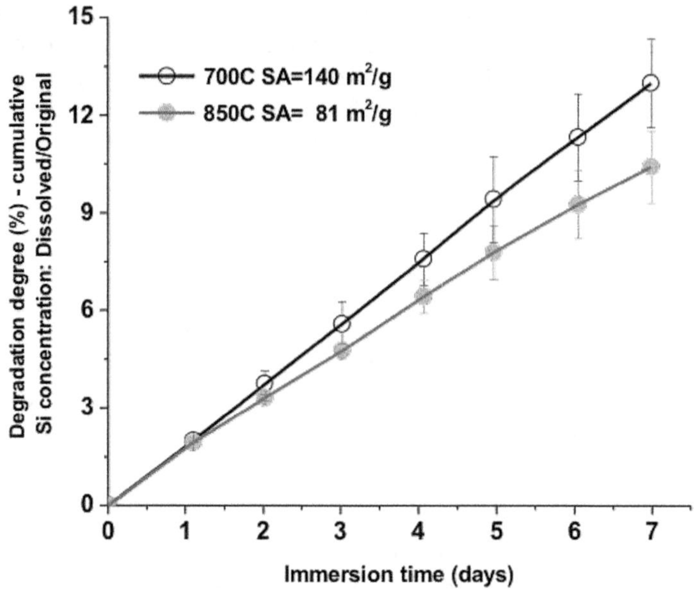

Figure 5.5 The effect of surface area (SA = 140 *vs.* 81 $m^2 g^{-1}$) on the degradation (%) of $30CaO–70SiO_2$ TAMP scaffold as a function of the time of immersion in simulated body fluid (SBF). The degradation is quantified by the normalized cumulative concentration of Si dissolved in SBF for two samples sintered at 700 and 850 °C.

5.4 Cell Response to TAMP Bioscaffolds

The TAMP silicates owe hard tissue regeneration properties to their progressive dissolution in body fluids that stimulate bone cell differentiation.[57] However, the interaction between biomaterials and cells of tissues and organs is a highly complex and poorly understood process, particularly between cells and bioactive glasses. Ample research over the past years has shown that the physico-chemical characteristics of biomaterials, including bioactive glasses, such as chemical composition, surface topology, chemical reactivity, pH, charge, hydrophobicity/hydrophilicity, and dissolution behavior influence and dictate their biological performance.

We know that cells are highly capable of sensing changes in their extracellular environment including the characteristics of the surfaces to which they adhere. Cells have trans-membrane surface receptors (*e.g.* integrins) that are organized in focal adhesions and hemi-desmosomes, which are complex cellular multi-protein signaling structures that interact with specific components of natural cell substrates (the extracellular matrix, ECM) such as collagens, fibronectin, or bone-cell specific ECM-proteins (*e.g.* osteopontin, osteonectin, bone sialoprotein, *etc.*). These receptors are linked *via* actin filaments to the cell interior and allow for signaling across the cell surface membrane in both directions (inside/out signaling).[58–60]

Arginine–glycine–aspartic acid amino acid sequence motifs (Arg-Gly-Asp in the three-letter, RGD in the single letter, amino acid codes) present in extracellular matrix proteins (*e.g.* in fibronectin) are the best-known signaling motifs that are recognized by integrin receptors. It is *via* these signaling receptors that cells 'sense' their adhesive environments, and render a substrate suitable for adhesion (*e.g.* mineralized bone extracts and bioactive glasses) or unsuitable (*e.g.* titanium, polymers, Teflon® and most ceramics). Furthermore, cells react to different characteristics of adhesive substrates, with clearly detectable morphological and physiological changes.[61–65]

It is now well established that biomaterials, upon contact with protein-rich body fluids ,such as blood and lymph, instantaneously absorb proteins, which coat their surface within seconds to minutes.[66,67] Thus cells do not actually contact the biomaterial itself, but rather interact with the molecular architecture of the surface-adsorbed protein layer. Research including our own (discussed below) has shown that the mode of protein adsorption on TAMP scaffolds is directly mediated by the above-mentioned physicochemical properties of the underlying biomaterial/scaffold.[59,64,65] Adhering cells detect protein type, protein-coat composition and protein conformation, which directly influence cellular behavior and cellular response. The scaffold morphology, composition and/or texture that may impact the cells are best investigated with cells in culture, allowing detailed microscopic and biochemical analyses of scaffold-induced cellular responses. We have primarily used MG63 human bone osteosarcoma cells (ATCC CRL-1427) and MC3T3-E1 subclone 4 mouse calvarial bone pre-osteoblasts (ATCC CRL-2593) (*i.e.* bone forming cells), which are model cell lines for bone research. Mouse osteoclasts (*i.e.* the bone degrading cells) that we isolated from mouse calvarial bone marrow, as well as epithelial and dermis cells, have also been used in our studies. MG63 and MC3T3-E1, as well as other cell types, readily adhere to both melt–quench and SG-derived TAMP bioactive glass scaffolds *in vitro*,[22,36,37,68–70] and efficiently colonize the surface as well as the inside of the scaffolds within a few days.[22] In the following sections, three examples of cellular responses to specific bioactive glass scaffold characteristics, *viz.* chemical composition, scaffold surface topology, and nanostructure, discovered in our research are described. Sections on challenges in working with highly porous bioactive glasses *in vitro* (Section 5.4.4) and on addressing future research questions follow (Section 5.6).

5.4.1 The Effect of Chemical Composition: Bioactive Glasses Substituted with Boron

Recent studies indicate that the addition of boron to bioactive glasses may further enhance bone formation.[71] Boron is an essential micronutrient acting as an ultra-trace element in concentrations <1 ppm. In plants, boron is required primarily for maintaining the integrity of cell walls while in mammals, it plays a crucial role in osteogenesis and maintenance of bone.[72,73] Under conditions of boron deficiency, development and regeneration of bone declines.[74–76] Yet, very little is known about how this ultratrace element exerts its beneficial health effects. Boron may interact with steroid hormones, and thus is involved in the prevention of calcium loss and bone de-mineralization.[77] Scientists began to evaluate the role of boron on the differentiation of osteoblasts and the formation of bone by adding boron to cell culture media, or Bioglass® implants. Although somewhat contradictory reports were published,[78,79] the beneficial effect of boron on osteoblast differentiation and bone formation seems compelling.[80–83] Therefore, we investigated the effect of boron by synthesizing TAMP bioactive glass scaffolds with and without a small amount of boron: normal $30CaO–70SiO_2$ and boron-substituted $30CaO–2B_2O_3–68SiO_2$ with similar structures. We found that MC3T3-E1 pre-osteoblast cells adhered significantly faster and more efficiently to boron-containing TAMP samples compared to boron-free samples. The difference became insignificant when cells grew for longer periods of time (4 and 24 hours). However, a different cellular morphology with more lamellipodial extensions, membrane ruffles and less pronounced actin stress fibers remained on boron-containing samples compared to a more roundish morphology with pronounced peripheral actin belts and pronounced stress fibers on non-boron containing TAMP samples.[84] In a related study we compared MC3T3-E1 cells on borosilicate microscopic cover glasses *vs.* on boron-free soda lime glass or tissue culture plastic substrates. The results showed a significant up-regulation of bone-specific proteins (including RunX2/Cbfa1, the master transcription factor responsible for bone-cell differentiation; bone sialoprotein 1 or osteopontin, osteocalcin, two bone-specific secreted proteins involved in matrix mineralization; collagen 1A, alkaline phosphatase, connexin 43, and three other osteoblast-relevant proteins) on boron-containing samples,[85] suggesting that doping of bioactive TAMP scaffolds with boron may further enhance their performance.

5.4.2 The Effect of Surface Topology on Attachment and Proliferation of Osteoblast Cells to Bioactive Glasses

Studies on titanium metal implants have shown that surface micro-architecture (roughness and texture) influences cell behavior.[86] However, little is known about the role of surface topology of glass on its use as an implant. In a preliminary study, Levy *et al.*[87] compared two glass samples and found greater cell proliferation on smoother samples. We systematically investigated the effect of surface roughness ($R_a \sim 0.01–1.1$ μm) on cell adhesion and proliferation on classical melt–quench prepared 45S5 Bioglass® samples *in vitro*.[67] MG63 osteosarcoma and MC3T3-E1 osteoblast precursor cells were seeded on the glass samples, and incubated for up to 6 days. The number, viability, morphology, and attachment of cells were investigated using fluorescence microscopy. The results showed that cell attachment (as indicated by cell spreading and number of focal adhesion sites) and proliferation rate decreased with increasing roughness of the bioactive glass surface. Although these findings provide important insight for improving surface characteristics of bioactive glass bone implants, they may have less relevance for TAMP scaffolds that have an intrinsically high surface roughness from their macropores. However, sol–gel glasses prepared without macropores may benefit from these considerations.

5.4.3 The Role of Bioactive Glass Nanostructure (Nanopore Size and Phase Separation) on Cell and Protein Attachment

Nanoporosity is known to impact the performance of implants and scaffolds, such as bioactive glass (BG) scaffolds, either by providing a higher concentration of bioactive chemical species from enhanced surface area, or due to inherent nanoscale topology, or both.[20,88–92] To delineate the role of these two characteristics we fabricated bioactive glass scaffolds with nearly identical surface areas ($81–83 \pm 2\ m^2\ g^{-1}$) but significantly different nanopore sizes (av. 3.7 nm, sample F; and av. 17.7 nm, sample E) by varying both the sintering temperature and the ammonia concentration during the solvent exchange phase of the SG fabrication process (Figure 5.6A). We then performed *in vitro* tests with MC3T3-E1 pre-osteoblast cells seeded and cultured on these two types of bioactive glass scaffolds. Within 12 hours post-seeding, cells attached to the surface of both sample types and began to proliferate within 24 hours. As indicated by cellular actin–cytoskeleton staining (green), cells on all specimens exhibited a well spread epithelioid-like morphology featuring prominent stress fibers as is typical for cells adhering to stiff substrates (Figure 5.6B). However, cell density on sample F (with smaller nanopore size) was significantly higher than on sample E (with larger nanopore size) 12 hours post-seeding (Figure 5.6C). The difference in cell density on the two sample types was less pronounced by 48 hours post-seeding, although a slightly higher density of cells was detected on samples with smaller nanopore size (Figure 5.6C). No obvious differences in actin organization were observed between the two scaffold types, although higher magnification may have revealed potential structural differences. At the same time, cells do not respond to non-porous samples as well as to nanoporous surfaces (see further in Section 5.5). Taken together, these results indicate that optimized nanoporosity (in our case ~4 nm compared to ~18 nm diameter) can boost bioactive glass scaffold bioactivity and cell adhesion, providing clear evidence of the beneficial effect of nanopore topography in tissue engineering-relevant bioactive glass scaffolds.[68]

Interestingly, we recently obtained additional comparable results with spinodally *versus* droplet-like phase-separated 45S5 bioactive glass scaffolds that were fabricated by cooling the glass melts from different temperatures.[93] We found that MC3T3-E1 pre-osteoblast cells attached significantly faster and more efficiently to samples with spinodal morphology compared to droplet-like morphology based on cell number and morphological

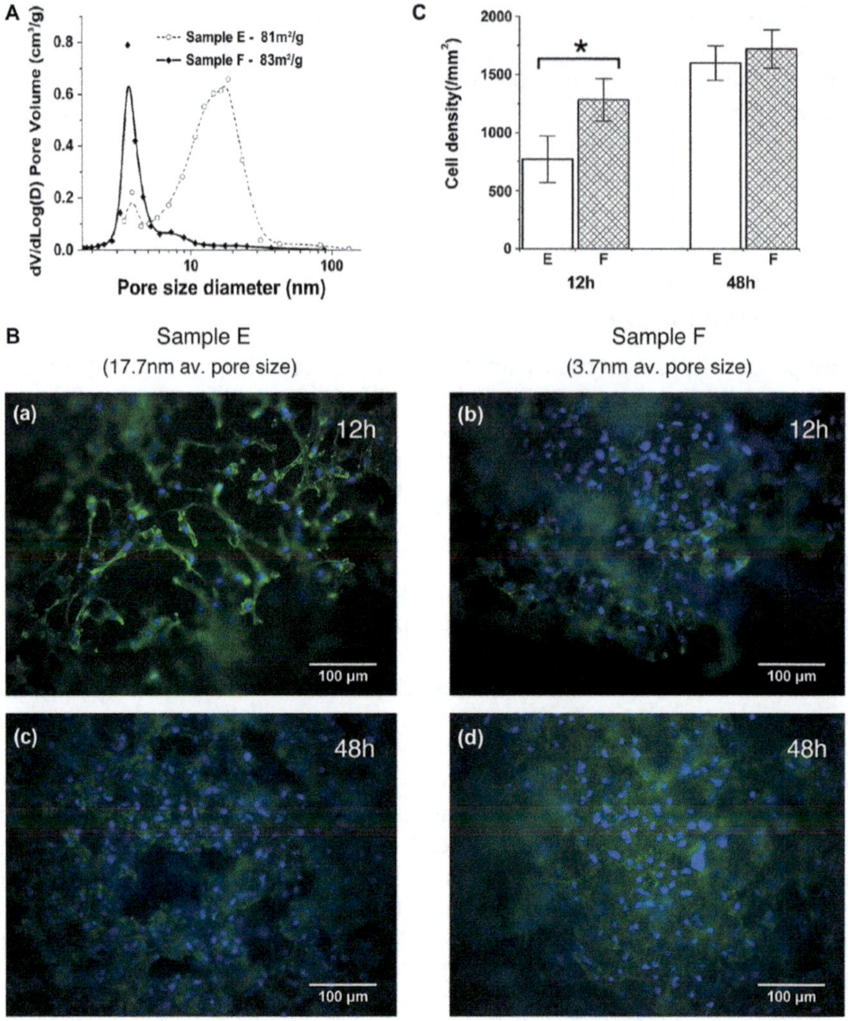

Figure 5.6 Proliferation and adhesion of bone cells on TAMP bioactive glasses with different nanoporosity. (A) Pore size distribution of samples E and F. (B) Representative micrographs of (a) samples E at 12 hours, (b) samples F at 12 hours, (c) samples E at 48 hours, and (d) samples F at 48 hours post cell seeding. MC3T3-E1 mouse calvarial bone pre-osteoblast cells were fixed with formaldehyde, F-actin was stained with Alexa 488-Phalloidin (green) to evaluate cell morphology, and cell nuclei were stained with DAPI (blue) to quantify cell density. Images were acquired using a 20× objective. (C) Cell density on samples E and F 12 and 48 hours post cell seeding. The error bars represent the standard deviation of cell density on three samples (* = statistically significant, $P<0.05$). (Reprinted from Wang *et al.*, *Tissue Eng., Part A*, 2013, ref. 70 with permission from Mary Ann Liebert, Inc.)

characteristics. A similar result was found for bovine serum albumin (BSA), a soluble globular model protein present in abundance in the serum component of cell culture media, which attached more efficiently to the surface of spinodally phase-separated glass. Raman spectroscopy indicated a different conformational state of attached protein molecules on the two glass varieties with significantly more β-sheet and β-turn topologies on spinodally phase-separated samples.[94]

What could be the underlying mechanism that provokes such a significantly different cellular adhesion response to chemically identical substrates that differ only in structure at ~10 nm or less? Recently, the impact of nanostructure on cell functions has been studied on various material systems.[36,95–101] In these studies, introducing nanostructures, such as nano carbon/polymer fibers, or nanopores to the scaffolds enhanced cell performance. Results obtained in these studies correlate well with our *in vitro* and *in vivo* observations (see Figure 5.6 and Section 5.5). However, the precise influence of nanostructure on cell performance is still unclear and under debate. For example, Woo *et al.* suggested that incorporation of nanostructures would lead to an increased surface area, promoting protein adsorption, and hence improved cell attachment.[96] Others have suggested that nanopore topography directly influences cellular functions, either by enhancing protein adsorption,[98] changing the conformation of certain cellular attachment proteins,[102] or by changing the surface energy.[97,99] A recent *in vitro* study investigating the influence of surface characteristics of poly(methyl methacrylate) on bone formation suggests that a distinct arrangement of nanoscale disorder can stimulate mesenchymal stem cells to produce bone mineral *in vitro*, even under conditions where all other parameters (*e.g.* the size and number of nanopores, comparable to our experiments) remained unchanged,[103] supporting our findings that nanotopography determines cellular performance.

Based on these, and our own findings, we have developed the following hypothesis. Figure 5.7 represents a schematic drawn to approximate scale of the surface of nanoporous TAMP scaffolds (~4 nm on the left, ~18 nm on the right), and potential conformations of adsorbed protein that cells may encounter upon adhesion to these scaffolds. Since the chemical composition of the scaffolds is identical, proteins may absorb in similar conformations on both scaffold types (represented by the '&' symbol in scenario A). So how may a variation in pore size (the only difference) then influence cellular behavior (manifested in faster or slower cell adhesion and faster and slower initial proliferation)? One possibility (shown in A, 'Early'), for example, is that adsorbed proteins fill the small nanopores, and thus a smooth, homogenous and favorable scaffold surface is presented to cells, resulting in more efficient cell adhesion and proliferation. The larger nanopores may be too big to be filled with adsorbent proteins (*e.g.* globular albumin), generating an inhomogeneous, 'checkered' less favorable substrate surface to which cells adhere less efficiently. Adherent cells are known to secrete over time their own extracellular matrix consisting of elongated, fibrous proteins

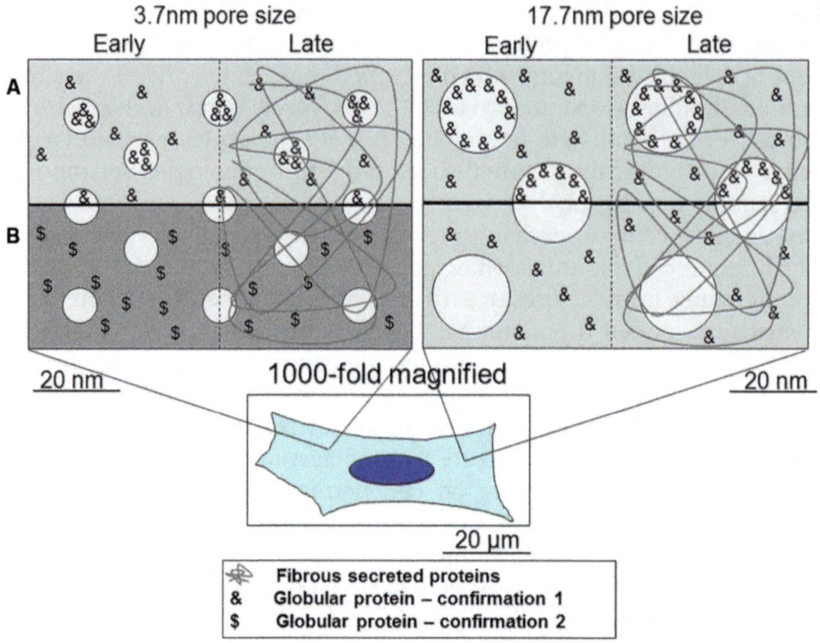

Figure 5.7 Schematic of TAMP scaffold surface topology drawn to approximate scale (2D-view; note that pores within the scaffolds are interconnected) and two hypothesized types of protein adsorption (A and B) triggering different observed cell responses (see text for further details).

(represented by crisscrossing curved lines in 'Late'). These then may also traverse and cover the larger pores transforming the scaffolds with larger pores into a similarly functioning substrate. This possibility correlates with our observation that after extended times (≥48 hours) cells grew with comparable efficiency on both scaffolds (Figure 5.6C).[36]

A second, more favorable possibility (shown in scenario B) depicts that the different physical properties of the two scaffold types cause a protein type (*e.g.* albumin) to adsorb in different conformations ('$' conformation on scaffolds with smaller nanopores, '&' conformation on scaffolds with larger nanopores). Cells prefer the protein $ conformation and thus adhere and proliferate more efficiently on the scaffolds with smaller pore size ('Early'). Later ('Late'), adsorbed proteins are again covered with cell-secreted fibrous matrix proteins and scaffold performance again becomes about equal. An earlier study, conducted on alumina, supports our second hypothesis (B), suggesting that the unfolding of the cellular adhesion protein, vitronectin, by introducing nanophase could expose an increased number of cell-adhesive epitopes that then are recognized by specific cell membrane receptors leading to enhanced cell attachment.[100] Similarly, our results investigating phase-separated 45S5 Bioglass® suggest that cell attachment to bioactive glasses is mediated by a protein layer and that the conformational state of this layer shapes cell attachment. In summary, our *in vitro* cell tests on bioactive glass scaffolds with different nanopore size, and on phase-separated samples indicate that initial cell attachment is significantly enhanced on bioactive scaffolds with an optimum nanopore size or spinodal decomposition (a glass topology on a similar nanoscale), providing compelling evidence that nanoscale topology can significantly affect the biological performance of engineered bioactive glass scaffolds.

5.4.4 Challenges of Investigating Cells on Porous Glasses

Macroporous glasses, appearing opaque from light scattering, are unsuitable for standard light microscopic analyses that depend on transmitted light (phase-contrast, differential interference contrast (DIC), and other contrast-enhancing light microscopy techniques). Furthermore, they have a highly contoured surface topology that makes detection with high resolution, high numerical aperture (NA) oil immersion objectives challenging as some cells attached to the scaffold appear out of focus. We partially overcame these issues by staining cells with fluorescent probes (Live/Dead Viability stains, DAPI, Hoechst, Phalloidin, *etc.*) or by using antibodies directed against specific proteins (vinculin, *etc.*) that were conjugated with fluorophores (Alexa 488, Alexa 568, Rhodamine, Texas Red, Cy3, *etc.*), a technique termed indirect immunofluorescence (see Figures 5.6B and 5.8A). Acquiring images with 40× oil, and 10×, 20× and 40× air, long-distance objectives, compared to higher resolution (but narrower focal plane) 60× and 100× oil immersion objectives, also may sometimes also help generate decent images of cells growing on scaffolds that otherwise would have been beyond acquisition (see, *e.g.* Figures 5.6C and 5.8A). Figure 5.8A shows cells growing on the surface of a TAMP scaffold (detected by staining the cell nuclei with DAPI) that were imaged at two different focal planes. The image on the left shows the cells growing on the scaffold surface in focus, while the image on the right shows the cells in focus growing deep inside macroporous depressions. Scanning electron microscopy (SEM) analysis of scaffolds in general allows for adequate detection of the scaffold and overall cellular morphology in great detail (Figure 5.8B), however it does not allow observation of protein distribution inside cells. Recently, for studies that only concerned nanoporosity, we crushed TAMP samples into fine powder, which was then pressed into disks. These disk-shaped scaffolds still maintain their specific nanoporosity, but feature a much smoother surface that is far more suitable for high-resolution fluorescence light microscopic analysis.

Another challenge, especially of porous bioactive glasses, is their tremendous ability to adsorb proteins including proteins present in the serum component of cell culture media (*e.g.* albumin). Such dominant adsorption can make Western blot analyses and other biochemical assays challenging, as matrix proteins, transcription factors and other bone-cell specific marker proteins are generally expressed in much lower amounts and thus may become difficult to resolve by SDS-PAGE gels. Figure 5.8C shows a Coomassie-stained SDS-PAGE gel of MC3T3-E1 cell proteins that grew on

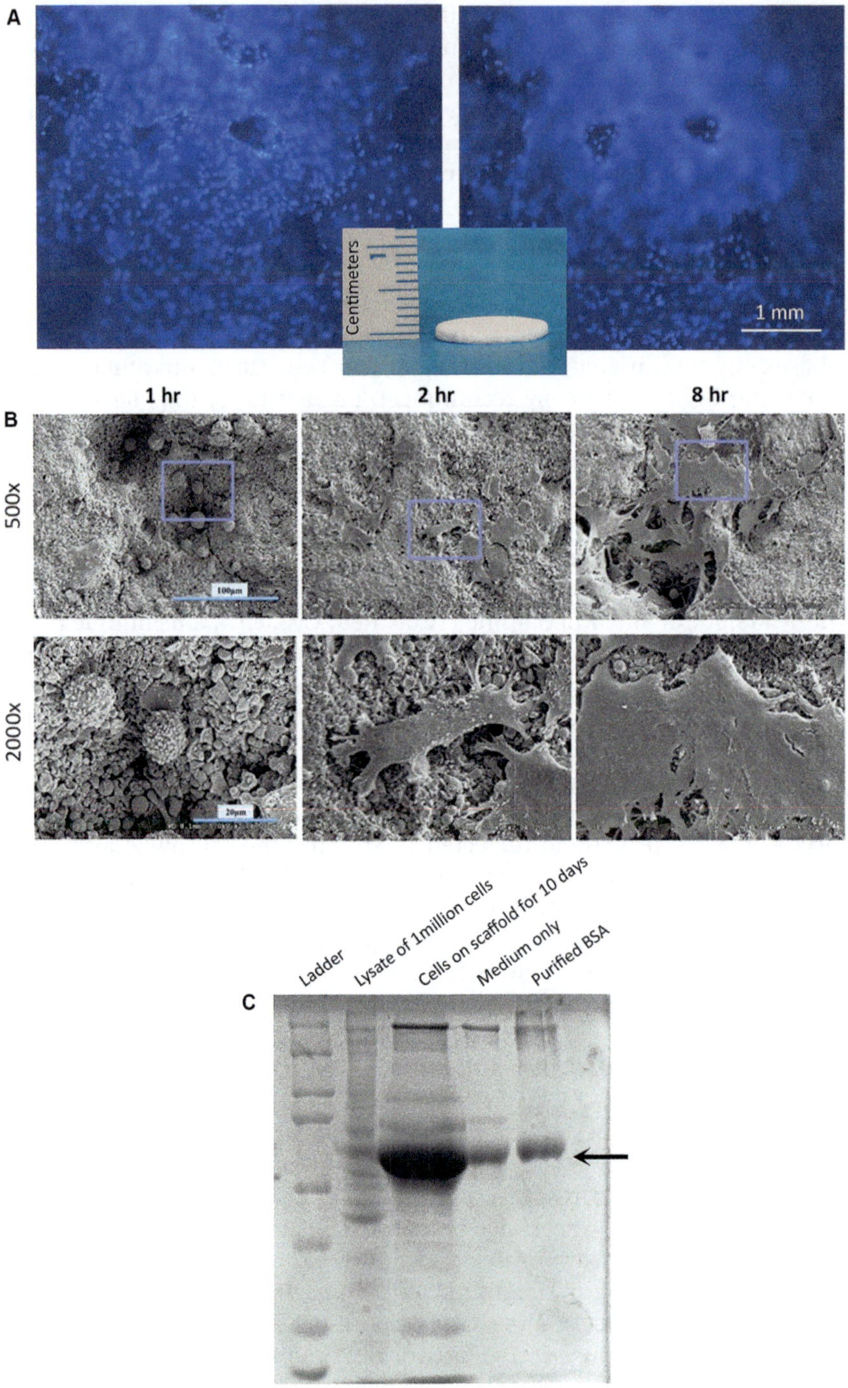
A
Centimeters
1 mm
1 hr
2 hr
8 hr
B
500x
2000x
100μm
20μm
Ladder
Lysate of 1million cells
Cells on scaffold for 10 days
Medium only
Purified BSA
C

sol–gel-fabricated TAMP scaffolds. Note the huge amount of albumin on the gel (labeled with an arrow) that totally obscured the resolution of smaller, faster-migrating cellular proteins (compare with cell lysate analyzed in the lane to the left derived from TC plastic-grown cells). We succeeded by normalizing cell lysates to a housekeeping protein that was larger than albumin (α-adaptin) and thus migrated more slowly. Consequently, its detection and quantification was not affected by the adsorbed albumin. TAMP scaffold protein adsorption can also be challenging when performing indirect immunofluorescence analyses as the secondary, fluorophore-conjugated antibodies readily adsorb to the scaffold, generating a high background fluorescence. Blocking permeabilized cell samples with large amounts of protein (5–10% bovine serum albumin or fetal calf serum) for extended times (overnight) before fluorescence analysis helped to partially overcome this issue. Quantitative cell assays may be challenging when cells have colonized inside the three-dimensional interconnected macroporosity of TAMP scaffolds, making consistent extraction of cells from the samples difficult. Crushing scaffolds into powder before adding solvent buffers has helped us perform these assays on a more quantitative basis.

Finally, a standard colorimetric assay to detect osteoblast differentiation, alizarin staining (a red stain that is based on the detection of calcified matrix as secreted by differentiated osteoblasts), cannot be performed on typical TAMP scaffolds that contain calcium. In this case, the entire scaffold is stained red, obscuring observation of cell differentiation. Overcoming these numerous, often unexpected and annoying challenges, requires more complicated analytical techniques to acquire seemingly simple results. Yet, the type of data that can be obtained (especially concerning nanostructure of scaffolds influencing cell response) makes the *in vitro* analyses of bioactive glass scaffold–cell interactions highly informative.

5.5 *In vivo* Experimental Animal Studies and Clinical Trial

A distinctive feature of TAMP bioscaffolds is the presence of interconnected nanopores that are absent in other common bioscaffolds, while interconnected macropores must exist in all bioscaffolds to facilitate tissue growth in 3D. The results in Section 5.3 and references therein have shown

Figure 5.8 Challenges of investigating cells on highly porous TAMP scaffolds. (A) Cell nuclei of MC3T3-E1 pre-osteoblasts growing on, and inside, a TAMP scaffold were stained with DAPI (blue) and imaged at low magnification in two different planes revealing the coarse surface topology of the TAMP material shown in the insert. (B) MC3T3-E1 cells growing for the indicated times on TAMP scaffolds were fixed and imaged by SEM. (C) A Coomassie-stained SDS-PAGE gel showing the large amount of albumin present in the serum component of cell culture medium that efficiently adsorbs to TAMP scaffolds (arrowed) and makes quantitative analyses, such as estimation of cell numbers, challenging.

that, by controlling nanoporosity, the biodegradation rate of a bioscaffold of a given composition can be controlled to match the tissue regeneration needs of a patient, at least under quasi-dynamic laboratory conditions. Notwithstanding, for the successful use of TAMP scaffolds in real patients, it is necessary to establish the role of nanopores under realistic testing conditions. In particular, we must establish how the larger changes in local chemistry from faster degradation of a nanoporous scaffold would affect the tissue in its vicinity? Then in view of the enhanced attachment of pre-osteoblast to TAMP substrates of specific nanopore size, as discussed in Section 5.4.3, it is important to determine whether nanoporosity may also produce a positive or negative effect on tissue response *in vivo*.

We sought to answer the above questions by performing *in vivo* experiments.[22] We implanted eight TAMP scaffolds under the skin (subcutaneously) of a New Zealand rabbit. These degraded rapidly with minimal inflammation and were replaced by normal connective tissue, as seen in Figure 5.9. Samples were harvested at different times (1, 3 and 5 weeks) and analyzed by fixation, staining (Stevenel's blue and Van Geison), and microscopic examination. Scaffolds are marked with an '*' in the figure. Only minimal inflammation (marked with yellow arrows in the 1 week image) is visible. Note how cells (stained blue) colonize the scaffolds from their edges. At three weeks, new blood vessels (marked with arrows in the 3 week image)

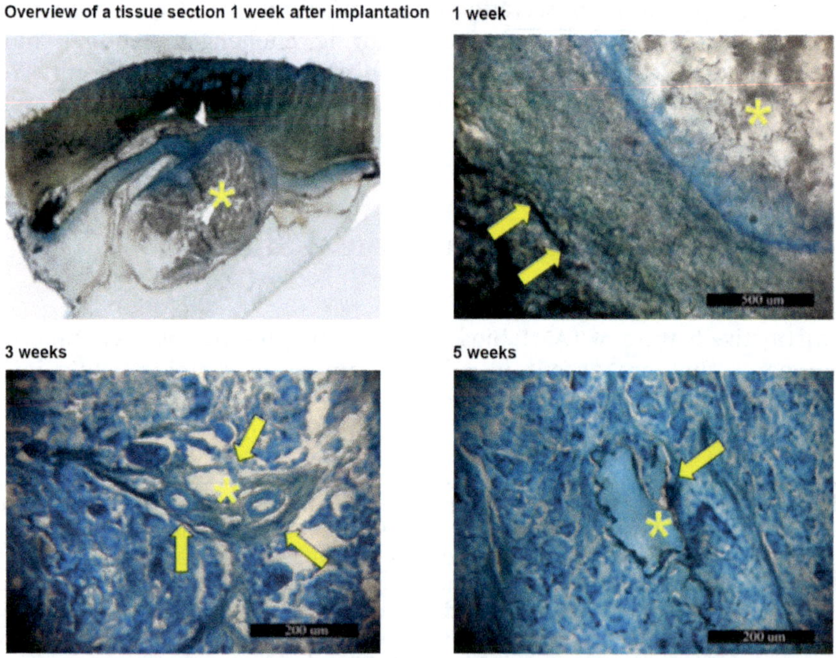

Figure 5.9 Histological analysis of TAMP scaffold and surrounding tissue after *in vivo* tests on New Zealand rabbit models for 1, 3 and 5 weeks. Scaffolds are marked with *. Top-left image shows overview at one week.

are formed in the area of the TAMP scaffold which begins to break down. At 5 weeks, only fragments of the scaffolds are left after being replaced by connective tissue with new vascularization. High cellular penetration into the scaffold is found throughout the whole region. Biodegradation of the nano–macro porous scaffold material and the observed high cellularity demonstrate the strong potential of this multi-porous material for enhancing the recruitment and proliferation of cells for new tissue formation, that gradually replace the dissolving bioactive glass. Three important conclusions are made from this study: (1) The $30CaO–70SiO_2$ TAMP scaffold is completely biocompatible. (2) It degrades *in vivo* at a rate that is comparable to the value from the laboratory test performed under quasi-dynamic conditions. Thus it will not be difficult to tailor bioscaffolds for a particular degradation rate. (3) This particular TAMP material promotes tissue regeneration of soft tissue that is similar to the surrounding tissue. Interestingly, we also found mineralized-like deposits in some samples. Evidently, this bioscaffold appears to induce regeneration of tissue appropriate for the local site, and therefore it has potential for a much broader range of applications than just hard tissue, like bone.

To establish the role of nanoporosity alone, it was important to fabricate the samples that had the same macroporosity and chemical composition, but differed only in nanoporosity. This was accomplished by preparing a nano–macro porous $30CaO–70SiO_2$ sample by sintering at 700 °C.[68] Some pieces of these samples were then further heat-treated at 840 °C, which decreased the surface area by more than an order of magnitude without altering the composition. Effectively, the latter treatment closed almost all the nanopores. These TAMP samples with and without nanopores were implanted subcutaneously in a New Zealand rabbit. The infiltration of cells, tissue formation and degradation of the scaffold were characterized at one and two weeks. At one week post implantation, cells penetrated to a depth of more than 200 μm into the macropores of the nanoporous sample, as seen in Figure 5.10. By comparison, there was no clear sign of cell penetration in the sample in which the nanopores were eliminated. At two weeks, cells also began to penetrate the macropores of the latter sample. However, the depth of penetration and the density of cells colonizing the inside of the macropores remained significantly below that observed for the former sample.

Furthermore, the nanopores seemed to enhance tissue integration with the scaffold. To reach this conclusion samples without macroporosity were implanted subcutaneously in a rabbit, with one sample having nanoporosity while another lacked any porosity.[68] Two weeks post-implantation, tissue fully integrated with the nanoporous BG scaffold, whereas the sample without nanopores consisted of unbonded interstitial regions indicative of a relatively poor scaffold integration. The origin of this difference in the tissue integration of the two types of samples can be traced to the fact that nanopores promote cell attachment as described above. As mentioned above, it is well known that cell attachment, migration, and growth are mediated by

Figure 5.10 Much faster infiltration of cells in (A) nano-macro porous than in (B) macroporous-only scaffolds. Yellow and green lines represent tissue-scaffold interface at 0 and 7 days, respectively.
(Reprinted from Wang *et al.*, *Tissue Eng., Part A*, 2013, ref. 70 with permission from Mary Ann Liebert, Inc.)

proteins that absorb from serum on the surface of biomaterials.[94] Furthermore, for calcium silicates the appropriate concentration of silicate and calcium ions released from the surface can stimulate the formation and growth of bone nodules.[16] The incorporation of nanopores enhances the surface area, which, in principle, can provide an increased number of sites for protein adsorption, as well as ion concentrations optimized for more efficient cell growth. These observations and reasons strongly support the beneficial effect of nanopores incorporated into BG scaffolds, in agreement with the *in vitro* cell response to nanoporous samples discussed in Section 5.4.3.

To specifically assess the hard tissue regeneration efficacy of 30CaO–70SiO_2 TAMP material, researchers at Alexandria University evaluated *in vivo* titanium implant osseointegration in the dog mandible after using TAMP as a graft material (unpublished data). The space between the implant and socket was packed with either the TAMP powder or autogenous bone chips from the same animal, the latter serving as a reference for the TAMP test sample (see Figure 5.11). At eight weeks the tissue was harvested *en bloc* to examine the tissue regenerated around the dental implant. The results showed high osseointegration along the serrated implant surface for both kinds of powders (Figure 5.11B and C). However, the TAMP powder induced remarkably more extensive trabecular pattern and larger marrow tissue, which is attributed to the role of the bioactive glass powder that degrades rapidly due to nanoporosity. It induces angiogenesis, as demonstrated by the rich blood supply that distinguishes the cancellous bone type with its high potential of remodeling activity. This characteristic is necessary to respond to functional changes and force adaptation.

Inspired by the encouraging *in vivo* results on bone formation and osseointegration in the dog model, recently a phase I/II randomized controlled clinical trial has been initiated in humans.[104] It utilizes a split mouth technique in order to evaluate the effect of TAMP bioactive glass in bone regeneration immediately following extraction of mandibular and/or

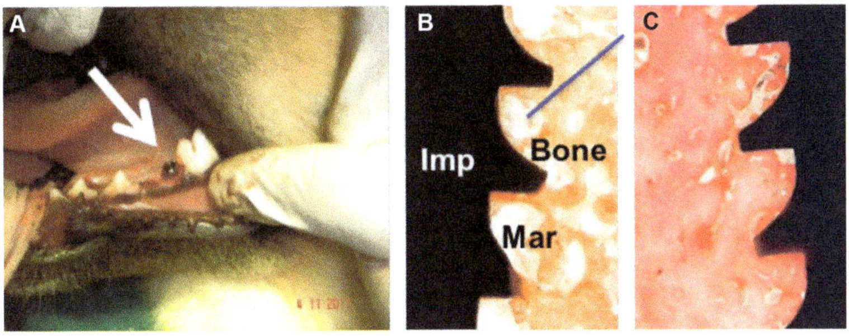

Figure 5.11 (A) Bone regeneration in canine mandible around titanium implant (Imp, shown in black) at 8 weeks. Microstructure of implant–tissue interface generated with (B) TAMP powder or (C) autogenous bone powder (40×).

maxillary premolars. Patients are followed up clinically and radiographically at 1 week, 2 weeks, and 1, 2, 3, and 6 months post-operatively. Core biopsies from the healed extraction sockets were obtained at 3 months from grafting. Preliminary histological analyses showed that the TAMP graft did not interfere with the normal healing cascade and enhanced cellularity and vascularity.

In conclusion, TAMP structures have demonstrated superior response for tissue regeneration. The results are highly promising, and provide a basis for additional required *in vivo* tests and clinical trials, before the product can be sold for use by surgeons and dentists.

5.6 Conclusions and Outlook

By exploiting a modified sol–gel preparation method, tailored amorphous multiporous bioscaffolds of calcium silicate compositions have been fabricated. The degradation rates of these materials can be optimized to suit the needs of a specific patient *via* independent control of nano and macro porosity. *In vitro* tests with cells and *in vivo* tests in animal models have established their biocompatibility and bioactivity for tissue regeneration. A clinical trial is underway to assess their potential for use in dental tissue regeneration.

In regard to the outlook for the fabrication of bioscaffolds by the modified SG route, more reproducible preparation conditions should be progressively implemented in terms of temperature and humidity control. One may also expect increasing applications involving 3D additive manufacturing and SG printing of different objects, including bioactive glass scaffolds with tailored porosity for tissue engineering and regenerative medicine.

Investigating bioactive glass–cell responses *in vitro* has revealed a large body of detailed information. Specifically, preparing TAMP bioactive glass scaffolds with small amounts of boron shed new light on the potential role of this ultra-trace element on bone cell differentiation. Future experiments should include a careful analysis of the dissolution of boron from borosilicate and boron-containing bioactive glasses, and of mechanistic aspects that may allow cells to sense boron in solution, as well as in growth-supportive substrates. It also would seem important to test other elements such as heavy metals that may have antimicrobial or other beneficial effects.

Clearly, as our initial results have demonstrated, the nanoscale features of bioactive glass substrates, which are one thousand-fold smaller than cells themselves, have a significant impact on cellular response. The cells are able to recognize such structures of a few nanometers, which is also the size range of single integrin receptors.[105] In future many more samples with a range of nanopore size (but with the same surface area) will need to be examined to establish the optimum pore size or nanostructure for the adhesion of a specific cell type. As protein adsorption to bioactive glass scaffolds is also influenced by nanostructure, coating/impregnating scaffolds with growth factors and adhesive peptides[106] should be explored for designing 'smart' bioactive glass scaffolds.

Finally, we note that the present TAMP materials can be useful also for other biomedical applications. For instance, there is considerable current interest in biocompatible, mesoporous drug delivery systems.[107] Due to their interconnected, multi-porous structure TAMP silicates will be particularly useful for multiple drug delivery implants, which will biodegrade in a prescribed time and therefore not need removal surgery.

Acknowledgements

The multinational collaboration that made this work possible was initiated with support from the US National Science Foundation's Materials World Network (DMR-0602975) and the International Materials Institute for New Functionality in Glass (IMI-NFG, DMR-0844014) programs. This overview provides only a sampling of the literature that is available on this topic and we apologize to authors whose original works have not been cited. We thank specifically Raina H. Jain, Jutta Y. Marzillier, Shaojie Wang, and Jui Chakraborty for performing parts of the work described in Section 5.4. Work in the laboratory of MMF is supported by funds from the National Institutes of Health (NIGMS, grant GM55725).

References

1. R. Langer and J. P. Vacanti, *Science*, 1993, **260**, 920.
2. L. G. Griffith and G. Naughton, *Science*, 2002, **295**, 1009.
3. C. Constance, *Science*, 2008, **320**, 437.
4. R. F. Service, *Science*, 2008, **322**, 1460.
5. G. Naik, *The Wall Street Journal*, Front page, April 10, 2014.
6. A. Khademhosseini, R. Langer, J. Borenstein and J. P. Vacanti, *Proc. Natl. Acad. Sci.*, 2006, **103**, 2480.

7. S. J. Hollister, *Nat. Mater.*, 2005, **4**, 518.
8. G. Orlando, K. J. Wood, R. J. Stratta, J. J. Yoo, A. Atala and S. Soker, *Transplantation*, 2011, **91**, 1310.
9. V. P. Shastri and A. Lendlein, *MRS Bull.*, 2010, **35**(8), 571.
10. A. Atala, *J. Endourol.*, 2000, **14**, 49.
11. T. H. Petersen, E. A. Calle, L. Zhao, E. J. Lee, L. Gui, M. B. Raredon, K. Gavrilov, T. Yi, Z. W. Zhuang, C. Breuer and E. Herzog, *Science*, 2010, **329**, 538.
12. Q. Fu, E. Saiz, M. N. Rahaman and A. P. Tomsia, *Mater. Sci. Eng.*, 2011, **7**, 1245.
13. E. A. Kaufman, P. Ducheyne and I. M. Shapiro, *Tissue Eng.*, 2000, **6**, 19.
14. L. L. Hench, *Biomaterials*, 1998, **19**, 419.
15. J. R. Jones, L. M. Ehrenfried and L. L. Hench, *Biomaterials*, 2006, **27**, 964.
16. A. Martinez, I. Izquierdo-Barba and M. Vallet-Regi, *Chem. Mater.*, 2000, **12**, 3080.
17. P. Saravanapavan, J. R. Jones, R. S. Pryce and L. L. Hench, *J. Biomed. Mater. Res., Part A*, 2003, **66**, 110.
18. L. L. Hench, *J. Mater. Sci.: Mater. Med.*, 2006, **17**, 967.
19. J. R. Jones, P. D. Lee and L. L. Hench, *Philos. Trans. R. Soc., A*, 2006, **364**, 263.
20. L. L. Hench and J. M. Polak, *Science*, 2002, **295**, 1014.
21. L. D. K. Buttery, S. Bourne, J. D. Xynos, H. Wood, F. J. Hughes, S. P. F. Hughes, V. Episkopou and J. M. Polak, *Tissue Eng.*, 2001, **7**, 89.
22. V. Karageorgiou and D. Kaplan, *Biomaterials*, 2005, **26**, 5474.
23. Q. Z. Z. Chen, I. D. Thompson and A. R. Boccaccini, *Biomaterials*, 2006, **27**, 414.
24. S. Wang, M. M. Falk, A. Rashad, M. M. Saad, A. C. Marques, R. M. Almeida, M. K. Marei and H. Jain, *J. Mater. Sci.: Mater. Med.*, 2011, **22**, 1195.
25. H. R. Ramay and M. Q. Zhang, *Biomaterials*, 2003, **24**, 3293.
26. Q. Fu, M. N. Rahaman, B. S. Bal, R. F. Brown and D. E. Day, *Acta Biomater.*, 2008, **4**, 1854.
27. C. Vitale-Brovarone, F. Baino and E. Verne, *J. Mater. Sci.: Mater. Med.*, 2009, **20**, 643.
28. S. Michna, W. Wu and J. A. Lewis, *Biomaterials*, 2005, **26**, 5632.
29. T. Livingston, P. Ducheyne and J. Garino, *J. Biomed. Mater. Res.*, 2002, **62**, 1.
30. W. Liang, M. N. Rahaman, D. E. Day, N. W. Marion, G. C. Riley and J. J. Mao, *J. Non-Cryst. Solids*, 2008, **354**, 1690.
31. Q. Fu, M. N. Rahaman, B. S. Bal and R. F. Brown, *J. Biomed. Mater. Res., Part A*, 2010, **93**, 1380.
32. R. F. Brown, D. E. Day, T. E. Day, S. Jung, M. N. Rahaman and Q. Fu, *Acta Biomater.*, 2008, **4**, 387.
33. A. K. Varshneya, *Fundamentals of Inorganic Glasses*, Academic Press, New York, 1994, p. 66.

34. S. Wang and H. Jain, *J. Am. Ceram. Soc.*, 2010, **93**, 3002.
35. H. S. Yun, S. E. Kim, Y. T. Hyun, S. J. Heo and J. W. Shin, *Chem. Mater.*, 2007, **19**, 6363.
36. J. R. Jones and L. L. Hench, *J. Mater. Sci.*, 2003, **38**, 3783.
37. A. C. Marques, H. Jain and R. M. Almeida, *Eur. J. Glass Sci. Technol.*, 2007, **48**, 65.
38. A. C. Marques, R. M. Almeida, A. Thiema, S. J. Wang, M. M. Falk and H. Jain, *J. Mater. Res.*, 2009, **24**, 3495.
39. Y. Vueva, A. Gama, A. V. Teixeira, R. M. Almeida, S. J. Wang, M. M. Falk and H. Jain, *J. Am. Ceram. Soc.*, 2010, **93**, 1945.
40. C. J. Brinker and G. W. Scherer, *Sol-Gel Science: The Physics and Chemistry of Sol-Gel Processing*, Academic Press, NY, 1990, ch. 3, vol. 5, p. 11.
41. H. Kozuka, *Sol-gel Processing, Handbook of Sol-Gel Science and Technology*, Kluwer Academic Publishers, London, 2005, vol. 1.
42. Y. Vueva, A. Teixeira and R. M. Almeida, (unpublished results).
43. X. M. Du and R. M. Almeida, *J. Mater. Res.*, 1996, **11**, 353.
44. Y. Li and R. M. Almeida, *J. Sol-Gel Sci. Technol.*, 2012, **61**, 332.
45. K. Nakanishi, *J. Porous Mater.*, 1997, **4**, 67.
46. J. R. Jones, O. Tsigkou, E. E. Coates, M. M. Stevens, J. M. Polak and L. L. Hench, *Biomaterials*, 2007, **28**, 1653.
47. A. C. Marques, H. Jain, C. Kiely and R. M. Almeida, *J. Sol-Gel Sci. Technol.*, 2009, **51**, 42.
48. R. M. Almeida, A. Gama and Y. Vueva, *J. Sol-Gel Sci. Technol.*, 2011, **57**, 336.
49. H. Jain, A. C. Marques and R. M. Almeida, *US Pat.* 8,277,829 B2, Oct. 2, 2012.
50. H. Maekawa, J. Esquena, S. Bishop, C. Solans and B. F. Chmelka, *Adv. Mater.*, 2003, **15**, 591.
51. A. K. Varshneya, *Fundamentals of Inorganic Glasses*, The Society of Glass Technology, Sheffield, U.K., 2006, ch. 4.
52. L. L. Hench, *J. Am. Ceram. Soc.*, 1998, **81**, 1705.
53. T. Kokubo and H. Takadama, *Biomaterials*, 2006, **27**, 2907.
54. L. L. Hench, J. R. Jones and P. Sepulveda, in *Future Strategies for Tissue and Organ Replacement*, ed. J. M. Polak, L. L. Hench and P. Kemp, Imperial College Press, London, 2002, ch. 1.
55. D. Zhang, H. Jain, M. Hupa and L. Hupa, *J. Am. Ceram. Soc.*, 2012, **95**, 2687.
56. L. Wu and J. Ding, *Biomaterials*, 2004, **25**, 5821.
57. I. D. Xynos, A. J. Edgar, L. D. Butter, L. L. Hench and J. M. Polak, *Biochem. Biophys. Res. Commun.*, 2000, **276**, 461.
58. C. Kim, F. Ye and M. H. Ginsberg, *Ann. Rev. Cell Dev. Biol.*, 2011, **27**, 321.
59. N. J. Anthis and I. D. Campbell, *Trends Biochem. Sci.*, 2011, **36**, 191.
60. P. Hu and B. H. Luo, *J. Cell. Physiol.*, 2013, **228**, 306.

61. B. Geiger, J. P. Spatz and A. D. Bershadsky, *Nat. Rev. Mol. Cell Biol.*, 2009, **10**, 21.
62. E. Dejana, S. Colella, G. Conforti, M. Abbadini, M. Gaboli and P. C. Marchisio, *J. Cell Biol.*, 1988, **107**, 1215.
63. T. Yeung, P. C. Georges, L. A. Flanagan, B. Marg, M. Ortiz, M. Funaki, N. Zahir, W. Ming, V. Weaver and P. A. Janmey, *Cell Motil. Cytoskeleton*, 2005, **60**, 24.
64. A. J. Engler, S. Sen, H. L. Sweeney and D. E. Discher, *Cell*, 2006, **126**, 677.
65. D. T. Butcher, T. Alliston and V. M. Weaver, *Nat. Rev. Cancer*, 2009, **9**, 108.
66. K. Wang, C. Zhou, Y. Hong and X. Zhang, *Interface Focus*, 2012, **2**, 259.
67. K. Magyari, L. Baia, O. Popescu, S. Simon and V. Simon, *Vib. Spectrosc.*, 2012, **62**, 172.
68. H. M. M. Moawad and H. Jain, *J. Mater. Sci.: Mater. Med.*, 2009, **20**, 1409.
69. R. H. Jain, S. Wang, H. M. M. Moawad, M. M. Falk and H. Jain, in *Engineering Biomaterials for Regenerative Medicine*, ed. S. Bhatia, S. Bryant, J. A. Burdick, J. M. Karp and K. Walline, Materials Research Society, Warrendale, PA, 2010, 1235-RR03-47.
70. S. Wang, T. J. Kowal, M. K. Marei, M. M. Falk and H. Jain, *Tissue Eng., Part A*, 2013, **19**, 1632.
71. A. Hoppe, V. Mouriño and A. R. Boccaccini, *Biomater. Sci.*, 2013, **1.3**, 254.
72. D. G. Blevins and K. M. Lukaszewski, *Annu. Rev. Plant Physiol. Plant Mol. Biol.*, 1998, **49**, 481.
73. M. T. Gallardo-Williams, R. R. Maronpot, C. H. Turner, C. S. Johnson, M. W. Harris, M. J. Jayo *et al.*, *Biol. Trace Elem. Res.*, 2003, **93**, 155.
74. M. R. Naghii, G. Torkaman and M. Mofid, *Biofactors*, 2006, **28**, 195.
75. A. A. Gorustovich, T. Steimetz, F. H. Nielsen and M. B. Guglielmotti, *Anat. Rec.*, 2008, **291**, 441.
76. F. H. Nielsen, *Nutr. Rev.*, 2008, **66**, 183.
77. F. H. Nielsen, C. D. Hunt, L. M. Mullen and J. R. Hunt, *FASEB J.*, 1987, **1**, 394.
78. W. C. A. Vrouwenvelder, C. G. Groot and K. de Groot, *Biomaterials*, 1994, **15**, 97.
79. L. L. Hench, *J. Am. Ceram. Soc.*, 1991, **74**, 1487.
80. S. S. Hakki, B. S. Bozkurt and E. E. Hakki, *J. Trace Elem. Med. Biol.*, 2010, **24**, 243.
81. N. W. Marion, W. Liang, G. C. Reilly, D. E. Day, M. N. Rahaman and J. J. Mao, *Mech. Adv. Mater. Struct.*, 2005, **12**, 239.
82. A. A. Gorustovich, J. M. Porto-Lopez, M. B. Guglielmotti and R. L. Cabrini, *Biomed. Mater.*, 2006, **1**, 100.
83. R. F. Brown, M. N. Rahaman, A. B. Dwilewicz, W. Huang, D. E. Day, Y. Li *et al.*, *J. Biomed. Mater. Res., Part A*, 2009, **88**, 392.

84. J. Chakraborty, J. Y. Marzillier, T. J. Kowal, D. Zhang, H. Jain and M. M. Falk, unpublished work.
85. R. H. Jain, J. Y. Marzillier, T. J. Kowal, S. Wang, H. Jain and M. M. Falk, in *Advances in Bioceramics and Porous Ceramics IV: Ceramic Engineering & Science Proceedings*, ed. R. Narayan and P. Colombo, 2011, vol. 32, p. 111.
86. B. Boyan and Z. Schwartz, *J. Am. Acad. Orthop. Surg.*, 2006, **6**, 157.
87. S. Levy, M. V. Dalen, S. Agonafer and W. Soboyejo, *J. Mater. Sci.: Mater. Med.*, 2007, **18**, 1573.
88. D. W. Hutmacher, *Biomaterials*, 2000, **21**, 2529.
89. K. Rezwan, Q. Z. Chen, J. J. Blaker and A. R. Boccaccini, *Biomaterials*, 2006, **27**, 3413.
90. T. M. Freyman, I. V. Yannas and L. J. Gibson, *Prog. Mater. Sci.*, 2001, **46**, 273.
91. L. G. Griffith, in *Reparative Medicine: Growing Tissues and Organs*, ed. J. D. Sipe, C. A. Kelley and L. A. McNicol, New York Academy of Sciences, New York, 2002, p. 83.
92. J. R. Jones, P. D. Lee and L. L. Hench, *Philos. Trans. R. Soc., A*, 2006, **364**, 263.
93. R. Golovchak, P. Thapar, A. Ingram, D. Savytskii and H. Jain, *Acta Biomater.*, 2014, **10**, 4878.
94. R. Golovchak, T. Kowal, T. Chokshi, J. Harmes, H. Jain and M. Falk, Role of Bioglass Phase Separation in Protein and Cell Attachment, to be published.
95. K. L. Elias, R. L. Price and T. J. Webster, *Biomaterials*, 2002, **23**, 3279.
96. K. M. Woo, V. J. Chen and P. X. Ma, *J. Biomed. Mater. Res., Part A*, 2003, **67A**, 531.
97. J. Huang, S. M. Best, W. Bonfield, R. A. Brooks, N. Rushton, S. N. Jayasinghe and M. J. Edirisinghe, *J. Mater. Sci.: Mater. Med.*, 2004, **15**, 441.
98. A. L. Teixeira, P. F. Nealey and C. J. Murphy, *J. Biomed. Mater. Res., Part A*, 2004, **71A**, 369.
99. M. J. Dalby, M. J. P. Biggs, N. Gadegaard, G. Kalna, C. D. W. Wilkinson and A. S. G. Curtis, *J. Cell. Biochem.*, 2007, **100**, 326.
100. S. K. Misra, D. Mohn, T. J. Brunner, W. J. Stark, S. E. Philip, I. Roy, V. Salih, J. C. Knowles and A. R. Boccaccini, *Biomaterials*, 2008, **29**, 1750.
101. T. Raimondo, S. Puckett and T. J. Webster, *Int. J. Nanomed.*, 2010, **5**, 647.
102. T. J. Webster, L. S. Schadler, R. W. Siegel and R. Bizios, *Tissue Eng.*, 2001, **7**, 291.
103. M. J. Dalby, N. Gadegaard, R. Tare, A. Andar, M. O. Riehle, P. Herzyk, C. D. W. Wilkinson and R. O. C. Oreffo, *Nat. Mater.*, 2007, **6**, 997.

104. N. El Shazley, A. Hamdy, H. A. El-Eneen, R. M. El Backly, M. M. Saad, W. Essam, H. Moussa, M. El Tantawi, H. Jain and M. K. Marei, *JDR Clin. Trans. Res.*, 2016, DOI: 10.1177/2380084416660672.
105. M. V. Nermut, N. M. Green, P. Eason, S. S. Yamada and K. M. Yamada, *EMBO J.*, 1988, 7, 4093.
106. C. Przybylowski, M. Ammar, C. LeBlon and S. S. Jedlicka, *J. Biomater. Nanobiotechnol.*, 2015, **6**, 146.
107. M. Vallet-Regí, F. Balas and D. Arcos, *Angew. Chem., Int. Ed*, 2007, **46**, 7548.

CHAPTER 6

Tailoring of Bioactive Glasses

LEENA HUPA* AND KAJ H. KARLSSON

Johan Gadolin Process Chemistry Centre, Åbo Akademi University, Turku, Finland
*Email: leena.hupa@abo.fi

6.1 Introduction

The discovery of bioactive glasses by Professor Hench and co-workers in the late 1960's is the outcome of a fascinating exploration of developing glasses as materials to be implanted inside the human body.[1–5] It was hypothesized that if the prosthetic material is compatible with the inorganic constituent of bone, hydroxyapatite (HA), it will not be surrounded by a fibrous capsule. This demand was expressed as the ability of the material to form an HA-like surface layer *in vivo*. After defining the challenge and the task, certain low-silicate glasses, so-called invert glasses, were identified to fulfil the criterion. Why silicate glasses? Interestingly, typical silicate glasses consist of elements abundant in the human body, thus minimizing any adverse effects on tissue. Bioactive glasses were developed *per se* to slowly be degraded by the body.[1–5] The gradual dissolution releases ions into the surrounding solution while favouring the formation of the desired HA layer. The first bioactive glasses tested consisted of the oxides Na_2O, CaO, P_2O_5 and SiO_2 in such ratios, which were known to support the formation of a $CaO–P_2O_5$ surface layer.[6] After testing several compositions, the glasses within certain oxide ranges were identified to develop the desired HA surface layer which gave a firm bond with bone and also with soft tissue.

RSC Smart Materials No. 23
Bioactive Glasses: Fundamentals, Technology and Applications
Edited by Aldo R. Boccaccini, Delia S. Brauer and Leena Hupa

Published by the Royal Society of Chemistry, www.rsc.org

Accordingly, the first bioactive glasses were tailored to interact with the tissue in a controlled, desired manner.

Commercial uses of bioactive glasses are mainly found in bone repair applications while recent research deals with utilization of the glasses in scaffolds for hard and soft tissue regeneration.[7–10] Since the discovery of the first bioactive glasses, the biological performance of the glasses has been characterized in detail. After implantation, a bioactive glass bonds to the tissue but the released ions from the dissolving glass may also promote the growth of new bone. Bioactive glasses are often divided into two classes according to their capability to affect bone growth and regeneration: the less bioactive glasses are osteoconductive if bone migrates along interfaces, while the highly bioactive glasses are both osteoconductive and osteoinductive since they release a critical concentration of ions which stimulate the cellular processes to form bone.[4,11–13] Recent development has demonstrated that the glass dissolution has several effects on biological processes. When released in suitable concentration the ions stimulate bone regeneration, enhance angiogenesis, prevent bacterial growth, aid in the treatment of diseases, and have potential also in soft tissue regeneration.[8,9,13–17] Glasses releasing inorganic ions in critical concentrations needed to activate the genes, are investigated as constituents of tissue engineering scaffolds which help the body's own regenerative mechanisms to restore a damaged tissue, also without the aid of integrated growth factors. Accordingly, controlled, desired dissolution of inorganic ions from the bioactive glasses during the degradation process is paramount for developing scaffolds.

This chapter discusses tailoring the bioactive glass composition to match the requirements of controlled dissolution kinetics and release rate of ions. When used as a structure-forming component in bioinspired implants and scaffolds, the glass composition must also meet several criteria essential for the structure fabrication. The amorphous structure of glass enables incorporation of almost any element in the composition, thus enabling optimization of the composition to induce specific effects on a particular property. However, since every element affects any property in a manner typical for the element, the glass composition must satisfy a set of properties of importance for the final use and the fabrication of the product. Finding a glass to suit a targeted site of implantation relies on understanding and interpreting how the requirements and constraints can be realized by adjusting the glass composition. Ideally, composition tailoring minimises the number of experiments needed to assess the effectiveness and safety of the bioactive glass device in the intended application. During the past years increasing interest has been paid to the phosphate and borate systems, but so far the information of the influence of composition on the properties of these glass families allows optimization only within rather narrow composition ranges. These two systems are therefore excluded from this chapter. Although the choice of composition is discussed below for melt-derived silicate glasses, the same tailoring principles are valid also for other bioactive glass types.

6.2 Development of the Bioactive Glasses 45S5 and S53P4

Most commercial clinical products of melt-derived bioactive glasses are based on the melt-derived silicate-based compositions 45S5 and S53P4. Monolithic products of 45S5 were commercialized in 1985 while the first products based on particles of the same composition are from 1993.[1,7] Products of S53P4 have been commercially available since 2006.[7]

The discovery of the first bioactive glass, Bioglass® 45S5 has resulted in a totally new branch of science dealing with development and utilisation of bioactive glasses and glass-ceramics as implants and components of medical devices. Professor Hench has documented the background and history of the bioactive glass 45S5 in several papers.[1,3,5] Today, the bioactive glass 45S5 (in wt%: 45 SiO_2, 24.5 Na_2O, 24.5 CaO and 6 P_2O_5) is the gold standard of all bioactive glass research and applications. Numerous research papers have been published based on glass 45S5 and it is still used as a basis when tailoring glasses to new potential applications. The recent research trend has been to dope the glass with certain therapeutic elements or to functionalize the surface with biomolecules for effects additional to the total tissue healing and regeneration process (see Chapters 16 and 17).

Glass S53P4 (in wt%: 53 SiO_2, 23 Na_2O, 20 CaO, 4 P_2O_5) has its birth in a series of 16 glass compositions tested in rat tibia.[18] The experimental compositions were selected to provide a basis for statistical evaluation of the influence of the oxide composition on the *in vivo* behaviour. Based on the good bone bonding properties, glass S53P4 was selected for further *in vivo* and clinical studies.[19–23] Also, the effects of glass S53P4 on growth and colonization of several microorganisms were studied in detail.[14–17,24,25] The clinical applications of the bioactive glasses 45S5 and S53P4 are discussed in Chapters 14, 15 and 19. While clinical utilization of commercial products is restricted to their approved and registered indications, experimental products based on 45S5 and S53P4 have been investigated frequently in several other applications. Since these compositions crystallize easily in hot-working, they are not suited for fabrication of amorphous products requiring traditional thermal treatments.[26] S53P4 can, to a limited degree, be sintered above the glass transition temperature,[27] while 45S5 crystallizes easily during thermal treatment. In contrast, glass 45S5 has frequently been utilized to fabricate porous glass-ceramic scaffolds.[28,29] However, the properties of glass-ceramics differ from parent glass, and tailoring of the properties of glass-ceramics cannot be based on the same principles as those for amorphous products. The easy crystallization of 45S5 and S53P4 also challenges the drawing of continuous fibres when using conventional fibre-drawing methods. Similarly, the utilization of these bioactive glasses as coatings enhancing the tissue attachment of a bioinert implant is limited by their unsuitable thermal properties.

Particles of bioactive glass 45S5 and S53P4 have been studied as constituents in composite scaffolds based on bioinert or biodegradable polymers and glasses. In these applications, the polymer enabled fabrication of a desired geometry and porosity while the bioactive glass induced bioactivity in the scaffold.[30,31] Bioactive glass particles have also been used to increase the mechanical strength of the biopolymer composite.[30] One composite approach has been to coat the polymer scaffold with glass particles for enhanced osteointegration.[32–34] Easy injection and filling of bone defects have been accomplished by thermoplastic biodegradable polymers mixed with bioactive glass particles.[35,36] In addition, bioactive glass particles have been added to cranial composite implants to enhance osteointegration and induce antibacterial effects.[37,38] When used in a composite device, the morphology and amount of a certain bioactive glass constituent must be tailored for controlled and desired performance in the implantation site. In dense composites, cracks due to poor adherence between the glass and the polymer have been found to accelerate degradation of the composite.[35,39]

6.3 Requirements for a Bioactive Glass

In general, the properties of bioactive glasses can be divided into two categories: the properties important to the final application, and the properties crucial for the fabrication of the desired bioactive glass device. Although composition tailoring balances between these property categories, the final choice is always based on the requirements the device must meet in the final use. The specific clinical requirements in bone tissue applications may be expressed as osteoconduction, osteostimulation, an antibacterial property, non-toxicity, safety, effectiveness, *etc.* Other specific measurements for the device include shape, surface condition, mechanical properties, handling properties during the surgery, *etc.* For devices based on amorphous bioactive glasses, several of the properties can at least to some extent be adjusted and controlled by the glass composition.

In most applications, controlled dissolution is the most important requirement for bioactive glasses. Ideally, the glass dissolves at a rate adapted to the tissue growth rate while the ions released from the glass activate the genes and thus enhance new tissue growth.[4,40–43] The mechanical performance of the device is a balance between the growth rate of new tissue and the dissolution rate of the glass. Thus, detailed knowledge of the reaction mechanisms and ion release kinetics of the glass in contact with aqueous solutions *in vitro* and *in vivo* provides the basis for tailoring the compositions for any particular clinical application. Fortunately, the dissolution reactions of glasses in aqueous solutions are highly composition dependent.

The *in vivo* interactions leading to bonding of the bioactive glass with bone tissue have been described as a sequence of eleven successive, but partly overlapping, steps.[3] These reactions are described in detail in Chapters 1 and 15 of this book. Briefly, the five first reactions taking place at glass surface are usually described as (1) exchange of alkalis from the surface with H^+ or H_3O^+ in the solution, (2) breaking of siloxane bonds in the glass surface leading to loss of soluble silica, (3) condensation and repolymerisation of an SiO_2 surface layer, (4) formation of an amorphous calcium phosphate-rich layer on top of the SiO_2 layer, and (5) crystallization of the amorphous calcium phosphate layer by incorporation of OH^-, CO_3^{2-}, or F^- from solution to form a mixed hydroxyl, carbonate, fluorapatite layer.[3] Usually, the interaction of the surface with proteins and cells leading to tissue bonding is assumed to start after these first reactions steps. The first five reactions described above take place also under *in vitro* conditions. Thus, these reactions are often used to study and compare the bioactivity of different glass compositions in buffered aqueous solutions, such as the Tris buffer or the so-called simulated body fluid. Today, the bioactive glasses are thought not only to bond to bone but also stimulate tissue regeneration. For this, the release of certain ions from the glasses must take place over critical time periods and in critical concentrations needed for activation and support of cellular processes.[4,13,40–43] Thus, enhanced understanding of the dissolution kinetics is one major challenge for tailoring novel compositions for future medical applications.

The properties describing the behaviour of the glass melt during processing into various product forms also depend on glass composition. The most important properties commonly used to control the forming processes are the viscosity of the melt and its change with temperature, and the crystallization characteristics of the glass in various thermal treatments (*cf.* Chapter 1 Melt-derived bioactive glasses). All glass melts are prone to crystallize in the temperature window between glass transformation and liquidus temperatures. Typically, the temperature range for crystal nucleation is lower than for crystal growth, but the overlapping of these temperatures depends on the glass composition. Liquidus temperature again gives the temperature at which the glass melt starts to crystallize upon cooling. Table 6.1 summarises the properties to be controlled and their mutual importance in forming melt-derived glasses into various products.

Particles and powdered glass only require that melt has been homogenous before the casting or quenching. The rapid cooling of the melt, *e.g.* into water, does not favour nucleation and also crystal growth is prevented. Manufacturing of small monoliths may interfere with crystallization. The forming must start and preferably be finished above the liquidus temperature to avoid crystals in the final product. Effective cooling and good heat transfer after forming prevent crystallization. Fibre drawing from the melt is usually performed in the narrow viscosity range of $10^{3.5}$–$10^{4.0}$ dPa s.[44] The temperature at the fibre drawing must be above the liquidus temperature of the melt. Sintering of porous implants and 3D tissue engineering scaffolds with interconnected porosity is controlled by the viscosity, around 10^{8}–$10^{8.8}$ dPa s, and the crystallization tendency of the glass.[44] The sintering must be finalized before nucleation and crystal growth commence. If fibres are drawn from marbles or preforms, crystal nuclei may form and grow during the heating to optimal fibre drawing viscosities. Finally, manufacture of glass-ceramics is based on controlled nucleation and growth at appropriate temperatures.

Table 6.1 High temperature properties to be controlled in forming operations of products based on melt-derived bioactive glasses.[a]

Product	Nucleation of crystals	Crystal growth	Liquidus temperature	Viscosity–temperature relation
Particles, powder	±	−	±	±
Small monoliths	+	+	+	+
Porous structures	+++	+++	−	+++
Fibres from melt	+	+	+++	+++
Fibres from preform	+++	+++	+++	+++
Glass-ceramics	+++	+++	−	+

[a] − Less important, ± must be taken into account, + important, +++ crucially important.

Apart from these criteria, the manufacturer must also control the melting process from initial melting reactions to fining and homogenizing of the glass melt. Volatilization losses may affect the final glass composition and thus the properties. All these properties depend on the glass batch and oxide compositions but for some properties temperature and furnace atmosphere also become important.

Composition–property relations can be used to estimate the behaviour of the glass in the target application in different product forms. One major challenge in tailoring is to have reliable predictions of the influence of composition on glass properties. In principle, the glass structure determines the properties but experimental data have been frequently used to give properties as functions of composition. The experimental models have been based on studies exploring the influence of systematic changes or statistical changes in the glass composition over the range of interest.

6.4 Properties from Glass Structure

The range of functional properties exhibited by glasses varies from optical and superconducting to chemical properties, *etc.* Optimal properties for a particular application are achieved by using various elements, or usually oxides, in carefully controlled ratios in the composition. A glass may consist of only one single oxide, *e.g.* SiO_2, but all commercial soda lime-silicate glasses are multi-oxide systems. Specific for glasses, is that each oxide or element has a certain effect on each property, albeit its concentration in the formula may be small. This specific character of composition dependence of properties allows tailoring of the best glass to meet the property requirements, both in the product fabrication but more importantly in the final use.

6.4.1 Glass Network and Bioactivity

Several properties depend on the structure of glasses. The classical glass structure theories, of the network-forming and modifying oxides according to Zachariasen and Warren, are used as the basis for theoretical estimation of properties.[45] These structure concepts allow prediction of most glass properties from the amount of the network formers (Si, B, P, *etc.*) in relation to the network modifiers (Na, K, Ca, *etc.*) and intermediates (Al, Mg, Zn, *etc.*) in the composition. Some sources for property predictions can be found in the literature, *e.g.* Volf.[46]

The basic building unit of a silicate glass network is the silicate ion $[SiO_4]^{4-}$, which is bonded in the shape of a tetrahedron sharing the O^{2-} corners with each other. The basic building unit can be illustrated by the symbol Q^n where n represents the number of shared corners. Adding modifying oxides, *e.g.* Na_2O and CaO, breaks some of the oxygen bonds and leads to oxygen ions, which are not shared by two tetrahedra. These oxygen ions are called non-bridging oxygens (NBO) while the oxygens shared by tetrahedra are bridging oxygens (BO). The network connectivity (NC) of a glass is defined as the average number of BO bound to a network former, *i.e.* Si in silicate glasses. The higher the NC, the more connected the network. Adding modifiers breaks Si–O bonds and leads to the formation of Q^3, Q^2, and Q^1 units, which share 3, 2 or 1 of their oxygen ions, respectively, with neighbouring units. Q^0 denotes an isolated silicate ion. Since bioactive glasses have low silica content, around 45 to 55 wt%, they mainly consist of Q^2 and Q^3 units.[47,48] Network connectivity is easily calculated from the glass composition by taking into account the molar amounts and nature of each oxide in the structure. The network connectivity in silicate glasses varies from $NC = 4$ in pure silica glass to $NC = 2$ in chain-like structures. Simplified, the higher the content of network modifier, the less cross-linked the structure. With decreasing network connectivity, the dissolution of ions in aqueous solutions increases, which is rarely a desired property for a glass but essential for bioactive glasses. $NC > 3$ is typical for bio-inactive compositions while $NC = 2–3$ suggests bioactivity.[49,50] In contrast, the low silica content of bioactive glasses leads to extensive depolymerisation of the network structure and thus makes them prone to crystallization during the manufacture. However, since the size and charge of network modifier affects the bonding strength to the basic network, the calculated NC does not give an accurate estimation of the properties and gives only an average value. The NC value can be used for overall estimation of the properties of a particular composition. More accurate data on the share of the different Q^n units in a glass have been obtained using solid-state NMR or Raman spectroscopies.[51–56]

6.4.2 Molecular Dynamics Simulations and Bioactivity

The various components in a homogeneous glass are assumed statistically distributed throughout the structure. Thus, the structure of the amorphous glass resembles the structure of a liquid. Atomistic computer simulations such as classical molecular dynamics (MD) simulations have been utilized to give fundamental information on the structure of binary and ternary glasses.[57]

The short-range and medium-range order structures giving the local environments in bioactive glasses containing four or more oxides have been studied with classical and *ab initio* MD simulations.[58,59] The influence of changes in the network modifiers in the structure of bioactive glass, usually 45S5, has also been studied using MD simulations.[60–66] These studies provide information on the changes in the network connectivity induced by composition changes, information on the environment and the distribution of the Q^n species in the glass. In general, the structural features from MD are used to explain some measured property, often strength or ion migration dynamics. Recent developments in MD simulations of the atomistic structures show that the method also has potential in predicting the properties for multicomponent systems and is thus an interesting tool for studying the dynamic behaviour also, *e.g.* network modifier ion migration to the surface of the bioactive glass. This information can then be used to predict dissolution of the ions from the glass surface into the interfacial solutions. Although providing an enhanced understanding of the structures of the glasses, the atomistic computer simulations require large computer capacity and as theoretical tools are not easily accessible to larger research societies. So far, predictions of the properties of the glasses from the structure are still not satisfying.

6.5 Property–Composition Models

6.5.1 *In vivo* Models

The properties of the homogeneous mixture depend on the amount of its components, *i.e.* the oxides in a silicate glass. Several properties are additive functions of the composition. The properties are often assumed using simple linear functions of composition using an equation of the type given in eqn (6.1)

$$x = a_0 + a_1 \cdot c_{Na_2O} + a_2 \cdot c_{CaO} + \cdots\ a_n \cdot c_{SiO_2} \tag{6.1}$$

where x is the property value, a is a coefficient for each oxide present, and c the concentration of each oxide present. The concentrations are expressed as molar or weight percentages. The models can also include non-linear terms; in such cases the property value markedly changes with composition. Several models express the properties of multi-oxide compositions as weight parts of oxides normalized to silica content.[67] Basically, describing the properties from the molar composition of the glasses is feasible since it would enable a direct comparison of composition changes with the glass structure. The choice of the glass compositions used to develop the models is vital for their usability. In contrast, using molar or weight percentage does not give any essential differences in the accuracy or applicability models, only in understanding. Composition–property models for several physical properties of silicate glasses can be found in classical literature on glasses.[46,68]

Albeit describing mostly conventional silicate glasses, some models are valid also for the composition ranges of bioactive glasses.

The first models describing bone contact, thermal expansion and glass transition temperature for multi-oxide bioactive glasses were based on regression analyses of experimental data for statistically chosen compositions.[69] The same principle was used to describe the *in vivo* bone bonding of glasses in the system Na_2O–CaO–B_2O_3–Al_2O_3–P_2O_5–SiO_2.[18] The *in vivo* bone bonding model was based on relative values based on SEM–EDX analysis of the composition of the interfacial area between the glass implant and the rabbit tibia after 8 weeks of implantation. Depending on the interactions, the glasses could be divided into five groups ranging from nearly inert compositions showing hardly any bone contact to compositions with thick surface layers of silica and calcium phosphate and strong bone bonding. Formation of the layered structure of silica and calcium phosphate—more exactly carbonated hydroxyapatite—on the glass surface was taken as an indication of the ability of the glass to bond to bone. Between these two extreme behaviour types a varying degree of silica-rich layer formation, with or without limited calcium phosphate formation but no bone bonding, was observed. These interaction types were given relative values from 1 to 6 according to the interactions between the glass and bone: 1 = inert composition leading to gap formation, 2 = poor contact, 3 = bone contact, 4 = good bone contact, and 6 = bone bonding. In the model development, the bone bonding ability was expressed by the relative value 6 instead of 5 to emphasize the important bone-bonding property. The composition dependence of these values, *i.e.* capability of the glass to bond to bone, was expressed as the reaction number, RN. The RN value given in eqn (6.2) enables a first estimation of the bone bonding capability of glasses in the composition range: (in wt%) 15–30 Na_2O, 10–25 CaO, 0–3 B_2O_3, 0–3 Al_2O_3, 0–8 P_2O_5, 38–65.5 SiO_2. The work also showed that $Al_2O_3 > 1.5$ wt% hindered bone bonding. Interestingly, the work also showed that although not accessible by histological analyses, bone bonding can be quantified from compositional analyses of the distribution of the inorganic components in the bone-to-glass interfaces.[18]

$$\begin{aligned} RN = {} & 88.3875 - 0.0116272[SiO_2]^2 + 0.980188[Na_2O] - 1.12306[CaO] \\ & - 1.20556[P_2O_5] - 0.560527[B_2O_3]^2 - 2.08689[Al_2O_3] \end{aligned} \tag{6.2}$$

The oxide compositions in eqn (6.2) are given in weight percentage.

Later, the strong tendency of the bioactive glasses containing low concentrations of SiO_2 and high concentrations of Na_2O and CaO to crystallize was suppressed by introducing K_2O and MgO into the formulations.[70] For establishing property–composition models, 30 statistically chosen compositions were selected in the range (wt%): 3–25 Na_2O, 0–15 K_2O, 0–4 MgO, 8–20 CaO, 0–3 B_2O_3, 0–6 P_2O_5, 39–70 SiO_2. Also, for this composition range, a relative *in vivo* bioactivity, the index of surface activity (ISA), was expressed based on the surface layer structures at the glass implants after 8 weeks in rabbit tibia.[70] The ISA number as suggested by SEM–EDX analyses were as follows: inert glass = 1, only silica-rich layer = 2, layered structure of silica and calcium phosphate = 3, bioactive = 4. Eqn (6.3) predicts the ISA number from the oxide composition (wt%) of the glass.

$$\mathrm{ISA} = -8.4 + 79.0\frac{[\mathrm{Na_2O}]+[\mathrm{K_2O}]}{[\mathrm{SiO_2}]} - 1.8\frac{([\mathrm{Na_2O}]+[\mathrm{K_2O}])^2}{[\mathrm{SiO_2}]} - 11.6\frac{[\mathrm{MgO}]}{[\mathrm{SiO_2}]} - 9.2\frac{[\mathrm{CaO}]}{[\mathrm{SiO_2}]} - 84.8\frac{[\mathrm{B_2O_3}]}{[\mathrm{SiO_2}]} + 24.1\frac{([\mathrm{B_2O_3}])^2}{[\mathrm{SiO_2}]} + 21.9\frac{[\mathrm{P_2O_5}]}{[\mathrm{SiO_2}]} - 4.5\frac{([\mathrm{P_2O_5}])^2}{[\mathrm{SiO_2}]} \quad (6.3)$$

Later *in vivo* and *in vitro* studies have verified that the models given by eqn (6.2) and (6.3) give reasonable estimations of the bioactivity.[71] Interestingly, one composition was selected from both series for further studies based on the positive properties measured in the *in vivo* experiments. Glass S53P4 from the first series (eqn (6.2)) is today commercially available for clinical applications, while the glass noted as 13-93 from the second series (eqn (6.3)) has been used in several studies to develop amorphous tissue engineering scaffolds or to draw continuous fibres. Both compositions were originally selected only to enable statistically valid fitting of properties from composition with regression analysis. Glasses S53P4 and 13-93 succeeded to meet the desired *in vivo* property requirements.

6.5.2 *In vitro* Models

In vivo models of the ability of the glass to bond to bone give a first estimation of the overall reactivity and layer formation in the body environment. As discussed above, the formation of the layered structure of silica and hydroxyapatite on the glass surface takes place also in buffered solutions, *e.g.* SBF. Changes in the pH or ion concentrations in the solution after different immersion times are then used to describe the reactions of the bioactive glasses. Similarly, development of the silica-rich and hydroxyapatite layers on the glass surface at selected time points are correlated with the bioactivity. The extent of the changes depends on the composition of the glass and immersion solution, the ratio of the surface area of the glass to volume of solution (S_a/V), the surface condition of the glass, and the flow of the solution in the experimental setup. These *in vitro* characteristics are then used for a first approximation of the behaviour of the bioactive glasses *in vivo*, especially when in contact with bone tissue.

In general, the *in vitro* behaviour of the glass is relatively easily measured and the observations can be used to compare the reactivity of a glass with the behaviour of well-established bioactive glass compositions. Several models have been developed to describe the interactions of bioactive glasses in SBF as functions of the glass oxide composition. These models describe changes in either glass sample or in the immersion solution after various immersion times: glass mass, silicate-rich layer formation, calcium phosphate layer formation, increase of the pH, *etc.*[72] Although the models do not give any direct measure of the bioactivity, they give indications of the likely reactivity of the glass and can thus be used to tailor glasses for desired *in vivo* behaviour. For example, the higher the pH increase of the immersion solution, the more the glass dissolves and the higher the bioactivity. Eqn (6.4)–(6.7) give models for calculating the pH of SBF from glass composition after 8, 24, 72 and 168 h.[72] The models were based on regression analysis of the measured property values for a series of 30 statistically selected glasses in the composition range (wt%): 5–25 SiO_2, 0–15 K_2O, 0–6 MgO, 15–25 CaO, 0–4 B_2O_3, 50–65 SiO_2. The oxide compositions in the equations are given in wt%.

$$pH_{8h} = 6.85159 + 0.021145([Na_2O] + [K_2O]) + 0.0172862([MgO] + [CaO]) \tag{6.4}$$

$$pH_{24h} = 7.01719 + 0.000646532([Na_2O] + [K_2O])^2 + 0.0195208([MgO] + [CaO]) + 0.00664098[B_2O_3]^2 \tag{6.5}$$

$$pH_{72h} = 4.41582 + 0.00197681([Na_2O] + [K_2O])^2 + 0.0673915([MgO] + [CaO]) + 0.0687312[B_2O_3] + 0.0590847[P_2O_5] + 0.000286703[SiO_2]^2 \tag{6.6}$$

$$pH_{168h} = 6.48241 + 0.00125778([Na_2O] + [K_2O])^2 + 0.0340536([MgO] + [CaO]) + 0.00856182[B_2O_3]^2 + 0.0360865[P_2O_5] \tag{6.7}$$

Figure 6.1 compares the calculated and measured pH of the bioactive glasses 45S5, S53P4 and 13-93 in similar experimental conditions.[72,73] The glass compositions are given in Table 6.2. As seen from Figure 6.1, the calculated pH values are in good agreement for S53P4 and 13-93 while higher deviations are seen for 45S5. Taking into account that the composition of 45S5 is outside the validity range of the models, the extrapolated values nevertheless successfully describe the overall changes in the pH of SBF in the immersion of glass 45S5.

In general, the bioactivity of the glass may be related to the thicknesses of the silica-rich and hydroxyapatite layers forming at the glass surface. However, models for evolution of layer thickness *in vitro* are ambiguous; the layers are partly mixed since the hydroxyapatite may form on top of, or inside, the silica-rich layer.[74] Also, the thickness of the layers may vary locally within wide ranges on the sample surface. In static SBF, the dissolution of the glass and the layer formation depends partly on saturation of the silicon ion species during immersion. Figure 6.2 shows that on glasses, which showed proper hydroxyapatite formation in SBF, the layer thickness was almost similar, independent of the glass composition.[75] The oxide compositions of the glasses are given in Table 6.2.

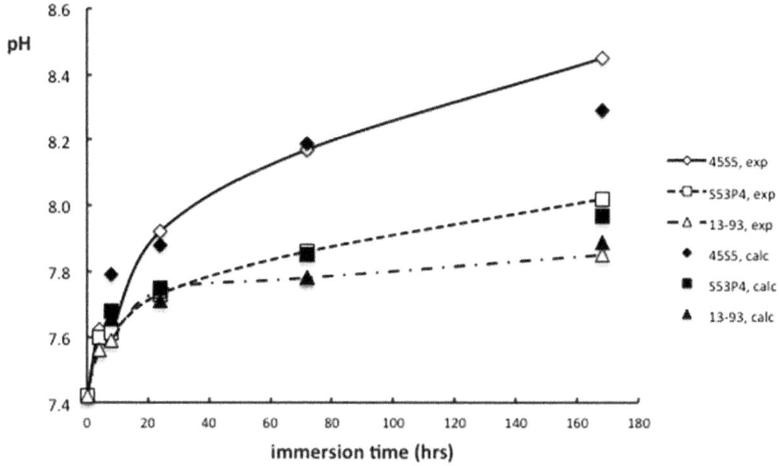

Figure 6.1 Measured (exp) and calculated (calc) pH values of SBF as a function of immersion for the bioactive glasses 45S5, S53P4 and 13-93.
Data from Zhang and Zhang *et al.*[72,73]

Table 6.2 Oxide composition of biocompatible glasses (wt%).

Glass	Na_2O	K_2O	MgO	CaO	B_2O_3	P_2O_5	SiO_2
45S5	24.5			24.5		6	45
S53P4	23			20		4	53
13-93	6	12	5	20		4	53
0106	5.9	12	5.3	22.6	0.2	4	50
0206	12.1	14		19.8	1.6	2.5	50
0306	24.6			21.6	1.3	2.5	50
1806	18.4		0.1	15	1.5		65

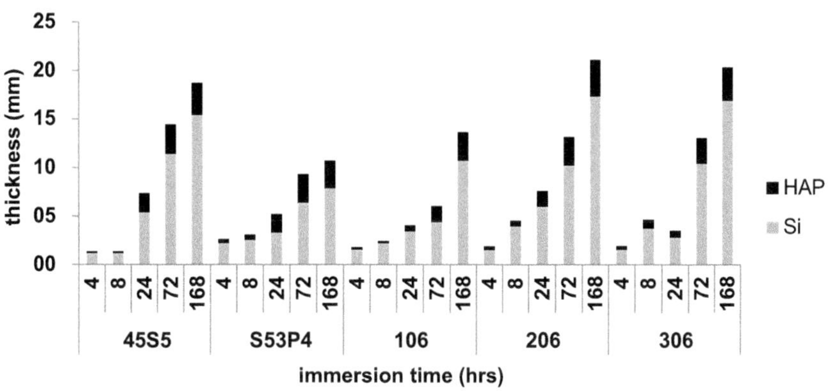

Figure 6.2 The average thickness of silica-rich and hydroxyapatite layers on bioactive glasses after different immersion times in SBF.
Data from Vedel *et al.*[75]

6.5.3 Dissolution in Dynamic Solutions

The utilization of bioactive glasses in bone tissue regeneration and especially in the emerging soft tissue applications are based on controlled release of biologically active inorganic ions from the glasses.[4,8] Thus, detailed understanding of the release rates of the inorganic ions from the glasses in the dynamic body environment provides valuable information for tailoring compositions for various applications. Specific studies concerning the dissolution of bioactive glasses have been mostly confined to static conditions. Plenty of data are available for the amounts of ions released from bioactive glasses after several immersion times in static SBF or Tris buffer solutions. All the results indicate that the dissolution mechanisms and rates are strongly composition dependent.[71–75]

A dynamic approach has been utilized to determine the ion dissolution profiles and concentrations of ions released from biocompatible glasses during the initial stages of contact with a fresh Tris buffer solution fed continuously through the glass samples.[71,76–78] Also, the pH of the solution was measured on-line to get additional information on the dissolution. When comparing the measured ion dissolution profiles with the bone-bonding ability of the glasses, four different ion release profiles (A–C), depending on the glass composition, were obtained.[71] Figure 6.3 shows ion dissolution profiles of glasses 45S5, 13-93, 1806, and a commercial aluminium borosilicate glass (E-glass). High initial release of Si, Ca and Na ions was typical (profile A, 45S5 in Figure 6.3) for very bioactive glasses, such as S53P4, 206 and 304. The pH of the Tris solution rapidly increased from the initial 7.4 to around 8 for profile A glasses. The second ion release, profile type B (glass 13-93 in Figure 6.3) with notable but clearly lower ion concentrations, was typical for glasses which also bond with bone but then form thinner surface layers, *e.g.* glasses 13-93 and 106. The profile B glasses increased the pH of the solution to around 7.8. Glasses giving dissolution profile C (glass 1806 in Figure 6.3) showed a sharp initial dissolution peak for alkaline and alkaline earth metals while the peaks for silicon and phosphorus were low. Further, the overall concentrations of all ions were significantly lower than for glasses exhibiting profiles A and B. Thus, profile C glasses are likely to only slowly dissolve *in vitro* and *in vivo*, as also verified by the slightly increased pH of the solution. Finally, profile D (E-glass in Figure 6.3) was obtained for compositions typically used as inert references in *in vivo* studies, *i.e.* float glass or aluminium borosilicate (E-glass) compositions. These glasses showed a sharp but minor initial dissolution peak (~2 min) for alkali metals, especially sodium, after which all measured concentrations decreased to near, or under, the level of quantification. The minor alkali exchange reaction did not give any marked changes in the pH of the solution thus suggesting that these glasses are almost inert in solutions mimicking the conditions in the body.

Table 6.3 shows the calculated values of bone bonding ability (RN number, eqn (6.2)), the relative bioactivity value (ISA number, eqn (6.3)) and the pH of

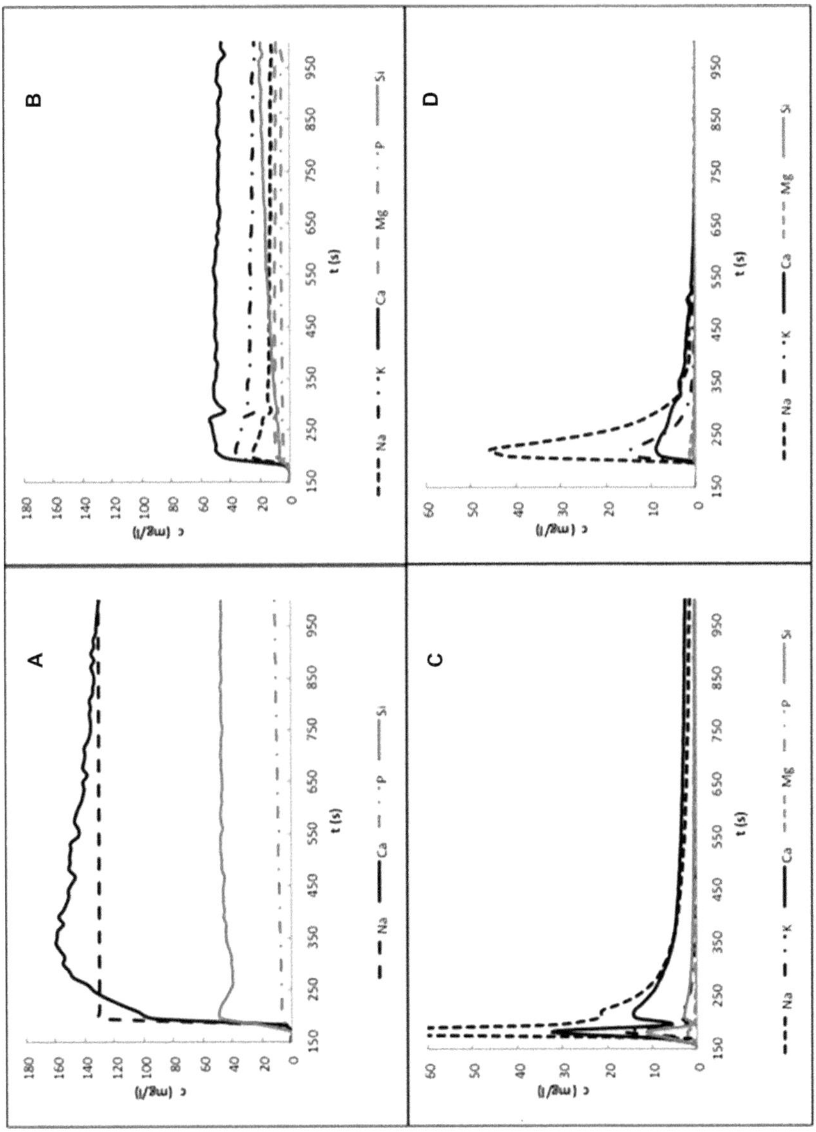

Figure 6.3 Typical initial profiles for glasses in a continuous flow of Tris buffer: A 45S5, B 13-93, C 1806, D E-glass. (Modified from data in Fagerlund *et al.*, Dissolution patterns of biocompatible glasses in 2-amino-2-hydroxymethyl-propane-1,3-diol (Tris) buffer, *Acta Biomater.*, **9**, 5400–5410, Copyright 2013 with permission from Elsevier.[71])

Table 6.3 Calculated and measured (ICP–OES) values of the *in vivo* and *in vitro* reactivity of the glasses in Table 6.2. RN describes the bone bonding ability and ISA the relative bioactivity in contact with bone. pH_{168h} is calculated for 168 h immersion of glass plates in SBF. The measured pH and ion concentration values are given at the time point 1000 s (≈15 min) for a fresh Tris buffer fed continuously through glass particles.[71] The profile gives the overall ion release profile type for the glasses.

	Calculated values			Measured pH and ion concentrations ($mg\ L^{-1}$) at 1000 s in continuous flow of Tris							
	Eqn (6.2)	Eqn (6.3)	Eqn (6.7)								
Glass	RN	ISA	pH_{168h}	pH	Na	K	Mg	Ca	P	Si	Profile
45S5	6.1	4.9[a]	8.3[a]	8.4	s.[b]	1	0.3	130	10	50	A
S53P4	5.9	4.7	8.0	7.9	110	1	0.2	100	10	40	A
13-93	5.2[a]	3.2	7.9	7.8	10	20	9	50	6	20	B
0106	5.6[a]	3.0[a]	8.0	7.8	20	50	20	90	3	30	B
0206	7.1[a]	3.7	8.1	8.0	70	90	a.[c]	120	8	50	A
0306	7.0	3.9[a]	8.1	8.0	s.[b]	0.3	0.3	130	7	50	A
1806	3.0[a]	1.3	7.4	7.5	1	a.[c]	a.[c]	3	0.3	0.4	C

[a]Composition outside the model validity range.
[b]s. = solution saturated.
[c]a. <limit of quantification.

SBF at 168 h (eqn (6.7)) for the glasses in Table 6.2. Table 6.3 shows also pH and the concentrations of the ions Na, K, Mg, Ca, B, P, and Si released into the Tris solution at the time point 1000 s (~15 min). No reliable values could be measured for the release of boron from the glasses. Glasses 45S5, 203 and 306 have RN>6 while a slightly lower value was calculated for S53P4. Also, the ISA number is high for these four compositions. In addition, the concentrations of all ions released from these A-profile glasses were high. The RN value and ISA number are slightly lower for 13-93 and 106. The measured concentrations of Ca and Si were lower but still notable for these profile B glasses. Finally, glass 1806 showed minor ion dissolution and also all the calculated values suggested that this glass only forms bone contact (RN = 3) or a silica-rich layer (ISA = 1.3). Interestingly, the calculated pH values of SBF after 168 h immersion of glass plates ($S_a/V = 0.4\ cm^{-1}$) in static conditions was, for all glasses, very close to the pH of the Tris buffer measured at 15 min for a bed of 300–500 μm glass particles in the dynamic conditions. The pH of the immersion solution also correlates with the thickness of the silica-rich layer (Figure 6.2). These results imply that the pH of SBF after 168 h immersion is a good indication of the bioactivity. In addition, eqn (6.7) seems to give a first approximation of the reaction layer formation ability of the glass without any *in vitro* or *in vivo* experiments.

Hydroxyapatite formation partly controls the layer formation in static SBF, while only minor hydroxyapatite precipitation takes place during the short initial dissolution in fresh Tris buffer flow. Nevertheless, the order of dissolution of the glasses is the same in both conditions. The *in vivo* results of the bone bonding of these glasses were in line with the ion release profiles.[71]

The initial ion dissolution profiles correlated with different *in vivo* reactivity types from rapid bone bonding (profile A) and bone bonding (profile B) to slowly dissolving (profile C) and inert (profile D) glasses.[71] However, more detailed data on the dissolution kinetics are needed for predicting the ion concentrations over time periods needed for tissue regeneration. So far, no models exist for calculating the dissolution rates of ions from glasses as functions of the glass composition.

6.5.4 Models for Hot-working Properties

The viscosity–temperature relationship of glass melts is highly composition dependent. Usually, η is described using the so-called Vogel–Fulcher–Tamman (VFT) equation, $\log \eta = -A + B/(T - T_0)$, in which the composition dependence is given by the constants A, B, and T_0, which are fitted with experimentally obtained viscosity data.[68]

Typically, the viscosity–temperature relations are given over the whole temperature range of interest from around the glass transformation temperature T_g ($\eta \approx 10^{14}$ dPa s at around 500 °C) up to practical melting temperatures corresponding to viscosity values of 100 dPa s (≈1300–1500 °C). However, the strong tendency of bioactive glasses to crystallize during heating at around 100–200 °C above the glass transition temperature prevents viscosity measurement.[79,80] Most silicate-based bioactive glasses also easily crystallize shortly below the liquidus temperature (≈1200–1300 °C).[79–81] In the series of 30 bioactive and biocompatible glasses, a total of 10 compositions crystallized at very low viscosities (<100 dPa s) when measured with a rotational viscometer.[79] Prediction over the whole temperature range from glass transition to practical melting temperatures is thus not meaningful. However, temperature–viscosity relationships can be estimated from the oxide composition for typical sintering ranges ($\eta = 10^{8}$–$10^{8.8}$ dPa s) using eqn (6.8) and (6.9).[79] The composition range of the models is the same as for eqn (6.4)–(6.7).

$$T_{\eta=10^{8.7}\ \mathrm{dPa\ s}}\,(^{\circ}\mathrm{C}) = -150.746 + 3.886[\mathrm{K_2O}] + 7.506[\mathrm{MgO}] + 9.429[\mathrm{CaO}] + 6.972[\mathrm{B_2O_3}] + 8.548[\mathrm{P_2O_5}] + 9.806[\mathrm{SiO_2}] \quad (6.8)$$

$$T_{\eta=10^{8.4}\ \mathrm{dPa\ s}}\,(^{\circ}\mathrm{C}) = 322.895 - 4.346[\mathrm{Na_2O}] + 2.643[\mathrm{MgO}] + 4.720[\mathrm{CaO}] + 4.202[\mathrm{P_2O_5}] + 5.201[\mathrm{SiO_2}] \quad (6.9)$$

The glass transformation temperature and the temperature at which the crystallization commenced in heating the glasses in the series of 30 compositions discussed above depended strongly on the composition of the primary crystals formed above the glass transformation temperature.[80] Eqn (6.10) predicts the crystallization temperature from the composition.[80]

Table 6.4 Measured and calculated values for viscosity–temperature points and crystallization temperature for glass compositions in Table 6.2.[79,80]

Glass	T ($\log\eta = 8.7$) (°C)		T ($\log\eta = 8.4$) (°C)		T (cryst.) (°C)	
	Meas	Calc	Meas	Calc	Meas	Calc
45S5	c[a]	573[b]	c[a]	591[b]	600[b]	561[b]
S53P4	620	592	649	610	645	617
13-93	693	676	708	697	825	820
0106	675	675	690	695	820	793
0206	610	613	634	634	620	670
0306	c[a]	574	c[a]	588	580	547
1806	645	639	660	652	780	715

[a]c – Glass crystallized.
[b]Composition outside model validity range.

The model does not take into account the crystal growth rate and thus, the crystallization tendency predicted is only indicative.

$$T_{\text{CRYSTALLIZATION}}\ (^{\circ}\text{C}) = 191.70 - 9.51[\text{Na}_2\text{O}] + 5.91[\text{MgO}] + 5.93[\text{CaO}] + 15.29[\text{B}_2\text{O}_3] + 12.92[\text{P}_2\text{O}_5] + 9.77[\text{SiO}_2] \quad (6.10)$$

Several experimental models have been suggested for predicting glass transformation and crystallization temperatures from composition using different approaches.[80,82,83] In general, the accuracy of the models is fairly good but also taking into account structural aspects gave better estimates of the actual values.[84] Also, some models which describe the properties using up to fourth degree equations have been suggested;[83] there is a risk, however, that they give oscillating behaviour and are thus not suitable for composition optimization and tailoring.

Table 6.4 compares the measured and calculated high temperature property values for the glasses given in Table 6.2. The compositions 0106, 206, 306 and 1806 were originally optimised using eqn (6.4)–(6.10) to possess various bioactivity and hot-working properties. The calculated viscosity values are in fairly good agreement with the measured values. Similarly, although the calculated and measured crystallization temperature values show large deviations for some compositions, they give a first approximation of the suitability of the glass to various hot-working procedures without losing bioactivity.

6.6 Tailoring of Glass Composition

In tailoring the composition for a certain clinical application, three different types of information are needed: (1) specification of the necessary and critical properties and their numerical or relative values, (2) knowledge of the composition–property relations, and (3) a procedure for choosing the most suitable composition. The overall pipeline for tailoring a product to fulfil the target use criteria for clinical testing is given in Figure 6.4.

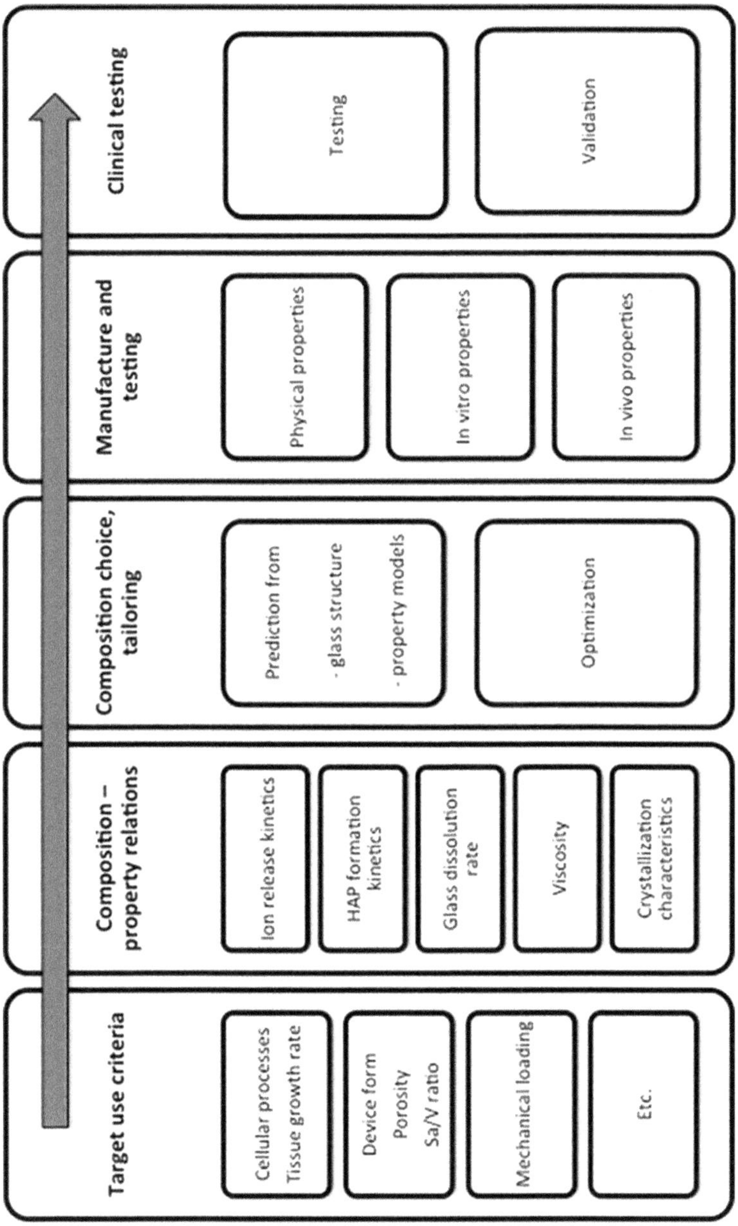

Figure 6.4 Pipeline for developing new medical devices based on bioactive glasses.

The necessary fabrication and clinical properties in a particular application are dictated by the desired product form and host response. These depend on the cellular processes involved, mechanical loading, tissue growth rate, fluid flow rate, *etc.* The most critical properties and their constraints are specified by the target application, the cellular processes involved, the desired implant or scaffold structure, its mechanical loading, *etc.*

Theoretical approaches for predicting the glass structure or models describing the properties as functions of the glass composition can then be used to predict the property values for a certain composition. Alternatively, the composition–property models can be used to optimize the composition to meet the predetermined property criteria.[67,75] The behaviour and properties of the tailored compositions must be verified experimentally before any clinical experiments. In general, tailoring of the glass composition to a particular application only diminishes the need for *in vitro* and *in vivo* testing.

Developing and tailoring new glass compositions or fabricating glasses into various product forms starts from the requirements of the implantation site. Essential criteria are that the glass must not induce any toxic effects and it should be biocompatible. These criteria were fulfilled already by the very first bioactive glasses, which consisted of the physiological elements Ca, P and Na in a silicate matrix. Besides the basic elements O, C, H and N, the human body comprises the elements Ca, P, K, S, Na, Cl, and Mg in concentrations higher than around 0.05 wt%.[85] All other elements present in the body are classified as trace elements of which several have essential functions in the human body. Si, the main element after oxygen in the melt-derived bioactive glasses, has an important role in the bone metabolism.[86] The bioactive glass network structure, and simultaneously the overall fabrication and dissolution properties, have been modified mainly by K and Mg.[70] Mg and several other elements present in small amounts, less than 0.01%, also have essential functions in the human body. Figure 6.5 shows

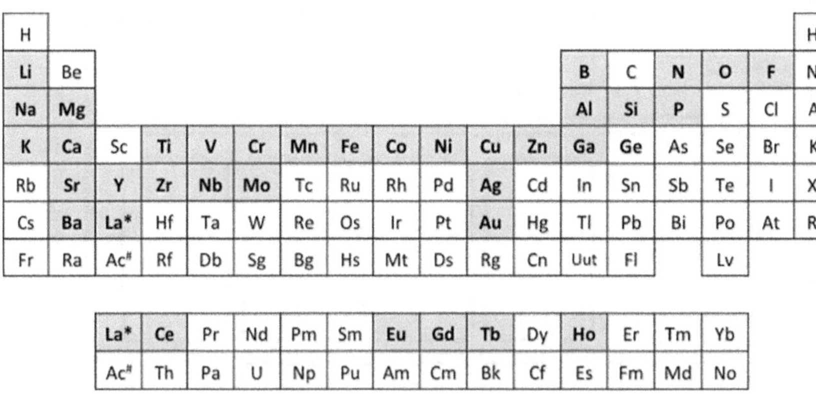

Figure 6.5 Periodic table of the elements showing in grey colour those used in bioactive and biocompatible glasses to achieve desired cellular effects or easier fabrication into various product forms.

the elements which have been used in bioactive and biocompatible glasses by several different research groups to enhance bone and soft tissue growth and vascularization, induce bacteriostatic effect, cure dental hypersensitivity, prevent caries, and aid in cancer treatment, *etc.* The elements inducing some specific cellular effect are often called therapeutic elements or bioinorganics. The therapeutic elements should be released from the glass into the surrounding solution in critical concentrations needed for activation and support of cellular processes. However, a trace element in the glass may induce toxic effects if released into the interfacial liquid in too high a concentration. Thus, doping the glass with a trace element should be carefully optimized to avoid any undesired effects especially if used in devices intended for long-term implantation. The amount of a particular therapeutic element in the glass formulation may vary from less than one wt% to total replacement of some of the elements in the typical Na_2O–CaO–P_2O_5–SiO_2 glasses. A low element content may have only a minor effect on the hot-working or overall dissolution properties of the glass while higher element content may induce large changes in the physical and chemical properties of the glass. So far, most models describing the properties of bioactive glasses take into account only the oxides of Na, K, Mg, Ca, B, Al, Si and P. The effects of other elements on the properties have been studied for several limited systems, mostly by substituting a trace metal oxide for an oxide in the formulations of the bioactive glasses 45S5 or S53P4. These studies provide information on the overall effect of the oxides on the properties and also data to be added to different databases for obtaining new statistical models including the effect of increasing the number of oxides on the properties.

Glass compositions can be tailored *via* different routes. One important goal with the tailoring is to avoid the laborious trial and error approach to decrease the amount of *in vitro* and particularly *in vivo* experiments. Intelligent guesses and experience with the effects of changes in the glass composition on the properties already minimise the experimental work. The models describing the properties as functions of compositions can be used to exclude the formulations which, according to calculated values, seem unsuitable to a specific target use. The models can also be included in computerized optimization routines to calculate a glass composition satisfying a set of desired predetermined properties.[67,75] Since melt-derived bioactive glasses typically consist of four oxides, at least three properties must be specified to find and test a glass composition. Several different combinations of oxides in a multicomponent system can meet a set of desired properties for the device. Then, the most suitable composition can be chosen by using one independent variable—a property that is critically important in the application—as the decisive criterion.

One major challenge in composition tailoring and optimization is to identify the critical properties for the device. In general, the anatomical site of implantation gives the limits of size, geometry, mechanical properties, and dissolution rate of the bioactive glass-based device. The size and geometry specify the fabrication method. Well-established principles of using viscosity–temperature relationships and crystallization characteristics to describe the hot-working methods give the numerical limits of the composition choice. For this part, the composition tailoring deals with chemical engineering, materials science, and glass technology. Mechanical strength is also a critical issue for medical devices. The brittle nature of glasses always challenges the utilization of glasses in load-bearing applications. Unlike most other glass properties, the actual strength value is controlled by the glass composition to a limited degree only. Thus, tailoring of glass compositions for some specific mechanical strength value is not feasible and the strength of the final device must be verified using appropriate methods.

Expressing the bioactivity or biocompatibility of glasses using numerical values is ambiguous. In general, the bioactive glasses are designed to dissolve at a rate matched to the tissue growth rate. The overall dissolution rate depends on the glass composition but also the flow rate, pH and saturation of the solution affect the release rate of the ions from the glass. Therefore, the following questions are relevant for specifying the criteria for the composition choice: What is the desired overall dissolution rate? What is the fluid flow rate in the implantation site? What is the desired concentration of the therapeutic ions released from the glass over a certain period of time? The materials scientist, chemical engineers, clinicians, biologists, *etc.* must together work out and refine the needs and property requirements in terms of measurable properties. Then, it will become possible to develop advanced tools and models for composition tailoring. As discussed above, one major challenge for the future in composition tailoring is to understand how the local conditions in the implantation site affect the glass dissolution and more specifically, the dissolution rate of the therapeutic ions. Special concern should be placed on predicting the amount of ions releasing from devices with a high surface area in sites with restricted flow conditions. Especially, some trace metals may then reach concentrations which become toxic for cellular processes. Several factors affect and control the final fate and dissolution rate of the glasses in the implantation site. Nevertheless, composition–property relationships in systems with the oxides of the physiological elements and bioactive glass constituents Na_2O, K_2O, MgO, CaO, B_2O, P_2O_5 and SiO_2 provide a solid base for smaller composition adjustments, doping with oxides of trace elements for some specific cellular response accompanying tissue healing and regeneration.

6.7 Conclusions and Outlook

Tailoring bioactive glasses for future medical devices is a multidisciplinary task with the goal to find the most suitable composition, which meets the complex criteria set by the cellular processes in the device application site and by the device fabrication. Today, several models are available for predicting the *in vivo*, *in vitro* and physical properties of bioactive glasses from oxide composition. These models can also be used for optimizing the composition, which meets certain specified property values. However, for more reliable prediction, accurate models, especially for the *in vivo* behaviour, are still needed. An interesting future task is to tailor novel glass compositions for scaffolds and devices for novel bone and soft tissue applications. For this, advanced models of the dissolution kinetics of the glasses in the dynamic body environment are needed. The surface area-specific release rates of specific ions would provide valuable information for predicting the biological performance of the glass at a certain implantation site. For a comprehensive understanding of the fate of the bioactive glasses in clinical applications, the long-term interactions of the glasses with cells, proteins, *etc.* is required.

In future, molecular dynamics simulations of the glass structure are likely to provide accurate theoretical predictions of the composition–property relationships, but so far empirical models give the easiest way to tailor glass compositions and also to minimise the amount of *in vitro* and *in vivo* experiments needed for assessing the suitability of a particular composition to the final clinical application. Successful tailoring of bioactive glasses for future medical devices—often biomimetic porous scaffold structures—requires multidisciplinary efforts by the materials scientists, chemical engineers, biochemists, and clinicians.

References

1. L. L. Hench, *J. Mater. Sci. Mater. Med.*, 2006, **17**, 967.
2. L. L. Hench, R. J. Splinter, W. C. Allen and T. T. Greenlee Jr., *J. Biomed. Mater. Res.*, 1971, **5**, 117.
3. L. L. Hench, *J. Am. Ceram. Soc.*, 1991, **74**, 1487.
4. L. L. Hench, *J. Eur. Ceram. Soc.*, 2009, **29**, 1257.
5. L. L. Hench, *New J. Glass Ceram.*, 2013, **3**, 67.
6. L. L. Hench and D. E. Clark, *J. Non-Cryst. Solids*, 1978, **28**, 83.
7. J. R. Jones, *Acta Biomater.*, 2013, **9**, 4457.
8. V. Miguez-Pacheco, L. L. Hench and A. R. Boccaccini, *Acta Biomater.*, 2015, **13**, 1.
9. M. N. Rahaman, D. E. Day, B. S. Bal, Q. Fu, S. B. Jung, L. F. Bonewald and A. P. Tomsia, *Acta Biomater.*, 2011, **7**, 2355.
10. L. L. Hench, *J. Am. Ceram. Soc.*, 1998, **81**, 1705.
11. H. Oonishi, L. L. Hench, J. Wilson, F. Sugihara, E. Tsuji, S. Kushitani and H. Iwaki, *J. Biomed. Mater. Res.*, 1999, **44**, 3.
12. H. Oonishi, L. L. Hench, J. Wilson, F. Sugihara, E. Tsuji, M. Matsuwura, S. Kin, T. Yamamoto and S. Mizokawa, *J. Biomed. Mater. Res.*, 2000, **51**, 37.
13. A. Hoppe, N. S. Güldal and A. R. Boccaccini, *Biomaterials*, 2011, **32**, 2757.
14. P. Stoor, E. Söderling and R. Grénman, *J. Biomed. Mater. Res., Part B*, 1999, **48**, 2869.
15. E. Munukka, O. Leppäranta, M. Vaahtio, T. Peltola, D. Zhang, L. Hupa, H. Ylänen, J. I. Salonen, M. K. Viljanen and E. Eerola, *J. Mater. Sci. Mater. Med.*, 2008, **19**, 27.

16. O. Leppäranta, M. Vaahtio, T. Peltola, D. Zhang, L. Hupa, M. Hupa, H. Ylänen, J. Salonen, M. K. Viljanen and E. Eerola, *J. Mater. Sci. Mater. Med.*, 2008, **19**, 547.
17. D. Zhang, O. Leppäranta, E. Munukka, H. Ylänen, M. K. Viljanen, E. Eerola, M. Hupa and L. Hupa, *J. Biomed. Mater. Res., Part A*, 2010, **93A**, 475.
18. Ö. H. Andersson, G. Liu, K. H. Karlsson, L. Niemi and J. Juhanoja, *J. Mater. Sci. Mater. Med.*, 1990, **1**, 219.
19. P. Virolainen, J. Heikkilä, A. Yli-Urpo, E. Vuorio and H. T. Aro, *J. Biomed. Mater. Res.*, 1997, **35**, 9.
20. M. Peltola, J. Suonpää, K. Aitasalo, M. Varpula, A. Yli-Urpo and R. P. Happonen, *Head Neck*, 1998, **20**, 315.
21. M. J. Peltola, J. T. Suonpää, H. S. Määttänen, M. J. Varpula, K. M. Aitasalo and A. Yli-Urpo, *J. Biomed. Mater.*, 2001, **58**, 54.
22. N. C. Lindfors, K. Tallroth and A. J. Aho, *J. Biomed. Mater. Res.*, 2002, **63**, 237.
23. N. C. Lindfors and A. J. Aho, *J. Mater. Sci. Mater. Med.*, 2003, **14**, 365.
24. P. Stoor, E. Söderling and J. I. Salonen, *Acta Odontol. Scand.*, 1998, **56**, 161.
25. P. Stoor, E. Söderling and R. Grénman, *J. Biomed. Mater. Res.*, 2001, **58**, 113.
26. J. Massera, S. Fagerlund, L. Hupa and M. Hupa, *J. Am. Ceram. Soc.*, 2012, **95**, 607.
27. S. Fagerlund, J. Massera, N. Moritz, L. Hupa and M. Hupa, *Acta Biomater.*, 2012, **8**, 2331.
28. Q. Z. Chen, I. D. Thompson and A. R. Boccaccini, *Biomaterials*, 2006, **11**, 2414.
29. S. Eqtesadi, A. Motellah, P. Miranda, A. Pajares, A. Lemos and J. M. F. Ferreira, *J. Eur. Ceram. Soc.*, 2014, **34**, 107.
30. K. Rezwan, Q. Z. Chen, J. J. Blaker and A. R. Boccaccini, *Biomaterials*, 2006, **27**, 3413.
31. J. Yao, S. Radin, P. S. Leboy and P. Ducheyne, *Biomaterials*, 2005, **26**, 1935.
32. J. A. Rother, A. R. Boccaccini, L. L. Hench, V. Maquet, S. Gautier and R. Jerome, *Biomaterials*, 2002, **23**, 3871.
33. S. M. R. Tuusa, M. J. Peltola, T. Tirri, L. V. J. Lassila and P. K. Vallittu, *J. Biomed. Mater. Res., Part B*, 2007, **82**, 149.
34. A. M. Ballo, L. V. Lassila, P. K. Vallittu and T. O. Närhi, *J. Mater. Sci. Mater. Med.*, 2007, **18**, 2025.
35. J. Rich, T. Jaakkola, T. Tirri, T. Närhi, A. Yli-Urpo and J. Seppälä, *Biomaterials*, 2002, **23**, 2143.
36. T. O. Närhi, J. A. Jansen, T. Jaakkola, A. de Rujiter, J. Rich, J. Seppälä and A. Yli-Urpo, *Biomaterials*, 2003, **24**, 1697.
37. S. Nganga, D. Zhang, N. Moritz, P. K. Vallittu and L. Hupa, *Dent. Mater.*, 2012, **28**, 1134.

38. K. M. J. Aitasalo, J. M. Piitulainen, J. Rekola and P. K. Vallittu, *Head Neck*, 2014, **36**, 722.
39. L. Varila, T. Lehtonen, J. Tuominen, M. Hupa and L. Hupa, *J. Mater. Sci. Mater. Med.*, 2012, **23**, 2425.
40. L. L. Hench, J. M. Polak, I. D. Xynos and L. D. K. Buttery, *Mater. Res. Innovations*, 2000, **3**, 313.
41. I. D. Xynos, M. V. J. Hukkanen, J. J. Batten, L. D. Buttery, L. L. Hench and J. M. Polak, *Calcif. Tissue Int.*, 2000, **67**, 321.
42. I. D. Xynos, A. J. Edgar, L. D. Buttery, L. L. Hench and J. M. Polak, *Biochem. Biophys. Res. Commun.*, 2000, **276**, 461.
43. I. D. Xynos, A. J. Edgar, L. D. K. Buttery, L. L. Hench and J. M. Polak, *J. Biomed. Mater. Res.*, 2001, **55**, 151.
44. F. T. Wallenberger and A. Smrček, *Int. J. Appl. Glass Sci.*, 2010, **1**, 151.
45. W. Vogel, *Structure and Crystallization of Glasses*, Pergamon Press, Oxford, Edition Leipzig, 1971.
46. M. B. Volf, *Mathematical Approach to Glass*, Glass Science and Technology, Elsevier, Amsterdam, 1988, vol. 9.
47. K. H. Karlsson, K. Fröberg and B. Skrifvars, *Glastech. Ber.*, 1990, **63**, 37.
48. M. W. G. Lockyer, D. Holland and R. Dupree, *J. Non-Cryst. Solids*, 1995, **188**, 207.
49. M. Edén, *J. Non-Cryst. Solids*, 2011, **357**, 1595.
50. R. G. Hill and D. S. Brauer, *J. Non-Cryst. Solids*, 2011, **357**, 3884.
51. D. S. Brauer, A. Al-Noaman, R. G. Hill and H. Doweidar, *Mat. Chem. Phys.*, 2011, **130**, 121.
52. R. A. Martin, H. L. Twyman, G. J. Rees, J. M. Smith, E. R. Barney, M. E. Smith, J. V. Hanna and R. J. Newport, *Phys. Chem. Chem. Phys.*, 2012, **14**, 12105.
53. K. Fujikura, N. Karpukhina, T. Kasuga, D. S. Brauer, R. G. Hill and R. V. Law, *J. Mater. Chem.*, 2012, **22**, 7395.
54. M. Magallanes-Perdomo, A. H. De Aza, I. Sobrados, J. Sanz and P. Pena, *Acta Biomater.*, 2012, **8**, 820.
55. R. Mathew, B. Stevensson, A. Tilocca and M. Edén, *J. Phys. Chem. B*, 2014, **118**, 833.
56. K. H. Karlsson, K. Fröberg and M. Perander, *J. Non-Cryst. Solids*, 1986, **84**, 183.
57. A. N. Cormack, J. Du and T. R. Zeitler, *Phys. Chem. Chem. Phys.*, 2002, **4**, 3193.
58. A. Tilocca and N. H. de Leeuw, *J. Phys. Chem. B*, 2006, **110**, 25810.
59. A. Tilocca, *Proc. R. Soc. A*, 2009, **465**, 1003.
60. J. K. Christie, J. Malik and A. Tilocca, *Phys. Chem. Chem. Phys.*, 2011, **12**, 17749.
61. G. Lusvardi, G. Malavasi, F. Tarsitano, L. Menabue, M. C. Menziani and A. Pedone, *J. Phys. Chem. B*, 2009, **113**, 10331.
62. A. Pedone, *J. Phys. Chem. C*, 2009, **113**, 20773.
63. A. Pedone, G. Malavasi and M. C. Menziani, *J. Phys. Chem. C*, 2009, **113**, 15723.

64. A. Tilocca, *Phys. Chem. Chem. Phys.*, 2014, **16**, 3874.
65. J. Du and Y. Xiang, *J. Non-Cryst. Solids*, 2016, **432**, 35.
66. A. Tilocca, *Phys. Chem. Chem. Phys.*, 2015, **17**, 2696.
67. T. Westerlund, L. Hatakka and K. H. Karlsson, *J. Am. Ceram. Soc.*, 1983, **66**, 574.
68. H. Scholze, *Glas*, Springer-Verlag, New York, 1977.
69. Ö. H. Andersson, K. H. Karlsson, K. Kangasniemi and A. Yli-Urpo, *Glastech. Ber.*, 1988, **61**, 300.
70. M. Brink, T. Turunen, R. Happonen and A. Yli-Urpo, *J. Biomed. Mater. Res.*, 1997, **37**, 114.
71. S. Fagerlund, L. Hupa and M. Hupa, *Acta Biomater.*, 2013, **9**, 5400.
72. D. Zhang, E. Vedel, L. Hupa, H. T. Aro and M. Hupa, *Glass Technol.: Eur. J. Glass Sci. Technol., Part A*, 2009, **50**, 1.
73. D. Zhang, In vitro characterization of bioactive glass, Academic dissertation, Uniprint, Turku, 2008.
74. L. Hupa, K. H. Karlsson, H. Aro and M. Hupa, *Glass Technol.: Eur. J. Glass Sci. Technol., Part A*, 2010, **51**, 89.
75. E. Vedel, D. Zhang, H. Arstila, L. Hupa and M. Hupa, *Glass Technol.: Eur. J. Glass Sci. Technol., Part A*, 2009, **50**, 9.
76. S. Fagerlund, P. Ek, L. Hupa and M. Hupa, *J. Am. Ceram. Soc.*, 2012, **95**, 3130.
77. L. Hupa, S. Fagerlund, J. Massera and L. Björkvik, *J. Non-Cryst. Solids*, 2016, **432**, 41.
78. M. Blochberger, L. Hupa and D. S. Brauer, *Biomed. Glasses*, 2015, **1**, 106.
79. E. Vedel, Predicting properties of bioactive glasses, Academic dissertation, Uniprint, Turku, 2008.
80. H. Arstila, E. Vedel, L. Hupa and M. Hupa, *Glass Technol.: Eur. J. Glass Sci. Technol., Part A*, 2008, **49**, 260.
81. F. Döhler, D. Groh, S. Chiba, J. Bierlich, J. Kobelke and D. S. Brauer, *J. Non-Cryst. Solids*, 2016, **432**, 130.
82. M. O'Donnell, *Acta Biomater.*, 2011, **7**, 2264.
83. C. Duée, F. Désanglois, I. Lebecq and C. Follet-Houttemane, *J. Non-Cryst. Solids*, 2012, **358**, 1083.
84. R. G. Hill and D. S. Brauer, *Acta Biomater.*, 2011, **7**, 3601.
85. J. Black, *Biological Performance of Mateirals: Fundamentals of Biocompatibility*, Marcel Decker, New York, 3rd edn, 1999.
86. E. M. Carlisle, *Science*, 1970, **167**, 279.

CHAPTER 7

Bioactive Glass Products Produced via *Sintering*

MOHAMED N. RAHAMAN,*[a] WENHAI HUANG[b] AND B. SONNY BAL[c]

[a] Department of Materials Science and Engineering, Missouri University of Science and Technology, Rolla, Missouri 65409, USA; [b] Institute of Bioengineering and Information Technology Materials, Tongji University, Shanghai 200092, China; [c] Department of Orthopaedic Surgery, University of Missouri, Columbia, Missouri 65212, USA
*Email: rahaman@mst.edu

7.1 Introduction

Although the first bioactive glass, a silicate composition designated 45S5, was developed more than 40 years ago,[1] research, development and applications of bioactive glass products continue to grow. Bioactive silicate glasses, such as 45S5 glass and glasses based on the 45S5 composition, have been the most widely researched and applied bioactive glasses.[2] However, bioactive borate and phosphate glasses developed in the last 10–20 years are also attracting interest.[3] Several biomedical applications require bioactive glass products in the form of spherical particles, granules, coatings, composites and three-dimensional scaffolds with the requisite geometry and microstructure. As these geometrical or microstructural requirements cannot be easily obtained using the widely-used process of casting or forming a molten glass, several bioactive glass products are formed by a route that involves a sintering step. Typically, glass particles are formed into a porous construct with the requisite geometry and heated to an appropriate

RSC Smart Materials No. 23
Bioactive Glasses: Fundamentals, Technology and Applications
Edited by Aldo R. Boccaccini, Delia S. Brauer and Leena Hupa

Published by the Royal Society of Chemistry, www.rsc.org

temperature, typically above the glass transition temperature but below the softening temperature of the glass (or below the onset temperature of crystallization). Less commonly, an amorphous sol–gel-derived material is used. Viscous flow of the glass (or amorphous material) leads to bonding of the particles, a reduction in the porosity, pore size and surface area of the object, and an increase in the strength of the construct. The driving force for the sintering process is a reduction in surface free energy (surface area) of the porous construct.

The sintering of glass (or amorphous materials) depends on a variety of variables, such as the particle or pore size of the as-formed construct, the isothermal sintering temperature and the rate of heating to the sintering temperature.[4] The viscosity of a glass depends on its composition and the temperature.[5] Consequently, the sintering temperature used for a glass will depend on its composition. Analytical and numerical models have been developed to analyze the sintering of glass with idealized or practical geometries. In general, experimental data for the sintering of glasses are in good agreement with the model predictions. Complications can arise if the glass is prone to crystallization in the temperature range required for sintering or is constrained from sintering freely due to the incorporation of a reinforcing phase in the formation of a composite material. Crystallization of the glass prior to appreciable densification or the incorporation of a high volume fraction of a reinforcing phase into the glass can severely reduce the ability to sinter the porous glass to a high density.

7.2 Glass Sintering Principles

7.2.1 Viscous Sintering

Analytical models have been developed to predict the sintering of idealized glass geometries by a viscous flow mechanism.[4] Matter transport is driven by the reduction in surface free energy (surface area) which is assumed to be governed by an energy balance concept first put forward by Frenkel.[6] In this concept, the rate of energy dissipation by viscous flow of the glass is assumed to be equal to the energy gained by the reduction in surface energy. Idealized geometries have been analyzed composed of: (1) two spherical particles of the same radius in contact (Frenkel model),[6] more applicable to the early stage of sintering, (2) a uniform grid-like geometry composed of orthogonal dense glass cylinders of the same radius (Scherer model),[7] more applicable to the intermediate stage of sintering, and (3) a random arrangement of spherical pores of the same radius in a dense glass phase (Mackenzie and Shuttleworth model),[8] more applicable to the late stage of sintering. In general, the predictions are not very sensitive to the geometry of the model except for densities less than ~0.2 of the theoretical density of the glass (which are not relevant to most sintering experiments) or greater than ~0.95 of the theoretical density.

For isothermal sintering (sintering at a given temperature), the predictions of the analytical models for the rate at which a porous glass object will densify can be expressed as[4]

$$\frac{1}{\rho}\frac{\mathrm{d}\rho}{\mathrm{d}t} = AF(\rho)\frac{\gamma_{sv}}{\eta r} \tag{7.1}$$

where ρ is the relative density (the bulk density expressed as a fraction of the density of the fully-dense glass), A is a numerical factor that depends on the geometry of the model, $F(\rho)$ is a function of the relative density, γ_{sv} is the specific surface energy (or surface tension) of the glass, η is the viscosity of the fully-dense glass and r is the pore radius of the material. The surface tension of glass does not change markedly with temperature or with the glass composition.[9] In comparison, the viscosity of glass depends strongly on the temperature and the composition of the glass.[5] The dependence of the viscosity of a glass on temperature is well described by the Vogel–Fulcher equation:

$$\eta = \eta_o \exp\left(\frac{C}{T - T_o}\right) \tag{7.2}$$

where T is the absolute temperature (in degrees Kelvin), and η_o, C and T_o are constants for the glass. For example, the viscosity can decrease by a few orders of magnitude with an increase in the temperature of only 100–200 degrees. Based on the strong dependence of the viscosity on temperature, eqn (7.1) predicts that for a given glass composition, a higher temperature and a finer pore size are beneficial for improving the rate of sintering. For glass objects composed of particles, the particle size scales as the pore size. Consequently, eqn (7.1) predicts that finer particles are beneficial for improving the rate of sintering. In general, these predictions have been well validated by experimental data.[4] The viscous sintering of glass has also been analyzed using finite element modeling (FEM).[10–12] Simulations of the sintering of two axisymmetric spherical particles showed good agreement with the predictions of the Scherer model and the Mackenzie and Shuttleworth model over a wide density range.

The sintering temperature used for a porous glass construct is typically above the glass transition temperature (T_g) but below its softening temperature (T_s). However, isothermal sintering analyzed in most theoretical models represents an ideal condition. Commonly, the as-formed glass article is heated at a given rate to the isothermal sintering temperature and, in some cases, appreciable densification can occur during this heating step. The time spent at a given temperature interval decreases with increasing heating rate, allowing a smaller amount of time for densification. Consequently, data for the measured density *versus* temperature typically shift to higher temperature with increasing heating rate.[13,14]

Although there is no thermodynamic barrier to the removal of all the pores in a glass, problems can arise in the removal of the last few percent of porosity if an insoluble gas is trapped in the isolated pores in the final stage of sintering (relative density greater than ~0.90).[4] For an insoluble gas (a gas than cannot diffuse out of the glass to the surface within the timescale of the sintering experiment), as the pore shrinks, the pressure of the gas in the closed pore increases. Eventually, the gas pressure becomes equal to the sintering pressure due to the driving force for sintering and further shrinkage of the pore stops. Typically, the amount of residual porosity is less than a few percent, decreases with smaller size of the glass particles, and is not a serious problem unless almost full densification (almost zero residual porosity) is a requirement.

7.2.2 Viscous Sintering with Crystallization

For glasses that are prone to crystallization during heating, the isothermal sintering temperature is commonly above T_g but below the onset temperature of crystallization (T_x). The occurrence of crystallization prior to appreciable densification can severely reduce the ability to sinter a porous glass to high density.[4,15] The viscosity of a crystalline phase is considerably greater than a glass phase of the same composition and, according to eqn (7.1), this will lead to a reduction in the densification rate. If present in a high enough volume fraction, the crystals can also form interconnected three-dimensional networks that constrain the glass phase from sintering. The magnitude of the reduction depends on the volume fraction of crystals formed and whether the crystallization occurs prior to appreciable densification or not. The formation of a few percent of crystals in the final stage of sintering has only a limited effect on the final density. In comparison, the formation of more than 10–15 vol% of crystals prior to the final stage of sintering can seriously limit the ability to reach a high final density.

For glasses that are prone to crystallization, the most favorable sintering trajectory for achieving high density is the completion of the densification process prior to significant crystallization. By considering the densification and crystallization processes separately, principles have been developed to qualitatively predict the influence of key process variables.[4] In general, for a given glass composition, finer particle (or pore) size and a faster heating rate in the lower region of the temperature–time schedule of sintering can improve the ability to sinter the glass. Finer particles favor rapid sintering (eqn (7.1)) without affecting the driving force for crystallization. Increasing the heating rate at lower temperature where the nucleation rate of crystals is fast can delay the onset of crystallization, leading to higher density for a given amount of crystallization. The application of an external pressure, as for example in hot pressing, is very effective but it is often uneconomical. The applied pressure enhances densification but generally has little effect on crystallization.

The bioactive silicate 45S5 glass (composition $24.5Na_2O$, $24.5CaO$, $45SiO_2$, $6P_2O_5$; wt%) is considered to be the gold standard for bioactive glasses.

However, it is often difficult to form 45S5 glass particles into strong porous three-dimensional scaffolds by sintering. The glass has a narrow window between T_g and the onset of crystallization (Figure 7.1a), making it prone to crystallization during sintering. This, coupled presumably with a high

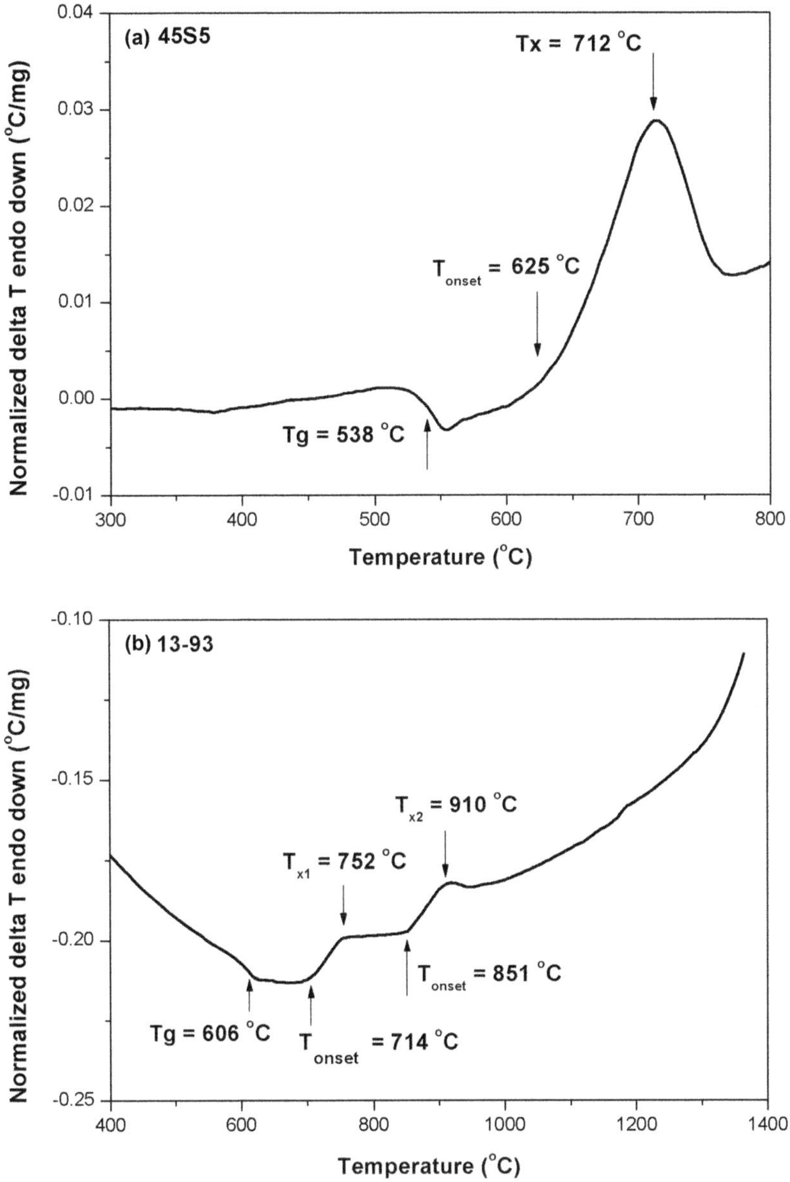

Figure 7.1 Differential thermal analysis (DTA) curves for (a) 45S5 glass particles (<150 μm) and (b) 13-93 glass particles (<45 μm) at a heating rate of 10 °C per minute. (T_g: glass transition temperature; T_x: crystallization temperature; T_{onset}: onset of crystallization.)

viscosity of the glass in the temperature range of sintering, leads to limited viscous flow sintering. Consequently, it is difficult to sinter the glass particles of the scaffolds into a dense glass phase.[16] In comparison, the bioactive silicate glass designated 13-93, with a modified 45S5 composition ($6Na_2O$, $12K_2O$, 5MgO, 20CaO, $53SiO_2$, $4P_2O_5$; wt%) has a wider window between T_g and the onset of crystallization (Figure 7.1b). The network modifiers in 13-93 glass limit the tendency of the glass to crystallize and the viscosity of the glass is presumably lower in the temperature range of sintering. Consequently, particles of 13-93 glass can be sintered to form a dense glass phase, resulting in scaffolds with higher strength than 45S5 scaffolds with a similar microstructure.[17–19]

7.2.3 Viscous Sintering of Composites

Bioactive glasses (and glass–ceramics) typically have a low fracture toughness which can limit their use in load-bearing medical and dental applications. One approach to improve their mechanical reliability is to incorporate a reinforcing phase (such as metal or glass particles) into the glass to form a composite. The formation of a glass matrix composite commonly consists of mixing the reinforcing phase with the powder matrix, forming the mixture into a green construct with the required geometry, and sintering the construct to produce a dense glass phase. Typically, the reinforcing particles or inclusions are dense and non-sinterable and they can be considered to be rigid. Sintering of a porous matrix around a rigid inclusion leads to the development of transient stresses that reduce the sintering stress (driving force for sintering due to reduction in surface free energy).[20] However, the transient stresses are smaller than the sintering stress and are relaxed by viscous creep of the glass matrix. The major effects of rigid particulate inclusions on sintering of a glass matrix are dependent on the volume fraction of the inclusions and the interactions of the inclusions with themselves and with the matrix phase.

For inclusion volume fractions below the percolation limit (the volume fraction when the inclusions just form a continuous touching network in three dimensions), experiments and theoretical models show that the sintering of the glass matrix in a composite is slower than the free matrix (without the reinforcing particles). However, the reduction in densification is small and the densification of the composite can be approximated by a rule of mixtures[20,21] which assumes that in the composite, each phase densifies in the same way as it would independently by itself. For particles of the same size in random arrangement, the percolation threshold is ~16 vol%. Above the percolation threshold, the ability to sinter the composite depends on the rigidity of the inclusion network.[22,23] High densities can be achieved if the glass wets the inclusions and prevents them from forming a rigid network. In comparison, if the inclusions form strong bonds (*e.g.*, they are not wetted by the glass matrix) or interlock (*e.g.*, due to surface roughness), densification of the glass matrix will be severely inhibited as the percolation threshold is approached. Densification would essentially cease at or above the percolation threshold.

7.3 Formation of Bioactive Glass Products by Sintering

The major bioactive glass products created by sintering are:

(1) Spherical particles or granules for use as a bioactive phase in polymer matrix composites, for implantation in non-loaded bone defects, or for conversion into porous or hollow calcium phosphate (or hydroxyapatite) microspheres for drug or growth factor delivery;
(2) Coatings to improve the biocompatibility of bioinert metallic and ceramic implants, impart bioactivity to the surface of the implants to improve their osseointegration, or impart antibacterial activity to the implants to treat or prevent infection;
(3) Composites composed of a particulate-reinforced glass matrix to improve the fracture toughness of the glass for load-bearing prosthetic devices;
(4) Porous three-dimensional scaffolds for use in engineered bone and tissue regeneration.

7.3.1 Spherical Particles and Granules

The formation of microspheres (typically a few tens of micrometers to a few hundred micrometers in diameter) involves heating crushed (irregular-shaped) glass particles of the required size in a flame or furnace.[24] Viscous flow due to surface tension leads to rounding of the sharply-curved regions and, eventually, to the formation of spherical particles to minimize the surface area of the particle. Typically, an optimal combination of the furnace (or flame) temperature and the residence time of the crushed particles in the furnace (or flame) is determined by trial-and-error. For particles of the same shape, the time taken to achieve a geometrically similar shape change by viscous flow sintering scales as the particle radius. Thus, larger crushed particles require a higher temperature or a longer time at a given temperature to achieve a spherical shape.

By sintering together bioactive glass particles (or microspheres), coarse porous bioactive glass granules, up to several hundred micrometers in diameter, are obtained. They provide an alternative to dense individual particles as implant materials to regenerate bone or to deliver drugs in non-loaded defect sites. Granules migrate more slowly out of the defect site than smaller individual particles and they can be more easily implanted in a defect with a complex geometry when compared to three-dimensional scaffolds with a specific geometry. The granules have a higher porosity (surface area) than a single dense particle of size equal to the granule size, which can enhance their ability to support tissue infiltration and integration. Granules can be formed from individual particles using methods such as tumbling and spraying.[25] After removal of any granulating agent (*e.g.*, by drying or thermal decomposition), the as-formed granules are partially sintered to bond the particles and provide the requisite strength.

The faster conversion of bioactive borate glasses to hydroxyapatite in an aqueous phosphate solution when compared to bioactive silicate glasses has been used in a novel process to form high-surface-area hollow or porous hydroxyapatite microspheres near room temperature.[26] Figure 7.2 shows spherical particles (150–250 μm) of a borate glass (composition $11Li_2O$, $15CaO$, $74B_2O_3$; wt%) formed from crushed particles that were allowed to fall under the force of gravity in a vertical tube furnace at 1100 °C. Hollow hydroxyapatite microspheres formed by reaction of the glass microspheres in an aqueous phosphate (K_2HPO_4) solution have been evaluated for the delivery of growth factors in healing bone defects.[27,28]

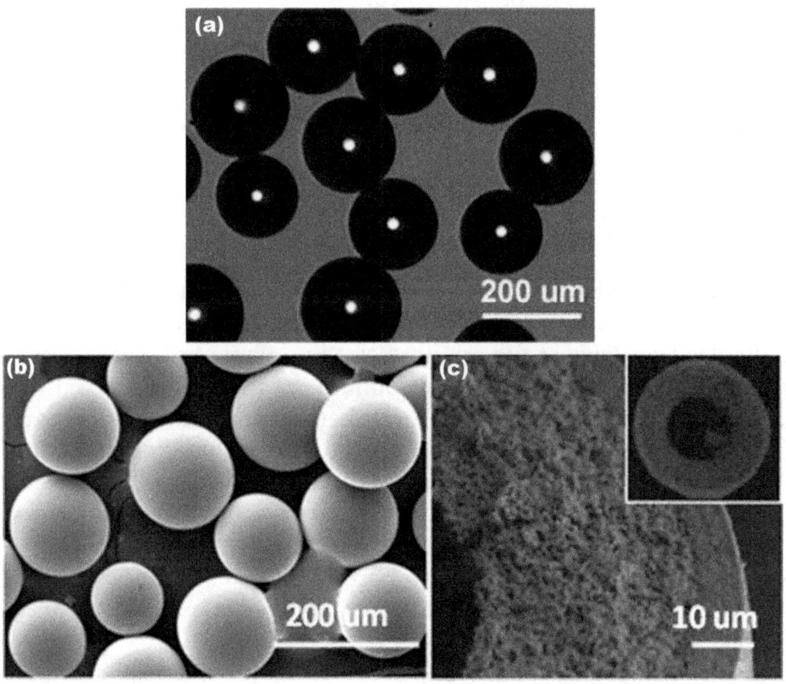

Figure 7.2 (a) Optical image of spherical bioactive borate glass particles (150–250 μm) formed by heating crushed particles in a furnace (the white circle in the center of the glass microspheres was due to reflection in the optical microscope); (b) Scanning electron microscope (SEM) image of hollow hydroxyapatite microspheres formed by converting the glass microspheres in an aqueous phosphate solution; (c) SEM image of the mesoporous shell wall of a hollow hydroxyapatite microsphere (inset: lower magnification image of the cross-section of a hollow hydroxyapatite microsphere).

7.3.2 Coatings

Prosthetic implants used in orthopedic surgery, such as Ti, Ti-6Al-4V, Co–Cr alloys, Al_2O_3 and stabilized ZrO_2, have desirable mechanical and biocompatible properties but their bioinertness limits their osseointegration. One approach that has been used to improve their osseointegration has been to coat them with a bioactive material.[29] A bioactive coating that can adhere strongly to the implant and bond strongly to host bone has the potential to improve stabilization of the implant and extend its life *in vivo*. While hydroxyapatite coatings deposited by plasma spraying have been widely studied and applied,[29] an alternative approach has been to coat the implant with a bioactive glass that converts to hydroxyapatite *in vivo*. Bioactive glass coatings on implant materials have been produced by plasma spraying,[29–34] immersion of the material in a molten glass,[35] and melting of a particulate layer on the substrate,[36,37] but these methods will not be discussed in this chapter because they do not involve the sintering route.

The production of a bioactive glass coating on an implant (or substrate) by the sintering route (sometimes referred to as "enameling") typically involves forming a suspension of glass particles (size <20–50 μm), coating the implant with a layer of the suspension, drying the coating and heating the coated implant under a controlled temperature–time schedule.[38–40] Deposition of the coating from a suspension is typically achieved by conventional methods such as casting, spraying, dip coating or spin coating. The sintering temperature, as described earlier, is above T_g but below T_s (or T_x). Thick coatings can be produced by repeating the process with the same starting suspension while coatings with a gradient composition can be produced by repeating the process with a different glass particle suspension.[38,40,41] An alternative coating method is the use of the solution sol–gel process.[42] Instead of using a suspension of glass particles, the implant is coated with a solution, typically composed of metal alkoxides (or a mixture of metal alkoxides and metal salts). After the coating step (for example, by spraying, dip coating or spin coating), hydrolysis and condensation reactions lead to the formation of a gel-like coating that is dried and sintered.

The desirable properties of a bioactive glass coating on an implant include: (1) a high enough density (typically <5–10% porosity) for adequate strength to transmit physiological stresses *in vivo*, (2) strong interfacial bonding with the surface of the implant to prevent delamination during fabrication or transmission of physiological stresses *in vivo*, and (3) ability to convert to hydroxyapatite at an acceptable rate to achieve osseointegration. The achievement of this desirable combination of properties by the sintering route depends largely on the bioactive glass composition and the sintering (temperature–time) schedule.

Sintering of glass particulate coatings on a rigid implant (or substrate) leads to densification of the coating by viscous flow. The coating adheres to the implant surface and, at the temperature required for sintering, the glass coating can react chemically with the surface of the implant. If the coating remains attached to the substrate during sintering and does not crack, shrinkage of the coating in the plane of the substrate is inhibited and a tensile stress develops in the plane of the coating.[43] All of the shrinkage occurs in the direction perpendicular to the plane of the coating (the thickness becomes smaller). When compared to a free (non-adherent) coating, the tensile stress reduces the densification rate of the adherent coating (increases the time to sinter the coating to a given final density) and can also lead to the growth of flaws.

The tensile stress in the coating does not limit its ability to sinter to the required density; it simply takes longer than a free (non-adherent) coating. However, other factors such as poor interfacial bonding (adhesion), thermal stresses due to thermal expansion mismatch between the glass and the implant, incompatible interfacial reactions, and deterioration of the glass and/or the implant at the sintering temperature can severely limit the ability to successfully form an adherent coating with the requisite combination of properties.

For a fixed interfacial strength, a critical thickness exists above which the interface can no longer support the tensile stress in the coating, resulting in interface failure (delamination).[44] The critical thickness can be a few tens of micrometers for coatings formed from glass particles. In comparison, the critical thickness for polymeric sol–gel coatings can often be <1 μm but it can be increased by aging the gel to achieve a viscosity near the gel point.

The thermal expansion of the glass and the implant should not be very different to avoid large thermal stresses in the coating that can lead to delamination and cracking. Preferably, the glass should have a slightly lower thermal expansion coefficient than the implant which will lead to a small compressive strength in the coating during cooling from the sintering temperature. The thermal expansion coefficients of bioactive silicate glasses 45S5 and 13-93 are 15.1×10^{-6} °C^{-1} and 12.5×10^{-6} °C^{-1}, respectively, which are larger than those (9.5–10.5×10^{-6} °C^{-1}) of Ti and Ti alloys such as Ti-6Al-4V that are commonly used in dental and medical applications. The thermal expansion coefficient of silicate glasses can be lowered by increasing their SiO_2 content but this reduces the bioactivity of the glass.[38,39] Bioactive silicate glasses with higher SiO_2 content have been used successfully to coat Ti alloys whereas glasses with lower SiO_2 content can be used to coat Co–Cr (thermal expansion coefficient $\sim14\times10^{-6}$ °C^{-1}). Bioactive borate glasses have a closer thermal expansion match to Ti alloys, are known to form strong hermetic seals to Ti alloys, are biocompatible and have a higher bioactive potential than bioactive silicate glasses.[45] Consequently, bioactive borate glasses can also be considered for coating Ti-based implants. An alternative approach is the use of a multilayered coating (Figure 7.3). The innermost layer should have a similar thermal expansion coefficient to the implant and should adhere strongly to the implant while the outermost layer should provide the desired bioactivity.[39,40]

The sintering (temperature–time) schedule should be designed to avoid degradation of the glass coating and the implant material, and to achieve the

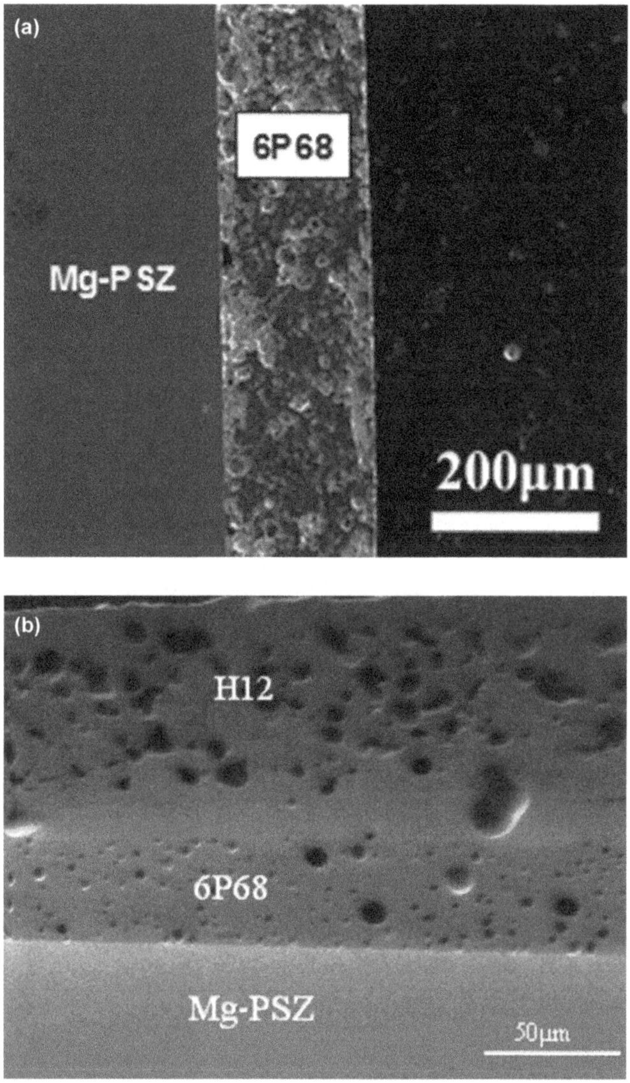

Figure 7.3 SEM images of magnesia-partially stabilized zirconia (Mg-PSZ) coated with (a) bioactive silicate glass (6P68) and (b) a bilayer composed of 6P68 glass and bioactive borate glass (H12).

optimal interfacial bonding. The isothermal sintering temperature, as described earlier, is typically above T_g but below T_s (or T_x). Crystallization of the glass prior to appreciable densification, should be avoided because it can limit the final density of the coating. Crystallization of the glass can also lead to poor adhesion to the implant. As Ti and Ti-6Al-4V implants undergo an α to β phase transition (855–950 °C for Ti, depending on its purity; 955–1010 °C for Ti-6Al-4V), the sintering temperature should be lower than the transition temperature to avoid degradation of their mechanical properties. Metal alloys used as prosthetic implants are also highly reactive and the glass–metal reactions that occur during sintering can be detrimental to the formation of a strong interfacial bond. The formation of a thin (~150 nm) titanium silicide interfacial layer between bioactive silicate glass and Ti-6Al-4V and a thin chromium oxide layer between bioactive silicate glass and Co–Cr under optimal sintering conditions were shown to result in strong interfacial bonding.[38,39,46] In comparison, sintering for longer times or at higher temperatures produced a thicker oxide interfacial layer (~1 μm) and weak interfacial bonding.

Ti and Ti-6Al-4V used as prosthetic implants in dental and orthopedic applications have been successfully coated with silicate or borate bioactive glass (Table 7.1). Zirconia partially stabilized with magnesia (Mg-PSZ) has been evaluated as bearings for total joint replacement because it does not suffer from the strength degradation of yttria-stabilized tetragonal zirconia polycrystals (Y-TZP). Because of the similarity in thermal expansion coefficient, Mg-PSZ has been successfully coated with a borate glass (designated H12) that was developed for coating Ti-6Al-4V. However, a gradient or bilayer coating, composed of an inner layer of bioactive silicate (6P68) glass adherent on the Mg-PSZ surface and an outer layer of H12 glass provided a better combination of interfacial strength and bioactivity.[40]

The use of bioactive glass coatings doped with small concentrations of metal ions (<3 wt%) that are known to enhance the osteogenic, angiogenic and antimicrobial activity of the coatings is also receiving interest. In one study, Ti plates were coated with a bioactive borate glass (composition $6Na_2O$, $8K_2O$, 8MgO, 16CaO, 6SrO, $18SiO_2$, $36B_2O_3$, $2P_2O_5$; mol%) doped with 1 wt% Ag_2O and evaluated in a rabbit tibial fracture model *in vivo*.[47] At 6 weeks postimplantation the Ag-doped bioactive glass-coated implants were

Table 7.1 Selected list of bioactive glasses used to successfully coat various implant materials by the sintering route.[a,b]

Implant material		Bioactive glass coating				
Composition	α (10^{-6} °C^{-1})	Designation	α (10^{-6} °C^{-1})	T_g (°C)	T_s (°C)	Ref.
Ti	9.5–10.5[1]	6P57	10.8[2]	557	609	38, 39
Ti-6Al-4V	9.5–10.5[1]	6P57	10.8[2]	557	609	38, 39
Ti-6Al-4V	10.8[3]	H12	9.7[3]	565 ± 5	590 ± 10	45
	10.8[3]	B18	10.1[3]	510 ± 5	547 ± 10	45
Co–Cr	~14	6P50	12.2[2]	522	560	39
Mg-PSZ	10.6[4]	6P68	8.9[4]	565	653	40
		H12	12.3[4]	575	607	40
		6P68 + H12				40

[a]T_g: glass transition temperature; T_s: glass softening temperature; α: thermal expansion coefficient, measured at the temperature or in temperature range indicated: 1: 400 °C; 2: 200–400 °C; 3: 100–500 °C; 4: 150–550 °C.

[b]Glass compositions (wt%): **6P50:** 15.5Na_2O, 4.2K_2O, 15.6CaO, 8.9MgO, 49.8SiO_2, 6.0P_2O_5; **6P57:** 11.0Na_2O, 3.0K_2O, 15.0CaO, 8.5MgO, 56.5SiO_2, 6.0P_2O_5; **6P68:** 8.3Na_2O, 2.2K_2O, 10.1CaO, 5.7MgO, 67.7SiO_2, 6.0P_2O_5; **H12:** 7.7Na_2O, 35.0CaO, 3.2Al_2O_3, 7.0SiO_2, 43.4B_2O_3, 3.7P_2O_5; **B18:** 12.0Na_2O, 30.0CaO, 5.4Al_2O_3, 6.0SiO_2, 44.3B_2O_3, 2.2P_2O_5.

effective for both stabilizing the fractured bone and eradicating methicillin-resistant *Staphylococcus aureus* (MRSA)-induced osteomyelitis.

7.3.2.1 Sol–Gel Coatings

Polymeric sol–gel coatings formed from a solution of metal alkoxides (or a solution of metal alkoxides and metal salts) have a lower sintering temperature than a particulate coating of the same composition due to their finer pore size and lower viscosity (eqn (7.1)). The coating can also be easily deposited by conventional methods such as dip coating and spray coating. Polymeric sol–gel coatings are more prone to cracking and delamination than particulate coatings due to much larger tensile stresses developed in the adherent coating during sintering and the lower strength of the coating. When compared to melt-derived glasses, difficulties can be encountered in preparing sol–gel-derived materials with the required composition. Sol–gel coatings with compositions in the system $CaO–SiO_2–P_2O_5$ have been used in most studies, such as the glasses designated 58S (38CaO, $58SiO_2$, $4P_2O_5$; mol%) and 77S (19CaO, $77SiO_2$, $4P_2O_5$; mol%).[48] Coatings have been deposited on implant materials such as stainless steel (316L), Co–Cr, Ti, Ti-6Al-4V and Al_2O_3 and evaluated for their ability to reduce corrosion of the metal and/or to impart bioactivity to the surface of the implants.[49–53] Silver-containing sol–gel compositions that can be used to form antibacterial coatings on implant materials have also been studied.[54]

7.3.3 Composites

Bioactive glasses and glass–ceramics have low fracture toughness which limits their use as implant materials in some load-bearing orthopedic and dental applications. One approach that has been used to improve their mechanical reliability is the incorporation of a reinforcing phase into the bioactive glass (or glass–ceramic) to form a composite material. While fibers or platelets can be used, the reinforcing phase commonly consists of particles for ease in processing the composites and to provide isotropic properties. Particles of a ductile metal have been used because previous studies have shown their ability to improve the strength, fracture toughness and fatigue properties of glasses. These ductile particles can improve the fracture toughness of the glass *via* energy dissipating mechanisms such as plastic deformation and crack bridging. Because Ti is well recognized for its biocompatibility and is widely used as an implant material, particles of this metal have been used most often as a reinforcement phase in bioactive glasses.[55–57] Another method has been the incorporation of Y_2O_3-stabilized ZrO_2 (3 mol% Y_2O_3) particles into the bioactive glass matrix.[58,59] These particles can improve the fracture toughness by a transformation toughening or crack deflection mechanism.

The production of the composite typically consists of mixing the particulate reinforcing phase with the bioactive glass powder, forming the mixture into the required shape, and sintering the green construct under an optimal temperature–time schedule in an appropriate atmosphere. As discussed earlier, for volume fractions of the reinforcing particulate phase below the percolation limit, the densification of the glass matrix in the composite is slower than in a free matrix (the glass without reinforcing particles) but the densification of the composite can be approximated by a rule of mixtures. In comparison, above the percolation limit, the densification is strongly dependent on the rigidity of the reinforcing particulate network.

In the production of composites composed of a glass or a glass–ceramic matrix, the starting glass powder should normally be amorphous and the composite should be densified by viscous flow sintering without appreciable crystallization of the glass matrix (less than a few vol% of crystals). The reinforcing particles reduce the densification rate of the glass matrix, requiring longer time or higher temperature for densification. On the other hand, the reinforcing particles can serve as a nucleating agent, lowering the onset temperature for crystallization and accelerating the crystallization of the glass matrix. Consequently, it is often necessary to experimentally determine an optimal temperature–time schedule for sintering to achieve densification without crystallization. Following the sintering step, the production of a glass–ceramic matrix (if required) is achieved by controlled thermal treatment of the dense composite. Reaction between the reinforcing particles and the glass matrix during sintering (or during the crystallization step to form a glass–ceramic) can lead to the formation of an interfacial layer. The matrix phase should bond strongly to the reinforcing particles for stress transfer and to prevent debonding at the interface. Complete debonding can result in the reinforcing particles serving as pseudo pores which can reduce the strength of the composite.

Bioverit® III is a bioactive phosphate-based glass–ceramic with a composition in the system P_2O_5–Al_2O_3–CaO–Na_2O which has been developed for medical applications such as orthopedic surgery, head and neck surgery, and stomatology.[60] Sieved particles (80–90 μm) showed a glass transition temperature $T_g = 493 \pm 2$ °C and a first onset crystallization temperature $T_{x1} = 597 \pm 2$ °C. Sintering of green compacts composed of the glass particles and 15 vol% Ti particles (40–50 μm) under the optimal temperature–time conditions (30 min at 540 °C) resulted in a composite with a final relative density of ~96%, little crystallization and a well-bonded interface between the Ti particles and the glass matrix.[56] After crystallization (3 h at 550 °C), the composite showed a higher fracture toughness (2.1 ± 0.1 MPa $m^{1/2}$) than the glass–ceramic (1.5 ± 0.1 MPa $m^{1/2}$). A related study was performed using reinforcing particles (17 wt%) of Y_2O_3-stabilized ZrO_3 (3 mol% Y_2O_3; particle size <38 μm) instead of Ti. Sintering under approximately similar conditions produced composites with a relative density of 93% and a higher fracture toughness (3.0 ± 0.1 MPa $m^{1/2}$) than the sintered glass (2.6 ± 0.1 MPa $m^{1/2}$).

7.3.4 Three-dimensional Scaffolds

Three-dimensional implants (scaffolds) composed of bioactive glass are being developed and applied mainly for bone repair. The scaffolds commonly have a porous microstructure with the requisite characteristics to accommodate bone infiltration. In general, interconnected pores with a mean diameter (or width) between neighboring pores of 100 μm or larger, and open porosity of more than ∼50% are considered to be the minimum requirements to permit tissue ingrowth and function in porous scaffolds.[2,3,61] As these pore characteristics cannot be easily obtained by casting a molten glass, bioactive glass scaffolds for bone repair are commonly fabricated by a sintering route. In the process, bioactive glass particles are formed into a weak porous construct with the desired shape and microstructure. Less commonly, the porous construct is formed using a sol–gel route.[42] The porous construct is then heated at an appropriate rate to the sintering temperature and held at that temperature for the required time (typically for a fraction of an hour to a few hours) to achieve the desired microstructure. The heating (temperature–time) schedule depends on the glass composition and the particle (or pore) size of the glass phase.

A variety of methods can be used to form porous bioactive glass constructs for the production of scaffolds by the sintering route.[3] The more commonly-used forming methods include: (1) pressing particles in a die or pouring particles or short fibers loosely into a die, (2) polymer foam replication, (3) foaming of sols, (4) freezing of suspensions, and (5) additive manufacturing (also referred to as rapid prototyping, solid freeform fabrication or three-dimensional printing) (Table 7.2). While the microstructure of the as-formed construct (prior to sintering) is dependent on the forming method, in general, it can be considered to consist of a bioactive glass matrix (porous or dense) and interconnected porosity composed of coarse pores. The bioactive glass matrix can consist of coarse glass particles, a combination of fine particles and fine inter-particle pores, or a sol–gel-derived amorphous phase containing fine pores. Commonly, the objective in the sintering step is to strengthen the glass matrix, by bonding the coarse dense particles at their contact points or eliminating the fine pores from the matrix, but to avoid appreciable shrinkage of the coarse pores that are typically present to support tissue infiltration. Figure 7.4 shows images of bioactive glass scaffolds that have been produced by sintering constructs formed by a variety of the methods described above.

One of the simplest methods for producing a scaffold is to thermally bond a loose, random packing of coarse particles or short fibers in a mold that has the required geometry.[62–65] (Figure 7.4a and b). The average size of the pores is determined by the average size of the particles and it can be approximated as one-third the average particle size. Common limitations include a fairly narrow porosity range (40–50%) and the potential for constricted connectivity between neighboring pores. The average pore size and porosity can be increased by the use of short fibers due to their lower packing density.

Table 7.2 Selected list of methods used to form bioactive glass scaffolds, and characteristics of the scaffolds produced by sintering.

Method	Glass	Porosity (%)	Pore size (μm)	Strength[a] (MPa)	Ref.
Thermal bonding of:					
Particles	13-93	40–45	100–300	22 ± 1	63
Short fibers	13-93	45–50	>100	5	65
Polymer foam replication	45S5	89–92	510–720	0.4 ± 0.1	16
	13-93	75–85	100–500	11 ± 1	17
	13-93B3	80–85	100–500	5 ± 0.5	67
Sol–gel foam	70S30C	82	500 (100)[a]	2.4	69
Unidirectional freezing of suspensions	13-93	53–57	90–110	25 ± 3	70
	13-93	50	50–150	47 ± 5	71
Rapid prototyping:					
Freeze extrusion fabrication	13-93	50	300	140 ± 70	73
Robocasting	6P53B	60	500	136 ± 22	74
	13-93B3	50	420	65 ± 11	77

[a]Macropore diameter = 500 μm; interconnected pore diameter = 100 μm.

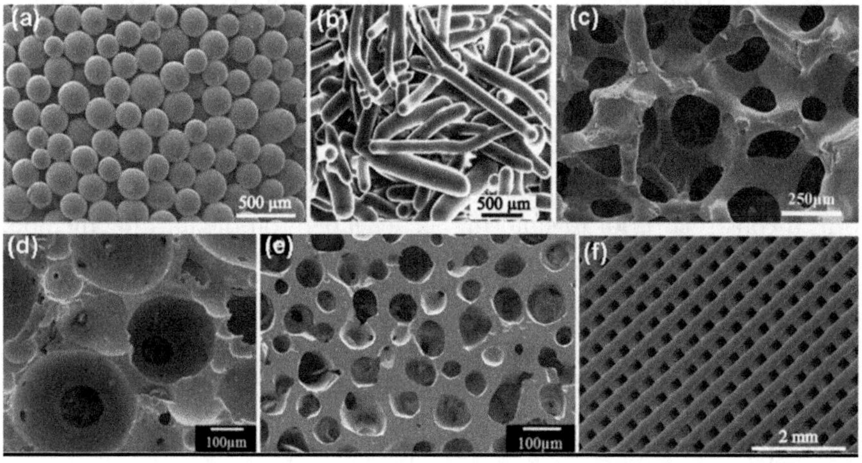

Figure 7.4 Microstructures of three-dimensional scaffolds prepared by a variety of methods: (a) thermal bonding (sintering) of particles (microspheres); (b) thermal bonding of short fibers; (c) polymer foam replication; (d) sol–gel foam; (e) unidirectional freezing of suspensions (plane perpendicular to the orientation direction); (f) robocasting. The scaffolds were formed from the bioactive glasses: (a) 16CaO–21Li_2O–63B_2O_3; (b), (c), (e) 13-93; (d) 100S; (f) 6P53B.

Another method is to press a mixture of bioactive glass particles and particles of a fugitive phase (for example, an organic material such as starch) in a die. Then the as-formed construct is heated to decompose the fugitive phase and sintered at a higher temperature to densify the glass phase. The size of the fugitive phase essentially determines the average pore size of the sintered scaffold. Constricted connectivity between neighboring pores still remains a problem in this method as well.

An extension of the thermal bonding method is the production of scaffolds composed of a mixture of two different glass compositions. Bioactive borate glasses commonly degrade faster than bioactive silicate glass (such as 45S5 or 13-93). Scaffolds composed of an optimized mixture of bioactive silicate and borate glasses can provide the requisite architecture to guide bone regeneration combined with a controllable degradation rate to stimulate bone and soft tissue healing. Scaffolds with varying ratios of borate to silicate glass were produced by thermally bonding a random mixture composed of short fibers of a bioactive borate glass (designated 13-93B3) and bioactive silicate 13-93 glass.[66] A difference in optimal sintering temperature between the borate glass and the silicate glass resulted in weak scaffolds. The addition of 5 wt% of fine borate 13-93B3 particles to the mixture resulted in better bonding and, thus, to the production of scaffolds with higher strength.

The polymer foam replication technique can provide a scaffold microstructure similar to that of dry human trabecular bone (Figure 7.4c). Scaffolds of silicate, borosilicate, and borate bioactive glass have been prepared with porosities in the range 60–90% using this method.[16,17,67] Crystallization of 45S5 glass prior to appreciable sintering, as mentioned earlier, leads to weak glass–ceramic scaffolds (compressive strength <1 MPa) (Table 7.2). In comparison, silicate 13-93 glass and borate 13-93B3 bioactive glass are less prone to crystallization during sintering and, consequently, scaffolds of these two glasses can be produced with higher compressive strength, in the range reported for trabecular bone (strength = 2–12 MPa).

Foaming of sol–gel systems, such as the bioactive silicate glass composition 58S described earlier in the production of coatings, has been used to produce scaffolds (Figure 7.4d) with a hierarchical structure composed of interconnected macropores (10–500 μm), due to the use of a foaming agent, and mesopores (2–50 nm) typical of sol–gel-derived glasses.[68,69] After foaming and gelation, the scaffold is partially sintered at an optimal temperature to stabilize and strengthen the structure, but the sintering temperature should not be too high so as to eliminate the mesopores. The benefit of the hierarchical structure is that the interconnected macropores can support tissue infiltration and vascularization whereas the mesopores can support cell adhesion, adsorption of biomolecules and resorption of the scaffolds at a desirable rate. However, the diameter of the necks between neighboring macropores can be much smaller than the diameter of the macropores themselves which can limit tissue infiltration and vascularization.

Unidirectional freezing of suspensions composed of fine bioactive glass particles (less than a few micrometers) in an aqueous or organic solvent has been used to prepare porous scaffolds with an oriented microstructure (Figure 7.4e).[70,71] A benefit of the oriented microstructure is that the scaffold can have a much higher strength in the pore orientation direction when compared to scaffolds with the same porosity but with a random pore arrangement. Unidirectional freezing of the suspension results in the nucleation and growth of crystalline columns (of ice or an organic material, depending on the solvent) in a preferred direction while the glass particles are expelled to form a matrix. Sublimation of the frozen liquid leads to oriented macropores while the glass matrix consists of glass particles and fine inter-particle pores. The diameter of the oriented macropores depends on the solvent, ranging from 10–40 μm for aqueous solvents to 100–150 μm for water–dioxane mixtures and camphene. The development of high strength (in the orientation direction) (Table 7.2) depends on the ability to sinter the glass matrix to high density. The use of fine glass particles and the formation of suspensions with optimal particle content and colloidal stability are beneficial for the attainment of a dense glass matrix.

Rapid prototyping techniques provide the ability to form constructs with customized external shape and predefined internal architecture (Figure 7.4f). The unprecedented control provided by these techniques can be used to form constructs with well-defined pore characteristics to support tissue infiltration, vascularization and diffusion of nutrients and with the precise external dimensions required to fill the defect site.[72] Because of their ability to provide unprecedented control of the scaffold geometry and architecture, rapid prototyping techniques are being used increasingly in the production of bioactive glass scaffolds. Similar to the other forming methods described above, the techniques are currently used as methods to form glass particles into a construct. A sintering step is invariably required to bond the glass particles by viscous flow in order to produce a scaffold with the requisite mechanical properties. Bioactive glass scaffolds produced by rapid prototyping techniques have shown compressive strength and elastic modulus values comparable to human cortical bone and which are far higher than those of scaffolds produced by other methods (Table 7.2). Strong porous scaffolds of bioactive silicate glasses (such as 13-93 and 6P53B) and bioactive borate glass (13-93B3) have been produced by freeze extrusion fabrication and robocasting techniques.[3,73–77] These scaffolds have shown the capacity to support bone infiltration in rat calvarial defects and to heal segmental defects in rat femurs (a loaded bone defect model) *in vivo*.[78,79]

7.4 Conclusions and Outlook

The sintering route provides an economical method for the production of bioactive glass products with the requisite geometry and microstructure for dental and medical applications which cannot be easily achieved by casting or forming a molten glass, the most widely used production method for glass. The ability to produce a bioactive glass product with the desired microstructure by the sintering route depends on several factors such as the glass composition, particle size, heating rate and sintering temperature. Additional complexities arise when the glass is constrained from sintering freely due to crystallization in the temperature range of sintering, the presence of an adherent substrate in the formation of bioactive coatings on bioinert implants, or the incorporation of reinforcing particles to form composites with improved mechanical properties. The principles of sintering that control the ability to produce bioactive glasses and glass–ceramics with the requisite geometry and microstructure were reviewed. Examples of the application of these principles to the production of adherent bioactive glass coatings on implant materials, particle reinforced bioactive glass and glass–ceramic matrix composites, and porous three-dimensional scaffolds were discussed. The principles and practical examples discussed in this chapter should provide a useful basis for creating bioactive glass and glass–ceramic implants with the requisite geometry and microstructure for a growing range of applications in medicine and dentistry.

Acknowledgements

Supported by the National Institutes of Health, National Institute of Dental and Craniofacial Research, Grant Number 1R15DE023987-01.

References

1. L. L. Hench, *J. Mater. Sci.: Mater. Med.*, 2006, **17**, 967.
2. J. R. Jones, *Acta Biomater.*, 2013, **9**, 4457.
3. M. N. Rahaman, D. E. Day, B. S. Bal, Q. Fu, S. B. Jung, L. F. Bonewald and A. P. Tomsia, *Acta Biomater.*, 2011, **7**, 2355.
4. M. N. Rahaman, *Sintering of Ceramics*, CRC Press, Boca Raton, FL, 2008.
5. A. K. Varshneya, *Fundamentals of Inorganic Glasses*, Academic Press, New York, 1994.
6. J. Frenkel, *J. Phys.*, 1945, **5**, 385.
7. G. W. Scherer, *J. Am. Ceram. Soc.*, 1977, **60**, 236.
8. J. K. Mackenzie and R. Shuttleworth, *Proc. Phys. Soc.*, 1949, **62**, 833.
9. N. M. Parikh, *J. Am. Ceram. Soc.*, 1958, **41**, 18.
10. A. Jagota and P. R. Dawson, *Acta Metall.*, 1988, **36**, 2551.
11. A. Jagota and P. R. Dawson, *J. Am. Ceram. Soc.*, 1990, **73**, 173.
12. H. Zhou and J. J. Derby, *J. Am. Ceram. Soc.*, 1998, **81**, 533.
13. P. C. Pan and R. Raj, *J. Am. Ceram. Soc.*, 1989, **72**, 1564.
14. P. C. Pan, W. M. Mobley and R. Raj, *J. Am. Ceram. Soc.*, 1989, **72**, 2361.
15. G. W. Scherer, *Mater. Res. Soc. Symp. Proc.*, 1990, **180**, 503.
16. Z. Q. Chen, I. D. Thompson and A. R. Boccaccini, *Biomaterials*, 2006, **27**, 2414.
17. Q. Fu, M. N. Rahaman, B. S. Bal, R. F. Brown and D. E. Day, *Acta Biomater.*, 2008, **4**, 1854.
18. X. Liu, M. N. Rahaman, Q. Fu and A. P. Tomsia, *Acta Biomater.*, 2012, **8**, 415.
19. T. S. Huang, N. D. Doiphode, M. N. Rahaman, M. C. Leu, B. S. Bal and D. E. Day, *Mater. Sci. Eng., C*, 2011, **31**, 1482.
20. G. W. Scherer, *J. Am. Ceram. Soc.*, 1987, **70**, 719.
21. M. N. Rahaman and L. C. De Jonghe, *J. Am. Ceram. Soc.*, 1987, **70**, C348.

22. R. E. Dutton and M. N. Rahaman, *J. Am. Ceram. Soc.*, 1992, **75**, 2146.
23. G. W. Scherer and A. Jagota, *Ceram. Trans.*, 1991, **19**, 1991.
24. D. E. Day, J. E. White, R. F. Brown and K. D. McMenamin, *Glass Technol.: Eur. J. Glass Sci. Technol., Part A*, 2003, **44**, 75.
25. M. N. Rahaman, *Ceramic Processing*, CRC Press, Boca Raton, FL, 2007.
26. H. Fu, M. N. Rahaman, D. E. Day and Q. Fu, *J. Am. Ceram. Soc.*, 2010, **93**, 3116.
27. H. Fu, M. N. Rahaman, D. E. Day and R. F. Brown, *J. Mater. Sci.: Mater. Med.*, 2011, **22**, 579.
28. X. Xiao, H. Fu, X. Liu, B. S. Bal and M. N. Rahaman, *Acta Biomater.*, 2013, **9**, 8374.
29. L. L. Hench and O. Andersson, *An Introduction to Bioceramics*, ed. L. L. Hench, Imperial College Press, UK, 2nd edn, 2013, ch. 22, pp. 349–354.
30. T. Kitsugi, T. Nakamura, M. Oka, Y. Senaha, T. Goto and T. Shibuya, *J. Biomed. Mater. Res.*, 1996, **30**, 261.
31. J. A. Helsen, J. Proost, J. Schrooten, G. Timmermans, E. Brauns and J. Vanderstraeten, *J. Eur. Ceram. Soc.*, 1997, **17**, 147.
32. E. Verné, M. Ferraris, A. Ventrella, L. Paracchini, A. Krajewski and A. Ravaglioli, *J. Eur. Ceram. Soc.*, 1998, **18**, 363.
33. E. Verné, M. Ferraris, C. Jana and L. Paracchini, *J. Eur. Ceram. Soc.*, 2000, **20**, 473.
34. J. Schrooten and J. A. Helsen, *Biomaterials*, 2000, **21**, 1461.
35. L. L. Hench and P. J. Buscemi, *US Pat.* 4,159,358, June 26, 1979.
36. M. Ferraris, E. Verné, P. Appendino, C. Moisescu, A. Krajewski, A. Ravaglioli and A. Piancastelli, *Biomaterials*, 2000, **21**, 765.
37. M. Bosetti, E. Verné, M. Ferraris, A. Ravaglioli and A. Cannas, *Biomaterials*, 2001, **22**, 987.
38. J. M. Gomez-Vega, E. Saiz and A. P. Tomsia, *J. Biomed. Mater. Res.*, 1999, **46**, 549.
39. S. Lopez-Esteban, E. Saiz, S. Fujino, T. Oku, K. Suganuma and A. P. Tomsia, *J. Eur. Ceram. Soc.*, 2003, **23**, 2921.
40. M. N. Rahaman, Y. Li, B. S. Bal and W. Huang, *J. Mater. Sci.: Mater. Med.*, 2008, **19**, 2325.
41. L. L. Hench and D. C. Greenspan, *US Pat.* 4,103,002, July 25, 1978.
42. C. J. Brinker and R. W. Scherer, *Sol–Gel Science*, Academic Press, New York, 1990.
43. G. W. Scherer and T. J. Garino, *J. Am. Ceram. Soc.*, 1985, **68**, 216.
44. M. Hu, M. Thouless and A. G. Evans, *Acta Metall.*, 1988, **36**, 1301.
45. L. Peddi, R. K. Brow and R. F. Brown, *J. Mater. Sci.: Mater. Med.*, 2008, **19**, 1345.
46. J. Gomez-Vega, E. Saiz, A. P. Tomsia, G. W. Marshall and S. J. Marshall, *Biomaterials*, 2000, **21**, 105.
47. W. Xiao, S.-H. Luo, X.-J. Wei, C.-Q. Zhang, W.-H. Huang, J.-K. Chen, Y. Cai, Y. Rui and M. N. Rahaman, *J. Mater. Res.*, 2012, **27**, 3147.
48. L. L. Hench, *Curr. Opin. Solid State Mater. Sci.*, 1997, **2**, 604.

49. P. Galliano, J. J. De Damborenea, M. J. Pascual and A. Durán, *J. Sol-Gel Sci. Technol.*, 1998, **13**, 723.
50. A. Durán, A. Conde, A. G. Coedo, T. Dorado, C. Garcia and S. Ceré, *J. Mater. Chem.*, 2004, **14**, 2282.
51. M. H. Fathi and A. D. Mohammadi, *Mater. Sci. Eng., A*, 2008, **474**, 128.
52. M. Hamadouche, A. Meunier, D. C. Greenspan, C. Blanchat, J. P. Zhong, G. P. La Torre and L. Sedel, *J. Biomed. Mater. Res.*, 2000, **52**, 422.
53. J. Liu and X. Miao, *Ceram. Int.*, 2004, **30**, 1781.
54. M. Bellantone, N. J. Coleman and L. L. Hench, *US Pat.* 6,482,444B1, November 19, 2002.
55. T. B. Troczynski and P. S. Nicholson, *J. Am. Ceram. Soc.*, 1991, **74**, 1803.
56. E. Verné, M. Ferraris and C. Jana, *J. Eur. Ceram. Soc.*, 1999, **19**, 2039.
57. B. A. Taylor, R. D. Rawlings and P. S. Rogers, *Bioceramics*, ed. O. H. Anderssson, R.-P. Happonen and A. Yli-Urpo, Butterworth-Heinemann, Oxford (UK), 1994, vol. 7, pp. 255–260.
58. E. Verné, R. Defilippi, G. Carl, C. Vitale Brovarone and P. Appendino, *J. Eur. Ceram. Soc.*, 2003, **23**, 675.
59. C. Fernandez, E. Verné, J. Vogel and G. Carl, *J. Eur. Ceram. Soc.*, 2003, **23**, 1031.
60. W. Höland and W. Vogel, *An Introduction to Bioceramics*, ed. L. L. Hench, Imperial College Press, UK, 2nd edn, 2013, ch. 16, pp. 215–228.
61. L.-C. Gerhardt and A. R. Boccaccini, *Materials*, 2010, **3**, 3867.
62. E. Pirhonen, L. Moimas and J. Haapanen, *Key Eng. Mater.*, 2003, **240–242**, 237.
63. Q. Fu, M. N. Rahaman, B. S. Bal, W. Huang and D. E. Day, *J. Biomed. Mater. Res., Part A*, 2007, **82**, 222.
64. W. Liang, M. N. Rahaman, D. E. Day, M. W. Marion, G. C. Riley and J. J. Mao, *J. Non-Cryst. Solids*, 2008, **354**, 1690.
65. M. N. Rahaman, D. E. Day, R. F. Brown, Q. Fu and S. B. Jung, *Ceram. Eng. Sci. Proc.*, 2008, **29**, 211.
66. Y. Gu, W. Huang, M. N. Rahaman and D. E. Day, *Acta Biomater.*, 2013, **9**, 9126.
67. Q. Fu, M. N. Rahaman, H. Fu and X. Liu, *J. Biomed. Mater. Res., Part A*, 2010, **95**, 164.
68. P. Sepulveda, J. R. Jones and L. L. Hench, *J. Biomed. Mater. Res.*, 2002, **59**, 340.
69. J. R. Jones, L. M. Ehrenfried and L. L. Hench, *Biomaterials*, 2006, **27**, 964.
70. Q. Fu, M. N. Rahaman, B. S. Bal and R. F. Brown, *J. Biomed. Mater. Res., Part A*, 2010, **93**, 1380.
71. X. Liu, M. N. Rahaman and Q. Fu, *Acta Biomater.*, 2011, **7**, 406.
72. E. Sachlos and J. T. Czernusnuszka, *Eur. Cells Mater.*, 2003, **5**, 29.
73. N. D. Doiphode, T. S. Huang, M. C. Leu, M. N. Rahaman and D. E. Day, *J. Mater. Sci.: Mater. Med.*, 2011, **22**, 515.

74. Q. Fu, E. Saiz and A. P. Tomsia, *Acta Biomater.*, 2011, **7**, 3547.
75. Q. Fu, E. Saiz and A. P. Tomsia, *Adv. Funct. Mater.*, 2011, **21**, 1058.
76. A. Deliormanli and M. N. Rahaman, *J. Eur. Ceram. Soc.*, 2012, **32**, 3637.
77. X. Liu, M. N. Rahaman, G. E. Hilmas and B. S. Bal, *Acta Biomater.*, 2013, **9**, 7025.
78. X. Liu, M. N. Rahaman, Y. Liu, B. S. Bal and L. F. Bonewald, *Acta Biomater.*, 2013, **9**, 7506.
79. L. Bi, B. Zobell, X. Liu, M. N. Rahaman and L. F. Bonewald, *Mater. Sci. Eng., C*, 2014, **42**, 816.

CHAPTER 8

Bioactive Nanoparticles, Nanofibers, and Polymeric Nanocomposites

ÁLVARO J. LEITE[a,b] AND JOÃO F. MANO*[a,b,†]

[a] 3B's Research Group – Biomaterials, Biodegradables and Biomimetics, University of Minho, Headquarters of the European Institute of Excellence of Tissue Engineering and Regenerative Medicine, Avepark – Parque de Ciência e Tecnologia, Zona Industrial da Gandra, 4805-017 Barco, Guimarães, Portugal; [b] ICVS/3B's – PT Government Associate Laboratory, Braga/Guimarães, Portugal
*Email: jmano@ua.pt

8.1 Introduction

Implants made of bioactive glasses lead to an adequate bonding with hard and soft living tissues.[1] Without this bonding, a fibrous tissue is likely to form around the implant, which allows micro-motion to occur, causing pain to the patient and leads to the failure of the biomaterial.[2] In orthopaedic applications, such anchorage is attributed to a bone-like apatite layer that develops on the surface of the bioactive glasses when they are in the body.[3] This characteristic has been referred to as bioactivity or bone bonding.[4] Usually, the precipitation of this essential apatite layer is assessed *in vitro* by the soaking of the biomaterial in simulated body fluid (SBF). This is an

†Present address: Department of Chemistry, CICECO, University of Aveiro, 3810-193 Aveiro, Portugal.

RSC Smart Materials No. 23
Bioactive Glasses: Fundamentals, Technology and Applications
Edited by Aldo R. Boccaccini, Delia S. Brauer and Leena Hupa

Published by the Royal Society of Chemistry, www.rsc.org

established test to predict the bonding with the surrounding biological tissue.[4] Regardless of their classical orthopaedic applicability, the bioactive glasses have also shown high potential for soft tissue regeneration (*e.g.* skin and cartilage). For this, their bioactivity is not limited to the formation of a bone-like apatite layer but also augmented by their capability to release ionic dissolution products that affect cellular processes involved in tissue repair.[5] A known example is the angiogenic potential of bioactive glasses in soft tissue repair strategies.[6,7]

Researchers have realized that the components of the biological tissues have nano-size dimensions and that important interactions between cells and biomaterials occur at the nano-size level.[8] Therefore, in recent decades, nanotechnology has become an emergent topic in tissue engineering, originating new materials with high potential for application. The suitability of the nanosize is also relevant in the case of inorganic biomaterials, such as ceramics and glasses. In this context, bioactive glass nanoparticles and bioactive glass nanofibers have been exploited. Nowadays, rapid technological advances are allowing the design and fabrication of a variety of nanostructured bioactive glasses for biomedical applications. The devices include tissue-engineered scaffolds, site-specific drug delivery systems, non-viral gene carriers, biosensors, screening systems, and devices for clinical diagnostics.

Several aspects of the nanometre scale differ from the conventional micrometre scale.[9,10] A substantial number of these are interrelated properties that determine the cell–biomaterial interactions such as nanotopography, surface area, hydrophilicity, surface energy, or chemical composition.[11,12] These features regulate the interfaces with proteins, modulating cell adhesion, spreading and proliferation, and ultimately, affecting the long-term functionality of implants.[10,13,14] Moreover, experimental evidence suggests that different cell lines can have specific and varying interactions with micrometre or nanometre topography.[15]

Nanomaterials have a high surface area to volume ratio. For instance, Gerhardt *et al.* described microsized Bioglass® with a median particle size of 4.3 μm and a specific surface area of 2.7 $m^2\ g^{-1}$, whereas nanoscaled Bioglass®, with diameters of 35–40 nm, presented a large specific surface area of 79 $m^2\ g^{-1}$.[16] This feature confers to nanoscale bioactive glasses a higher surface energy and surface reactivity when compared with the microscale counterparts. Therefore, nanoscale bioactive glass produces a faster ion release rate as well as increased surface area for protein adsorption, which is predicted to increase the bioactivity.[17–19] As a result, it is expected that nanoscale bioactive glass can promote increased cell adhesion, proliferation, and differentiation.[14,20–23] Studies have evaluated the *in vitro* osteogenic properties of nanoscaled Bioglass® compared with microsized Bioglass®.[24,25] They found that Bioglass® nanoparticles induced an early precipitation of apatite in SBF. In fact, compared with microsized particles, Bioglass® glass nanoparticles promoted higher human osteoblast-like cell (MG-63) adhesion, and induced osteogenic gene expression of the alkaline phosphatase (ALP), runt-related transcription factor two (RUNX2) and collagen I (COL1), at an earlier stage and to a greater extent.[25] Gong *et al.* reported similar results with a different composition of nanoscale bioactive glass.[26] They found that the ionic dissolution of bioactive glass nanoparticles improved cell proliferation and activation of osteogenic genes, when compared with micro bioactive glass particles.[26] These studies demonstrated that bioactive glass nanoparticles hold a higher capacity for apatite formation and stronger osteogenic ability when compared with microscale bioactive glass particles.

Even inside the nanolimits, different sizes lead to different outcomes. Ajita *et al.* compared the influence of three sizes of nano bioactive glasses (approx. 74.7 nm, 43.25 nm, and 37 nm) in mouse mesenchymal stem cells.[27] They showed that the sizes were non-cytotoxic at a concentration of 20 $mg\,mL^{-1}$ but observed increased proliferation in cells treated with the smallest size.[27] Moreover, the dissolution rate of ions also varied depending on the size of the nanoparticles. They found that the particles with smaller size performed better because of the sustained activation of extracellular regulated kinases (ERKs) and upregulation of cyclin genes, that modulated the cell cycle from the G0/G1 phase to the S and G2/M phases.[27] This study emphasized the nanosize dependency on the cell proliferative action.

Bioactive glasses are well known for their use in bone tissue engineering due to their high bioactivity. Besides, studies also indicated that bioactive glasses have angiogenic potential.[6,7] Following this rationale, Mao *et al.* studied the effect of ionic extracts of two compositions of nanoscale bioactive glasses on human umbilical vein endothelial cells (HUVECs).[5] The extracts accelerated cell migration and upregulated the expression of the vascular endothelial growth factor, basic fibroblast growth factor, their receptors, and endothelial nitric oxide synthase, resulting in capillary tube formation *in vitro*. They also found that the enhanced angiogenic response correlated with increased levels of Ca and Si ions. These biological effects might be due to the increased level of Si released, with the released Ca ions playing a supporting role. These results indicated that nanoscale glasses are an approach for stimulating neovascularization in tissue engineering applications.

The safety of nanoscale bioactive glasses is also an area of concern. Studies demonstrated that nanoscale bioactive glass showed an equal performance when compared with commercial micro powder bioactive glasses. Rismanchian *et al.* compared the cytotoxicity of a nanopowder bioactive glass with a commercial micropowder bioactive glass named NovaBone®, an FDA-approved bone graft.[28] They concluded that the cytotoxicity of nano-sized bioactive glasses at concentrations $\leq 2\ mg\,mL^{-1}$ was similar to NovaBone® at 24 h and 48 h. However, it is more cytotoxic at concentrations $\geq 5\ mg\,mL^{-1}$ in the first 48 h of application.[28] Also, Tavakoli *et al.* evaluated the genotoxicity of nano-sized bioactive glasses and NovaBone® using gingival fibroblast cells.[29] They showed that concentrations less than 4 $mg\,mL^{-1}$ had no genotoxic effect without any statistical difference between nano-sized bioactive glasses and NovaBone®.[29]

In this chapter, we will describe the main features of bioactive glass nanoparticles and bioactive glass nanofibers and give a brief description of their methods of production. We will also present the state of the art, and the applications of nanocomposites constituted of bioactive glass nanoparticles and bioactive glass nanofibers. Finally, a summary of the addressed topics is presented and areas for future research are discussed.

8.2 Bioactive Glass Nanoparticles

Bioactive glass nanoparticles (BGNPs) are usually composed of silicates or phosphosilicates combined with different proportions of glass modifiers such as sodium oxide (Na_2O) and calcium oxide (CaO), among others. As a bioactive glass, they exhibit high bioactivity and excellent bone bonding properties. Typical bioactive compositions are binary systems, *e.g.*, SiO_2–CaO; ternary systems, *e.g.*, SiO_2–CaO–P_2O_5; or quaternary systems, *e.g.*, SiO_2–CaO–P_2O_5–Na_2O.[30,31]

The composition of BGNPs could determine the ability of the material to support the proliferation and function of cells.[32,33] Moorthi *et al.* synthesized ternary BGNPs (SiO_2–CaO–P_2O_5) with different amounts of calcium oxide and silica (70 : 25 : 5 and 64 : 31 : 5 mol%), and investigated their role in osteoblast proliferation.[34] Whereas both BGNPs were non-cytotoxic to MG-63 cells, they found the formulation with higher calcium content to be more effective on osteoblast proliferation and promoted more cells to enter into the G2/M cell cycle phase. Additionally, they also stated a pronounced expression of cyclin proteins.[34] These results suggest that the adaptation of the ionic constituents of BGNPs can lead to different biological properties.

It is estimated that the controlled release of ions from bioactive glasses can stimulate the interaction with tissues, *e.g.*, new bone growth.[32] The rate of ion release is predominantly dependent on the composition, particle size and surface area, which in turn affect osteogenesis and angiogenesis. In this respect, BGNPs became a vehicle for the local delivery of specific ions to control cellular functions. Ion-doped BGNPs are being designed for the controlled release of specific metal ions, which can lead to increased osteoblast activity (*e.g.*, in strontium-doped BGNPs), to have beneficial anti-inflammatory effects (*e.g.*, in zinc-doped BGNPs) or to possess angiogenic effects (*e.g.*, in Co-doped and Cu-doped BGNPs).

Copper-doped BGNPs (SiO_2–CaO–P_2O_5–CuO) and silver-doped BGNPs (SiO_2–CaO–P_2O_5–Ag_2O) possess antibacterial properties, preventing bacterial colonization.[35,36] Goh *et al.* described the ion release profiles of these BGNPs. They observed a faster initial release of Ag, which made silver-doped BGNPs a rapid bacteria-killing agent. Meanwhile, due to a slower release, copper-doped BGNPs might be suitable candidates for long-term antibacterial protection.[35] Additionally, titanium dioxide can be a powerful antibacterial substance as TiO_2 shows high oxidizing power under irradiation with UV light. Therefore, Rajendran *et al.* prepared titanium-doped BGNPs (SiO_2–CaO–P_2O_5–TiO_2).[37] They reported effective antibacterial inhibition against *Escherichia coli* and *Staphylococcus aureus*. Moreover, the cytotoxicity assay revealed that the nanoparticles were non-cytotoxic and showed improved cell viability in MG-63 cells at a concentration of 125 $\mu g\,mL^{-1}$.[37] These particles could reduce the occurrence of bacterial infections in biomaterial implants.

Zinc-doped BGNPs (SiO_2–CaO–ZnO) had been shown to promote odontogenic differentiation and angiogenesis. The zinc ion is an essential trace element in bone and is considered for osteoporosis treatments.[38,39] Zhang *et al.* evaluated the effect of zinc-doped BGNPs on human dental pulp cells.[40] Zinc-doped BGNPs enhanced ALP activity, improved the formation of mineralized nodules, and upregulated mRNA expression of odontogenic differentiation marker genes, in a time- and dose-dependent manner.[40]

Strontium-doped BGNPs are known to have beneficial effects on osteogenic stimulation and *in vivo* bone formation. In this regard, Strobel *et al.* assessed the osteogenic potential of BGNPs containing strontium (SiO_2–CaO–P_2O_5–Na_2O–K_2O–MgO–SrO) on human bone marrow stromal cells (hBMSCs).[41] Cell growth and morphology indicated the high cytocompatibility of the nanoparticulate bioactive glass. The strontium-doped particles enhanced the expression of osteocalcin, collagen type I and vascular endothelial growth factor.[41] However is also important to notice that increasing strontium content can delay the bioactive behaviour of BGNPs, due to the inhibitory effect of strontium on apatite mineralization.[42]

Despite the acceptable non-cytotoxicity and non-genotoxicity of BGNPs, concerns have arisen about their internalization from the cells and the consequent long-term fate in the body. A localized increase in intracellular silica and calcium concentrations could cause a marked effect on cell metabolism or inflammatory response. As particles dissolve intracellularly, they might also break up into finer particles that can escape the endosomal pathway and enter the cell cytoplasm or even the nucleus. In this way, studies have explored this subject by investigating the response of human mesenchymal stem cells (MSCs) and adipose tissue-derived stem cells (ADSCs) to the BGNPs.[43,44] The results confirmed cellular uptake and localization of the nanoparticles in cell cytoplasm and cell endosomes and partial particle dissolution inside the cells. It was hypothesized that the dispersion of the particles, and finer control of particle size, dramatically increased the amount of the particles taken up by the cells. However, the uptake mechanism remained unclear. More important, the particles were non-cytotoxic at lower concentrations and held minimal effects on primary MSCs and ADSC viability at higher concentrations (100 and 200 $\mu g\,mL^{-1}$). However, BGNPs had an irrelevant effect on cell proliferation compared with the control. Also, BGNPs induced osteogenic differentiation in the presence of osteogenic supplements. These minimal effects on cell metabolism and proliferation combined with the large numbers of particles detected inside the cells showed acceptable safety. Additionally, these results can be used to explore the potential of BGNPs as injectable target particles to be internalized by cells for the sustained local delivery of inorganic therapeutic ions.[43,44]

BGNPs can be produced by different techniques such as sol–gel, flame synthesis and microemulsion. The following concise description of each method is presented and can be complemented with contents discussed elsewhere.[45]

8.2.1 Sol–Gel

The most common sol–gel technique for BGNPs production is an adapted procedure of the Stöber method.[46] In this approach, the glass compositional precursors are first hydrolyzed to a sol under acidic conditions and then condensed and precipitated to gel particles in alkaline solution. A typical silicate precursor is tetraethyl orthosilicate (TEOS), whereas common precursors for calcium and phosphate groups are calcium nitrate tetrahydrate and triethylphosphate, respectively. In brief, the procedure starts with the hydrolysis of TEOS in water or a water/alcohol acidic mixture (creating Si–OH groups). Then, addition of a base (usually ammonia) as a catalyst allows the condensation process through the release of water molecules and formation of Si–O–Si bonds, and a colloidal sol is generated. The primary nanoparticles present in the sol become coarse and bond together, forming a silica-based gel network of assembled nanoparticles. This gelation step proceeds from minutes to a few hours after the addition of pre-hydrolysed TEOS to the ammonia solution. A further drying step is commonly used to remove the excess water and prevent the secondary aggregation of the gel particles *via* linkage by water molecules. The gel is calcined above 680 °C to remove the organic phase and residual precursors, which decomposes the gel network, and affords the final BGNPs (Figure 8.1A). Experimental conditions which could be manipulated are the pH, the volume of solvents, ratio of reagents, and the inclusion of surfactants which influences the processes of hydrolysis and condensation that determines the chemical composition, the final size and agglomeration of the BGNPs. Moreover, the synthesis and calcination temperature has a considerable effect on the crystallization process of the BGNPs. The sol–gel route permits easier control of the chemical composition and morphology.[43,47–53] For instance, spherical or rice grain-shaped nanoparticles have been obtained by this method.[54]

8.2.2 Flame Spray Synthesis

In this process, the metal-organic precursors present in a liquid mixture are dispersed by oxygen over a nozzle forming a spray.[45,55–57] The spray is ignited and, as it burns at temperatures above 1000 °C, the organic constituents of the liquid precursor combust to water and carbon dioxide while metal constituents oxidize to form the nanoparticles. The gas phase synthesis method starts with the formation of molecular nuclei followed by condensation and coalescence, inducing subsequent growth of nanoparticles in high-temperature regions during the process.[45,55–57] The most decisive factor for the final particle size is the mean residence time of the

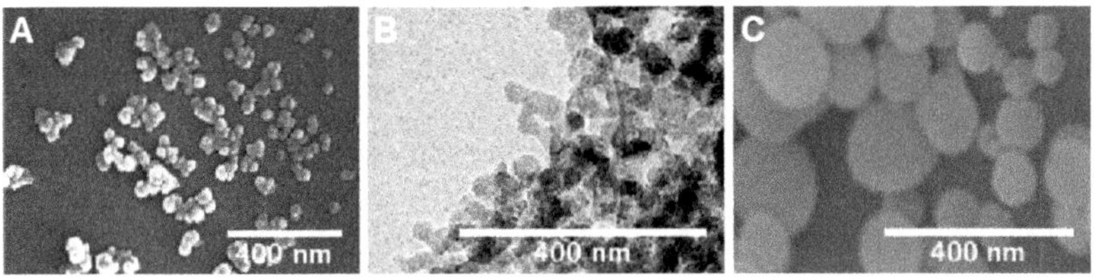

Figure 8.1 A: ESEM micrographs of BGNPs produced by Sol–gel process (SiO_2–CaO–P_2O_5). B: TEM image of a flame-made BGNPs sample showing a high degree of agglomeration (SiO_2–CaO–P_2O_5–Na_2O). C: SEM images of BGNPs produced by microemulsion (SiO_2–CaO–P_2O_5). [A] Adapted from Z. Hong *et al.*, Preparation of bioactive glass ceramic nanoparticles by combination of sol-gel and coprecipitation method, *J. Non-Cryst. Solids*, **355**, 368–372, Copyright 2009 with permission from Elsevier,[48] [B] Reprinted from ref. 55 with permission from the Royal Society of Chemistry, [C] H. Wang *et al.*, Preparation and characterization of the system SiO_2-CaO-P_2O_5 bioactive glasses by microemulsion approach, *J. Wuhan Univ. Technol., Mater. Sci. Ed.*, **28**, with permission from Springer.[58]

particles in the high-temperature regions. High cooling rates (>1000 K s^{-1}) and short residence times (1 ms) enable the formation of nanoparticles (Figure 8.1B). The process can produce a variety of nanoparticles with different compositions and with a high chemical homogeneity.[45,55–57]

8.2.3 Microemulsion

In water-in-oil microemulsions, nanosize water droplets are dispersed in the hydrocarbon continuous phase and surrounded by a monolayer of surfactant molecules. The size of the aqueous droplets is usually in the range of 5–20 nm in diameter. These aqueous droplets act as minuscule reactors in which reactions take place when droplets containing reactants collide. Precursor particles of hydroxide or oxalate are first formed in a microemulsion system. Then, the desired oxide system is obtained after drying and calcination of the precursor powder at a defined temperature (Figure 8.1C). Microemulsion techniques can give rise to nanoscale particles with low agglomeration. However, the main disadvantages of the microemulsion technique are the small production yields and the use of a large amount of oil and surfactants.[58]

8.3 Bioactive Glass Nanofibers

Bioactive glass nanofiber (BGNFs) structures have excellent potential to act as cell supporters for extracellular matrix production of hard tissues and to promote tissue growth.[59–61] It is known that the surface topography of hydroxyapatite crystals can influence fibronectin, vitronectin and osteoblast adhesion, thus it is expected that the profile of nanofibers could also affect cell behaviour.[62] Moreover, the structure of nanofibrous bioactive glasses presents two benefits. On the one hand, BGNFs with ultrathin diameter show highly specific surface areas. On the other hand, due to their length, the nanofibers can be assembled into 3D structures, such as membranes, with an interconnected microporous network.[60] BGNFs matrices can provide an appropriate environment for cell adhesion, proliferation, differentiation, and mineralization.[63] BGNFs can also exhibit the previously described features related with BGNPs, such as typical compositions, improved bioactivity, exceptional bone bonding, promotion of osteogenic and angiogenic activity and ion doping effects.[60,61,64–66]

8.3.1 Electrospinning

BGNFs can be prepared by electrospinning of the sol–gel precursor (Figure 8.2A).[60,63,65,67–69] In the last decades of the 19th century, electrospinning was first described as a process to fabricate thin fibers and today is a well-known technique in the biomedical field.[70] To form bare bioactive glass fibers using electrospinning, the sol solution is mixed with a coadjuvant polymer to adjust the viscosity to a level suitable for electrospinning. Some polymers used include polyvinylbutyral (PVB), polyvinyl alcohol (PVA), polyvinyl pyrrolidone, pluronic P123 and (PEO).[60,63,65,67–69] The solution is forced through a needle, forming a drop at the tip. Then, a high voltage is applied to extrude nanofibers. When the electric field strength overcomes the surface tension of the droplet, a solution jet is initiated and accelerated towards the collection target. As the jet travels through the air, the solvent evaporates, and a non-woven fabric is formed on the target. The non-woven fabric is calcined at temperatures high enough to eliminate the coadjuvant polymer and other organic sources and nitrates, but low enough to avoid

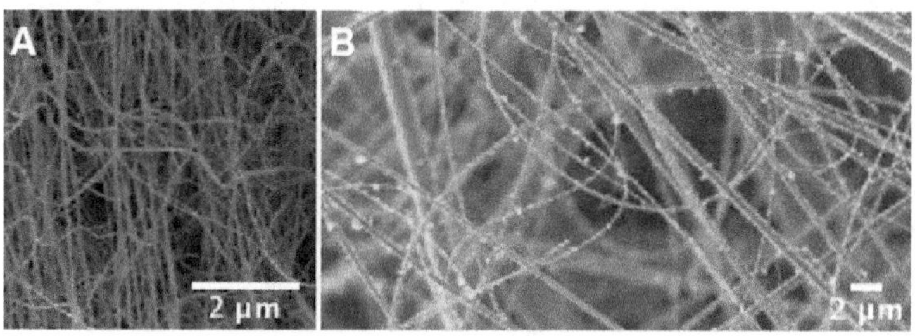

Figure 8.2 A: SEM image of the BGNFs produced by electrospinning of sol–gel precursors (SiO_2–CaO–P_2O_5). B: FESEM micrograph showing the typical appearance of BGNFs produced by laser spinning (SiO_2–CaO–P_2O_5–Al_2O_3–Na_2O–K_2O). [A] Adapted from ref. 60 with permission from John Wiley and Sons. Copyright © 2006 WILEY-VCH Verlag GmbH & Co. KGaA, Weinheim, [B] adapted from ref. 71 with permission from John Wiley and Sons. Copyright © 2009 WILEY-VCH Verlag GmbH & Co. KGaA, Weinheim.

crystallization. The resulting nanofibers can be further modified to increase their bioactivity.

Kim *et al.* reported, for the first time, the use of an electrospinning process based on sol–gel to generate continuous bioactive glass nanofibers with variable diameters (Figure 8.2A).[60] Before electrospinning, PVB was added to adjust the rheological properties of the sol to be suitable for fiber generation during the electrospinning process. Electrospun fibers were then heated at 700 °C.[60] The excellent bioactivity of the nanofiber was confirmed by the rapid induction of bone-like minerals onto the nanofiber surface. Their results showed that the ability for apatite formation on nanofibers in SBF was higher compared with that on the traditional bioactive glasses. They also studied the effect of the glass sol concentration on the diameter of the glass nanofibers, which reduced with the decrease in the sol concentration. However, the diameter of electrospun fibers was dependent on other processing parameters besides sol concentration, such as the strength of the applied electric field. Studies with bone marrow-derived cells showed that cells could attach and proliferate on the nanofiber mesh, and differentiate into osteoblastic cells, which showed the excellent osteogenic potential of BGNFs.[60] BGNFs have also been shown to be suitable for drug delivery matrices. Hsu *et al.*, produced BGNFs with high drug loading efficiency that reduced the burst release of gentamicin sulphate and prolonged its release over 10 days.[63]

To increase the bioactivity potential and the nanostructure complexity of BGNFs, Hong *et al.* synthesized nanoporous BGNFs with different pore diameters.[69] Experiments demonstrate that the prepared bioactive glass fibers were highly homogenous, and the nanopores in the fibers endowed BGNFs with high bioactivity. Moreover, drug loading and release experiments showed that, following the decrease of pore size, the drug loading capacity increased and higher amounts of drug could be released in a controlled manner.[69] Hollow BGNFs were also developed by Durgalakshmi *et al.*[68] The authors fabricated hollow BGNFs using the sol–gel electrospinning technique with the assistance of polyvinyl pyrrolidone (PVP). The mechanism of the formation of the hollow fibers was attributed to the rapid solvent evaporation during electrospinning and phase separation between the PVP and the solvent. Fibers presented a sustained release of 85% of ibuprofen within release times of 18 h. This showed the superior performance of this system in drug loading and releasing kinetics. The porous BGNFs could also be advantageous for adsorbing bone marrow aspirate.[68]

8.3.2 Laser Spinning

The techniques based on sol–gel precursors offer greater flexibility to vary composition, but they rely on the use of chemical solvents. To overcome this drawback, Quintero *et al.* first produced BGNFs using a laser spinning technology, avoiding post heat treatment or chemical additives.[71] Laser spinning involves the melting of a small volume of the supplied ceramic material at high temperatures using a high-power laser. Instantaneously, a high velocity gas jet blows the molten material. The molten ceramic is initially a hanging drop, but under the action of the gas jet it is quickly stretched and cooled. This process occurs at high rates and, as a result, the elongation of the molten material can produce nanofibers in a fraction of a second. The relative movement between the plate of supplied material and the laser beam provides a cut that continuously feeds molten material to the process. Thus, Quintero and co-workers obtained dense nanofibers with proper consistency and flexibility. The bioactivity was confirmed by apatite precipitation obtained after five days of immersion in SBF. The laser spinning technique has been demonstrated to produce fibers from a broad range of compositions.[71]

Despite the processability of simple BGNFs, the most common approach to fabricating BGNFs is to combine BGNPs and polymers through electrospinning giving composites. A series of composites with various morphologies, such as nanofibrous membranes and 3D porous scaffolds, are being developed which suggests their versatility and potential applications in bone tissue engineering.[72] Some representative examples will be addressed in the next section.

8.4 Polymer Nanocomposites

The development of composite materials is an attractive field in tissue engineering. The beneficial properties of two or more types of materials can be compounded to tailor a biomaterial to respond to the mechanical and physiological demands of the host tissue. Composite materials could be classified based on the matrix (*e.g.*, metals, ceramics and polymers) or the reinforcement dimensions (*e.g.*, particulates, short fibers, continuous fibers and nanofibers). This chapter will give relevance to synthetic and natural polymeric matrices and to particles and fibers as nano reinforcements (BGNPs and BGNFs).

In biomedical applications, namely in bone tissue engineering, the bioactivity of a biomaterial is an important feature. However, several attractive biomaterials are not bioactive *per se*.[73] A strategy to instruct the desirable bioactive behaviour is by the inclusion of nanoscale bioactive glasses in the polymer matrix (Table 8.1). The bioactivity of the resulting composite can be adjusted by the volume fraction of the nano bioactive glasses while keeping the suitable properties of the polymer, such as the capacity to deform under loads.[74,75]

BGNPs were first used as coatings for metal and ceramic implants to avoid the formation of a surrounding fibrous capsule.[76] Currently, BGNPs are being incorporated within polymeric systems, where the polymer acts as a continuous medium for their immobilization. This combination created nanocomposite materials for orthopaedic implants, bone filler materials, injectable bone materials, 3D biocompatible scaffolds, and membranes.[74,75,77–80] An overview of these nanocomposites will be emphasized within this section.

Table 8.1 Summary of polymeric nanocomposites based on nanosized bioactive glass.

Polymer	Nanosized bioactive glass	Nanocomposite aspect	Production technique	Ref.
Alginate	SiO_2–CaO–P_2O_5	Scaffold	Freeze-drying	123
Alginate/Gelatin	SiO_2–CaO–P_2O_5–Na_2O	Films	Casting	132
Chitin	SiO_2–CaO–P_2O_5	Scaffold	Freeze-drying	110, 111
Chitosan	SiO_2–CaO	Membrane	Solvent casting	77
	SiO_2–CaO–P_2O_5	Hydrogel	Mixing	80, 114
		Membrane	Stamping	117
			Solvent casting	116
		Scaffold	Freeze-drying	118, 119, 122
	SiO_2–CaO–P_2O_5–MgO	Membrane	Solvent casting	115
Collagen	SiO_2–CaO–P_2O_5	Membrane	Electrospinning	126
	SiO_2–CaO–P_2O_5–Na_2O	Films	Compression moulding	125
	Not reported	Scaffold	Freeze-drying	127
Collagen/hyaluronic Acid/ phosphatidylserine	SiO_2–CaO–P_2O_5	Scaffold	Freeze-drying	133, 134
Gelatin	SiO_2–CaO–P_2O_5	Scaffold	Freeze-drying	128, 129
Gelatin/chitosan	SiO_2–CaO–P_2O_5	Scaffold	Freeze-drying	131
Gellan gum	SiO_2–CaO	Scaffold	Freeze-drying	130
	SiO_2–CaO–P_2O_5–Na_2O	Scaffold	Freeze-drying	130
Bis-GMA/TEGDMA bisphenol-*A*-glycidyldimethacrylate	SiO_2–CaO–P_2O_5–Na_2O–Bi_2O_2	Resin	Infiltration by centrifugation	106
P(3HB) (poly(3-hydroxybutyrate))	SiO_2–CaO–P_2O_5	Scaffold	Particle leaching	98, 99
	SiO_2–CaO–P_2O_5–Na_2O	Films	Solvent casting	19, 97

Table 8.1 (*Continued*)

Polymer	Nanosized bioactive glass	Nanocomposite aspect	Production technique	Ref.
PCL (poly(ε-caprolactone))	SiO_2–CaO	Bulk	Melt Blended and injection moulding	105
	SiO_2–CaO–P_2O_5	Film	Solvent casting	72, 83
		Membrane	Electrospinning	72, 104
	SiO_2–CaO–P_2O_5–Na_2O	Scaffold	Particle leaching	99
PLGA (poly(D,L-lactide-*co*-glycolide))	SiO_2–CaO–P_2O_5	Microsphere	Emulsion solvent evaporation method	102
	SiO_2–CaO–P_2O_5–Na_2O	Fiber	Electrospinning	88
PLLA (poly(L-lactic acid))	SiO_2–CaO–P_2O_5	Fiber	Electrospinning	61
		Film	Solvent casting	78
		Scaffold	Freeze-drying	79, 85
	SiO_2–CaO–P_2O_5–Na_2O	Film	Solvent casting	16
Polyurethane/polyvinyl alcohol	SiO_2–CaO–P_2O_5	Scaffold	Freeze-drying	107
PHBV ((hydroxybutyrate-2-*co*-2-hydroxyvalerate))	SiO_2–CaO–P_2O_5	Scaffold	Particle leaching	142
Poly(vinyl pyrrolidone)/Pluronic P123	Not reported	Fibers	Electrospinning	64
Poly(caprolactone)/chitosan	Not reported	Fiber	Electrospinning	137
Poly(vinyl alcohol)/chitosan	SiO_2–CaO–P_2O_5	Scaffold	Drying	139
Poly(ethylene oxide)/chitosan	SiO_2–CaO–P_2O_5–Na_2O	Fiber	Electrospinning	141

Apart from the bioactive benefits, other general properties can be observed when nanoscale bioactive glasses are incorporated into polymer matrices.

8.4.1 Mechanical Properties

It is expected that nanoscale bioactive glasses will cause a significant improvement in the mechanical properties of composite scaffolds, namely the elastic modulus and, in the case of strong interfacial bonding, tensile compressive strength.[11,19,33,73,79,81–83] However, this feature results in a synergistic gain. In one way, the use of polymers overcomes the brittle behaviour of bulk bioactive glasses, which is a major impairment for their use in biomedical applications. In fact, the bending strength of most bioactive glasses is between 40–60 MPa, which is unsuitable for load-bearing applications[84] but in another way, their nano size enhances the reinforcement of the polymers. As previously stated, due to the greater specific surface area, the nano filler will lead to higher interface effects resulting in improved mechanical properties of the materials when compared with microstructured composites.[24] Moreover, strategies to prevent the agglomeration of the nanoscale reinforcements enables a finer homogeneous dispersion of the nano fillers in the polymer matrix, which also contributes to enhanced mechanical properties.[78,79,85]

8.4.2 Degradation Properties

The addition of nano bioactive glasses to bioresorbable polymers can alter the polymer degradation behaviour. This occurs by facilitating the rapid exchange of protons in water for alkali elements in the bioactive glass. The resulting modification of the surface and bulk properties of the composites increases the hydrophilicity and water absorption of the polymer matrix which alters the scaffold degradation kinetics.[19,86] Moreover, bioactive glasses can produce an alkaline environment.[87] Thus, when a composite incorporates nano bioactive glasses, this increase of pH can counteract the acidic degradation of some polymers, *e.g.*, polylactic acid. However, this feature can have undesirable implications. Hild *et al.* studied the viability of human oral bacteria cultured on electrospun poly(lactide-*co*-glycolide) (PLGA) and BGNPs/PLGA composites.[88] The question arises whether an antibacterial effect is obtained more from pure PLGA or BGNPs/PLGA composites, as both PLGA and BGNPs have proved to be antimicrobial. They found that the evolution of colony-forming units (CFU) of human oral bacteria related to the pH shift of the medium, induced by the scaffolds. In addition, the increase of pH caused by BGNPs in the scaffold was insufficient to show a reduction in CFU. On the contrary, pure PLGA demonstrated antibacterial properties that should be taken into consideration when designing biomaterials for dental applications. It can be concluded that even if advantageous in some cases, the buffering effect of BGNPs in this dental application prevented the disinfection. This finding showed that the biomaterial design has to be adapted to the targeted function.

8.4.3 Biological Properties

Regarding the biological properties, nano bioactive glasses can create nanostructured features in the composite (*e.g.*, nanotopography on scaffold surface), which should lead to increased interface effects, and it should contribute to improved bioactivity. Thus, incorporating nanoparticles in biopolymers can increase the surface energy of the nanocomposite, which modulates protein absorption.[89] This might improve cell attachment and subsequent cell behaviour.[73] In fact, higher surface energy enhances the interactions between an implant surface and its biological environment.[90]

Moreover, in bone applications, the insertion of nanoparticles in a polymeric matrix is thought to mimic the bone structure. Natural bone can be considered as a nanostructured composite as it contains hydroxyapatite nanocrystals combined with a polymeric phase of collagen, handling the desirable mechanical properties of bone. In this context, Webster *et al.* have reported a significant increase in protein adsorption and osteoblast adhesion on nanoscale ceramic materials compared with micron-sized ceramic materials.[91] Also, mimicking the nano features of bone on the surface of synthetic bone implants, has been shown to increase cell adhesion, proliferation, and bone formation.[23] Additionally, cells in their natural environment are surrounded by the extracellular matrix (ECM), which is also a nanostructure formed by biomolecules configured in different geometrical arrangements.[92,93] Biopolymer matrix composites containing nanoscale bioactive glasses have been discussed elsewhere, and it was shown that the use of BGNPs led to composites that mimicked more closely the structure of natural bone.[45]

8.4.4 Synthetic Polymer Nanocomposites

Synthetic polymers have been widely studied for implantable biomedical applications. However, synthetic polymers have intrinsically poor cellular affinity and low bone bioactivity (*e.g.*, poly(lactic acid), poly(lactide-*co*-glycolide) and poly(ε-caprolactone)).[94–96] Nevertheless, they have many beneficial properties, such as low rigidity, shape-forming ability, degradability, and drug-encapsulating capacity. Hence, the combination of synthetic polymers with nanoscale bioactive glasses has been proposed to produce biomaterials for tissue regeneration.

8.4.4.1 Poly(3-hydroxybutyrate)

Poly(3-hydroxybutyrate) (P(3HB)) belongs to the biodegradable polyhydroxyalkanoate family, is a biodegradable hydrophobic polyester, and has been applied in a range of biomedical applications.[86]

Misra *et al.* obtained quaternary BGNPs (SiO_2–Na_2O–CaO–P_2O_5) and conjugated them with P(3HB) to produce membranes.[19,97] They found that nanoparticles had a stiffening effect and that the composites were highly bioactive due to the distribution of a uniform apatite layer over the entire surface. They also showed that weight loss and water uptake increased with the increase of BGNPs content. In addition, they also pointed out that the systematic addition of BGNPs induced nanostructured topography on the surface of the composites. This considerably changed the morphology of the surface, and improved protein adsorption when comparing pure polymer membranes with composite membranes containing microsized bioactive glass. Furthermore, they also showed that MG-63 cells, cultured in osteogenic and non-osteogenic media, were able to attach, proliferate and differentiate, showing that the nanocomposites were cytocompatible and suitable for bone tissue engineering applications.[19,97]

Hajiali *et al.* reinforced P(3HB) with different weight ratios of BGNPs to achieve superior mechanical and physiological properties.[98,99] The nanocomposite scaffolds were prepared by a particle leaching process. The analysis showed that apatite was formed on the scaffold surfaces and that the nanocomposite exhibited high bioactivity when compared with pure P(3HB) scaffolds. The results showed a favourable interaction between the polymer matrix and the BGNPs that improved the mechanical properties, and that by increasing the BGNPs content, the tensile strength of the scaffolds improved significantly. The cell studies with MG-63 cell line demonstrated that a particular concentration of BGNPs (7.5 wt%) could improve cell attachment, proliferation and osteoconductivity, when compared with pure P(3HB) scaffolds.[98,99]

8.4.4.2 Poly(lactic acid)

Bioactive glasses and poly(lactic acid) (PLA) composites are known for their high osteoconductivity, osteoinductivity, and biodegradability. Moreover, these composites have attracted interest for load-bearing applications because they combine bioactive behaviour with mechanical strength and processability.[79,100]

Gerhardt *et al.* conjugated BGNPs and PLA to fabricate both composite membranes by solvent casting, and scaffolds by particulate leaching techniques. The nanocomposite scaffolds showed compressive strength values from 0.4 to 1.6 MPa that were in the range of the values of trabecular bone.[16,101] On the composite membranes containing 20 wt% of BGNPs, human fibroblasts produced five times higher vascular endothelial growth factor (VEGF) than on pure PLA membranes. Furthermore, after eight weeks of implantation in rats, the nanocomposite scaffolds exhibited pro-angiogenic properties. They were infiltrated with newly-formed tissue and demonstrated a higher percentage of blood vessel to tissue (11.6–15.1%) in comparison with neat PLA scaffolds (8.5%).[16]

Composite scaffolds were also prepared by a thermal-induced phase separation method.[79] *In vitro* tests revealed that these composites were bioactive. They found that increasing the amount of BGNPs in the PLA matrix altered the morphology and porosity of the scaffolds. The addition of

BGNPs to the PLA matrix induced an increase of both compressive modulus and strength of the scaffolds. Furthermore, the authors found that BGNPs increased the water uptake of the scaffolds, especially at a low BGNP loading, and significantly affected the degradation rate of the PLA matrix.[79]

An attractive strategy to ameliorate the interaction of BGNPs with the polymer was achieved by Liu *et al.* These authors functionalized BGNPs by grafting low molecular weight PLA onto their surfaces, and combined them with a PLA matrix. The resulting nanocomposite scaffolds were produced *via* a particle-leaching technique.[78] Therefore, the grafting modification induced an increase in both tensile strength and tensile modulus of the composites by improving the interphase compatibility. The *in vitro* bioactivity tests comparing pure PLA scaffolds and BGNPs/PLA nanocomposite demonstrated that the nanocomposite had a higher capacity to induce the formation of an apatite surface layer. Furthermore, cell culture studies revealed that BGNPs enhanced the attachment and proliferation of bone stromal cells onto the surface of the scaffolds. This strategy can thus give rise to composites with stronger structures and shows the potential of BGNPs for surface functionalization.[78]

Moreover, a nanofiber composite of BGNFs in PLA matrices was developed by Kim *et al.*[61] The nanocomposites showed a morphology of uniformly dispersed nanofibers within the PLA matrix. The nanocomposites demonstrated excellent bioactivity, inducing apatite precipitation within 24 h of immersion in SBF. Also, as the amount of BGNFs increased, the *in vitro* bioactivity was enhanced. Furthermore, osteoblasts attached and grew well on nanocomposites, and secreted collagen in initial culturing periods. The nanocomposites significantly improved the differentiation of cells when compared to pure PLA. Moreover, the mineralization produced by the cells was higher on the nanocomposites. Such studies demonstrate again that the development of a nanocomposite constituted of bioactive nanofibers and a degradable polymer is a promising strategy for bone regeneration.[61]

*8.4.4.3 Poly(lactide-*co*-glycolide)*

Biomaterials made of biodegradable poly(lactide-*co*-glycolide) (PLGA) are known to decrease the pH in the vicinity of the implants. Bioactive glass is being investigated as a counteracting agent, buffering the acidic degradation products.

Microsphere nanocomposites made of BGNPs and lysozyme/PLGA were obtained by an emulsion solvent evaporation method.[102] The SEM results showed that the BGNPs distributed uniformly in the PLGA microspheres. The lysozyme exhibited a release up to one month, and more than 90% of the released lysozyme was preserved in a bioactive form. Liu *et al.* also showed that the microspheres incorporated with BGNPs provided permanent protein stability against the acidic environment resulting from PLGA degradation. This nanocomposite showed potential for sustained protein release.[102]

8.4.4.4 Poly(ε-caprolactone)

Poly(ε-caprolactone) (PCL) is an FDA-approved biomaterial with good processability, toughness, flexibility, and biocompatibility. However, PCL has limitations in regenerative applications because of its low stiffness, hydrophobic nature, and lack of bioactivity.[103]

Nanocomposite thin membranes were produced by blending of PCL and BGNFs.[104] The BGNFs distributed well within the PCL matrix, conferring roughness to the membrane surface. *In vitro* SBF studies showed a layer of apatite precipitate that almost covers the surface of the nanocomposite membranes. Osteoblastic cells (MC3T3) spread and grew actively, with many cytoplasmic extensions, and showed improved proliferation behaviour on the nanocomposites compared to the pure PCL membrane.[104] Ji *et al.* also inserted BGNPs in PCL to prepare a composite by a melt-blended method.[105] They homogeneously dispersed the BGNFs in the PCL matrix up to a concentration of 40 wt%. The elastic modulus of the nanocomposites improved remarkably from 198 ± 13 MPa to 851 ± 43 MPa. Meanwhile, the tensile strength was retained in the range 19–21.5 MPa. The addition of BGNFs also improved the hydrophilicity and degradation behaviour of the nanocomposites. Moreover, the nanocomposites showed *in vitro* bioactivity after being immersed in SBF. These composites, with excellent mechanical properties and bioactivity, could be potential candidates for bone tissue regeneration and remodelling.[105] In a similar approach, the biocompatibility and mechanical properties of PCL/BGNFs composites were compared with composites containing bioactive glass microparticulates.[72] BGNFs were uniformly distributed in the PCL matrix because of their similar shape and size, in contrast to the bioactive glass microparticles. Furthermore, *in vivo* test results revealed the suitable biocompatibility of the composite and demonstrated its bone-forming capability when implanted in a calvarial bone defect.[72]

8.4.4.5 Dental Resins

Bioactive nanocomposites made of commercial light-curable dental resins were also studied. Tauböck *et al.* introduced bismuth-doped BGNPs into a matrix made of bisphenol-A-glycidyl dimethacrylate (Bis-GMA) and triethylene glycol dimethacrylate (TEGDMA).[106] This resin is a commercial dentin bonding agent under the trademark Heliobond®. Despite being a widely used material in restorative dentistry, it lacks bioactive features. The BGNPs were doped with bismuth oxide to render the final composite visible on X-ray. The results showed that the viscosity of the dental resin increased with nanoparticle loading but remained suitable for a flowable dental material. Upon nanocomposite polymerisation, the nanoparticles increased the water uptake and pH values. However, the addition of nanoparticles up to 20 wt% had no significant influence on the microhardness. Apatite could precipitate on the nanocomposites, but not on unfilled resin surfaces. This study showed a direct application of the benefits of BGNPs in clinical practice.[106]

8.4.4.6 Polyurethane/Polyvinyl Alcohol

Bioactive nanocomposite films and foams were obtained by incorporating BGNPs in polyurethane/polyvinyl alcohol (PU/PVA) blends by drying or freeze-drying.[107] The films presented a deformation of 350%, and the foams exhibited pore structure and mechanical properties adequate to support cell growth and proliferation. The incorporation of 10% BGNPs in the PU/PVA foam reinforced the structure, improving the compressive strength and the elastic modulus of the materials. Additionally, to assess the bioactivity of the composites, an SBF immersion test revealed that apatite formed on the nanocomposites. To confirm the biocompatibility of the BGNPs/PU/PVA composites, cells were put in contact with the decomposition products. After 24 h, an acceptable percentage of cells remained viable. These nanocomposites could be easily shaped during surgical implantation, being suitable for uses as flexible scaffolds in several tissue engineering applications.[107]

8.4.4.7 Poly(vinyl pyrrolidone)/Pluronic P123

Xia *et al.* produced composite nanofibers by electrospinning of bioactive glass sol–gel precursors in the presence of poly(vinyl pyrrolidone) (PVP) and pluronic P123.[64] They showed that the average diameter of these BGNFs composites could be controlled in the range from 85 to 400 nm. They concluded that the addition of PVP resulted in sufficient chain entanglement and smoothed bioactive glass composite nanofibers. Moreover, the pluronic led to a further decrease of the diameter and held the balance between the electrostatic repulsive force and surface tension of the electrospinning solution. The biomineralization study indicated that composites had an excellent ability to induce apatite deposition. They found that the apatite deposited first at the cross points of the fibers and gradually grew to form globes, which connected with each other after prolonged SBF soaking time.[64]

8.4.5 Natural Polymer Nanocomposites

Besides the synthetic polymers discussed above, natural polymers can also be used as matrices to prepare nanocomposites. Naturally-derived polymers have been widely proposed for tissue regeneration as they are similar to macromolecules present in the biological environment and produce degradation products that are recognizable and metabolically processed by the body.[108]

8.4.5.1 Chitin

Chitin is a non-toxic, biodegradable polymer that has been studied for a variety of biomedical applications. Chitin is an appealing material because it is structurally similar to the extracellular matrix of many tissues and can be easily processed.[109]

Nanocomposite scaffolds that combined chitin with BGNPs have been prepared by a freeze-drying technique.[110,111] Chitin was first added to a saturated mixture of $CaCl_2 \cdot 2H_2O$/methanol and stirred until a transparent solution was obtained. Then the solution was filtered to remove undissolved chitin and dialysed to remove the calcium ions. The chitin hydrogel was mixed with BGNPs and freeze-dried to get the composite scaffolds. The resulting composite scaffolds had enhanced porosity and the BGNPs were distributed homogeneously on the porous walls. Also, the scaffolds were found to have improved swelling, bioactivity and degradation compared to the control scaffolds without BGNPs. The composite scaffolds were non-toxic to MG-63 and human primary osteoblast (hPOB) cells and supported cell attachment, spreading and proliferation. Mineralized bone nodules and calcium deposits were also formed, which proved that hPOB cells could mineralize in these scaffolds. The maturation and mineralization of the hPOB cells in the absence of osteogenic supplements proved the beneficial properties of the nanocomposite.[111] These nanocomposite scaffolds could be promising candidates for tissue engineering applications.

8.4.5.2 Chitosan

Chitosan (CHT) is a natural polysaccharide obtained from the deacetylation of chitin and is one of the most widely used biopolymers for biomedical research.[112,113] The possibility of dissolving chitosan in acidic aqueous solutions offers high processability potential as compared to chitin. Chitosan is considered a functional material because of its high biocompatibility, biodegradability to harmless products and non-antigenicity.[108] However, continuous efforts are being made to improve the bioactivity and mechanical properties of CHT.

Nanocomposite hydrogels that combined CHT and BGNPs are also being addressed. Oliveira *et al.* developed a combinatorial study of this system.[114] They tested up to 30 conditions to identify the most predominant combination that affects the pre-osteoblasts' response in the composites. The addition of BGNPs in amounts higher than 12.5% (w/w of CHT) led to a significant increase in the storage modulus of the hydrogels. The authors showed that the MC3T3 pre-osteoblast cell proliferation and spreading were favoured with this amount of BGNPs. The elastic modulus measured for this formulation meets the values reported for the granulation tissue occurring during bone regeneration, where fibroblasts produce collagen.[114] Moreover, Couto *et al.* combined a CHT–glycerophosphate salt formulation with BGNPs to prepare bioactive thermo-responsive composite hydrogels (Figure 8.3A).[80] The system gelled at around 36.8 °C, being suitable for an intracorporeal application. The nano size and spherical shape of the particles guaranteed the efficient injection of the composite into bone defects through small-gauge needles. The composites were bioactive as they induced the formation of apatite deposits and their density augmented with the increase of BGNPs content. Such systems could thus have potential to be applied as thermo-responsive hydrogels for orthopaedic reconstruction.

Composite membranes have also been developed for guided tissue regeneration. Membranes are used as barriers to protect the faster-growing soft tissue cells from the defect site and to regenerate periodontal ligament, cementum, and bone (Figure 8.3B).[77,115,116] CHT was blended with different formulations of BGNPs, namely ternary nanoparticles and magnesium-doped nanoparticles.[77,115] The composite membranes can be easily

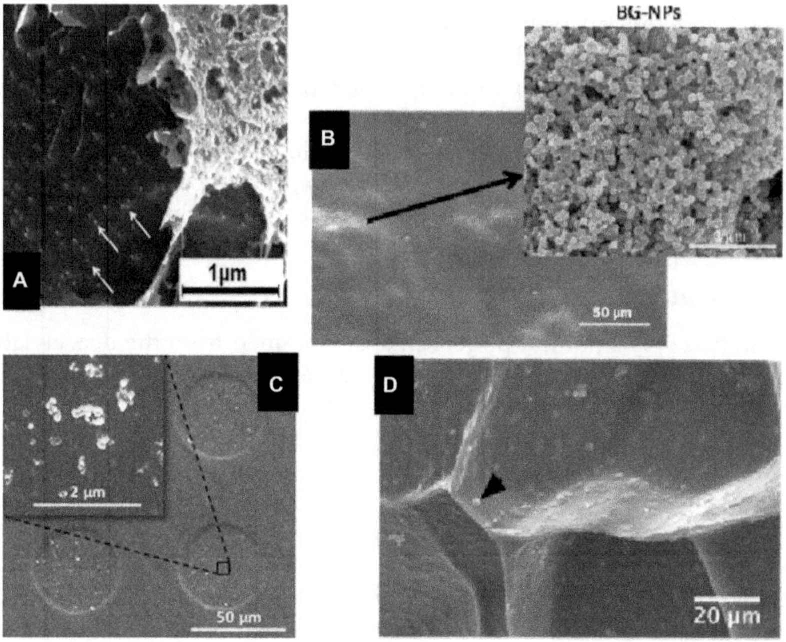

Figure 8.3 Different BGNPs/CHT nanocomposites. A: CryoSEM image of the surface of a nanocomposite hydrogel, where the arrows indicate small agglomerates of BGNPs deposited on the hydrogel walls; in the upper right-hand side there is part of a cluster of nanoparticles involved with chitosan. B: SEM images of CHT/BGNPs nanocomposite membranes, the inset shows the morphology of the BGNPs. C: SEM image of patterned nanocomposite membrane. D: SEM image of nanocomposite scaffolds, the arrow indicates a cluster of BGNPs. [A] Reprinted from D. S. Couto *et al.*, Development of bioactive and biodegradable chitosan-based injectable systems containing bioactive glass nanoparticles, *Acta Biomater.*, **5**, 115–123, Copyright 2009 with permission from Elsevier,[80] [B] Reproduced with permission from G. M. Luz, L. Boesel, A. del Campo and J. F. Mano, Micropatterning of Bioactive Glass Nanoparticles on Chitosan Membranes for Spatial Controlled Biomineralization, *Langmuir*, 2012, **28**,[17] 6970. Copyright 2012 American Chemical Society,[117] [C] Reprinted from J. Mota *et al.*, Chitosan/bioactive glass nanoparticle composite membranes for periodontal regeneration, *Acta Biomater.*, **8**, 4173–4180, Copyright 2015 with permission from Elsevier,[116] [D] Reprinted from A. J. Leite *et al.*, Synthesis and characterization of bioactive biodegradable chitosan composite spheres with shape memory capability, *J. Non-Cryst. Solids*, **432**, 158–166, Copyright 2016 with permission from Elsevier.[119]

produced by solvent casting. The nanoparticles had a homogeneous dispersion in the polymer. The membranes prepared with the BGNPs formulations presented an increased stiffness, showed adequate extensibility in wet conditions and early degradation was absent. Moreover, it was found that the magnesium-doped BGNPs increased the hydrophilicity of the composites when compared with composites loaded with non-doped BGNPs, showing that the properties of the nanocomposite could be tailored by adjusting the BGNPs formulation. Upon immersion in SBF the composite membranes induced apatite deposition when compared with pure CHT membranes. Additionally, Caridade *et al.* followed the biomineralization of CHT/BGNPs composite membranes by dynamical mechanical analyses (DMA).[24] The nanostructured membranes showed an increase of the Young's modulus associated with the onset of the apatite formation. Thus, the membranes had enhanced performance in the mechanical properties and higher bioactivity than both bare CHT membranes and CHT membranes containing micro bioactive glass particles.[24] This evidenced the bioactive improvement by the addition of BGNPs in the chitosan membrane. Studies with osteoblast-like cells (SaOs-2) showed that the nanocomposite membranes were non-cytotoxic. Moreover, the composite with magnesium-doped BGNPs stimulated a more appropriated osteoblastic response toward cellular differentiation and mineralization. It was also noticed that supplementing with BGNPs enhanced proliferation and the metabolic activity of human periodontal ligament cells (hPDLCs) and human bone marrow stromal cells (hBMSCs). Furthermore, BGNPs promoted greater cell matrix mineralization by both types of cells.[77,115,116] Still regarding membranes as applications, Luz *et al.* used an innovative approach to fabricate BGNPs micropatterned chitosan membranes.[117] The authors used soft lithography to imprint highly localized bioactive motifs onto CHT freestanding membranes (Figure 8.3C). The bioactive character of the BGNPs spots allowed a highly restricted nucleation and growth of apatite. Results showed that fibroblast L929 cells replicated the initial bioactive pattern, preferring the environment created by the BGNPs rather than migrating to the exposed CHT regions. With this approach, it was proved that it is possible to spatially control the bioactive behaviour and cellular interactions in a polymeric matrix. This approach could potentially be used for skin, vascular, articular, and bone tissue engineering applications. Additionally, this system can also be used under co-culture systems or to develop substrates that are able to confine cells in specific regions.[117]

Besides hydrogels and membranes, CHT nanocomposite scaffolds containing BGNPs have also been synthesized by freeze-drying techniques.[118] Such composite scaffolds have demonstrated adequate swelling and degradation properties. *In vitro* studies showed apatite deposition on the surface of the nanocomposite scaffolds, indicating their bioactive ability. The investigation of the behaviour of MG-63 cells indicated that cells attached to the pore walls of the scaffolds and showed initial signs of spreading. Another interesting feature of CHT scaffolds is their shape memory ability. Shape memory polymers can be deformed and fixed into a temporary shape, which remains stable unless it is exposed to an external stimulus that triggers the polymer to recover its original form. Therefore, nanocomposite scaffolds and spheres were produced to combine the shape memory properties of CHT and the biomineralization ability of BGNPs for applications in bone regeneration (Figure 8.3C).[119,120] The addition of BGNPs improved the bioactivity of the nanocomposite scaffold as seen by the precipitation of the apatite layer upon immersion in SBF, which evidenced their osteoconductive potential. Furthermore, the introduction of BGNPs in the chitosan matrix increased the stiffness of the constructs. CHT/BGNPs exhibit better shape memory properties than CHT, and the composite scaffolds presented good shape memory properties characterized by high recovery ratios and high fixity ratio. The applicability of such structures was demonstrated by a suitable geometrical accommodation of a previously compressed nanocomposite in a bone defect. The intrinsic shape memory behaviour showed the potential of these nanocomposites for bone tissue engineering therapies in particular as a device to be implanted using minimally invasive procedures.[119,120]

8.4.5.3 Alginate

Alginate is a natural polysaccharide extracted from brown seaweeds. In the presence of divalent cations, alginate can form stable hydrogels. It is highly hydrophilic, biocompatible, relatively economical and widely utilized in the food and pharmaceutical industries.[121,122] Srinivasan *et al.* evaluated the influence of adding BGNPs to alginate for periodontal tissue regeneration.[123] For this purpose, they fabricated composite scaffolds by freeze-drying. The results indicated reduced swelling, which might be due to the strong interaction between alginate and BGNPs and to the reduction in the pore size. Moreover, the detected enhanced biomineralization and protein adsorption could be due to the exposed BGNPs on the scaffold surfaces that increases the binding sites on the material surface. In addition, BGNPs helped the attachment and proliferation of human periodontal ligament fibroblast (hPDLF) and MG-63 cells. This effect was attributed to the increase in surface area and surface roughness due to the incorporation of BGNPs. The presence of BGNPs enhanced ALP activity of hPDLF cells with a maximal activity at seven days due to the release of ions and dissolution products from the bioactive glass followed by a decrease in this expression after that. This decrease after seven days could be attributed to the completion of osteoblastic differentiation.[123] Thus, results suggest that these biocompatible composite scaffolds could be useful for periodontal tissue regeneration.

8.4.5.4 Collagen

The use of collagen-based biomaterials in tissue engineering and regenerative medicine applications has been growing intensively over the past decades. Collagen plays a significant role in the formation of tissues and organs and co-operates in various functional expressions of cells. Particularly, type I collagen is the major organic constituent of the bone matrix (approximately 35 wt%) and one of the most used biomaterials.[124]

The angiogenic potential of collagen nanocomposites containing BGNPs using the quail chorioallantoic membrane (CAM) was evaluated as an alternative to the traditional mammalian models of angiogenesis.[125] Interestingly, 24 h post-implantation, collagen films with 10 wt% of BGNPs stimulated angiogenesis by an increase of 40% in the number of blood vessel branch points. In contrast, composite films containing 20 wt% BGNPs were found to inhibit angiogenesis. This experimental study evidenced that the addition of a particular concentration of BGNPs to collagen films induces an early angiogenic response.[125] Kim *et al.* also developed BGNF–collagen nanocomposite both as a thin membrane and as a scaffold.[126] SEM revealed a similar composite microstructure on both membranes and porous scaffolds with uniformly distributed BGNF in the collagen matrix. Those authors also observed that the nanocomposites assisted the adhesion and growth of human osteoblast-like cells.[126]

Collagen–BGNFs composite scaffolds were also fabricated by a gelation and freeze-drying process, and their bioactive properties were evaluated.[127] The composite scaffolds rapidly formed a layer of apatite minerals on their surfaces when incubated in SBF, showing excellent bioactivity *in vitro*. The MG-63 cells on the collagen–BGNFs scaffolds expressed significantly higher ALP activity, bone sialoprotein (BSP) and osteocalcin (OCN) levels than on collagen alone. The composite scaffolds provided a suitable environment for attachment to the cytoskeleton and promoted the differentiation and mineralization of MG-63 osteoblast-like cells. Moreover, the bone regeneration ability of the composite scaffold was examined using a rat calvarial defect model *in vivo*. The results revealed that the composite is biodegradable, and the bone regeneration ability *in vivo* of composite scaffolds was better than on the pure collagen formulation.[127]

8.4.5.5 Gelatin

Gelatin is a denatured protein obtained by hydrolysis of animal collagen and has a range of uses in the pharmaceutical, food, and cosmetic industries. It also decreases the concerns of immunogenicity and pathogen transmission associated with collagen. Gelatin was used as a polymer to increase the flexibility of the biomaterials but it has poor mechanical properties.

Considering the advantages of gelatin and bioactive glasses, Koudehi *et al.* designed a BGNPs–gelatin conduit for peripheral nerve regeneration.[128] The guidance channel was examined in the right sciatic nerve of a male Wistar rat. Twenty rats were randomly divided into two experimental groups, one with the nanocomposite and the other consisting of normal rats. The results indicated that after three months, nerve regeneration of the nanocomposite group was statistically equivalent to the standard group. They suggested that the nanocomposite could be a suitable candidate for peripheral nerve repair as a biocompatible, biodegradable and novel biomaterial. The bone regeneration capacity of BGNPs–gelatin scaffold was also evaluated in a bone critical-sized defect.[129] The radiographic evaluation indicated that the nanocomposite scaffolds could successfully bridge the critical-sized defect. X-ray analysis also proposed that bioactive glass scaffolds supported healthy bone formation *via* intramembranous formation. Such a study is representative of the development of nanocomposite scaffolds that contributed significantly to growth and healing of bone.[129]

8.4.5.6 Gellan Gum

Composites were prepared to verify the possibility to improve the mechanical properties and the bioactivity of gellan gum using BGNPs.[130] Although the BGNPs were agglomerated within the gellan gum matrix, they contributed significantly to an improved microstructure of the original gellan gum hydrogels as they increased the roughness of the pore walls, which can be assumed to be beneficial for cell attachment. Despite the Young's modulus and the point of failure appearing at much higher loads when gellan gum was reinforced, their mechanical properties were below the desired value that would allow load-bearing applications and also lower than the highest values achieved reported for hydrophobic biopolymers reinforced with BGNPs. A further improvement of the mechanical properties could be obtained by a better dispersion of the nanoparticles in the matrix. Nevertheless, by incorporating the bioactive glass particles, the composite material acquired the ability to form an apatite layer when soaked in SBF. Furthermore, human adipose-derived stem cells remained viable and could adhere and spread within the fabricated spongy-like nanocomposite, which is an important result considering their use in bone-tissue engineering. Thus, hydrogels based on gellan gum and bioactive glass could be promising biomaterials to be used in the orthopaedic field.[130]

8.4.5.7 Gelatin/Chitosan

Peter *et al.* introduced ternary BGNPs into gelatin–CHT matrix, which upon freeze-drying resulted in porous composite scaffolds.[131] They showed that the degradation and swelling behaviour of the scaffolds decreased, whereas protein adsorption increased with the addition of BGNPs. Biomineralization studies showed a higher amount of mineral deposits on the nanocomposite scaffold, which increases with the incubation time in SBF. Cell studies with MG-63 cells indicated that the composite scaffolds with BGNPs provided a more suitable environment for cell attachment and spreading. Therefore, the developed nanocomposite scaffold is a potential candidate for periodontal/alveolar bone tissue applications.[131]

8.4.5.8 Alginate/Gelatin

The combination of alginate dialdehyde and gelatin (ADA–GEL) intends to overcome the limitations of the alginate, such as low viscosity and dissolution in physiological media. Most relevant, alginate discourages cell interactions that are crucial features in tissue engineering. Thus, ADA–GEL has been shown to promote cell adhesion and proliferation. Rottensteiner *et al.* combined ADA–GEL with quaternary BGNPs to produce hydrogel films by solvent casting.[132] *In vivo* implantation did not show a significant immune reaction and revealed an apparent degradation of the films after four weeks, which would ideally match the beginning of bone formation after this time. Furthermore, ongoing vascularization could be seen after four weeks. Despite this, no difference was reported between the composite and the pure ADA–GEL. This result reveals that the addition BGNPs might contribute to the bioactivity of the gel whereas the benefits of ADA–GEL were maintained.[132]

8.4.5.9 Collagen, Hyaluronic Acid and Phosphatidylserine

Wang *et al.* developed a new porous bioactive nanocomposite composed of sol–gel-derived bioactive glass nanoparticles, collagen, hyaluronic acid, and phosphatidylserine (BGNPs–COL–HYA–PS) by freeze-drying.[133] They also synthesized a cross-linked bioactive nanocomposite using 1-ethyl-3-(3-dimethylaminopropyl) carbodiimide and *N*-hydroxysuccinimide (EDC–NHS cross-linked BGNPs–COL–HYA–PS). It was reported that biomineralization, degradation, and mechanical strength of the cross-linked composite scaffolds were superior in comparison with scaffolds without the cross-linking process. Further *in vitro* cell culture studies demonstrated that MC3T3 cells attached and spread on the surface of cross-linked BGNPs–COL–HYA–PS scaffolds indicating the biocompatibility of the nanocomposite.[133] Following this work, Xie *et al.* have investigated the *in vivo* bone regeneration ability of the EDC–NHS cross-linked BGNPs–COL–HYA–PS composite scaffolds using a rabbit radius defect model.[134] X-ray and histological studies showed the ability of bone regeneration for plain nanocomposites and nanocomposites combined with growth factors (bone morphogenetic protein, BMP). The bone defect was covered with new bone only in the nanocomposites grafted with BMP at eight weeks. Moreover, the nanocomposite combined with BMP showed ameliorated ectopic bone formation compared with the composites without BMP.[134]

8.4.6 Blends of Synthetic and Natural Polymeric Nanocomposites

The blending of synthetic and natural polymers can result in the development of new polymeric material that retains the mechanical strength, processability and durability of the synthetic component and the biocompatibility and biological functionality of a natural polymer.[135,136]

8.4.6.1 Poly(caprolactone)/Chitosan

Shalumon *et al.* carried out a comparative study between nano hydroxyapatite (nHAp) incorporated poly(caprolactone)–chitosan (PCL–CHT) and

BGNPs incorporated PCL–CHT nanofibrous scaffolds by electrospinning.[137] Protein adsorption studies showed that PCL–CHT incorporated nHAp and BGNPs adsorbed more proteins on their surface than other samples. Cell attachment and proliferation studies using human periodontal ligament fibroblast (hPLFs) and MG-63 cells showed that BGNPs incorporated samples are slightly superior compared to nHAp incorporated scaffolds. Cell viability tests confirmed that the scaffolds are cytocompatible. ALP activity confirmed the osteoblastic characteristics of hPDLFs. Also, the presence of nHAp and BGNPs enhanced the ALP activity of hPDLF. These studies indicate that BGNPs incorporated electrospun scaffolds were good candidates for orthopaedic and periodontal tissue engineering applications.[137]

8.4.6.2 Poly(vinyl alcohol)/Chitosan

Poly(vinyl alcohol) (PVA) is a water-soluble polymer and has frequently been explored as an implant material in biomedical applications because of its excellent mechanical strength, biocompatibility and non-toxicity.[138–140] However, PVA has some reduced integration to living tissues due to its limited biodegradability and bioactivity. Therefore, it is promising to blend PVA with biopolymers such as chitosan (CHT) to overcome the limited biological performance of synthetic polymers and to enhance the mechanical characteristics of biopolymers. Mansur *et al.* reported the development and characterization of macroporous scaffolds of PVA–CHT–BGNPs.[139] The polymeric blends were synthesized by blending PVA with CHT by foaming a mixture of polymer solution and then reacting with bioactive glass reagents using the sol–gel route. The developed nanocomposites have presented suitable mechanical, morphological and cell viability properties for potential biomedical applications.[139]

8.4.6.3 Poly(ethylene oxide)/Chitosan

Talebian *et al.* employed the electrospinning technique to fabricate a nanofibrous nanocomposite membrane from CHT–PEO solution incorporating BGNPs particles.[141] They investigated various properties of the nanocomposite membrane including mechanical properties, wettability, and biomineralization. The measurements of tensile strength and water contact angles showed that the incorporation of BGNPs into the nanofibers improves the mechanical properties and hydrophilicity of the scaffolds. The results evidenced that BGNPs containing nanofibers could induce apatite formation on the surface of the composite after 14 days immersion in SBF. The cell adhesion results showed that hMSCs were viable on the CHT–PEO–BGNPs nanofiber scaffolds. In addition, BGNPs enhanced the ALP activity of hMSCs cultured on composite scaffolds compared with that on pure CHT–PEO scaffolds. Their results suggest that a CHT–PEO–BGNPs nanofibrous composite could be a potential candidate for application in tissue engineering due to its improved mechanical and biological properties as compared with pure CHT–PEO nanofibers.[141]

8.5 Conclusions and Outlook

The combination of tissue engineering and materials science/technology has created new high performance biomaterials for tissue regeneration.

Bioactive glasses, in the form of nanoparticles, nanofibers or nanocomposites, have undergone significant progress, in particular in their synthesis, processing and structural characterization, in an attempt to understand their interactions with cells and their applications in tissue engineering. Moreover, they have been found to enhance and direct the proliferation and differentiation of odontoblasts, cementoblasts, osteoblasts and fibroblasts.

Although a large number of studies have been made to regenerate tissues using nanoscale bioactive glasses, these efforts are still mainly limited to laboratory investigations for *in vitro* and *in vivo* studies. Moreover, the conjugation of nanostructured biomaterials with stem cells and growth factors to achieve the desired goal of tissue regeneration is showing promising results, but more research is needed, especially in the area of soft tissue reconstruction.

Significant progress is also being made in the introduction of nanoscale bioactive glass as fillers in polymers, to use the different materials' properties synergistically and facilitate tissue regeneration. A current area of extensive research is the generation of nano size features in biomaterials to mimic certain extracellular matrix structures to enhance the biological function of the existing polymeric biomaterials. Systems that integrate macroscale shapes, microscale pores, and nanoscale architecture have been rapidly designed to optimize biomaterials function. However, more research is required to exploit the novel properties of these composites, in different morphologies, for a variety of requests for hard and soft tissue regeneration. Promising research areas are the incorporation of biomolecules, such as growth factors, and the manipulation of the geometrical arrangement of BGNPs and BGNFs within the polymeric matrix. This would allow for better control and tailoring of the mechanical and biological properties of the nanocomposite materials.

Further progress in the development of new nanoscale bioactive glasses and their clever combination with polymers and bioactive molecules has the potential to produce a biomimetic material that fulfils the demands of hard and soft tissue engineering applications.

Acknowledgements

Álvaro J. Leite acknowledges the Portuguese Foundation for Science and Technology (FCT) for his doctoral grant (SFRH/BD/73174/2010).

References

1. J. R. Jones, Review of bioactive glass: From Hench to hybrids, *Acta Biomater.*, 2013, **9**(1), 4457–4486.

2. L. L. Hench, Bioceramics – from Concept to Clinic, *Am. Ceram. Soc. Bull.*, 1993, **72**(4), 93–98.
3. T. Kokubo, Bioactive glass ceramics: properties and applications, *Biomaterials*, 1991, **12**(2), 155–163.
4. T. Kokubo and H. Takadama, How useful is SBF in predicting *in vivo* bone bioactivity? *Biomaterials*, 2006, **27**(15), 2907–2915.
5. C. Mao, X. Chen, G. Miao and C. Lin, Angiogenesis stimulated by novel nanoscale bioactive glasses, *Biomed. Mater.*, 2015, **10**(2), 025005.
6. A. A. Gorustovich, J. A. Roether and A. R. Boccaccini, Effect of Bioactive Glasses on Angiogenesis: A Review of *In Vitro* and *In Vivo* Evidences, *Tissue Eng. Part B*, 2009, **16**(2), 199–207.
7. Z. K. Hong, G. M. Luz, P. J. Hampel, M. S. Jin, A. X. Liu, X. S. Chen *et al.*, Mono-dispersed bioactive glass nanospheres: Preparation and effects on biomechanics of mammalian cells, *J. Biomed. Mater. Res., Part A*, 2010, **95A**(3), 747–754.
8. S. M. Amit, S. Xinfeng and G. M. Antonios, Nanocomposite Scaffolds for Tissue Engineering, *Tissue Engineering and Artificial Organs*, Electrical Engineering Handbook, CRC Press, 2006, p. 1–11.
9. E. Engel, A. Michiardi, M. Navarro, D. Lacroix and J. A. Planell, Nanotechnology in regenerative medicine: the materials side, *Trends Biotechnol.*, 2008, **26**(1), 39–47.
10. C. T. Laurencin, S. G. Kumbar and S. P. Nukavarapu, Nanotechnology and orthopedics: a personal perspective, *Wiley Interdiscip. Rev.: Nanomed. Nanobiotechnol.*, 2009, **1**(1), 6–10.
11. G. Wei and P. X. Ma, Nanostructured Biomaterials for Regeneration, *Adv. Funct. Mater.*, 2008, **18**(22), 3566–3582.
12. N. M. Alves, I. Pashkuleva, R. L. Reis and J. F. Mano, Controlling Cell Behavior Through the Design of Polymer Surfaces, *Small*, 2010, **6**(20), 2208–2220.
13. L. Zhang and T. J. Webster, Nanotechnology and nanomaterials: Promises for improved tissue regeneration, *Nano Today*, 2009, **4**(1), 66–80.
14. T. J. Webster, C. Ergun, R. H. Doremus, R. W. Siegel and R. Bizios, Enhanced functions of osteoblasts on nanophase ceramics, *Biomaterials*, 2000, **21**(17), 1803–1810.
15. K. von der Mark, J. Park, S. Bauer and P. Schmuki, Nanoscale engineering of biomimetic surfaces: cues from the extracellular matrix, *Cell Tissue Res.*, 2010, **339**(1), 131–153.
16. L. C. Gerhardt, K. L. Widdows, M. M. Erol, C. W. Burch, J. A. Sanz-Herrera, I. Ochoa *et al.*, The pro-angiogenic properties of multi-functional bioactive glass composite scaffolds, *Biomaterials*, 2011, **32**(17), 4096–4108.
17. M. Mackovic, A. Hoppe, R. Detsch, D. Mohn, W. J. Stark, E. Spiecker *et al.*, Bioactive glass (type 45S5) nanoparticles: *in vitro* reactivity on nanoscale and biocompatibility, *J. Nanopart. Res.*, 2012, **14**(7), 1–22.

18. A. A. Rocha de Oliveira, D. A. de Souza, L. L. Silveira Dias, S. M. de Carvalho, H. S. Mansur and MM Pereira, Synthesis, characterization and cytocompatibility of spherical bioactive glass nanoparticles for potential hard tissue engineering applications, *Biomed. Mater*, 2013, **8**(2), 025011.
19. S. K. Misra, D. Mohn, T. J. Brunner, W. J. Stark, S. E. Philip, I. Roy *et al.*, Comparison of nanoscale and microscale bioactive glass on the properties of P(3HB)/Bioglass® composites, *Biomaterials*, 2008, **29**(12), 1750–1761.
20. S. M. Carvalho, A. A. R. Oliveira, C. A. Jardim, C. B. S. Melo, D. A. Gomes, M. de Fátima Leite *et al.*, Characterization and induction of cementoblast cell proliferation by bioactive glass nanoparticles, *J. Tissue Eng. Regener. Med.*, 2012, **6**(10), 813–821.
21. M. Vollenweider, T. J. Brunner, S. Knecht, R. N. Grass, M. Zehnder, T. Imfeld *et al.*, Remineralization of human dentin using ultrafine bioactive glass particles, *Acta Biomater.*, 2007, **3**(6), 936–943.
22. G. Mendonça, D. B. S. Mendonça, F. J. L. Aragão and L. F. Cooper, Advancing dental implant surface technology – From micron- to nanotopography, *Biomaterials*, 2008, **29**(28), 3822–3835.
23. E. Palin, H. N. Liu and T. J. Webster, Mimicking the nanofeatures of bone increases bone-forming cell adhesion and proliferation, *Nanotechnology*, 2005, **16**(9), 1828–1835.
24. S. G. Caridade, E. G. Merino, N. M. Alves, Z. Bermudez Vde, A. R. Boccaccini and J. F. Mano, Chitosan membranes containing micro or nano-size bioactive glass particles: evolution of biomineralization followed by *in situ* dynamic mechanical analysis, *J.Mech. Behav. Biomed. Mater*, 2013, **20**, 173–183.
25. S. Liu, W. Gong, Y. Dong, Q. Hu, X. Chen and X. Gao, The effect of submicron bioactive glass particles on *in vitro* osteogenesis, *RSC Adv.*, 2015, **5**(49), 38830–38836.
26. W. Y. Gong, Y. M. Dong, X. F. Chen and B. Karabucak, Nano-sized 58S bioactive glass enhances proliferation and osteogenic genes expression of osteoblast-like cells, *Chin. J. Dent. Res.*, 2012, **15**(2), 145–152.
27. J. Ajita, S. Saravanan and N. Selvamurugan, Effect of size of bioactive glass nanoparticles on mesenchymal stem cell proliferation for dental and orthopedic applications, *Mater. Sci. Eng., C*, 2015, **53**, 142–149.
28. M. Rismanchian, N. Khodaeian, L. Bahramian, M. Fathi and H. Sadeghi-Aliabadi, In-vitro Comparison of Cytotoxicity of Two Bioactive Glasses in Micropowder and Nanopowder forms, *Iran. J. Pharm. Res.*, 2013, **12**(3), 437–443.
29. M. Tavakoli, E. Bateni, M. Rismanchian, M. Fathi, A. Doostmohammadi, A. Rabiei *et al.*, Genotoxicity effects of nano bioactive glass and Novabone bioglass on gingival fibroblasts using single cell gel electrophoresis (comet assay): An *in vitro* study, *Dent. Res. J.l*, 2012, **9**(3), 314–320.

30. P. Saravanapavan, J. R. Jones, R. S. Pryce and L. L. Hench, Bioactivity of gel-glass powders in the CaO-SiO2 system: A comparison with ternary (CaO-P2O5-SiO2) and quaternary glasses (SiO2-CaO-P2O5-Na2O), *J. Biomed. Mater. Res., Part A*, 2003, **66A**(1), 110–119.
31. R. L. Siqueira, O. Peitl and E. D. Zanotto, Gel-derived SiO2-CaO-Na2O-P2O5 bioactive powders: Synthesis and *in vitro* bioactivity, *Mater. Sci. Eng., C*, 2011, **31**(5), 983–991.
32. A. Hoppe, N. S. Güldal and A. R. Boccaccini, A review of the biological response to ionic dissolution products from bioactive glasses and glass-ceramics, *Biomaterials*, 2011, **32**(11), 2757–2774.
33. M. N. Rahaman, D. E. Day, B. Sonny Bal, Q. Fu, S. B. Jung, L. F. Bonewald *et al.*, Bioactive glass in tissue engineering, *Acta Biomater.*, 2011, **7**(6), 2355–2373.
34. A. Moorthi, P. R. Parihar, S. Saravanan, M. Vairamani and N. Selvamurugan, Effects of silica and calcium levels in nanobioglass ceramic particles on osteoblast proliferation, *Mater. Sci. Eng., C*, 2014, **43**, 458–464.
35. Y.-F. Goh, A. Z. Alshemary, M. Akram, M. R. A. Kadir and R. Hussain, Bioactive Glass: An In-Vitro Comparative Study of Doping with Nanoscale Copper and Silver Particles, *Int. J. Appl. Glass Sci.*, 2014, **5**(3), 255–266.
36. A. M. El-Kady, A. F. Ali, R. A. Rizk and M. M. Ahmed, Synthesis, characterization and microbiological response of silver doped bioactive glass nanoparticles, *Ceram. Int.*, 2012, **38**(1), 177–188.
37. V. Rajendran, M. Prabhu and R. Suriyaprabha, Synthesis of TiO2-doped mesoporous nanobioactive glass particles and their cytocompatibility against osteoblast cell line, *J. Mater. Sci.*, 2015, **50**(15), 5145–5156.
38. B. Pemmer, A. Roschger, A. Wastl, J. G. Hofstaetter, P. Wobrauschek, R. Simon *et al.*, Spatial distribution of the trace elements zinc, strontium and lead in human bone tissue, *Bone*, 2013, **57**(1), 184–193.
39. H. Rico and L. F. Villa, Zinc, a New Coherent Therapy for Osteoporosis? *Calcif. Tissue Int.*, 2000, **67**(5), 422–423.
40. J. Zhang, Y.-D. Park, W.-J. Bae, A. El-Fiqi, S.-H. Shin, E.-J. Lee *et al.*, Effects of bioactive cements incorporating zinc-bioglass nanoparticles on odontogenic and angiogenic potential of human dental pulp cells, *J. Biomater. Appl.*, 2015, **29**(7), 954–964.
41. L. A. Strobel, N. Hild, D. Mohn, W. J. Stark, A. Hoppe, U. Gbureck *et al.*, Novel strontium-doped bioactive glass nanoparticles enhance proliferation and osteogenic differentiation of human bone marrow stromal cells, *J. Nanopart. Res.*, 2013, **15**(7), 1–9.
42. A. Hoppe, B. Sarker, R. Detsch, N. Hild, D. Mohn, W. J. Stark *et al.*, *In vitro* reactivity of Sr-containing bioactive glass (type 1393) nanoparticles, *J. Non-Cryst. Solids*, 2014, **387**, 41–46.
43. O. Tsigkou, S. Labbaf, M. M. Stevens, A. E. Porter and J. R. Jones, Monodispersed Bioactive Glass Submicron Particles and Their Effect

on Bone Marrow and Adipose Tissue-Derived Stem Cells, *Adv. Healthcare Mater.*, 2014, **3**(1), 115–125.

44. S. Labbaf, O. Tsigkou, K. H. Müller, M. M. Stevens, A. E. Porter and J. R. Jones, Spherical bioactive glass particles and their interaction with human mesenchymal stem cells *in vitro*, *Biomaterials*, 2011, **32**(4), 1010–1018.
45. A. R. Boccaccini, M. Erol, W. J. Stark, D. Mohn, Z. K. Hong and J. F. Mano, Polymer/bioactive glass nanocomposites for biomedical applications: A review, *Compos. Sci. Technol.*, 2010, **70**(13), 1764–1776.
46. W. Stöber, A. Fink and E. Bohn, Controlled growth of monodisperse silica spheres in the micron size range, *J. Colloid Interface Sci.*, 1968, **26**(1), 62–69.
47. A. Lukowiak, J. Lao, J. Lacroix and J.-M. Nedelec, Bioactive glass nanoparticles obtained through sol-gel chemistry, *Chem. Commun.*, 2013, **49**(59), 6620–6622.
48. Z. Hong, A. Liu, L. Chen, X. Chen and X. Jing, Preparation of bioactive glass ceramic nanoparticles by combination of sol–gel and coprecipitation method, *J. Non-Cryst. Solids*, 2009, **355**(6), 368–372.
49. W. Xia and J. Chang, Preparation and characterization of nano-bioactive-glasses (NBG) by a quick alkali-mediated sol–gel method, *Mater. Lett.*, 2007, **61**(14–15), 3251–3253.
50. G. M. Luz and J. F. Mano, Preparation and characterization of bioactive glass nanoparticles prepared by sol-gel for biomedical applications, *Nanotechnology*, 2011, **22**(49), 494014.
51. G. M. Luz, Á. J. Leite, A. I. Neto, W. Song and J. F. Mano, Wettable arrays onto superhydrophobic surfaces for bioactivity testing of inorganic nanoparticles, *Mater. Lett.*, 2011, **65**(2), 296–299.
52. Z. Hong, G. M. Luz, P. J. Hampel, M. Jin, A. Liu, X. Chen *et al.*, Monodispersed bioactive glass nanospheres: Preparation and effects on biomechanics of mammalian cells, *J. Biomed. Mater. Res., Part A*, 2010, **95A**(3), 747–754.
53. X. Chen, B. Lei, Y. Wang and N. Zhao, Morphological control and *in vitro* bioactivity of nanoscale bioactive glasses, *J. Non-Cryst. Solids*, 2009, **355**(13), 791–796.
54. Z. Hong, E. G. Merino, R. L. Reis and J. F. Mano, Novel Rice-shaped Bioactive Ceramic Nanoparticles, *Adv. Eng. Mater.*, 2009, **11**(5), B25–B29.
55. T. J. Brunner, R. N. Grass and W. J. Stark, Glass and bioglass nanopowders by flame synthesis, *Chem. Commun.*, 2006, **13**, 1384–1386.
56. T. J. Brunner, W. J. Stark and A. R. Boccaccini, Nanoscale Bioactive Silicate Glasses in Biomedical Applications, *Nanotechnologies for the Life Sciences*, Wiley-VCH Verlag GmbH & Co. KGaA, 2007.
57. S. Loher, W. J. Stark, M. Maciejewski, A. Baiker, S. E. Pratsinis, D. Reichardt *et al.*, Fluoro-apatite and Calcium Phosphate Nanoparticles by Flame Synthesis, *Chem. Mater.*, 2005, **17**(1), 36–42.

58. N. R. Zhao, Y. J. Wang, X. F. Chen, Y. X. Yang, K. Wei and G. Wu, Preparation of bioactive nanoparticles in the system CaO-P(2)O(5)-SiO(2) using microemulsions, *ASBM6: Adv. Biomater. VI*, 2005, **288–289**, 179–182.
59. M. A. De Diego, N. J. Coleman and L. L. Hench, Tensile properties of bioactive fibers for tissue engineering applications, *J. Biomed. Mater. Res.*, 2000, **53**(3), 199–203.
60. H. W. Kim, H. E. Kim and J. C. Knowles, Production and Potential of Bioactive Glass Nanofibers as a Next-Generation Biomaterial, *Adv. Funct. Mater.*, 2006, **16**(12), 1529–1535.
61. H. W. Kim, H. H. Lee and G. S. Chun, Bioactivity and osteoblast responses of novel biomedical nanocomposites of bioactive glass nanofiber filled poly(lactic acid), *J. Biomed. Mater. Res., Part A*, 2008, **85A**(3), 651–663.
62. T. J. Webster, C. Ergun, R. H. Doremus, R. W. Siegel and R. Bizios, Specific proteins mediate enhanced osteoblast adhesion on nanophase ceramics, *J. Biomed. Mater. Res.*, 2000, **51**(3), 475–483.
63. F.-Y. Hsu, R.-C. Weng, H.-M. Lin, Y.-H. Lin, M.-R. Lu, J.-L. Yu *et al.*, A biomimetic extracellular matrix composed of mesoporous bioactive glass as a bone graft material, *Microporous Mesoporous Mater.*, 2015, **212**, 56–65.
64. X. Wei, Z. Daming and C. Jiang, Fabrication and *in vitro* biomineralization of bioactive glass (BG) nanofibres, *Nanotechnology*, 2007, **18**(13), 135601.
65. A. M. Deliormanli, Preparation and *in vitro* characterization of electrospun 45S5 bioactive glass nanofibers, *Ceram. Int.*, 2015, **41**(1), 417–425.
66. S. Zhao, L. Li, H. Wang, Y. Zhang, X. Cheng, N. Zhou *et al.*, Wound dressings composed of copper-doped borate bioactive glass microfibers stimulate angiogenesis and heal full-thickness skin defects in a rodent model, *Biomaterials*, 2015, **53**, 379–391.
67. H. Lu, T. Zhang, X. P. Wang and Q. F. Fang, Electrospun submicron bioactive glass fibers for bone tissue scaffold, *J. Mater. Sci.: Mater. Med.*, 2009, **20**(3), 793–798.
68. D. Durgalakshmi and S. Balakumar, Phase separation induced shell thickness variations in electrospun hollow Bioglass 45S5 fiber mats for drug delivery applications, *Phys. Chem. Chem. Phys.*, 2015, **17**(23), 15316–15323.
69. Y. Hong, X. Chen, X. Jing, H. Fan, B. Guo, Z. Gu *et al.*, Preparation, Bioactivity, and Drug Release of Hierarchical Nanoporous Bioactive Glass Ultrathin Fibers, *Adv. Mater.*, 2010, **22**(6), 754–758.
70. N. Tucker, J. J. Stanger, M. P. Staiger, H. Razzaq and K. Hofman, The History of the Science and Technology of Electrospinning from 1600 to 1995, *J. Eng. Fibers Fabr.*, 2012, **7**, 63–73.
71. F. Quintero, J. Pou, R. Comesaña, F. Lusquiños, A. Riveiro, A. B. Mann *et al.*, Laser Spinning of Bioactive Glass Nanofibers, *Adv. Funct. Mater.*, 2009, **19**(19), 3084–3090.

72. J.-H. Jo, E.-J. Lee, D.-S. Shin, H.-E. Kim, H.-W. Kim, Y.-H. Koh *et al.*, *In vitro/in vivo* biocompatibility and mechanical properties of bioactive glass nanofiber and poly(ε-caprolactone) composite materials, *J. Biomed. Mater. Res., Part B*, 2009, **91B**(1), 213–220.
73. N. M. Alves, I. B. Leonor, H. S. Azevedo, R. L. Reis and J. F. Mano, Designing biomaterials based on biomineralization of bone, *J. Mater. Chem.*, 2010, **20**(15), 2911–2921.
74. K. Rezwan, Q. Z. Chen, J. J. Blaker and A. R. Boccaccini, Biodegradable and bioactive porous polymer/inorganic composite scaffolds for bone tissue engineering, *Biomaterials*, 2006, **27**(18), 3413–3431.
75. V. Guarino, F. Causa and L. Ambrosio, Bioactive scaffolds for bone and ligament tissue, *Expert Rev. Med. Devices*, 2007, **4**(3), 405–418.
76. L. L. Hench, R. J. Splinter, W. C. Allen and T. K. Greenlee, Bonding mechanisms at the interface of ceramic prosthetic materials, *J. Biomed. Mater. Res.*, 1971, **5**(6), 117–141.
77. G. M. Luz and J. F. Mano, New Composite Membranes Containing Bioactive Glass-Ceramic Nanoparticles and Chitosan for Biomedical Applications, *Adv. Mater. Forum V, Pts 1 and 2*, 2010, **636–637**, 31–35.
78. A. X. Liu, Z. K. Hong, X. L. Zhuang, X. S. Chen, Y. Cui, Y. Liu *et al.*, Surface modification of bioactive glass nanoparticles and the mechanical and biological properties of poly(L-lactide) composites, *Acta Biomater.*, 2008, **4**(4), 1005–1015.
79. Z. K. Hong, R. L. Reis and J. F. Mano, Preparation and *in vitro* characterization of scaffolds of poly(L-lactic acid) containing bioactive glass ceramic nanoparticles, *Acta Biomater.*, 2008, **4**(5), 1297–1306.
80. D. Couto, Z. Hong and J. Mano, Development of bioactive and biodegradable chitosan-based injectable systems containing bioactive glass nanoparticles, *Acta Biomater.*, 2009, **5**(1), 115–123.
81. Q. Z. Z. Chen, I. D. Thompson and A. R. Boccaccini, 45S5 Bioglass (R)-derived glass-ceramic scaffolds for bone tissue engineering, *Biomaterials*, 2006, **27**(11), 2414–2425.
82. F. Baino and C. Vitale-Brovarone, Three-dimensional glass-derived scaffolds for bone tissue engineering: Current trends and forecasts for the future, *J. Biomed. Mater. Res., Part A*, 2011, **97A**(4), 514–535.
83. B. Lei, K. H. Shin, D. Y. Noh, I. H. Jo, Y. H. Koh, H. E. Kim *et al.*, Sol-gel derived nanoscale bioactive glass (NBG) particles reinforced poly(epsilon-caprolactone) composites for bone tissue engineering, *Mater. Sci. Eng., C*, 2013, **33**(3), 1102–1108.
84. J. A. Burdick and R. L. Mauck, *Biomaterials for Tissue Engineering Applications: A Review of the Past and Future Trends*, Springer, 2010.
85. Z. Hong, R. L. Reis and J. F. Mano, Preparation and *in vitro* characterization of novel bioactive glass ceramic nanoparticles, *J. Biomed. Mater. Res., Part A*, 2009, **88A**(2), 304–313.
86. S. K. Misra, S. P. Valappil, I. Roy and A. R. Boccaccini, Polyhydroxyalkanoate (PHA)/Inorganic Phase Composites for Tissue Engineering Applications, *Biomacromolecules*, 2006, **7**(8), 2249–2258.

87. J. Rodenas-Rochina, J. L. Gomez Ribelles and M. Lebourg, Comparative study of PCL-HAp and PCL-bioglass composite scaffolds for bone tissue engineering, *J. Mater. Sci.: Mater. Med.*, 2013, **24**(5), 1293–1308.
88. N. Hild, P. N. Tawakoli, J. G. Halter, B. Sauer, W. Buchalla, W. J. Stark *et al.*, pH-dependent antibacterial effects on oral microorganisms through pure PLGA implants and composites with nanosized bioactive glass, *Acta Biomater.*, 2013, **9**(11), 9118–9125.
89. S. B. Kennedy, N. R. Washburn, C. G. Simon Jr and E. J. Amis, Combinatorial screen of the effect of surface energy on fibronectin-mediated osteoblast adhesion, spreading and proliferation, *Biomaterials*, 2006, **27**(20), 3817–3824.
90. G. Zhao, A. L. Raines, M. Wieland, Z. Schwartz and B. D. Boyan, Requirement for both micron- and submicron scale structure for synergistic responses of osteoblasts to substrate surface energy and topography, *Biomaterials*, 2007, **28**(18), 2821–2829.
91. T. J. Webster, R. W. Siegel and R. Bizios, Osteoblast adhesion on nanophase ceramics, *Biomaterials*, 1999, **20**(13), 1221–1227.
92. J. F. Mano, *Biomimetic Approaches for Biomaterials Development*, Wiley, 2013.
93. H. Ylänen, *Bioactive Glasses: Materials, Properties and Applications*, Elsevier Science, 2011.
94. C. M. Alves, Y. Yang, D. Marton, D. L. Carnes, J. L. Ong, V. L. Sylvia *et al.*, Plasma surface modification of poly(D,L-lactic acid) as a tool to enhance protein adsorption and the attachment of different cell types, *J. Biomed. Mater. Res. Part B*, 2008, **87B**(1), 59–66.
95. H. Shen, X. Hu, F. Yang, J. Bei and S. Wang, Combining oxygen plasma treatment with anchorage of cationized gelatin for enhancing cell affinity of poly(lactide-*co*-glycolide), *Biomaterials*, 2007, **28**(29), 4219–4230.
96. H.-S. Yu, J.-H. Jang, T.-I. Kim, H.-H. Lee and H.-W. Kim, Apatite-mineralized polycaprolactone nanofibrous web as a bone tissue regeneration substrate, *J. Biomed. Mater. Res., Part A*, 2009, **88A**(3), 747–754.
97. S. K. Misra, T. Ansari, D. Mohn, S. P. Valappil, T. J. Brunner, W. J. Stark *et al.*, Effect of nanoparticulate bioactive glass particles on bioactivity and cytocompatibility of poly(3-hydroxybutyrate) composites, *J. R. Soc., Interface*, 2010, **7**(44), 453–465.
98. H. Hajiali, S. Karbasi, M. Hosseinalipour and H. R. Rezaie, Effects of bioglass nanoparticles on bioactivity and mechanical property of poly(3-hydroxybutirate) scaffolds, *Sci. Iran.*, 2013, **20**(6), 2306–2313.
99. H. Hajiali, M. Hosseinalipour, S. Karbasi and M. A. Shokrgozar, The influence of bioglass nanoparticles on the biodegradation and biocompatibility of poly (3-hydroxybutyrate) scaffolds, *Int. J. Artif. Organs*, 2012, **35**(11), 1015–1024.
100. J. J. Blaker, V. Maquet, R. Jérôme, A. R. Boccaccini and S. N. Nazhat, Mechanical properties of highly porous PDLLA/Bioglass® composite

foams as scaffolds for bone tissue engineering, *Acta Biomater.*, 2005, **1**(6), 643–652.
101. A. Nazarian, D. von Stechow, D. Zurakowski, R. Müller and B. D. Snyder, Bone Volume Fraction Explains the Variation in Strength and Stiffness of Cancellous Bone Affected by Metastatic Cancer and Osteoporosis, *Calcif. Tissue Int.*, 2008, **83**(6), 368–379.
102. H. Liu, S. Shi, J. Cao, L. Ji, Y. He and J. Xi, Preparation and evaluation of a novel bioactive glass/lysozyme/PLGA composite microsphere, *Drug Dev. Ind. Pharm.*, 2015, **41**(3), 458–463.
103. M. A. Woodruff and D. W. Hutmacher, The return of a forgotten polymer—Polycaprolactone in the 21st century, *Prog. Polym. Sci.*, 2010, **35**(10), 1217–1256.
104. H. H. Lee, H. S. Yu, J. H. Jang and H. W. Kim, Bioactivity improvement of poly(epsilon-caprolactone) membrane with the addition of nanofibrous bioactive glass, *Acta Biomater.*, 2008, **4**(3), 622–629.
105. L. Ji, W. Wang, D. Jin, S. Zhou and X. Song, *In vitro* bioactivity and mechanical properties of bioactive glass nanoparticles/polycaprolactone composites, *Mater. Sci. Eng., C*, 2015, **46**, 1–9.
106. T. T. Tauböck, M. Zehnder, T. Schweizer, W. J. Stark, T. Attin and D. Mohn, Functionalizing a dentin bonding resin to become bioactive, *Dent. Mater.*, 2014, **30**(8), 868–875.
107. A. A. Rocha de Oliveira, S. M. de Carvalho, M. F. Leite, R. L. Orefice and M. M. Pereira, Development of biodegradable polyurethane and bioactive glass nanoparticles scaffolds for bone tissue engineering applications, *J. Biomed. Mater. Res., Part B*, 2012, **100B**(5), 1387–1396.
108. J. F. Mano, G. A. Silva, H. S. Azevedo, P. B. Malafaya, R. A. Sousa, S. S. Silva *et al.*, Natural origin biodegradable systems in tissue engineering and regenerative medicine: present status and some moving trends, *J. R. Soc., Interface*, 2007, **4**(17), 999–1030.
109. T. H. Silva, A. Alves, B. M. Ferreira, J. M. Oliveira, L. L. Reys, R. J. F. Ferreira *et al.*, Materials of marine origin: a review on polymers and ceramics of biomedical interest, *Int. Mater. Rev.*, 2012, **57**(5), 276–306.
110. M. Peter, P. T. S. Kumar, N. S. Binulal, S. V. Nair, H. Tamura and R. Jayakumar, Development of novel alpha-chitin/nanobioactive glass ceramic composite scaffolds for tissue engineering applications, *Carbohydr. Polym.*, 2009, **78**(4), 926–931.
111. S. Sowmya, P. T. S. Kumar, K. P. Chennazhi, S. V. Nair, H. Tamura and R. Jayakumar, Biocompatible β-chitin hydrogel/nanobioactive glass ceramic nanocomposite scaffolds for periodontal bone regeneration, *Trends Biomater. Artif. Organs*, 2011, **25**(1), 1–11.
112. R. C. F. Cheung, T. B. Ng, J. H. Wong and W. Y. Chan, Chitosan: An Update on Potential Biomedical and Pharmaceutical Applications, *Mar. Drugs*, 2015, **13**(8), 5156–5186.
113. S. Kalia and L. Avérous, *Biopolymers: Biomedical and Environmental Applications*, Wiley, 2011.

114. M. B. Oliveira, G. M. Luz and J. F. Mano, A combinatorial study of nanocomposite hydrogels: on-chip mechanical/viscoelastic and pre-osteoblast interaction characterization, *J. Mater. Chem. B*, 2014, **2**(34), 5627–5638.
115. G. M. Luz and J. F. Mano, Chitosan/bioactive glass nanoparticles composites for biomedical applications, *Biomed. Mater.*, 2012, **7**(5), 054104.
116. J. Mota, N. Yu, S. G. Caridade, G. M. Luz, M. E. Gomes, R. L. Reis *et al.*, Chitosan/bioactive glass nanoparticle composite membranes for periodontal regeneration, *Acta Biomater.*, 2012, **8**(11), 4173–4180.
117. G. M. Luz, L. Boesel, A. del Campo and J. F. Mano, Micropatterning of Bioactive Glass Nanoparticles on Chitosan Membranes for Spatial Controlled Biomineralization, *Langmuir*, 2012, **28**(17), 6970–6977.
118. M. Peter, N. S. Binulal, S. Soumya, S. V. Nair, T. Furuike and H. Tamura *et al.* Nanocomposite scaffolds of bioactive glass ceramic nanoparticles disseminated chitosan matrix for tissue engineering applications, *Carbohydr. Polym.*, 2010, **79**(2), 284–289.
119. Á. J. Leite, S. G. Caridade, J. F. Mano, Synthesis and characterization of bioactive biodegradable chitosan composite spheres with shape memory capability, *J. Non-Cryst. Solids*, 2016, **432**, 158–166.
120. C. O. Correia, Á. J. Leite and J. F. Mano, Chitosan/bioactive glass nanoparticles scaffolds with shape memory properties, *Carbohydr. Polym.*, 2015, **123**, 39–45.
121. P. de Vos, M. M. Faas, M. Spasojevic and J. Sikkema, Encapsulation for preservation of functionality and targeted delivery of bioactive food components, *Int. Dairy J.*, 2010, **20**(4), 292–302.
122. K. Y. Lee and D. J. Mooney, Alginate: Properties and biomedical applications, *Prog. Polym. Sci.*, 2012, **37**(1), 106–126.
123. S. Srinivasan, R. Jayasree, K. P. Chennazhi, S. V. Nair and R. Jayakumar, Biocompatible alginate/nano bioactive glass ceramic composite scaffolds for periodontal tissue regeneration, *Carbohydr. Polym.*, 2012, **87**(1), 274–283.
124. C. H. Lee, A. Singla and Y. Lee, Biomedical applications of collagen, *Int. J. Pharm.*, 2001, **221**(1–2), 1–22.
125. G. E. Vargas, L. A. Haro Durand, V. Cadena, M. Romero, R. Vera Mesones, M. Mačković *et al.*, Effect of nano-sized bioactive glass particles on the angiogenic properties of collagen based composites, *J. Mater. Sci.: Mater. Med.*, 2013, **24**(5), 1261–1269.
126. H. W. Kim, J. H. Song and H. E. Kim, Bioactive glass nanofiber-collagen nanocomposite as a novel bone regeneration matrix, *J. Biomed. Mater. Res., Part A*, 2006, **79A**(3), 698–705.
127. H. Fu-Yin, L. Meng-Ru, W. Ru-Chun and L. Hsiu-Mei, Hierarchically biomimetic scaffold of a collagen–mesoporous bioactive glass nanofiber composite for bone tissue engineering, *Biomed. Mater.*, 2015, **10**(2), 025007.
128. M. F. Koudehi, A. A. I. Fooladi, K. Mansoori, Z. Jamalpoor, A. Amiri and M. R. Nourani, Preparation and evaluation of novel nano-bioglass/gelatin

conduit for peripheral nerve regeneration, *J. Mater. Sci.: Mater. Med.*, 2014, **25**(2), 363–373.

129. F. Hafezi, F. Hosseinnejad, A. A. I. Fooladi, S. M. Mafi, A. Amiri and M. R. Nourani, Transplantation of nano-bioglass/gelatin scaffold in a non-autogenous setting for bone regeneration in a rabbit ulna, *J. Mater. Sci.: Mater. Med.*, 2012, **23**(11), 2783–2792.
130. A. Gantar, L. P. da Silva, J. M. Oliveira, A. P. Marques, V. M. Correlo, S. Novak *et al.*, Nanoparticulate bioactive-glass-reinforced gellan-gum hydrogels for bone-tissue engineering, *Mater. Sci. Eng., C*, 2014, **43**, 27–36.
131. M. Peter, N. S. Binulal, S. V. Nair, N. Selvamurugan, H. Tamura and R. Jayakumar, Novel biodegradable chitosan-gelatin/nano-bioactive glass ceramic composite scaffolds for alveolar bone tissue engineering, *Chem. Eng. J.*, 2010, **158**(2), 353–361.
132. U. Rottensteiner, B. Sarker, D. Heusinger, D. Dafinova, S. N. Rath, J. P. Beier *et al.*, *In vitro* and *in vivo* Biocompatibility of Alginate Dialdehyde/Gelatin Hydrogels with and without Nanoscaled Bioactive Glass for Bone Tissue Engineering Applications, *Materials*, 2014, 7(3), 1957–1974.
133. Y. J. Wang, C. R. Yang, X. F. Chen and N. R. Zhao, Development and characterization of novel biomimetic composite scaffolds based on bioglass-collagen-hyaluronic acid-phosphatidylserine for tissue engineering applications, *Macromol. Mater. Eng.*, 2006, **291**(3), 254–262.
134. E. Xie, Y. Y. Hu, X. F. Chen, J. P. Bai, L. Ren and Z. Zhang, A novel nanocomposite and its application in repairing bone defects, *2008 3rd IEEE International Conference on Nano/Micro Engineered and Molecular Systems*, 2008, vols. 1–3, pp. 943–946.
135. M. Jawaid and H. P. S. Abdul Khalil, Cellulosic/synthetic fibre reinforced polymer hybrid composites: A review, *Carbohydr. Polym.*, 2011, **86**(1), 1–18.
136. J. Gunn and M. Zhang, Polyblend nanofibers for biomedical applications: perspectives and challenges, *Trends Biotechnol.*, 2010, **28**(4), 189–197.
137. K. T. Shalumon, S. Sowmya, D. Sathish, K. P. Chennazhi, S. V. Nair and R. Jayakumar, Effect of Incorporation of Nanoscale Bioactive Glass and Hydroxyapatite in PCL/Chitosan Nanofibers for Bone and Periodontal Tissue Engineering, *J. Biomed. Nanotechnol.*, 2013, **9**(3), 430–440.
138. E. Chiellini, A. Corti, S. D'Antone and R. Solaro, Biodegradation of poly (vinyl alcohol) based materials, *Prog. Polym. Sci.*, 2003, **28**(6), 963–1014.
139. H. S. Mansur and H. S. Costa, Nanostructured poly(vinyl alcohol)/bioactive glass and poly(vinyl alcohol)/chitosan/bioactive glass hybrid scaffolds for biomedical applications, *Chem. Eng. J.*, 2008, **137**(1), 72–83.
140. W. Y. Chuang, T. H. Young, C. H. Yao and W. Y. Chiu, Properties of the poly(vinyl alcohol)/chitosan blend and its effect on the culture of fibroblast *in vitro*, *Biomaterials*, 1999, **20**(16), 1479–1487.

141. S. Talebian, M. Mehrali, S. Mohan, Hr Balaji raghavendran, M Mehrali, HM Khanlou *et al.*, Chitosan (PEO)/bioactive glass hybrid nanofibers for bone tissue engineering, *RSC Adv.*, 2014, **4**(90), 49144–49152.
142. H. D. Zheng, Y. J. Wang, C. R. Yang, X. F. Chen and N. R. Zhao, Investigation on the porous biomaterial for bone reconstruction with addition of bio-mimetic Nano-Sized inorganic particles, *High-Performance Ceram. IV, Pts 1–3*, 2007, **336–338**, 1534–1537.

CHAPTER 9

Surface Functionalization of Bioactive Glasses: Reactive Groups, Biomolecules and Drugs on Bioactive Surfaces for Smart and Functional Biomaterials

SARA FERRARIS AND ENRICA VERNÈ*

Department of Applied Science and Technology, Institute of Materials Physics and Engineering, Politecnico di Torino, Italy
*Email: enrica.verne@polito.it

9.1 Bioactive Glasses and their Surface Reactivity

Bioactive glasses and glass-ceramics are a class of biomaterials of particular interest for bone reconstruction due to their ability to react with physiological fluids and to form a mineralized layer of hydroxycarbonatoapatite (HCA) on their surface. This layer grows through a complex mechanism that has been thoroughly studied both *in vitro* and *in vivo* in different aqueous

RSC Smart Materials No. 23
Bioactive Glasses: Fundamentals, Technology and Applications
Edited by Aldo R. Boccaccini, Delia S. Brauer and Leena Hupa

Published by the Royal Society of Chemistry, www.rsc.org

media and can be summarized and schematized in a series of sequential steps:[1,2]

1. Ion exchange between the alkaline cations leached from the glass surface (mainly Na^+) and the H^+ (more properly H_3O^+) ions coming from the surrounding solution;
2. Breaking of $\equiv$Si–O–Si$\equiv$ bonds, formation of $\equiv$Si–OH bonds (silanols) at the glass surface and release of soluble $Si(OH)_4$;
3. Polycondensation of silanols ($\equiv$Si–OH) to form a silica rich layer (silica gel);
4. Adsorption of Ca^{2+}, PO_4^{3-} and CO_3^{2-} from the solution into the silica gel layer and formation of an amorphous CaO–P_2O_5-rich layer;
5. Crystallization of the CaO–P_2O_5-rich layer to form hydroxy-carbonatoapatite (HCA);
6. Absorption of proteins from the biological fluid;
7. Attachment of macrophages;
8. Attachment of stem cells;
9. Osteoblast differentiation and proliferation;
10. Matrix production;
11. Matrix crystallization;
12. Bone formation.

The first five steps can take place *in vitro*, in inorganic media and have been thoroughly investigated in order to study the bioactivity of bioactive glasses in physiological fluids (*e.g.* simulated body fluid, SBF).[1–5]

In the biological environment the interaction of the glass surface with proteins and cells can also take place and the surface reactivity has a fundamental role for the interaction with tissues and, in turn, for the integration of biomaterial.[1,2,6,7] The *in vivo* completion of steps 6–12 leads to the growth of new healthy bone on the bioactive surface.

A schematization of the above-described bioactivity reactions is reported in Figure 9.1.

It has been underlined by several authors that, in real conditions, protein absorption is not an isolated step after HCA crystallization: in a protein-containing medium this step evolves dynamically together with the silica gel formation, the adsorption of Ca^{2+}, PO_4^{3-} and CO_3^{2-} and the HCA crystallization, radically affecting their concurrent development.[2,8]

Surface topography and chemistry play a determinant role on the material reactivity, on its ability to adsorb water and proteins and, in turn, on the possibility to interact with cells and tissues. The surface reactivity of bioactive glasses is strictly correlated to their composition,[1,2,9] surface area[10] and to the medium in which the analysis is performed.[11,12]

While for melt derived bioactive glasses reactivity is mainly driven by chemical composition (*e.g.* 60% SiO_2 is the superior limit to have bioactive behavior), for sol–gel bioactive glasses the presence of large surface area and surface nano-porosity, together with high surface concentration of silanols

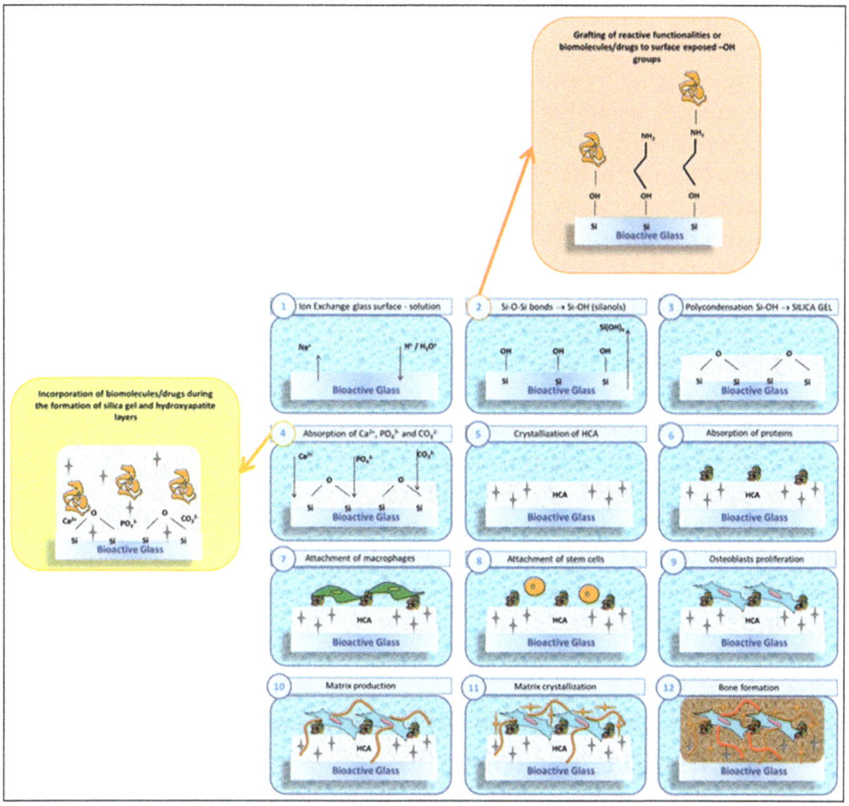

Figure 9.1 Bioactivity reactions and associated surface functionalization strategies.

increase reactivity and allow a significant bioactive behavior even if the silica content rises up to 88%.[1,13] In the case of mesoporous bioactive glasses, ordered porous structures with high pore volume, narrow pore size and large surface area further increase surface reactivity and open up the opportunity for the realization of drug/biomolecule delivery systems.[13]

The surface reactivity of bioactive glasses and glass-ceramics is not only the mechanism by which these materials can interact with biological tissues, but also an opportunity to modify their surface for a more specific interaction, as will be described in Section 9.2 of the present chapter.

As schematized in Figure 9.1, the second step of the bioactivity mechanism promotes the exposure of –OH groups on the bioactive glass (or glass-ceramic) surface. These groups are useful reactive sites for the surface anchoring of various molecules.[14–20] In order to obtain reactive –OH groups, prone to surface functionalization, it is of great importance to activate the material surface by a careful selection of the medium and of the activation conditions, in order to avoid the fast condensation of silanols to silica gel, as observed by the authors on various bioactive glasses and glass-ceramics with different degrees of bioactivity.[14,15,19] On the other hand, biomolecules can be also entrapped in the bioactive glass surface reaction layer during the early stages of its formation, exploiting the step of silica gel formation and its enrichment with ions and biomolecules. For example, several authors[21,22] have reported the possibility to incorporate antibiotics on the bioactive glass surfaces. More recently, Miola *et al.*[23] described different incorporation mechanisms of antibiotics on a bioactive glass surface activated by different pre-treatments in aqueous solution, working over the bioactivity process. Depending on the duration of the pre-treatment the glass surface is activated to a different stage of the bioactivity mechanism and thus exposes different reaction layers. On the basis of the exposed chemical species, three different interaction mechanisms have been suggested: (i) direct bonding with the exposed silanols on the glass surface, (ii) drug diffusion into the silica gel layer in competition with the direct bonding with the residual exposed silanols, (iii) precipitation of the antibiotic into the silica gel layer, assisted by the interaction with calcium ions involved in the HCA crystallization.

In the case of sol–gel bioactive glasses, due to the low temperature of processing, biomolecules and drugs can be incorporated within the network during the glass preparation.[13] Finally, as far as mesoporous bioactive glasses are concerned, biomolecules and drugs can be absorbed into the mesopores following a size selectivity rule.[13]

9.2 Introduction of Specific Reactive Functionalities and/or Spacer Molecules

As described in the previous Section, the presence of reactive sites is crucial for further anchoring of biomolecules on the surface of bioactive glasses and glass-ceramics. Upon exposition to water-based media these materials naturally expose –OH functionalities that can be exploited for functionalization purposes.

The optimization of hydroxyl group exposure and activation has been widely investigated by the authors,[14,15,19] considering washings and soaking in aqueous solution with different pH (acetone, water, H_2SO_4, NaOH and SBF) and also by Schickle *et al.*[24] who investigated the effect of different acids (nitric acid, hydrochloric acid, sulfuric acid and chromosulfuric acid) on both surface area and –OH exposition. However, the majority of research works concerning surface modification of bioactive glasses simply refer to the glass surface hydroxyls groups without further specification of their exposition/activation strategy.

Different functionalization strategies can be used in order to improve bone integration of an implant, such as grafting of proteins from the extracellular matrix (ECM proteins), their peptides, bone morphogenetic proteins (BMPs) or angiogenic growth factor (VEGF), collagen, fibronectin and alkaline phosphatase. These biomolecules have been anchored on different substrates *via* silanization, tresyl chloride activation, plasma surface activation and the electrospray technique.

The nucleation of Au nanoparticles in the glass has also been reported in order to favor the surface grafting of specific reactive groups (–SH and $-NH_2$) by means of a self assembled monolayer (SAM) mechanism as a preliminary activation step for further surface functionalization of bioactive glasses.[25,26]

Finally, the introduction of specific ions (*e.g.* Ga^{3+}) able to act as preferential sites for molecular complexation has been reported in order to obtain bioactive glass–biomolecule coupling.[27]

Some attempts to introduce functional groups on bioactive glass surfaces have been described as preliminary steps for further biomolecular grafting. Specific functional groups (*e.g.* $-NH_2$) can be introduced by different strategies on the surface of bioactive glasses in order to obtain covalent grafting of biomolecules or drugs. Among the various strategies reported in the literature, silanization is worth of mentioning. Silanization is a consolidated and widely applied technique for the effective and stable introduction of reactive functionalities on inert glass surface for various applications, such as the improvement of adhesion between the phases of composite materials or coatings, the anchorage of specific molecules, the development of sensors and the modification of surface properties (*e.g.* wettability, organic molecule absorption). Silanes have also been widely employed for the introduction of reactive functional groups on the surface of bioactive glasses. In particular 3-aminopropyltriethoxysilane (APTES) is the most widely used silane for the introduction of amino groups ($-NH_2$) on the surface of bioactive glasses.[24,28–37] Its grafting has been obtained from aqueous,[28,32,35,36] toluene[24,29,34,37] or ethanol[14,15,31] solutions. The use of glutaraldehyde has sometimes been associated with APTES use[32,35] as well as the introduction of silane molecules directly into the bioactive glass synthetic route in the case of sol–gel-derived bioactive glasses.[33]

Silanization of mesoporous silica has also been considered in order to improve the drug loading of these materials.[38]

Phenyltriethoxysilane, (3-mercaptopropyl)triethoxysilane, (3-chloropropyl)-triethoxysilane and *n*-butyltrimethoxysilane have also been considered for the introduction of phenyl, hydroxypropyl, mercaptopropylbutyl and chloropropyl functionalities.[37] Silanization has been carried out on Bioglass® scaffolds.[39] Amino and mercapto alchols, cysteine, and cysteamine have been also proposed for the introduction of $-NH_2$ and –SH functionalities on the surface of bioactive glasses.[25,26,40]

The main aim of the introduction of reactive sites on the surface of bioactive glasses is the further grafting of biomolecules or drugs in order to improve the biological response of the artificial substrates.[24,25,28,29,31,33,35–37,40] However, their employment has been suggested also for the improvement of the glass particles' dispersibility,[30] to evaluate and improve protein absorption,[32] improve chemical bonding with a matrix in the realization of composites (*e.g.* collagen[34]) and regulate the degradation kinetics of polymeric substrates coupled with the glass.[41]

The main strategies considered in the literature for surface activation and further functionalization are reported schematically in Table 9.1.

Table 9.1 Some example of surface modifications in bioactive glasses and glass-ceramics: reactive sites, functional groups and biomolecules.

Substrate	Activation	Reactive group/spacer	Biomolecule/drug	Application	Reference
45S5 Bioglass® scaffolds and pellets		NH_2 (3-aminopropyl-triethoxysilane)		Further protein grafting for bone substitution applications	28
Mesoporous bioactive glasses		NH_2 (3-aminopropyl-triethoxysilane) COOH (triethoxysilylpropyl succinic anhydride)		Tissue engineering and drug storage	29
Nano sol–gel bioactive glass (60 SiO_2, 36 CaO, 4 P_2O_5, mol%)		NH_2 (3-aminopropyl-triethoxysilane)		Improve particle dispersibility	39
Ga-containing bioactive glass (47.5 SiO_2, 24.1 Na_2O, 26.6 CaO, 2.6 P_2O_5, 1.0 Ga_2O_3) and Ga-free parent glass		OH (tetraethoxysilane) NH_2 (3-aminopropyl-triethoxysilane)		Improve ability to bind drugs, proteins and peptides	31
Sol–gel bioactive glass (45 SiO_2, 24.5 Na_2O, 24.5 CaO, 6 P_2O_5, mol%)		NH_2 (3-aminopropyl-triethoxysilane)		Evaluate/improve protein absorption	32
Nano-mesoporous bioactive glass (85 SiO_2, 15 CaO, mol%)		NH_2 (3-aminopropyl-triethoxysilane)		Improve chemical bonding with collagen for collagen hydrogel-bioactive glass composites	34
Cu-containing sol–gel bioactive glass (80 SiO_2, 15 CaO, 5 $P_2O_5 \times$ CuO, $x = 0/1$, mol%)		NH_2 (2-amino-ethan-1-ol, 6-amino-hexan-1-ol) SH (2-mercapto-ethan-1-ol, 6-mercapto-hexan-1-ol)		Further drug grafting	40

Sol–gel bioactive glass with Au nanoparticles (80 SiO_2, 15 CaO, 5 P_2O_5, 1 Au_2O, mol%)	Au nanoparticles nucleation	SH (2-mercapto ethanol) NH_2 (2-amino ethanol) Cysteine Cysteamine		Realization of chemical bonds with different strengths for protein grafting/drug release	25
45S5 Bioglass® scaffolds		NH_2 (3-aminopropyl-triethoxysilane)		Further protein grafting	28
Mesoporous silica		NH_2 (3-aminopropyl-triethoxysilane)	Ibuprofen	Improve drug loading	38
Sol–gel bioactive glass (80 SiO_2, 15 CaO, 5 P_2O_5, mol%)		NH_2 (3-aminopropyl-triethoxysilane)	5-Aminofluorescein	Controlled release of drugs at the tumor site	33
Sol–gel bioactive glass with Au nanoparticles (80 SiO_2, 15 CaO, 5 P_2O_5, *x* Au_2O, mol%)	Au nanoparticles nucleation	NH_2 (Cysteamine)	Soybean peroxidase	Improve bioactivity and reduce oxidative stress associated with implantable biomaterials	26
Apatite–wollastonite glass-ceramic		*N*-β-(Aminoethyl)-γ-aminopropyl trimethoxy silane	Lysine	Improve cytocompatibility	50
Starch–poly(ethylene-vinylalchol) blend–45S5 Bioglass® powder			Bovine serum albumin α-Amylase	Induce osteoconductive properties in the polymer and regulate its degradation kinetics	41
Starch–poly(ethylene-vinylalchol) blend–45S5 Bioglass® powder			Bovine serum albumin α-Amylase	Biomolecule release	51
45S5 Bioglass® granules			Emdogain® (commercial mixture of enamel matrix proteins)	Improve proliferation and differentiation of pre-osteoblastic cells	53
40% Tricalcium phosphate (β-TCP) 60% 45S5 Bioglass® composite	Chromosufuric acid treatment for OH exposition	NH_2 (3-aminopropyl-triethoxysilane)	Recombinant human bone morphogenetic protein 2 (rhBMP-2)	Stimulation of bone regeneration	24

Table 9.1 (*Continued*)

Substrate	Activation	Reactive group/spacer	Biomolecule/drug	Application	Reference
45S5 Bioglass® porous scaffold		NH_2 (3-aminopropyl-triethoxysilane)	Collagen	Improving cell adhesion and bone regeneration	36
Mesoporous bioactive glass (85 SiO_2, 10 CaO, 5 P_2O_5 mol%)		Phenyl Hydroxypropyl Aminopropyl Mercaptopropyl	Ipriflavone	Reduce osteoporosis	37
Sol–gel bioactive glasses and mesoporous sol–gel bioactive glasses with/without Ga	Gallium introduction as preferential site for molecular anchoring		Curcumin	Release of an antioxidant, anti-inflammatory and anticancer agent	27
Mesoporous bioactive glass (80 SiO_2, 15 CaO, 5 P_2O_5, mol%)			Dexamethasone Bovine serum albumin	Delivery of chemotherapeutic drugs and proteins in bone regeneration applications	51
Silica-based bioactive glasses with different reactivity (47.5 SiO_2, 2.5 P_2O_5, 30 CaO, 10 K_2O, 10 Na_2O; 45 SiO_2, 3 P_2O_5, 7 MgO, 26 CaO, 4 K_2O, 15 Na_2O; 57 SiO_2, 34 CaO, 6 Na_2O, 3 Al_2O_3, mol%)	Washing/soaking in aqueous solutions with different pH (acetone, water, H_2SO_4, NaOH and SBF) for OH exposition/activation	NH_2 (3-aminopropyl-triethoxysilane)	Human bone morphogenetic protein 2 (BMP-2)	Bone regeneration	14
Silica-based bioactive glasses and glass ceramics with different reactivity (45 SiO_2, 3 P_2O_5, 7 MgO, 26 CaO, 4 K_2O, 15 Na_2O; 57 SiO_2, 34 CaO, 6 Na_2O, 3 Al_2O_3, mol%)	Washing in acetone and water for OH exposition/activation	NH_2 (3-aminopropyl-triethoxysilane)	Alkaline phosphatase (ALP)	Improvement of bioactivity and specific bone stimulation ability	15

Silica-based bioactive glasses with different reactivity (45 SiO_2, 3 P_2O_5, 7 MgO, 26 CaO, 4 K_2O, 15 Na_2O; 57 SiO_2, 34 CaO, 6 Na_2O, 3 Al_2O_3, mol%)	Washing in acetone and water for OH exposition/activation	NH_2 (3-aminopropyl-triethoxysilane)	Alkaline phosphatase (ALP)	Improvement of bioactivity and specific bone stimulation ability	55
Biocompatible glass-ceramic scaffolds (57 SiO_2, 34 CaO, 6 Na_2O, 3 Al_2O_3, mol%)	Washing in acetone and water for OH exposition/activation	NH_2 (3-aminopropyl-triethoxysilane)	Alkaline phosphatase (ALP)	Improvement of bioactivity and specific bone stimulation ability	56
45S5 Bioglass® pellets	Washing in acetone and water for OH exposition/activation	NH_2 (3-aminopropyl-triethoxysilane)	Alkaline phosphatase (ALP)	Improvement of bioactivity and specific bone stimulation ability	54
Mesoporous bioactive glasses (M58S, M77S)			Gentamicin	Localized drug delivery	57, 58
Mesoporous bioactive glasses with different Ca content			Tetracycline	Localized drug delivery	59
Mesoporous bioactive glass			Ibuprofen	Localized drug delivery	60
Silica-based bioactive glasses with different reactivity (45 SiO_2, 3 P_2O_5, 7 MgO, 26 CaO, 4 K_2O, 15 Na_2O; 57 SiO_2, 34 CaO, 6 Na_2O, 3 Al_2O_3, mol%)	Washing in acetone and water for OH exposition/activation		Gallic acid	Induction of antioxidant and anticancer activity	18
Silica based bioactive glasses with different reactivity (45 SiO_2, 3 P_2O_5, 7 MgO, 26 CaO, 4 K_2O, 15 Na_2O; 57 SiO_2, 34 CaO, 6 Na_2O, 3 Al_2O_3, mol%)	Washing in acetone and water for OH exposition/activation		Natural polyphenols from grape skin	Induction of antioxidant and anticancer activity	16

Table 9.1 (*Continued*)

Substrate	Activation	Reactive group/spacer	Biomolecule/drug	Application	Reference
Ferrimagnetic glass-ceramic (24.7 SiO_2, 13.5 Na_2O, 13.5 CaO, 3.3 P_2O_5, 14 FeO, 31 Fe_2O_3, wt%)	Soaking in water for OH exposition/activation		Gallic acid	Induction of anticancer activity	20
Ferrimagnetic glass-ceramic (24.7 SiO_2, 13.5 Na_2O, 13.5 CaO, 3.3 P_2O_5, 14 FeO, 31 Fe_2O_3, wt%)	Soaking in water for OH exposition/activation		Doxorubicin Cisplatinum	Controlled and localized delivery of chemotherapeutic drugs	19
Mesoporous bioactive glass nanospheres			Alendronate	Reduce osteosarcoma cells viability and inhibit osteoclasts activity	61
Silica–zirconia mixed oxides			Alendronate, zolendronate	Localized drug delivery	62

9.3 Surface Grafting of Biomolecules and Drugs

Despite wide research on the surface functionalization of polymers and metals[42–49] few works can be enumerated concerning the introduction of biomolecules and drugs on the surface of bioactive glasses.

On the other hand, surface functionalization of bioactive glasses is a promising strategy in order to combine the typical properties of bioactive glasses and glass-ceramics (*e.g.* versatility, biocompatibility, reactivity, bone bonding and cell stimulating ability) with the ones of specific molecules (*e.g.* therapeutic action, specific cell stimulation) for the preparation of functional biomaterials for tailored applications.

Various researches report the grafting of biomolecules to bioactive glasses and glass-ceramics in order to improve bioactivity, cell adhesion and differentiation for bone regeneration purposes. Lysine,[50] soybean peroxidase,[25] bovine serum albumin,[37,51,52] α-amylase,[37,51] Emdogain® (commercial mixture of enamel matrix proteins),[53] bone morphogenetic proteins, BMPs,[14,24] collagen,[36] alkaline phosphatase, ALP[15,54–56] and ipriflavone[37] have been considered for this purpose.

Antibiotics and anti-inflammatory agents, such as gentamicin,[57,58] tetracycline[59] and ibuprofen[60] have been loaded into mesoporous bioactive glasses in order to obtain localized drug delivery from bioactive materials. A crucial role of Ca in the glass composition has been evidenced in the ability of mesoporous bioactive glasses to load and release tetracycline.[59]

Another interesting application of surface-modified bioactive glasses is the localized treatment of cancer. Various biomolecules and drugs have been proposed in order to reach this aim: 5-aminofluorescein, as a model for doxorubicin,[33] doxorubicin and cisplatin,[19] gallic acid,[17,18,20] curcumin,[27] dexamethasone[52] and alendronate.[61] Bisphosphonates (alendronate and zolendronate) have been loaded into silica–zirconia mixed oxides.[62] By varying the zirconium content surface properties can be tailored in order to control drug uptake and release.

Recently, an increasing interest in natural biomolecules has been registered due to their potential health benefits and low toxicity.[63–66] A few papers report on their coupling with synthetic substrates and some of them consider bioactive glasses and glass-ceramics as possible carriers. In particular gallic acid,[17,18] natural polyphenols,[16] soybean peroxidase[26] and curcumin[27] have been grafted to these materials in order to induce antioxidant activity.

The grafting of biomolecules has been obtained by direct anchoring to –OH groups characteristic of bioactive glasses and glass-ceramic surfaces, or exposed by specific surface activation treatments,[15,16,18–20,53–56] or to previously inserted functional groups (*e.g.* $-NH_2$), as described in Section 9.2.[14,15,24,26,33,36,37,50,54–56] The effect of Bioglass® coatings and of vascular endothelial growth factor (VEGF) release on *in vitro* response and *in vivo* angiogenesis and bone regeneration has been studied.[67] In this context, Vernè *et al.* successfully grafted alkaline phosphatase to different bioactive glasses maintaining the enzymatic activity and improving both glass bioactivity and the ability to stimulate osteoblast cells.[68]

The introduction of a further spacer molecule has been proposed in order to obtain controlled release under specific conditions (*e.g.* maleic anhydride for selective release at acidic pH typical of tumor sites).[33]

The first conjugation of a biomolecule with silane, with the silane carrying the bioactive principle, has been proposed for *N*-β-(aminoethyl)-γ-aminopropyl trimethoxy silane and lysine grafting to apatite–wollastonite glass-ceramic.[50]

Moreover, the incorporation of biomolecules during the calcium phosphate/hydroxyapatite precipitation in SBF has been performed, introducing reactive molecules in SBF solution and exploiting the peculiar reactivity of bioactive glasses and glass-ceramics, described in Section 9.1. Bovine serum albumin, α-amylase and dexamethasone have been coupled with bioactive glasses and glass-ceramics by this route.[41,51,52]

The various strategies that can be employed in order to exploit the surface reactivity of bioactive glasses and glass-ceramics are schematized in Figure 9.1, directly in surface reactivity sequence.

The possibility to obtain different bonding strength between the biomolecule/drug and the glass/glass-ceramic surface has been evidenced by the introduction of various functional groups ($-NH_2$ or $-SH$)[25] or by the covalent/non covalent grafting obtained by the use, or not, of 1,1′-carbonyldiimidazole.[24] Higher release and bioactivity has been highlighted for non-covalent grafting of proteins, compared to covalent grafting.[24]

The main research works concerning biomolecule/drug coupling with bioactive glasses and glass-ceramics are schematically reported in Table 9.1.

9.4 Conclusions and Outlook

The scientific community has recently focused its efforts on the study of surface properties of biomaterials, since their interaction with biological tissues starts mainly at their surface. Aiming at optimizing the performances of biomaterials in contact with living tissues, several challenging approaches to surface modification have been investigated, including biological functionalization, which introduces specific biomolecules on the materials surface to stimulate specific cellular responses.

Surface functionalization of bioactive glasses takes its inspiration from surface modification techniques already exploited for different purposes on pure silica, common glasses or metallic and polymeric substrates. However, the literature concerning surface functionalization of bioactive glasses still comprises few reports. In the scientific literature some examples of the introduction of reactive sites (alternatives to hydroxyls) as well as the grafting of model and complex biomolecules have been reported. In the specific case of bioactive glasses the surface functionalization procedures have been adapted to combine the peculiar properties of the substrate (such as bioactivity and release of ions able to stimulate cells) with those resulting from the grafted molecule. Moreover, for these reactive substrates, surface reactivity represents both a peculiar feature to be combined with grafted biomolecules and at the same time a tool for surface modification. In fact, the surface reactivity of bioactive glasses allows the exposition of free reactive hydroxyl groups that can be further used for biomolecules grafting, introducing biological signaling, beside the peculiar inorganic signals of the bioactive glass. This approach will open the way to a variety of surface grafting strategies that, in association with the specific therapeutic role of the ions released from the bioactive glasses, will allow further challenging applications for these very versatile materials. Surface functionalization of bioactive glasses is actually a poorly explored field in scientific research but represents a promising and challenging strategy for the development of innovative smart biomaterials. In fact the result of this procedure is a smart multifunctional surface that can be designed in order to match various biological requirements associated with different applications and clinical needs.

References

1. W. Cao and L. L. Hench, *Ceram. Int.*, 1996, **22**, 493.
2. L. L. Hench, N. Roki and M. B. Fenn, *J. Mol. Struct.*, 2014, **1073**, 24.
3. T. Kokubo and H. Takadama, *Biomaterials*, 2006, **27**, 2907.
4. C. Wu and Y. Xiao, *Bone Tissue Regener. Insights*, 2009, **2**, 25.
5. S. Fujibayashi, M. Neo, H. M. Kim, T. Kokubo and T. Nakamura, *Biomaterials*, 2003, **24**, 1349.
6. D. G. Castner and B. D. Ratner, *Surf. Sci.*, 2002, **500**, 28.
7. B. Kasemo, *Surf. Sci.*, 2002, **500**, 656.
8. C. Knabe, P. Ducheyne, in *Comprehensive Biomaterials*, ed. P. Ducheyne, K. Healy, D. Hutmacher, D. W. Grainger and J. P. Kirkpatrick, Elsevier, Oxford, UK, 2011, ch. 1.114, vol. 1, pp. 245–258.
9. S. Ferraris, M. Miola, in *Surface Tailoring of Inorganic Materials for Biomedical Applications*, ed. L. Rimondini, C. L. Bianchi and E. Vernè, Bentham e-Books, 2012, ch. 16, pp. 392–405.
10. S. Naseri, W. C. Lepry, W. Li, K. E. Waters, A. R. Boccaccini and S. N. Nazhat, *J. Non-Cryst. Solids*, 2016, **432**, 47.
11. L. Varila, S. Fagerlund, T. Lehtonen, J. Tuominen and L. Hupa, *J. Eur. Ceram. Soc.*, 2012, **32**, 2757.
12. D. Bellucci, G. Bolelli, V. Cannillo, A. Cattini and A. Sola, *Mater. Charact.*, 2011, **62**, 1021.
13. D. Arcos and M. Vallet-Regì, *Acta Mater.*, 2013, **61**, 890–911.
14. E. Vernè, C. Vitale-Brovarone, E. Bui, C. L. Bianchi and A. R. Boccaccini, *J. Biomed. Mater. Res., Part A*, 2009, **90**, 981.
15. E. Vernè, S. Ferraris, C. Vitale-Brovarone, S. Spriano, C. L. Bianchi, A. Naldoni, M. Morra and C. Cassinelli, *Acta Biomater.*, 2010, **6**, 229.
16. X. Zhang, S. Ferraris, E. Prenesti and E. Vernè, *Appl. Surf. Sci.*, 2013, **287**, 341.
17. X. Zhang, S. Ferraris, E. Vernè, A. Venturello, E. Prenesti, *Proceedings of the 10th IASTED International Conference on Biomedical Engineering, BioMed*, Innsbruck, Austria, 2013, 604–608.

18. X. Zhang, S. Ferraris, E. Prenesti and E. Vernè, *Appl. Surf. Sci.*, 2013, **287**, 329.
19. E. Vernè, M. Miola, S. Ferraris, C. L. Bianchi, A. Naldoni, G. Maina and O. Bretcanu, *Adv. Biomater.*, 2010, **12**, B309.
20. S. Ferraris, X. Zhang, E. Prenesti, I. Corazzari, F. Turci, M. Tomatis and E. Vernè, *J. Non-Cryst. Solids*, 2016, **432**, 167.
21. M. Otsuka, M. Sawada, Y. Matsuda *et al.*, *Biomaterials*, 1997, **18**, 1559.
22. H. Li and J. Chang, *J. Controlled Release*, 2005, **107**(3), 463.
23. M. Miola, C. Vitale-Brovarone, C. Mattu and E. Vernè, *J. Biomater. Appl.*, 2012, **28**, 308.
24. K. Schickle, K. Zurlinden, C. Bergmann, M. Lindner, A. Kirsten, M. Laub, R. Telle, H. Jennissen and H. Fischer, *J. Mater. Sci.: Mater. Med.*, 2011, **22**, 763.
25. V. Aina, T. Marchis, E. Laurenti, E. Diana, G. Lusvardi, G. Malavasi, L. Menabue, G. Cerrato and C. Morterra, *Langmuir*, 2010, **26**, 18600.
26. V. Aina, D. Ghigo, T. Marchis, G. Cerrato, E. Laurenti, C. Morterra, G. Malavasi, G. Lusvardi, L. Menabue and L. Bergandi, *J. Mater. Chem.*, 2011, **21**, 10970.
27. G. Malavasi, E. Ferrari, G. Lusvardi, V. Aina, F. Fantini, C. Morterra, F. Pignedoli, M. Saladini and L. Menabue, *J. Mater. Chem.*, 2011, **21**, 5027.
28. Q. Z. Chen, K. Rezwan, V. Francon, D. Armitage, S. N. Nazhat, F. H. Jones and A. R. Boccaccini, *Acta Biomater.*, 2007, **3**, 551.
29. J. Sun, Y. Li, L. Li, W. Zhao, L. Li, J. Gao, M. Ruan and J. Shi, *J. Non-Cryst. Solids*, 2008, **354**, 3799.
30. X. Chen, C. Guo and N. Zhao, *Appl. Surf. Sci.*, 2008, **255**, 466.
31. G. Lusvardi, G. Malavasi, L. Menabue and S. Shruti, *Mater. Sci. Eng., C*, 2013, **33**, 3190.
32. C. Gruian, E. Vanea, S. Simon and V. Simon, *Biochim. Biophys. Acta*, 2012, **1824**, 873.
33. V. Aina, G. Malavasi, C. Magistris, G. Cerrato, G. Martra, G. Viscardi, L. Menabue and G. Lusvardi, *J. Mater. Sci.: Mater. Med.*, 2014, **25**, 2243.
34. A. El-Fiqi, J. H. Lee, E. J. Lee and H. W. Kim, *Acta Biomater.*, 2013, **9**, 9508.
35. Q. Z. Chen, K. Rezwan, D. Armitage, S. N. Nazhat and A. R. Boccaccini, *J. Mater. Sci.: Mater. Med.*, 2006, **17**, 979.
36. Q. Z. Chen, I. Ahmed, J. C. Knowles, S. N. Nazhat, A. R. Boccaccini and K. Rezwan, *J. Biomed. Mater. Res., Part A*, 2008, **86**, 987.
37. A. Lopez-Noriega, D. Arcos and M. Vallet-Regi, *Chem. – Eur. J.*, 2010, **16**, 10879.
38. M. Manzano *et al.*, *Chem. Eng. J.*, 2008, **137**, 30–37.
39. Q. Z. Chen *et al.*, *Acta Biomater.*, 2007, **3**, 551.
40. A. Bonici, G. Lusvardi, G. Malavasi, L. Menabue and A. Piva, *J. Eur. Ceram. Soc.*, 2012, **32**, 2777.
41. H. S. Azevedo, I. B. Leonor, C. M. Alves and R. L. Reis, *Mater. Sci. Eng., C*, 2005, **25**, 169.

42. L. T. De Jonge, S. C. G. Leeuwenburgh, J. G. C. Wolke and J. A. Jansen, *Pharm. Res.*, 2008, **25**, 2357.
43. T. Hanawa, *Jpn. J. Dent. Sci. Rev.*, 2010, **46**, 93.
44. S. J. Xiao, G. Kenausis and M. Textor, in *Titanium in Medicine*, ed. D. M. Brunette, P. Tengvall, M. Textor and P. Thomsen, Springer-Verlag, Berlin/Heidelberg/New York, 2001, pp. 417–453.
45. S. Spriano, S. Ferraris, C. L. Bianchi, C. Cassinelli, P. Torricelli, M. Fini, L. Rimondini and R. Giardino, in *Titanium Alloys: Preparation, Properties and Applications*, ed. P. N. Sanchez, Nova Science Publishers Inc., 2010, ch. 5, pp. 269–294.
46. V. Luginbuehl, L. Meinel, H. P. Merkle and B. Gander, *Eur. J. Pharm. Biopharm.*, 2004, **58**, 197.
47. S. H. Lee and H. Shin, *Adv. Drug Delivery Rev.*, 2007, **59**, 339.
48. R. Bosco, E. R. Urquia Edreira, J. G. C. Wolke, S. C. G. Leeuwenburgh, J. J. J. P. van den Beucken and J. A. Jansen, *Surf. Coat. Technol.*, 2013, **233**, 91.
49. S. Yoshida, K. Hagiwara, T. Hasebe and A. Hotta, *Surf. Coat. Technol.*, 2013, **233**, 99.
50. Q. Long, D. L. Zhou, X. Zhang and J. B. Zhou, *Front. Mater. Sci.*, 2014, **8**, 157.
51. I. B. Leonor, C. M. Alves, H. S. Azevedo and R. L. Reis, *Mater. Sci. Eng., C*, 2009, **29**, 913.
52. M. Zhu, J. Zhang, C. Tao, X. He and Y. Zhu, *Mater. Lett.*, 2014, **115**, 194.
53. S. Hattar, A. Asselin, D. Greenspan, M. Oboeuf, A. Berdal and J. M. Sautier, *Biomaterials*, 2005, **26**, 839.
54. E. Vernè, S. Ferraris, C. Cassinelli and A. R. Boccaccini, *Surf. Coat. Technol.*, 2015, **264**, 132.
55. E. Vernè, S. Ferraris, C. Vitale-Brovarone, A. Cochis and L. Rimondini, *Appl. Surf. Sci.*, 2014, **313**, 372.
56. S. Ferraris, C. Vitale-Brovarone, O. Bretcanu, C. Cassinelli and E. Vernè, *Appl. Surf. Sci.*, 2013, **271**, 412.
57. W. Xia and J. Chang, *J. Non-Cryst. Solids*, 2008, **354**, 1338–1341.
58. W. Xia and J. Chang, *J. Controlled Release*, 2006, **110**, 522–530.
59. L. Zhao, X. Yan, X. Zhou, L. Zhou, H. Wang, J. Tang and C. Yu, *Microporous Mesoporous Mater.*, 2008, **109**, 210–215.
60. F. Baino *et al.*, *J. Appl. Biomater. Funct. Mater.*, 2012, **10**, 12–21.
61. E. Boanini *et al.*, *Materials*, 2016, **9**, 135.
62. M. Colilla, M. Manzano, I. Izquiedo-Barba, M. Vallet-Regì, C. Boissiere and C. Sanchez, *Chem. Mater.*, 2010, **22**, 1821–1830.
63. S. Quideau, D. Deffieux, C. Douat-Casassus and L. Pouysegu, *Angew. Chem., Int. Ed.*, 2011, **50**, 586.
64. L. Bravo, *Nutr. Rev.*, 1998, **56**, 317.
65. S. C. Gupta, A. K. Tyagi, P. Deshmukh-Taskar, M. Hinojosa, S. Prasad and B. B. Aggarwal, *Arch. Biochem. Biophys.*, 2014, **559**, 91.
66. C. J. Weng and G. C. Yen, *Cancer Treat. Rev.*, 2012, **38**, 76.
67. L. C. Gerhardt *et al.*, *Biomaterials*, 2011, **32**, 4096.
68. E. Vernè *et al.*, *Acta Biomater.*, 2010, **6**, 229.

CHAPTER 10

Bioactive Glass Particulate-incorporated Polymer Composites

SHIVA NASERI,[a] ALDO R. BOCCACCINI[b] AND SHOWAN N. NAZHAT*[a]

[a] Department of Mining and Materials Engineering, McGill University, Montreal, QC, Canada; [b] Institute of Biomaterials, University of Erlangen-Nuremberg, 91058 Erlangen, Germany
*Email: showan.nazhat@mcgill.ca

10.1 Introduction

Tissue engineering (TE) is a multidisciplinary field that utilizes the expertise and knowledge of various subjects to develop and engineer systems intended to partially or fully repair damaged tissues or enhance their functions.[1] One approach in TE is to provide a supportive system (template or scaffold) for cells and growth factors to heal and repair the damaged tissue. A combination of biomaterials and fabrication techniques can determine the optimal supportive system for a desired application.[2,3]

According to their interaction with living tissues, biomaterials can be categorized either as bioinert (chemically inert forming a fibrous tissue layer), bioresorbable (dissolves, degrades, and is replaced by the surrounding tissue) or bioactive.[4,5] In the early 1970's, Hench and co-workers

RSC Smart Materials No. 23
Bioactive Glasses: Fundamentals, Technology and Applications
Edited by Aldo R. Boccaccini, Delia S. Brauer and Leena Hupa

Published by the Royal Society of Chemistry, www.rsc.org

defined a bioactive material as "one that elicits a specific biological response at the interface of the material, which results in the formation of a bond between the tissues and the material".[4] Specific compositions of silicate-based glasses in the system of SiO_2–Na_2O–CaO–P_2O_5 are classified as bioactive glasses.[6] The most well known and clinically used composition is 45S5 Bioglass® of composition: 45 SiO_2–24.5 Na_2O–24.5 CaO–6 P_2O_5 (wt%).[6,7] Indeed, silicate-based bioactive glasses are the most widely used and studied,[7] however, more recently, both borate- and phosphate-based glasses have been researched. Borate-based glasses have been investigated as they have shown higher carbonate hydroxyapatite (CHA) conversion rates than silicate-based glasses.[8] In addition, soluble phosphate-based glasses (PGs) have been utilized to provide a desirable scaffold for TE applications.[9] PGs have been studied for a wide range of biomedical applications since their degradation rates can be controlled through their chemistry.[7] The flexibility in ionic doping allows for tuneable dissolution properties of the glass. Indeed, numerous PGs, with a wide range of different compositions and degradation rates have been fabricated[7] and investigated for potential applications in bone repair and regeneration,[10] the repair of damaged nerves,[11] angiogenesis and cell differentiation[12] and the engineering of human craniofacial muscle.[13]

Biocompatibility and mechanical properties of biomaterials are determinant factors for TE applications. Bioactive glasses are biocompatible, biodegradable, osteoconductive (the ability to form bone on their surface[14]), and osteoinductive (the ability to induce osteogenic differentiation to form bone cells[14]).[15] However, bioactive glasses are brittle, which limits their applications.[16,17] In order to overcome some of the drawbacks of bioactive glasses, one strategy has been to form a new complex through their combination with polymers that fulfill the properties potentially suitable for TE purposes. By learning from nature one can develop novel multifunctional composites that possess favourable material properties. For example, bioinert or degradable polymers can be combined with bioactive glasses to improve their mechanical, bioactive and biological properties. However, the composition ratio and fabrication process should be carefully chosen to avoid any potential negative effects on the interfacial properties, the bioactivity of the glasses or biocompatibility of the final composites. There are several processing techniques to fabricate either dense or three-dimensional (3D) structures. In this chapter, the main processing techniques for fabricating bioactive and soluble glass particulate filler composites, both porous and non-porous, are reviewed. In particular, it focuses on composites based on classical degradable polymers, such as polyesters and polyhydroxyalkanoates as well as some naturally derived polymers. Furthermore, the characteristics of each processing technique are discussed. Table 10.1 lists some of the advantages and disadvantages of these approaches.

Table 10.1 Summary of fabrication techniques and their advantages/disadvantages.

	Fabrication techniques	Advantages	Disadvantages
Dense composites	Melt blending and injection moulding	– Ability to fabricate different shapes.	– High temperature process.
	Melt blending and Compression moulding	– Easy processing method.	– Complex structure is not possible. – Damage owing to excessive shear.
	Solvent casting	– Easy processing method.	– Use of organic solvent. – Complex structures are not possible. – Uneven shrinkage may occur. – Porosity.
Porous composites	Particulate leaching	– Pore size, pore shape and porosity can be controlled. – Graded porosity is possible.	– Poor interconnectivity. – Particulates may not be leached out completely. – Restricted porogen shapes (*e.g.* cubic salt).
	Gas foaming	– No need of organic solvent. – No solvent (gas) residuals in the system.	– CO_2 may not dissolve all types of polymers.
	Thermally induced phase separation	– High level of porosity is possible. – High pores interconnectivity. – Pore size and shape can be tuned. – Tubular and anisotropic pores are possible.	– Long process due to the sublimation. – May use organic solvent. – Shrinkage can occur.
	Solid freeform fabrication	– Possibility of scale up. – Ability to fabricate complex structure and fine features. – High control of porosity, pore size and pore shape based on the specific application.	– Expensive technique. – Need a professional technician. – May use organic solvent. – Binder requirements. – Limited materials. – UV curing agent toxicity.
	Microsphere sintering	– Ability to fabricate complex structures. – Possibility of graded porosity. – Control on porosity.	– Poor interconnectivity.
	Coated scaffolds	– Simple technique.	– Possibility to block pores.
	Electrospinning	– Versatile and simple method to produce fibrous structure. – Ability to tune morphologies and structures.	– Use of organic solvents. – Particulate charge.

10.2 Techniques for the Processing of Dense Composites Incorporating Bioactive Glass

10.2.1 Melt Blending and Injection Moulding

Melt extrusion is a conventional method to produce polymer composites. Thermoplastic polymers can be reshaped and softened above either their glass transition (in amorphous polymers) or melting (in semi-crystalline polymers) temperatures and solidified upon cooling. In this method a single or combination of polymers are melt mixed with bioactive glass particles in a heated rotating screw extruder followed by injection into a mould of a desired shape and eventually solidified upon cooling of the mould.[18] For example, melt blending followed by thermal injection moulding has recently been applied to fabricate a polycaprolactone (PCL) polymer–nano bioactive glass composite for bone tissue regeneration which showed enhanced mechanical properties in comparison with bioactive glass and neat PCL.[19] It was shown that adding 0 to 40 wt% nano-bioactive glass resulted in the formation of a composite with elastic modulus in the range of 198 ± 13 to 851 ± 43 MPa, tensile strength of 19–21.5 MPa, and high bioactivity in simulated body fluid (SBF).[19] However, a drawback of this technique is that biomolecules and proteins cannot be added during the fabrication process as the elevated temperatures can cause them to denature and degrade. Much research has studied the cellular responses to different composites containing bioactive glasses. Vergnol *et al.* investigated dose-dependent cellular (MG63 osteoblastic cells) responses to 45S5 Bioglass® (content 0, 20, 30 and 50 wt%) in poly(L,DL) lactide matrices.[20] There was no significant difference in cell viability between 20 and 30 wt% bioactive glass, however, a composite containing 50 wt% bioactive glass showed less cell viability and poor *in vitro* cytocompatibility,[20] which may be due to a higher initial ion release rate.[21]

10.2.2 Melt Blending and Compression Moulding

Compression moulding is a relatively simple technique for fabricating a composite incorporating bioactive glass particles. One approach in this method is to apply high pressure and temperature to pre-extruded composite pellets to produce a compact structure. However, the main drawback of this technique is the difficulty of fabricating large and thick structures owing to excessive time required to obtain evenly pressed and heated composites, and complex structures owing to limited flow of materials. Another disadvantage is that particles may not be homogenously distributed in case of dry-state blending. Rich *et al.* investigated the effect of S53P4 (53 SiO_2–23 Na_2O–20 CaO–4 P_2O_5 (wt%)) bioactive glass content (40, 60 and 70 wt%) and particle size (<45 and 90–135 μm) on the properties of poly(ε-caprolactone-*co*-DL-lactide) (P(CL/DL-LA)) composites fabricated by compression moulding.[22] The results showed that higher bioactive glass loading and smaller particle size leads to more rapid bioactivity rates. It was concluded that the incorporation of the smaller sized particles at lower content was more beneficial than the incorporation of larger sized particles at higher content,[22] attributable to the higher surface areas and therefore greater exposure to SBF. Additionally, degradation was determined by evaluating the water uptake and molecular weight change in SBF.[22] Due to the hydrophobic nature of PCL, the neat copolymer had less than 1% water uptake after 6 months immersion in SBF.[22] It was also shown that the incorporation of bioactive glass particles into the copolymer increased water absorption and decreased the polymer molecular weight due to the larger amount of interfacial surfaces between filler and matrix, with the smaller sized particles dramatically increasing the polymer degradation rate.[22] Blaker *et al.* studied the effect of thermal processing on composites of poly(α-hydroxyesters) and bioactive glasses.[23] It was shown that the compression moulding of poly-DL-lactide (PDLLA) at high temperature in the presence of 45S5 Bioglass® filler, led to premature polymer degradation and reduction in its molecular weight as well as poor interfacial adhesion between matrix and filler. As a consequence, the composites resulted in a reduction in mechanical stability.[23] Simpson *et al.* investigated the effect of different fillers, including calcium carbonate, hydroxyapatite, bioactive glass 45S5 and ICIE4 (49.46 SiO_2–6.6 Na_2O–42.87 CaO–1.07 P_2O_5 (mol%)), on the properties of poly(lactide-*co*-glycolide) (PLGA) matrix composites.[24] PLGA granules with 25 vol% bioactive fillers were melt-blended and composites were produced by either compression or rudimentary injection moulding. Thermal analysis showed that the PLGA degradation temperature was significantly decreased upon addition of 45S5 and ICIE4 bioactive glasses. Additionally, the PLGA molecular weight decreased due to processing in the presence of these bioactive glasses. It was shown that the main reduction occurs during the melt-blending (extrusion) step and further processing (injection or moulding) leads to greater reduction in the molecular weight. As a conclusion, high temperature processing techniques may not be optimal for the fabrication of bioactive glass composites with poly(α-hydroxyester) matrices.[23,24]

10.2.3 Solvent Casting

Another commonly used composite fabrication method is the solvent casting technique. In this method, a polymer is dissolved in an organic solvent, mixed with bioactive glass particles and cast into a predefined mould, where a solid structure is finally formed after the solvent is evaporated.[17] Ease of fabrication and low temperature processing are some of its advantages. However, a challenge in this process is the reliance on organic solvents that could impact the efficacy of biomolecules and be toxic to cells. Moreover, the glass particles tend to agglomerate and sediment in the polymer solution leading to their inhomogeneous distribution within the composite matrix. Another drawback is process scalability as solvent removal is very time consuming and may result in residual solvent within the final composite structure which is toxic.[17,25] As an example of the application of solvent casting, a non-porous membrane composite of chitosan and 45S5 Bioglass® in nano- and micro-particle sizes has been fabricated for bone regeneration applications.[26] In addition, the effect of the bioactive glass composition has been studied in PCL matrix composites with 50 wt% 45S5 silicate-based glass and 13-93 B3 (53 B_2O_3–12 K_2O–20 CaO–6 Na_2O–5 MgO–4 P_2O_5 (wt%)) borate-based glass.[27] It was shown that the borate-based glass filler was fully converted to CHA in 3 days, whereas only half of the silicate-based glass was converted to CHA during the same period.[27] Furthermore, borate-based glass showed increased degradation and ion release rates compared to silicate-based glass due to the formation of a silica rich layer in the latter. Shah Mohammadi *et al.* fabricated a composite of PCL and mixed PG fillers (Si- and Fe-doped PGs).[28] PCL/PG particle composite films were initially prepared by the solvent casting method, then five composite films were compression moulded. Weight loss and mechanical properties of composites were examined to investigate the effect of PG composition on the PCL matrix. It was shown that the incorporation of higher Si-doped PGs resulted in an increase in weight loss and Young's modulus but also a decrease in flexural strength and strain at failure. It was proposed that higher Si content caused a reaction between the PG fillers and the PCL matrix during high temperature processing and resulted in a reduction in PCL molecular weight.[28] The incorporation of bioactive glass in poly(3-hydroxybutyrate) (P(3HB)) through solvent casting resulted in a decrease in the polymer crystallinity and the composite Young's modulus, and an increase in the composite hardness.[29] The reduction of Young's modulus was attributed to the weak interfacial bond between the polymer and filler particles.[29] Tamjid *et al.* investigated the effect of 45S5 Bioglass® particle size on the characteristics of PCL matrix composites prepared using the solvent casting method.[30] The effect of bioactive glass particles (5 wt%) of three different particles sizes: 6 μm, 250 nm and less than 100 nm were investigated. *In vitro* bioactivity was evaluated by SBF conditioning. It was shown that the 250 nm particle size resulted in greater increase in the bioactivity of PCL film in comparison with the 6 μm particle size.[30] This bioactivity enhancement was proposed to occur due to the higher surface area in contact with the SBF solution and better nanotopography. However, composites with less than 100 nm particle size showed lower bioactivity compared to 250 nm particle size, which was attributed to nanoparticles agglomeration and partial crystallization during the mechanical milling.[30]

10.2.4 Coatings

The coating of dense structures with bioactive glass particles can enhance the bioactivity, mechanical and surface properties, as well as the fixation of sutures and dental and orthopaedic implants for biomedical applications.[31–34] Various techniques, such as slurry dipping, electrochemical, electrophoretic and laser depositions have been applied to coat dense structures.[31,35,36] For example, a sol–gel-derived silver-doped bioactive glass was used to coat resorbable and non-resorbable sutures by applying the slurry dipping technique.[31] It was shown that adding silver ions to conventional sutures enhanced the bioactivity and antimicrobial properties, making the product highly desirable for wound healing and tissue engineering applications.[31] As another example, polymethyl methacrylate (PMMA) was coated with bioactive glasses (S53P4) to enhance adhesion between implant and skull for craniofacial bone reconstruction.[34]

10.3 Techniques for the Processing of Porous Composite Scaffolds Incorporating Bioactive Glass

10.3.1 Particulate Leaching

Particulate leaching is a common approach to produce porous scaffolds. The addition of soluble/extractable templating agents in a dense composite produced through either solvent casting or moulding, which then leach out in an aqueous solvent results in a porous structure.[25,37] The main advantage of the particulate leaching technique is that it offers control over porosity, where the leaching agent can be used to tune the pore size and shape, thus enabling the fabrication of graded porosity structures. For example, in many biomedical applications, scaffolds with interconnected pores are required for vascularization. However, it is difficult to obtain interconnected pores through this technique especially at low porosity and with a low particulate volume fraction.[37] Another drawback is the potential for residual porogens in the scaffold.[25] A composite scaffold of P(3HB) and 45S5 Bioglass® was fabricated for bone tissue engineering applications through solvent casting particulate leaching (sugar cubes) technique, resulting in 85% interconnected porosity.[38] It was shown that the addition of bioactive glass promoted the bioactivity of the composite, along with cytocompatiblity and antibacterial properties.[38] Alternatively, composites can be initially processed through compression moulding followed by particulate leaching. For example, a porous composite of poly(3-hydroxybutyrate-*co*-3-hydroxyvalerate) and bioactive glass 58S (60 SiO_2–36 CaO–4 P_2O_5 (mol%)) was fabricated through the addition of NaCl particulates (150–300 μm particle size range). Three different scaffolds with 0, 10 and 20 wt% of bioactive glass were produced for further characterizations.[39] It was shown that an increase in filler content did not alter the porosity, though there was an increases in compressive yield strength and bioactivity.[39] It has been shown that the addition of bioactive glass to zein (a biocompatible protein from corn with antimicrobial properties[40]) enables formation of a porous bioactive scaffold, on which hydroxyapatite can be deposited;[41] a zein–45S5 Bioglass® composite scaffold was recently fabricated by employing compression moulding and particulate leaching (NaCl salt of 72–175 μm particle size range).[41]

10.3.2 Gas Foaming

In order to overcome toxicity of organic solvents, a supercritical liquid was proposed for use as a solvent. Here, a saturated polymer–gas solution is prepared at high pressure and temperature then, upon sudden gas release at low pressure and temperature, a solid porous structure will be formed.[42,43] Carbon dioxide is non-toxic and a gas at room temperature, which acts as a supercritical fluid (having the properties of both liquid and gas) above a critical pressure (72.9 atm) and temperature (304.25 K). Additionally, as opposed to solvents, it can be completely removed from the scaffold.[44] Recently, a composite hydrogel of chitosan and sol–gel bioactive glass was fabricated by using carbon dioxide gas to make a porous foam.[45] It was shown that the compressive modulus was increased with bioactive glass incorporation, though at higher filler content, there was a decrease in the average pore size due to particle agglomeration and the blockage of pores.[45] PG fillers have also been used in polymeric composites in order to develop fully degradable scaffolds for TE applications. As an example, dense and porous composites of polylactic acid (PLA) and PG particles in the ternary system (P_2O_5–CaO–Na_2O) have been developed using both compression moulding and gas foaming techniques.[42] In this study, the effect of PG filler content up to 20 wt% (0, 5, 10 and 20 wt%) on the PLA matrix was investigated. It was shown that an increase in PG filler content led to an increase in storage modulus in dense structures, as well as an increase in degradation rate. However, the filler content did not significantly affect the compressive modulus of the foamed composites. Additionally, the proliferation rate of seeded human foetal bone cells was initially higher when seeded on the neat PLA foams when compared to composite foams. It was proposed, this was due to the rapid degradation of the PG filler that initially hinders cellular attachment to the foam.[42]

10.3.3 Thermally Induced Phase Separation

A 3D structure with high porosity (more than 95%) can be produced by employing the thermally induced phase separation (TIPS) technique.[46] In this technique, a homogenous polymer solution is prepared by dissolution in an organic solvent, and mixed with bioactive glass particles. Liquid nitrogen is then used to quench the polymer–bioactive glass solution. The frozen solvent is sublimated in a freeze-dryer or cooling bath under vacuum. Sublimation of frozen solvent leads to the formation of tubular and interconnected pores with high porosity.[46] Various processing parameters of the TIPS method have been studied to optimize and tune the final foam properties.[46] These parameters, including polymer concentration, polymer molecular weight and cooling rate, can affect the pore morphology, mechanical properties, degradation rate and bioactivity of the final composite.[46,47] Furthermore, the mixture can be sonicated to achieve a homogenous distribution of bioactive glass particles within the polymer solution.

Several bioactive glass composites of PDLLA and PLGA matrices and 45S5 Bioglass® particles (with 10, 25 and 50 wt%) have been fabricated through TIPS.[48] It was shown that PDLLA/45S5 Bioglass® resulted in slightly higher pore volume than PLGA/45S5 Bioglass® composites. In addition, composites with higher filler content had more irregular pore structure, but resulted in higher compressive modulus, bioactivity and degradation rate.[48] Tamjid *et al.* produced a highly porous PCL/45S5 Bioglass® composite using the TIPS technique.[30] Compression tests indicated that the addition of 5 wt% bioactive glass particles (6 μm) increased the elastic modulus up to 427 ± 4 kPa compared to 128 ± 3 kPa for unmodified PCL. The addition of particles of smaller size (250 nm) resulted in higher elastic modulus values reaching 819 ± 9 kPa.[30] However, no significant difference was reported between glass particles of 6 μm and less than 100 nm,[30] which may have been due to the agglomeration and inhomogeneous distribution of the nanosized particles. It is worth mentioning that as a result of mechanical milling, nanosized particles can turn partially crystalline on the surface, which may affect the particle surface properties and cause less interfacial adhesion between glass particles and polymer matrix compared to amorphous glass particles.[30] Dziadek *et al.* investigated a composite of PCL and 21 vol% bioactive glass filler using two different fabrication techniques including particulate leaching of solvent-cast composites and solid–liquid phase separation.[49] It was shown that these two techniques resulted in formation of composites with different porosity levels, pore sizes, and shapes. When the solvent-cast particulate leaching method was employed, a porous composite was formed with interconnected, irregular and well-distributed pores, where the pore sizes varied in the range of 10 to 300 μm (Figure 10.1a). In contrast, when TIPS was applied, oriented elongated pores with smaller pore size (10 to 150 μm) were formed (Figure 10.1b).[49] The mechanical properties of highly porous (~93%) PDLLA/45S5 Bioglass® composites (2 and 15 vol% filler) were investigated through quasi-static and dynamic mechanical analysis in

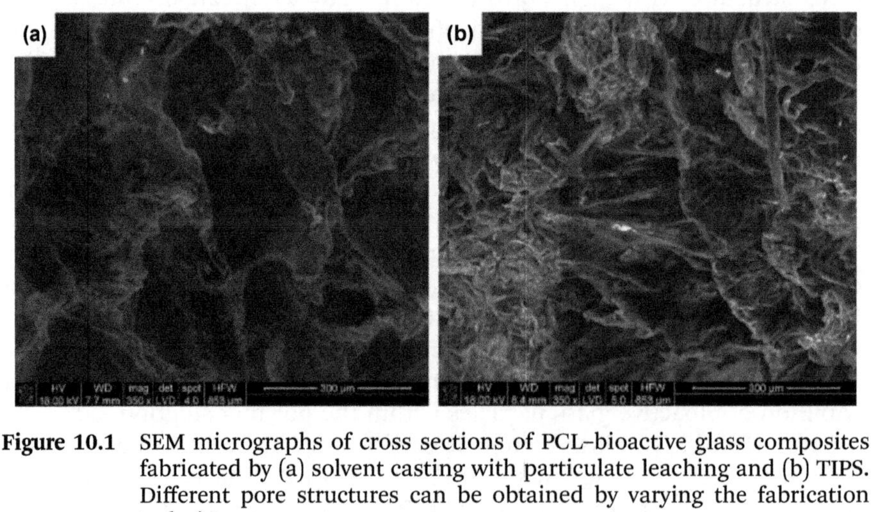

Figure 10.1 SEM micrographs of cross sections of PCL–bioactive glass composites fabricated by (a) solvent casting with particulate leaching and (b) TIPS. Different pore structures can be obtained by varying the fabrication technique.
Reproduced from ref. 49 with permission from John Wiley and Sons.

compression mode.[50] The results showed that the mechanical properties were affected by the composite anisotropic pore structures defined by continuous tubular macro-pores and interconnected by a micro-porous network. Additionally, no significant differences between neat (unmodified) polymer and composites were observed. It was proposed that the high porosity of the composites was a dominant factor when comparing with the effect of the filler.[50] Longer-term degradation of PDLLA/45S5 Bioglass® composites fabricated through TIPS has also been examined.[51] It was shown that the degradation process consisted of three main stages: (1) water absorption and plasticization with weight loss due to glass particles dissolution, as well as reduction in wet mechanical stability; (2) a small increase in wet mechanical properties; and (3) a dramatic increase in weight loss and disruption of the pore structure.[51]

Freeze-drying has also been used to develop naturally derived (*e.g.* collagen, chitosan and silk fibroin) hydrogel-based porous composites with bioactive glass resulting in a 3D structure with high porosity (~85%).[52] In this method, the hydrogel is frozen and freeze-dried at low temperature and pressure to provide a 3D porous structure. The pore size and structure has a direct relationship with the ice crystals formed after freezing. Temperature forms one of the parameters that can be varied to adjust pore size, where at lower temperatures smaller ice crystals can be formed resulting in sponges with smaller pore size.[52] In addition, various crosslinking approaches have been used to increase the mechanical properties of the sponges.[53] However, the toxicity of chemical agents is an important issue and should be considered. Another challenge in this method is making a composite with a homogenous bioactive glass filler distribution in the matrix.[52] Nevertheless, the freeze-drying method is a well-used technique to produce collagen–bioactive glass composites.[53] For example, Xu *et al.* studied a composite hydrogel of collagen–phosphatidylserine and bioactive glasses (58S, of 70–100 nm particle size range).[54] It was shown that lower porosity (84.75, 75.40, and 70.67%) and higher compression strength (0.4827, 1.5469, and 1.9724 MPa) values were obtained at higher bioactive glass contents (50, 65 and 80 wt%, respectively). Additionally, biocompatibility of composites was verified via *in vitro* study.[54]

10.3.4 Solid Freeform Fabrication

Recently, the solid freeform fabrication (SFF) or rapid prototyping (RP) technique has been extensively studied in order to fabricate complex structures of 3D scaffolds for tissue engineering applications.[55,56] In this technique, computer-aided design (CAD) is utilized to form a 3D structure fabricated layer-by-layer. In contrast to most of the common fabrication techniques, this benefits from a great potential scale-up and can reproduce complex structures and compositions. In addition, scaffold properties such as interconnectivity, shape, orientation and pore size can be customized.[55] There are different types of SFF technique, such as using *stereolithography apparatus*, using UV light to fabricate photo polymerisable liquid materials,[57] *selective laser sintering*, using a CO_2 laser to sinter powdered polymer materials,[58] *fused deposition modelling*, using heating and extrusion approaches,[59] and *3D printing/direct ink writing*.[60] Among these techniques, 3D printing is the most studied approach for biomedical applications. It was first introduced at the Massachusetts Institute of Technology by employing an inkjet printing technique and using a liquid binder solution.[61] Each layer is formed by selective deposition of binder solution along the powder bed of the desired material, and this procedure is repeated layer-by-layer until the final structure is formed.[62] Various materials have been utilized to fabricate 3D bioactive composites. For example, a multifunctional composite containing magnetic Fe_3O_4 nanoparticles, mesoporous bioactive glasses and PCL was developed for bone regeneration, local anticancer drug release and magnetic hyperthermia.[63] It was shown that a uniform structure with 60% porosity can be fabricated by employing a 3D plotter.[63]

10.3.5 Microsphere Sintering

Utilization of microsphere technology along with a sintering process can enable the fabrication of 3D porous composites of polymers and bioactive glasses.[64] In this method, polymer–bioactive glass microspheres are first synthesized by preparing an emulsion and stirring until the solvent is completely evaporated. The polymer microspheres are then packed and sintered in a desired mould above the glass transition temperature in order to fabricate a 3D porous structure.[65] Upon cooling, a fixed 3D structure can be formed.[65] Composite structural properties can be modified by varying the heating time and microsphere size.[65] This technique enables control over porosity grade, and the fabrication of more complex structures.[17] However, the main drawback is the lack of interconnectivity between pores, which is a significant factor in tissue engineering of scaffolds for the circulation of wastes and nutrients, ingrowth of blood vessels and adhesion of cells for new ingrown tissue.[66] A 3D composite composed of PLGA and 45S5 Bioglass® was produced using this technique for bone tissue engineering applications.[67] It was shown that the composite was bioactive and biodegradable, with enhanced mechanical properties in comparison with neat PLGA scaffolds. In addition, adhesion and the proliferation and differentiation of seeded human osteoblast-like cells was demonstrated.[67] A similar composite of PLGA and 30 wt% 45S5 Bioglass® promoted the osteogenesis of marrow stromal cells.[68]

10.3.6 Coated Porous Scaffold

Another method to obtain a composite scaffold is through the coating of 3D porous polymers.[25,69] Porous polymer foams can be coated with bioactive glass particles of different sizes through slurry dipping in order to promote

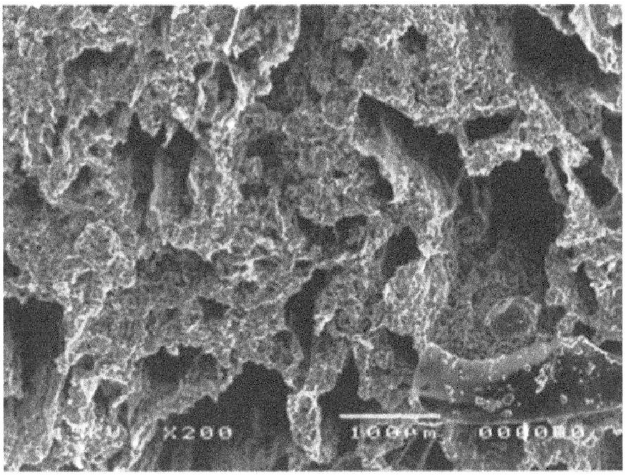

Figure 10.2 SEM micrograph of cross-sections of PDLLA foam coated by 45S5 Bioglass® using slurry dipping technique; homogenous distribution of Bioglass® particles within polymeric foam can be observed. J. A. Roether, Novel bioresorbable and bioactive composites based on bioactive glass and polylactide foams for bone tissue engineering, *J. Mater. Sci.*, 2002, **13**, with permission of Springer.[70]

the scaffold bioactivity.[25] A macroporous PDLLA foam was coated with 45S5 Bioglass® of less than 5 μm size range by employing a slurry of 42 wt% bioactive glass in distilled water.[70] A uniform and homogenous distribution of bioactive glass particles was achieved resulting in enhanced bioactivity (Figure 10.2).[70] The same approach has been used to fabricate a calcium phosphate with an interconnected porous structure and 80% porosity coated with 38 wt% 58S bioactive glass–PLGA. The resultant material showed improved mechanical properties with compressive modulus values suitable for low load-bearing applications and increased bioactivity.[71] On the other hand, it has been shown that the coating of a polymeric structure with bioactive glass can adversely affect its mechanical stability. Stamboulis *et al.* demonstrated a decrease in tensile strength when coating a polyglactin 910 (Vicryl®) suture with 45S5 Bioglass®.[72] However, upon conditioning in SBF, the coated suture indicated a lower rate of decrease in tensile strength attributable to the hydroxyapatite formation.[72]

10.3.7 Electrospinning

Fibrous structures are attractive for biomedical applications since they can potentially be designed to mimic the native extracellular matrix architecture. One approach to fabricate fibrous composites is through electrospinning. In order to develop composites, bioactive glass particulates can be incorporated into fibrous polymers to adjust their biological and mechanical properties. Electrospinning is a process that uses electrical forces to form fibres with a wide range of diameters from nanometres to several micrometres.[73,74] All electrospinning devices contain three main parts: a voltage power supply, a spinneret and a collector (often a thin foil sheet or a rotating mandrel).[73] In this technique, the solution is highly charged and then moved toward a collector of opposite polarity to enforce the repulsive electrical forces and overcome the surface tension forces which would result in formation of fibres.

Along with the inherent polymer properties, other factors such as fibre length, diameter, orientation, volume fraction and interaction between fibres and particles play significant roles in the composite reinforcement efficacy.[75,76] It was shown that the manufacturing parameters including solution viscosity, flow rate, applied voltage and distance between needle and collector significantly affect the fibre quality.[76,77] A number of studies have investigated different composites of natural and synthetic polymers with inorganic particulates produced by electrospinning. El-Fiqi *et al.* developed an electrospun fibrous composite of mesoporous bioactive glass nanoparticles (0, 2.5, 5, and 10 wt%) incorporated into PCL–gelatin fibres for bone tissue engineering applications.[78] Upon addition of bioactive glass nanoparticles, there was an increase in both elastic modulus and tensile strength and a decrease in elongation to failure. Additionally, there was an improvement in bioactivity and HA deposition. Furthermore, it was demonstrated that therapeutic drugs can be loaded into the mesoporous bioactive glass to fabricate a fibrous composite allowing for potential drug delivery.[78] Noh *et al.* created an electrospun composite of bioactive glass nanoparticles (up to 20 wt%) in a PLA matrix.[79] The addition of nanoparticle fillers up to 10% showed good processability and fibrous morphology without the formation of beads. *In vitro* bioactivity demonstrated that HA formation was enhanced with nanofiller content (Figure 10.3).[79]

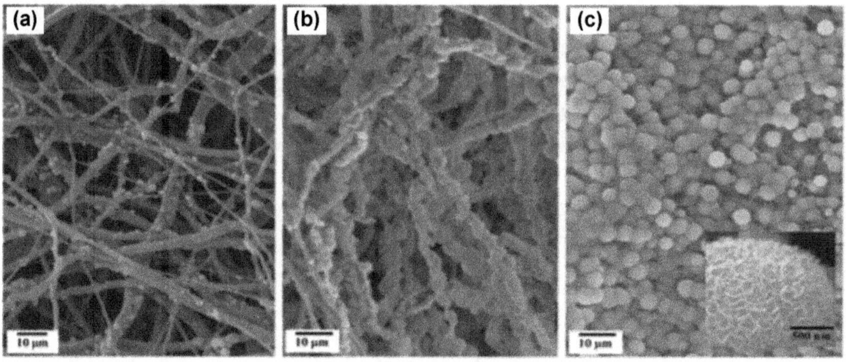

Figure 10.3 SEM micrographs of bioactive glass nanoparticulates (10 wt%)–PCL fibrous composite after (a) 3 days, (b) 7 days, and (c) 14 days immersion in SBF.
Reproduced from K.-T. Noh *et al.*, Composite nanofiber of bioactive glass nanofiller incorporated poly(lactic acid) for bone regeneration, *Mater. Lett.*, **64**, 802–805, Copyright 2010 with permission from Elsevier.[79]

10.4 Hydrogel-based Composite with Bioactive Glasses

The high water contents of biodegradable hydrogels provide interesting characteristics such as suitability for therapeutic incorporation.[80] These characteristics make hydrogels attractive to be employed in many fields such as drug delivery[81] and tissue engineering.[82] Depending on its application, a suitable hydrogel is achieved by adjusting its crosslinked structure (chemically or physically), which affects its mechanical, degradation, swelling, and diffusion properties. Different crosslinked structures can be achieved by varying the fabrication parameters. Hydrogel based composites can be based either on natural polymers[83] such as collagen, hyaluronic acid, alginate, gelatin and chitosan, or on synthetic polymers[83] such as polyethylene glycol (PEG) and poly acrylic acid (PAA). Hydrogel materials have the ability to be used as an injectable composites or as a 3D scaffolds.

There are different fabrication techniques to develop hydrogel-based composites with bioactive glasses including the plastic compression technique, and bioprinting. The plastic compression technique, introduced by Brown *et al.*[84] relies on rapidly expelling the fluid content of a highly hydrated collagen hydrogel through the application of a compressive load to generate dense biomimetic collagen gels. Marelli *et al.* developed a composite of collagen type I and 45S5 Bioglass® using the plastic compression technique.[85] Incorporation of bioactive glass was shown to enhance collagen mineralization with potential applications in bone TE.[85] In a follow up study, a pre-osteoblastic cell-seeded dense collagen–nano-bioactive glass composite was fabricated by plastic compression.[86] Addition of nano-bioactive glasses demonstrated an increase in mineralization, mechanical properties and regulated osteoblastic function within the scaffold.[86] Marelli *et al.* then developed an injectable dense collagen via using gel-aspiration ejection (GAE), where the highly-hydrated hydrogel solution is guided through negative pressure into a needle and eventually, by reversal of pressure, it is ejected as a dense collagen.[87] It was shown that collagen fibrillar densities (CFDs) and microstructures of hydrogels could be tuned by this method. Recently, Miri *et al.* used a GAE device to develop a rapidly mineralizable, injectable dense collagen–45S5 Bioglass® (25 wt%) hybrid for bone TE applications.[88] These subcutaneously-injected gels demonstrated that the incorporation of bioactive glasses into the dense collagen enhanced mineralization, angiogenesis and the formation of woven bone-like tissue.[88]

Bioprinting is another technique that has been used to develop composite hydrogels. Luo *et al.* fabricated a hierarchically organised alginate–mesoporous bioactive glass composite via the 3D printing technique.[89] It was shown that the biofabrication allows for developing a composite with hierarchal pore size including nano-channels, micropores and macropores.[89]

10.5 Conclusions and Outlook

Bioactive glasses are suitable materials for tissue engineering applications. They have been used clinically in mineralized tissue repair in dental and orthopaedic applications due to their CHA formation. Composite strategies have been widely used to fabricate novel materials and structures for specific high technology applications. Bioactive glasses can also be combined with inert, biocompatible and biodegradable polymers to generate composites of enhanced mechanical properties, as well as tailored, bioactivity and degradation properties for tissue regeneration. These composites can be either dense or porous scaffolds and have been generated using a wide range of fabrication routes. It is predicted that future research will continue to expand on the development of novel bioactive glass particulate composites with distinctive characteristics for various potential applications.

References

1. R. Langer and J. P. Vacanti, Tissue engineering, *Science*, 1993, **260**(5110), 920–926.
2. E. Eisenbarth, Biomaterials for tissue engineering, *Adv. Eng. Mater.*, 2007, **9**(12), 1051–1060.
3. D. W. Hutmacher, Scaffold design and fabrication technologies for engineering tissues – State of the art and future perspectives, *J. Biomater. Sci., Polym. Ed.*, 2001, **12**(1), 107–124.
4. L. L. Hench, R. J. Splinter, W. C. Allen and T. K. Greenlee, Bonding mechanisms at the interface of ceramic prosthetic materials, *J. Biomed. Mater. Res.*, 1971, **5**(6), 117–141.
5. L. L. Hench, Bioceramics – from Concept to Clinic, *J. Am. Ceram. Soc.*, 1991, **74**(7), 1487–1510.
6. M. N. Rahaman, D. E. Day, B. Sonny Bal, Q. Fu, S. B. Jung, L. F. Bonewald *et al.*, Bioactive glass in tissue engineering, *Acta Biomater.*, 2011, 7(6), 2355–2373.
7. E. A. Abou Neel, D. M. Pickup, S. P. Valappil, R. J. Newport and J. C. Knowles, Bioactive functional materials: a perspective on phosphate-based glasses, *J. Mater. Chem.*, 2009, **19**(6), 690–701.
8. Q. Fu, M. N. Rahaman, H. Fu and X. Liu, Silicate, borosilicate, and borate bioactive glass scaffolds with controllable degradation rate for bone tissue engineering applications. I. Preparation and in vitro degradation, *J. Biomed. Mater. Res., Part A*, 2010, **95**(1), 164–171.
9. A. M. Young, S. M. Ho, E. A. Abou Neel, I. Ahmed, J. E. Barralet, J. C. Knowles *et al.*, Chemical characterization of a degradable polymeric bone adhesive containing hydrolysable fillers and interpretation of anomalous mechanical properties, *Acta Biomater.*, 2009, **5**(6), 2072–2083.
10. V. Salih, K. Franks, M. James, G. W. Hastings, J. C. Knowles and I. Olsen, Development of soluble glasses for biomedical use Part II: The biological

response of human osteoblast cell lines to phosphate-based soluble glasses, *J. Mater. Sci.: Mater. Med.*, 2000, **11**(10), 615–620.
11. D. V. Lenihan, A. J. Carter, T. Gilchrist, D. M. Healy, I. A. Miller, L. M. Myles *et al.*, Biodegradable controlled release glass in the repair of peripheral nerve injuries, *J. Hand Surg.*, 1998, **23 B**(5), 588–593.
12. C. Sandino, S. Checa, P. J. Prendergast and D. Lacroix, Simulation of angiogenesis and cell differentiation in a CaP scaffold subjected to compressive strains using a lattice modeling approach, *Biomaterials*, 2010, **31**(8), 2446–2452.
13. R. Shah, A. C. M. Sinanan, J. C. Knowles, N. P. Hunt and M. P. Lewis, Craniofacial muscle engineering using a 3-dimensional phosphate glass fibre construct, *Biomaterials*, 2005, **26**(13), 1497–1505.
14. T. Albrektsson and C. Johansson, Osteoinduction, osteoconduction and osseointegration, *Eur. Spine J.*, 2001, **10**(Suppl 2), S96–S101.
15. H. Yuan, J. D. De Bruijn, X. Zhang, C. A. Van Blitterswijk and K. De Groot, Bone induction by porous glass ceramic made from Bioglass® (45S5), *J. Biomed. Mater. Res.*, 2001, **58**(3), 270–276.
16. M. N. Rahaman, R. F. Brown, B. S. Bal and D. E. Day, Bioactive Glasses for Nonbearing Applications in Total Joint Replacement, *Semin. Arthroplasty*, 2006, **17**(3–4), 102–112.
17. K. Rezwan, Q. Z. Chen, J. J. Blaker and A. R. Boccaccini, Biodegradable and bioactive porous polymer/inorganic composite scaffolds for bone tissue engineering, *Biomaterials*, 2006, **27**(18), 3413–3431.
18. M. Wang, Developing bioactive composite materials for tissue replacement, *Biomaterials*, 2003, **24**(13), 2133–2151.
19. L. Ji, W. Wang, D. Jin, S. Zhou and X. Song, In vitro bioactivity and mechanical properties of bioactive glass nanoparticles/polycaprolactone composites, *Mater. Sci. Eng., C*, 2015, **46**, 1–9.
20. G. Vergnol, N. Ginsac, P. Rivory, S. Meille, J. M. Chenal, S. Balvay *et al.*, In vitro and in vivo evaluation of a polylactic acid-bioactive glass composite for bone fixation devices, *J. Biomed. Mater. Res. Part B Appl. Biomater.*, 2015, **104**(1), 180–191.
21. X. B. Yang, D. Webb, J. Blaker, A. R. Boccaccini, V. Maquet, C. Cooper *et al.*, Evaluation of human bone marrow stromal cell growth on biodegradable polymer/Bioglass® composites, *Biochem. Biophys. Res. Commun.*, 2006, **342**(4), 1098–1107.
22. J. Rich, T. Jaakkola, T. Tirri, T. Närhi, A. Yli-Urpo and J. Seppälä, In vitro evaluation of poly(ε-caprolactone-co-DL-lactide)/bioactive glass composites, *Biomaterials*, 2002, **23**(10), 2143–2150.
23. J. J. Blaker, A. Bismarck, A. R. Boccaccini, A. M. Young and S. N. Nazhat, Premature degradation of poly(α-hydroxyesters) during thermal processing of Bioglass®-containing composites, *Acta Biomater.*, 2010, **6**(3), 756–762.
24. R. L. Simpson, S. N. Nazhat, J. J. Blaker, A. Bismarck, R. Hill, A. R. Boccaccini *et al.*, A comparative study of the effects of different bioactive fillers in PLGA matrix composites and their suitability as bone

substitute materials: A thermo-mechanical and in vitro investigation, *J. Mech. Behav. Biomed. Mater.*, 2015, **50**, 277–289.
25. E. Zeimaran, S. Pourshahrestani, I. Djordjevic, B. Pingguan-Murphy, N. A. Kadri and M. R. Towler, Bioactive glass reinforced elastomer composites for skeletal regeneration: A review, *Mater. Sci. Eng., C*, 2015, **53**, 175–188.
26. S. G. Caridade, E. G. Merino, N. M. Alves, V. D. Z. Bermudez, A. R. Boccaccini and J. F. Mano, Chitosan membranes containing micro or nano-size bioactive glass particles: Evolution of biomineralization followed by in situ dynamic mechanical analysis, *J. Mech. Behav. Biomed. Mater.*, 2013, **20**, 173–183.
27. A. Mohammadkhah, L. M. Marquardt, S. E. Sakiyama-Elbert, D. E. Day and A. B. Harkins, Fabrication and characterization of poly-(ε)-caprolactone and bioactive glass composites for tissue engineering applications, *Mater. Sci. Eng., Proc. Conf.*, 2015, **49**(0), 632–639.
28. M. S. Mohammadi, I. Ahmed, B. Marelli, C. Rudd, M. N. Bureau and S. N. Nazhat, Modulation of polycaprolactone composite properties through incorporation of mixed phosphate glass formulations, *Acta Biomater.*, 2010, **6**(8), 3157–3168.
29. S. K. Misra, S. N. Nazhat, S. P. Valappil, M. Moshrefi-Torbati, R. J. K. Wood, I. Roy *et al.*, Fabrication and characterization of biodegradable poly(3-hydroxybutyrate) composite containing bioglass, *Biomacromolecules*, 2007, **8**(7), 2112–2119.
30. E. Tamjid, R. Bagheri, M. Vossoughi and A. Simchi, Effect of particle size on the in vitro bioactivity, hydrophilicity and mechanical properties of bioactive glass-reinforced polycaprolactone composites, *Mater. Sci. Eng., Proc. Conf.*, 2011, **31**(7), 1526–1533.
31. J. J. Blaker, S. N. Nazhat and A. R. Boccaccini, Development and characterisation of silver-doped bioactive glass-coated sutures for tissue engineering and wound healing applications, *Biomaterials*, 2004, **25**(7–8), 1319–1329.
32. M. Xuereb, J. Camilleri and N. J. Attard, Systematic review of current dental implant coating materials and novel coating techniques, *Int. J. Prosthodont.*, 2015, **28**(1), 51–59.
33. A. Simchi, E. Tamjid, F. Pishbin and A. R. Boccaccini, Recent progress in inorganic and composite coatings with bactericidal capability for orthopaedic applications, *Nanomedicine*, 2011, 7(1), 22–39.
34. M. J. Peltola, P. K. Vallittu, V. Vuorinen, A. A. J. Aho, A. Puntala and K. M. J. Aitasalo, Novel composite implant in craniofacial bone reconstruction, *Eur. Arch. Oto-Rhino-Laryngol.*, 2012, **269**(2), 623–628.
35. A. R. Boccaccini, S. Keim, R. Ma, Y. Li and I. Zhitomirsky, Electrophoretic deposition of biomaterials, *J. R. Soc., Interface*, 2010, 7(SUPPL. 5), S581–S613.
36. B. Saravanakumar, M. Prabhu, V. Rajendran, J. Gopal and U. K. Mudali, Electrochemical deposition of 58SiO2-33CaO-9P2O5 nanobioactive glass

particles on Ti-6Al-4V alloy for biomedical applications, *Int. J. Appl. Ceram. Technol.*, 2015, **12**(1), 95–105.
37. A. R. Boccaccini and J. J. Blaker, Bioactive composite materials for tissue engineering scaffolds, *Expert Rev. Med. Devices*, 2005, **2**(3), 303–317.
38. S. K. Misra, T. I. Ansari, S. P. Valappil, D. Mohn, S. E. Philip, W. J. Stark *et al.*, Poly(3-hydroxybutyrate) multifunctional composite scaffolds for tissue engineering applications, *Biomaterials*, 2010, **31**(10), 2806–2815.
39. H. Li, R. Du and J. Chang, Fabrication, characterization, and in vitro degradation of composite scaffolds based on PHBV and bioactive glass, *J. Biomater. Appl.*, 2005, **20**(2), 137–155.
40. P. Zhou, Y. Xia, J. Wang, C. Liang, L. Yu, W. Tang *et al.*, Antibacterial properties and bioactivity of HACC- and HACC-Zein-modified mesoporous bioactive glass scaffolds, *J. Mater. Chem. B.*, 2013, **1**(5), 685–692.
41. S. Naseri, J. Hum, W. C. Lepry, A. K. Miri, S. N. Nazhat and A. R. Boccaccini, Fabrication and characterization of zein-bioactive glass scaffolds, *Bioinspired, Biomimetic Nanobiomater.*, 2015, **4**(1), 73–78.
42. G. Georgiou, L. Mathieu, D. P. Pioletti, P. E. Bourban, J. A. E. Manson, J. C. Knowles *et al.*, Polylactic acid-phosphate glass composite foams as scaffolds for bone tissue engineering, *J. Biomed. Mater. Res. Part B Appl. Biomater.*, 2007, **80b**(2), 322–331.
43. D. J. Mooney, D. F. Baldwin, N. P. Suh, J. P. Vacanti and R. Langer, Novel approach to fabricate porous sponges of poly(d,l-lactic-co-glycolic acid) without the use of organic solvents, *Biomaterials*, 1996, **17**(14), 1417–1422.
44. I. Armentano, M. Dottori, E. Fortunati, S. Mattioli and J. M. Kenny, Biodegradable polymer matrix nanocomposites for tissue engineering: A review, *Polym. Degrad. Stab.*, 2010, **95**(11), 2126–2146.
45. A. Fathi, M. Barnard, R. Ravarian and F. Dehghani Fabrication of Porous Chitosan/Bioactive Glass Hydrogels in an Aqueous Media Using CO2 as a Gas Foaming Agent, *10th International Symposium on Supercritical Fluids (ISSF)*, 2012, San Francisco, USA: Elsevier.
46. A. R. Boccaccini and V. Maquet, Bioresorbable and bioactive polymer/Bioglass (R) composites with tailored pore structure for tissue engineering applications, *Compos. Sci. Technol.*, 2003, **63**(16), 2417–2429.
47. C. Schugens, V. Maquet, C. Grandfils, R. Jerome and P. Teyssie, Biodegradable and macroporous polylactide implants for cell transplantation: 1. Preparation of macroporous polylactide supports by solid-liquid phase separation, *Polymer*, 1996, **37**(6), 1027–1038.
48. V. Maquet, A. R. Boccaccini, L. Pravata, I. Notingher and R. Jérôme, Porous poly(α-hydroxyacid)/Bioglass® composite scaffolds for bone tissue engineering. I: Preparation and in vitro characterisation, *Biomaterials*, 2004, **25**(18), 4185–4194.
49. M. Dziadek, J. Pawlik, E. Menaszek, E. Stodolak-Zych and K. Cholewa-Kowalska, Effect of the preparation methods on architecture, crystallinity, hydrolytic degradation, bioactivity, and biocompatibility of

PCL/bioglass composite scaffolds, *J. Biomed. Mater. Res. Part B Appl. Biomater.*, 2015, **103**(8), 1580–1593.
50. J. J. Blaker, V. Maquet, R. Jerome, A. R. Boccaccini and S. N. Nazhat, Mechanical properties of highly porous PDLLA/Bioglass (R) composite foams as scaffolds for bone tissue engineering, *Acta Biomater.*, 2005, **1**(6), 643–652.
51. J. J. Blaker, S. N. Nazhat, V. Maquet and A. R. Boccaccini, Long-term in vitro degradation of PDLLA/Bioglass (R) bone scaffolds in acellular simulated body fluid, *Acta Biomater.*, 2011, **7**(2), 829–840.
52. Y. Wang, C. Yang, X. Chen and N. Zhao, Development and characterization of novel biomimetic composite scaffolds based on bioglass-collagen-hyaluronic acid-phosphatidylserine for tissue engineering applications, *Macromol. Mater. Eng.*, 2006, **291**(3), 254–262.
53. B. Sarker, J. Hum, S. N. Nazhat and A. R. Boccaccini, Combining Collagen and Bioactive Glasses for Bone Tissue Engineering: A Review, *Adv. Healthcare Mater.*, 2015, **4**(2), 176–194.
54. C. Xu, P. Su, X. Chen, Y. Meng, W. Yu, A. P. Xiang *et al.*, Biocompatibility and osteogenesis of biomimetic Bioglass-Collagen-Phosphatidylserine composite scaffolds for bone tissue engineering, *Biomaterials*, 2011, **32**(4), 1051–1058.
55. D. W. Hutmacher and S. Cool, Concepts of scaffold-based tissue engineering-the rationale to use solid free-form fabrication techniques, *J. Cell. Mol. Med.*, 2007, **11**(4), 654–669.
56. J. J. Beaman, *Solid Freeform Fabrication: A New Direction in Manufacturing: With Research and Applications in Thermal Laser Processing*, Kluwer Academic Publishers, Dordrecht, Boston, 1997, p. 330.
57. A. Naumann, J. Aigner, R. Staudenmaier, M. Seemann, R. Bruening, K. H. Englmeier *et al.*, Clinical aspects and strategy for biomaterial engineering of an auricle based on three-dimensional stereolithography, *Eur. Arch. Oto-Rhino-Laryngol.*, 2003, **260**(10), 568–575.
58. B. K. Paul and S. Baskaran, Issues in fabricating manufacturing tooling using powder-based additive freeform fabrication, *J. Mater. Process. Technol.*, 1996, **61**(1–2), 168–172.
59. A. Bellini, L. Shor and S. I. Guceri, New developments in fused deposition modeling of ceramics, *Rapid Prototyping J.*, 2005, **11**(4), 214–220.
60. A. Curodeau, E. Sachs and S. Caldarise, Design and fabrication of cast orthopedic implants with freeform surface textures from 3-D printed ceramic shell, *J. Biomed. Mater. Res.*, 2000, **53**(5), 525–535.
61. D. W. Hutmacher, M. Sittinger and M. V. Risbud, Scaffold-based tissue engineering: Rationale for computer-aided design and solid free-form fabrication systems, *Trends Biotechnol.*, 2004, **22**(7), 354–362.
62. H. N. Chia and B. M. Wu, Recent advances in 3D printing of biomaterials, *J. Biol. Eng.*, 2015, **9**, 1.
63. J. Zhang, S. Zhao, M. Zhu, Y. Zhu, Y. Zhang, Z. Liu *et al.*, 3D-printed magnetic Fe3O4/MBG/PCL composite scaffolds with multifunctionality

of bone regeneration, local anticancer drug delivery and hyperthermia, *J. Mater. Chem. B.*, 2014, **2**(43), 7583–7595.

64. P. X. Ma and J. W. Choi, Biodegradable polymer scaffolds with well-defined interconnected spherical pore network, *Tissue Eng.*, 2001, **7**(1), 23–33.
65. M. Borden, M. Attawia, Y. Khan and C. T. Laurencin, Tissue engineered microsphere-based matrices for bone repair: Design and evaluation, *Biomaterials*, 2002, **23**(2), 551–559.
66. S. W. Choi, Y. Zhang and Y. Xia, Three-dimensional scaffolds for tissue engineering: The importance of uniformity in pore size and structure, *Langmuir*, 2010, **26**(24), 19001–19006.
67. H. H. Lu, S. F. El-Amin, K. D. Scott and C. T. Laurencin, Three-dimensional, bioactive, biodegradable, polymer-bioactive glass composite scaffolds with improved mechanical properties support collagen synthesis and mineralization of human osteoblast-like cells in vitro, *J. Biomed. Mater. Res. A*, 2003, **64**(3), 465–474.
68. J. Yao, S. Radin, S. Leboy and P. Ducheyne, The effect of bioactive glass content on synthesis and bioactivity of composite poly (lactic-co-glycolic acid)/bioactive glass substrate for tissue engineering, *Biomaterials*, 2005, **26**(14), 1935–1943.
69. A. R. Boccaccini, I. Notingher, V. Maquet and R. Jerome, Bioresorbable and bioactive composite materials based on polylactide foams filled with and coated by Bioglass particles for tissue engineering applications, *J. Mater. Sci.: Mater. Med.*, 2003, **14**(5), 443–450.
70. J. A. Roether, J. E. Gough, A. R. Boccaccini, L. L. Hench, V. Maquet and R. Jérôme, Novel bioresorbable and bioactive composites based on bioactive glass and polylactide foams for bone tissue engineering, *J. Mater. Sci.: Mater. Med.*, 2002, **13**(12), 1207–1214.
71. X. Miao, L. P. Tan, L. S. Tan and X. Huang, Porous calcium phosphate ceramics modified with PLGA-bioactive glass, *Mater. Sci. Eng. C.*, 2007, **27**(2), 274–279.
72. A. Stamboulis, L. L. Hench and A. R. Boccaccini, Mechanical properties of biodegradable polymer sutures coated with bioactive glass, *J. Mater. Sci.: Mater. Med.*, 2002, **13**(9), 843–848.
73. N. Bhardwaj and S. C. Kundu, Electrospinning: A fascinating fiber fabrication technique, *Biotechnol. Adv.*, 2010, **28**(3), 325–347.
74. J. H. Jang, O. Castano and H. W. Kim, Electrospun materials as potential platforms for bone tissue engineering, *Adv. Drug Delivery Rev.*, 2009, **61**(12), 1065–1083.
75. P. K. Vallittu, T. O. Närhi and L. Hupa, Fiber glass–bioactive glass composite for bone replacing and bone anchoring implants, *Dent. Mater.*, 2015, **31**(4), 371–381.
76. Q. P. Pham, U. Sharma and A. G. Mikos, Electrospinning of polymeric nanofibers for tissue engineering applications: a review, *Tissue Eng.*, 2006, **12**(5), 1197–1211.

77. K. Fujihara, M. Kotaki and S. Ramakrishna, Guided bone regeneration membrane made of polycaprolactone/calcium carbonate composite nano-fibers, *Biomaterials*, 2005, **26**(19), 4139–4147.
78. A. El-Fiqi, J. H. Kim and H. W. Kim, Osteoinductive Fibrous Scaffolds of Biopolymer/Mesoporous Bioactive Glass Nanocarriers with Excellent Bioactivity and Long-Term Delivery of Osteogenic Drug, *ACS Appl. Mater. Interfaces*, 2015, **7**(2), 1140–1152.
79. K. T. Noh, H. Y. Lee, U. S. Shin and H. W. Kim, Composite nanofiber of bioactive glass nanofiller incorporated poly(lactic acid) for bone regeneration, *Mater. Lett.*, 2010, **64**(7), 802–805.
80. N. A. Peppas, J. Z. Hilt, A. Khademhosseini and R. Langer, Hydrogels in biology and medicine: From molecular principles to bionanotechnology, *Adv. Mater.*, 2006, **18**(11), 1345–1360.
81. E. J. Oh, K. Park, K. S. Kim, J. Kim, J. A. Yang, J. H. Kong *et al.*, Target specific and long-acting delivery of protein, peptide, and nucleotide therapeutics using hyaluronic acid derivatives, *J. Controlled Release*, 2010, **141**(1), 2–12.
82. D. A. Wang, C. G. Williams, F. Yang, N. Cher, H. Lee and J. H. Elisseeff, Bioresponsive phosphoester hydrogels for bone tissue engineering, *Tissue Eng.*, 2005, **11**(1–2), 201–213.
83. S. Utech and A. R. Boccaccini, A review of hydrogel-based composites for biomedical applications: enhancement of hydrogel properties by addition of rigid inorganic fillers, *J. Mater. Sci.*, 2016, **51**(1), 271–310.
84. R. A. Brown, M. Wiseman, C. B. Chuo, U. Cheema and S. N. Nazhat, Ultrarapid engineering of biomimetic materials and tissues: Fabrication of nano- and microstructures by plastic compression, *Adv. Funct. Mater.*, 2005, **15**(11), 1762–1770.
85. B. Marelli, C. E. Ghezzi, J. E. Barralet, A. R. Boccaccini and S. N. Nazhat, Three-Dimensional Mineralization of Dense Nanofibrillar Collagen-Bioglass Hybrid Scaffolds, *Biomacromolecules*, 2010, **11**(6), 1470–1479.
86. B. Marelli, C. E. Ghezzi, D. Mohn, W. J. Stark, J. E. Barralet, A. R. Boccaccini *et al.*, Accelerated mineralization of dense collagen-nano bioactive glass hybrid gels increases scaffold stiffness and regulates osteoblastic function, *Biomaterials*, 2011, **32**(34), 8915–8926.
87. B. Marelli, C. E. Ghezzi, M. James-Bhasin and S. N. Nazhat, Fabrication of injectable, cellular, anisotropic collagen tissue equivalents with modular fibrillar densities, *Biomaterials*, 2015, **37**, 183–193.
88. A. K. Miri, N. Muja, N. O. Kamranpour, W. C. Lepry, A. R. Boccaccini, S. A. Clarke *et al.*, Ectopic bone formation in rapidly fabricated acellular injectable dense collagen-Bioglass hybrid scaffolds via gel aspiration-ejection, *Biomaterials*, 2016, **85**, 128–141.
89. Y. Luo, C. Wu, A. Lode and M. Gelinsky, Hierarchical mesoporous bioactive glass/alginate composite scaffolds fabricated by three-dimensional plotting for bone tissue engineering, *Biofabrication*, 2013, **5**(1).

CHAPTER 11

Phosphate Glass Fibres and Their Composites

KAZI M. ZAKIR HOSSAIN,[a] REDA M. FELFEL,[a,b] DAVID M. GRANT[a] AND IFTY AHMED*[a]

[a] Advanced Materials Research Group, Faculty of Engineering, University of Nottingham, UK; [b] Physics Department, Faculty of Science, Mansoura University, Mansoura 35516, Egypt
*Email: Ifty.Ahmed@nottingham.ac.uk

11.1 Introduction

The interest in use of phosphate glasses (PGs) for biomedical applications has been steadily increasing due to their unique properties, such as their easily controllable degradation profiles coupled with ion release rates, cytocompatibility and mechanical properties.[1,2] The chemical compositions of PGs can be made to resemble the mineral content of bone, which makes them extremely promising candidates for use as resorbable biomaterials and in resorbable implantable devices.[3] PGs have mainly been investigated for applications where only a temporary presence of the implant material is required, thus negating secondary surgical procedures for their removal and enabling the targeted tissue of repair to fully replace the implant *in situ*. In addition, PGs have been employed in various geometries such as in the form of particles,[4] fibres[5,6] and microspheres[7] to enhance their applicability for varying biomedical applications.

RSC Smart Materials No. 23
Bioactive Glasses: Fundamentals, Technology and Applications
Edited by Aldo R. Boccaccini, Delia S. Brauer and Leena Hupa

Published by the Royal Society of Chemistry, www.rsc.org

11.2 Phosphate Glass Fibres (PGFs)

Moving from the bulk glass to fibrous structures, PGFs possess unique advantageous features, such as excellent mechanical properties thus enabling a greater reinforcing capability of (bio)polymer matrices, and have also been explored as cell guides and in soft tissue repair applications. PGs have been fabricated into continuous fibres *via* both melt-drawing and preform rod manufacturing processes and relatively good mechanical properties have been achieved (for example, tensile modulus ~74 GPa[8] and tensile strength ~1.2 GPa[9]). One of the main areas for exploration of phosphate glass fibres (PGFs) has been the manufacture of fully bioresorbable composites as fracture fixation devices[10] and they have also been heavily investigated for the fabrication of scaffolds for tissue engineering and regenerative medicine applications.[5]

11.2.1 Manufacture of Phosphate Glass Fibres

PGFs are commonly produced *via* a melt or pre-form drawing process depending on the fragility index of the glass materials. For instance, comparatively less fragile glasses can be melt drawn and pulled from a solid pre-from to manufacture continuous fibres, whilst highly fragile glasses require rapid quenching of upward drawn fibres processed from a melt.[11] The 'drawing point viscosity' of the phosphate glasses is also very important during fibre production (typically assumed to be ~100 Pa s),[12] as these types of glasses are very fragile and prone to crystallise quickly if appropriate conditions are not met. Therefore, the fibre drawing temperature is maintained within a relatively narrow window (which falls below the liquidus melt temperature) so that the melt viscosity can enable continuous fibres to be drawn whilst maintaining their amorphous nature.

11.2.1.1 Melt Spinning

Glass fibres (Figure 11.1) are generally produced *via* melt spinning techniques which involve melting glass frits in platinum–rhodium crucibles which encompass bushing tips with small holes for the molten glass to flow through gravitationally. Continuous glass fibres are drawn down from the holes and wound onto rotating drums (see Figure 11.2a). Composition selection of glass materials is also important in order to provide sufficient P–O–P bonds within the molten glass to enable production of continuous fibre. At the same time, stable P–O–P bonds impart the required strength to the fibre to withstand the tensile stresses applied during the drawing process at high temperatures. Fibre properties and dimensions can be controlled *via* process variables such as, the diameter of the bushing exit holes, melt temperature (hence viscosity of the glass), mass flow and drawing speed. In addition, heat treatment (also known as annealing) of the fibres produced is sometimes necessary to relieve internal stresses created within the fibres

Figure 11.1 SEM images of typical melt-drawn PGFs.
Reproduced from ref. 3 with permission from the Royal Society of Chemistry.

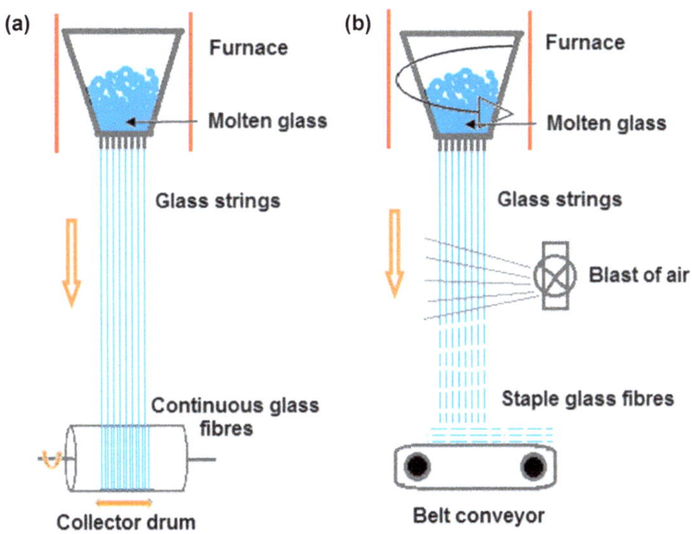

Figure 11.2 Schematic diagram representing the glass manufacturing processes: (a) melt-drawn continuous glass fibres and (b) melt-blown staple glass fibres.

during the fibre drawing process by reforming P–O–P bonds into more stable configurations.[13] On the other hand, short fibres can also be produced *via* a melt-blowing technique which is similar to the melt spinning process, except that an external force is applied to the molten glass exit holes and the fibres are cut to required lengths using mechanical means or by air jets (see Figure 11.2b).

11.2.1.2 Pre-form Drawing

Glass fibre drawing using the pre-form technique involves two steps: in the first step, a glass pre-form in the form of a rod is produced. This rod can be produced *via* a glass extrusion process or can be produced *via* melting the glass and pouring into a pre-form shaped mould.[14] In the second step, this glass pre-form is then heated in a furnace above its T_g, maintaining the viscosity at around 4.5–5 Pascal second (Pa s), and allowed to form a gob at the end of the pre-form which falls away from the tip due to gravity. The fibrous strand developed is then wound around and collected on a rotating drum. The fibre properties produced *via* this technique can also be altered by changing the drawing parameters such as the pre-form heating zone length, heat supplied and the pre-form feed and fibre drawing rates. Recently, Ahmed *et al.*[15] investigated the manufacturing process of two different metal ion (Ti^{2+} and Fe^{3+})-doped phosphate core/clad glass fibres, a process previously limited to manufacturing optical glass fibres. The manufacturing process of core/clad glass fibres included the following steps: (a) preparation of glass billets with varying compositions (*but similar thermal transition and viscosity profiles*) *via* casting in a graphite mould, (b) co-extrusion of the stacked glass billets by placing the cladding glass underneath the core glass to produce the core/clad pre-form (see Figure 11.3a), and (c) drawing the core/clad fibres (Figure 11.3b) from the core/clad pre-form utilising a fibre drawing tower.

11.2.2 Properties of Phosphate Glass Fibres

Producing fibres from phosphate glass compositions initially proved to be quite challenging, as various parameters, such as the melt temperature, melt viscosity, O : P ratio and drawing speed for manufacture of continuous fibres from varying formulations, were ascertained, mainly through trial and error experimentation. Properties of glass fibres, such as thermal, mechanical, and dissolution properties differ significantly from those of the bulk glass. This variation in properties depends on several factors such as fibre

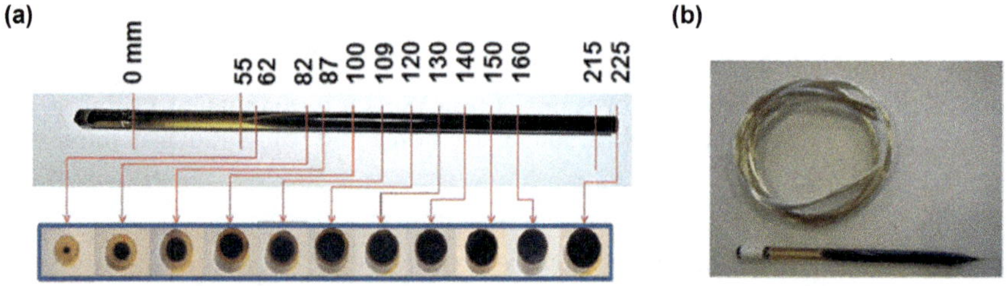

Figure 11.3 (a) Cross-sections of an extruded core/clad preform (the values indicate the cut distance of the preform into discs) and (b) an extruded core/clad preform and resultant fibres.
© 2015 I. Ahmed *et al.*[15]

manufacturing methodology, drawing parameters (temperature, speed, viscosity) and fibre diameter. In addition, the properties of PGFs also depend on the bonds present within the molten glass which provide the required strength to withstand the applied stresses during the fibre drawing process. The type of chain-forming bonds and the cross-linked networks within the glass structure can be controlled by varying the compositions produced in order to tailor the fibre properties required.[3]

The main glass-forming oxide for phosphate glasses is phosphorus pentoxide (P_2O_5) which consists of tetrahedral phosphate anions ($PO_4{}^{-3}$). The addition of alkaline and alkaline earth metal ions results in depolymerisation of the phosphate network *via* creation of a non-bridging oxygen (O^-M^+) by breaking the bridging oxygen (P–O bonds). Moreover, the field strength created *via* the formation of strong O^-M^+ bonds due to addition of modifying ions within the glass structure improves their chemical durability.[16] Some metal oxides have high resistance to hydration; for example, addition of Al_2O_3, Fe_2O_3 and TiO_2 within PG systems have been shown to significantly decrease the dissolution rates of these glasses.[17] Trivalent ions such as iron (Fe^{3+}) and (Ti^{3+}) have been found to have a greater influence on the solubility of phosphate glasses than divalent or monovalent ions, which has been suggested to be due to the strengthening of the phosphate network *via* their cross-linking effect.

Phosphate glass fibres with various compositions, including binary, ternary and quaternary formulations have been investigated for their biocompatibility and degradation profiles.[5,18] Ahmed *et al.*[1,5] reported that the degradation rates of phosphate glass fibres (within ternary systems P_2O_5–CaO–Na_2O) increased significantly compared to bulk glass of the same composition. Moreover, the degradation rates for glass fibres increased with decreasing fibre diameter, which was due to the vast increase in surface area compared to the bulk glass. Furthermore, varying the additions of mono-, di- or trivalent cations (such as Na_2O, CaO, MgO, Fe_2O_3, *etc.*) within the glass structure enables further control over their degradation rates in aqueous media. For example, addition of Fe_2O_3 content (from 1 to 5 mole%) within the iron phosphate glass system (P_2O_5–CaO–Na_2O–Fe_2O_3) revealed a significant reduction in the glass dissolution rates (dissolution rate ~0.00045 and ~0.00004 $mg\,cm^{-2}\,h^{-1}$ for the addition of 1 and 5 mol% Fe_2O_3 in the glass system, respectively.[5]

Mechanical properties of glass fibres depend not only on the bonds present within the glass but also its molecular organisation which is related to the chemical composition.[19,20] The chemical bonds in bulk phosphate glass generally exist as an isotropic structure, however during fibre production the glass network transforms into anisotropic structure and the PO_4 tetrahedra align in the direction of the pull.[21,22] The diameter of glass fibres is also known to strongly influence the mechanical properties of glass fibres.[23] For example, Lin *et al.*[24] reported that tensile strength and modulus properties of calcium–iron phosphate glass fibres increased from 587 MPa to 1045 MPa and 43 GPa to 64 GPa, respectively, as the iron content increased from 3.6 to

Table 11.1 Mechanical properties of phosphate glass fibres.

Glass composition (mol%)	Fibre diameter (μm)	Tensile strength (MPa)	Tensile modulus (GPa)	Ref.
$63.7P_2O_5$–$32.7CaO$–$3.6Fe_2O_3$	~20	587	43	Lin *et al.*[24]
$56.9P_2O_5$–$26.2CaO$–$16.9Fe_2O_3$	~20	1045	64	Lin *et al.*[24]
$50P_2O_5$–$50CaO$	20–25	474	44	Ahmed *et al.*[25]
$50P_2O_5$–$40CaO$–$5Na_2O$–$5Fe_2O_3$	20–25	456	51.5	Ahmed *et al.*[6]
$45P_2O_5$–$16CaO$–$15Na_2O$–$24MgO$–$0B_2O_3$	~20	530	53.4	Sharmin *et al.*[9]
$45P_2O_5$–$16CaO$–$10Na_2O$–$24MgO$–$5B_2O_3$	~20	1050	59.6	Sharmin *et al.*[9]
$45P_2O_5$–$16CaO$–$5Na_2O$–$24MgO$–$10B_2O_3$	~20	1200	60	Sharmin *et al.*[9]
$45P_2O_5$–$16CaO$–$11Na_2O$–$24MgO$–$4Fe_2O_3$	~26	569	57	Liu *et al.*[26]
$40P_2O_5$–$16CaO$–$16Na_2O$–$24MgO$–$4Fe_2O_3$	~15	370	62	Felfel *et al.*[27]
$40P_2O_5$–$16CaO$–$20Na_2O$–$24MgO$	~15	484	44	Cozien-Cazuc *et al.*[28]

16.9 mol%. More recently, Sharmin *et al.*[9] reported that addition of 5 and 10 mol% boron within phosphate glass fibre structures resulted in a significant increase in tensile strength with values of 1050 MPa and 1200 MPa reported, respectively, whilst the control glass fibres without boron exhibited tensile strengths around 530 MPa. They also suggested that incorporation of boron had a significant effect on improving the fibre drawing properties (for example, easing fibre formation) which was attributed to the extension of the phosphate chains *via* boron addition. By simple modification of the glass chemical composition a range of mechanical properties can be achieved (see Table 11.1).

11.2.3 Biomedical Applications of Phosphate Glass Fibres

As mentioned above, phosphate glasses in the form of fibrous structures containing various modifiers such as iron, zinc, copper, and titanium have been explored for a range of biomedical applications. The iron ions released from iron–phosphate glass could potentially participate in redox reactions for certain types of proteins (such as, cytochrome, myoglobin, *etc.*) and could also promote cell attachment and differentiation. For example, Ahmed *et al.*[5] investigated the use of iron–phosphate glass fibres as potential cell delivery vehicles for transplantation purposes and used them to orientate muscle precursor cells along the axis of the fibres to form myotubes (see Figure 11.4). The chemical durability of these glass fibres played a vital role in improving their biocompatibility and showed that by incorporating modifier ions, such as iron oxide, the dissolution rates of the glass fibres could be reduced by orders of magnitude. It has also been suggested that iron–phosphate glass fibres containing 4–5 mol% Fe_2O_3 were favourable for cell attachment and

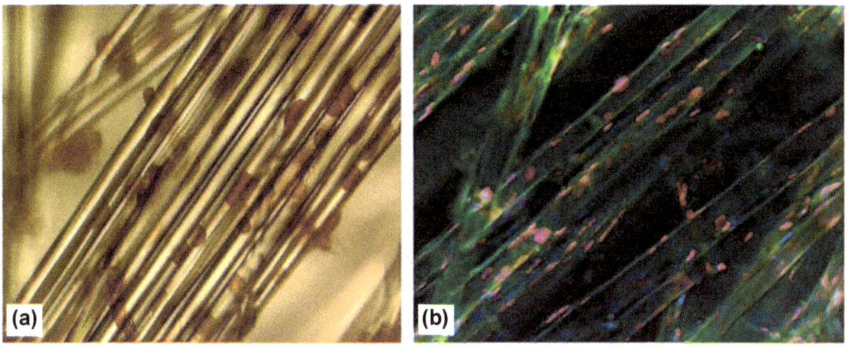

Figure 11.4 (a) Attached muscle precursor cells (MPCs) on iron–phosphate glass fibres and (b) MSCs differentiation on PG fibres: Desmin (seen in green) is a cytoplasmic marker of all skeletal muscle cells. Myogenin (seen as red) is a nuclear marker of differentiation. The blue is DAPI (4,6-diamidino-2-phenylindole) which stains all nuclei.
Adapted from I. Ahmed *et al.*, Processing, characterisation and biocompatibility of iron–phosphate glass fibres for tissue engineering, *Biomaterials*, **25**, 3223–3232, Copyright 2004 with permission from Elsevier.[5]

differentiation.[5] In addition, iron-doped PGFs with diameters of 20–25 μm were suggested to be of suitable size for attachment of human osteoblast and fibroblast cells compared to 10–15 μm fibres due to their higher surface curvature.[29] Fibres with various diameters differ in their surface curvature; usually fibres with smaller diameter possess higher surface curvature which could positively influence cell attachment and spreading.

Zn ions are well known to stimulate protein synthesis in osteoblast cells and also to increase ATPase and ALP activity.[30] Zinc-doped PGFs in the form of 3D-scaffolds were investigated by Shah *et al.*[31] for the construction of muscle organoid units. Skeletal muscle cells were found at the interstices of the phosphate fibres and revealed increased cell numbers on the fibre constructs. Ti-doped phosphate glass fibres and their role in neuronal polarisation and axonal growth direction have also been investigated by Vitale-Brovarone *et al.*[32] who suggested that the aligned configuration of the fibres provided the directional cue for growing dorsal root ganglia (DRG) neurons along the fibre axis (see Figure 11.5a). Active proliferation of neonatal olfactory bulb ensheathing cells (NOBEC) extending along the PGF surfaces was observed as shown in Figure 11.5b.

Release of copper ions from PGFs was investigated by Abou Neel *et al.*[33] for its potential antibacterial properties against *Staphylococci*. It has been suggested that incorporation of 10 mol% CuO released sufficient Cu^{2+} ions as the glass degraded to prevent bacterial colonisation and reduce the number of viable bacteria in the local environment. Copper ions have also been reported to have other advantages, such as stimulation of angiogenesis and also proliferation of human endothelial cells. As such, Cu-doped glasses could also potentially be used for wound healing applications.[34]

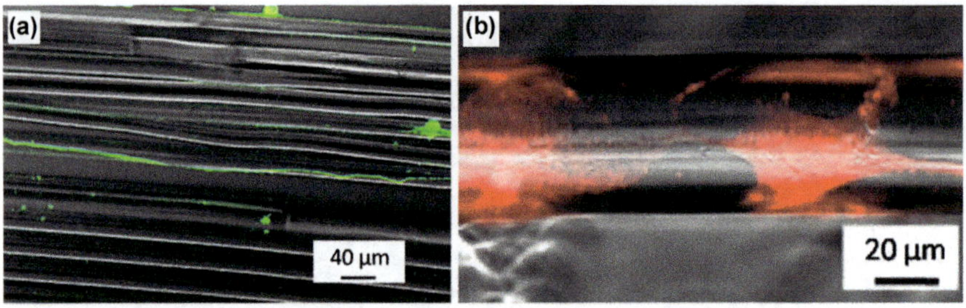

Figure 11.5 Confocal microscope images of (a) DRG neurons on glass fibres presenting long neuritis extended along the fibre axis direction and (b) NOBEC cells showing active proliferation on phosphate glass fibres.
Adapted from C. Vitale-Brovarone *et al.*, Phosphate glass fibres and their role in neuronal polarization and axonal growth direction, *Acta Biomater.*, **8**, 1125–1136, Copyright 2012 with permission from Elsevier.[32]

11.3 Phosphate Glass Fibre-reinforced Composites

The principal limitation to widespread use of bioresorbable polymers for biomedical applications is usually due to their mechanical properties, although the mechanical characteristics for some bioresorbable polymers are known to be sufficient for non-load-bearing applications such as in the cranial (skull) bones and for maxillofacial fractures. Fractures at load-bearing sites (*e.g.* femur and tibia) require sufficiently robust fixation devices to avoid implant failure until the fractured bone has healed. Therefore, reinforcement of these by polymers has been considered to produce composite implants with enhanced mechanical properties to match cortical bone. In order to produce fully bioresorbable implants, the reinforcement phase should also be made from biocompatible and bioresorbable materials. These composite devices could also eliminate potential inflammatory responses associated with the use of resorbable polymers alone, which have mainly been ascribed to the acidic degradation products from these polymers (such as lactic acid in the case of PLA).[35–37] These acidic breakdown products could potentially be buffered by the degradation by-products from the reinforcement phase. The pH of the media during glass degradation is important for composite stability and degrading glass compositions which cause an acidic pH may not be ideal for biomedical applications. However, controlling pH during degradation of PGs is just a matter of altering their composition. This was demonstrated by Ahmed *et al.*,[2] who showed that PGs could maintain a neutral pH level which would assist in buffering the polymer degradation by-products (see Figure 11.6). Moreover, the amount of polymer matrix within the composite would decrease at the expense of the reinforcement, leading to less polymer remaining to be eliminated.

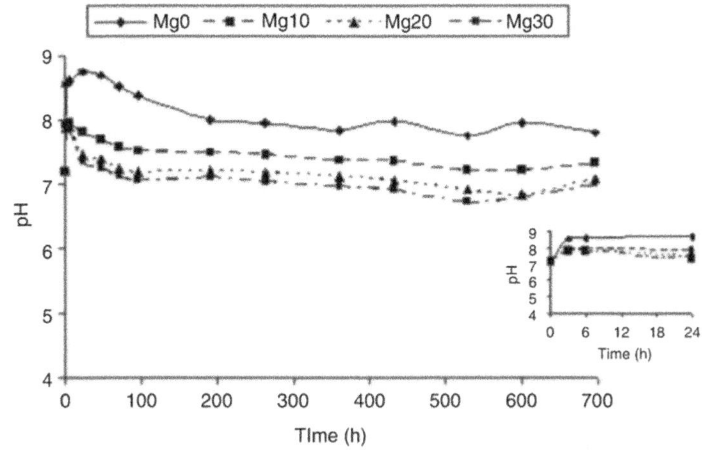

Figure 11.6 pH values obtained for $40P_2O_5$–25CaO–(35 − x)Na_2O–xMgO glasses, x varied from 0 to 30 mol%. pH remained relatively neutral over time. Adapted with permission from ref. 2.

Composites are generally comprised of a stiff and strong dispersed phase (reinforcement) and continuous phase (matrix). Properties of the composites depend on properties of the constituents as well as the geometry, size, distribution and volume fraction of the reinforcement phase. Furthermore, the bonding strength (adhesion) between the phases plays a crucial role in their properties. Increase in reinforcement (fibres–matrix bonding) allows for effective load transfer leading to higher strength profiles for the composite.[38]

Biocomposites are composite materials that can be implanted inside the body to replace the function of living tissue.[39] They may be grouped into fully, partially and non-bioresorbable composites based on the ability of the constituents to degrade and resorb within the body.[40] The matrix and reinforcement have to be bioresorbable for a fully bioresorbable composite (*e.g.* PCL–PGF, PLA–PGF[25,41]) whilst both can be biostable for non-resorbable composites (*e.g.* ultra-high molecular weight polyethylene–hydroxyapatite (UHMWPE–HA)[42]). For partially bioresorbable composites, one of the composite components (usually the reinforcement) is biostable (*e.g.* PLA–Bioglass, PLA–carbon fibre[43–45]).

Fibre-reinforced composites (FRC) could potentially provide a wide range of mechanical properties to match both cortical and cancellous bone.[46,47] For instance, the anisotropic nature of the bone (*i.e.* the longitudinal mechanical properties of bone are greater than those in the transverse direction) can be achieved by continuous unidirectional fibre-reinforced composites.[46] The orientation of fibres could be aligned or be present in a random format (as in chopped-strand fibre mats) to influence the mechanical performance of the composites produced.[48,49] Properties of fibre-reinforced composites are controlled by fibre volume fraction, type, length, distribution and strength of fibre–matrix interface.[46,47] Various types of fibres have been utilised in the orthopaedic field such as carbon,[50] Bioglass[45,51–56] and phosphate glass fibres.[25,41,57–63]

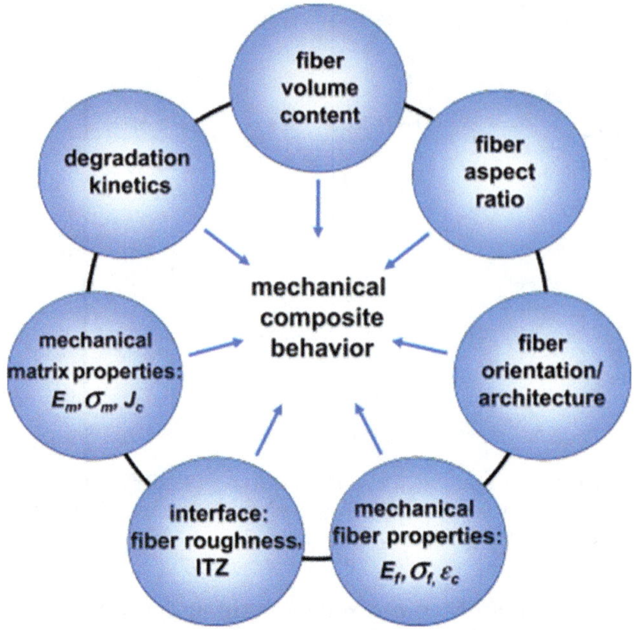

Figure 11.7 Schematic diagram of parameters which control the mechanical properties of fibre-reinforced composites.
Adapted from R. Krüger and J. Groll, Fiber reinforced calcium phosphate cements – On the way to degradable load bearing bone substitutes? *Biomaterials*, **33**, 5887–5900, Copyright 2012 with permission from Elsevier.[66]

Aspect ratio (length : diameter ratio) of the fillers also has a significant influence on the modulus/stiffness of a particulate composite.[64,65] Mechanical properties of composites are known to increase as the aspect ratio of the particles increases. Therefore, it is expected that the reinforcing efficiency of fibres is significantly greater than particles. The parameters which affect the mechanical behaviour of FRCs are summarised in Figure 11.7.

Bioresorbable fibres such as phosphate glass fibres (PGF) have been investigated to manufacture fully bioresorbable composites due to their unique resorption profiles. Composite plates based on PLA and PGF have been produced with different fibre content and lay-up geometries. Flexural properties for unidirectional composites with fibre volume fraction (V_f) ~30% and ~55% of 115 MPa and 170 MPa strength and 16 GPa and 15 GPa modulus, respectively, have been achieved.[57,58] Furthermore, PLA reinforced with random PGF mats (V_f ~ 14%) achieved properties of 90 MPa and 5 GPa for bending strength and modulus, respectively.[41]

11.3.1 Manufacturing PGF Composites

11.3.1.1 Compression Moulding

Random mat (RM) and unidirectional (UD) composites have been prepared *via* a laminate film stacking process. PLA or PCL films were stacked alternately with RM or UD aligned fibres (see Figure 11.8) in a mould cavity between two metallic plates. The stack was then heated in the press for 15 mins above melting temperature of the polymer matrix (210 and 120 °C for PLA and PCL, respectively) and pressed for 15 mins at 38 bar. The plates were transferred to a second press for cooling to room temperature at 38 bar for 15 mins.[25,67]

11.3.1.2 In situ *Polymerisation*

PCL-based composites have also been prepared *via* an *in situ* polymerisation process. The catalyst $Sn(Oct)_2$ was mixed with ε-caprolactone at the molar ratio of 1 : 1000 and then injected into a dried polytetrafluoroethylene (PTFE) mould (see Figure 11.9) at room temperature. The amount of unidirectional PGF was determined based on the target fibre volume fraction of the composites and placed in the mould before injection of reaction mixture. Parts of

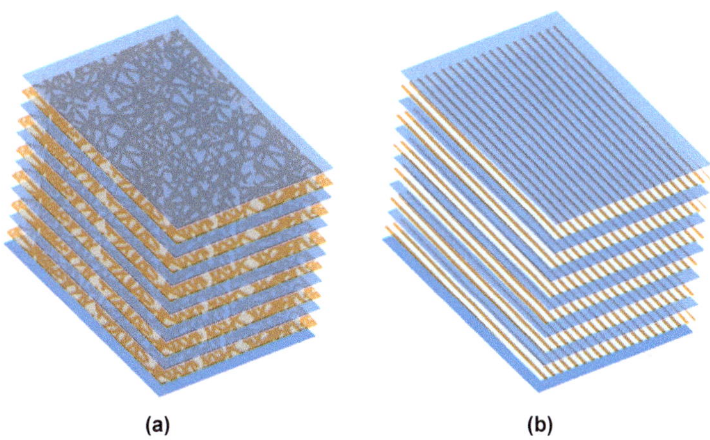

Figure 11.8 Schematic diagram of film stacking for producing (a) RM and (b) UD composites.

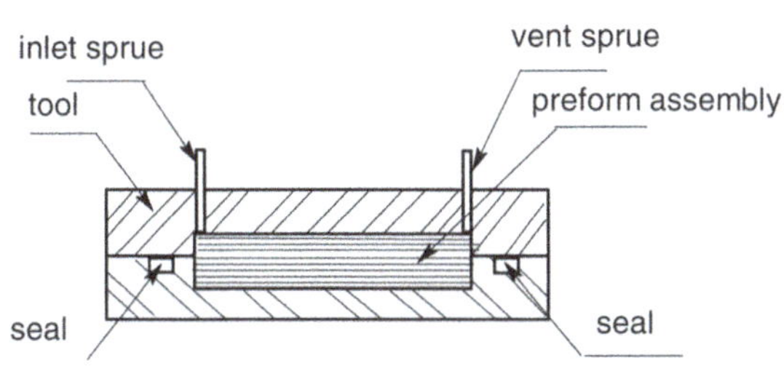

Figure 11.9 A schematic diagram for *in situ* polymerisation process used in preparation of PCL–PGF composites.
Adapted from G. Jiang *et al.*, Preparation of poly(ε-caprolactone)/continuous bioglass fibre composite using monomer transfer moulding for bone implant, *Biomaterials*, **26**, 2281–2288, Copyright 2005 permission from Elsevier.[45]

the mould were carefully sealed and then kept at 110 °C for 24 hours to complete the polymerisation reaction.[45]

11.3.2 Properties of PGF-based composites

The flexural properties of PGF-reinforced composites investigated over recent years have been summarised in Figures 11.10 and 11.11. Flexural strength and modulus for both RM and UD composites increased as the fibre volume fraction (V_f) increased, especially for the 40% and 45% V_f RM and UD composites, as expected. The range of values obtained spanned 15 to 370 MPa for strength and 1 to 25 GPa for modulus (well exceeding the mechanical properties of human bone which ranges between 45 and 270 MPa and 5 and 23 GPa for strength and modulus, respectively). When the fibre content exceeded 35%, the strength and modulus for the composites remained approximately constant or decreased slightly due to breaking of fibres and lack of fibre impregnation within the polymer matrix. UD composites showed significantly higher flexural properties compared to RM composites, as expected, based on the rule of mixtures and due to high efficiency of aligned fibres to transfer stress across the fibre–matrix interface in comparison with random chopped fibres. Sizing the fibres with coupling

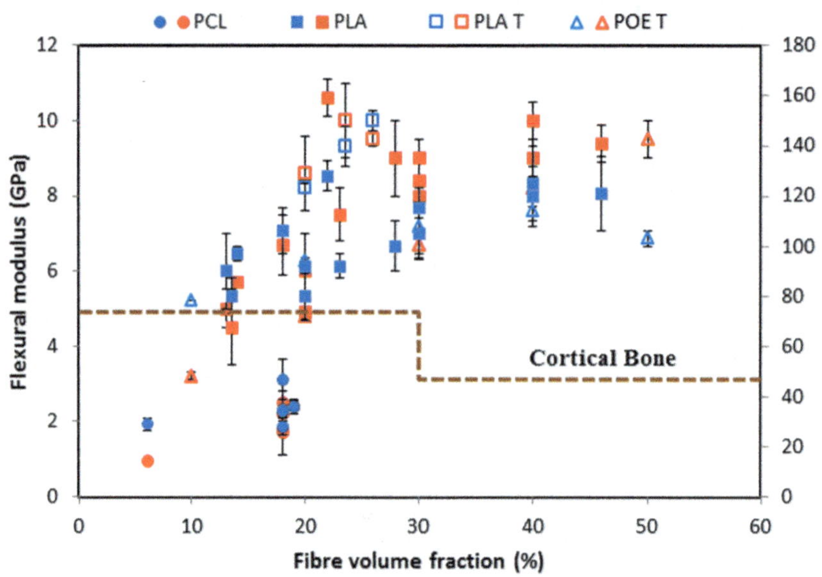

Figure 11.10 Flexural properties of compression moulded PGF RM composites against fibre volume fraction. Symbols in blue and red represents flexural strength and modulus, respectively, and unfilled symbols refer to composites containing treated PGF with coupling agents. PCL: polycaprolactone, PLA: polylactic acid, POE: poly(ortho ester) and the additional T refers to treated fibres with coupling agent.[41,61,68–72]

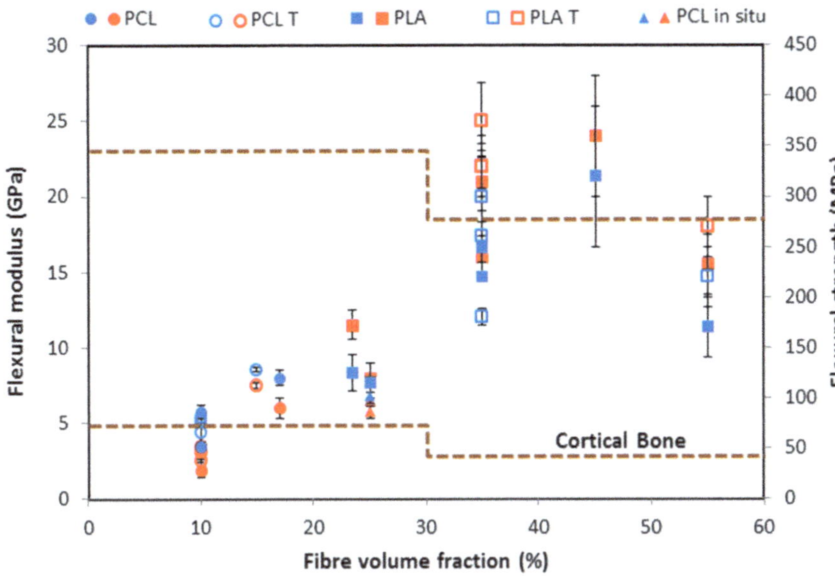

Figure 11.11 Flexural modulus and strength of PLA and PCL reinforced with UD PGF *versus* fibre volume fraction. Symbols in blue and red represents flexural strength and modulus respectively and unfilled symbols refer to composites containing treated PGF with coupling agents. PCL: polycaprolactone, PLA: polylactic acid and the additional T refers to treated fibres with coupling agent.[58,61,71,73–77]

agents caused a significant increase in flexural properties for RM and UD composites. This was attributed to enhancement of the fibre–matrix interface. As seen from Figure 11.11, the flexural properties of UD composites containing more than 15% fibre volume fraction surpassed the lower limits of cortical bone mechanical properties. The huge variation of flexural properties for composites at constant fibre volume fraction is ascribed to differences in mechanical properties of the PGFs used (with varying compositions), as highlighted in Table 11.1.

Ahmed *et al.*[41] produced PLA–PGF composites containing 14% fibre volume fraction with initial flexural properties of 90 MPa for strength and 5 GPa for modulus. The values decreased to 40 MPa and 1 GPa, respectively, after immersion in deionised water at 37 °C for 6 weeks. Further studies by Ahmed *et al.*[61] investigated degradation and mechanical retention for PLA–PGF composites with a range of fibre volume fractions (from 20% to 45%) and fibre orientations. The initial flexural strength and modulus for RM composites increased to ~120 MPa and ~10 GPa, respectively, with increasing fibre volume fraction to ~45%. Flexural properties for UD composites were ~130 MPa for strength and ~11.5 GPa for modulus for 24% fibre volume fraction composites. These composite mechanical properties decreased by between ~50–80% of their initial strength and modulus after 14 days immersion in deionised water at 37 °C, which was attributed to degradation of the fibre–matrix interface and the PGF fibres. The overall mass loss for the composites also increased with increasing fibre volume fraction. Most of the studies investigating particulate and fibre-reinforced composites based on PG have reported a rapid loss of mechanical properties when exposed to media.

Annealed and non-annealed binary calcium–phosphate ($50P_2O_5$–50CaO) glass fibres have also been used to reinforce PCL. The composites contained two different volume fractions of non-woven random mat fibres, 6.4 and 17%, and revealed flexural modulus and strength of up to 2.5 GPa and 30 MPa, respectively, which were in the range of human cancellous bone. A good correlation between degradation and ion release profiles was observed which also revealed a porous structure within the PCL matrix at the end of the degradation period of 5 weeks as a result of the glass fibres resorbing and leaving behind continuous channels.[25]

Borbely *et al.*[63] explored the degradation and ion release profiles of PCL–PGF composites over 7 days at 37 °C. The weight of all composites initially increased followed by loss in both distilled water and Hanks Buffered Saline Solution (HBSS). The change in weight was the result of water uptake and mass loss due to hydrolytic degradation over time. Initially, water uptake was the dominant factor over weight loss as a result of the hydrophilic nature of PG. However, the weight loss at the end of the study in water was higher than that in HBSS as a result of the buffering effect of HBSS. The SEM micrographs of the composites showed a distinct precipitation of calcium phosphate on their surfaces and the authors reported that these composites could potentially be used as bioactive root fillings.

Mohammadi *et al.*[78] prepared PCL–PGF random mat composites with 18% fibre volume fraction *via* compression moulding using two formulations of PGF: $50P_2O_5$–40CaO–$10Fe_2O_3$ (Fe_{10}) and $50P_2O_5$–40CaO–5Fe–$5SiO_2$ (Fe_5Si_5). PCL–Fe_5Si_5 showed approximately 55% decrease in flexural strength and modulus after 28 days of degradation at 37 °C, whilst no statistically significant change was seen in flexural strength and modulus for PCL–Fe_{10} composites after 7 and 28 days. They also investigated the change in storage modulus for PCL–PGF composites over the temperature range from −90 to 37 °C; this increased from 3.7 to 4.5 and 5.5 GPa by incorporation of Fe_{10} and Fe_5Si_5 fibres into the PCL matrix. PCL–Fe_5Si_5 samples revealed greater weight loss and ion release in comparison to PCL–Fe_{10} and neat PCL. They also reported that the PCL–Fe_5Si_5 composites initiated the formation of brushite precipitates on their surface after 14 and 28 days of conditioning in simulated body fluid (SBF) at 37 °C. In contrast, no precipitates were observed on the surfaces of PCL–Fe_{10} and PCL alone until the 28 day interval.

Rates of degradation and ion release of PLA and PCL composites reinforced with PGF can mainly be controlled by the dissolution rates of the glass formulations used. Moreover, a good correlation has been observed between mass loss and ion release profiles.[25,79] Felfel *et al.*[80,81] investigated the effect of chemical formulation of PGF on degradation rate of PLA composites. Two glass formulations (P50: $50P_2O_5$–40CaO–$5Na_2O$–$5Fe_2O_3$ in mol) and (P40: $40P_2O_5$–24MgO–16CaO–$16Na_2O$–$4Fe_2O_3$ in mol%) were used in this study. They found that P50 composites lost 16% of their weight after 9 weeks of degradation at 37 °C, whilst P40 composites lost approximately 1% during the same period. Rates of cations and anions released from the P50 composites were also significantly greater than for the P40 composites due to higher chemical durability of the P40 formulation. Lower phosphate content and shorter chains, excess of Q1 structure and existence of divalent cations, capable of crosslinking phosphate chains, were the main reasons suggested for the enhanced chemical durability of the P40 fibres. Scanning electron micrographs (see Figure 11.12) of the composites throughout the degradation period revealed that P50 fibres had almost fully degraded and leaving behind porous PLA structure while P40 fibres remained intact.

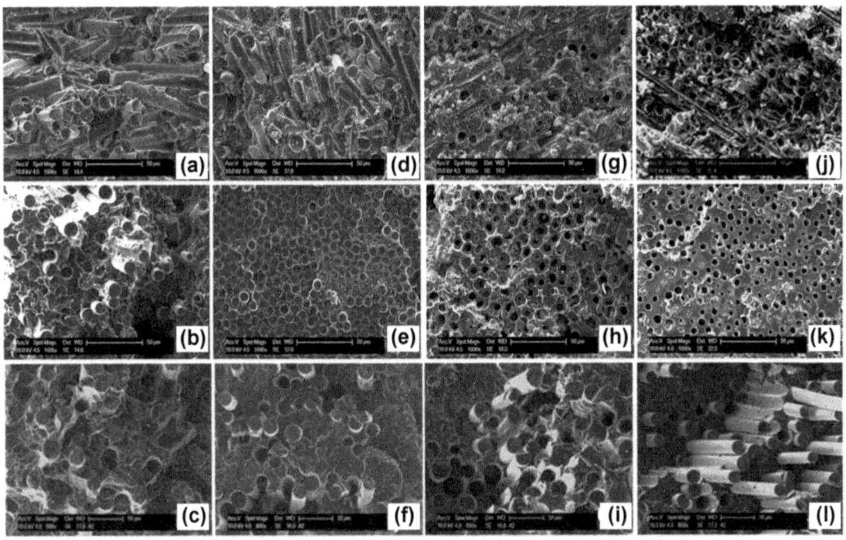

Figure 11.12 SEM micrographs of a freeze-fractured surface for (a) P50 RM, (b) P50 UD and (c) P40 UD composite rods before degradation, (d), (e) and (f), respectively, after 2 weeks of degradation in PBS at 37 °C, (g), (h) and (i), respectively, after 4 weeks of degradation in PBS at 37 °C, and (j), (k) and (l), respectively, after 9 weeks of degradation in PBS at 37 °C. P40 fibres remained intact after 9 weeks of degradation, while P50 fibres had degraded and left behind porous PLA structure. Scale bars for all micrographs represent 50 μm. (a), (b), (d), (e), (g), (h), (j) and (k) Adapted from R. M. Felfel *et al.*, *In vitro* degradation, flexural, compressive and shear properties of fully bioresorbable composite rods, *J. Mech. Behav. Biomed. Mater.*, **4**, 1462–1472, Copyright 2011 with permission from Elsevier[80] and (c), (f) (i) and (l) Adapted from R. M. Felfel *et al.*, Cytocompatibility, degradation, mechanical property retention and ion release profiles for phosphate glass fibre reinforced composite rods, *Mater. Sci. Eng., C*, **33**, 1914–1924, Copyright 2013 with permission from Elsevier.[81]

11.3.2.1 The Fibre–Matrix Interface

PGF-based composites commonly lose substantial percentages of their initial mechanical property values after immersion in aqueous media due to loss of interfacial integrity. Several studies have thus investigated improving the bonding between PGF and biodegradable polymer matrices such as PCL and PLA using coupling agents.[8,68,73–76,82] In the absence of coupling agents, the quality of the fibre–matrix interface is mainly attributed to mechanical adhesion. Treatment of the fibres with coupling agents could potentially introduce chemical bonding with the polymer matrix.

Hasan *et al.*[68,73,83] used varying chemical agents to investigate the fibre–matrix interface in order to enhance mechanical property retention by the composites. They trialed 3-phosphonopropionic acid (PPA), glycerol 2-phosphate disodium salt (GP), etidronic acid (EA), 3-aminopropyltriethoxy silane (APS), sorbitol–sodium-ended PLA oligomers (S/Na–PLA), hexamethylene diisocyanate (HDI) and amino phosphonic acid (APA). Covalent bonding was detected between phosphate glass and APS, EA and HDI using IR spectroscopy, whilst GP, PPA, S/Na–PLA and APA were suggested to be linked *via* hydrogen bonds. Interfacial shear strength (IFSS) values of between 7 and 22 MPa were achieved after treatment of glasses with APS, EA, HDI and S/Na–PLA, whereas other agents showed little increase in IFSS after immersion in aqueous media. Initial flexural strength of PLA composites reinforced with RM and UD PGF treated with APS exhibited 15% and 30% increase, respectively, compared with composites containing non-treated fibres. However, their strength profiles decreased gradually to be similar to the control composites after one week immersion in PBS at 37 °C. Cytocompatibility of the composites containing treated fibres with different coupling agents was examined using primary human osteoblasts over 21 days. All composites exhibited comparable cytocompatibility to the tissue culture plastic control. Haque *et al.*[8,82] also found that the IFSS for PGF treated with Na–PLA and GP increased from 9.5 MPa (for non-coated fibres) to 14.4 and 17.5 MPa, respectively. However, the effects of these sizing agents also decreased after 3 days of immersion in PBS at 37 °C to level with the IFSS values of the non-coated single fibre composites.

Onal *et al.*[84] also investigated degradation mechanisms of PGF-reinforced PCL composites fabricated using an *in situ* polymerisation method. PGF fibres were treated with a coupling agent (3-aminopropyltriethoxysilane) and the percentage moisture content was observed to increase from 0.38 to 2.42% as the fibre volume fraction within the composites increased from 0 to 20%. The moisture diffusion coefficient also increased by 40% with inclusion of 20% fibre content.

Liu *et al.*[85] investigated coating PGFs with magnesium (Mg) using physical vapour deposition (PVD) in order to enhance adhesion between PGF and PCL and consequently control degradation and mechanical retention properties of the composites. Magnesium coating yielded roughened fibre surfaces (see Figure 11.13) which could potentially induce a mechanical interlock between the fibres and polymer matrix. Tensile properties of the

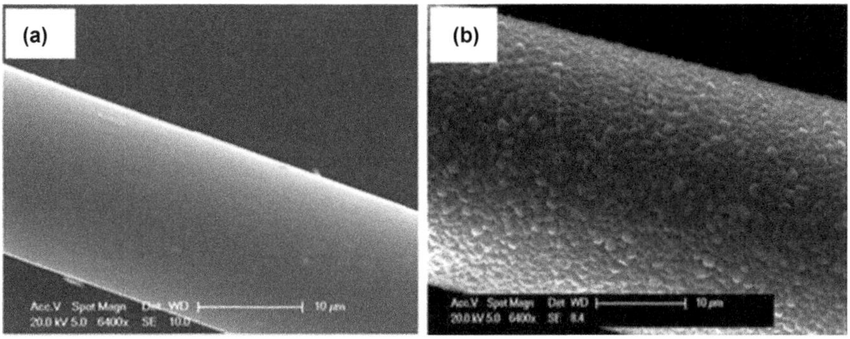

Figure 11.13 SEM of (a) non-coated PGF and (b) PGF coated with magnesium using PVD.
Adapted from ref. 85. Copyright © 2013 Xiaoling Liu *et al.*

coated fibres decreased initially after coating with magnesium which was suggested to be due to surface damage due to bombardment of Mg particles during the coating process. The IFSS of the PCL–PGF interface increased from 5.2 to 8.9 MPa as the thickness of magnesium coating increased to 4 μm. However, the IFSS then decreased to similar levels as the non-coated counterparts when the Mg coating had reached 9.5 μm. Mg-coated PCL–PGF random mat composites showed significantly lower water uptake, mass loss and rate of ion release compared to the non-coated fibre-reinforced composites (control group). The tensile strength and modulus of these Mg-coated fibre composites were 17 and 47%, respectively, higher than the control (non-coated fibre) composites. In addition, the coated fibre composites retained higher tensile and flexural properties than control composites after 10 days of degradation in PBS at 37 °C. Both coated and non-coated composites revealed good cytocompatibility and no significant differences were observed. Similar findings were obtained for Mg-coated unidirectional (UD) composites. Initial tensile properties for UD composites increased more than 50% as a result of Mg coating. They concluded that the Mg coating inhibited fibre degradation and enhanced the fibre–matrix adhesion, which led to the improvement in mechanical retention and degradation properties achieved.

11.4 Biomedical Applications of PGF Composites

11.4.1 Bone Repair

Han *et al.*[86] manufactured resorbable PLA–PGF composite plates and demonstrated the effects of drilled screw holes on their mechanical and degradation properties (see Figure 11.14). Flexural strength and modulus of the UD composites increased gradually from ~200 to 320 MPa and from 10 GPa to 23 GPa, respectively, as the fibre volume fraction increased from 25 to 45%. Moreover, use of a larger number of thin UD layers for the same fibre fraction had a positive effect on mechanical properties for the

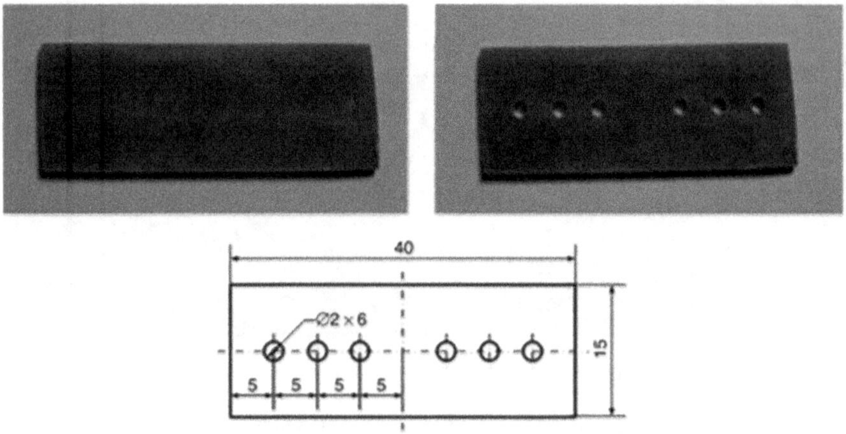

Figure 11.14 PLA–PGF composite plates before and after drilling screw holes. Screw holes have 2 mm diameter and the span between holes is 5 mm. Adapted with permission from ref. 86.

composites. For instance, composite containing 8 UD thin fibre layers revealed 60% higher flexural properties than the composite composed of one thick layer of UD fibres. This was attributed to the use of multiple, thinner UD mats providing better fibre wet-out, impregnation and dispersion across the composite thickness which led to a stronger fibre–matrix interface resulting in enhanced stress transfer. Drilling screw holes into composite plates revealed a decrease in flexural strength by 6–20% in comparison with composite plates without holes. Interestingly, they also found that gamma sterilisation had no significant influence on the flexural properties of PLA–PGF composite plates with and without screw holes, which was ascribed to stability of the glass fibres against gamma radiation and that the fibres were the dominating factor in the properties of the composites. Drilling screw holes into laminate composites caused different types of damage such as delamination, cracking, debonding, fibre tearing and polymer softening and deformation around the screw hole. It was also found that the damage was more substantial for UD composites than the RM counterparts. Consequently, use of RM layers at the top and bottom of UD composites revealed a significant decrease in screw-hole damage. Weight loss profiles for both PLA–PGF UD and RM composites increased initially at day 3 followed by a gradual decrease which was ascribed to water uptake. Furthermore, composites containing 45% fibre volume fraction showed significantly higher mass loss compared to other RM and UD composites. Composite plates with screw holes exhibited slightly faster degradation compared to plates without holes due to increase in the surface area.[86]

A novel method introduced by Felfel *et al.*[87–90] for reshaping PLA–PGF composite plates into rods and screws (*i.e. via* forging or thermomechanical deformation) was applied during their rubbery state (*i.e.* above T_g of PLA) and below the crystallisation temperature (T_c) in order to prevent an increase in

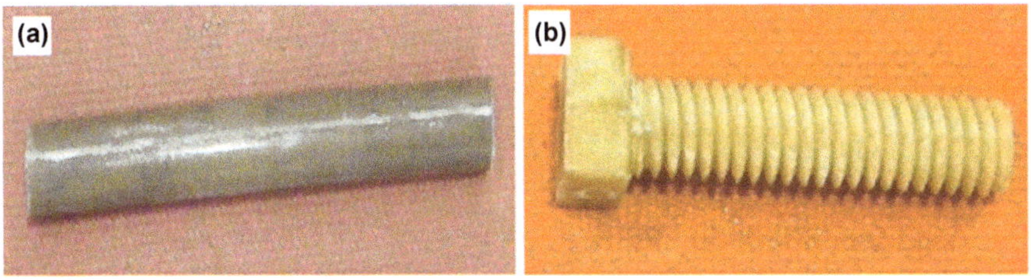

Figure 11.15 Images of (a) composite rod and (b) composite screw prepared *via* the forging process. (a) R. M. Felfel, Initial mechanical properties of phosphate-glass fibre-reinforced rods for use as resorbable intramedullary nails, *J. Mater. Sci.*, 2012, **47**, with permission by Springer[27] and (b) adapted from R. M. Felfel *et al.*, Bioresorbable screws reinforced with phosphate glass fibre: Manufacturing and mechanical property characterisation, *J. Mech. Behav. Biomed. Mater.*, **17**, 76–88, Copyright 2013 with permission from Elsevier.[88]

matrix crystallinity which could lower the degradation rate of the implant.[91–93] The forging process window was selected as 70–90 °C and this technique had numerous advantages: low processing temperatures, low cost, small void ratio, appropriate for different designs, shapes and materials, exclusion of aggressive thermal degradation and improvement of the matrix mechanical properties. The composite plates were forged at ~90 °C into rods and screws (Figure 11.15). The forging technique improved the mechanical properties which were attributed to both the orientation of polymer chains in the matrix and the short chopped random mat fibres rearranging along the principal axes of the rods. PLA–PGF rods and screws showed comparable initial mechanical properties to cortical bone. Initial flexural, shear and compressive strengths of the rods varied within ranges of 100–250 MPa, 60–90 MPa, 130–420 MPa, respectively, and the flexural modulus ranged from 7 to 25 GPa depending on fibre volume fraction and architecture.

An *in vivo* study conducted for PCL–PGF composites for 26 weeks used a rat calvarium model in order to provide information on the biocompatibility of the composite (which was designed for craniomaxillofacial reconstruction). No clinical complications were observed and all animals recovered well after surgery. A lack of inflammatory response was also observed from histological assessment. No bone was detected at the dural face of the implants up to the 4 week time point. Mineralised new bone was present at 8 weeks post-surgery for PCL, PCL–PGF and PCL–bioglass composites. Monolithic PCL did not support new bone formation, whilst PCL composite discs (see Figure 11.16a) showed a gradual increase in the amount of mineralised bone from 20 to 35% over time. At the 26 week time point, PCL–PGF composites revealed a significantly greater quantity of new bone formation compared with PCL alone and PCL reinforced with 45S5 bioglass fibres. Extensive bone growth was seen at 26 weeks which was examined using histological examination and micro-CT (see Figure 11.16b and c).[60]

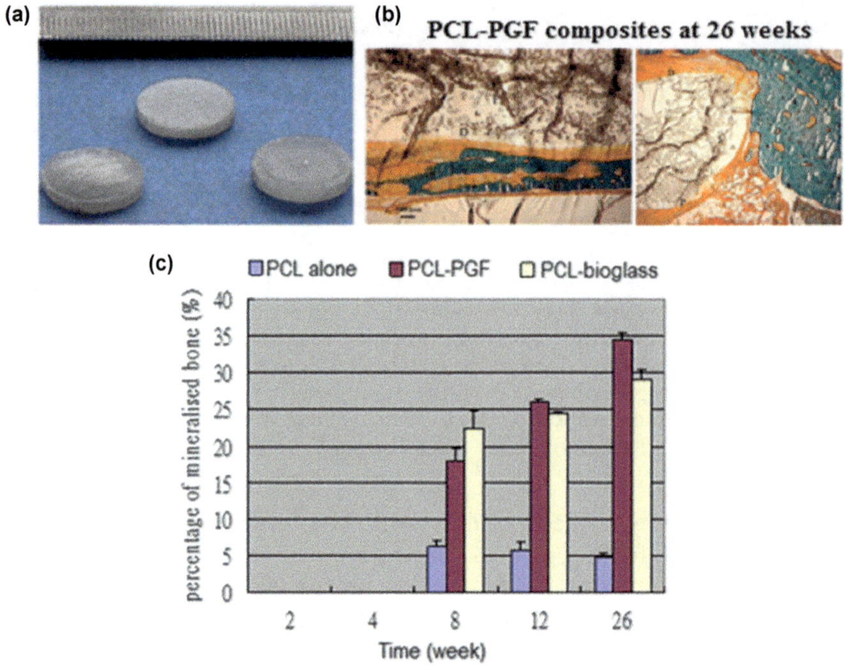

Figure 11.16 (a) PCL–phosphate glass discs before implantation (8 mm diameter), (b) examples of bone associated with the dural face of implants after 26 weeks implantation and (c) percentage of bone associated with the dural face of the implants for PCL, PCL–PGF and PCL-bioglass composites.
Adapted from ref. 60 with permission from John Wiley and Sons. Copyright © 2010 Wiley Periodicals, Inc.

11.4.2 Dental Applications

PCL–PGF composites have also been investigated as root canal obturation materials and showed potential due to their capability of sealing in aqueous environments and releasing certain ions at controlled rates based on composition of PGF.[94] The amount of ions released was found to be inversely proportional to content of iron oxide within the glass composition. PCL–PGF composites revealed better adaptation in root canal than the conventional gutta-percha (GP) and adhered firmly to the canal wall (see Figure 11.17a). Strong adhesion with dentine was attributed to formation of precipitate along the entire canal wall. In contrast, a clear gap was detected between the dentine–GP interface (see Figure 11.17b).[94]

11.4.3 Soft Tissue Repair

Nazhat *et al.*[95] produced spiral collagen scaffolds *via* incorporation of soluble UD PGF into dense collagen gels. They initially prepared sheets (50 μm thickness) of collagen–PGF UD composites *via* an unconfined plastic

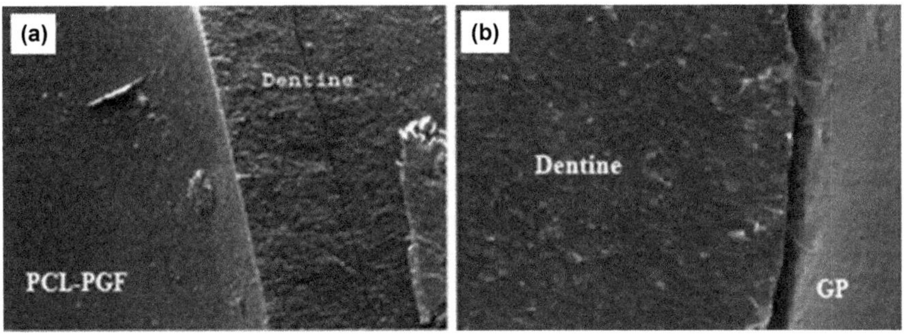

Figure 11.17 SEM micrographs for filling material–dentine interface; (a) good adhesion can be seen between PCL–PGF composite and canal wall (magnification 900×) and (b) a clear gap is present between gutta-percha (GP) and dentine at lower magnification (400×).
Adapted from A. Alani *et al.*, Ion release characteristics, precipitate formation and sealing ability of a phosphate glass-polycaprolactone-based composite for use as a root canal obturation material, *Dent. Mater.*, **25**, 00-410, Copyright 2009 with permission from Elsevier.[94]

compression method and then rolled these composite sheets to form three-dimensional constructs. The mechanical properties of these scaffolds could be tuned by the amount of PGFs incorporated within. Tensile strength and modulus of the scaffolds increased to 2.3 and 165 MPa, respectively, by inclusion of 30 wt% of PGF. Controlled degradation of PGF could enhance cell ingrowth, perfusion and integration of the scaffolds with the surrounding tissue. Joo *et al.*[96] investigated the spiral collagen–PGF scaffolds as a potential candidates for axonal outgrowth following spinal cord injuries. They implanted cylindrical scaffolds into transected spinal cord of rats. The collagen–PGF scaffolds exhibited better performance compared to collagen alone and no inflammatory responses were observed for either group. They concluded that these PGF-incorporated constructs had potential for functional recovery of injured, or even totally transected, spinal cord in rats.

11.5 Conclusions and Outlook

Different aspects of phosphate glass fibres and their composites, such as manufacturing process, chemical composition and processing technique-related properties (for example, dissolution, ion release and mechanical properties) for biomedical applications have been reviewed.

Compared to bulk glass, phosphate glass fibres possess many advantageous features, such as excellent mechanical properties, higher surface area, can enable greater reinforcing capability of polymer matrices and have also been explored as cell guides for soft tissue repair applications.

In addition, by altering their chemical compositions and adding of modifier ions within the phosphate glass fibre structures, the strength of the chain-forming bonds and the cross-linked networks within the glass structure could be controlled in order to manipulate the fibre degradation and mechanical properties required.

Applications of phosphate glass fibre alone spanning cell delivery vehicles, angiogenesis, nerve guides, antibacterial implants and scaffolds for regeneration of soft–hard tissue interfaces have all been investigated.

Phosphate glass fibre-reinforced composite devices in the form of plates, rods, screws and scaffolds have been shown to have vast potential with a wide range of mechanical properties for load-bearing bone and for soft tissue repair applications. The fibre volume fraction, type, length, orientation (aligned or random) and strength of fibre–matrix interface have a significant influence on the mechanical performance of the composites produced.

Though phosphate glass fibres and their composites have been investigated for varying biomedical applications due to their biocompatibilty, favourable degradation and mechanical properties, there still remain considerable developments to be achieved (for example, industrial fibre scale-up) in order to fully exploit these biomaterials for commercialisation.

Abbreviations

PG	Phosphate glasses
PGFs	Phosphate glass fibres
PLA	Polylactic acid
PCL	Polycaprolactone
FRC	Fibre-reinforced composites
UD	Unidirectional
RM	Random mat
V_f	Volume fraction
PBS	Phosphate buffered saline
SBF	Simulated body fluid
HBSS	Hanks buffered saline solution

References

1. I. Ahmed, M. Lewis, I. Olsen and J. C. Knowles, Phosphate glasses for tissue engineering: Part 1. Processing and characterisation of a ternary-based P2O5–CaO–Na2O glass system, *Biomaterials*, 2004, **25**, 491–499.
2. I. Ahmed, A. Parsons, A. Jones, G. Walker, C. Scotchford and C. Rudd, Cytocompatibility and Effect of Increasing MgO Content in a Range of Quaternary Invert Phosphate-based Glasses, *J. Biomater. Appl.*, 2010, **24**, 555–575.
3. J. C. Knowles, Phosphate based glasses for biomedical applications, *J. Mater. Chem.*, 2003, **13**, 2395–2401.
4. H.-J. Moon, K.-N. Kim, K.-M. Kim, S.-H. Choi, C.-K. Kim, K.-D. Kim, R. LeGeros and Y.-K. Lee, Effect of calcium phosphate glass on bone formation in calvarial defects of Sprague-Dawley rats, *J. Mater. Sci.: Mater. Med.*, 2006, **17**, 807–813.

5. I. Ahmed, C. A. Collins, M. P. Lewis, I. Olsen and J. C. Knowles, Processing, characterisation and biocompatibility of iron-phosphate glass fibres for tissue engineering, *Biomaterials*, 2004, **25**, 3223–3232.
6. I. Ahmed, P. S. Cronin, E. A. Abou Neel, A. J. Parsons, J. C. Knowles and C. D. Rudd, Retention of mechanical properties and cytocompatibility of a phosphate-based glass fiber/polylactic acid composite, *J. Biomed. Mater. Res., Part B*, 2009, **89**, 18–27.
7. N. J. Lakhkar, J.-H. Park, N. J. Mordan, V. Salih, I. B. Wall, H.-W. Kim, S. P. King, J. V. Hanna, R. A. Martin, O. Addison, J. F. W. Mosselmans and J. C. Knowles, Titanium phosphate glass microspheres for bone tissue engineering, *Acta Biomater.*, 2012, **8**, 4181–4190.
8. P. Haque, A. J. Parsons, I. A. Barker, I. Ahmed, D. J. Irvine, G. S. Walker and C. D. Rudd, Interfacial properties of phosphate glass fibres/PLA composites: Effect of the end functionalities of oligomeric PLA coupling agents, *Compos. Sci. Technol.*, 2010, **70**, 1854–1860.
9. N. Sharmin, A. J. Parsons, C. D. Rudd and I. Ahmed, Effect of boron oxide addition on fibre drawing, mechanical properties and dissolution behaviour of phosphate-based glass fibres with fixed 40, 45 and 50 mol% P(2)O(5), *J. Biomater. Appl.*, 2014, **29**, 639–653.
10. I. Ahmed, I. Jones, A. Parsons, J. Bernard, J. Farmer, C. Scotchford, G. Walker and C. Rudd, Composites for bone repair: phosphate glass fibre reinforced PLA with varying fibre architecture, *J. Mater. Sci.: Mater. Med.*, 2011, **22**, 1825–1834.
11. F. T. Wallenberger and N. E. Weston, Glass Fibers from High and Low Viscosity Melts, *Mater. Res. Soc. Symp. Proc.*, 2002, **702**, 165–172.
12. K. M. Z. Hossain, H. Muhammad Sami, F. Reda and A. Ifty, Development of phosphate-based glass fibers for biomedical applications, in *Hot Topics in Biomaterials*, Future Science Ltd, 2014, pp. 104–115.
13. J. Choueka, J. L. Charvet, H. Alexander, Y. O. Oh, G. Joseph, N. C. Blumenthal and W. C. LaCourse, Effect of annealing temperature on the degradation of reinforcing fibers for absorbable implants, *J. Biomed. Mater. Res.*, 1995, **29**, 1309–1315.
14. D. Furniss and A. B. Seddon, Towards monomode proportioned fibreoptic preforms by extrusion, *J. Non-Cryst. Solids*, 1999, **256–257**, 232–236.
15. I. Ahmed, S. S. Shaharuddin, N. Sharmin, D. Furniss and C. Rudd, Core/Clad Phosphate Glass Fibres Containing Iron and/or Titanium, *Biomed. Glasses*, 2015, **1**, 20–30.
16. S. Lee, A. Obata and T. Kasuga, Ion release from SrO-CaO-TiO_2-P_2O5 glasses in Tris buffer solution, *J. Ceram. Soc. Jpn.*, 2009, **117**, 935–938.
17. M. Navarro, M. P. Ginebra and J. A. Planell, Cellular response to calcium phosphate glasses with controlled solubility, *J. Biomed. Mater. Res. A*, 2003, **67**, 1009–1015.
18. E. A. Abou Neel, I. Ahmed, J. J. Blaker, A. Bismarck, A. R. Boccaccini, M. P. Lewis, S. N. Nazhat and J. C. Knowles, Effect of iron on the surface, degradation and ion release properties of phosphate-based glass fibres, *Acta Biomater.*, 2005, **1**, 553–563.

19. Y. Nishikubo, S. Yoshida, T. Sugawara and J. Matsuoka, Intrinsic strength of sodium borosilicate glass fibers by using a two-point bending technique, *IOP Conf. Ser.: Mater. Sci. Eng.*, 2011, **18**, 112019.
20. T. Rouxel, Elastic properties of glasses: a multiscale approach, *C. R. Mec.*, 2006, **334**, 743–753.
21. M. Goldstein and T. H. Davies, Glass Fibers with Oriented Chain Molecules, *J. Am. Ceram. Soc.*, 1955, **38**, 223–226.
22. S. Inaba, H. Hosono and S. Ito, Entropic shrinkage of an oxide glass, *Nat. Mater.*, 2015, **14**, 312–317.
23. W. H. Otto, Relationship of Tensile Strength of Glass Fibers to Diameter, *J. Am. Ceram. Soc.*, 1955, **38**, 122–125.
24. S. T. Lin, S. L. Krebs, S. Kadiyala, K. W. Leong, W. C. LaCourse and B. Kumar, Development of bioabsorbable glass fibres, *Biomaterials*, 1994, **15**, 1057–1061.
25. I. Ahmed, A. J. Parsons, G. Palmer, J. C. Knowles, G. S. Walker and C. D. Rudd, Weight loss, ion release and initial mechanical properties of a binary calcium phosphate glass fibre/PCL composite, *Acta Biomater.*, 2008, **4**, 1307–1314.
26. X. Liu, D. M. Grant, A. J. Parsons, L. T. Harper, C. D. Rudd and I. Ahmed, Magnesium Coated Bioresorbable Phosphate Glass Fibres: Investigation of the Interface between Fibre and Polyester Matrices, *BioMed Res. Int.*, 2013, **2013**, 10.
27. R. Felfel, I. Ahmed, A. Parsons, L. Harper and C. Rudd, Initial mechanical properties of phosphate-glass fibre-reinforced rods for use as resorbable intramedullary nails, *J. Mater. Sci.*, 2012, **47**, 4884–4894.
28. S. Cozien-Cazuc, A. J. Parsons, G. S. Walker, I. A. Jones and C. D. Rudd, Effects of aqueous aging on the mechanical properties of P40Na20-Ca16Mg24 phosphate glass fibres, *J. Mater. Sci.*, 2008, **43**, 4834–4839.
29. M. Bitar, V. Salih, J. C. Knowles and M. P. Lewis, Iron-phosphate glass fiber scaffolds for the hard–soft interface regeneration: The effect of fiber diameter and flow culture condition on cell survival and differentiation, *J. Biomed. Mater. Res., Part A*, 2008, **87A**, 1017–1026.
30. M. Bitar, V. Salih, J. C. Knowles and M. P. Lewis, Iron-phosphate glass fiber scaffolds for the hard–soft interface regeneration: The effect of fiber diameter and flow culture condition on cell survival and differentiation, *J. Biomed. Mater. Res., Part A*, 2008, **87A**, 1017–1026.
31. R. Shah, A. C. M. Sinanan, J. C. Knowles, N. P. Hunt and M. P. Lewis, Craniofacial muscle engineering using a 3-dimensional phosphate glass fibre construct, *Biomaterials*, 2005, **26**, 1497–1505.
32. C. Vitale-Brovarone, G. Novajra, J. Lousteau, D. Milanese, S. Raimondo and M. Fornaro, Phosphate glass fibres and their role in neuronal polarization and axonal growth direction, *Acta Biomater.*, 2012, **8**, 1125–1136.
33. E. A. Abou Neel, I. Ahmed, J. Pratten, S. N. Nazhat and J. C. Knowles, Characterisation of antibacterial copper releasing degradable phosphate glass fibres, *Biomaterials*, 2005, **26**, 2247–2254.

34. S. Zhao, L. Li, H. Wang, Y. Zhang, X. Cheng, N. Zhou, M. N. Rahaman, Z. Liu, W. Huang and C. Zhang, Wound dressings composed of copper-doped borate bioactive glass microfibers stimulate angiogenesis and heal full-thickness skin defects in a rodent model, *Biomaterials*, 2015, **53**, 379–391.
35. J. E. Bergsma, W. C. de Bruijn, F. R. Rozema, R. R. M. Bos and G. Boering, Late degradation tissue response to poly(L-lactide) bone plates and screws, *Biomaterials*, 1995, **16**, 25–31.
36. M. Bühler, P.-E. Bourban and J.-A. E. Månson, Cellular composites based on continuous fibres and bioresorbable polymers, *Composites, Part A*, 2008, **39**, 1779–1786.
37. A. Weiler, R. F. G. Hoffmann, A. C. Stahelin, H.-J. Helling and N. P. Sudkamp, Biodegradable implants in sports medicine: The biological base, *Arthroscopy: The Journal of Arthroscopic & Related Surgery*, 2000, **16**, 305–321.
38. S. Fu, X. Feng, B. Lauke and Y.-W. Mai, Effects of particle size, particle/matrix interface adhesion and particle loading on mechanical properties of particulate-polymer composites, *Composites, Part B*, 2008, **39**, 933–961.
39. D. F. Williams, *The Williams dictionary of biomaterials*, Liverpool University Press, Liverpool, 1999.
40. S. Ramakrishna, J. Mayer, E. Wintermantel and K. W. Leong, Biomedical applications of polymer-composite materials: a review, *Compos. Sci. Technol.*, 2001, **61**, 1189–1224.
41. I. Ahmed, P. S. Cronin, E. A. Abou Neel, A. J. Parsons, J. C. Knowles and C. D. Rudd, Retention of mechanical properties and cytocompatibility of a phosphate-based glass fiber/polylactic acid composite, *J. Biomed. Mater. Res., Part B*, 2009, **89B**, 18–27.
42. W. Bonfield, Hydroxyapatite-Reinforced Polyethylene as an Analogous Material for Bone Replacementa, *Ann. N. Y. Acad. Sci.*, 1988, **523**, 173–177.
43. Y. Z. Wan, Y. L. Wang, X. H. Xu and Q. Y. Li, In vitro degradation behavior of carbon fiber-reinforced PLA composites and influence of interfacial adhesion strength, *J. Appl. Polym. Sci.*, 2001, **82**, 150–158.
44. C. Zimmerman Mark, H. Alexander, J. R. Parsons and P. K. Bajpai, in *High-Tech Fibrous Materials*, American Chemical Society, 1991, vol. 457, pp. 132–148.
45. G. Jiang, M. E. Evans, I. A. Jones, C. D. Rudd, C. A. Scotchford and G. S. Walker, Preparation of poly([epsilon]-caprolactone)/continuous bioglass fibre composite using monomer transfer moulding for bone implant, *Biomaterials*, 2005, **26**, 2281–2288.
46. A. Z. Kharazi, M. H. Fathi and F. Bahmany, Design of a textile composite bone plate using 3D-finite element method, *Mater. Des.*, 2010, **31**, 1468–1474.
47. S. L. Evans and P. J. Gregson, Composite technology in load-bearing orthopaedic implants, *Biomaterials*, 1998, **19**, 1329–1342.
48. D. Hull and T. W. Clyne, *An Introduction to Composite Materials*, Cambridge, 1981.

49. P. A. Smith and J. A. Yeomans, in *Material Science and Engineering*, ed. R. D. Rawlings, 2009, vol. II, pp. 133–154.
50. S. Saidpour, Assessment of Carbon Fibre Composite Fracture Fixation Plate Using Finite Element Analysis, *Ann. Biomed. Eng.*, 2006, **34**, 1157–1163.
51. E. Pirhonen and P. Törmälä, Coating of bioactive glass 13-93 fibres with biomedical polymers, *J. Mater. Sci.*, 2006, **41**, 2031–2036.
52. A. D. D. Matilde, J. C. Nichola and L. H. Larry, Tensile properties of bioactive fibers for tissue engineering applications, *J. Biomed. Mater. Res.*, 2000, **53**, 199–203.
53. H. W. Kim, H. E. Kim and J. C. Knowles, Production and Potential of Bioactive Glass Nanofibers as a Next-Generation Biomaterial, *Adv. Funct. Mater.*, 2006, **16**, 1529–1535.
54. E. Pirhonen, H. Niiranen, T. Niemelä, M. Brink and P. Törmälä, Manufacturing, mechanical characterization, and *in vitro* performance of bioactive glass 13-93 fibers, *J. Biomed. Mater. Res., Part B*, 2006, **77B**, 227–233.
55. A. R. Boccaccini and V. Maquet, Bioresorbable and bioactive polymer/Bioglass® composites with tailored pore structure for tissue engineering applications, *Compos. Sci. Technol.*, 2003, **63**, 2417–2429.
56. A. R. Boccaccini, M. Erol, W. J. Stark, D. Mohn, Z. Hong and J. F. Mano, Polymer/bioactive glass nanocomposites for biomedical applications: A review, *Compos. Sci. Technol.*, 2010, **70**, 1764–1776.
57. D. Brauer, C. Rüssel, S. Vogt, J. Weisser and M. Schnabelrauch, Degradable phosphate glass fiber reinforced polymer matrices: mechanical properties and cell response, *J. Mater. Sci.: Mater. Med.*, 2008, **19**, 121–127.
58. A. J. Parsons, I. Ahmed, P. Haque, B. Fitzpatrick, M. I. K. Niazi, G. S. Walker and C. D. Rudd, Phosphate Glass Fibre Composites for Bone Repair, *J. Bionic Eng.*, 2009, **6**, 318–323.
59. S. Lin, S. Krebs, A. Nazre and R. King, presented in part at the 39th International SAMPE Symposium, April 11–14, 1994.
60. C. A. Scotchford, M. Shataheri, P. S. Chen, M. Evans, A. J. Parsons, G. A. Aitchison, C. Efeoglu, J. L. Burke, A. Vikram, S. E. Fisher and C. D. Rudd, Repair of calvarial defects in rats by prefabricated, degradable, long fibre composite implants, *J. Biomed. Mater. Res. A*, 2010, **96**, 230–238.
61. I. Ahmed, I. Jones, A. Parsons, J. Bernard, J. Farmer, C. Scotchford, G. Walker and C. Rudd, Composites for bone repair: phosphate glass fibre reinforced PLA with varying fibre architecture, *J. Mater. Sci.: Mater. Med.*, 2011, **22**, 1825–1834.
62. J. L. Charvet, J. A. Cordes and H. Alexander, Mechanical and fracture behavior of a fiber-reinforced bioabsorbable material for orthopaedic applications, *J. Mater. Sci.: Mater. Med.*, 2000, **11**, 101–109.
63. P. Borbely, K. Gulabivala and J. C. Knowles, Degradation properties and ion release characteristics of Resilon and phosphate glass/polycaprolactone composites, *Int. Endod. J.*, 2008, **41**, 1093–1100.

64. W. S. DePolo and D. G. Baird, Particulate reinforced PC/PBT composites. I. Effect of particle size (nanotalc versus fine talc particles) on dimensional stability and properties, *Polym. Compos.*, 2009, **30**, 188–199.
65. S. K. De and J. R. White, *Short Fibre-Polymer Composites*, Woodhead Publishing, 1996.
66. R. Krüger and J. Groll, Fiber reinforced calcium phosphate cements – On the way to degradable load bearing bone substitutes? *Biomaterials*, 2012, **33**, 5887–5900.
67. R. M. Felfel, I. Ahmed, A. J. Parsons, P. Haque, G. S. Walker and C. D. Rudd, Investigation of Crystallinity, Molecular Weight Change, and Mechanical Properties of PLA/PBG Bioresorbable Composites as Bone Fracture Fixation Plates, *J. Biomater. Appl.*, 2012, **26**, 765–789.
68. M. Hasan, I. Ahmed, A. Parsons, G. Walker and C. Scotchford, Cytocompatibility and Mechanical Properties of Short Phosphate Glass Fibre Reinforced Polylactic Acid (PLA) Composites: Effect of Coupling Agent Mediated Interface, *J. Funct. Biomater.*, 2012, **3**, 706.
69. X. Liu, M. S. Hasan, D. M. Grant, L. T. Harper, A. J. Parsons, G. Palmer, C. D. Rudd and I. Ahmed, Mechanical, degradation and cytocompatibility properties of magnesium coated phosphate glass fibre reinforced polycaprolactone composites, *J. Biomater. Appl.*, 2014, **29**, 675–687.
70. L. T. Harper, I. Ahmed, R. M. Felfel and C. Qian, Finite element modelling of the flexural performance of resorbable phosphate glass fibre reinforced PLA composite bone plates, *J. Mech. Behav. Biomed. Mater.*, 2012, **15**, 13–23.
71. A. J. Parsons, I. Ahmed, N. Han, R. Felfel and C. D. Rudd, Mimicking Bone Structure and Function with Structural Composite Materials, *J. Bionic Eng.*, 2010, **7**, S1–S10.
72. K. P. Andriano, A. U. Daniels and J. Heller, Biocompatibility and mechanical properties of a totally absorbable composite material for orthopaedic fixation devices, *J. Appl. Biomater.*, 1992, **3**, 197–206.
73. M. S. Hasan, I. Ahmed, A. J. Parsons, G. S. Walker and C. A. Scotchford, The influence of coupling agents on mechanical property retention and long-term cytocompatibility of phosphate glass fibre reinforced PLA composites, *J. Mech. Behav. Biomed. Mater.*, 2013, **28**, 1–14.
74. R. A. Khan, A. J. Parsons, I. A. Jones, G. S. Walker and C. D. Rudd, Effectiveness of 3-Aminopropyl-Triethoxy-Silane as a Coupling Agent for Phosphate Glass Fiber-Reinforced Poly(caprolactone)-based Composites for Fracture Fixation Devices, *J. Thermoplast. Compos. Mater.*, 2011, **24**, 517–534.
75. R. A. Khan, A. J. Parsons, I. A. Jones, G. S. Walker and C. D. Rudd, Surface treatment of phosphate glass fibers using 2-hydroxyethyl methacrylate: Fabrication of poly(caprolactone)-based composites, *J. Appl. Polym. Sci.*, 2009, **111**, 246–254.
76. R. A. Khan, A. J. Parsons, I. A. Jones, G. S. Walker and C. D. Rudd, Degradation and Interfacial Properties of Iron Phosphate Glass

Fiber-Reinforced PCL-Based Composite for Synthetic Bone Replacement Materials, *Polym.-Plast. Technol. Eng.*, 2010, **49**, 1265–1274.
77. X. Liu, D. M. Grant, G. Palmer, A. J. Parsons, C. D. Rudd and I. Ahmed, Magnesium coated phosphate glass fibers for unidirectional reinforcement of polycaprolactone composites, *J. Biomed. Mater. Res., Part B*, 2014, DOI: 10.1002/jbm.b.33324, n/a–n/a.
78. M. S. Mohammadi, I. Ahmed, N. Muja, S. Almeida, C. D. Rudd, M. N. Bureau and S. N. Nazhat, Effect of Si and Fe doping on calcium phosphate glass fibre reinforced polycaprolactone bone analogous composites, *Acta Biomater.*, 2012, **8**, 1616–1626.
79. E. El-Meliegy, M. M. Farag and J. C. Knowles, Dissolution and drug release profiles of phosphate glasses doped with high valency oxides, *J. Mater. Sci.: Mater. Med.*, 2016, **27**, 1–10.
80. R. M. Felfel, I. Ahmed, A. J. Parsons, G. S. Walker and C. D. Rudd, In vitro degradation, flexural, compressive and shear properties of fully bioresorbable composite rods, *J. Mech. Behav. Biomed. Mater.*, 2011, **4**, 1462–1472.
81. R. M. Felfel, I. Ahmed, A. J. Parsons, G. Palmer, V. Sottile and C. D. Rudd, Cytocompatibility, degradation, mechanical property retention and ion release profiles for phosphate glass fibre reinforced composite rods, *Mater. Sci. Eng.: C*, 2013, **33**, 1914–1924.
82. P. Haque, I. A. Barker, A. Parsons, K. J. Thurecht, I. Ahmed, G. S. Walker, C. D. Rudd and D. J. Irvine, Influence of compatibilizing agent molecular structure on the mechanical properties of phosphate glass fiber-reinforced PLA composites, *J. Polym. Sci., Part A: Polym. Chem.*, 2010, **48**, 3082–3094.
83. M. S. Hasan, I. Ahmed, A. J. Parsons, C. D. Rudd, G. S. Walker and C. A. Scotchford, Investigating the use of coupling agents to improve the interfacial properties between a resorbable phosphate glass and polylactic acid matrix, *J. Biomater. Appl.*, 2013, **28**, 354–366.
84. L. Onal, S. Cozien-Cazuc, I. A. Jones and C. D. Rudd, Water absorption properties of phosphate glass fiber-reinforced poly-ϵ-caprolactone composites for craniofacial bone repair, *J. Appl. Polym. Sci.*, 2008, **107**, 3750–3755.
85. X. Liu, D. M. Grant, A. J. Parsons, L. T. Harper, C. D. Rudd and I. Ahmed, Magnesium Coated Bioresorbable Phosphate Glass Fibres: Investigation of the Interface between Fibre and Polyester Matrices, *BioMed Res. Int.*, 2013, **2013**, 10.
86. N. Han, I. Ahmed, A. J. Parsons, L. Harper, C. A. Scotchford, B. E. Scammell and C. D. Rudd, Influence of screw holes and gamma sterilization on properties of phosphate glass fiber-reinforced composite bone plates, *J. Biomater. Appl.*, 2013, **27**, 990–1002.
87. R. Felfel, I. Ahmed, A. Parsons, G. Walker and C. Rudd, Flexural, Shear and Compressive Properties of Bioresorbable PLA-PBG Fibre Reinforced Composite Intramedullary Rods, *World J. Eng.*, 2010, 323–324.

88. R. M. Felfel, I. Ahmed, A. J. Parsons and C. D. Rudd, Bioresorbable screws reinforced with phosphate glass fibre: Manufacturing and mechanical property characterisation, *J. Mech. Behav. Biomed. Mater.*, 2013, **17**, 76–88.
89. R. M. Felfel, I. Ahmed, A. J. Parsons and C. D. Rudd, Bioresorbable composite screws manufactured via forging process: Pull-out, shear, flexural and degradation characteristics, *J. Mech. Behav. Biomed. Mater.*, 2013, **18**, 108–122.
90. R. M. Felfel, I. Ahmed, A. J. Parsons, G. S. Walker and C. D. Rudd, In vitro degradation, flexural, compressive and shear properties of fully bioresorbable composite rods, *J. Mech. Behav. Biomed. Mater.*, 2011, **4**, 1462–1472.
91. H. Cai, V. Dave, R. A. Gross and S. P. McCarthy, Effects of physical aging, crystallinity, and orientation on the enzymatic degradation of poly(lactic acid), *J. Polym. Sci., Part B: Polym. Phys.*, 1996, **34**, 2701–2708.
92. Y. Tokiwa, B. Calabia, C. Ugwu and S. Aiba, Biodegradability of Plastics, *Int. J. Mol. Sci.*, 2009, **10**, 3722–3742.
93. Q. Zhou and M. Xanthos, Nanoclay and crystallinity effects on the hydrolytic degradation of polylactides, *Polym. Degrad. Stab.*, 2008, **93**, 1450–1459.
94. A. Alani, J. C. Knowles, W. Chrzanowski, Y. L. Ng and K. Gulabivala, Ion release characteristics, precipitate formation and sealing ability of a phosphate glass-polycaprolactone-based composite for use as a root canal obturation material, *Dent. Mater.*, 2009, **25**, 400–410.
95. S. N. Nazhat, E. A. Abou Neel, A. Kidane, I. Ahmed, C. Hope, M. Kershaw, P. D. Lee, E. Stride, N. Saffari, J. C. Knowles and R. A. Brown, Controlled Microchannelling in Dense Collagen Scaffolds by Soluble Phosphate Glass Fibers, *Biomacromolecules*, 2007, **8**, 543–551.
96. N.-Y. Joo, J. C. Knowles, G.-S. Lee, J.-W. Kim, H.-W. Kim, Y.-J. Son and J. K. Hyun, Effects of phosphate glass fiber–collagen scaffolds on functional recovery of completely transected rat spinal cords, *Acta Biomater.*, 2012, **8**, 1802–1812.

CHAPTER 12

Organic–Inorganic Hybrid Biomaterials

GOWSIHAN POOLOGASUNDARAMPILLAI*[a] AND ANTHONY L. B. MAÇON*[b]

[a] The University of Manchester, School of Materials, Research Complex at Harwell, Rutherford Appleton Laboratory, Didcot, Oxon OX11 0FA, UK; [b] Frontier Research Institute for Materials Science, Nagoya Institute of Technology, Gokiso-cho, Showa-ku, Nagoya 466-8555, Japan
*Email: g.poolo@manchester.ac.uk; alb.macon@nitech.ac.jp

12.1 Introduction

12.1.1 Lessons From Nature

We only have to look at nature to realise the importance of combining materials with properties that complement each other. Over millions of years of evolution nature has produced structures, such as bone, tooth and nacre that are durable and mechanically strong whilst also performing multitude of other biological functions. Nature does this by assembling organic and inorganic (O–I) molecular building-blocks, from the atomic scale to the macroscale, into several orders of hierarchical organisation. Bone, for example, has seven levels of hierarchy and at the nanoscale it is composed of ~3 nm thick platelets of inorganic hydroxyl carbonate apatite (HCA) mineral, which are bonded through an amorphous interfibril matrix to collagen fibrils, which are 1.5 nm thick. From an engineering point of view, the high degree of mixing of the two nanometric phases, within the mineralised fibrils, is responsible for the high strength and Young's

RSC Smart Materials No. 23
Bioactive Glasses: Fundamentals, Technology and Applications
Edited by Aldo R. Boccaccini, Delia S. Brauer and Leena Hupa

Published by the Royal Society of Chemistry, www.rsc.org

modulus while the interfibrillar matrix that binds the mineralised fibrils is responsible for the high fracture toughness by accommodating large amount of shear stress during loading.[1] Such complex structure in bone has resulted in mechanical properties which are greater than the sum of its constituents and their homogeneous mixtures.[2] It is therefore a combination of the nanoscale and tailored interfacial bonding between the organic and inorganic that give bone its unique and outstanding properties.

Aiming for the design of synthetic materials with such intricate structure could be the key towards ideal biomaterials with a high degree of tailorability. For instance, when targeting the regeneration of a defect caused by a trauma or disease located in hard tissues, such as bone, the rate of success of the implant greatly depends on its ability to match the mechanical properties of the tissue. Thus, by mimicking bone in its composition (organic–inorganic) and structure (high interfacial bonding), it could be possible to synthetically reproduce its outstanding and adaptive nature. In addition, and in line with the principle of materials for regenerative medicine, organic–inorganic materials with a high degree of mixing and small domain sizes could also potentially degrade uniformly as new tissue repopulates the defect and therefore allow a smooth transition from an implant to new tissue within the original defect site.

Such materials can be designed using the sol–gel process. This "soft-chemistry" approach allows for the introduction of an organic component, usually polymeric, while the inorganic network forms *via* the condensation of the sol–gel precursors, yielding materials with organic and inorganic constituents interacting at the atomic level with a high degree of mixing. Thus, these materials, called hybrids, are defined as an interpenetrating molecular mixture of organic and inorganic constituents. From this definition, it is important to mention that nanocomposites should not be classified as hybrids. Even though one of the phases, in a nanocomposite has nanometric dimension (1–100 nm), the two phases may not have interpenetrating networks, which therefore implies that clear boundaries between the phases exist. Excellent books and review articles on O–I hybrids have been published elsewhere.[3–7]

12.1.2 Silica Hybrids and Their Classes

Silica-based hybrids are of particular interest due to the beneficial effect of silicic acid, a degradation product of sol–gel silica glasses, on the metabolism of bone cells.[8–11] Furthermore, the kinetics of hydrolysis and condensation reactions of the silica sol–gel precursor is reasonably slow allowing the addition of an organic phase before the gelation occurs.[12,13] In addition, a wide variety of organosilane coupling agents are commercially available and can be used to functionalise polymers with alkoxysilane moieties (more details in Section 12.2). When dissimilar materials that may be hydrophobic and hydrophilic are mixed together, thermodynamic equilibrium will drive the two phases apart leading to microphase separation.[13,14] In the case of O–I hybrids, the majority, if not all, of the atoms are surface atoms; here the interface between the organic and inorganic components plays a major role in the structure formation and final properties. As in bone, it is not only important to optimise the interfacial bonding to obtain the desired properties but also to obtain homogenous materials.

O–I hybrids can be broadly divided into Class I and Class II depending on the bonding strength between the different components (Figure 12.1).[15] In Class I hybrids, the organic and inorganic constituents are weakly bonded with dynamic interactions, such as ionic, hydrogen bonding, van der Waals' and electrostatic forces, while in Class II hybrids, strong covalent or ionic–covalent bonds exist between the organic and inorganic components. With increasing interfacial bonding strength more homogeneous distribution of the components in the O–I hybrids is observed. This leads to enhanced optical and mechanical properties.[16,17]

This chapter focuses on the synthesis and properties of Class II hybrids towards the design of biomaterials for application in regenerative medicine, where the organic component of the hybrids is a polymer or oligomer. Class II hybrids can be synthesised from polymers that contain alkoxysilane precursors in their structure. Thus, upon addition of these functional polymers to the sol during the sol–gel process their alkoxysilane moieties can hydrolyse and co-condense with the growing silica network and therefore covalently bond the inorganic network. Polymers can therefore be classified into two structural classes (only linear polymers are discussed here): (1) polymers, such as silanised polyesters and polyethers presenting sol–gel functionalities as terminal end-groups (2) polymers, such as silanised polysaccharides, polypeptides or polyacrylates presenting sol–gel functionalities as pendant groups

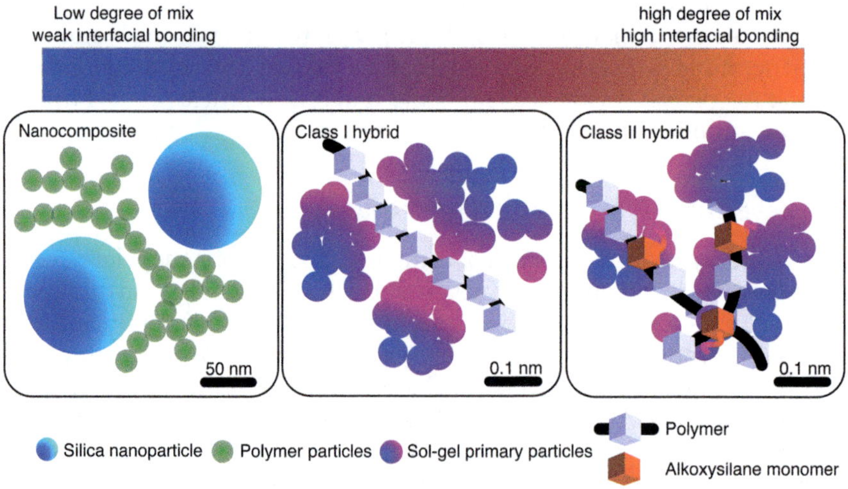

Figure 12.1 Schematic showing the degree of mixing of the organic and inorganic components within a nanocomposite and hybrids of Class I and Class II.

along the backbone of the polymers. The ratio of sol–gel precursors covalently bonded to the polymer to the number of repeating units in the polymer is defined as the cross-linking density (CLD). CLD can be varied independently of the molecular weight of the polymer when the polymer has functional pendant groups on its backbone. However, the CLD in organosilane-terminated polymers is directly linked to the molecular weight of the polymer. Thus, any variation in the molecular weight of the polymer leads to a variation in the CLD. For instance, a two-fold decrease in molecular weight of the polymer translates to a two-fold decrease of the number of repeating units in the polymer, which in turn increases the CLD by two.

12.2 Polymer Source and Silanisation Reactions

Apart from polydimethylsiloxane (PDMS) which has a backbone made of silicon bridging oxygen, the incorporation of alkoxysilane moieties within the structure of a polymer has to be the result of either a functionalisation of the polymer with an organosilane coupling agent, or the bottom-up design of the polymer with alkoxysilane functionalities *via* radical or ring-opening polymerisation. This is an important design criterion if true Class II hybrids are to be synthesised.[15]

12.2.1 Incorporation of Alkoxysilane Moieties

Polymer silanisation consists of incorporating alkoxysilane moieties within the structure of a pre-formed or synthesised polymer, such as a polysaccharide,[18–20] polyester,[21–25] polyether[26–30] or polypeptide.[31–38] To do so, organosilane coupling agents are used which possess an organic functional group that can react with an available reactive site on the polymer, yielding a polymer with a covalently-grafted sol–gel precursor. Thus, the alkoxysilane moieties present on the polymer can hydrolyse and co-condense with the growing silica network during the sol-gel process generating silicate Class II hybrids.[15] The grafting reaction is usually carried out before the sol–gel process and the choice of organosilane coupling agent depends on the available reactive site (end-groups or pendant groups) that the polymer possesses. However, three coupling agents, (3-isocyanatopropyl)triethoxysilane (ICPTES),[24–26,28–30] (3-aminopropyl)triethoxysilane (APTES)[21,22,27,39] and (3-glycidyloxypropyl)-trimethoxysilane (GPTMS)[18,31,33–36,38] have been extensively used for the silanisation of pre-formed polymers towards the synthesis of hybrids for biomedical applications, owing to the low selectivity and high reactivity of their organic moiety (Figure 12.2).

The organic moiety of ICPTES is terminated by an isocyanate motif (R–N=C=O) in which the central carbon is highly reactive towards nucleophiles, such as hydroxyls (R–OH) forming a urethane motif, due to the delocalisation of the electron density towards the nitrogen and oxygen atoms.[40] The reaction can be auto-catalysed, where a urethane group can increase the reactivity of the central carbon of the isocyanate motif *via* hydrogen bonding,

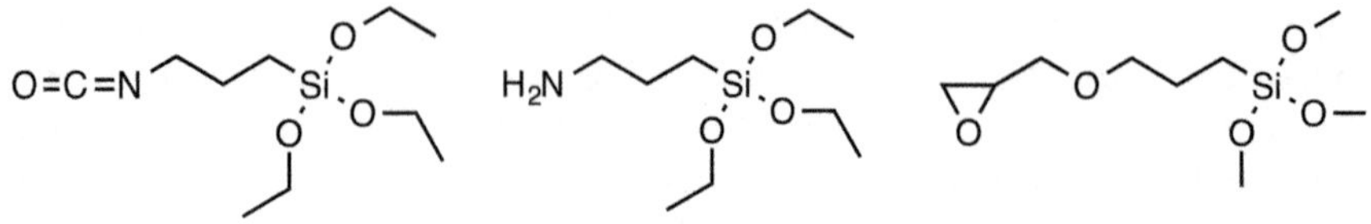

Figure 12.2 Chemical structures of organosilane coupling agents: (3-isocyanatopropyl) triethoxysilane (ICPTES), (3-aminopropyl) triethoxysilane (APTES), (3-glycidyloxypropyl) trimethoxysilane (GPTMS).

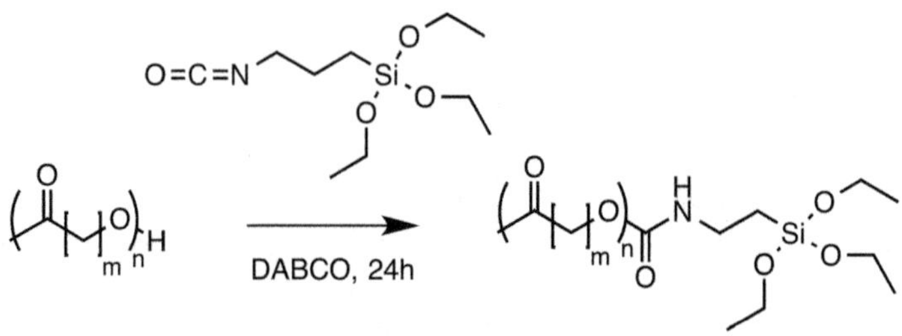

Figure 12.3 Schematic representing the silanisation of a polyester using (3-isocyanatopropyl) triethoxysilane (ICPTES) and catalysed by 2,2′-diazabicyclo[2.2.2]octane (DABCO). *m* represents the length of the alkane chain between two ester groups ($m = 5$ for polycaprolactone) and *n* the degree of polymerisation of the polymer.

or catalysed by a tertiary amine, such as 2,2′-diazabicyclo[2.2.2]octane (DABCO).[40] ICPTES has been mainly used for the silanisation of hydroxyl-terminated polyesters (polycaprolactone (PCL) and poly(lactic acid) (PLA))[21–25] and polyethers (polyethylene glycol (PEG) and polytetramethylene oxide (PTMO)).[26,27,29,30] A well-documented use of ICPTES for the synthesis of hybrids is for the silanisation of PCL, as reported by Tian *et al.*[24] and Rhee *et al.*[25] (Figure 12.3). However, silanisation reactions using ICPTES must be carried out in the absence of water which limits the use of this coupling agent to polymers that are soluble in organic solvents, and preferably in aprotic solvents. When an isocyanate motif reacts with water, an unstable carbamic acid is formed which subsequently decomposes into a primary amine and gaseous carbon dioxide.[40]

APTES is characterised by the presence of a primary amine at the extremity of its organic moiety, which can be used to functionalise carboxylic residues of polymers through the formation of amide bonds (Figure 12.4). However, to be quantitative the amidation reaction must occur after activation of the carboxylic acid.[41] One convenient and straightforward pathway is to form an *O*-acylisourea intermediate through the reaction of the acid with a compound containing a carbodiimide motif, such as *N*-hydroxysuccinimide or *N,N'*-dicyclohexylcarbodiimide.[21,22] Due to the high delocalisation of the electron density around the central carbon, similar to the isocyanate, the activated carboxylic group is highly reactive towards primary amines, where the *O*-acylisourea intermediate acts as an active leaving group. Here, the

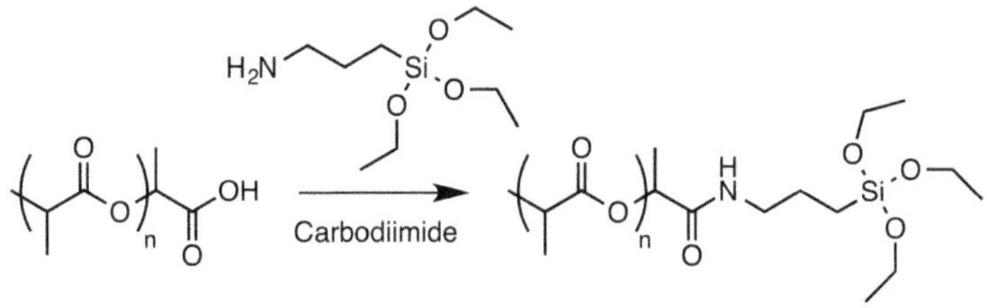

Figure 12.4 Schematic representing the silanisation of poly(lactic acid) using carbodiimide as catalyst and (3-aminopropyl) triethoxysilane (APTES) as organosilane coupling agent. *n* represents the degree of polymerisation of the polymer.

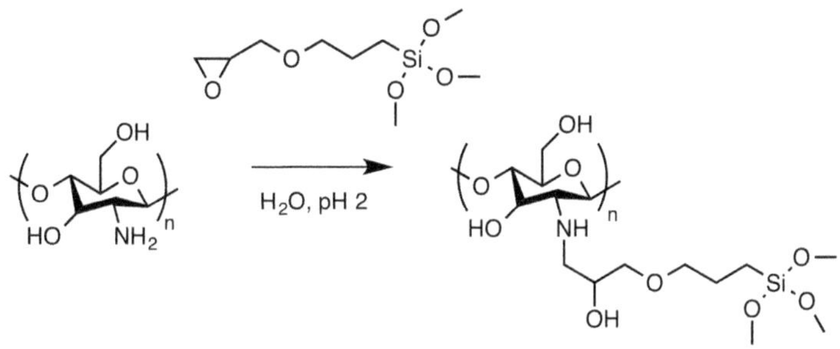

Figure 12.5 Schematic representing the silanisation of chitosan using (3-glycidyloxypropyl) trimethoxysilane (GPTMS) as organosilane coupling agent. *n* represents the degree of polymerisation of the polymer.

work of Kasuga *et al.* must be acknowledged for the silanisation of carboxyl-terminated poly(lactic acid) towards the synthesis of hybrids.[42,43]

The organic moiety of GPTMS has an epoxide, which is a cyclic ether under strain due to its 3-membered ring conformation, making it much more reactive than conventional ethers.[44] Thus, GPTMS can react with weak nucleophiles, such as amine, hydroxyl or carboxylic acid *via* nucleophilic substitution reaction by attacking the electrophilic asymmetric carbon on the epoxide, breaking the ring and relieving the strain.[44] GPTMS was assumed to be one of the most versatile organosilane cross-linkers and has been used for the silanisation of a large number of biocompatible polymers which are only soluble in water, such as gelatine, chitosan and poly(γ-glutamic acid), as shown in Figure 12.5.[45–49] All these polymers contain pendant nucleophilic groups that can potentially attack the epoxide of GPTMS. Despite its popularity, little analytical proof or indirect characterisation methods were used to confirm that the cross-linking reaction occurred with respect to the experimental conditions used. For example, Ren *et al.*, who were one of the pioneers in the silanisation of biopolymers, analysed the amino acid fragments obtained after silanisation of gelatine and showed that GPTMS reacts quantitatively and preferentially with the aspartic acid carboxylic acid (–COOH) residue.[33] However, gelatine fragmented *via* the aminolysis of the peptide bonds when reaction was performed in 6 N HCl at 110 °C for 22 h. These harsh conditions would inevitably cause the hydrolysis of the ester bond formed by the reaction of the epoxide of GPTMS with the carboxylic group of the aspartic acid, as esters are much more prone to hydrolysis than amides. Thus, the validity of these data is questionable.

In addition, several recent studies reported the low stability of the epoxide group on GPTMS when used as a coupling agent in water.[18,50,51] They showed that the epoxide reacts directly with water, resulting in the formation of diols (by ring opening), which considerably reduces its reactivity towards nucleophiles. Gabrielli *et al.* used simple nucleophiles (propanol, propylamine, propanthiol and propanoic acid) at pH 6 and found that only propanoic acid (with a carboxylic acid group) could effectively attack the epoxide ring to form an ester, however the reaction was not quantitative (*i.e.*, was of low yield).[50,51] Connell *et. al.* showed by quantitative 2D NMR and ^{15}N Magic Angle Spinning (MAS) NMR that chitosan (with a primary amine group) can react with GPTMS at pH 2 in a 24 h reaction, while at pH 6 no significant coupling was observed.[18] However, the reaction was once again not quantitative with a yield of ~20%. In addition, ring opening was observed at all pH values investigated (pH = 2, 4 and 6) with an increase in the rate of hydrolysis as the pH decreased.

The other major limitation in carrying out the silanisation in an aqueous environment is the simultaneous hydrolysis and condensation (formation of Si–O–Si bonds) of the alkoxysilane residue of GPTMS, cross-linking the polymer chains together or with unreacted GPTMS.[50] Thus, the isolation of silanised polymer is challenging and could lead to increased phase separation between the growing silica network and the polymer during the sol–gel process.[35]

To this day, no experimental evidence of quantitative functionalisation of water-soluble polymers with alkoxysilane sol–gel precursors has been reported.

12.2.2 Bottom-up Design of Polymers

In contrast to polymers requiring a silanisation step before the sol–gel process, such as these mentioned in the previous section, polymers synthesised *via* radical polymerisation can have built-in alkoxysilane moieties and therefore do not require any post-functionalisation.[52–55]

The organic moiety of the alkoxysilane coupling agent can be terminated with a vinyl group, such as in 3-(alkoxysilyl)propyl methacrylate (Figure 12.6),[56] acrylate[57] or acrylamide,[58] allowing them to react with radicals during a polymerisation process and therefore directly incorporate alkoxysilane moieties within the structure of the polymer. Alkoxysilane-containing monomers can be polymerised with co-monomers with various residual groups for adding functionality. In particular the work of Ravarian *et al.*, Lee *et al.* and Otsuki *et al.* must be acknowledged for their initial contributions on the synthesis of hybrids using polymethacrylates.[59–62] They

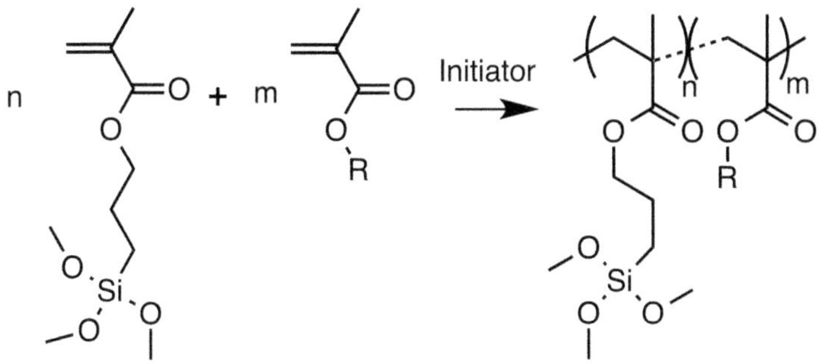

Figure 12.6 Schematic representing the free radical polymerisation of (trimethoxysilyl)-propyl methacrylate (TMSPMA)

reported the copolymerisation of 3-(methoxysilyl)propyl methacrylate with methylmethacrylate or hydroxyethylmethacrylate by free radical polymerisation.

However, great care must be taken when using alkoxysilane containing polymethacrylate as an organic source for the synthesis of hybrids as their backbone cannot be hydrolytically or enzymatically degraded.[63] This implies that the polymer chains released upon degradation of the hybrids *in vivo* must clear through the porous structure of the renal glomerulus to avoid accumulation in the spleen and liver.[64] Thus, polymethacrylate must be synthesised with a hydrodynamic radius smaller than 8 nm in order to comply with the pore size of the kidney.[64] This is particularly challenging using conventional polymerisation techniques as broad size distribution is inherently obtained.[65] However, with the emergence of controlled polymerisation techniques, which allow the design of polymers with predefined molecular weight and narrow polydispersity, these design criteria can be easily fulfilled.[66] In addition, many reports have already been published describing successful polymerisation of well-defined polymethacrylates with alkoxysilane moieties incorporated in their structure.[52–55,59]

12.3 Inorganic Bioactive Source

Synthesis of hybrids is challenging but it promises to mimic the complexity in nature's hierarchical structures. Their design is primarily centred on a biomimetic approach where a high degree of mixing and strong interfacial bonding between the organic and inorganic constituents enhances mechanical properties and molecular level homogeneity in the structure which leads to congruent degradation.[67] However, the coexistence of organic and inorganic entities in a material is not enough to guarantee its success as a biomaterial. Bioactive glasses (inorganic constituents in the hybrids) owe their osteogenic property to their ability to release ions that act as chemical cues while also providing a suitable surface for the formation of new tissue.[68] For example, and when targeting the regeneration of bone, silicic acid has been proven to be essential for the formation of calcified tissue;[8–11] calcium promotes the proliferation and differentiation of preosteoblast cells that lay down new bone;[69,70] magnesium can increase cell adhesion and stimulates the formation of new bone.[71,72]

It is therefore important that the design of hybrids is not only centred on the selection of an appropriate organic phase but should also focus on the incorporation of these inorganic cues to actively promote the regeneration of tissues.

Metal oxides are incorporated within sol–gel silica networks by mixing together silica sol and mineral salts, such as calcium nitrate.[73] However, the metal ions are incorporated within the silica network as metal oxide only after heat stabilisation at high temperatures above 400 °C. Lin *et al.* demonstrated that, using an acid catalysed route, metallic salts stay solubilised within the silica gel where they have minimal interactions with the silanol groups due to the low pH of the sol.[74] Upon drying, the salt precipitates within the mesopores (Figure 12.7). It is not until the heat stabilisation step, when the calcination of the counterions takes place, that the metallic ions diffuse into the silica network. In the case of calcium nitrate, the stabilization occurs above 400 °C,[74] which is incompatible with the organic part in hybrids, regardless of the Class of hybrid synthesised.

Therefore, if metallic salts are used in the synthesis of hybrids, it is likely that they remain as salt particles within the structure. Thus, due to the interconnected mesoporous structure of sol–gel bioactive glasses and hybrids, the salt is likely to be released in a non-controlled manner, with ions undergoing a burst release into the surrounding physiological media upon contact with biological tissue, which could lead to an acute inflammatory response.[75] Hence, alternative metal oxide precursors to those conventionally used in the synthesis of inorganic glasses should be used when attempting to incorporate metallic ions within the structure of Class II hybrids.

12.3.1 Metal Alkoxide

One approach for incorporating metallic ions into the structure of hybrids is to use ionic metal alkoxide as the source.[76] Metal alkoxides are formed from the conjugated base of an alcohol, or deprotonated alcohol (RO–), that acts as ligand for metals. Metal alkoxide precursors can be obtained through the reduction of metals where the alcohol acts as a weak acid or from the reaction with a strong base, such as metal hydride. Similar to a silicon alkoxide network former, such as tetraethylorthosilicate (TEOS), metal alkoxides are not stable in the presence of water and are often much more reactive than TEOS. Thus in an aqueous environment, they undergo rapid hydrolysis forming metallic ions and alcohols. The latter can be easily removed or dried-off at temperatures low enough such that thermal degradation of the organic phase can be avoided. In addition, upon hydrolysis, the alcohol re-protonates causing the pH of the sol to increase, which usually comes with a change in the kinetics of condensation of the silica network when the synthesis is acid catalysed. This increase in pH favours the

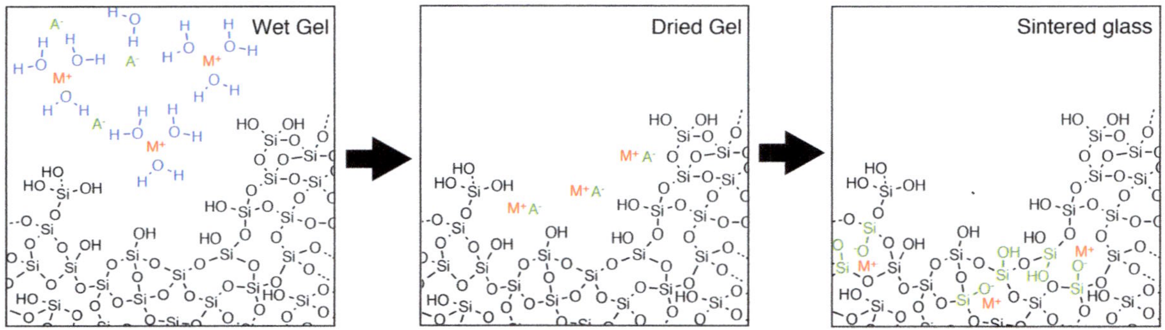

Figure 12.7 Schematic demonstrating the mechanism by which metal salts are incorporated into the silica network. In the wet gel the cations stay solubilised within the pore liquor. Upon drying (Dried gel) the salt precipitates within the mesopores. Once sintered, the metal ions diffuse in the silica network forming metal oxides.

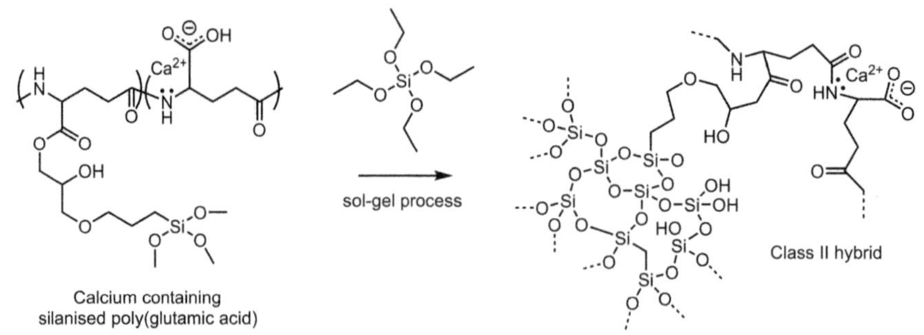

Figure 12.8 Calcium chelation on the silanised poly(γ-glutamic acid) and subsequent reaction with TEOS in the sol–gel process.

deprotonation of the silanols which can therefore act as weak chelators for the metallic ions. Thus, a burst release of metallic ions into physiological media can potentially be prevented.

The use of metal alkoxide to improve the bioactive properties of Class II hybrids has already been reported using silanised poly(γ-glutamic acid),[77] gelatine[34] and poly(ethylene glycol)[27] as organic sources and calcium ethoxide as the metal alkoxide precursor.

12.3.2 Polymers with Cation Chelating Property

Another approach to incorporating metal ions within the structure of Class II hybrids is by using the organic phase as an ionic carrier. Polymers, such as polypeptides or polyacrylates, can be synthesised with a wide range of functional pendant groups, along their backbone. Several of these pendant groups have good affinity for metal ions, described as the chelating effect. Chelators are usually composed of co-existing weak base and weak acid, such as secondary amine and carboxylic acid, which can coordinate metal ions. The strength of the coordination depends on the number of coordination sites available on the metal ion, where at least two sites are necessary to form a chelate.

This strategy has been used by Vaillant *et al.* and Obata *et al.* for the design of Class II hybrids towards the regeneration of bone tissue using poly(γ-glutamic acid) as organic phase silanised with GPTMS.[35,38] The amino group of the polypeptide backbone coupled with the pendant carboxylic acid residue of glutamic acid was used to chelate calcium ions as shown in Figure 12.8.[78]

12.4 Structure and Properties of O–I Hybrid Materials

The versatile soft chemistry approach allows endless combinations of organic and inorganic moieties resulting in hybrids with infinite numbers of novel composition–structure–property relationships. Optimisation of the sol–gel reaction conditions, such as the choice of precursors and composition (components ratio) and their respective solubilities and interactions in the sol and gel phases, will allow the creation of an amorphous hybrid material with molecularly interpenetrating organic and inorganic networks (Figure 12.1). However, this is a tall order, for when dissimilar materials are mixed together, for example hydrophobic polyesters with hydrophilic silica, they will often phase separate at the microscale.[24,25,79,80] This will compromise the final properties. In this section, focus will be on the structure and properties of Class II organic–bioactive inorganic-based hybrids for biomedical applications. In particular, emphasis will be on the effect of the interfacial region between the organic and inorganic and how this influences the structure–property relationships.

12.4.1 Organic–Inorganic Interfacial Region

At the O–I interfacial regions there can be weak or strong interactions. The interface between the organic and inorganic materials has to be tailored to overcome serious problems associated with phase separation in the preparation of hybrid materials.

12.4.1.1 Polyester–Silica Hybrids

O–I hybrids using polyesters, such as polycaprolactone (PCL), can covalently bond to the inorganic network through their alkoxysilane-functionalised end-groups (for more details see Section 12.3).[80–86] Tian *et al.* showed through small angle X-ray scattering (SAXS) and dynamic mechanical analysis that PCL end-group-functionalised with triethoxysilanes had more intimate mixing in the O–I hybrids than the hydroxyl end-groups.[83] SAXS is a powerful technique to study long-range order within disordered materials, such as sol–gel O–I hybrid materials. Depending on the experimental conditions used, nanoscale phase separation leading to cluster formation, nanoparticle size distributions and their distances can be studied. In the case of the PCL–silica O–I hybrids the scattering maximum in the SAXS profiles and the correlation length corresponding to this maximum were identified to relate to PCL and silica-rich domains within a microphase-separated structure. Correlation length decreased from 9.5 nm to 4.2 nm when PCL was end-capped with triethoxysilanes. This suggests that intermixing was improved with triethoxysilane end-capped PCL which is more reactive with TEOS than –OH end-groups on PCL. It is more likely that this correlation length corresponds to the primary silica particles within a microphase-separated structure.

As well as end-group functionalisation, using low molecular weight PCL leads to increased homogeneity. TEM images in Figure 12.9 show the extent of microphase separation within PCL–silica hybrids synthesised with relatively low (a) and high (b) molecular weight PCL.[80] The microphase separation is much more severe at relatively high molecular weight. The silica-rich

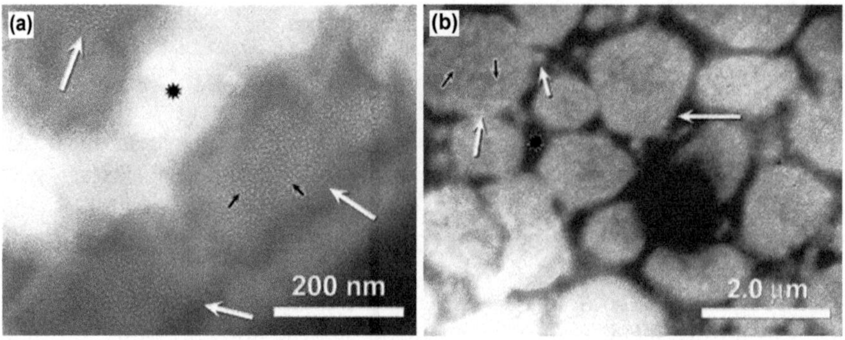

Figure 12.9 TEM micrographs of (a) low and (b) high molecular weight poly(ε-caprolactone)–silica hybrids. Silica-rich domains (white arrows) surrounded by (black asterisk) numerous small silica particles (black arrows) with PCL (dark gray color) in between the particles.
[Reprinted from S.-H. Rhee, Effect of molecular weight of poly(ε-caprolactone) on interpenetrating network structure, apatite-forming ability, and degradability of poly(ε-caprolactone)/silica nano-hybrid materials, *Biomaterials*, **24**, 1721–1727, Copyright 2003 with permission from Elsevier.[80]]

domains increase in size from 100's of nanometres to the micron scale for the low and high molecular weight PCL, respectively. SAXS study showed that the silica nanoparticles within the silica-rich domains were also observed to increase in size from 7 nm, for the low molecular weight PCL, to 160 nm, for the high molecular weight PCL, suggesting limited interactions when high molecular weight PCL is used. PCL being hydrophobic, it phase separates as the silica network forms during gelation. This leads to sub-optimal mechanical properties and uncontrolled degradation of the organic and inorganic species, where the organic species could freely leach out within a short period on soaking in a liquid.[85]

12.4.1.2 Polypeptide–Silica Hybrids

Polypeptides, such as collagen, gelatine and poly(γ-glutamic acid), are water-soluble and have multiple reactive functional groups on their backbone. This could be used advantageously to produce O–I hybrids with more intimate interactions between the organic and inorganic moieties.[31,32,36,38,77] Mahony *et al.* synthesised Class II gelatin–silica hybrids by functionalising gelatin with the organosilane glycidoxypropyl trimethoxysilane (GPTMS).[36] Using tapping mode atomic force microscopy (AFM) it was found that the silica-rich domains (dark region) were in the 10's of nanometres (Figure 12.10). In the un-functionalised, *i.e.* Class I hybrids with weak interactions, the characteristic size of the phase domains was 90.6 ± 9.5 nm and with the introduction of functionalisation the domain size decreased to 44.3 ± 6.1 nm. In comparison to polyesters which had silica-rich domains that were of the orders of microns, this indicates that gelatine–silica hybrids

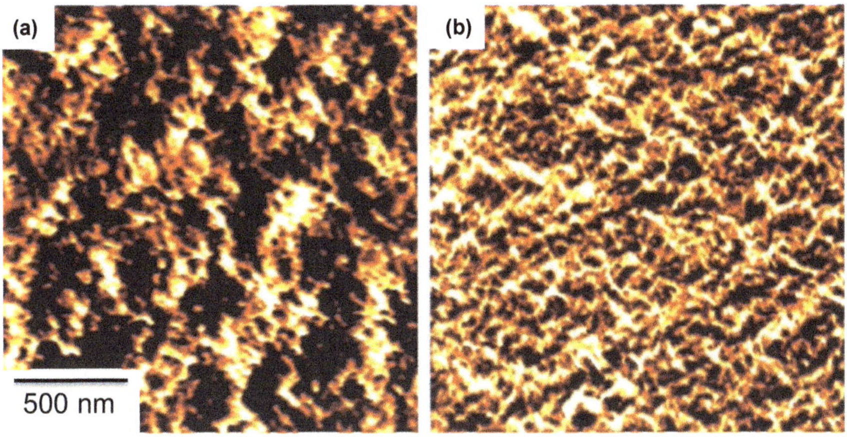

Figure 12.10 Tapping mode AFM phase images for gelatine–silica hybrids (a) Class I and (b) Class II. Light regions signify the organic while the darker regions represent the inorganic phase.
[Reproduced from Mahoney *et al.*, *Adv. Funct. Mater.*, 2010, **20**, 3835–3845 with permission from John Wiley and Sons. Copyright © 2010 WILEY-VCH Verlag GmbH & Co. KGaA, Weinheim.]

with or without functionalisation have more homogeneous phase mixing than polyesters. Furthermore, by varying the degree of functionalisation in the gelatine–silica hybrids, tailorable mechanical properties and, importantly, congruent degradation of Si and gelatine were observed.[36]

12.4.1.3 Polyacrylate–Silica Hybrids

O–I hybrid materials produced with polymers that were designed with built-in alkoxysilane moieties, such as 3-(trimethoxysilyl)propyl methacrylate, through a radical polymerisation have the most homogeneous type of interpenetrating networks.[56,58,87] Maçon *et al.* co-condensed poly(TMSPMA) and TEOS to produce O–I hybrids that had phase domains as small as 11 nm[56] The mechanical properties of these hybrids measured using nanoindentation revealed that the increase in elastic modulus and the deformation mechanism were a function of the inorganic : organic ratio suggesting a true hybrid had been synthesised,[15] although the polymer poly(TMSPMA) is non-degradable. In this study, the molecular weight was chosen to be less than 30 kDa where the hydrodynamic radius was ~5 nm, which can be passed through the kidneys.[64]

12.5 Conclusions and Outlook

In this chapter we have described in detail the different types of organosilanes which can be used to functionalise polyesters, polysaccharides and polypeptides as well as organosilane monomers for the synthesis of polymethacrylate polymers. We have also discussed in detail the different types of metal precursors that can be introduced into the sol–gel process to synthesise bioactive O–I hybrid materials. We then introduced some bioactive and biodegradable Class II O–I hybrid materials and discussed the importance of strong interactions between organic and inorganic components to avoid phase separation and to achieve true hybridisation.

The versatile soft chemistry approach that allows the synthesis of O–I hybrid materials with endless composition–property relationships is very promising for biomedical applications. The relatively simple chemistry and inexpensive setup to fabricate these hybrids means it is easy to adopt in laboratories around the world. However, complete understanding of organic–inorganic interactions in the hybrids is of paramount importance. With advanced characterisation methods that can resolve structures at atomic resolution combined with cryo-fixing of dynamic states we now have the possibility to study the nature of the organic–inorganic interfacial regions with unprecedented detail allowing development of multifunctional O–I hybrid materials.

References

1. R. Wang and H. S. Gupta, *Annu. Rev. Mater. Res.*, 2011, **41**, 41–73.
2. U. G. K. Wegst *et al.*, *Nat. Mater.*, 2015, **14**, 23–36.
3. C. Sanchez, K. J. Shea and S. Kitagawa, *Chem. Soc. Rev.*, 2011, **40**, 471–472.
4. C. Sanchez and F. Ribot, *New J. Chem.*, 1994, **18**, 1007.
5. G. Kickelbick, *Hybrid Materials: Strategies, Syntheses, Characterization and Applications*, Wiley-VCH, Weinheim, 2007.
6. E. Ruiz-Hitzky, K. Ariga, and Y. M. Lvov, *Bio-inorganic Hybrid Nanomaterials: Strategies, Syntheses, Characterization and Applications*, Wiley-VCH Verlag GmbH & Co. KGaA, Weinheim, 2007.
7. M. Vallet-Regí, M. Colilla and B. Gonzàlez, *Chem. Soc. Rev.*, 2011, **40**, 596–607.
8. J. Henstock, L. Canham and S. Anderson, *Acta Biomater.*, 2015, **11**, 17–26.
9. Z. Mladenovi *et al.*, *Acta Biomater.*, 2014, **10**, 406–418.
10. H. Li and J. Chang, *Acta Biomater.*, 2013, **9**, 6981–6991.
11. D. Reffitt *et al.*, *Bone*, 2003, **32**, 127–135.
12. M. Landry *et al.*, *J. Polym. Sci., Part B: Polym. Phys.*, 1995, **33**, 637–655.
13. D. Rodrigues, B. Risch and G. Wilkes, *Chem. Mater.*, 1997, **9**, 2709–2719.
14. M. Kleman and O. Laverntovich, *Soft Matter Physics: An Introduction*, 2003, Springer-Verlag, New York.
15. M. B. Novak, *Adv. Mater.*, 1993, **5**, 422–433.
16. F. Mammeri *et al.*, *J. Mater. Chem.*, 2005, **15**, 3787–3811.
17. C.-C. Chang and W.-C. Chen, *Chem. Mater.*, 2002, **14**, 4242–4248.
18. L. Connell *et al.*, *J. Mater. Chem. B*, 2014, **2**, 668.
19. Y. Shirosaki *et al.*, *J. Biomed. Mater. Res., Part A*, 2014, **1**, 289–299.
20. D. Wang *et al.*, *J. Mater. Chem. B*, 2015, **3**, 7560.
21. H. Maeda, T. Kasuga and L. Hench, *Biomaterials*, 2006, **27**, 1216–1222.

22. H. Maeda and T. Kasuga, *J. Biomed. Mater. Res., Part A*, 2008, **85A**, 742–746.
23. L. Mazzocchetti, S. Sandri and M. Scandola, *Biomacromolecules*, 2007, **8**, 672–678.
24. D. Tian, P. Dubois and R. Jérôme, *J. Polym. Sci., Part A: Polym. Chem.*, 1997, **35**, 2295–2309.
25. S.-H. Rhee, J.-Y. Choi and H.-M. Kim, *Biomaterials*, 2002, **23**, 4915–4921.
26. B. Granqvist *et al.*, *Colloid polym. Sci.*, 2004, **282**, 495–501.
27. A. Li *et al.*, *J. Mater. Chem. B*, 2015, **3**, 1379.
28. N. Miyata *et al.*, *Biomaterials*, 2002, **23**, 3033–3040.
29. M.-Y. Koh *et al.*, *J. Ceram. Soc. Jpn.*, 2007, **115**, 732–737.
30. J. Wen *et al.*, *Chem. Mater.*, 1997, **9**, 1968–1971.
31. G. Poologasundarampillai *et al.*, *J. Mater. Chem.*, 2010, **20**, 8952–8961.
32. G. Poologasundarampillai *et al.*, *Soft Matter*, 2012, **8**, 4822–4832.
33. L. Ren *et al.*, *J. Sol–Gel Sci. Technol.*, 2001, **21**, 115–121.
34. X. Dieudonnée *et al.*, *Chem. Commun.*, 2014, **20**, 8701–8704.
35. A. Obata *et al.*, *RSC Adv.*, 2014, **4**, 52491.
36. O. Mahony *et al.*, *Adv. Funct. Mater.*, 2010, **20**, 3835–3845.
37. L. Ren, *J. Sol–Gel Sci. Technol.*, 2001, **21**, 115–121.
38. E. M. Valliant *et al.*, *Acta Biomater.*, 2013, **9**, 7662–7671.
39. C. Airoldi and J. Monteiro, *J. Appl. Polym. Sci.*, 2000, **77**, 797–804.
40. E. Delebecq *et al.*, *Chem. Rev.*, 2013, **113**, 80–118.
41. A. Williams and I. Ibrahim, *Chem. Rev.*, 1981, **81**, 589–636.
42. H. Maeda, T. Kasuga and L. L. Hench, *Biomaterials*, 2006, **27**, 1216–1222.
43. H. Maeda and T. Kasuga, *J. Biomed. Mater. Res., Part A*, 2008, **85A**, 742–746.
44. R. Parker and N. Issacs, *Chem. Rev.*, 1959, **59**, 737–799.
45. Y. Shirosaki *et al.*, *Chem. Eng. J.*, 2008, **137**, 122–128.
46. Y. Shirosaki *et al.*, *J. Biomed. Mater. Res., Part A*, 2014, **1**, 289–299.
47. Y.-L. Liu, Y.-H. Su and J.-Y. Lai, *Polymer*, 2004, **45**, 6831–6837.
48. Y. Shirosaki *et al.*, *Acta Biomater.*, 2009, **5**, 366.
49. Y. Shirosaki *et al.*, *J. Ceram. Soc. Jpn*, 2010, **118**, 989–992.
50. L. Gabrielli *et al.*, *Chem. – Eur. J.*, 2013, **19**, 7856–7864.
51. L. Gabrielli *et al.*, *RSC Adv.*, 2014, **4**, 1841–1848.
52. I. Varma, A. Tomar and R. Anaud, *J. Appl. Polym. Sci.*, 1987, **33**, 1377–1388.
53. H. Ozaki, A. Hirao and S. Nakahama, *Macromolecules*, 1992, **25**, 1391–1395.
54. J. Du and Y. Chen, *Macromolecules*, 2004, **37**, 6322–6328.
55. V. Mellon *et al.*, *Macromolecules*, 2005, **38**, 1591–1598.
56. A. L. B. Maçon *et al.*, *Phys. Chem. Chem. Phys.*, 2015, **17**, 29124–29133.
57. C. Gamys *et al.*, *Macromol. Chem. Phys.*, 2012, **213**, 10–18.
58. A. L. B. Maçon *et al.*, *Macromol. Rapid Commun.*, 2015, **36**, 2060–2064.
59. R. Ravarian *et al.*, *J. Mater. Chem. B*, 2013, **1**, 1835–1845.
60. R. Ravarian *et al.*, *ACS Nano*, 2013, **7**, 8469–8483.

61. C. Ohtsuki, T. Miyazaki and M. Tanihara, *Mater. Sci. Eng., C*, 2002, **22**, 27–34.
62. K.-H. Lee and S.-H. Rhee, *Biomaterials*, 2009, **30**, 3444–3449.
63. H. Oonichi *et al.*, *Acta Orthopaedica*, 2011, **82**, 553–558.
64. M. Fox, F. Szoka and J. Fréchet, *Acc. Chem. Res.*, 2009, **42**, 1141–1151.
65. P. Flory, *Principles of Polymer Chemistry*, Cornell University Press, 1953.
66. W. Braunecker and K. Matyjaszewski, *Prog. Polym. Sci.*, 2007, **32**, 93–146.
67. J. R. Jones, *Acta Biomater.*, 2013, **9**, 4457–4486.
68. A. Hoppe, N. Güldal and A. Boccaccini, *Biomaterials*, 2011, **32**, 2757–2774.
69. D. Xynos *et al.*, *Calcif. Tissue Int.*, 2000, **67**, 321–329.
70. I. D. Xynos *et al.*, *J. Biomed. Mater. Res.*, 2001, **55**, 151–157.
71. M. Takeichi and T. Osada, *Exp. Cell Res.*, 1972, **74**, 51–60.
72. F. Wolf and A. Cittadini, *Front. Biosci.*, 1999, **4**, 607–617.
73. P. Sepulveda, J. Jones and L. Hench, *J. Biomed. Mater. Res., Part A*, 2002, **61**, 301–311.
74. S. Lin *et al.*, *J. Mater. Chem.*, 2009, **19**, 1276–1282.
75. S. Maeno *et al.*, *Biomaterials*, 2005, **26**, 4847–4855.
76. B. Yu *et al.*, *Langmuir*, 2012, **28**, 17465–17476.
77. G. Poologasundarampillai *et al.*, *Chem. - Eur. J.*, 2014, **20**, 8149–8160.
78. Takashi *et al.*, *J. Microbiol. Biotechnol.*, 2010, **20**, 1436–1439.
79. D. Tian, S. Blacher and R. Jerome, *Polymer*, 1999, **40**, 951–957.
80. S.-H. Rhee, *Biomaterials*, 2003, **24**, 1721–1727.
81. D. Tian, P. Dubois and R. Jerome, *J. Polym. Chem.*, 1997, **35**, 2295–2309.
82. D. Tian *et al.*, *J. Sol–Gel Sci. Technol.*, 1998, **13**, 415–419.
83. D. Tian, S. Blacher and R. Jerome, *Polym. Bull.*, 1999, **40**, 951.
84. D. Tian *et al.*, *Langmuir*, 1998, **14**, 1905–1910.
85. S.-H. Rhee, J.-Y. Choi and H.-M. Kim, *Biomaterials*, 2002, **23**, 4915–4921.
86. S.-H. Rhee, *Biomaterials*, 2004, **25**, 1167–1175.
87. J. J. Chung, J. R. Jones and T. K. Georgiou, *Macromol. Rapid Commun.*, 2015, **36**, 1806–1809.

CHAPTER 13

Cell Interactions with Calcium Phosphate Glasses

OSCAR CASTAÑO,[a,b,d] SOLEDAD PÉREZ,[a,b] MIGUEL A. MATEOS-TIMONEDA[a,b,c] AND ELISABETH ENGEL*[a,b,c]

[a] Biomaterials for Regenerative Therapies, Institute for Bioengineering of Catalonia (IBEC), 08028, Barcelona, Spain; [b] CIBER de Bioingeniería, Biomateriales y Nanomedicina (CIBER-BBN), 08028 Barcelona, Spain; [c] Dpt. Materials Science and Metallurgical Engineering, Universitat Politècnica de Catalunya (UPC), 08028 Barcelona, Spain; [d] Dpt. Materials Science and Metallurgical Engineering, Universitat de Barcelona (UB), 08028 Barcelona, Spain
*Email: eengel@ibecbarcelona.eu

13.1 Biomaterials on the Regeneration Cosmos

Nowadays there is a wide range of biomaterials and processing techniques that have allowed the preparation of scaffolds, which are currently available for numerous applications in regenerative medicine. However, in general, available synthetic materials lack appropriate signals to stimulate cellular events such as adhesion, migration and differentiation into the desired cell lineage. This limitation is currently overcome by the incorporation of various morphogens and growth factors into the scaffolds in order to activate a specific cell behavior. The addition of biological molecules, whose clinical effect is still under discussion, is another factor that hampers the arrival of these treatments for the patients.

RSC Smart Materials No. 23
Bioactive Glasses: Fundamentals, Technology and Applications
Edited by Aldo R. Boccaccini, Delia S. Brauer and Leena Hupa

Published by the Royal Society of Chemistry, www.rsc.org

Despite considerable advances in the understanding of physical and chemical material properties, only few so-called smart biomaterials have found their way into clinical application so far.

From a clinical point of view an instructive material/implant may be defined as a material that actively participates in the regeneration of the damaged tissue and responds and reacts to stimuli from its environment in a valuable way. These instructive biomaterials effectively combine two features that, in our experience, actively participate in the regeneration of the damaged tissue and respond and react to stimuli from its environment: bioactivity and biodegradability. Bioactive materials are required in order to recruit cells and to stimulate their migration, growth and differentiation towards each specific tissue. Biodegradable materials are required to provide the structural support and the chemical cues to activate the regeneration process, while combining their gradual biodegradation into non-toxic products with the progressive release of the signals required to guide tissue repair. These bioactive and biodegradable materials should end up totally replaced by the natural regenerated tissue.

Taking into account these two features, CaP glasses and glass-ceramics can be considered as biodegradable and bioactive.

13.2 Introduction to Calcium Phosphate Glasses

Calcium phosphate glasses were introduced as biomaterials in the 1980's by Burnie and Gilchrist.[1] In contrast to silicate-based bioactive glasses, phosphate-based glasses may dissolve completely, and their degradation time can vary from a few days to several months.[2] They offer interesting solutions as a reinforcing phase for composite materials (*cf.* Chapter 11), drug delivery systems, and cavity fillers for maxillofacial surgery, as well as being good candidates for the fabrication of scaffolds for tissue engineering. They are based on $[PO_4]$ tetrahedra that form long chains or rings leading the arrangement into a vitreous network[2] (*cf.* Chapter 3). Because vitreous P_2O_5 has demonstrated a high solubility, different glass systems have been investigated to improve their chemical stability. In the system P_2O_5–CaO–Na_2O particularly, TiO_2 addition improves both the chemical stability and the mechanical properties of the glasses.[3] As Ti^{4+} has a small ionic radius and a large electrical charge, it cross-links the phosphate chains very effectively and the vitreous network is more stable, allowing for variation of solubility, degradation and the mechanical properties of glasses. Hence, degradation rate and mechanical performance can be tuned according to the chemical formulation, making these materials an interesting ion release system. Recently, our group developed a new approach, namely organically-modified glasses (ORMOGLASSES), which involve phosphorus, calcium, sodium and silicon (or titanium) ions and act as an efficient ionic release system as do inorganic glasses.[4] They demonstrate a better interaction with polymers and a versatility to be adapted to different shapes when processed as hybrid devices. It is well known that ionic dissolution products from inorganic materials are crucial in *in vivo* and *in vitro* performance and in their effect on several cellular events. In fact, some ions, such as Ca^{2+} and PO_4^{3-} and other trace elements, namely Sr^{2+}, Cu^{2+}, Zn^{2+}, Mg^{2+}, Fe^{3+}, K^{+} and Co^{2+}, are present in the human body and have anabolic effects in the metabolism of tissues, such as bone and skin.[5–7] Therefore, it is believed that the release of specific ions from the biomaterial after exposure to a physiological environment favorably affects the behavior of human cells and enhances bioactivity of the scaffolds. Indeed, recent studies showed that glasses in the system P_2O_5–CaO–Na_2O–TiO_2 promote vessel formation both *in vitro* and *in vivo*.[8,9] This phenomenon is mainly due to the release of Ca^{2+} and their specific angiogenic promoting effect.

The high solubility of some compositions, initially harmful and rejected by tissues, can have a great potential for the biomedical field. This aspect offers high biodegradability and allows for acting as initial supports for the earlier stages of bone healing and gradually decreasing mechanical resistance at the same time that the bone tissue is renewed. In other words, this material is resorbable.

Glasses can be processed through basically two main scalable methods: melting and sol–gel processes. The melting processes is a top-down process that involves high temperatures (>1300 °C) and a quench to freeze the disordered atomic structured of the melt to get an amorphous glass. Other thermal treatments at lower temperatures should also perform in order to relax the initial structure.[3] Other variations can be developed to shape the glass, for example, into fibers using a melt spinning technique.[10] Sol–gel is a bottom-up process that involves the use of inorganic or organic precursors, which, after different processes of hydrolysis and condensation, are shaped into powder, nanoparticles, fibers, coatings, *etc.* To date, calcium phosphate glasses have only been prepared in the form of powder,[11] microspheres[12] and nanospheres[13] by this method. However, hybrid calcium phosphate glasses, such as organically modified glasses (ormoglasses), can be fabricated by the sol–gel method in the form of nanoparticles (Table 13.1).[14]

Composites have also been developed, based on solvent casting fabrication and 3D printing, allowing the production of highly porous scaffolds

Table 13.1 Different glass types.

Type of glass	Main feature	Ref.
Silicate glass (Bioglass®)	Osteogenesis at the micro- and nanoscale Slow degradation	[15]Chen *et al.*, *Biomaterials*, 2006
Calcium phosphate glasses	Angiogenesis and osteogenesis at the microscale Potentially fast degradation	[16]Navarro *et al.*, *J. Biomed. Mater. Res., Part A*, 2006
Glass-ceramics	Osteogenesis Medium degradation	[17]Navarro *et al.*, *Biomaterials*, 2004
siloxanes and ormoglasses	Angiogenesis and osteogenesis at the nanoscale Very fast degradation	[14]Oliveira *et al.*, *Acta Biomater.*, 2016

that will help cells to colonize the scaffolds and the ingrowth of blood vessels. Microparticles of CaP around 40 μm in size can be embedded on these matrices giving bioactivity to the polymers. Figure 13.1 shows a rapid prototyping PLA scaffold with CaP microparticles. Thus, nanofibers containing sol–gel nanoparticles allow mimicking of the structure of collagen fibers in bone and other tissues with the glass nanoparticles embedded on the fibers. Figure 13.2 shows SEM images of electrospun PLA nanofibers with and without CaP nanoparticles.

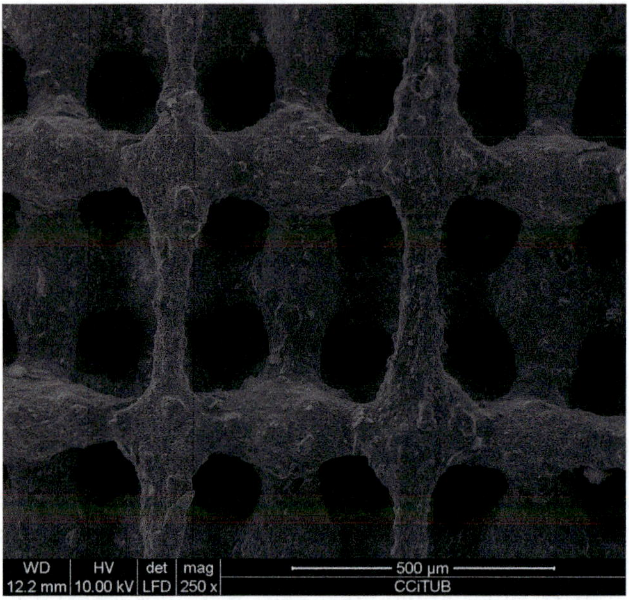

Figure 13.1 SEM images of PLA scaffolds with CaP glass microparticles of 40 μm or less produced by rapid prototyping.

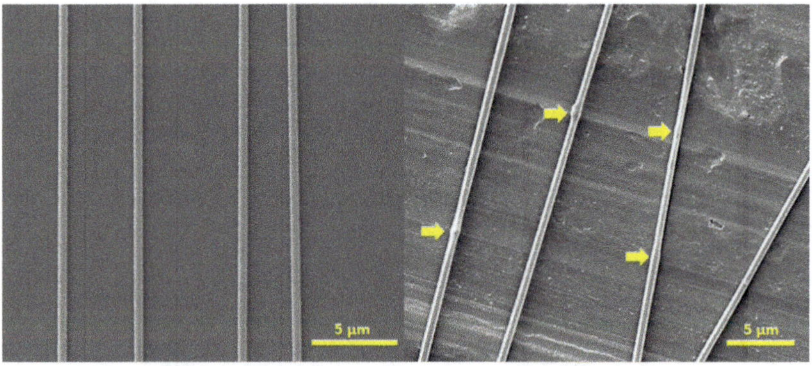

Figure 13.2 Nanofiber scaffolds (left) without and (right) with CaP glass nanoparticles produced by the sol–gel method. Arrows show the presence of the nanoparticles on the fibers.

Throughout this chapter, we will review the studies performed in order to further understand the different cell types' interactions with CaP-based glasses. Preparation techniques as well as combinations with polymers to obtain either composite materials or hybrids, such as ormoglasses (a new hybrid material that opens opportunities due to the versatility in its fabrication) will be the focus of the chapter.

13.3 CaP Glass and Its Biological Interactions

13.3.1 Mesenchymal and Bone Cells

Bone has been the target tissue for CaP glasses since the very beginning. It is the main reservoir of calcium and phosphate ions in the body and plays a pivotal role in calcium homeostasis. During bone remodeling, a process that occurs throughout our lives, osteoclasts resorb the damaged or old bone, and it is believed that Ca is released in high concentrations. Then pre-osteoblasts are recruited and arrived at the resorbed area and start to produce new extracellular matrix (mainly collagen I). Once the osteoid has been produced, osteoblasts start the process of bone mineralization, using calcium and phosphates that will be deposited on the collagen I fibers. This activity makes clear the interest to use glasses and ceramics for bone regeneration, as they are composed from the main inorganic elements of mineralized bone.

Many studies have been performed using bone cells in contact or cultured with extracts from glasses. Mainly all these studies have tested the effect of the different ions on cells in order to elucidate the effect of degradation and ion release.

Many ions have been correlated with different effects, as reviewed by Hoppe *et al.*[6] Since the inorganic phase of bone contains Ca and P, these ions have been the ones that have attracted more attention. Ca is demonstrated to favor osteoblast proliferation, differentiation and extracellular matrix (ECM) mineralization,[18] and to activate the calcium sensing receptor (CaSR) in osteoblasts[19] and bone marrow mesenchymal stem cells (BMSCs).[20] The role of P has not been as clearly correlated, but some indicators have phosphate ions acting as an extracellular 'pool' responsible for the release of Cbfa-1, an important bone marker, from bone cells.[21] Thus, phosphorylation of ERK1/2 and Fra-1 in osteoblast was reported.[22]

TiO_2 has been shown to be an efficient oxide for the improvement, not only of the chemical durability of the glasses, but also of their mechanical properties. The use of MG63 cell line cultured with four different glasses that varied the amount of TiO_2 (at 0, 1, 3 and 5 mol%) showed improved cell proliferation behavior and gene expression, in particular in 3 and 5 mol% TiO_2-containing glasses.[23] The improvement is most probably due to synergistic effects involving all the ions released and to the increased structural stability rather than to the isolated effect of titanium ions. In other studies on composites using 5 mol% TiO_2 in the system P_2O_5–CaO–Na_2O–TiO_2, cells clearly showed a preference of the glass microparticles towards the polylactic acid film[24] indicating a certain capability of cell recruitment (Figure 13.1).

Even though MG63 is not the most appropriate cell model owing to its tumoral origin, other studies based on primary cells have demonstrated the chemoattractive effect of these glasses.[8]

Another study based on a macroporous chitosan composite containing either a CaP glass (57.0 CaO–28.0 P_2O_5–12.0 TiO_2–3.0 Na_2O; in mol%) or hydroxyapatatite (HA) showed a difference in MG63 cell adhesion and viability on the CaP glass-containing samples compared to the HA-containing ones. The slow degradation of HA had no effect on cell proliferation and differentiation, while the glass–chitosan composite induced higher alkaline phosphatase (ALP) and osteocalcin (OC) production after 7 and 11 days of culture.[25]

Ceramic scaffolds were produced with different percentages of H_2O_2 solution (10% v/v : 40 and 60 wt%) to produce different porosities.[26] The 60% showed a higher porosity and a higher degree of interconnectivity than the 40%. SAOS-2 osteoblast-like cells were cultured in contact with the 60% scaffolds to test biocompatibility. Proliferation of cells at different time points (24 h, 3 days and 6 days) showed a higher proliferation rate on the porous scaffolds compared to control tissue culture plate. Compared to results obtained with extracts, no differences were observed except for the extraction after more days (12 and 22 days) that were correlated with the higher amounts of ions present in the extracts. Ti concentration did not change with time while Ca and P concentrations double and triple, respectively, after 22 days of extraction, supporting the hypothesis that these ions do have an effect on cell behavior. Again, even though hSAOS-2 are not the best model cells owing to their tumoral source, further studies using primary cells demonstrated that amounts of extracellular calcium on rBMSc induced migration, proliferation and differentiation towards osteoblasts as well as mineralization *in vitro*.[20] Figure 13.3 shows the adhesion of rBMSc

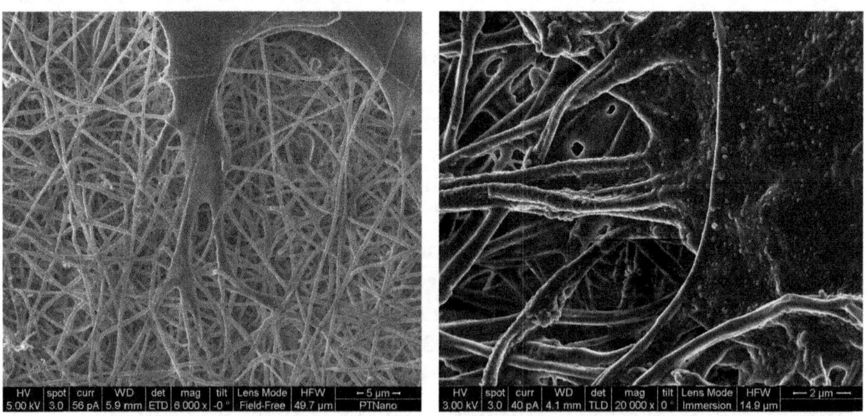

Figure 13.3 Mesenchymal stem cells obtained from rat bone marrow cultured on electrospun nanofibers containing 25% of sol–gel CaP nanoparticles. Filopodia follow and attach to fibers making it difficult to differentiate filopodia from fibers.

cultured on nanofibers with and without CaP nanoparticles. The fibers induced the organization of the cells, mainly the filopodia. The presence of CaP induced higher proliferation of rBMSc (unpublished data).

Previous studies on fetal rat calvarial cells and murine clonal osteoblastic 2T3 cells treated with Ca ions at concentrations up to 5 mM showed an increase in cell proliferation, differentiation and mineralization nodule formation.[27] The effect of increasing extracellular calcium was correlated to its effect on bone cells through the activation of the CaSR. This receptor had been postulated already as the molecular mechanism through which several cells recognize and respond to small, physiologically relevant changes in extracellular Ca^{2+}. Raising Ca^{2+} concentration inhibits bone resorption and stimulates bone formation *in vitro* in bone explants and exerts similar actions on isolated osteoblasts and osteoclasts.[28] Elevating Ca^{2+} also promotes the chemotaxis and proliferation of osteoblast precursors, thereby potentially expanding the pool of mature osteoblasts at sites of new bone formation.[29]

Another type of mesenchymal stem cells are human adipose tissue-derived mesenchymal stromal cells (hAMSC). These cells have the same properties as their homologues from the bone marrow source, such as similar phenotype and differentiation potential, but they overcome some of their limitations: hAMSC are more abundant, offer less traumatic extraction and seem to have low predisposition to produce tumors.[30] These cells were used on PLA–CaP scaffolds and *in vivo*. To monitor their performance cells were labeled with two promoters, one Photinus pyralis luciferase (PLuc) as a reporter of endothelial cell differentiation and hypoxia state and the other for the platelet endothelial cell adhesion molecule-1 (PECAM-1) gene, a frequently-used marker for endothelial cells. A non-invasive bioluminescence imaging (BLI) platform was used to monitor the cells. Results indicated that PLA–CaP had a superior proangiogenic capacity than pure PLA. Even though both materials supported cell growth; after 5 days the cells on PLA–CaP decreased, however the hAMSC went through differentiation to the endothelial lineage significantly.

13.3.2 Fibroblast Cells

Fibroblast cells have been widely used to assay cytotoxicity and biocompatibility of biomaterials. No uses have already been assayed with CaP glasses for regenerating soft tissues, so all the examples that we will show are related to cytotoxicity assay and biocompatibility evaluation.

A previous work from Navarro *et al.* produced glass in the system P_2O_5–CaO–Na_2–TiO_2. The molar composition was 44.5% P_2O_5, 44.5% CaO, $(11-x)$% Na_2O, and x% TiO_2, where $x=0$ for G0 and $x=5$ for G5. In this study, cultures in contact and with extracts of human skin fibroblasts obtained from primary cultures were performed. The results showed a better proliferation with the G0 samples compared to the G5. However, in cultures with extracts at different time points showed no large differences. It was noted that a protein layer was deposited on the G0 samples while it was not existent on the G5. The authors correlated the amount of Ca^{2+} released to the production of proteins, even though the measurement of the ions by ICP did not match very well.[31]

Another study conducted by Bitar *et al.*[32] used human oral fibroblasts and hand flexor tendon fibroblasts that were co-cultured on glasses with different degradation rates ($(P_2O_5)_{45}$–$(CaO)_x$–$(Na_2O)_{55-x}$, where $x = 30$–48). Results showed that glass compositions with high CaO content (46 and 48 mol%) supported high numbers of adherent and viable cells and maintain cellular function. This result clearly indicates that our findings on the effect of Ca ions on cells' behavior through the CaSR[8,20,33] can be tuned with CaP-based glasses. Even though this study did not analyze the concentration of Ca ions during glass degradation, we have observed that fibroblasts are sensitive to changes in extracellular Ca^{2+} concentrations, compared to endothelial progenitor cells or mesenchymal stem cells. Concentrations up to 5 mM are too high and can cause a proliferation inhibitory effect (unpublished data), but that correlates well with other studies.

13.3.3 Endothelial and Vascular Cells

Almost all our tissues are vascularized with only a few exceptions, such as cartilage being avascular. This is the reason why angiogenesis and vascularization is nowadays considered the bottleneck in tissue regeneration. The lack of blood vessels in scaffolds or tissue constructs produced *in vitro* will render them non-viable when implanted in the body.

Angiogenesis occurs in nature when a pro-angiogenic stimulus, such as hypoxia, induces the disruption of the nearby vessels to start to sprout towards the stimulus. This is a multistep process that culminates with the formation of new capillaries on the required tissue. It has been demonstrated that Bioglass can enhance angiogenesis. Several works have revealed that fibroblasts cultured on scaffolds with Bioglass produced VEGF *in vitro* and *in vivo*.[34] Other cells, such as human endothelial cells, fibroblasts and microvascular endothelial cells, also produced basic fibroblast growth factor (bFGF) as reviewed by Gorustovich *et al.*[35] A mechanism based on the induction of hypoxia caused by the release of ions such as cobalt has been proposed.[36]

With CaP-based glasses, a mechanism based on Ca ions and the presence of CaSR in endothelial progenitor cells (EPCs) has been demonstrated by our group.[20] Aguirre *et al.* demonstrated for the first time the presence of the CaSR on these cells and how calcium chloride had an effect on tube-like formation, migration and EPCs maturation.

There are few papers where CaP glasses have been cultured with vascular cells. Aguirre *et al.* demonstrated the induction of angiogenesis *in vitro* on composite scaffolds of PLA and CaP microparticles made by solvent casting. EPCs from rat bone marrow showed a greater migration towards the microparticles, as well as an enhanced capacity to create capillary-like tubes

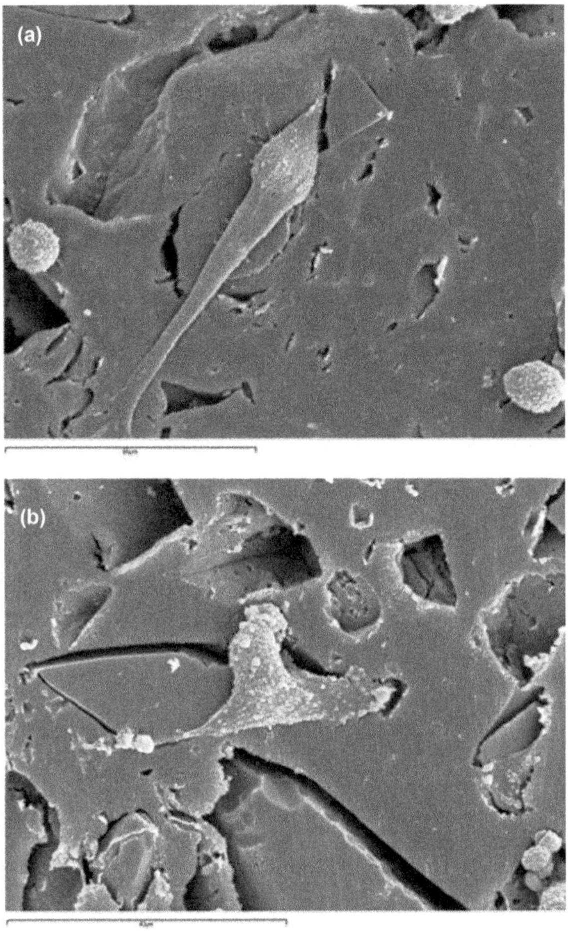

Figure 13.4 Endothelial progenitor cells show a great adhesion capability for the cells towards the glass microparticles on the composite scaffolds of PLA and CaP.
(Reproduced from ref. 24 with permission from John Wiley and Sons. Copyright © 2007 Wiley Periodicals, Inc.)

and express vascular endothelial growth factor (VEGF) and its corresponding receptor (VEGFR).[8] Figure 13.4 shows how EPCs attached preferentially to CaP microparticles than to the PLA scaffold, something that had been previously observed by Charles-Harris *et al.*[37] These results correlated faultlessly with the *in vitro* model stablished by the same authors just adding calcium chloride to the cultured EPCs. Figure 13.5 shows how EPCs cultured on solvent-cast PLA–CaP composite scaffolds support tube-like formation, in contrast to those on pure PLA.

An *in vivo* study using the same solvent casting PLA and CaP glass microparticles scaffolds confirmed the *in vitro* results. Scaffolds of PLA with

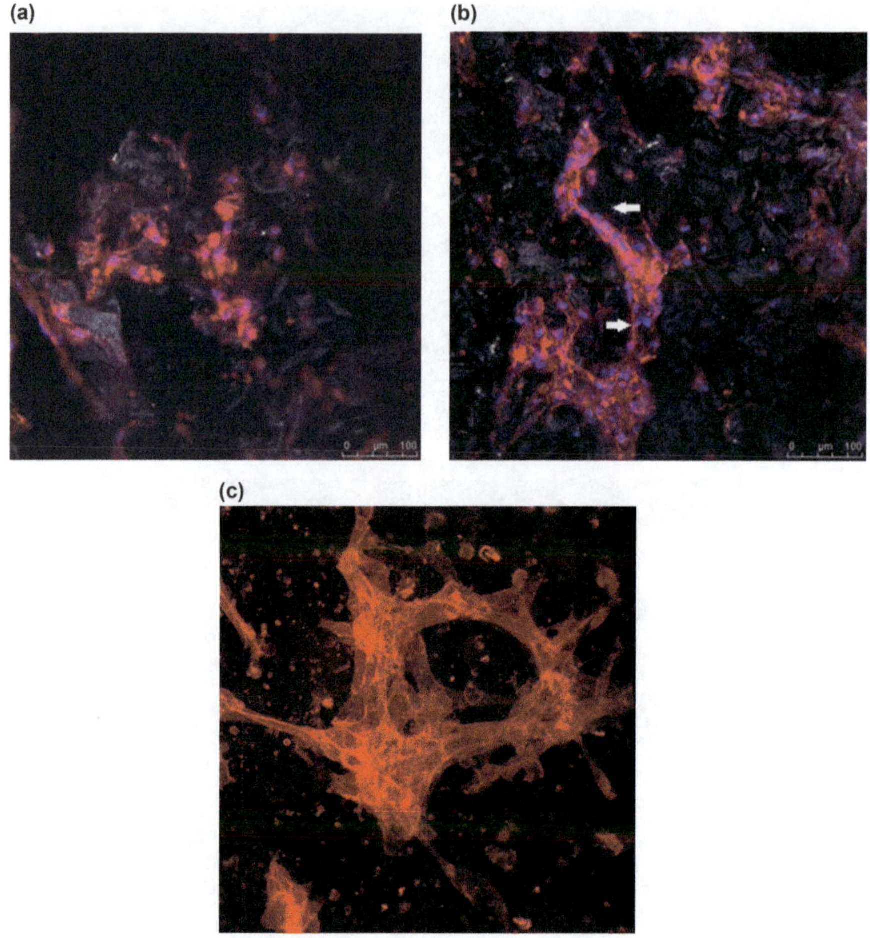

Figure 13.5 Confocal microscopy images of endothelial progenitor cells cultured on solvent-cast composite scaffolds: (a) cells individually attached to the scaffold, (b) tube-like formation organization of the EPCs highlighted by white arrow heads and (c) magnified view of the new vessels formed on the PLA composite scaffolds.

and without CaP glass microparticles were implanted subcutaneously on the back of mice for 28 days. PLA–CaP scaffolds showed substantial colonization from the host cells and a high number of blood vessels were quantified by histology on those scaffolds.[9]

In another *in vitro* and *in vivo* study of angiogenesis in bone formation, authors used silicate-based calcium phosphate glasses. Even though silicate-based glasses are not the focus of this review, we include this example as the amount of Si controlled the degradation rate of the glass. High amounts of Si in the Si-based glasses degraded slower than glasses with lower amounts of Si. They demonstrated that by decreasing amount of Si in the composition, better angiogenic performances could obtained, together *in vitro* and *in vivo*.[38] The study emphasizes the hypothesis that cells on the bone marrow (precursors and adults) have a receptor, the CaSR, that is able to detect the rise in Ca ion concentration in the external microenvironment and that concentrations around 10 mM have a strong effect on promoting angiogenesis.

However, other *in vivo* studies using hybrid fibers with nanoparticles of CaP have failed to correlate the increase of extracellular Ca ions with an increase of angiogenesis with an expression of CaSR.[14] Even though, the capability to promote angiogenesis is out of discussion, as demonstrated with the solvent-cast scaffolds of PLA and CaP glass microparticles in an *in vivo* model.[9]

13.4 Conclusions and Outlook

The cell response to calcium phosphate glasses has been significantly less studied than that to silicate bioactive glasses, such as Bioglass 45S5. As a result, fewer applications have been envisioned for this glass type. However, the properties of these glasses related to degradability and bioactivity have gained increasing recognition. The quantity of publications relating to bioactive and biodegradable glasses has increased greatly in the last 10 years. A search in PubMed showed more than 69 papers on Larry Hench's glasses for 2015. This number is a lot lower for CaP-based glasses, but we believe that they will increase in the next years and more uses and applications will be foreseen.

The interest in these glasses relates to the possibility of using a synthetic biomaterial that has great versatility. Different preparation methods can be applied in order to obtain diverse forms and structures: 3D printed scaffolds, solvent casting or freeze-drying, matrices of electrospun fibres, nanoparticles, microparticles or microspheres, allowing for a very varied manner of application. Thus, some of the bioactive properties demonstrated by these glasses, such as osteogenesis, angiogenesis, cell recruitment, to name a few, allow us to compare their performance to other advanced therapies based on cells and/or growth factors. The low cost production, easier regulatory issues and lack of undesired side effects point to this product becoming a highly competitive biomaterial in regenerative medicine applications.

Acknowledgements

The authors would like to acknowledge some of the group members that have contributed their data to this chapter. First, Prof. Josep A. Planell who was a pioneer in this kind of biomaterial and whose work and all the highly creative ideas have taken us to this point. Also the students and postdocs. that have been involved in this research, such as Dr Melba Navarro, Dr Tiziano Serra, Dr Aitor Aguirre, Dr Arlyng Gonzalez, Dr Nadege Sachot, Claudia Navarro and Joan Martí. Masters and undergraduate students in the group have also collaborated in the work.

We would also like to acknowledge our funding agencies and projects for their support: European project NMP3-LA-2008-214402, ANGIOSCAFF and European ERANET project (PI11/03030, NANGIOFRAC) and the Spanish Ministry of Economy and Competitiveness (Project MAT2012-38793). The crowdfunding project Dermoglass (all the people who have trusted us and have given money for this project) and the CaixaImpulse program (La Caixa).

References

1. T. Gilchrist and J. Burnie, *Biomaterials*, 1981, **2**, 244.
2. R. K. Brow, *J. Non. Cryst. Solids*, 2000, **263**, 1–28.
3. M. Navarro, M. Ginebra, J. Clément, M. Salvador, A. Gloria and J. A. Planell, *J. Am. Ceram. Soc.*, 2003, **86**, 1345–1352; T. Kasuga, Y. Hosoi, M. Nogami and M. Niinomi, *J. Am. Ceram. Soc.*, 2001, **84**, 450–452; D. Brauer, N. Karpukhina, R. V. Law and R. G. Hill, *J. Non-Cryst. Solids*, 2010, **356**, 2626–2633; E. A. Abou Neel, W. Chrzanowski, S. P. Valappil, L. A. O'Dell, D. M. Pickup, M. E. Smith, R. J. Newport and J. C. Knowles, *J. Non-Cryst. Solids*, 2009, **355**, 991–1000.
4. N. Sachot, O. Castaño, M. A. Mateos-Timoneda, E. Engel and J. A. Planell, *J. R. Soc., Interface*, 2013, **10**, 20130684.
5. S. Q. Lin, H. W. Xu, J. Xiao, Y. L. Liu, Y. Zhang, L. Cai, X. K. Li and Y. Tan, *J. Heal. Sci.*, 2009, **55**, 910–922.
6. A. Hoppe, N. S. Güldal and A. R. Boccaccini, *Biomaterials*, 2011, **32**, 2757–2774.
7. B. Lansdown, B. Sampson and A. Rowe, *J. Anat.*, 1999, **195**(Pt 3), 375–386.
8. A. Aguirre, A. Gonzalez, M. Navarro, O. Castano, J. A. Planell and E. Engel, *Eur. Cells Mater.*, 2012, **24**, 90–106.
9. O. F. Vila, J. R. Bagó, M. Navarro, M. Alieva, E. Aguilar, E. Engel, J. Planell, N. Rubio and J. Blanco, *J. Biomed. Mater. Res., Part A*, 2013, **101 A**, 932–941.
10. P. Haque, I. Ahmed, A. Parsons, R. Felfel, G. Walker and C. Rudd, *J. Non. Cryst. Solids*, 2013, **375**, 99–109.
11. D. Carta, D. M. Pickup, R. J. Newport, J. C. Knowles, M. E. Smith and K. O. Drake, *Phys. Chem. Glas.*, 2005, **46**, 365–371.
12. N. Lakhkar, J. Park, N. Mordan and V. Salih, *Acta Biomater.*, 2012.
13. F. Foroutan, J. V. Jokerst, S. S. Gambhir, O. Vermesh, H.-W. Kim and J. C. Knowles, *ACS Nano*, 2015, **9**, 1868–1877.
14. H. Oliveira, S. Catros, C. Boiziau, R. Siadous, J. Marti-Munoz, R. Bareille, S. Rey, O. Castano, J. Planell, J. Amédée and E. Engel, *Acta Biomater.*, 2016, **29**, 435–445.
15. Q. Z. Chen, I. D. Thompson and A. R. Boccaccini, *Biomaterials*, 2006, **27**, 2414–2425.
16. E. S. Sanzana, M. Navarro, M.-P. Ginebra, J. A. Planell, A. C. Ojeda and H. A. Montecinos, *J. Biomed. Mater. Res., Part A*, 2013, 1–7.
17. M. Navarro, S. del Valle, S. Martínez, S. Zeppetelli, L. Ambrosio, J. A. Planell and M. P. Ginebra, *Biomaterials*, 2004, **25**, 4233–4241.

18. S. Maeno, Y. Niki, H. Matsumoto, H. Morioka, T. Yatabe, A. Funayama, Y. Toyama, T. Taguchi and J. Tanaka, *Biomaterials*, 2005, **26**, 4847–4855.
19. E. M. Brown, N. Chattopadhyay and S. Yano, *J. Musculoskelet. Neuronal Interact.*, 2004, **4**, 412–413.
20. A. Aguirre, A. Gonzalez, J. A. Planell and E. Engel, *Biochem. Biophys. Res. Commun.*, 2010, **393**, 156–161.
21. T. Fujita, N. Izumo, R. Fukuyama, T. Meguro, H. Nakamuta, T. Kohno and M. Koida, *Biochem. Biophys. Res. Commun.*, 2001, **280**, 348–352.
22. M. Julien, S. Khoshniat, A. Lacreusette, M. Gatius, A. Bozec, E. F. Wagner, Y. Wittrant, M. Masson, P. Weiss, L. Beck, D. Magne and J. Guicheux, *J. Bone Miner. Res.*, 2009, **24**, 1856–1868.
23. E. A. Abou Neel, T. Mizoguchi, M. Ito, M. Bitar, V. Salih and J. C. Knowles, *Biomaterials*, 2007, **28**, 2967–2977.
24. M. Navarro, E. Engel, J. A. Planell, I. Amaral, M. Barbosa and M. P. Ginebra, *J. Biomed. Mater. Res., Part A*, 2008, **85**, 477–486.
25. Y. Zhang and M. Zhang, *J. Mater. Sci.: Mater. Med.*, 2004, **15**, 255–260.
26. M. Navarro, S. Del Valle, S. Martinez, S. Zeppetelli, L. Ambrosio, J. A. Planell and M. P. Ginebra, *Biomaterials*, 2004, **25**, 4233–4241.
27. M. M. Dvorak, A. Siddiqua, D. T. Ward, D. H. Carter, S. L. Dallas, E. F. Nemeth and D. Riccardi, *Proc. Natl. Acad. Sci. U. S. A.*, 2004, **101**, 5140–5145.
28. M. Zaidi, V. S. Shankar, R. Tunwell, O. A. Adebanjo, J. Mackrill, M. Pazianas, D. O. Connells, B. J. Simon, B. R. Rifkin, A. R. Venkitaraman, I. C. L. Huang and F. A. Lail, *J. Clin. Invest.*, 1995, **96**, 1582–1590.
29. L. D. Quarles, J. E. H. Ii, S. R. Siddhanti, R. Guo and T. K. Hinson, *J. Bone Miner. Res.*, 1997, **12**, 393–402.
30. M. Vilalta, I. R. Dégano, J. Bagó, D. Gould, M. Santos, M. García-arranz, R. Ayats, C. Fuster, Y. Chernajovsky, D. García-olmo, N. Rubio and J. Blanco, *Gene Ther.*, 2009, **16**, 547–557.
31. M. Navarro, M.-P. Ginebra and J. a Planell, *J. Biomed. Mater. Res., Part A*, 2003, **67**, 1009–1015.
32. M. Bitar, V. Salih, V. Mudera, J. C. Knowles and M. P. Lewis, *Biomaterials*, 2004, **25**, 2283–2292.
33. A. Gonzalez-Vazquez, J. A. Planell and E. Engel, *Acta Biomater.*, 2014, **10**, 2824–2833.
34. R. M. Day, *Tissue Eng.*, 2005, **11**, 768–777.
35. A. A. Gorustovich, D.D.S., C. Perio., Ph.D., J. A. Roether, M.Sc. and P. D. and A. R. Boccaccini, Ph.D., *Tissue Eng., Part B*, 2010, **16**, 199–207.
36. E. Quinlan, S. Partap, M. M. Azevedo, G. Jell, M. M. Stevens and F. J. O. Brien, *Biomaterials*, 2015, **52**, 358–366.
37. M. Charles-Harris, M. A. Koch, M. Navarro, D. Lacroix, E. Engel and J. A. Planell, *J. Mater. Sci.: Mater. Med.*, 2008, **19**, 1503–1513.
38. O. Castaño, N. Sachot, E. Xuriguera, E. Engel, J. A. Planell, J.-H. Park, G.-Z. Jin, T.-H. Kim, J.-H. Kim and H.-W. Kim, *Appl. Mater. Interfaces*, 2014, **6**, 7512–7522.

CHAPTER 14

Bioactive Glasses in Infection Treatment

NINA LINDFORS,*[a] CARLO ROMANO,[b] SARA SCARPONI,[b] DRAGO LORENZO,[c] BORTOLIN MONICA,[c] JANEK FRANTZÉN,[d] PIETER D. DE VEIJ MESTDAGH,[e] DAVID R. COLNOT,[e] PEPIJN A. BORGGREVEN[e] AND JASPER J. QUAK[e]

[a] Department of Musculoskeletal and Plastic Surgery, Helsinki University Central Hospital, Helsinki University, Helsinki, Finland; [b] Center of Reconstructive and Osteoarticular Infection Surgery, Istituto Ortopedico IRCCS Galeazzi, Via R. Galeazzi 4, 20161, Milano, Italy; [c] Laboratory of Clinical Chemistry and Microbiology Analysis, IRCCS Galeazzi Hospital, Department of Biomedical Sciences for Health University of Milan, Istituto Ortopedico IRCCS Galeazzi, Via R. Galeazzi 4, 20161, Milano, Italy; [d] Clinical Neurosciences, Department of Neurosurgery, Turku University Hospital, Turku University, Turku, Finland; [e] Diakonessenhuis Utrecht, Department of Otolaryngology, Head and Neck Surgery, The Netherlands
*Email: nina.c.lindfors@hus.fi

14.1 Introduction

Chronic bone infection is characterised as a destructive infection of the bone structure leading to bone necrosis. The infection can involve any bone in the body and may result from trauma, surgical procedures or the haematogenous spreading of bacteria. The disease is most often associated with prolonged hospitalisation, long-term antimicrobial treatment, numerous complex and lengthy surgical procedures, high risk for complications,

RSC Smart Materials No. 23
Bioactive Glasses: Fundamentals, Technology and Applications
Edited by Aldo R. Boccaccini, Delia S. Brauer and Leena Hupa

Published by the Royal Society of Chemistry, www.rsc.org

possible recurrence or chronic persistence, or both, and elevated social and economic costs. Subjects with specific co-morbidities, including diabetes, peripheral vasculopathies, renal failures, and others, or with personal habits such as smoking or drug or alcohol abuse are at higher risk for bone and joint infection and the worse prognosis. On the other hand, the frequent occurrence of bone and joint infections due to multi-resistant and biofilm-producer bacteria, associated with the low ability of antibacterials to reach effective concentrations in bone and joint tissues, limits the effectiveness of systemic antibiotic treatments. In fact, the frequent use of antibiotics in recent decades has led to increased antimicrobial resistance worldwide. According to estimates by the World Health Organization in 2011, antibacterial resistance may account for 25 000 deaths per year and overall societal costs of approximately €1.5 billion annually in the European Union.[1]

Bioactive glasses are bone substitutes with osteoconductive, osteo- and angiostimulative, and antibacterial properties.[2–4] The osteoconductive and osteostimulative properties of the glass promote the bone formation and growth needed to reconstruct the bone defect. The well-documented antibacterial properties of bioactive glass S53P4 (BAG-S53P4), resulting from a rise in pH and osmotic pressure around the implanted glass, have been observed for both planctonic bacteria and bacteria in biofilm, whether Gram-positive or Gram-negative.[4,5] These properties and the virtual absence of bacterial resistance make bioactive glasses well suited for infection treatment of bone defects.

14.2 Chronic Osteomyelitis: Basic Principles and Limits of Current Treatments

In 1852, the French surgeon Edouard Chassaignac was probably the first to introduce the term osteomyelitis by defining the disease as an inflammatory process accompanied by bone destruction and caused by an infectious microorganism.[6]

Bone infection tends to relapse and become chronic. In fact, the prevalence of chronic osteomyelitis in children treated an acute haematogenous bone infection ranges from 5 to 25%.[7–9] While the development of chronic osteomyelitis in children often follows acute onset, its clinical presentation in adults may be subacute or chronic *ab initio,* with poor symptoms and no clear distinction between acute and chronic clinical presentation. Given the absence of a clear cut-off period between acute and chronic osteomyelitis in many clinical situations, chronic osteomyelitis is diagnosed with the fulfillment of at least one of the following criteria: clinical and/or radiographic signs of osteomyelitis lasting for more than six weeks, radiographic evidence of bone destruction, sequestra, sclerosis or osteolysis, the persistence or recurrence of infection after adequate antibiotic treatment, bone infection associated with the presence of a foreign body or implant, bone infection associated with inadequate local vascular supply, and bone infection due to microorganisms able to induce specific bone destruction (*e.g.*, *Mycobacterium tuberculosis*).

Chronic osteomyelitis may be caused by haematogenous and/or post-traumatic or postoperative bacterial colonisation of bones and surrounding tissues.[10] In the past, osteomyelitis resulted mainly from direct bacterial penetration into the bone or adjacent tissues caused by open wounds or low-energy trauma, or due to the haematogenous spreading of the microorganisms

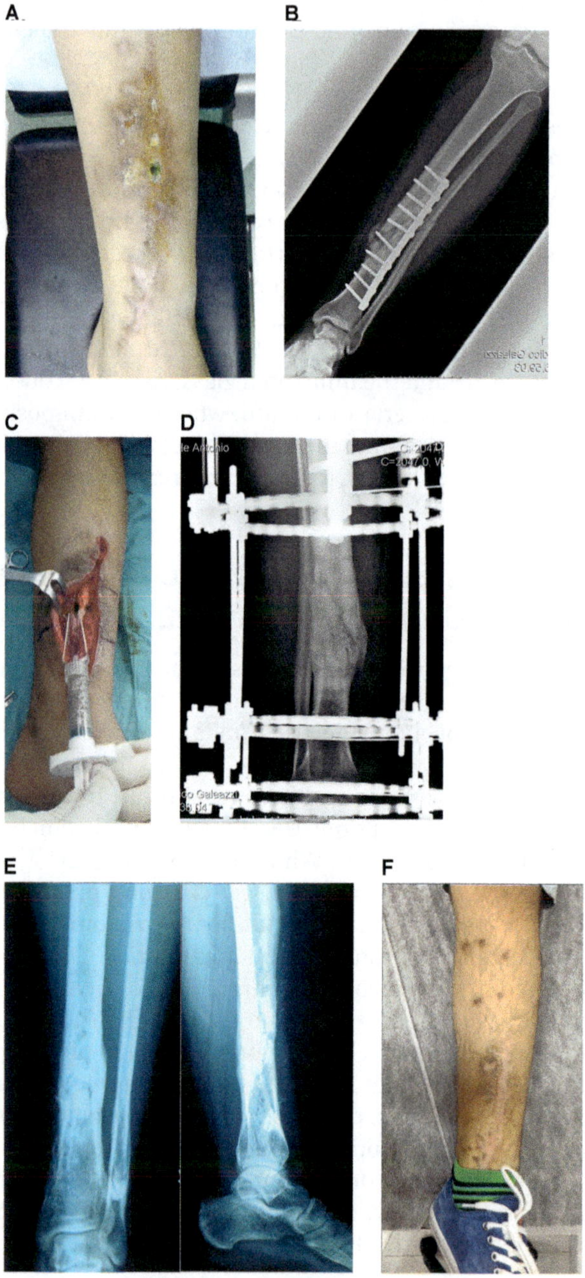

from septic foci localised in other organs and in apparatus. During the last century, however, the industrialised world has seen a progressive increase in bone and joint infections due to high-energy traumas (wars, traffic, sports, *etc.*) or secondary to surgical procedures. In addition, studies show that today, more osteomyelitis and septic arthritis are causally related to metabolic disorders (diabetes, renal insufficiency, *etc.*), peripheral vasculopathies and neuropathies, health-damaging life habits (smoking, drug or alcohol abuse), inherited or acquired immunodeficiencies and advanced age.[11,12]

Overall, estimates of the annual incidence of bone and joint infections in Europe range from 30 to 60 per 100 000 habitants, with a prevalence of several tens of thousands of patients.

Despite improvements in medical and surgical management, chronic osteomyelitis remains a difficult-to-treat condition, is associated with high economic and social costs, and is prone to relapse, complications and possible dramatic sequalae.

Complete eradication of chronic osteomyelitis is still considered a challenging goal depending, among other variables, on the host type[11] and anatomo-pathological presentation of the disease,[12] as well as microbiological, pharmacological and surgical aspects.

Staphylococcus aureus and other Gram-positive pathogens are the pathogens most often involved in chronic osteomyelitis, but Gram-negative and mixed florae (often multi-resistant and strong biofilm producers), may also be involved.[10] Given the limited ability of antibiotics to penetrate bacterial biofilms and poorly vascularised or devitalised infected bone and surrounding tissues, systemic antibiotic treatment alone is often ineffective, and nearly all cases require complex surgical interventions aimed at completely removing all foreign materials, infected or necrotic bone and any severely affected soft tissue.[13,14]

However, radical debridement often results in significant bone loss that may complicate reconstructive surgery and pose difficult reconstructive surgical dilemmas.

In fact, surgeons have traditionally used various surgical techniques to manage bone loss due to the septic process *per se*, to related inflammatory reactions and to necessary surgical debridement, depending on the site, size and shape of the defect.

Bone defect sizes in chronic osteomyelitis may range from a few mm^3 to several cm^3, and vary from the common "cavitary" defect to epiphyseal (involving the joint) or segmental defects (Figure 14.1A–F).[10]

Figure 14.1 (A) Preoperative clinical presentation and (B) radiographic finding of a 19-year-old male 12 months after internal osteosynthesis; septic nonunion or Cierny–Mader Stage IV osteomyelitis. Type B host (smoker). Persistence of infection: infection persisted despite three previous surgeries. Methicillin-resistant *Staph. aureus*. (C) Plate removal, surgical debridement, bioglass implant in the septic focus and application of a circular external fixator; (D) radiographic control six months after surgery. (E and F) Radiographic and clinical findings two years after surgery. Complete bone healing and no infection recurrence.

While large epiphyseal and segmental defects usually require highly specialised techniques (bone transport with external fixators, local or free soft-tissue flaps, *etc.*) and implants, the most common cavitary defects are generally treated by filling the defect with local antibiotic delivery carriers, such as antibiotic-loaded polymethylmethacrylate (PMMA),[15,16] bone grafts or calcium-based bone substitutes.[17–19] Studies have shown that under such circumstances and in the presence of low serum concentrations, local antibiotics delivery does in fact provide high local antibiotic levels, thereby reducing the risk for systemic toxicity and side effects.[20]

However, antibiotic-loaded PMMA has some drawbacks, including possible thermal damage to the antibiotic and to the surrounding tissues as the cement cures, thereby requiring a second intervention to remove the PMMA and the possible presence of small-colony variant multi-resistant strains and/or of biofilm on the bone cement itself due to inadequate concentrations of the slow-release antibiotic.[21]

On the other hand, antibiotic-loaded bone grafts require harvesting and processing, have limited availability and risk potentially transmitting viral disease, whereas calcium-based bone substitutes may induce local seromas and can be safely used with a limited range of antibiotics.[20,22,23] Furthermore, all of these solutions require adding an effective antibiotic to the local carrier, though in many cases the pathogen(s) may be unidentified prior to surgery due to difficult sampling and the growth of causative bacteria in chronic biofilm-related infections such as osteomyelitis.

In such complex situations bioactive glasses may promise a novel and attractive solution that offers some advantage over existing technologies.

In particular, due to its previously demonstrated osteoconductive[24–26] and angiogenetic properties and broad antibacterial[4,27] and antibiofilm capability,[5,28] BAG-S53P4 has recently been proposed as the treatment of choice for cavitary defects and less severe segmental defects in chronic osteomyelitis.[29,30]

14.3 Bioactive Glass in Bone Infection Treatment

The use of bioactive glasses in bone infection treatment began in the 1990's.

In frontal sinusitis, BAG-S53P4 has proved to be a reliable frontal sinus obliteration material, providing favourable conditions for total bony sinus obliteration.[31] In mastoiditis, BAG-S53P4 granules not only function safely but successfully as an obliteration material in problematic open radical cavities. BAG-S53P4 also appears to tolerate chronic infection in the mastoid and to prevent postoperative infections, and is an effective material in mastoid obliteration surgery after previous failed efforts to achieve a dry, safe ear.[32,33]

BAG-S53P4 has also been used to treat infection in chronic osteomyelitis. Since the first patient with osteomyelitis in the distal tibia received BAG treatment in 2007 in Oulu, Finland, three different studies have reported that the use of BAG-S53P4 without local antibiotics is both safe and effective in treating bone defects in chronic osteomyelitis.[29,30,34] In particular, an initial report on a limited series of 11 osteomyelitis patients explored the potential use of BAGs to treat chronic osteomyelitis with promising results;[29] a more recent study of 27 patients found that BAG-S53P4 was as effective as two different calcium-based antibiotic-loaded bone substitutes in treating chronic osteomyelitis. In addition, patients treated with BAG-S53P4 showed significantly lower prolonged wound serum leakage (3.7%) than did the two antibiotic-loaded calcium-based bone substitutes (29.6% and 27.2%); they also tended to shorten the patient's hospital stay.[34]

More recently, a multinational, multicenter cohort study of 116 patients with clinically and radiologically verified chronic osteomyelitis reported treatments with surgical debridement and bone void filling with BAG-S53P4 without local antibiotics. The study included cavitary defects as well as segmental bone defects with a bone gap of less than 1 cm. Overall, the results from six European centres showed a 90% success rate, thus confirming that BAG-S53P4 can serve as a bone substitute without local antibiotics in the treatment of chronic osteomyelitis and yield excellent results.

The main benefits of BAG-S53P4 in the treatment of chronic osteomyelitis include:

- osteoconductive properties,
- direct antibacterial and antibiofilm actions requiring no prior pathogen identification or use of local antibiotics,
- efficacy against multi-resistant strains,
- the possibility to perform the treatment as a single-stage procedure, unlike the current standard, which often consists of a two-stage procedure involving the first stage which involves the use of antibiotic-containing PMMA beads. Avoiding additional surgery reduces the patient burden and the risk for additional complications, while leaving the potential financial advantage for health services open to further investigation.

In conclusion, available data show that the antibacterial, osteostimulative and osteoconductive bone substitute BAG-S53P4 is suitable as a bone void filler in the treatment of chronic osteomyelitis and treatment can be performed as a single-stage procedure with excellent results. This makes the treatment protocol cost-effective, with the added trend of shortening the patients' hospital stays.

14.4 Antibacterial Properties of Bioactive Glasses

14.4.1 Bioactive Glasses as Antimicrobial Agents

Bioactive glasses (BAGs) are bone substitutes with osteoconductive and antibacterial properties. The process leading to bone bonding has been described as a sequence of reactions occurring on the surface of the glass ultimately resulting in the formation of a hydroxyapatite (HA) layer.[35] The antibacterial effect of BAGs appears to result from marked changes in the local physiological milieu, such as the high pH and osmotic pressure associated with the conversion of the glass to HA,[36] which creates an environment hostile to microbial growth.[35]

Soon after the implantation of BAG and its exposure to bodily fluids, surface reactions lead to the deposition of a layer of calcium phosphate. Sodium, silica, calcium and phosphate ions are released from the surface and raise the local pH and osmotic pressure. Thereafter, a layer of silica gel forms on the surface of the glass, and on this layer, amorphous calcium phosphates precipitate. These amorphous structures then crystallise to natural HA, which begins to activate osteoblasts forming new bone.[37–39] The continuous reactions and layer formation finally causes the glass to be absorbed.

The surface reactions are not only advantageous for the formation of new bone, but they also mediate the antibacterial properties of BAGs.[4,27,40–43] First of all, antibacterial properties appear to result from the rise in local pH resulting from the exchange of sodium ions with protons in bodily fluids.[42] The dissolution of sodium ions from the glass can initially raise the local pH of bodily fluids to 11 within 8 h, and maintain that pH level for a period of 48 h or longer.[36] pH neutralisation eliminates the antibacterial effect.[44] The extent of the change in pH varies depending on the composition of the glass, the ratio of surface area to volume, the sample dosage and the agitation rate of the system.[45–51] An additional factor contributing to antimicrobial properties could be the release of silica and calcium and phosphate ions, which disturbs the membrane potential of bacteria and raises the osmotic pressure.[37–39]

In a recent study, a SEM analysis served to gain further insight into the mechanism of action of BAGs by evaluating bacterial morphology changes occurring after the incubation of bacteria with BAG S53P4.[52] Such an analysis revealed a significant change in the morphology of *Staphylococcus epidermidis, Acinetobacter baumannii* and *Klebsiella pneumoniae* cells after treatment with BAG, including cell shrinkage, with a consequent reduction in cell dimensions, as well as damage to bacterial membranes. These findings correlate closely with the double mechanism of action of BAGs (mentioned above) due to osmolarity and variations in pH. The concentration of solutes within the bacterial cytoplasm is normally higher than in the surrounding environment, creating positive pressure on the cell membrane. A sudden increase in the concentration of external solutes, such as that which occurs after BAG immersion in bodily fluids, causes a rapid water efflux and a pressure drop across the cell membrane, thereby altering cell size, cell shape and membrane stress levels.[53] Similarly, shifting to a highly basic environment is stressful for bacteria, which respond to alkaline stress by changing morphology, ultrastructure, protein synthesis and gene expression.[54]

A relevant issue is the effect that such extreme local conditions (*i.e.*, large local pH and osmolarity changes resulting from BAG dissolution) may have on host cells and tissues. However, BAGs have been used in clinics to treat frontal sinus, mastoid, benign bone tumour, trauma and spine surgery for more than 20 years with no observable signs of toxic reaction.[31,33,55–57]

14.4.2 Broad Spectrum Antimicrobial Efficacy

BAGs have shown a considerable ability to inhibit bacterial growth *in vitro* and to eradicate infections *in vivo*. Different authors have pointed out the utility of the antibacterial activity of BAGs in dentistry[44,58–60] and orthopaedic surgery, especially in treating osteomyelitis,[29,34,61,62] thus demonstrating high efficacy against aerobic and anaerobic clinical isolates.

Though BAGs have often been combined with antibacterial ions (*e.g.*, silver and copper) in effort to develop antibacterial materials with promising results,[63,64] some BAGs with no antibacterial ions also kill bacteria effectively (Table 14.1) thanks to the dramatic changes created by the local physiological environment.

As Table 14.1 shows, published studies on the antibacterial properties of BAGs are characterised by broad heterogeneity in species of bacteria as well as the compositions, sizes and concentrations of the BAGs used. The wide variability among studies may hinder the specific factors influencing bactericidal capacity and may make it difficult to draw general conclusions.

BAGs showed efficacy against several clinically important bacterial strains, including Gram-positive and Gram-negative species, aerobic and anaerobic bacteria. The effect varied between species, but the effectiveness of BAGs against Gram-postitive and Gram-negative bacteria showed no significant difference.[4,27]

BAGs with different chemical compositions (varying percentages of SiO_2, Na_2O, CaO and P_2O_5) were used across several studies. All glasses inhibited bacterial growth, though concentrations and times depended on the specific BAG used. Zehnder *et al.*[59] found that the bactericidal effect of BAGs correlates with high levels of silicon ions in the supernatant, while Munukka *et al.*[27] showed that the sol–gel-derived BAG with a high CaO content had a greater antibacterial effect on aerobic bacteria than BAGs with a lower CaO content.

BAG particle sizes may range from nanometers to a maximum of 3 mm. The particle dimensions appears to influence the antimicrobial activity of BAGs. Reducing the particle size and the resultant increase in surface area, which enhances ion release, has served as one approach to raising the antibacterial activity of BAGs.[60,65]

14.4.3 Bioactive Glasses as an Enemy of Bacterial Biofilm

Bacteria prefer biofilms as the most common medium of growth in the environment, but this kind of bacterial colonisation can be very harmful during an infection. Biofilms can be up to one thousand-fold more resistant to antibiotic treatment than the same organism growing planktonically.

Table 14.1 Summary of the main *in vitro* studies of the antibacterial effect of BAGs.

BAG	Concentration ($mg\,ml^{-1}$)	Particle size (μm)	Bacterial species	Reference
S53P4	1667	<45	*A. actinomycetemcomitans, P. gingivalis, A. naeslundii, S. mutans, S. sanguis*	Stoor *et al.*[58]
45S5	1000, 2000 500	90–710 355–500	*S. sanguis, S. mutans, A. viscous, A. actinomycetemcomitans, P. gingivalis, F. nucleatum, P. intermedia*	Allan *et al.*[44]
45S5, $76SiO_2$–22CaO–$2P_2O$	0.05–1.10	90–710	*E. coli, P. aeruginosa, S. aureus*	Bellantone *et al.*[78]
45S5	10, 50, 100	<45	*S. aureus, S. epidermidis, E. coli*	Hu *et al.*[79]
S53P4	300	<45	*E. faecalis*	Zehnder *et al.*[59]
S53P4, $CaPSiO_2$	50, 100	<45	*S. epidermidis*	Vaahtio *et al.*[80]
45S5	0.1	0.02–0.06 100	*E. faecalis*	Waltimo *et al.*[60]
S53P4, 13-93, CaPSiO II, E-glass	50, 100, 200, 400	<45	*B. fragilis, B. thetaiotaomicron, B. adolescentis, C. difficile, C. perfringens, C. septicum, E. lentum, F. necrophorum, F. nucelatum, P. anaerobious, P. gingivalis, P. intermedia, P. melanogenica, P. acnes, P. propionicus, S. epidermidis, V. parvula*	Leppäranta *et al.*[4]
MBG0103, MBG0118, MBG0123, S53P4, 13-93, H2-02, CaPSiO, CaPSiO II	6.25, 12.5, 25, 50, 100	<45	*Acinetobacter sp., B. cereus, C. ulcerans, E. aerogenes, E. amnigenus, E. faecalis, E. faecium, E. coli, F. meningosepticum, H. influenzae, K. pneumoniae, L. monocytogenes, Micrococcus sp., M. catarrhalis, N. meningitidis, P. multocida, P. mirabilis, P. aeruginosa, S. typhimurium, S. sonnei, S. aureus, S. epidermidis, S. agalactiae, S. mutans, S. pneumoniae, S. pyogenes, S. sanguis, Y. enterocolitica*	Munukka *et al.*[27]
3-04, 13-93, H2-02, 18-04, 23-04	100 6.25, 12.5, 25, 50, 100	<45	*Acinetobacter sp., H. influenzae, M. catarrhalis, E. aerogenes, E. coli, E. faecalis, P. aeruginosa, S. epidermidis, N. meningitidis, Y. enterocolitica, F. meningosepticum, P. multocida, P. mirabilis, S. pyogenes, S. sanguis, S. pneumoniae*	Zhang *et al.*[36]
S53P4				
58S, 63S, 72S	6.25, 12.5, 25, 50, 100	0.02–0.09	*E. coli, S. aureus, S. typhimurium, P. aeruginosa*	Mortazavi *et al.*[65]
S53P4	400, 800	500–800	*S. aureus, S. epidermidis, A. Baumannii, P. Aeruginosa*	Drago *et al.*[61]
S53P4	400	<45	*S. epidermidis, K. Pneumoniae, A. baumannii*	Drago *et al.*[52]

The administration of antimicrobial agents has been a useful approach to eliminating biofilms, but the prolonged persistence of these agents in the environment could raise toxicity towards non-target organisms and, thus, resistance among microorganisms within biofilms. This characteristic has encouraged the development of more environmentally-friendly compounds to combat the issue, and some hypothesise that some bioactive compounds could prove useful in combating biofilm formation and infections.

Recent, studies have shown that BAGs are active not only against planktonic bacteria, but also in destroying and inhibiting biofilms,[5,28,66,67] making them ideal bone substitutes in biofilm-related infections such as prosthetic joint infections (PJIs).

14.4.4 Bioactive Glasses: Wide Antimicrobial Spectrum But No Resistance Induction

The widespread use of antibiotics is one of the main drivers of natural selection among microorganisms to develop resistance, which seriously hampers the treatment of bacterial-driven infections, including bone and prosthetic joint infections (PJIs). Despite advances in antibiotic therapies and surgical techniques, treating bone infections remains challenging. Debridement in conjunction with local administration of antibiotics (*e.g.*, antibiotic-loaded PMMA bone cement) is considered the standard of care. Although single or combinations of different antibiotics have been loaded into bone cements,[68–71] inadequate release kinetics can lead to the development of multi-resistant bacterial strains, often rendering commercially available antibiotic-loaded bone substitutes ineffective.[72] Thus in this scenario, the development of valuable alternatives to antibiotics is mandatory in the fight against drug resistance.

Because the antimicrobial working mechanism of BAGs differs entirely from that of antibiotics, they could be more reliable in the long term, with the growing prevalence of antibiotic-resistant bacteria.[73–75]

Thus far, only one study has evaluated the bacterial resistance of BAGs. The study aimed to evaluate the antimicrobial activity of BAG-S53P4 against multidrug-resistant (MDR) microorganisms involved in bone infections as well as the capability of BAGs to select for resistance in such microorganisms.[52] The development of resistance during prolonged exposure to BAG-S53P4 was evaluated *in vitro*. The selection of resistance took place over ten steps in an effort to mimic a long period of incubation with BAG-S53P4.[76] Repeated exposure failed to change MIC values, which were equal to basal values and remained stable after exposure to subcultures without antibiotics, thus demonstrating that exposure did not compromise susceptibility. Moreover, no induction of resistance to BAG-S53P4 occurred during incubation in bactericidal concentrations. Together, these observations suggest that the MDR bacteria tested are unable to adapt to the hostile environment created by BAG-S53P4.

14.4.5 Biofilm and Multidrug Resistance (MDR): A Harmful Combination

At present, the correlation between biofilm production and multidrug resistance remains unclear. Strains capable of forming a biofilm might be selected under antibiotic pressure or, conversely, might develop resistance to multiple drugs within biofilm communities. In either event, the high capacity of biofilm-forming strains to colonise, coupled with their resistance to multiple drugs, will contribute to a microorganism's survival and further dissemination.[77]

A very recent study evaluated the effectiveness of BAG-S53P4 against the biofilm produced *in vitro* by MDR strains isolated from PJIs.[67] The evaluation was carried out using confocal laser scanning microscope (CLSM) analysis and showed that BAG-S53P4 can reduce the biofilm produced by MDR strains on titanium substrates *in vitro*, probably by interfering with the viability of the bacterial cell.

In the scenario described above, the use of BAG-S53P4, which can reduce biofilm formation while bypassing drug resistance, could represent the *ideal bone* substitute for treating bone infections, even in the presence of multi-drug-resistant strains.

14.5 Bioactive Glass in Spinal Infections

14.5.1 Introduction

Infections in the spine can be iatrogenous (*e.g.*, postoperative infections due to any surgical procedure or biopsy, with or without hardware, resulting in pyogenic vertebral osteomyelitis). More commonly, this rare condition is caused by haematogenous seeding or spreading from a nearby infection. In general, the treatment of vertebral osteomyelitis is managed conservatively with antibiotics.[81]

The treatment of deep infections in the instrumented spine is more complicated, often requiring the removal of the implanted hardware, which can lead to a high incidence of long-term patient morbidity.[82,83]

The presence of an epidural abscess formation or larger involvement of the spine that comprises stability often requires surgical debridement and reconstruction. The use of metal implants or bone grafts can increase the patient's infectious burden because pathogens are able to produce biofilms on the surface of the implanted materials, thus complicating treatment outcomes.

14.5.2 Bioactive Glass in Spinal Infections

The first clinical publication on the use of BAG-S53P4 in treating osteomyelitis of the lower extremities and spine demonstrated its effectiveness as a single-stage procedure yielding favourable outcomes on 10 of 11 patients that lasted for a mean 24 months (range: 10–38 months).[29]

A second publication from the same author described three patients with severe spondylodiscitis caused by *Mycobacterium tuberculosis*, *Candida tropicalis* or *Staphylococcus aureus* treated surgically, due to neural compromise and instability of the spine, with fusion and BAG-S53P4 bioactive glass—as well as autograft in patients 2 and 3.

The follow-up was four years for patient 1, one year and eight months for patient 2, and two years and two months for patient 3. With no relapses or complications observed during that period, all three patients fused and showed excellent clinical results (Figures 14.2 and 14.3).[84]

BAG-S53P4 granules have served in centres around Europe and Latin America in treating spondylodiscitis of the whole spine. Although these results have not yet been published, the clinical results are encouraging (personal communication).

Based on these publications and personal experience, thorough debridement of all infected tissue is crucial. Any defect must be properly filled with bioactive glass granules. The size of the granules can be tailored to certain needs. To achieve maximal bactericidal effect a smaller granule size is preferable as it provides a larger reaction surface. *In vivo* studies have shown that the antibacterial properties of BAG differ depending on its composition. Comparing the bactericidal effect of BAG-S53P4 to that of other glasses has shown it to be the most effective with the fastest killing and growth

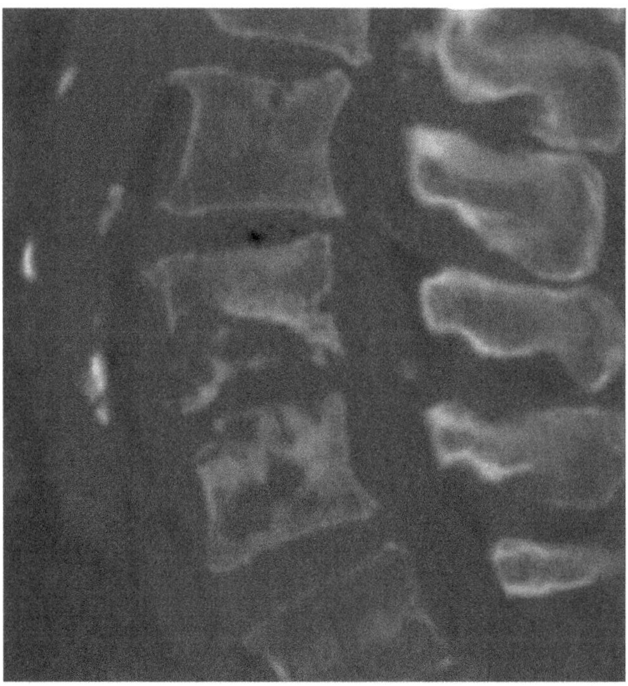

Figure 14.2 Preoperative MRI of patient showing osteomyelitis in the spine.

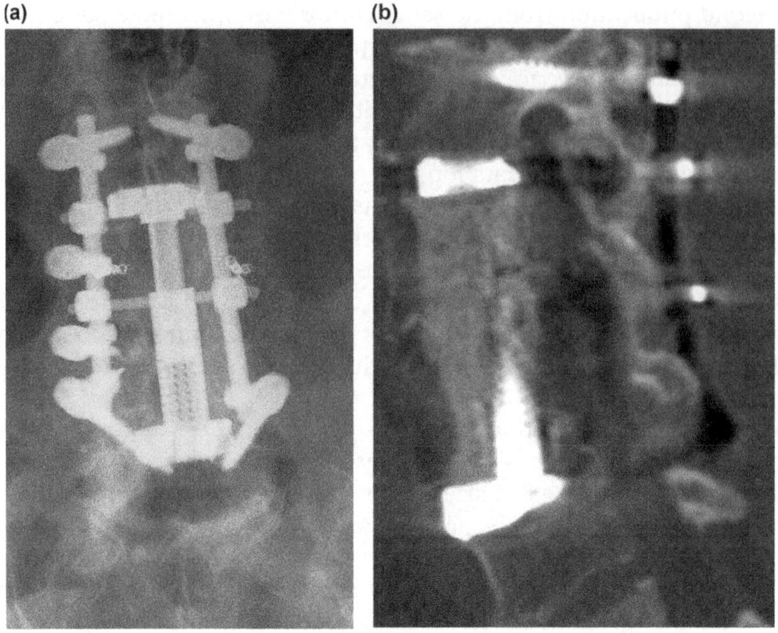

Figure 14.3 (a) Postoperative X-ray and (b) CT showing complete fusion in BAG bone subtituted area at two years.

inhibitory effect for 29 aerobic and 17 anaerobic pathogens tested, including multi-resistant pathogens.[4,27,51,85]

In conclusion, the antibacterial properties of BAG-S53P4 make it a suitable bone substitute for the treatment of severe spondylodiscitis. Nevertheless, further studies in this field would be welcomed.

14.6 Evaluation of S53P4 Bioactive Glass as a Filler Material in Mastoid Obliteration for Chronic Otitis Media and Cholesteatoma

14.6.1 Introduction

Canal wall up (CWU) tympanoplasty is considered the preferred technique in treating chronic otitis media (COM) that fails to respond to conservative treatment. Recurrent disease, however, can require multiple surgical procedures which can eventually lead to canal wall down (CWD) surgery.[86]

Obliteration of the mastoid cavity can prove useful in both CWU and CWD procedures. In CWU surgery, obliteration of the mastoid cavity appears to prevent the recurrence of both chronic infection of the mastoid and cholesteatoma by closing the possible route of a retraction pocket.[87–89] In (revision) CWD surgery, obliteration can reduce or help prevent problems associated with a large open mastoid cavity, such as the accumulation of cerumen, vertigo, hearing loss and water intolerance.[87,88,90,91]

Multiple obliteration techniques have been attempted in recent decades including numerous flap variations as well as autologous filler materials such as bone paté, fat, and cartilage.[87,88,90] Bone paté and flaps that incorporate muscle, however, have been shown to lose volume over time.[87,88,92,93] Cartilage holds its volume but is difficult to harvest in large enough quantities to fill large mastoid cavities.[88] Therefore clinicians have tested a multitude of allogenous and synthetic obliteration materials.

Filler materials such as demineralised bone matrix and hydroxyapatite granules have shown variable long-term results.[94–97] Both bone paté and hydroxyapatite cement reportedly yield low postoperative infection rates.[98–100]

A promising alternative filler material for mastoid obliteration is BAG-S53P4. Together with the observed antibacterial effect, BAG-S53P4 dissolves slowly, thereby retaining its volume over time.[24,25]

In 2006, BAG-S53P4 received the CE mark for use in mastoid sinus obliteration, nasal septum perforation reconstruction, and nasal cavity narrowing. In recent studies, BAG-S53P4 use has shown promising results in obliteration of the mastoid cavity.[33,101,102]

14.6.2 BAG-S53P4 in Mastoid Obliteration

A retrospective study of BAG-S53P4 use in mastoid obliteration has been conducted at Diakonessenhuis Utrecht, the Netherlands. The inclusion criterion was the use of BAG-S53P4 as an obliteration material in the mastoid cavity. Three different indications were defined for surgery in patients who had their mastoid obliterated with BAG-S53P4: cholesteatoma, COM without cholesteatoma, and cavity problems. The following outcome parameters were analyzed: procedural safety, functional outcome (hearing levels and incidence of otorrhea), and recurrent and residual rate of cholesteatoma. Follow-up for cholesteatomatous ears consisted of regular clinical otoscopy and either an MRI scan or a follow-up exam after one year. Having a dry ear was defined as presenting a dry ear continuously after surgery (or after the initial post-surgical 'wet period') or presenting with otorrhea (whether treated or untreated with systemic antibiotics) only once in the year prior to the most recent follow-up visit. Hearing levels appear as pure tone averages (PTAs) for four frequencies (500, 1000, 2000, 4000 Hz). Audiometric data were taken from the last preoperative audiogram and most recent postoperative audiogram (the first of which usually took place six weeks after surgery). Hearing results were analysed with the availability of both pre- and postoperative hearing tests. The cut-off for procedural safety analysis was a minimum follow-up of 12 weeks. For cholesteatoma and infection recurrence, we used a minimum follow-up period of at least one year.

Patients underwent either (revision) CWD or CWU tympanoplasty, depending on intraoperative findings (extent and localisation of the disease), hearing level before surgery and expected hearing level after surgery, as well as the preferences discussed with the patient preoperatively. The mastoid cavity was filled with a standard sachet of 2.5 ml S53P4 granules after closure of the antrum with cartilage. Fibrin glue (Tissucol, produced by Baxter, Deerfield, IL US) served to cover the BAG-S53P4 granules. A periosteal flap then served to additionally cover the granules and to close the mastoid. Patients received intravenous amoxicillin – clavulanic acid preoperatively, which was continued postoperatively for six days. Additionally, in patients undergoing combined approach tympanoplasty, a passive drain was left under the periosteal flap for seven days.

14.6.2.1 Results

A total of 117 patients (118 ears) underwent mastoid obliteration with BAG-S53P4 (Table 14.2).

Of these, 45 patients received treatment for chronic otitis media (one patient had surgery on both ears), 59 patients were treated for cholesteatoma, and 13 patients for cavity problems. Altogether 46 patients (47 ears) received treatment with combined-approach tympanoplasty and 71 patients underwent (revision) CWD mastoidectomy.

The duration of follow-up varied from four weeks to 37.6 months. A summary of the results appears in Table 14.3.

In conclusion, BAG-S53P4 bioactive glass granules are feasible for use as a filler material in mastoid obliteration. Obliteration of the mastoid using BAG-S53P4 granules contributes to the elimination of chronic suppurative

Table 14.2 Patient characteristics.

Surgery indication, type of surgery	Patient characteristics			
	Number	Males (%)	Mean age (years)	Previous surgery (mean number)
COM[a]	46	50	48.9	1.4
CWU[b]	34	47	49.2	1.4
CWD[c]	12	58	47.8	1.2
Cholesteatoma	59	61	34.5	1.1
CWU	13	53	27.2	1.0
CWD	46	63	36.5	1.2
Cavity problems	13	62	47.2	2.3
Total	118	57	41.5	1.4

[a]COM: chronic otitis media;
[b]CWU: canal wall up;
[c]CWD: canal wall down.

Table 14.3 Outcome and complications by indication and type of surgery.

Indication, type of surgery	Functional outcome			Complications	
	Dry ear (%)	PTA[d] pre-op (mean)	PTA post-op (mean)	Wound infection (%)	Recurrence (%)
COM[a]	95	48.30	49.11	13	
CWU	95	45.92	44.73	18	
CWD	100	59.37	69.58	0	
Cholesteatoma	90	39.70	39.78	0	10
CWU[b]	90	28.50	26.87	0	20
CWD[c]	90	42.73	43.27	0	0
Cavity problems	100	54.37	43.87	0	
Total	93	44.53	43.72	5	

[a]COM: chronic otitis media;
[b]CWU: canal wall up;
[c]CWD: canal wall down;
[d]PTA: pure tone average.

otitis media. Furthermore, the results are in line with those of other studies for which obliteration of the mastoid is the preferred treatment for (post-)cholesteatomatous ears. The results of this study suggest that BAG-S53P4 granules have a role to play in the treatment of COM with, and without, cholesteatoma, COM resistant to both conservative and surgical therapy, and in problematic mastoid cavities after CWD tympanoplasty.

References

1. E. Leung, D. E. Weil, M. Raviglione and H. Nakatani, *Bull. W. H. O.*, 2011, **89**, 390.
2. L. L. Hench and J. Wilson, *Science*, 1984, **226**, 630.
3. A. A. Gorustovich, J. A. Roether and A. R. Boccaccini, *Tissue Eng., Part B*, 2010, **16**, 199.
4. O. Leppäranta, M. Vaahtio, R. Peltola, D. Ziang, L. Hupa, M. Hupa, H. Ylänen, J. I. Salonen, M. K. Viljanen and E. Eerola, *J. Mater. Sci.: Mater. Med.*, 2008, **19**, 547.
5. L. Drago, C. Vassena, S. Fenu, E. De Vecchi, V. Signori, R. De Francesco and C. l. Romanò, *Future Microbiol.*, 2014, **9**, 593.
6. E. Chassaignac, *Bull. Mem. Soc. Chir.*, 1852, 431.
7. V. Q. Dich, J. D. Nelson and K. C. Haltalin, *Am. J. Dis. Child.*, 1975, **129**, 1273.
8. D. Shivarathre, H. George, N. Kaimal and L. James, *Acta Orthop. Belg.*, 2009, **75**, 81.
9. M. J. Blyth, R. Kincaid, M. A. Craigen and G. C. Bennet, *J. Bone Jt. Surg., Br. Vol.*, 2001, **83**, 99.

10. C. L. Romanò, D. Romanò, N. Logoluso and L. Drago, *Eur. Orthop. Traumatol.*, 2011, **1**, 207.
11. G. Cierny 3rd, J. T. Mader and J. J. Penninck, *Clin. Orthop. Relat. Res.*, 2003, **414**, 7.
12. E. J. McPherson, C. Woodson, O. Holtom, N. Roidis, C. Shufelt and M. Patzakis, *Clin. Orthop. Relat. Res.*, 2003, **403**, 8.
13. D. W. Haas and M. P. McAndrew, *Am. J. Med.*, 1996, **101**, 550.
14. D. P. Lew and F. A. Waldvogel, *Lancet*, 2004, **364**, 369.
15. J. Bridgens, S. Davies, L. Tilley, P. Norman and I. Stockley, *J. Bone Jt. Surg., Br. Vol.*, 2008, **90-B**, 643.
16. K. Anagnostakos and J. Kelm, *Biomed. Mater. Res., Part B*, 2009, **90**, 467.
17. J. C. Webb and R. F. Spencer, *J. Bone Jt. Surg., Br. Vol.*, 2007, **89-B**, 851.
18. B. Parsons and E. Strauss, *Am. J. Surg.*, 2004, **188**(1A Suppl), 57.
19. L. Gordon and E. J. Chiu, *J. Bone Jt. Surg., Am.*, 1988, **70**, 377.
20. C. G. Zalavras, M. J. Patzakis and P. Holtom, *Clin. Orthop. Relat. Res.*, 2004, **427**, 86.
21. K. Anagnostakos, P. Hitzler, D. Pape, D. Kohn and J. Kelm, *Acta Orthop.*, 2008, **79**, 302.
22. A. D. Hanssen, *Clin. Orthop. Relat. Res.*, 2005, **437**, 91.
23. A. C. McLaren, *Clin. Orthop. Relat. Res.*, 2004, **427**, 101.
24. P. Virolainen, J. Heikkilä, A. Yli-Urpo, E. Vuorio and H. T. Aro, *J. Biomed. Mater. Res.*, 1997, **35**(1), 9.
25. V. V. Välimäki and H. T. Aro, *Scand. J. Surg.*, 2006, **95**(2), 95.
26. J. T. Heikkilä, J. Kukkonen, A. J. Aho, S. Moisander, T. Kyyrönen and K. Mattila, *J. Mater. Sci.: Mater. Med.*, 2011, **22**(4), 1073.
27. E. Munukka, O. Leppäranta, M. Korkeamäki, M. Vaahto, T. Peltola, D. Zhang, L. Hupa, H. Ylänen, J. I. Salonen, M. K. Virtanen and E. Eerola, *J. Mater. Sci.: Mater. Med.*, 2008, **19**, 27.
28. D. Coraca-Huber, M. Fille, J. Hausdorfer, D. Putzer and M. Nogler, *J. Orthop. Res.*, 2014, **32**, 175.
29. N. C. Lindfors, P. Hyvönen, M. Nyyssönen, M. Kirjavainen, J. Kankare, E. Gullichsen and J. Salo, *Bone*, 2010, **47**, 212.
30. L. Drago, D. Romanò, E. De Vecchi, C. Vassena, N. Logoluso, R. Mattina and C. L. Romanò, *BMC Infect. Dis.*, 2013, **10**, 584.
31. M. Peltola, K. Aitasalo, J. Suonpää, M. Varpula and A. Yli-Urpo, *Head Neck*, 2006, **28**, 834.
32. J. Sarin, R. Grénman, K. Aitasalo and J. Pulkkinen, *Ann. Otol., Rhinol., Laryngol.*, 2012, **121**, 563.
33. J. T. Silvola, *Otolaryngol. - Head Neck Surg.*, 2012, **147**(1), 119.
34. C. L. Romanò, N. Logoluso, E. Meani, D. Romanò, E. De Vecchi, C. Vassena and L. A. Drago, *Bone Joint J.*, 2014, **96-B**, 845.
35. C. Liu, C. W. Chen and P. Ducheyne, *Biomed. Mater.*, 2008, **3**(3), 034111.
36. D. Zhang, E. Munukka, I. Hupa, H. O. Ylanen, M. K. Viljanen and M. Hupa, *Key Eng. Mater.*, 2007, **173**, 330.

37. J. T. Heikkila, in *Bioactive Glasses: Materials, Properties and Applications*, ed. H. Ylanen, Woodhead Publishing, Cambridge, 2011, p. 198.
38. L. Hupa, Melt-derived bioactive glasses, in *Bioactive Glasses: Materials, Properties and Applications*, ed. H. Ylanen, Woodhead Publishing, Cambridge, 2011, p. 3.
39. J. R. Jones, *Acta Biomater.*, 2013, **9**, 4457.
40. A. Leu and J. K. Leach, *Pharm. Res.*, 2008, **25**, 1222.
41. A. Leu, S. M. Stieger, P. Dayton, K. W. Ferrara and J. K. Leach, *Tissue Eng., Part A*, 2009, **15**, 877.
42. D. Zhang, O. Leppäranta, E. Munukka, H. Ylänen, M. K. Viljanen, E. Eerola, M. Hupa and L. Hupa, *J. Biomed. Mater. Res., Part A*, 2010, **93**, 475.
43. M. N. Rahaman, B. S. Bal and W. Huang, *Mater. Sci. Eng., C*, 2014, **41**, 224.
44. I. Allan, H. Newman and M. Wilson, *Biomaterials*, 2001, **22**, 1683.
45. D. C. Greenspan, I. P. Zhong and G. P. La Torre, *Bioceramics*, 1994, **7**, 55.
46. O. Peitl, E. D. Zanotto and L. L. Hench, *J. Non-Cryst. Solids*, 2001, **292**, 115.
47. J. R. Jones, P. Sepulveda and L. L. Hench, *J. Biomed. Mater. Res.*, 2001, **58**, 720.
48. A. Ràmila and M. Vallet-Regì, *Biomaterials*, 2001, **22**, 2301.
49. M. Cerruti, D. Greenspan and K. Powers, *Biomaterials*, 2005, **26**, 1665.
50. M. G. Cerruti, D. Greenspan and K. Powers, *Biomaterials*, 2005, **26**, 4903.
51. D. Zhang, M. Hupa and L. Hupa, *Acta Biomater.*, 2008, **4**, 1498.
52. L. Drago, E. De Vecchi, M. Bortolin, M. Toscano, R. Mattina and C. L. Romanò, *Future Microbiol.*, 2015, **10**, 1293.
53. T. Pilizota and J. W. Shaevitz, *Biophys. J*, 2013, **104**, 2733.
54. S. Ran, Z. He and J. Liang, *Arch. Oral. Biol.*, 2013, **58**, 1667.
55. N. C. Lindfors, I. Koski, J. T. Heikkilä, K. Mattila and A. J. Aho, *J. Biomed. Mater. Res., Part B*, 2010, **94**, 157.
56. J. Frantzén, J. Rantakokko, H. T. Aro, J. Heinänen, S. Kajander, E. Gullichsen, E. Kotilainen and N. C. Lindfors, *J. Spinal Disord. Tech.*, 2011, **24**, 455.
57. J. Rantakokko, J. P. Frantzén, J. Heinänen, S. Kajander, E. Kotilainen, E. Gullichsen and N. C. Lindfors, *Scand. J. Surg.*, 2012, **101**, 66.
58. P. Stoor, E. Söderling and J. I. Salonen, *Acta Odontol. Scand.*, 1998, **56**, 161.
59. M. Zehnder, T. Waltimo, B. Sener and E. Söderling, *Oral Surg., Oral Med., Oral Pathol. Oral Radiol. Endod.*, 2006, **101**, 530.
60. T. Waltimo, T. J. Brunner, M. Vollenweider, W. J. Stark and M. Zehnder, *J. Dent. Res.*, 2007, **86**, 754.
61. L. Drago, D. Romanò, E. De Vecchi, C. Vassena, N. Logoluso, R. Mattina and C. L. Romanò, *BMC Infect. Dis.*, 2013, **13**, 584.

62. J. McAndrew, C. Efrimescu, E. Sheehan and D. Niall, *Iran. J. Med. Sci.*, 2013, **182**, 509.
63. C. Wu, Y. Zhou, M. Xu, P. Han, L. Chen, J. Chang and Y. Xiao, *Biomaterials*, 2013, **34**, 422.
64. Y. Y. Wang, X. Chatzistavrou, D. Faulk, S. Badylak, L. Zheng, S. Papagerakis, L. Ge, H. Liu and P. Papagerakis, *Eur. Cells Mater.*, 2015, **29**, 342.
65. V. Mortazavi, M. M. Nahrkhalaji, M. H. Fathi, S. B. Mousavi and B. N. Esfahani, *J. Biomed. Mater. Res., Part A*, 2010, **94**, 160.
66. R. Pérez-Tanoira, M. García-Pedrazuela, T. Hyyrynen, A. Soininen, A. Aarnisalo, M. T. Nieminen, V. M. Tiainen, Y. T. Konttinen and T. J. Kinnari, *J. Mater. Sci.: Mater. Med.*, 2015, **26**, 239.
67. M. Bortolin, E. De Vecchi, C. L. Romanò, M. Toscano, R. Mattina and L. Drago, *J. Antimicrob. Chemother.*, 2016, **71**, 123.
68. H. van de Belt, D. Neut, W. Schenk and J. Horn, *Acta Orthop. Scand.*, 2001, **72**, 557.
69. J. Gallo, M. Kolár, A. V. Florschütz, R. Novotný, R. Pantůcek and M. Kesselová, *Biomed. Pap.*, 2005, **149**, 153.
70. W. A. Jiranek, A. D. Hanssen and A. S. Greenwald, *J. Bone Jt. Surg., Am. Vol.*, 2006, **887**, 2487.
71. M. Shi, J. D. Kretlow, A. Nguyen, S. Young, L. S. Laggett, M. E. Wong, F. K. Kasper and A. G. Mikos, *Biomaterials*, 2010, **31**, 4146.
72. A. C. Matos, L. M. Gonçalves, P. Rijo, M. A. Vaz, A. J. Almeida and A. F. Bettencourt, *Mater. Sci. Eng., C*, 2014, **38**, 218.
73. B. Parsons and E. Strauss, *Am. J. Surg.*, 2004, **188**(1A Suppl), 57.
74. J. H. Calhoun, M. M. Manring and M. Shirtliff, *Semin. Plast Surg.*, 2009, **23**, 59.
75. M. Bassetti and E. Righi, *Hematology Am. Soc. Hematol. Educ. Program*, 2013, **2013**, 428.
76. L. Drago, E. De Vecchi, L. Nicola, A. Colombo and M. R. Gismondo, *J. Antimicrob. Chemother.*, 2004, **54**, 542.
77. G. K. Badave and D. Kulkarni, *J. Clin. Diagn. Res.*, 2015, **9**, DC08.
78. M. Bellantone, H. D. Williams and L. L. Hench, *Antimicrob. Agents Chemother.*, 2002, **46**, 1940.
79. S. Hu, J. Chang, M. Liu and C. Ning, *J. Mater. Sci.: Mater. Med.*, 2005, **20**, 281.
80. M. Vaahtio, E. Munukka, O. Leppäranta, D. Zhang, E. Eerola, H. Ylänen and T. Peltola, *Key Eng. Mater.*, 2006, **309–311**, 349.
81. M. C. McHendy, K. A. Easley and G. A. Locker, *Clin. Infect. Dis.*, 2002, **34**, 1342.
82. J. L. Stambough and D. Beringer, *J. Spinal. Disord.*, 1992, **5**, 277.
83. L. Labler, M. Keel, O. Trentz and M. Heinzelmann, *Eur. Spine J.*, 2006, **15**, 1388.
84. J. Kankare and N. Lindfors, *Scand. J. Surg.*, 2016, DOI: 10.1177/1457496915626834.
85. M. M. Echezarreta-Lopez and M. Landin, *Int. J. Pharm.*, 2013, **453**, 641.

86. M. Verhoeff, E. L. van der Veen, M. M. Rovers, E. A. M. Sanders and A. G. M. Schilder, *Int. J. Pediatr. Otorhinolaryngol.*, 2006, **70**(1), 1.
87. R. P. Mehta and J. P. Harris, *Otolaryngol. Clin. North Am.*, 2006, **39**(6), 1129.
88. M. Yung and A. Bennett, *Curr. Opin. Otolaryngol. Head Neck Surg.*, 2013, **21**(5), 455.
89. U. Mercke, *Am. J. Otol.*, 1987, **8**(6), 534.
90. J. B. Nadol, *Otolaryngol. Clin. North Am.*, 2006, **39**(4), 723.
91. M. Yung, P. Tassone, I. Moumoulidis and S. Vivekanandan, *J. Laryngol. Otol.*, 2011, **125**(3), 221.
92. F. H. Linthicum, *Laryngoscope*, 2002, **112**(10), 1777.
93. M. K. Kang, J. K. Ahn, T. W. Gu and C. S. Han, *Otolaryngol. Head Neck Surg.*, 2009, **140**(5), 687.
94. B. Black, *Am. J. Otol.*, 1998, **19**, 551.
95. M. Bagot d'Arc, G. Daculsi and N. Emam, *Ann. Otol., Rhinol., Laryngol.*, 2004, **113**(9), 711.
96. J. B. Roberson Jr, T. P. Mason and K. R. Stidham, *Otol. Neurotol.*, 2003, **24**(2), 132.
97. D. Bernardeschi, Y. Nguyen, I. Mosnier, M. Smail, E. Ferrary and O. Sterkers, *Eur. Arch. Otorhinolaryngol*, 2014, **271**(1), 59.
98. B. J. Gantz, E. P. Wilkinson and M. R. Hansen, *Laryngoscope*, 2005, **115**(10), 1734.
99. S. Mahendran and M. W. Yung, *Otol. Neurotol.*, 2004, **25**(1), 19.
100. J. S. Ridenour, D. S. Poe and D. W. Roberson., *Otolaryngol. Head Neck Surg.*, 2008, **139**(5), 641.
101. J. Sarin, R. Grénman, K. Aitasalo and J. Pulkkinen, *Ann. Otol., Rhinol., Laryngol.*, 2012, **121**(9), 563.
102. P. Stoor, J. Pulkkinen and R. Grénman, *Ann. Otol., Rhinol., Laryngol.*, 2010, **119**(6), 377.

CHAPTER 15

Bioactive Glasses for Soft Tissue Engineering Applications

VALENTINA MIGUEZ-PACHECO,[a] ALEJANDRO A. GORUSTOVICH,[b] ALDO R. BOCCACCINI[a] AND JUDITH A. ROETHER*[c]

[a] Institute of Biomaterials, University of Erlangen-Nuremberg, 91058 Erlangen, Germany; [b] Interdisciplinary Materials Group-IESIING-UCASAL, INTECIN UBA-CONICET, A4400EDD Salta, Argentina; [c] Institute of Polymer Materials, University of Erlangen-Nuremberg, 91058 Erlangen, Germany
*Email: Judith.roether@ww.uni-erlangen.de

15.1 Introduction

It is a well-known fact that bioactive glasses (BGs) bond very strongly to soft and hard tissues, nevertheless the interactions between BGs (as well as BG-containing composites) and mainly *hard* tissues have been widely investigated.[1–4] The pioneering work of Wilson *et al.* more than 30 years ago analysed the effect of bioactive glass particles on a wide range of soft tissue cells from various species and performed *in vivo* studies in rats and rabbits demonstrating for the first time that some bioactive glasses can bond to both soft tissues and hard tissues.[5] In recent years, increasing attention has been devoted to investigating the *in vitro* and *in vivo* performance of BGs in the context of soft tissue repair, including skin, nerve, intestinal, lung, laryngeal, urinary tract and cardiac tissue, as well as laryngeal and nasal

RSC Smart Materials No. 23
Bioactive Glasses: Fundamentals, Technology and Applications
Edited by Aldo R. Boccaccini, Delia S. Brauer and Leena Hupa

Published by the Royal Society of Chemistry, www.rsc.org

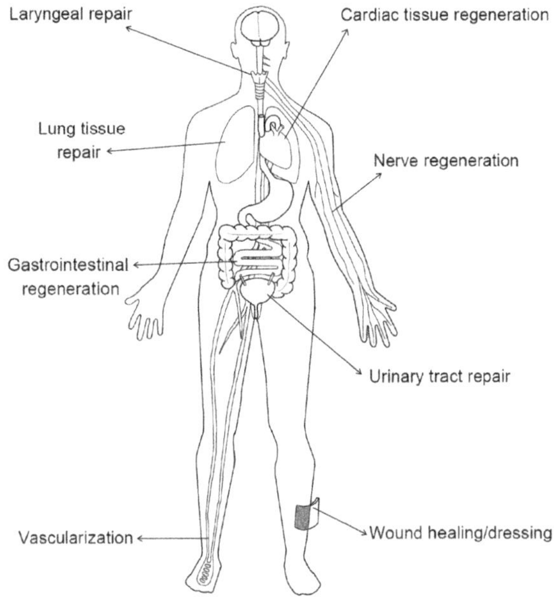

Figure 15.1 Main application areas of BGs for soft tissue engineering. Reproduced from ref. 10 with permission of Elsevier. Copyright © 2015 Elsevier.

ridge repair.[6–10] Figure 15.1 illustrates the main application areas of BGs for soft tissue engineering. The reasons for the increased interest in the use of bioactive glasses for soft tissue engineering applications investigated so far are on the one hand related to their angiogenic activity, which is a critical factor in tissue regeneration processes and on the other hand, due to the antibacterial properties of bioactive glasses which reduce the need for antibiotics to guard from infections during wound healing processes. Recent research has indicated that the release of ionic species from bioactive glasses is beneficial in a number of tissue repair processes, thus the doping with therapeutic ions (such as copper, zinc, gallium, *etc.*) can be strategically used to accelerate wound healing and tissue repair processes[11] (see also Chapter 16).

This chapter will give an overview of the main research efforts in the applications of BGs in soft tissue engineering, providing up-to-date coverage of the topic. The second section will cover the main mechanisms that arise after BG implantation *in vivo* with focus on the interaction of BGs with tissues. The third section of the chapter will focus on the latest research efforts in soft tissue repair. One important aspect of BGs is their angiogenic potential, which is of crucial importance for the repair of critical-size defects in both soft and hard tissue as vascularization of the graft is required. Experimental evidence in the context of soft tissues will be reviewed in the fourth section of this chapter, considering both *in vitro* and *in vivo* investigations, to illustrate that BGs can serve as angiogenic agents, inducing increased vascularization and contributing to the successful regeneration of new tissue. In the last section, the patents filed for the application of bioactive glasses for soft tissue repair will be summarized; these represent the potential translation of research results into biomedical products.

15.2 Interactions Between BGs and Soft Tissues

The successful binding of BGs and tissues relies primarily on a series of dissolution–precipitation reactions that occur on the surface of BGs when in contact with relevant body fluids, as summarized in Figure 15.2 for the case of bone. These reactions result in a change in the local concentration of several ions that leads to the formation of a reaction bilayer composed of an underlying hydrated silica layer and a hydroxyl-carbonated apatite (HCA) layer on top. This bilayer has a very high surface area, and its formation paves the way for the adsorption and desorption of biological moieties, crucial processes in the initial stages of tissue regeneration.[8] Of particular interest is the fact that during the formation of the HCA layer many biologically active molecules are bonded to it, most notably collagen type I (Coll I) produced by surrounding cell populations. Coll I is the most abundant type of collagen found in the body and is a crucial structural component of the extra cellular matrix (ECM) of many tissues. HCA crystallites and collagen fibres are strongly bonded chemically and this is a

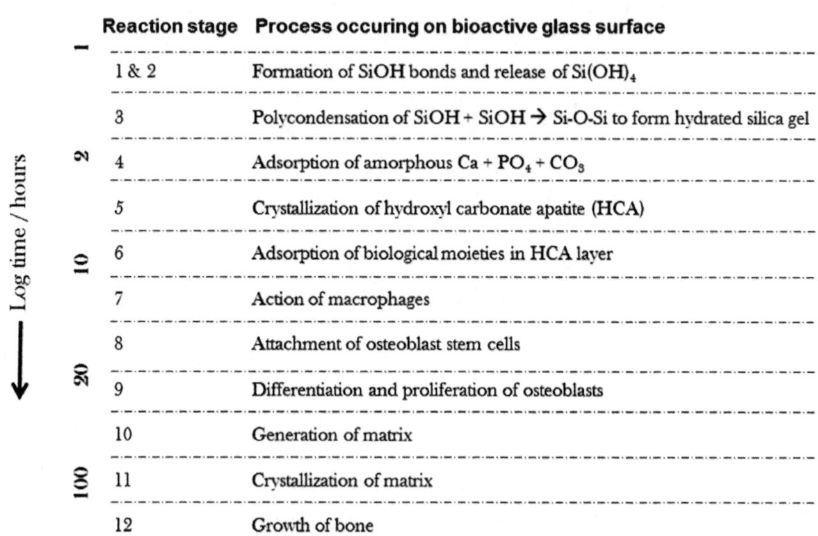

Figure 15.2 A sequence of interfacial reactions involved in forming a bond between tissue and bioactive glass with the focus on the BG–bone interface. Reproduced with permission from ref. 1 with permission from DeGruyter Open. Copyright © 2015 DeGruyter Open.

direct result of the series of reactions on the BG surface, the kinetics of which govern ECM protein adsorption and subsequent cell attachment.

Collagen fibrils present ideal sites for cell attachment via integrin binding, which is the main pathway of communication between intracellular and extracellular environments, and is crucial for controlling tissue regeneration through the regulation of cell proliferation, migration and differentiation, as well as other cellular processes.[6]

As mentioned above, the formation of an interfacial bilayer and subsequent adsorption of ECM molecules is a crucial process that allows for the successful bonding between tissues and BGs. While this has been highly investigated for the interaction of BGs with bone, the type of surface reactions described above are also relevant in the case of soft tissues/BG bonding. This bonding should be stable in the long term, avoiding micromotion which could result in the production of small particles that could cause an inflammatory reaction and eventual rejection of the implanted graft. Additionally, the layered structure on the surface of the BGs after implantation results in a graded stress transfer between the BG and tissue, which prevents the deterioration of the joints between implant and tissue due to stress shielding. Moreover, it helps to avoid fibrous tissue encapsulation of the implant, which can sometimes lead to implant failure. The high surface area and gradient stress factor across the reaction bilayer results in an even spread of the difference in elastic moduli between the bulk BG and the soft tissue and leads to a favourable stress transfer from the bonding of collagen fibrils. This successful stress transfer is considered to be responsible for the short- and long-term success of BG implants.

The specific applications for BGs in contact with soft tissues have been extensively covered in recent reviews[9,10] but this promising area of research is ever expanding and so the present chapter includes a thorough review of recent findings complementing previous publications[9,10] to provide an up-to-date overview of the field.

15.3 Applications of Bioactive Glasses in Soft Tissue Engineering

In recent years, an increasing number of research teams have started investigating the applicability of bioactive glasses for soft tissue engineering (Figure 15.1). In particular, wound healing and nerve regeneration have been the focus of research but, as detailed in the following sections, there are isolated studies that show the great potential bioactive glasses offer in other applications in soft tissue engineering.

15.3.1 Skin Repair and Wound Healing

The healing of a wound involves the complex interaction between a number of elements including cytokines, growth factors, proteases, eicosanoids, kinins, nitric oxide and cellular elements among others.[11] Wound healing is known to occur in four overlapping stages, *i.e.* hemostasis, inflammation, proliferation, and maturation, and is governed by a variety of local and systemic factors.[12] Bioactive glasses are being investigated as a means of aiding hemostasis and preventing wound infection. A composition based on 45S5 Bioglass® is marketed in China under the tradename DERMGLAS™ to prevent ulceration of the skin and accelerate tissue repair.[13]

In general, the early stages of wound closure can be accelerated using BGs, which reduces the risk of infections. Some diseases and increasing age of the patient lead to slower or disrupted wound healing capacity. Zhou *et al.* investigated micron-sized fibres produced by a traditional melting route of 45S5 BG and borate glasses, which were used as wound gauzes for treating full thickness wounds in a rat model.[13] The authors demonstrated that the presence of boron has a profound effect as it led to increased blood vessel formation and resulted in faster reduction in wound size compared to 45S5 BG or the control. In normal conditions, such approaches can lead to accelerated wound healing in the initial stages, which reduces the possibility of infection. However, if wound healing is compromised by conditions such as diabetes, cancer or others, wounds tend not to heal in a timely manner and can enter a state of chronic inflammation and become infected. There are several potential uses for BG in the repair of skin wounds where wound healing might be compromised.

One important aspect in this regard is the antibacterial activity of BGs as well as their hemostatic activity. In this respect, the treatment of those injured in a war zone with hemostatic agents needs to be considered where severe blood loss can lead to infections and even morbidity. A hemostatic agent is defined as one which, upon contact with blood and oxygen, will promote or accelerate blood coagulation thus stemming hemorrhages. A recent study considered the wound healing effect of borate- and silicate-based BGs comparing the relative efficacy *in vitro* of 3 different formulations including 45S5 BG, borate-based 1393B3 and 1605 BG, which contains boron (B) but is also doped with copper (Cu) and zinc (Zn) oxides.[14] The results of the study indicate that the presence of B might lead to increased cell growth rates, while the presence of Zn and Cu affects glass conversion, biocompatibility and bioactivity.

Borate-based fibres which were doped with between 0 and 3 wt% of Cu were investigated by Zhao *et al.* with respect to their effect on human umbilical vein endothelial cells (HUVECs) and fibroblasts.[15] All compositions tested were found to increase proliferation of fibroblasts and HUVECs as well as promoting HUVEC migration, tubule formation and secretion of vascular endothelial growth factor (VEGF). When the Cu-doped glasses were used on full-thickness skin defects in rodents they showed significantly increased capacity to stimulate angiogenesis in comparison to undoped fibres and to the untreated defects (control) at 7 and 14 days post-surgery.[15]

Silver- and fluoride-containing BGs fabricated using a sol–gel route have been studied for treating burns, in particular the effect of the glasses on multi-resistant bacterial strains isolated from patients with burns.[16] The results of this study showed that fluoride (F) was not effective in reducing the bacterial populations used in the work (namely *K. pneumoniae, S. aureus, E. coli*, and *P. aeruginosa*) while the addition of 1 wt% of silver (Ag) ions to the BG inhibited bacterial growth without being toxic to fibroblasts. Higher concentrations of Ag ions were found to have a cytotoxic effect.[16]

Mesoporous BG scaffolds containing varying concentrations of Ga_2O_3 fabricated by evaporation-induced self-assembly were studied to investigate antibacterial activity and hemostasis.[17] The results of this study showed that addition of 1% of gallium (Ga) led to improved stuctural properties such as mesopore size, surface area, pore volume and calcium and silicon (Si) ion release. The addition of Ga also activated blood coagulation and thrombus generation slightly, led to increased platelet adhesion, and induced a significantly higher antibacterial activity against *S. aureus* and *E. coli* compared with mesoporous BG without Ga.[17]

The reason for the increased fibroblast proliferation on BGs was recently investigated by a team of researchers who developed a fibroblast sheet, activated by BG, which was then applied to a full skin lesion of a nude mouse. It was found that the migration ability of fibroblasts could be stimulated by BG, which directly affected the differentiation of fibroblasts into myofibroblasts.[18]

Another recently published study developed a multifunctional integrated strategy for dermal tissue reconstruction to treat deep wounds.[19] The scaffold consisted of a porous BG–chitosan (CHI)–silk fibroin (SF) composite, which was prepared by freeze-drying a SF solution containing BG and CHI. Biocompatibility and capability for tissue repair were investigated both *in vitro* and *in vivo*. A vascularisation evaluation indicated that the scaffold possessed the ability to support newly formed and mature blood vessels, which may lead to accelerated wound healing.[19]

It should be pointed out that the addition of BGs in any form to a polymer matrix forming organic–inorganic composites has been shown to lead to a beneficial development in wound management as the inorganic particles possess a buffering effect, hence wound healing can be improved compared with pure polymer wound dressings, which have often been associated with inflammation, blisters, scar hyperplasia and severe wound contraction due to acidic byproducts of their degradation, hydrophobicity and lack of bioactivity of the polymer.[20]

15.3.2 Nerve Tissue Regeneration

Damage to the sensory or motor nerves as a result of disease or trauma is severely debilitating to patients and can be very difficult to treat. Peripheral neuropathies can be caused by trauma and conditions such as diabetes, kidney disorders, hypothyroidism, rheumatic disorders, *etc.* Nerves damaged by trauma are often treated using autografts or allografts, however the setbacks associated with this approach, most notably due to limited availability as well as donor site morbidity, have been well documented and hence only small gaps of 30 mm or less can be successfully treated.[20]

Using tissue engineering strategies, a range of electrically conductive and non-conductive materials are being considered as well as the addition of biomolecular cues to successfully treat damaged nerves using tubular scaffolds or nerve guidance conduits.[21] However, some limitations of this approach in terms of poor functionality of the repaired tissue have been reported.[22] Recently, the use of microsized phosphate glass fibres was introduced and has been hypothesized to be able to direct neurite growth.[23] Phosphate glass fibres were placed on a collagen sheet which was then manually rolled up to form a nerve conduit. *In vivo* experiments using 30 mm scaffolds with fibre-filled and unfilled collagen sheets on transected sciatic nerves in rats showed that the fibre-filled conduits exhibited a large number of axons along the length of the scaffold whereas the collagen tube contained none. However, the functional capacity of both conduits was similar, hence the role of phosphate glass fibres was limited.[23]

Composite conduits consisting of nanosized bioactive glass (nBG) and gelatin have been investigated by Koudehi *et al.*[24] The conduits were fabricated by dipping mandrels into an aqueous gelatin–nBG solution (70 : 30 wt%), which were subsequently subjected to a freeze-drying stage to fabricate a porous structure, which was crosslinked using gluteraldehyde. Both *in vitro* and *in vivo* studies using male Wistar rats indicated biocompatibility of the developed structures and an increased rate of growth of myelinated axon fibres.[24] In order to develop a material that may be used to promote axon regeneration as a nerve guidance conduit, three BG–polymer sheet composites were generated.[25] The composite materials were of 50 : 50 wt% poly-(ε-caprolactone) (PCL) and either 45S5 BG microparticles, 1393B3 borate glass microparticles, or a 25 : 25 blend of 45S5 and 1393B3 microparticles. Table 15.1 shows the chemical composition of the glasses used. To evaluate the effect that the addition of BG particles (<20 μm) to PCL had on neurite extension, the whole dorsal root ganglion (DRG) neurons outgrowth was studied on the composite sheets and compared to commonly used 100% PCL. None of the added glasses negatively affected neurite extension compared to PCL alone. In fact, the neurites growing on the 45S5 BG-containing composite showed a significant increase in length. According

Table 15.1 Selected bioactive glass compositions considered for soft tissue engineering applications.

BG denomination	BG composition (mol%)
45S5	46.1 SiO_2, 26.9 CaO, 24.4 Na_2O, 2.5 P_2O_5
58S	60 SiO_2, 36 CaO, 4 P_2O_5
70S30C	70 SiO_2, 30 CaO
S53P4	53 SiO_2, 23 Na_2O, 20 CaO, 4 P_2O_5
1393	6 Na_2O, 8 K_2O, 8 MgO, 22 CaO, 2 P_2O_5, 54 SiO_2
1393B3	6 Na_2O, 7.9 K_2O, 7.7 MgO, 22.1 CaO, 1.7 P_2O_5, 54.6 B_2O_3
1605	6.5 Na_2O, 8.5 K_2O, 8.3 MgO, 23.9 CaO, 0.8 ZnO, 1.9 P_2O_5, 0.33 CuO, 49.7 B_2O_3

to previous reports, the addition of 45S5 glass particles significantly improved regeneration in an *in vivo* sciatic nerve injury model.[26] Interestingly however, when composite sheets were pre-reacted for 24 h prior to DRG seeding, the positive effect seen with 45S5 particles was decreased to normal levels while neurite extension increased when cultured on bioactive 1393B3 borate glass composites. It might be hypothesized that the conversion to HA leads to a more growth-permissive substrate, either through improved adhesion or by some as-yet-unknown mechanism.

A recent study by Ahn *et al.*[27] investigated the feasibility of using aligned phosphate glass fibres which had aminated carbon nanotubes (CNTs) chemically tethered onto the surface as nerve conduits, which were then placed into poly(DL-lactic acid) tubes. Phosphate glass fibres and CNT–phosphate glass fibres were implanted into a 10 mm gap of a rat sciatic nerve for sixteen weeks. The number of regenerating axons along the scaffold was determined, as well as the cross-sectional area of re-innervated muscles. The presence of CNTs was found to significantly improve the electrophysiological findings. The *in vivo* study showed that CNT-interfaced phosphate glass fibre scaffold could be of benefit in the regeneration of a transected rat sciatic nerve at the interface between the nerve conduit and peripheral neural tissues.[27] Using an electroconductive material, such as CNT, might thus offer some advantages.

Various biomedical applications of Zn-containing BGs in the context of tissue regeneration have been proposed in the literature, as comprehensively discussed elsewhere,[28] highlighting the potential of Zn incorporation for the development of different products such as scaffolds for bone TE, bone filling granules, bone cements and coatings for implants. To verify the compatibility of rigid supports with neuronal cells for biomechanical application, Sabbatini *et al.*[29] have analyzed the biocompatibility profile of undifferentiated and differentiated neuronal cell line SKNBE on 45S5 BG doped with different quantities of ZnO (5%, 10% and 20%; wt%). BGs doped with a low concentration of Zn favor cell adhesion and proliferation of undifferentiated SKNBE neuronal cells, while high Zn concentration exerted an inhibiting action.[29] The discussed studies in this section add to the emerging number of *in vitro* and *in vivo* results which confirm nerve repair and regeneration as one of the areas of promising applications of BGs in their silicate, phophate and borate compositions.

15.3.3 Other Applications

A number of other applications of BGs or BG-containing composites are emerging in soft tissue engineering apart from those mentioned in the previous sections. Such novel appliation fields for BGs include muscle, urinary tract, and intestinal tissue among others.[9,10] Two areas of research recently explored are in ophthalmology and nasal ridge repair, which are briefly discussed in this section.

(i) Oculo-orbital fracture restoration and other applications related to ophthalmology

Bioactive glasses have been considered as ocular implants, as artificial cornea and retina as well as more recently orbital implants.[30,31] A few studies have investigated the usage of BGs as well as bioactive glass-ceramics for the restoration of orbital floor fractures. For the most part, the use of slow-resorbing glass-ceramics such as Ceravital® has raised concerns due to the relatively slow resorbability of the material which could lead to suboptimal tissue healing, and hence research into this type of bioactive material is being discontinued.[31]

A recent study by Stoor *et al.*[32] considered the application of anatomically drop-shaped BG devices of S53P4 composition (see Table 15.1) which were placed in 20 patients who had either suffered an isolated blow-out fracture of the orbital floor or a combined zytomatico-orbito-maxillary complex fracture. The results of the long term study spanning almost 3 years were positive with no sign of major complications or resorption of the implant over the time period. Studies have also indicated that porous bioceramics can stimulate proliferation and adhesion of keratocytes, and hence could be exploited as skirt keratoprostheses.[31,33] Mesoporous BG with a drug-release capacity could offer new possibilities in the treatment of ocular tumors. Recent detailed reviews on the possibilities of bioactive glasses and ceramics in this particular medical area have been published.[31,33]

(ii) Nasal ridge perforation repair

Nasal septum repair surgery is the most likely method to be used in cases of perforations that fall beyond a critical size range, thus impeding spontaneous full closure of the wound. These surgeries require the use of different surgical techniques and materials, known as obturators, to bridge the tissue gap and onto which the nasal mucosa can grow.[34] Silicone and acrylic acid were amongst the first materials used to close nasal septum perforations, but they were found to be poorly tolerated over long periods of time and approaches were then focused on using tissue grafts, whether autologous, allogenic or xenogeneic. These approaches presented much more satisfactory results and have been shown to lead to more consistent healing rates. However, they also present drawbacks, namely donor site morbidity, graft resorption, risk of graft rejection and the spread of infections.[35]

Bioactive glasses (BGs) have been used to reconstruct facial bone defects and in the treatment of frontal sinusitis and atropic rhinitis, and more recently to repair nasal septal perforations. In a study by P. Stoor *et al.*,[36] a group of 39 patients were operated upon, of which 23 with medium-sized (10–20 mm) injuries were treated using BG S53P4 disks as an interpositional graft in conjunction with turbinate flaps; the rest of the patients[16] with small-sized (<10 mm) injuries were treated with mucoperiosteal flaps only. The perforations were found to be completely healed in 22 of the 23 patients treated with BG disks and in all of the patients treated without interpositional grafts. It is worth noting, however, that 2 patients from the latter group required secondary surgeries where BG disks were used to achieve full closure of the defect. The authors conclude that BG disks were well tolerated and are suitable as interpositional implants for the treatment of medium to large septal defects which would otherwise present complications during healing. The BG composition used in this study was used in another similar study by Stoor *et al.*[37] In this case, in addition to repairing septal defects in human patients, S53P4 BG disks and granules were subjected to growth inhibition and adhesion tests, respectively, of bacteria normally found in human airways. These *in vitro* studies showed that BG did not favour bacterial attachment and growth and, as such, posed no significant risks when used as interpositional implants. Following surgical implantation in 11 patients with different septal defects, it was found that the implants were well tolerated and the perforations were sealed satisfactorily in 10 out of 11 patients.

15.4 Angiogenic Activity of Bioactive Glasses

15.4.1 *In vitro* Evidence in Bone and Soft Tissues

Previous *in vitro* studies have demonstrated that increases in angiogenic indicators have been achieved through both direct and indirect contact of relevant cells with BG particles or scaffolds and with their ionic dissolution products (IDPs).[38–65] This section will review some key relevant studies in this important field of application of BGs.

Day *et al.*[38,40] showed a higher concentration of vascular endothelial growth factor (VEGF) in the culture medium of human (CCD-18Co) and rat fibroblasts (208F) grown on surfaces coated with colloidal suspensions (0.01 and 0.1 w/v) of particles (<5 μm) of 45S5 Bioglass®. Similar results were observed when mouse fibroblasts (L929) were cultured on the surfaces of discs from a composite material based on PLGA with 0.01, 0.1, and 1% w/v particles (<5 μm) of 45S5 Bioglass®.[39] On the other hand, Keshaw *et al.*[41,42] quantified higher VEGF secretion by human fibroblasts (CCD-18Co) either grown in the presence of microporous spheres of PLGA or encapsulated in alginate beads with 10% (w/w) or 0.01% and 0.1% (w/v) of particles (~4 μm) of 45S5 Bioglass®, respectively. The authors of the studies observed an increase in the proliferation of human dermal microvascular endothelial cells (HDMEC) treated with culture medium from the incubation of human fibroblasts (CCD-18Co) encapsulated in alginate beads with 0.1% (w/v) of glass particles (of ~4 μm in size) of 45S5 Bioglass®.[41] These results are consistent with those of Day *et al.*,[40] who described an increase of 61.5% in the number of HDMEC after exposure for 24 h to a culture medium obtained from the incubation of human fibroblasts (CCD-18Co) grown on polystyrene plates coated with 0.1% (w/v) of microparticles (<5 μm) of 45S5 Bioglass®. In addition, Gerhardt *et al.*[43] found that human fibroblasts (CCD-18Co) grown on films of poly(DL-lactic acid) (PDLLA) filled with 5% and 20% of nanoparticles (35–40 nm) or 20% of microparticles (0.1–25 μm) of 45S5 Bioglass® secrete up to 5 times more VEGF than those grown on PDLLA films without bioactive glass particles.

Regarding the effects of 45S5 Bioglass® on the secretion of growth factors with pro-angiogenic activity, previous studies have shown an increase in the *in vitro* production of basic fibroblast growth factor (bFGF) in human fibroblasts (CCD-18Co) grown on polystyrene plates coated with 0.1–2% (w/v) of microparticles (<5 μm) of 45S5 Bioglass®.[40,44]

Leach *et al.*[45] showed that scaffolds based on poly(lactic-*co*-glycolic acid) (PLGA) coated with 0.5 mg of microparticles of 45S5 Bioglass® can stimulate the *in vitro* proliferation of human microvascular endothelial cells (HMVEC). Leu and Leach[46] also described an increase in cell proliferation and VEGF production in HMVEC exposed to IDPs released from collagen sponges containing 0.12 and 1.2 mg of 45S5 Bioglass® microparticles in their structure. Similarly, Deb *et al.*[47] observed an increase in the proliferation of human umbilical vein endothelial cells (HUVECs) directly cultured on scaffolds composed of 45S5 Bioglass®. In accordance with these studies, Handel *et al.* showed that minimal medium containing 1% of crushed 45S5 Bioglass®-based scaffolds induced a significant proliferation of HUVECs.[48]

In a recent study, 45S5 BG microparticles (90% <34.86 μm) were incorporated into alginate (ALG) hydrogel to prepare bioactive BG–ALG composite hydrogel beads with different BG content (0, 0.1, 0.2, or 0.4% w/v) as cell carriers for bone regeneration and tissue engineering applications.[49]

BG–ALG beads showed a favorable *in vitro* bioactivity by inducing apatite formation on the surface of the composite hydrogels. In addition, ionic extracts of 0.2% BG–ALG beads stimulated proliferation and differentiation of rat bone marrow mesenchymal stem cells (rBMSCs) as well as angiogenesis of HUVECs. Furthermore, BG–ALG beads enhanced the cell proliferation and stimulated osteogenic differentiation of the encapsulated MC3T3-E1 cells as compared with ALG beads. The positive regulation effect of BG–ALG bead extracts on cells may be caused by silicon ions (0.2–0.9 mmol L^{-1}), since silicon ions were only released by the BG-ALG samples and not by the pore ALG beads.[49]

The *in vitro* ability of BG S53P4 (the bioactive glass BonAlive™, see Table 15.1 for composition) of three different size ranges to stimulate the expression and secretion of the angiogenic growth factor VEGF from human fibroblasts has also been investigated.[50] All granules were seen to be highly cytocompatible, as assessed by cell morphology, LHD activity, and mitochondrial activity. Both the size of the granules and the glass concentration affected the secretion of VEGF. The VEGF expression increased with BG concentration for small granules (0.5–0.8 mm) and decreased for larger granule sizes (2.0–3.15 mm). The highest VEGF expression was observed for granules of 1–2 mm at a concentration of 0.1% w/v. The S53P4 BG granule sizes used in this study are in clinical use, which indicates the clinical relevance of the results.[50]

It has been established that boron (B) may play an important part in angiogenesis as well as osteogenesis, hence the controlled and localized release of B ions from BGs is expected to provide a promising therapeutic alternative for regenerative medicine of vascularized tissues. In this context, Haro Durand *et al.* found that HUVECs possess greater proliferative and migratory response and greater ability to form *in vitro* tubules and to secrete pro-angiogenic cytokines (IL-6 and bFGF) when stimulated with the ionic dissolution products from microparticles of 45S5 BG enriched with 2 wt% B_2O_3 (Figure 15.3).[51]

The pro-angiogenic action of Cu-doped BG microparticles has also been reported.[52] Controlled copper ion release from phosphate-based glasses (PGs) improved HUVECs survival in a reduced nutrient environment (RNE) with growth factor concentrations which approximate conditions in the core of large volume scaffolds where nutrient diffusion is limited.[52] HUVECs death induced by the RNE was delayed by an optimal dose of 4 ppm copper ions released from the 10 mol% CuO PG composition, suggesting the potential of these materials to extend the survival of transplanted endothelial cells in large volume scaffolds, which was associated with a down-regulation of apoptosis.[52] However, another study that examined the formation of capillary-like networks by SVEC4-10 endothelial cells seeded in a three-dimensional type I collagen hydrogel matrix mixed with PG particles showed that the network length was initially reduced in a concentration-dependent fashion by 10% CuO PG and, to a lesser extent, by 0% CuO PG, but reached values identical to the non-PG control by day 5 in culture.[53] This reduction was attributed to a PG-mediated decrease in cell metabolic activity while cell proliferation, as well as network connectivity and branching, were independent of PG content.[53]

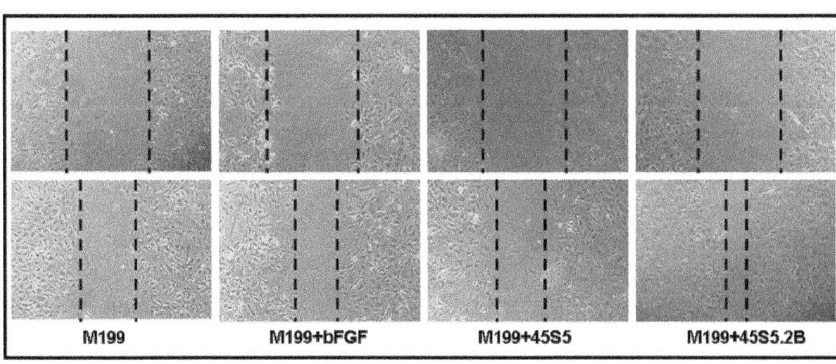

Figure 15.3 Endothelial wound healing assay. Note the increased migratory response of HUVECs when stimulated with ionic dissolution products from microparticles of 45S5 BG enriched with 2 wt% B_2O_3 (M199 + 45S5.2B). (Reproduced from ref. 51 with permission of Royal Society of Chemistry, copyright 2014 Royal Society of Chemistry).

Stähli *et al.*[54] have also examined the effect of copper-doped 45S5 BG on 3D capillary-like network formation by SVEC4-10 endothelial cells in type I collagen matrices.[54] Higher amounts of silicon release and calcium consumption resulting from both BG and Cu–BG dissolution were associated with a reduced endothelial cells network length, connectivity and branching as well as proliferation and metabolic activity, whereas a Cu^{2+} ion dose equivalent to the highest Cu–BG concentration (6.4 mg mL^{-1} Cu–BG) had no effect on endothelial cells. In addition, the lowest Cu–BG concentration investigated in their study (0.4 mg mL^{-1}) exhibited no adverse effects on endothelial cells.[54]

In vitro cell culture have also shown that the ionic dissolution products of borate bioactive glass (1393B3) microfibres doped with varying amounts of Cu (0.5–3 wt% CuO) were not toxic to HUVECs and fibroblasts, promoted HUVECs migration, tubule formation and secretion of VEGF, and stimulated the expression of angiogenic-related genes of the fibroblasts to a higher extent than the cells cultured in the ionic dissolution products of the undoped microfibres.[15] These results are consistent with those obtained using Cu-doped borate-based 1605 fibres (see Table 15.1 for composition).[14] Evaluation on human skin fibroblast cells (CCL-110) demonstrated that pre-soaked Cu-doped borate-based 1605 fibres can significantly stimulate cell growth, as evidenced by a higher cell proliferation rate and migration ability under a dynamic flow rate with a continuous supply of fresh media.[14]

Novel nanocomposite biomaterials made of 45S5 BG nanoparticles (nBG) and alginate cross-linked with Cu^{2+} or Ca^{2+} were developed with potential applications for the preparation of multifunctional scaffolds for bone tissue engineering.[55] *In vitro* results indicated that the bioactive ions released from both nanocomposite biomaterials were able to stimulate the differentiation of rat bone marrow-derived mesenchymal stem cells towards the osteogenic lineage. In addition, the typical endothelial cell property of forming tubes in Matrigel was observed for HUVECs when in contact with the novel biomaterials, particularly the nanocomposite made of nBG and alginate cross-linked with Cu^{2+}, which indicates their angiogenic properties.[55]

It has already been reported that Cu^{2+}-doped 45S5 BG scaffolds exhibit high acellular bioactivity as proven by rapid formation (after 3 days of immersion in SBF) of a carbonated HA layer on BG scaffold surfaces.[56] Moreover, the released Cu^{2+} levels in SBF are within the therapeutic range, indicating a potential angiogenic effect of such Cu^{2+}-releasing scaffolds.[56] In another study, Rath *et al.*[57] investigated the *in vitro* biocompatibility and bioactivity of Cu^{2+}-doped 45S5 BG-derived scaffolds in either BMSC-only culture (bone marrow-derived mesenchymal stem cells) or co-culture of BMSC and human dermal microvascular endothelial cells (HDMEC). Results from their study have shown that Cu^{2+} contents of 0.1 wt% and 1 wt% in the BG scaffold act on BMSCs to facilitate high VEGF secretion into the medium. Additionally, the scaffolds showed no toxicity to BMSCs and there were healthy live cells found throughout the scaffold. The secreted VEGF and the Cu^{2+} ions have an influence on the functional properties of HDMECs. The HDMECs retained their surface antigens such as CD 31, vWF, and VEGFR2 (KDR, VEGF receptor 2) and produced tube-like structures even in cell culture only in the presence of Cu^{2+} and BMSCs.[57] This approach is advantageous for the successful application of the scaffold in bone regenerative medicine, since it enables controlled VEGF release mediated by cells, which can be adopted physiologically to the required local conditions, avoiding growth factor overdose. Hence, Cu^{2+}-doped 45S5 BG scaffolds in combination with BMSCs were represented as superior candidates for bone tissue engineering applications with enhanced angiogenic potential.[57] Another study was reported by Wu *et al.*,[59] in which Cu^{2+}-doped (at 0, 1, 2 and 5 mol%) mesoporous BG scaffolds were investigated, having combined angiogenic and osteo-stimulative properties. The authors explained the results of increased angiogenesis attributed to a hypoxia-like tissue reaction due to which both the scaffolds and their ionic extracts could stimulate hypoxia-inducible factor (HIF)-1α and vascular endothelial growth factor (VEGF) expression in human bone marrow stromal cells (hBMSCs). It was also shown that Cu^{2+} has an added osteogenic ability only at high concentrations. However, with the highest Cu^{2+} concentration, there was increased cytotoxicity and hence decreased cell viability.[59]

In addition to Cu^{2+}, cobalt (Co^{2+}) ions are also known to stimulate angiogenesis via inducing hypoxic conditions and hence are interesting agents to be used in conjunction with BGs in bone tissue engineering applications.[60–65] Recently, Hoppe *et al.*[63] reported the biocompatibility of Co-releasing 13-93 BG composition (for composition see Table 15.1) and BG-derived scaffolds containing 1 wt% CoO towards osteoblast-like cells (MG-63) and human dermal micro endothelial cells (hDMECs) while addition of 5 wt% of CoO was cytotoxic to both cells types.[63] Similarly, Azevedo *et al.*[64] reported that human mesenchymal stem cell viability was unaffected by the hypoxia-mimicking BG except when the cells were treated with the BG containing the highest Co^{2+} concentration. It was also shown that several genes crucial for angiogenesis were overexpressed, including VEGF at both the transcript and protein levels, in a concentration-dependent manner. This simple strategy to stabilize the hypoxia-sensing transcription factor (HIF-1a) through the controlled release of ions from BGs is able to induce the activation of a variety of genes crucial for tissue regeneration and could be of great potential for use in bone and soft tissue regenerative therapies.[64] Another study confirmed that a cobalt bioactive glass–collagen glycosaminoglycan scaffold-conditioned medium significantly enhanced the *in vitro* osteoblast cell proliferation, tubule formation, production and expression of VEGF in endothelial cells (HUVECs).[65]

15.4.2 *In vivo* Evidence in Bone and Soft Tissues

An increasing number of previous and recent investigations demonstrate that BGs may improve *in vivo* vascularization.[38,43,45,66–74] For example, an early study has shown that a biodegradable polyglycolic acid (PGA) mesh coated with 45S5 Bioglass® particles implanted subcutaneously into rats became infiltrated by a significantly increased number of blood vessels compared with uncoated control scaffolds.[38] More recently, the angiogenic properties of micron-sized (mBG) and nano-sized (nBG) 45S5 BG-filled poly(D,L-lactic acid)(PDLLA) composites were investigated in a rat model.[43] After 8 weeks of implantation, mBG- and nBG-containing scaffolds were well infiltrated with newly formed tissue and demonstrated higher vascularization and percentages of blood vessel to tissue than neat PDLLA scaffolds.[43] In accordance with previous results, 10 wt% nBG-containing collagen films stimulated angiogenesis in the quail chorioallantoic membrane (CAM) by increasing by 41% the number of blood vessel branch points. In contrast, composite films containing 20 wt% nBG were found to inhibit angiogenesis.[66] However, these results do not agree with those by Handel *et al.* who showed that in their native, non-biofunctionalized state, neither Bioglass®-based nor biologically inert fibrous polypropylene control scaffolds showed angiogenic properties.[67] However, significant vascularization was induced by human adipose tissue-derived stem cells (hASC)-seeded scaffolds (Bioglass® and polypropylene) in the chick CAM angiogenesis assay.[67]

Related *in vivo* experiments showed that approximately 0.5 mg of 45S5 Bioglass® particles coated on a VEGF-releasing PLGA porous scaffold were capable of enhancing neovascularization in a critical-sized cranial bone defect in rats.[45] In addition, greater neovascularization and bone regeneration were found in irradiated critical-sized calvarial defects filled with collagen sponges loaded with 1.2 mg of Bioglass® in comparison to controls at 2 weeks post-implantation in rats.[68] These results, showing a positive angiogenic response to 45SS BG, are consistent with those of Arkudas *et al.* for the implantation of sintered 45S5 Bioglass® granula matrix in a rat arteriovenous loop (AVL) model.[69] At 3 weeks, all constructs were filled with cell-rich, highly vascularized connective tissue around the vascular axis (Figure 15.4).[69]

In addition to 45S5 Bioglass®, BGs and silicate glass-ceramics of other compositions are under investigation for their ability to promote angiogenesis.[14,15,50–54,56–65,67–74] It has been described that the ionic dissolution products from a BG in the 45S5 system doped with 2 wt% B_2O_3 increased angiogenesis, quantitatively evidenced by the greater expression of integrin $\alpha_v\beta_3$ and higher vascular density in the embryonic quail chorioallantoic membrane (CAM).[70] The response observed at 2 and 5 days post-treatment was equivalent to that achieved by applying 10 $\mu g\,mL^{-1}$ of basic fibroblast growth factor.[70]

Furthermore, *in vivo* results revealed that 3-D printed Sr-containing mesoporous bioactive glass scaffolds had osteogenic capability and stimulated new blood vessel formation in critical-sized rat calvarial defects within 8 weeks.[71]

Figure 15.4 The arteriovenous loop placed in a Teflon isolation chamber filled with sintered 45S5 Bioglass® granulated matrix and fibrin gel with a fibrinogen concentration of 10 mg mL^{-1} and a thrombin concentration of 2 I.U. mL^{-1}. (A) arteriovenous loop (AVL) placed on the first half of the matrix; (B) chamber filled with the complex matrix; (C) dense network of newly formed vessels originating from the AVL as demonstrated by microcomputed tomography (Reproduced from ref. 69 with permission of Mary Ann Liebert Inc. Copyright 2013 Mary Ann Liebert, Inc.).

Several studies indicate that Cu-doped borate glass microfibres have a promising capacity to stimulate angiogenesis.[15,72] When used to treat full-thickness skin defects in rodents, the wound dressings composed of borate bioactive glass (13-93B3) microfibres doped with 3 wt% CuO showed a significantly better capacity to stimulate angiogenesis than the undoped fibres and the untreated defects (control) at 7 and 14 days post-surgery. The defects treated with the Cu-doped and undoped fibres showed improved collagen deposition, maturity and orientation when compared to the untreated defects, the improvement shown by the Cu-doped fibres was not markedly better than the undoped fibres at 14 days post-surgery.[15] These results are consistent with those of Lin *et al.*[72] showing that subcutaneous implantation of Cu-containing borate bioactive glass (13-93B3) microfibres in rats significantly enhanced the growth of capillaries and small blood vessels when compared to silicate 45S5 bioactive glass microfibres and sham implant controls.

A recent study provided additional evidence that Cu-doped bioactive borate glasses promote angiogenesis.[73] Quantitative histomorphometry of subcutaneous tissue implanted with plugs of 45S5, 13-93B3, and 13-93B3Cu glass microfibres (composition shown in Table 15.1) revealed microvascular densities that were 1.6, 2.3, and 2.7 times higher, respectively, than the sham control values whereas 13-93, 13-93B3, and 13-93B3Cu glass beads caused the microvascular density to increase 1.3, 1.6, and 2.5 fold, respectively, relative to sham controls. Moreover, a marginally significant increased expression of VEGF mRNA in tissues with 13-93B3Cu glass beads was observed, an outcome that supported the hypothesis that copper-doped borate glass could promote VEGF expression followed by angiogenesis for enhanced wound healing.[73] It has already been reported that, when implanted in rat calvarial defects *in vivo*, porous scaffolds of a borosilicate bioactive glass (composition 6 Na_2O, 8 K_2O, 8 MgO, 22 CaO, 36 B_2O_3, 18 SiO_2, 2 P_2O_5; mol%) doped with 3 wt% CuO significantly enhanced both blood vessel formation and bone regeneration in the defects at 8 weeks post-implantation.[58] These results show that doping BGs with Cu is a promising approach for enhancing angiogenesis and osteogenesis in the healing of osseous defects.[58,74]

Collectively, these findings suggest that BGs are effective vehicles for delivery of pro-angiogenic ions as an alternative to expensive growth factors for bone and soft tissue repair or regeneration. Chapter 16 of this book further discusses the effect of different biologically active ions added to BGs.

15.5 Existing Patents Involving Bioactive Glasses for Soft Tissue Repair

Hundreds of patents have been filed since the discovery of the "grandfather bioactive glass composition" 45S5 Bioglass® in 1969. These are for applications in hard tissue engineering, e.g. for bone and teeth. A search of the existing patents for soft tissue engineering applications has come up with around 20 patents, most of them filed fairly recently.[75–88] Most of the patents are intended for dermal repair and wound healing, as reflected in the research field outlined in Section 3.1. A summary of the patents proposing the use of bioactive glasses for soft tissue engineering applications is included in Table 15.2.

A brief discussion of the relevant patents is provided in this section.

Table 15.2 Summary of patents filed on the application of bioactive glasses in soft tissue engineering.

Biomedical application	Material system	Year	Ref.
Skin/hemostasis	BG–chitosan composite (80 : 20 wt%)	2014	75
Skin/hemostasis	BG mesoporous	2015	77
Skin/cosmetic	BG particles (micronsized)	2015	78
Skin/wound or burn healing	Lactoferrite–BG particle composite	2010	79
Skin/wound or burn healing	BG particles with/without silver	2001	80
Skin/wound or burn healing	BG particles with topical antibiotic	1998	81
Skin/wound or burn healing	non-interlinked particles in a carrier	2013	82
Skin/wound or burn healing	hydrocolloid dressing with salicylic acid, cellulose derivatives, BG, water	2013	83
Percutaneous fixation	Particulate BAG	1999	84
Gastrointestinal ulcer	Liquid BAG in capsule	2005/2014/ 2014/2013	85–88

15.5.1 Hemostatic

A search for patents concerning BG-based products designed as hemostatic agents came up with two products:

1. A porous BG and chitosan composite[75] in which a combination of 80 wt% BG and 20 wt% chitosan is claimed to show shorter hemostatic time, or least amount of bleeding, compared to a commercial™ hemostatic powder under the same conditions.
2. A mesoporous BG of 58S composition (see Table 15.1 for composition) with good hemostatic effects that, in addition, exhibits wide antibacterial action and fungicidal effects promoting tissue regeneration.[76]

15.5.2 Cosmetic

Cosmetic dermal fillers together with other aesthetic injectables, such as botulinum toxin, and intense pulsed light (IPL) being used as anti-wrinkle treatments. A recent patent[77] describes a novel composite of medical grade hyaluronic acid and micron-sized particulate BG (composition not disclosed) claiming to show improved anti-wrinkle effects compared to the aforementioned treatment options.

15.5.3 Wound/Burn Healing

Several products have been developed with the purpose of enhancing and accelerating wound or burn healing as well as improving their appearance, mostly composed of particulate BG in a carrier. One patent[78] is for a composite material of lactoferrite and BG (composition not disclosed) which, in addition to enhancing wound healing, also claims to have antibacterial, antiviral and hemostatic effects. Another patent[79] pertains to compositions of a particulate BG containing silver (48–50 wt% SiO_2, 10–32 wt% CaO, 10–32 wt% Na_2O, 2–10% P_2O_5 and 0–8% Ag_2O) intended for wound healing, with another patent[80] describing a method for wound dressing involving amounts and composition of a particulate BG used in combination with a topical antibiotic. In another patent[81] a particular BG composition (not disclosed) of non-interlinked particles in a carrier is briefly described. A final patent describing a hydrocolloid dressing consisting of salicylic acid, cellulose derivatives, BG and water amongst other components, claims to promote formation of granulation tissue and to be safe, non-toxic and free of side effects.[82]

15.5.4 Percutaneous Fixation

Percutaneous implants are placed in patients in order to facilitate access to certain organs in the body, and are prone to bacterial infections due to the formation of a fibrous, non-adherent tissue layer around the device. A patent filed by Marotta *et al.*[84] outlines the production of a particulate BG (composition not disclosed) coating to promote a dermal adhesion area between the implant and dermal region in the patient. Such a coating would present BG particles with a spacing of about 20 μm.

15.5.5 Gastrointestinal Ulcers

A product consisting of a soft capsule filled with powder or liquid BG (composition not disclosed) is described in a patent filed by Zhong,[85] with the intent to deliver said BG to gastroduodenal ulcers, where it would neutralize the acid environment in order to avoid further erosion of the mucous gastrointestinal walls and to promote tissue repair and regeneration. More recently, the treatment described above has been filed as a US, a European and a WO patent by the same group.[85–87]

15.6 Conclusions and Future Scope

Bioactive glasses of varying compositions based mainly on silicate or borate glasses have been investigated for a range of applications exploiting their ability to stimulate soft tissue regeneration processes as well as to enhance angiogenesis. Although clear experimental evidence both *in vivo* and *in vitro* confirms the potential of bioactive glasses for various different soft tissue engineering applications, so far the biochemical mechanisms which regulate these processes are still not completely understood. Individual ionic species have been shown to influence the efficacy of bioactive glasses in contact with soft tissues, for example silicate, phosphate or borate matrices doped with ions, including copper, cobalt, lithium, gallium, boron and other biologically active bioinorganics, which then act as biochemical cues represents a useful approach to accelerate soft tissue regeneration by inducing angiogenesis. In the long run, the use of BGs in regenerative medicine of soft tissues could be an alternative to the application of expensive growth factors or other complex biological moieties which are frequently difficult to handle and challenging to be incorporated in biomedical devices. Bioactive glasses can be manufactured into many useful geometries and/or incorporated into a polymer (flexible) matrix to be able to mimic the structure and mechanical properties of specific soft tissues as closely as possible. The overwhelming evidence in the literature, part of which was captured in this chapter, is that bioactive glasses are strong contenders for the regeneration of soft tissues and anticipate the need for more research efforts focussing on *in vivo* and *in vitro* studies, which are starting to appear in the literature, such as the study by Li *et al.*[89] which shows that 45S5 BG has a pronounced effects on connexin 43 (Cx43) gap junction channels as well as gap junction-related behavior of endothelial cells. The authors conclude from their results that BG, on the one hand, enhances cell-to-cell communication and, on the other hand, protects endothelial cells. These relevant studies should lead to the gathering of more evidence to ultimately establish the exact mechanism by which BG affects soft tissue repair, which will help in the development of optimized BG scaffolds for soft tissue engineering applications in the not too distant future.

References

1. L. L. Hench, Opening paper 2015-Some comments on Bioglass: Four Eras of Discovery and Development, *Biomed. Glasses*, 2015, **1**, 1–11.
2. L. L. Hench, H. R. Stanley, A. E. Clark, M. Hall and J. Wilson, Dental Applications of Bioglass® Implants, *Bioceram. Proc.*, 1991, 231–238.
3. L. L. Hench, N. Roki and M. B. Fenn, Bioactive glasses: Importance of structure and properties in bone regeneration, *J. Mol. Struct.*, 2014, **1073**, 24–30.
4. W. Li, P. Nooeaid, J. A. Roether, D. W. Schubert and A. R. Boccaccini, Preparation and characterization of vancomycin releasing PHBV coated 45S5 Bioglass®-based glass–ceramic scaffolds for bone tissue engineering, *ECerS*, 2014, **34**, 505–514.
5. J. Wilson, G. H. Pigott, F. J. Schoen and L. L. Hench, Toxicology and biocompatibility of bioglasses, *J. Biomed. Mater. Res.*, 1981, **15**, 805–817.
6. V. Miguez-Pacheco, D. Greenspan, L. L. Hench and A. R. Boccaccini, Bioactive glasses in soft tissue repair, *Am. Ceram. Soc. Bull.*, 2015, **94**, 27–31.
7. L. L. Hench, Chronology of Bioactive Glass Development and Clinical Applications, *New J. Glass Ceram.*, 2013, **03**, 67–73. Scientific Research Publishing.
8. L. L. Hench and D. Greenspan, Interactions between Bioactive Glass and Collagen: A Review and New Perspectives, *J. Aust. Ceram. Soc.*, 2013, **49**, 1–40.
9. F. Baino, G. Novajra, V. Miguez-Pacheco, A. R. Boccaccini and C. Vitale-Brovarone, Bioactive glasses: Special applications outside the skeletal system, *J. Non-Cryst. Solids*, 2015, **432**, 15–30.
10. V. Miguez-Pacheco, L. L. Hench and A. R. Boccaccini, Bioactive glasses beyond bone and teeth: Emerging applications in contact with soft tissues, *Acta Biomater.*, 2015, **13**, 1–15.
11. G. Broughton 2nd, J. E. Janis and C. E. Attinger, The basic sciences of wound healing, *Plast. Reconstr. Surg.*, 2006, **117**, 12S–34S.
12. S. Guo and L. A. Dipietro, Factors Affecting Wound Healing, *J. Dent. Res.*, 2010, **89**, 219–229.
13. J. Zhou, H. Wang, S. Zhao, N. Zhou, L. Li, W. Huang, D. Wang and C. Zhang, In vivo and in vitro studies of borate based glass micro-fibers for dermal repairing, *Mater. Sci. Eng., C*, 2016, **60**, 437–445.
14. Q. Yang, S. Chen, H. Shi, H. Xiao and Y. Ma, In vitro study of improved wound-healing effect of bioactive borate-based glass nano-/micro-fibers, *Mater. Sci. Eng., C*, 2015, **55**, 105–117.
15. S. Zhao, L. Li, H. Wang, Y. Zhang, X. Cheng, N. Zhou, M. N. Rahaman, Z. Liu, W. Huang and C. Zhang, Wound dressings composed of copper-doped borate bioactive glass microfibers stimulate angiogenesis and

heal full-thickness skin defects in a rodent model, *Biomaterials*, 2015, **53**, 379–391.
16. M. Gholipourmalekabadi, M. Sameni, A. Hashemi, F. Zamani, A. Rostami and M. Mozafari, Silver- and fluoride-containing mesoporous bioactive glasses versus commonly used antibiotics: Activity against multi-resistant bacterial strains isolated from patietns with burns, *Burns*, 2016, **42**, 131–140.
17. S. Pourshahrestani, E. Zeimaran, N. Adib Kadri, N. Gargiulo, S. Samuel, S. Vasudevaraj Naveen, T. Kamarulc and M. R. Towler, Gallium-containing mesoporous bioactive glass with potent hemostatic activity and antibacterial efficacy, *J. Mater. Chem. B*, 2016, **4**, 71–86.
18. H. Yu, J. Peng, Y. Xu, J. Chang and H. Li, Bioglass Activated Skin Tissue Engineering Constructs for Wound Healing, *ACS Appl. Mater. Interfaces*, 2016, **8**, 703–715.
19. D. Li, G. Jiao, W. Zhang, X. Chen, R. Ning and C. Du, Hybrid scaffolding strategy for dermal tissue reconstruction: a bioactive glass/chitosan/silk fibroin composite, *RSC Adv.*, 2016, **6**, 19887–19896.
20. B. J. Pfister, T. Gordon, J. R. Loverde, A. S. Kochar, S. E. Mackinnon and D. Kacy Cullen, Biomedical engineering strategies for peripheral nerve repair: Surgical applications, state of the art, and future challenges (Review), *CRC Crit. Rev. Bioeng.*, 2011, **39**, 81–124.
21. T. Y. Nguyen and H. Liu, A review of current advances in biomaterials for neural tissue regeneration, *Recent Pat. Biomed. Eng.*, 2013, **6**, 29–39.
22. X. Jiang, S. H. Lim, H. Q. Mao and S. Y. Chew, Current applications and future perspectives of artificial nerve conduits, *Exp. Neurol.*, 2010, **223**, 86–101.
23. Y.-P. Kim, G.-S. Lee, J.-W. Kim, M. S. Kim, H.-S. Ahn, J.-Y. Lim, H.-W. Kim, Y.-J. Son, J. C. Knowles and J. K. Hyun, Phosphate glass fibres promote neurite outgrowth and early regeneration in a peripheral nerve injury model, *J. Tissue Eng. Regener. Med.*, 2015, **9**, 236–246.
24. M. F. Koudehi, A. A. Fooladi, K. Mansoori, Z. Jamalpoor, A. Amiri and M. R. Nourani, Preparation and evaluation of novel nano-bioglass/gelatin conduit for peripheral nerve regeneration, *J. Mater. Sci.: Mater. Med.*, 2014, **25**, 363–373.
25. A. Mohammadkhah, L. M. Marquardt, S. E. Sakiyama-Elbert, D. E. Day and A. B. Harkins, Fabrication and characterization of poly-(ε)-caprolactone and bioactive glass composites for tissue engineering applications, *Mater. Sci. Eng., C*, 2015, **49**, 632–639.
26. S. Bunting, L. Di Silvio, S. Deb and S. Hall, Bioresorbable glass fibres facilitate peripheral nerve regeneration, *J. Hand. Surg. Br.*, 2005, **30**, 242–247.
27. H.-S. Ahn, J.-Y. Hwang, M. S. Kim, J.-Y. Lee, J.-W. Kim, H.-S. Kim, U. S. Shin, J. C. Knowles, H.-W. Kim and J. K. Hyun, Carbon-nanotube-interfaced glass fiber scaffold for regeneration of transected sciatic nerve, *Acta Biomater.*, 2015, **13**, 324–334.

28. P. Balasubramanian, L. A. Strobel, K. Ulrich and A. R. Boccaccini, Zinc-containing bioactive glasses for bone regeneration, dental and orthopedic applications, *Biomed. Glasses*, 2015, **1**, 51–69.
29. M. Sabbatini, F. Boccafoschi, M. Bosetti and M. Cannas, Adhesion and differentiation of neuronal cells on Zn-doped bioactive glasses, *J. Biomater. Appl.*, 2014, **28**(5), 708–718.
30. D. Liang, J. Chen, Y. Li, Z. Chen, B. Dan and H. Zhou, Tissue interaction with bioglass ceramic implanted in the rabbit cornea, *Yan Ke Xue Bao*, 2001, **17**, 198–201.
31. F. Baino, S. Perero, S. Ferraris, M. Miola, C. Balagna, E. Verné, C. Vitale-Brovarone, A. Coggiola, D. Dolcino and M. Ferraris, Biomaterials for orbital implants and ocular prostheses: overview and future prospects, *Acta Biomater.*, 2014, **10**, 1064–1087.
32. P. Stoor, K. Meismäki, C. Lindqvist and R. Knotio, The use of anatomically drop-shaped bioactive glass implants in the reconstruction of orbital floor fractures – a prospective long-term follow-up study, *J. Craniomaxillofac. Surg.*, 2015, **43**, 969–975.
33. F. Baino, How can bioactive glasses be useful in ocular surgery? *J. Biomed. Mater. Res., Part A*, 2015, **103**, 1259–1275.
34. M. Re, L. Paolucci, R. Romeo and V. Mallardi, Surgical treatment of nasal septal perforations. Our experience, *Acta Otorhinol*, 2006, **26**, 102–109.
35. L. Kogan, P. Gilbey, A. Samet and Y. Talmon, Nasal septal perforation repair using oral mucosal flaps, *Isr. Med. Assoc. J.*, 2007, **9**, 373–375.
36. P. Stoor and R. Grénman, Bioactive glass and turbinate flaps in the repair of nasal septal perforations, *Ann. Otol., Rhinol., Laryngol.*, 2004, **113**, 655–661.
37. P. Stoor, E. Söderling and R. Grénman, Bioactive glass S53P4 in repair of septal perforations and its interactions with the respiratory infection-associated microorganisms Haemophilus influenzae and Streptococcus pneumoniae, *J. Biomed. Mater. Res.*, 2001, **58**, 113–120.
38. R. M. Day, A. R. Boccaccini, S. Shurey, J. A. Roether, A. Forbes, L. L. Hench and S. M. Gabe, Assessment of polyglycolic acid mesh and bioactive glass for soft-tissue engineering scaffolds, *Biomaterials*, 2004, **25**, 5857–5866.
39. R. M. Day, V. Maquet, A. R. Boccaccini, R. Jerome and A. Forbes, In vitro and in vivo analysis of macroporous biodegradable poly(d,l-lactide-co-glycolide) scaffolds containing bioactive glass, *J. Biomed. Mater. Res.*, 2005, **75A**, 778–787.
40. R. M. Day, Bioactive Glass Stimulates the Secretion of Angiogenic Growth Factors and Angiogenesis in Vitro, *Tissue Eng.*, 2005, **11**, 768–777.
41. H. Keshaw, A. Forbes and R. M. Day, Release of angiogenic growth factors from cells encapsulated in alginate beads with bioactive glass, *Biomaterials*, 2005, **26**, 4171–4179.
42. H. Keshaw, G. Georgiou, J. J. Blaker, A. Forbes, J. C. Knowles and R. M. Day, Assessment of polymer/bioactive glass-composite

microporous spheres for tissue regeneration applications, *Tissue Eng., Part A*, 2009, **15**, 1451–1461.
43. L. C. Gerhardt, K. L. Widdows, M. M. Erol, C. W. Burch, J. A. Sanz-Herrera, I. Ochoa, R. Stämpfli, I. S. Roqan, S. Gabe, T. Ansari and A. R. Boccaccini, The pro-angiogenic properties of multi-functional bioactive glass composite scaffolds, *Biomaterials*, 2011, **32**, 4096–4108.
44. S. R. Moosvi and R. M. Day, Bioactive glass modulation of intestinal epithelial cell restitution, *Acta Biomater.*, 2009, **5**, 76–83.
45. J. K. Leach, D. Kaigler, Z. Wang, P. H. Krebsbach and D. J. Mooney, Coating of VEGF-releasing scaffolds with bioactive glass for angiogenesis and bone regeneration, *Biomaterials*, 2006, **27**, 3249–3255.
46. A. Leu and J. K. Leach, Proangiogenic potential of a collagen/bioactive glass substrate, *Pharm. Res.*, 2008, **25**, 1222–1229.
47. S. Deb, R. Mandegaran and L. Di Silvio, A porous scaffold for bone tissue engineering/45S5 Bioglass® derived porous scaffolds for co-culturing osteoblasts and endothelial cells, *J. Mater. Sci.: Mater. Med.*, 2010, **21**, 893–905.
48. M. Handel, T. R. Hammer, P. Nooeaid, A. R. Boccaccini and D. Hoefer, 45S5-Bioglass®-Based 3D-Scaffolds Seeded with Human Adipose Tissue-Derived Stem Cells Induce In Vivo Vascularization in the CAM Angiogenesis Assay, *Tissue Eng., Part A*, 2013, **19**, 2703–2712.
49. Q. Zeng, Y. Han, H. Li and J. Chang, Bioglass/alginate composite hydrogel beads as cell carriers for bone regeneration, *J. Biomed. Mater. Res., Part B*, 2014, **102**, 42–51.
50. R. Detsch, P. Stoor, A. Grünewald, J. A. Roether, N. C. Lindfors and A. R. Boccaccini, Increase in VEGF secretion from human fibroblast cells by bioactive glass S53P4 to stimulate angiogenesis in bone, *J. Biomed. Mater. Res., Part A*, 2014, **102A**, 4055–4061.
51. L. A. Haro Durand, A. Góngora, J. M. Porto López, A. R. Boccaccini, M. P. Zago, A. Baldi and A. Gorustovich, In vitro endothelial cell response to ionic dissolution products from boron-doped bioactive glass in the SiO_2–CaO–P_2O_5–Na_2O system, *J. Mater. Chem. B*, 2014, **2**, 7620–7630.
52. C. Stähli, N. Muja and S. N. Nazhat, Controlled copper ion release from phosphate-based glasses improves human umbilical vein endothelial cell survival in a reduced nutrient environment, *Tissue Eng., Part A*, 2013, **19**, 548–557.
53. C. Stähli, M. James-Bhasin and S. N. Nazhat, Three-dimensional endothelial cell morphogenesis under controlled ion release from copper-doped phosphate glass, *J. Controlled Release*, 2015, **200**, 222–232.
54. C. Stähli, M. James-Bhasin, A. Hoppe, A. R. Boccaccini and S. N. Nazhat, Effect of ion release from Cu-doped 45S5 Bioglass® on 3D endothelial cell morphogenesis, *Acta Biomater.*, 2015, **19**, 15–22.
55. J. P. Cattalini, A. Hoppe, F. Pishbin, J. A. Roether, A. R. Boccaccini, S. Lucangioli and V. Mouriño, Novel nanocomposite biomaterials with

controlled copper/calcium release capability for bone tissue engineering multifunctional scaffolds, *J. R. Soc., Interface*, 2015, **12**, 05–09.
56. A. Hoppe, R. Meszaros, C. Stähli, S. Romeis, J. Schmidt, W. Peukert, B. Marelli, S. Nazhat, L. Wondraczek, J. Lao, E. Jallot and A. R. Boccaccini, In vitro reactivity of Cu doped 45S5 Bioglass® derived scaffolds for bone tissue engineering, *J. Mater. Chem. B*, 2013, **1**, 5659–5674.
57. S. N. Rath, A. Brandl, D. Hiller, A. Hoppe, U. Gbureck, R. E. Horch, A. R. Boccaccini and U. Kneser, Bioactive copper-doped glass scaffolds can stimulate endothelial cells in co-culture in combination with mesenchymal stem cells, *PLoS One*, 2014, **9**, e113319.
58. H. Wang, S. Zhao, J. Zhou, Y. Shen, W. Huang, C. Zhang, M. N. Rahaman and D. Wang, Evaluation of borate bioactive glass scaffolds as a controlled delivery system for copper ions in stimulating osteogenesis and angiogenesis in bone healing, *J. Mater. Chem. B*, 2014, **2**, 8547–8557.
59. C. Wu, Y. Zhou, M. Xu, P. Han, L. Chen, J. Chang and Y. Xiao, Copper-containing mesoporous bioactive glass scaffolds with multifunctional properties of angiogenesis capacity, osteostimulation and antibacterial activity, *Biomaterials*, 2013, **34**, 422–433.
60. M. M. Azevedo, G. Jell, M. D. O'Donnell, R. V. Law, R. G. Hill and M. M. Stevens, Synthesis and characterization of hypoxia-mimicking bioactive glasses for skeletal regeneration, *J. Mater. Chem.*, 2010, **20**, 8854–8864.
61. C. Wu, Y. Zhou, W. Fan, P. Han, J. Chang, J. Yuen, M. Zhang and Y. Xiao, Hypoxia-mimicking mesoporous bioactive glass scaffolds with controllable cobalt ion release for bone tissue engineering, *Biomaterials*, 2012, **33**, 2076–2085.
62. A. Hoppe, B. Jokic, D. Janackovic, T. Fey, P. Greil, S. Romeis, J. Schmidt, W. Peukert, J. Lao, E. Jallot and A. R. Boccaccini, Cobalt-releasing 1393 bioactive glass-derived scaffolds for bone tissue engineering applications, *ACS Appl. Mater. Interfaces*, 2014, **6**, 2865–2877.
63. A. Hoppe, A. Brandl, O. Bleiziffer, A. Arkudas, R. E. Horch, B. Jokic, D. Janackovic and A. R. Boccaccini, In vitro cell response to Co-containing 1393 bioactive glass, *Mater. Sci. Eng., C*, 2015, **57**, 157–163.
64. M. M. Azevedo, O. Tsigkou, R. Nair, J. R. Jones, G. Jell and M. M. Stevens, Hypoxia Inducible Factor-Stabilizing Bioactive Glasses for Directing Mesenchymal Stem Cell Behavior, *Tissue Eng., Part A*, 2015, **21**, 382–389.
65. E. Quinlan, S. Partap, M. M. Azevedo, G. Jell, M. M. Stevens and F. J. O'Brien, Hypoxia-mimicking bioactive glass/collagen glycosaminoglycan composite scaffolds to enhance angiogenesis and bone repair, *Biomaterials*, 2015, **52**, 358–366.
66. G. E. Vargas, L. A. Haro Durand, V. Cadena, M. Romero, R. V. Mesones, M. Mačković, S. Spallek, E. Spiecker, A. R. Boccaccini and A. Gorustovich, Effect of nano-sized bioactive glass particles on the angiogenic properties of collagen based composites, *J. Mater. Sci.: Mater. Med.*, 2013, **24**, 1261–1269.

67. M. Handel, T. R. Hammer, P. Nooeaid, A. R. Boccaccini and D. Hoefer, 45S5-Bioglass®-Based 3D-Scaffolds Seeded with Human Adipose Tissue-Derived Stem Cells Induce In Vivo Vascularization in the CAM Angiogenesis Assay, *Tissue Eng., Part A*, 2013, **19**, 2703–2712.
68. A. Leu, S. M. Stieger, P. Dayton, K. W. Ferrara and J. K. Leach, Angiogenic response to bioactive glass promotes bone healing in an irradiated calvarial defect, *Tissue Eng., Part A*, 2009, **15**, 877–883.
69. A. Arkudas, A. Balzer, G. Buehrer, I. Arnold, A. Hoppe, R. Detsch, P. Newby, T. Fey, P. Greil, R. E. Horch, A. R. Boccaccini and U. Kneser, Evaluation of angiogenesis of bioactive glass in the arteriovenous loop model, *Tissue Eng., Part A*, 2013, **19**, 479–486.
70. L. A. Haro Durand, G. E. Vargas, N. M. Romero, R. Vera-Mesones, J. M. Porto-López, A. R. Boccaccini, M. P. Zago, A. Baldi and A. Gorustovich, Angiogenic effects of ionic dissolution products released from a boron-doped 45S5 bioactive glass, *J. Mater. Chem. B*, 2015, **3**, 1142–1148.
71. S. Zhao, J. Zhang, M. Zhu, Y. Zhang, Z. Liu, C. Tao, Y. Zhu and C. Zhang, Three-dimensional printed strontium-containing mesoporous bioactive glass scaffolds for repairing rat critical-sized calvarial defects, *Acta Biomater.*, 2015, **12**, 270–280.
72. Y. Lin, R. F. Brown, S. B. Jung and D. E. Day, Angiogenic effects of borate glass microfibers in a rodent model, *J. Biomed. Mater. Res., Part A*, 2014, **102**, 4491–4499.
73. R. J. Watters, R. F. Brown and D. E. Day, Angiogenic Effect of Bioactive Borate Glass Microfibers and Beads in the Hairless Mouse, *Biomed. Glasses*, 2015, **1**, 173–184.
74. H. Wang, S. Zhao, W. Xiao, J. Xue, Y. Shen, J. Zhou, W. Huang, M. N. Rahaman, C. Zhang and D. Wang, Influence of Cu doping in borosilicate bioactive glass and the properties of its derived scaffolds, *Mater. Sci. Eng., C*, 2016, **58**, 194–203.
75. J. Chang, X. Feng and C. Hou Mesoporous bioactive glass and composite porous chitosan hemostatic material and preparation method. CN101991875, 2014.
76. X. Feng and J. Chang, Synthesis of a Well-Ordered Mesoporous 58S Bioactive Glass by a Simple Method, *Int. J. Appl. Ceram. Technol.*, 2011, **8**(3), 547–552.
77. D. Zhihong, Mesoporous bioactive glass/lasiosphaera composite material, preparation and applications, CN104906629, 2015.
78. G. D. Hack and S. Asculai, Methods and compositions for enhancing and extending the cosmetic effects of non-surgical dermal interventions, WO2015153377, 2015.
79. G. Roberts, R. Daniels, X. Zhao and I. Thompson, Composite material, US2010143490, 2010.
80. S.-L. Yang, Compositions and methods for repair of osseous defects and accelerated wound healing, US2001024662, 2001.
81. D. C. Greenspan and J. K. West, Composition and method for acceleration of wound and burn healing, US5834008, 1998.

82. D. C. Greenspan, J. K. West, J. L. Meyers, S. Lee and M. Diamond Anti-inflammatory and antimicrobial uses for bioactive glass, EP2338448, 2013.
83. H. Li, Anti-inflammatory hydrocolloid dressing capable of promoting wound healing, CN103143054, 2013.
84. J. S. Marotta, G. LaTorre, C. D. Batich and L. L. Hench, Percutaneous biofixed medical implants, US5990380, 1999.
85. J. Zhong, Application of bioactive glass in preparing medicine for treating gastroduodenal ulcer, CN1636576, 2005.
86. J. Zhong, Y. Wang, J. Chang and Y. Xu, pH buffered bioactive glass, and its preparation and use, US 20140227365 A1, 2014.
87. J. Zhong, Y. Wang, J. Chang and Y. Xu, Bioactive ph buffer glass and preparation process and use thereof, EP 2759288 A4, 2014.
88. J. Zhong, Y. Wang, J. Chang and Y. Xu, Bioactive ph buffer glass and preparation process and use thereof, WO/2013/041030, 2013.
89. H. Li, J. He, H. Yu, C. R. Green and J. Chang, Bioglass promotes wound healing by affecting gap junction connexin 43 mediated endothelial cell behavior, *Biomaterials*, 2016, **84**, 64–75.

CHAPTER 16

Bioactive Glasses as Carriers of Therapeutic Ions and the Biological Implications

ALEXANDER HOPPE† AND ALDO R. BOCCACCINI*

Institute of Biomaterials, Department of Materials Science and Engineering, University of Erlangen-Nuremberg, Cauerstrasse 6, 91058 Erlangen, Germany
*Email: aldo.boccaccini@ww.uni-erlangen.de

16.1 Introduction

Apart from the bioactive properties of Bioactive Glasses (BGs) (the ability to bond to hard as well as to soft tissue) one of the key features of BGs is related to their ionic dissolution products which can stimulate gene expression in stem cells regulating their osteogenic differentiation.[1] Hence, BGs are considered 3rd generation biomaterials[2] which have the ability to induce specific intrinsic cell responses while bonding to hard tissue as well as to soft tissue. There is also experimental evidence showing that BGs are able to stimulate angiogenesis *in vitro* as well as *in vivo*.[3–5] Hereby, the role of ions released from BGs in the biological context has been emphasized by L. L. Hench who summarized and discussed a series of results supporting the hypothesis that "ionic dissolution products released from BGs stimulate the genes of cells towards a path of regeneration and self-repair".[6]

†Current affiliation: Johnson Matthey Advanced Glass Technologies. 6222NZ Maastricht, Fregatweg 38, The Netherlands.

RSC Smart Materials No. 23
Bioactive Glasses: Fundamentals, Technology and Applications
Edited by Aldo R. Boccaccini, Delia S. Brauer and Leena Hupa

Published by the Royal Society of Chemistry, www.rsc.org

Whilst this statement has been widely accepted by the biomaterials scientific community, there has been a large amount of research work generated in the last decade investigating the exact mechanisms behind the biological effect of BG dissolution products and resulting cell interactions which are, however, not yet fully understood.[1] In fact, it is apparent that gaining further knowledge about the role of inorganic ions (and combinations thereof) on cell behaviour is key to understanding the *in vitro* and *in vivo* behaviour of biomaterials which would clear the path toward developing smart implants and scaffolds with tailored (multiple) biological functionalities for individualised biomedical applications.

More specifically, it has been observed that ionic dissolution products from silicate-based BGs (*e.g.* Si, Ca, P) can stimulate the expression of osteogenic genes which should result in enhanced bone formation.[7] Beyond that, during the last decade significant progress has been made to enhance the biological impact of BGs by incorporating inorganic therapeutics and metallic ions (also termed bioinorganics) in silicate (or phosphate and borate) glasses leading to novel bioactive glass compositions which exhibit the capability of realising, in controlled manner, such biologically active ions to induce specific acellular effects relevant for a broad variety of biomedical applications.[1,8,9]

Trace elements such as Sr, Cu, Zn or Co present in the human body are known for their anabolic effects in bone metabolism,[10–12] and hence they have been proposed as active elements to be incorporated into scaffold materials for bone tissue engineering applications. The subsequent release of these ions after exposure to a physiological environment, if controlled to an optimal concentration, favours the behaviour of human cells and enhances the bioactivity of the scaffolds related to both osteogenesis and angiogenesis. However, considering that concentrations of metallic ions which are too high are known to be toxic,[13] their controlled release kinetics from a suitable carrier system is crucial.[8] Since the biodegradability of BGs can be controlled by their chemical composition, BGs are a natural choice as carrier matrices for therapeutic metal ions. The controlled crystallisation of selected compositions of BGs can also be exploited to control the release rate of ionic dissolution products.

The number of research studies published on the biological effects of ions released from BGs has continuously increased in the last decade. Figure 16.1 shows the number of published works found in the Scopus® database using the keywords "bioactive glass" + "ion release" (papers published between 1974 and 2014), which demonstrates the increasing interest in this particular research topic related to the biological impact of BGs.

In this chapter, the biological responses to ionic dissolution products released from bioactive glasses are summarized and discussed providing the scope for future investigations related to novel bioactive glass-derived materials. The content of this chapter can be considered to complement and update the previous review articles by the authors published in 2011 [1] and 2013.[8]

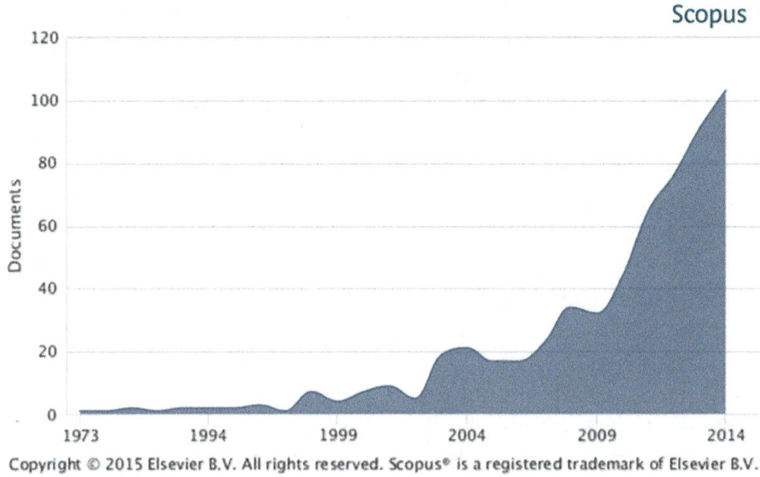

Figure 16.1 Number of published papers identified by keywords "bioactive glass" + "ion release" from 1974 to 2014 according to the Scopus® database.

16.2 Bioinorganics and Metallic Ions for Biomedical Applications

Specific metallic ions, known as therapeutic inorganic ions (TIIs, also termed bioinorganics), are essential in human metabolism and have been reported to play a critical role in osteogenesis and angiogenesis.[11,14–16] Hence, TIIs have been identified as promising bioactive agents for applications in biomedicine.[17] In particular, metal ions are known to act as enzyme co-factors and therefore to influence signalling pathways and stimulate metabolic effects occurring during hard and soft tissue formation.[11,18] Considering these effects and the fact that biomaterials bearing TIIs are more cost-effective, easy to process and less risky compared to gene therapy or the use of growth factors, TIIs are being increasingly considered as part of therapeutic approaches in the fields of hard and soft tissue engineering.[19] An added value of some TIIs is their antibacterial capability which may lead to antibiotic-free alternatives to combat infections.[9]

A broad variety of TIIs has been proposed as biologically active agents to be used in the biomedical field,[17] including calcium,[20–23] phosphorous,[24] silicon,[25,26] strontium,[27–31] zinc[32,33] as well as boron,[34,35] vanadium,[12,36,37] cobalt,[38–41] copper,[42,43] niobium,[44,45] lithium,[46,47] cerium[48,49] and magnesium.[50–54] For more details on the mechanisms of action and physiological role of specific metallic ions and inorganic species with therapeutic effects on bone formation and angiogenesis the reader is referred to comprehensive reviews available in the open literature.[1,9,14,16,19,55]

In this section some of the key therapeutic effects of selected TIIs are highlighted summarising their importance in bone physiology and framing

Table 16.1 Role of therapeutic inorganic ions (bioinorganics) in human bone metabolism and angiogenesis. Modified and updated after Hoppe *et al.*[1]

Ion	Biological response *in vivo/in vitro*
Si	- essential element for formation and calcification of hard tissue[25,79]
	- aqueous Si may induce hydroxyapatite precipitation[80]
	- $Si(OH)_4$ stimulates collagen I formation and osteoblastic differentiation[26]
Ca	- favours osteoblast proliferation, differentiation and the mineralization of extracellular matrix (ECM)[21]
	- activates Ca-sensing receptors in osteoblast cells, increases expression of IGF-I or IGF-II[22,23]
P	- stimulates expression of matrix la protein (MGP), a key regulator in bone formation[24]
Cu	- is essential for VEGF expression[42]
	- exhibits synergetic stimulating effects on angiogenesis when combined with angiogenic growth factor FGF-2[66]
	- stimulates proliferation and VEGF expression in endothelial cells[67]
	- induces neovascularisation *in vivo*[81]
	- induces osteogenic differentiation of mesenchymal cells[68]
Co	- mimics hypoxia, hence stabilises HIF-1 factor[76]
	- *In vivo*: $CoCl_2$ pre-treated bone marrow stromal cells (BMSCs) induced higher degree of vascularisation and enhanced osteogenesis within the implants[41]
B	- stimulates RNA synthesis in fibroblast cells[77,82]
	- dietary applied boron stimulates bone formation[35]
Zn	- anti-inflammatory agent and stimulates bone formation *in vitro* by activating protein synthesis in osteoblasts[33]
	- increases ATPase activity, regulates transcription of osteoblastic differentiation genes, *e.g.* collagen I, ALP, osteopontin and osteocalcin[64]
Mg	- stimulates new bone formation[54]
	- increases bone cell adhesion (probably due to interactions with integrins)[54,59]
Sr	- shows beneficial effects on bone cells and bone formation *in vivo*[29,31]
	- used as drug for treating osteoporosis[61]
Li	- enhances re-myelination of peripheral nerves[46]
	- stimulates bone formation and increases bone mass[47]
Ce	- cerium nanoparticles act pro-angiogenically *via* stabilising HIF-1α and stimulation of endothelial cell growth and tube formation[49,83]
B	- stimulates bone formation in rats[35]
	- boric acid increases bone strength[84]

the scope for their potential use as additives in bioactive glass compositions used in hard and soft tissue engineering. Table 16.1 gives an overview of biological responses to single inorganic ions focusing on their impact on osteogenesis and angiogenesis. Indeed, in many of the reported studies it is the synergetic effect of more than one TII which is found to provide beneficial effects.

Silicon is an essential element for metabolic processes, being associated with the formation and calcification of hard tissue whereby high Si contents have been detected in early stages of bone matrix calcification.[25] Additionally, dietary Si intake has been shown to increase the bone mineral density (BMD) in men and premenopausal women.[56] Regarding its physiological role, Nielsen *et al.*[57] suggested that Si has a biochemical function in bone growth processes affecting the bone collagen turnover. Furthermore, orthosilicic acid ($Si(OH)_4$) at a physiological concentration of 10 μmol stimulates collagen I formation in human osteoblast cells (HOC) also inducing osteoblastic differentiation.[26] The role of silicon in the biomaterials research field has been recently highlighted in the literature.[58]

Magnesium is another essential element for bone metabolism and it has been shown to have stimulating effects on new bone formation.[51–54] The beneficial effect of Mg is related to its interaction with integrins of osteoblast cells which are responsible for cell adhesion and stability.[54,59] Furthermore, Rude *et al.*[51] observed that Mg depletion results in impaired bone growth, increased bone resorption and loss in trabecular bone underlining the significant role that Mg plays in bone metabolism.

Because of the chemical analogy to Ca (same main group, similar atomic radius, $Ca^{2+} = 1.0$ Å, $Sr^{2+} = 1.16$ Å), Sr can accumulate in bone by exchanging with Ca in the hydroxyapatite crystal lattice.[60] The therapeutic potential of Sr in bone metabolism has been investigated by Marie *et al.* who showed that Sr exhibits beneficial effects on bone cells and bone formation *in vivo*.[29,31] Sr has also been shown to be a promising agent in treating osteoporosis.[61] Additionally, Sr applied as strontium renalate is a well-known drug for treatment of osteoporosis which enhances bone healing, indicated by increased callus-resistant bones.[62] Zinc is well known for its anti-inflammatory effects[63] and has also been shown to stimulate bone formation *in vitro* by activating protein synthesis in osteoblast cells and increasing ATPase activity in bone tissue.[33] Also, the inhibitory effect of Zn on bone resorption through obstructing the formation of osteoclast cells in mouse marrow cultures has been reported.[33] Zinc was also identified as a regulation agent in transcription of osteoblastic differentiation genes, such as collagen I, ALP, osteopontin and osteocalcin. It is assumed that Zn can be considered a Runx2 stimulating agent being able to directly stimulate bone formation through increasing Runx2-targeted osteoblast differentiation gene transcription.[64]

Cu is well recognized for its role in angiogenesis.[65–67] For example, significant distributions of cellular Cu have been found in human endothelial cells when they were induced to undergo angiogenesis revealing the importance of Cu as angiogenic agent.[65] Copper is furthermore associated with angiogenesis growth factor FGF-2 and has been shown to promote synergetic stimulating effects on angiogenesis *in vitro*.[66] In addition, Cu has been shown to stimulate proliferation of human endothelial cells[67] and to enhance osteogenic differentiation of mesenchymal stem cells (MSC).[68] Cu^{2+} ions at a concentration of 10^{-6} mol L^{-1} have been shown to inhibit osteoclast activity.[69]

Regarding angiogenesis, cobalt ions are of interest in the context of bone physiology[38–41] since Co is known to induce hypoxia conditions and to stabilize the hypoxia-inducible factor 1 (HIF-1).[70,71] Hypoxia conditions, in turn, activate several pro-regenerative processes in the human body[72] *via* regulation of HIF-1. HIF-1 activation has been shown to result in accelerated bone ingrowth whereby the HIF-1α pathway has been identified as being critical for angiogenesis and skeletal regeneration[73] and important for stem cell differentiation and fracture repair.[74,75] These potential stimulating effects of Co^{2+} ions were tested *in vitro* and *in vivo* whereby cobalt promoted angiogenesis *via* activation of hypoxia-inducible factor 1 (HIF-1) in a rat remnant kidney *in vivo* model when applied by subcutaneous injections.[41] Furthermore, in a rat bladder *in vivo* model Co was shown to enhance hypoxia response, cell growth and angiogenesis indicated by stimulated expression of HIF-1α and vascular endothelial growth factor (VEGF).[38] Furthermore, it has been reported that bone marrow-derived stem cells (BMSCs) pre-treated with $CoCl_2$ induced a higher degree of vascularisation and enhanced osteogenesis within collagen scaffolds implanted *in vivo*.[76]

Boron has also been described as a stimulating agent for bone formation[34,35,77] and has been recognised for its angiogenic effects.[78] Recently, cerium has been investigated in the context of biomaterials for tissue regeneration and has been shown to stimulate keratinocytes and fibroblasts and also to enhance angiogenesis.[48,49]

The biological effects of the metallic ions described above can be exploited in order to functionalise biomaterials, *e.g.* bioactive glasses in the context of the topic of this chapter. However, the balance between therapeutic effects and toxicity is dose-dependent and thus a controlled release mechanism is required in order to avoid overdosing. Hereby, bioactive glasses are considered suitable inorganic carriers since the glass degradation profile can be tailored by adjusting the glass chemistry and in some cases by the degree of crystallisation (in glass-ceramics). The suitability of BGs as inorganic carrier materials for TIIs is discussed in the following sections.

16.3 Bioactive Glasses as Carriers for Therapeutic Ions

BGs exhibit enhanced bioactivity which, in this context, is defined as the ability to bond to bone *via* formation of surface carbonated hydroxyapatite,[85] as well as osteostimulating and pro-angiogenic effects which makes them highly attractive biomaterials for bone and soft tissue engineering applications.[4,86,87] Furthermore, BGs are biodegradable, with varying degrees of degradability depending on the composition, and may be used as carriers for TIIs. BG systems have been proposed as carriers for TIIs, including melt-derived silicate glasses,[1] sol–gel-based mesoporous glasses,[88] phosphate glasses,[89] borate glasses[89–92] and borosilicate glasses.[92–95] The degradation behaviour and ion release profiles are the critical parameters to consider for specific applications and which differ for different types of BG. Usually, for common BG systems the degradation behaviour can be tailored through the glass composition. Moreover, processing technologies for BGs can also be adopted and optimised enabling shaping of the bioactive glass into particles, including nanoparticles,[96] scaffolds, fibers, dense discs and rods.[97] The type of processing, either by melting or by the sol–gel technique, has also a profound effect on ion releasing capability of the glass. In general, borate glasses are more reactive and show higher degradation rates compared to standard bioactive glass compositions based on SiO_2 as a network former.[98] The group of phosphate glasses, in turn, is more soluble compared to silicate as well as to borosilicate glasses. The main aspect that differentiates phosphate glasses from silicate glasses is the linear degradation behaviour with sustained ion release through the entire glass degradation time, which can range from days to weeks. Moreover, a variety of different phosphate species are released from phosphate-based glasses including orthophosphate (PO_4^{3-}), linear species such as pyrophosphate ($P_2O_7^{4-}$) and tripolyphosphate ($P_3O_{10}^{5-}$) and cyclic species such as cyclic trimetaphosphate ($P_3O_9^{3-}$).[89] The degradation rate of phosphate glasses can be readily adjusted by changing the composition, for example by addition of oxides such as TiO_2.[99] The release of therapeutic ions from phosphate glasses can also be controlled in a linear manner as demonstrated by Stähli *et al.*, for example,[100] for Cu release from a PO_5–CaO–NaO glass system, see Figure 16.2. In this study, the authors show that the Cu release rate is proportional to the amount of CuO incorporated in the glass matrix.[100]

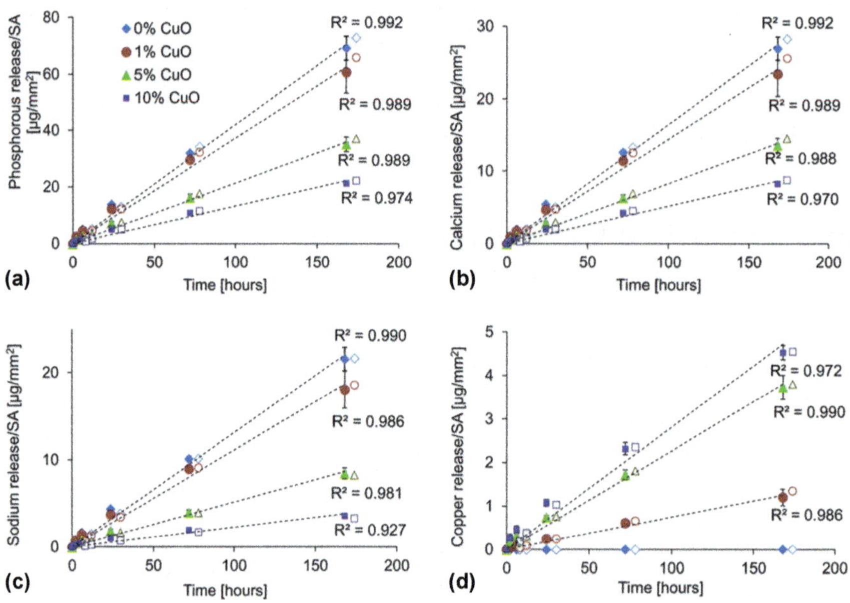

Figure 16.2 Ion release from Cu-containing phosphate glass as function of time. Linear Cu^{2+} release can be controlled *via* tuning the glass composition. [Reprinted from C. Stahli *et al.*, Characterization of aqueous interactions of copper-doped phosphate-based glasses by vapour sorption, *Acta Biomater.*, **10**, 3317–3326, Copyright 2014[100] with permission from Elsevier Ltd.]

Silicate glasses, on the other hand, form a carbonated hydroxyapatite (CHA) layer on the surface when exposed to biological fluids which acts as a diffusion barrier affecting the release of ions. Hence, silicate-based glasses typically show a diffusion-dominated non-linear ion release behaviour as shown in Figure 16.3 for the release of Co ions from a 1393 BG

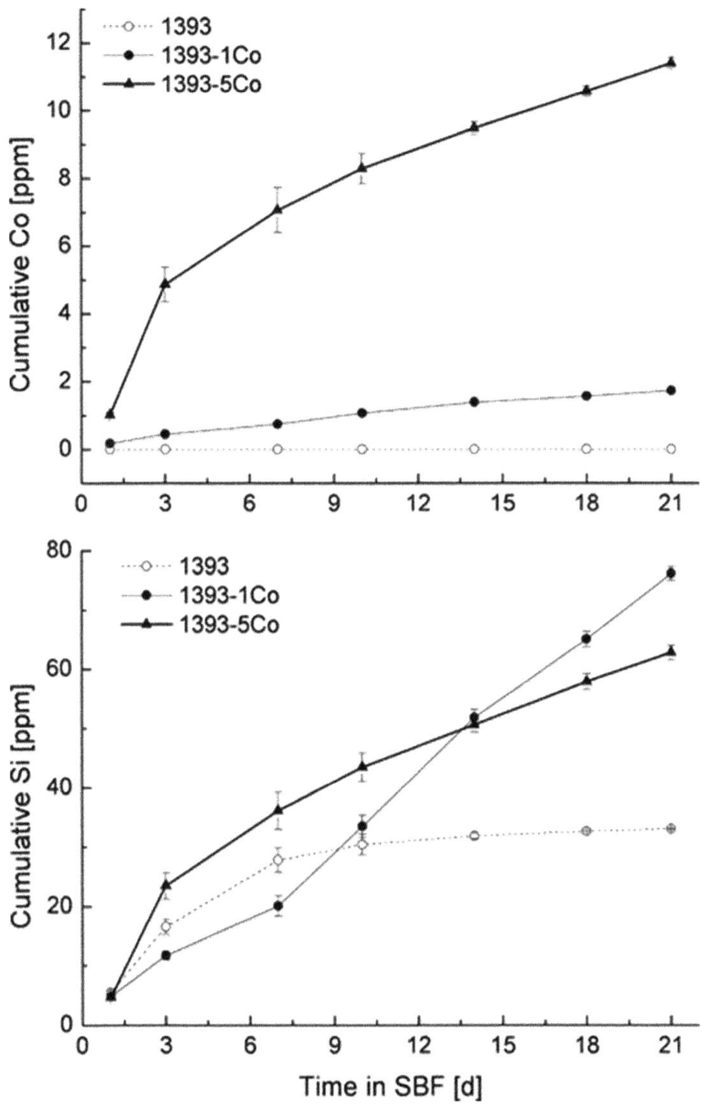

Figure 16.3 Controlled Co and Si release from 1393-type silicate bioactive glass-derived scaffolds in simulated body fluid (SBF) showing diffusion-controlled ion release profiles.
Reprinted with permission from A. Hoppe *et al.*, Cobalt-Releasing 1393 Bioactive Glass-Derived Scaffolds for Bone Tissue Engineering Applications, *ACS Appl. Mater. Interfaces*, 2014, **6**(4), 2865–2877. Copyright 2016 American Chemical Society.

(wt%: $53SiO_2$–$6Na_2O$–$12K_2O$–5MgO–20CaO–$4P_2O_5$).[101] Similar behaviour was observed for Cu-releasing 45S5 BG-derived scaffolds.[102]

Furthermore, it has been shown that metallic ions released from BG during reaction in biological fluids are incorporated in the calcium phosphate-rich surface layer, which is relevant for understanding the cell–material interaction of these glasses.[101,102]

It should be noted that the degradation behaviour and also the ion release kinetics are dependent on the conditions of the degradation test, *e.g.* being either static or dynamic.[103] For instance, literature reports have shown that amorphous BG scaffolds (mol%: $70SiO_2$–30CaO) reach a degradation stop after 3 days in simulated body fluid (SBF) under static conditions due to saturation of the solution in Si.[103]

In addition to the biological effect of TIIs, their influence on the surface chemistry and bioactivity should be taken into account when designing ion-doped BG compositions. Indeed, the incorporation of metallic ions has an influence on the *in vitro* bioactivity, which has been discussed in the literature.[104] When doped into the glass network, metallic ions can affect the material's bioactivity in two ways: first, by stabilising or disrupting the glass network and hence influencing the BG reactivity/bioactivity or, secondly, by affecting the hydroxyapatite formation on the BG surface once released from the BG matrix. Sr or Mg, for example, have been shown to have inhibitory effects on HAp formation on BG surfaces.[1,105] Moreover, HAp formation can be delayed when CoO, ZnO or MgO are incorporated in the glass network, as revealed by SBF immersion studies.[75] This effect, however, could be advantageous for BG applications in soft tissue and cartilage regeneration when no tissue calcification is desired and a reduction of BG bioactivity is needed.

Overall, when analysing the available literature, it becomes evident that BGs are suitable inorganic carries for therapeutic ions, whereby the ion release kinetics can be adjusted *via* tuning the glass compositions. In most cases, the intrinsic BG bioactivity can be retained and a desired compositional window can be found, suitable for incorporation of specific therapeutic ions and also for maintaining the characteristic glass properties.

16.4 Osteogenic Response to Bioactive Glasses (BGs)

16.4.1 Effect of Ca, Si and P from Standard Silicate Glass Compositions

One key property of BGs (besides the so-called acellular bioactivity, which is the ability to facilitate strong bonds to hard as well as soft tissue) is the stimulation of bone cells *via* upregulation of osteogenic genes resulting in enhanced bone regeneration.[85] These effects have been intensively investigated during the last 15 years[2,7,106] and to date the research has been expanding the area of "genetically designed" bioactive glasses, the so-called 3rd generation biomaterials.[2,6]

One early key finding from the studies of Xynos *et al.*[107] in 2001 was that dissolution products released from Bioglass® (45S5 composition, wt%: $45SiO_2$, $24.5Na_2O$, 24.5CaO, $6P_2O_5$) were able to regulate gene expression in human osteoblastic cells (HOC). Since then, the research efforts have increased to consider molecular interactions of ionic dissolution products of BGs and their physiological environment in order to gain a greater understanding of these mechanisms and develop "smart" BGs with tailored properties for specific tissue engineering applications. Several genes known to play a role in osteoblast metabolism, proliferation and cell–cell and matrix–cell adhesion can be upregulated up to 5-fold when human osteoblastic cells are cultured in 45S5 Bioglass®-conditioned medium.[7] Jell *et al.*,[108] for instance, showed six- and three-fold upregulation of the osteogenic markers sialoprotein (BSP) and alkaline phosphatase (ALP) in osteoblasts treated with BG-conditioned culture medium resulting in enhanced osteogenic differentiation.

In other studies it was confirmed that the osteogenic stimulation is not exclusive for 45S5 Bioglass® but was also observed for 1393 BG composition (wt%: $53SiO_2$–$6Na_2O$–$12K_2O$–5MgO–20CaO–$4P_2O_5$). 1393 BG-derived scaffolds have been shown to support osteoblast growth and differentiation *in vitro* as well as bone tissue formation *in vivo*.[109–111] The osteogenic potential of 1393 BG was also observed *in vivo* by implanting particles in bone defects of rat tibia and by investigating gene expression at the defect area.[112] 1393 BGs have been shown to stimulate mRNA expression of genes for both bone formation and bone resorption indicating their effect not only on osteoblasts but also on osteoclasts function, hence improving bone turnover.[112]

The osteogenic potential of BGs extends also to sol–gel-derived glasses which has been verified for a number of compositions, such as 77S BG (mol%: $80SiO_2$–16CaO–$4P_2O_5$) which has been shown to induce osteogenic differentiation of bone marrow stromal cells into osteoblast-like cells and to promote mineralisation.[113] Similarly, ionic dissolution products from sol–gel-derived 58S composition (mol%: $60SiO_2$–36CaO–$4P_2O_5$) have been shown to enhance proliferation of osteoblast cells[114] and to upregulate the expression of a number of genes including IGF-I, gpl30 or MAPK3/ERK1.[115] Interestingly, also sol–gel-derived phosphate-free glasses, such as 70S30C (mol%: $70SiO_2$–30CaO), have been observed to enhance osteoblast maturation and differentiation as well as production of bone-like mineral by osteoblasts suggesting that P may be not necessarily required for *in vitro* mineralization of the extracellular matrix.[116]

The research results introduced above represent just a selection from a large number of *in vitro* and *in vivo* studies which have confirmed the osteogenic effects of bioactive silicate glasses in relation to their dissolution products and which have been comprehensively reviewed in the literature.[1,6,117]

However, the exact mechanisms of the interaction between BG ionic dissolution products and cells are not fully understood and remain the focus of considerable research efforts in the biomaterial community. In this context, elucidating the right time-dependent critical concentrations of the therapeutic ions needed to achieve the optimal required biological effects, as well as synergetic effects, of different TIIs is at the centre of the research interest.

Hench has stated that concentrations of Ca and Si of 60–88 ppm and 17–21 ppm, respectively, are critical in order to stimulate osteogenesis in stem cells.[6] Other studies have highlighted the importance of Si. For example, it has been shown that despite higher Ca concentrations, no increased osteoblast activity was observed in the presence of biphasic calcium phosphates (BCP) compared to treatment with a 60S BG composition releasing Si.[118] Indeed, as discussed previously, this effect corresponds to the essential role of Si in metabolic processes associated with the formation and calcification of hard tissues.[25,79] Relevant for angiogenesis, Si concentrations of 0.1–1.8 $\mu g\, mL^{-1}$ have been shown to stimulate the proliferation of human umbilical vein endothelial cells (HUVECs) and to upregulate the expression of angiogenic factors VEGF and bFGF.[119] On the other hand, the importance of Si in osteogenesis and bone tissue formation and its relevance in the evolution of biomaterials has recently been emphasized and reviewed.[58]

Overall, osteogenic stimulation by BGs might also be attributed to Ca, in addition to Si.[120] Even though extensive Ca concentrations (in the range of 88–109 ppm) were reported to reduce osteoblast cells (Saos-2) proliferation when exposed to products released from "MBG85" (mol%: $85SiO_2$–10CaO–$5P_2O_5$) BG,[121] Ca-sensing receptors are important signalling targets in bone regeneration and bone turnover.[22,122] When investigated individually, and in combination, Ca and Si ions were confirmed to upregulate osteogenic genes in human stem cells.[123] Considering the importance of Ca in the bone regeneration process, Ca-surface enrichment was proposed as a strategy in order to enhance the BG acellular bioactivity which could also enhance the biological performance of BGs.[124] Despite the fact that P-free bioactive silicate glasses have also exhibited osteogenic potential (as mentioned above), phosphorus remains an important agent to be considered for bone tissue formation, taking into account also the wide variety of phosphate-based glasses being investigated for bone tissue engineering.[89] An overview of the biological effects of different phosphate species can be found in the literature.[89]

While the osteostimulating effect of BGs has been widely reported in the literature, little is known about the interaction of BG dissolution products with osteoclast cells which is also relevant for bone tissue engineering as it was recently highlighted in the literature.[125] Indeed, in one of the few studies available, soluble Si has been shown to inhibit osteoclast phenotype gene expression and osteoclast formation and activity.[126] Future studies should therefore expand the current research focus to consider various cell types including osteoclasts. In addition, the individual effect of single ions should be investigated in model systems in order to elucidate the full impact of single TIIs, including Si or Ca, on bone formation, vascularisation and bone remodelling processes, which will generate experimental data to understand, in a quantitative manner, the expected biological effects of combinations of TIIs in tailored concentrations.

16.4.2 Osteogenic Agents Released from BGs

16.4.2.1 Sr-releasing Bioactive Glasses (Sr-BGs)

As described in Section 16.2, Sr is a well-known agent for the treatment of osteoporosis and hence it has been used as a TII in BGs. Melt-derived Sr-BGs have been investigated showing stimulating effects on osteoblast cells while at the same time inhibiting osteoclast function.[127,128] The results indicate that Sr released from BGs can be used to regulate bone turnover and remodelling during the bone healing and regeneration process. A recent study investigated Sr-doped bioactive glass and the effect Sr has on osteogenesis using rat bone marrow-derived mesenchymal stromal cells showing that Sr-substituted BGs promoted osteogenesis in a differentiating bone cell culture model.[129] Interestingly, the authors also detected a reduction in cell viability after addition of BG powder to the cell culture medium which underlines the sensitivity of cell culture model experiments even though the well-known 45S5 BG was used as reference material.

Besides its use in scaffolds for bone regeneration, Sr is an interesting agent also for periodontal therapies as it has been shown to increase proliferation and alkaline phosphate (ALP) activity of periodontal ligament cells (PDLCs) as well as to upregulate osteogenesis/cementogenesis related genes when released from sol–gel-based mesoporous glass.[130] Also Sr-releasing borate glasses have been reported which show a linear Sr ion release profile which could be interesting for applications requiring rapid ion release kinetics.[131]

16.4.2.2 Mg-releasing Bioactive Glasses (Mg-BGs)

Mg-containing BGs have been extensively investigated *in vitro* and *in vivo*. As recently summarised by Diba *et al.*,[132] there is large body of research evidence demonstrating that Mg-BGs improve the attachment, proliferation and osteoblastic differentiation of relevant cell types (osteoblast progenitor cells or stem cells) involved in bone regeneration. For instance, the osteogenic markers collagen I, alkaline phosphatase (ALP) as well as the osteogenic markers Runx2 and osteocalcin were upregulated after treatment of osteoblast-like cells with dissolution extracts from Mg-BGs.[133] Similarly, increased osteoblast proliferation and differentiation [134] as well as enhanced ALP activity in human fetal osteoblastic cells (hFOB 1.19)[135] have been observed when cells were exposed to Mg-BGs.

16.4.2.3 Zn-releasing Bioactive Glasses (Zn-BGs)

Zn-releasing bioactive glass scaffolds have been investigated in relation to their possible osteogenic effects.[136–140] A recent review article summarises fundamentals and applications of Zn-containing glasses in bone regeneration strategies.[141]

Exemplarily, sol–gel-derived glasses containing 5 mol% ZnO were shown to increase ALP activity and proliferation of osteoblast cells.[140] In related research, Oki *et al.* observed increased ALP activity of human fetal osteoblastic cells (hFOB 1.19) when seeded on Zn-containing bioactive glass (5 mol%) discs compared to polystyrene control.[139] The evidence for stimulated osteogenesis induced by Zn released from BGs, however, is not fully concrete. Haimi *et al.*[137] investigated the effects of Zn-doped BG scaffolds (mol%: 10–12 Na_2O, 10–12 K_2O, 4–6 MgO, 10–18 CaO, 1–4 P_2O_5, 1–2 B_2O_3, 0–1 TiO_2, 50–56 SiO_2, 0–5 ZnO) on human adipose stem cells proliferation and osteogenic differentiation revealing no significant effect of Zn-doping on cell activity when cells were seeded on the Zn-doped BG scaffolds. The authors suggested that the possible stimulatory effect of Zn is inhibited through the decreased degradation profile of the bioactive glass caused by the Zn addition. Similar results were reported by Lusvardi *et al.*,[138] who found that Zn-doping provides no significant effect on adhesion and proliferation of osteoblast-like cells (MC3T3-E1) compared to un-doped control glasses, probably due to the slowed degradation of the BG caused by Zn addition resulting in a negligible amount of Zn ions released. Clearly, future studies are required to unveil the actual potential of Zn as TIIs in BGs.

16.4.2.4 Other Therapeutic Ions with Osteogenic Properties

Alternative elements, such as Mn[142] and Nb,[45] have also been proposed as therapeutic ions for bone regeneration since they are known for their stimulating effects on the skeleton and muscles,[11,143] and on osteoblast cells,[45] respectively. Also Li-containing BGs have been investigated in relation to their biocompatibility with osteoblast-like cells and periodontal ligament cells.[144,145] However, research devoted to this group of TIIs is still in its infancy and therefore an amount of future dedicated research in the field can be anticipated.

16.5 Angiogenic Potential of BGs

16.5.1 Angiogenesis in Bone Tissue Engineering

In addition to exhibiting enhanced osteogenic potential, biomaterials for bone tissue engineering should ideally also have angiogenic effects. Vascularisation of the engineered bone construct is essential for successful bone healing since the vasculature of the hard tissue plays a key role in bone remodelling through acting as a reservoir and conduit for bone cells, growth factors and providing key signals involved in bone metabolism.[146] Hence, the successful clinical application of engineered bone constructs depends on achieving a highly functional vascularised network which will result in enhanced bone metabolism and bone formation. In the literature, several approaches for induction of vascularisation in engineering bone tissue have been proposed and include:[147]

(i) Delivery of angiogenic growth factors (GFs, *e.g.* VEGF, bFGF);
(ii) Micro fabrication of blood vessels incorporated into a 3D scaffold;
(iii) Micro surgical techniques whereby the engineered (bone) tissue construct is connected to the host vascular system to induce vascularisation, (*e.g.* in an AV-loop[148] pre-vascularisation model), or
(iv) Pre-vascularisation *in vitro* using endothelial cells (ECs) and/or co-cultures of ECs and osteoblast cells (OCs) cultured in natural or synthetic scaffolds to form a pre-vascular network (micro capillary structures) which will connect to the host vasculature.

The techniques mentioned above show different degrees of complexity including challenging *in vitro* pre-vascularisation with multiple cell types, difficult surgical procedures and, in some cases, excessive use of growth factors which can be critical due to their potentially negative effects and relatively high costs.

More straightforward biomaterials approaches for stimulating vascularisation should target the enhancement of angiogenic properties of the scaffold material itself. Hence, research efforts are increasingly devoted to investigations on the angiogenic effects of biomaterials.[147] One approach includes the loading and subsequent release of angiogenic therapeutic agents from inorganic scaffold materials. In this context, specific TIIs identified as being angiogenic, can be released from BGs for promoting vascularisation. Such approaches are described in this section.

16.5.2 Angiogenic Potential of Standard Silicate Bioactive Glasses

Additionally to their osteostimulating properties, the angiogenic potential of silicate BGs has been reported in the literature.[3–5,149–151] Numerous *in vitro* as well as *in vivo* studies indicate the potential of BGs to stimulate angiogenesis through inducing the proliferation of endothelial cells, up-regulation of relevant genes and enhancement of growth factor release.[4] Day *et al.*[151] reported increased neovascularisation into Bioglass®-coated polymer meshes after these were subcutaneously implanted in rats. Furthermore, it was shown[3] that fibroblast cells (CCD-18Co) cultured in contact with Bioglass® particles secreted increased amounts of vascular endothelial growth factor (VEGF) and basic fibroblast growth factor (bFGF). A comprehensive review (up to 2010) of *in vitro* and *in vivo* evidence supporting the angiogenic potential of bioactive glasses can be found in the literature[4] and more recent relevant reports are discussed in Chapter 15 of this book.

However, it should be noted that the angiogenic effect of BGs seems to be dependent on the morphology and the form in which the BG is applied. It has been shown that the angiogenic effect of bioactive glass is more pronounced in bioactive glass-based scaffolds like BG-loaded collagen sponges,[149] discs,[152] meshes,[151] tubes,[153] and porous glass-ceramic scaffolds[4,154,155] than in composite structures which incorporate and fully embed BGs particles in polymer matrices such as microsphere composites[156] or foams.[157,158] Hence, the influence of the morphology, size and structure in which BGs are present, sometimes in combination with other materials, needs to be taken into account, in addition to the chemical composition, when evaluating BG angiogenic potential.

16.5.3 BGs Doped with Angiogenic Agents

16.5.3.1 Copper-releasing BGs

Even though there is indication of the osteogenic effects of copper, its most recognised function is related to angiogenesis. Several approaches have been made in order to incorporate Cu into biomaterial scaffolds to implement angiogenic characteristics, including Cu addition in calcium phosphates[159,160] and as a cross-linking agent in alginates used as coating on bioactive glass scaffolds.[161] Cu ions released from calcium phosphates were shown to enhance angiogenesis[159] whereas Cu-containing mesoporous bioactive glass scaffolds have been reported to stimulate angiogenesis as well as osteogenesis.[162] Furthermore, Cu ions have been incorporated in phosphate-based glasses,[163] melt-derived 45S5 bioactive glass-derived scaffolds,[102] in sol–gel mesoporous glass-derived scaffolds[162,164,165] as well as in borosilicate and borate glasses.[92,166] All the above-mentioned bioactive glass compositions have been shown to be suitable carriers for controlled Cu release, as discussed in Section 16.2.

Regarding the biological performance of Cu-containing BGs the studies by Wu *et al.*[162] revealed that Cu released from sol–gel-derived BG scaffolds stimulated hypoxia inducible factor (HIF)-1 and vascular endothelial growth factor (VEGF), two important transcription agents in angiogenesis, in human bone marrow stromal cells (hBMSCs). Related research has also shown that sol–gel-derived nano-sized BG particles with a diameter of ~50 nm with Cu release capability exhibit angiogenic and antibacterial properties.[165]

Moreover, Cu released from 45S5 bioactive glass-based scaffolds has been investigated in a co-culture study.[167] It was shown that Cu^{2+} ions released from 45S5 BG-derived scaffolds stimulate BMSCs to secrete VEGF which, in turn, can be used in a co-culture approach in order to promote activation of specific surface markers and formation of tubular-like structures in endothelial cell culture.[167]

The effect of Cu ions released from 45S5 BG on endothelial cell morphology in 3D has been studied in detail.[168] Network length, connectivity and branching as well as proliferation and metabolic activity of endothelial cells were reduced in a dose-dependent manner by BG and Cu-BG ionic release products indicating no direct correlation of pro-angiogenic properties of BG and Cu with endothelial cell activity. While this result is in contrast to earlier literature reports, as discussed in Sections 16.2 and 16.5.2, it reveals the complex nature of interactions of TIIs released from BGs with specific cell types. Stähli *et al.*[168] suggested a maximum glass concentration for non-harmful angiogenic stimulation of endothelial cells which will help in designing experiments for future investigations.

In the same 3D *in vitro* model utilising endothelial cells, Stähli *et al.*[169] investigated the effect of Cu ions released from phosphate glasses (PGs) on the 3D morphology of endothelial cells with similar outcome, *i.e.* revealing a reduction of the endothelial cell network length in a concentration-dependent fashion by CuO-containing phosphate glasses.

Synergetic effects of Cu and Si ions on angiogenesis were recently investigated in a co-culture cell model of human umbilical vein endothelial cells (HUVECs) and human dermal fibroblast (HDFs) cells showing that extracts from Cu-containing silicate ceramic (Cu–$CaSiO_3$) possessed strong stimulatory effects on upregulation of angiogenic growth factors.[170] However, further experimental work is needed which should consider the design of novel specific BG compositions to enable deciphering the synergetic effect of Cu with Si, P, Ca and other typical ions present in BG compositions.

16.5.3.2 Cobalt-releasing BGs

As described in Section 16.2 cobalt has usually been described as an ion inducing hypoxia (low O_2 pressure environment) condition hence stimulating tissue regenerative processes. Co has been proposed as a bioinorganic therapeutic agent to be included in hypoxia-mimicking biomaterials. Creating hypoxia conditions is suggested to be a strategy for activating pro- and anti-angiogenic genes[171] and, in this context, Co is known to activate relevant HIF-1 pathways for regenerative processes by mimicking hypoxia conditions.[41,72]

In this context, Co-containing glasses, co-doped with Zn and Mg, have been proposed as "hypoxia-mimicking" biomaterials for use in bone tissue engineering.[75] Melt-derived Co-containing silicate BGs as carriers for Co^{2+} ions have been reported by Azevedo *et al.*[75] who demonstrated that Co ion concentrations of ~13 ppm were released from BGs in TRIS buffer after 21 days without refreshing the solution.

In a previous work by the authors[101] Co-releasing melt-derived bioactive silicate-based glass scaffolds of 1393 BG composition were developed for use in bone regeneration applications. The results showed that 1393 BG scaffolds can be used for the controlled release of Co^{2+} ions with release rates (after the initial burst release) of between 0.3 ppm day^{-1} and 0.1 ppm day^{-1}, which can be adjusted through the glass composition. Corresponding cell biology studies with endothelial and osteoblast-like cells revealed that inclusion of 1 wt% CoO was not harmful towards osteoblast-like cells and endothelial cells while addition of 5 wt% of CoO was cytotoxic to both cells types.[172] The additions of 1 wt% CoO and 5 wt% CoO led to the release of ~2 ppm and ~12 ppm Co ions, respectively.[101] In conclusion, the release of ~2 ppm Co from 1393-1Co BG may be considered non-critical regarding potential cytotoxicity whereas ~12 ppm Co released from 1393-5Co BG was shown to be cytotoxic under the investigated experimental conditions.

Sol–gel-based Co-containing BGs have been processed to fabricate 3D scaffolds using a foam replica technique.[173] It was shown that Co ions released from BG-derived scaffolds significantly enhanced VEGF protein secretion and HIF-1 expression as well as the expression of bone-related genes in bone marrow-derived stem cells (BMSCs) while supporting BMSCs attachment and proliferation.[173] Wu *et al.*[173] reported controlled release of maximum Co levels of ~20 ppm from Co-containing sol–gel-derived scaffolds. Even though the observed Co^{2+} concentration of ~20 ppm did not cause significant cytotoxicity, a reduction of the bone marrow-derived stem cells' (BMSCs) viability was observed. The therapeutic window of concentration of Co ions released from BGs is rather narrow and potential cytotoxicity risks should be considered very carefully when designing hypoxia mimicking biomaterials based on Co release. Cell biology studies on Co ions, for instance, revealed that 12 ppm, 24 ppm and 48 ppm of $CoCl_2$ stimulated migration, proliferation, and tubule-like structure formation of umbilical cord blood-derived CD133(+) cells, hence indicating the angiogenic potential of Co inducing hypoxic conditions.[174] Similarly, related reports confirmed that treating human microvascular endothelial cells (HMEC-1) with 12 ppm $CoCl_2$ resulted in binding of HIF-1 hence mediating transcriptional responses to hypoxia.[175] By contrast, in other studies, Co levels higher than 10 ppm have been indicated to be potentially cytotoxic as treatment of osteoblast-like cells with 10 ppm of Co^{2+} resulted in a 40% decrease in cell number.[176]

16.5.3.3 Boron-containing BGs

Boron-containing glasses have been identified as an important group of BGs for use in tissue engineering applications. Borate glasses have been shown to exhibit osteogenic as well as angiogenic properties.[78,92,177] In an *in vivo* rat study, for instance, it has been shown that borate glass-derived (1393B3 composition supplemented with 0.4 wt% copper) microfiber meshes promoted extensive angiogenesis compared to silica glass microfibers when implanted subcutaneously.[178] Also, the 45S5 BG composition containing boron has been investigated for its angiogenic properties. In a recent study, boron-containing 45S5 BG was evaluated in an embryonic chorioallantoic membrane (CAM) model which revealed stimulated angiogenesis promoted by 2 wt% B_2O_3 in the glass network.[78] The stimulating effect of the boron-containing 45S5 BG was comparable to the results obtained with the addition of 10 g mL^{-1} of basic fibroblast growth factor in the cell culture medium.[78]

16.6 Antibacterial Properties of BGs

Bioactive glasses exhibit intrinsic antibacterial properties and are being considered for the treatment of infected bone defects. A recently published review article by Rahaman *et al.*[179] summarises the fundamentals and application of silicate and borosilicate glasses for treatment of infected prosthetic joints. The BG S53P4 (wt%: $53SiO_2$, $23Na_2O$, 20CaO, $4P_2O_5$) is being successfully applied in a clinical setting for the treatment of osteomyelitis.[180]

While some BG compositions show antibacterial properties which are mainly related to pH rise in the surroundings of the BG implants or granules,[181] BGs can be doped with antibacterial agents like Ag, Cu, Zn and Ga. The combination of the antibacterial actions of these ions and the osteostimulating and angiogenic effects of BGs as discussed above enables the development of multifunctional biomaterials for orthopaedic application which could lead to replacing the common use of antibiotics (*e.g.* delivered by a bone cement like PMMA) which is the current standard therapy for this indication. Table 16.2 shows examples of typical ion-doped BGs developed for antibacterial applications in the context of bone replacement and repair.

BGs with antibacterial capability investigated include Ga-releasing phosphate glasses,[185,186] Ag-releasing silicate glasses[187] and Cu-containing sol–gel-derived glasses.[164] Soda lime glasses with high CaO content (~20 wt%) were also reported to exhibit antibacterial effect against Gram-positive and Gram-negative bacteria as well as yeasts.[188]

Moreover, Ag-coated Zn-containing BGs have been tested with *Escherichia coli*, *Staphylococcus epidermidis* and *Candida albicans* showing antibacterial and antifouling properties in the zone inhibition test.[189] Beside angiogenic stimulation, Cu ions are also known for their antibacterial properties. Hence, antibacterial Cu-containing mesoporous BG coatings on porous hydroxyapatite were proposed to increase the biocompatibility of implants for orbital floor repair.[190]

Zn-containing mesoporous bioactive glass scaffolds in the system $80SiO_2$–15CaO–$5P_2O_5$ (mol%) containing 4.0 mol% or 7.0 mol% of ZnO were shown

Table 16.2 Selected studies on BGs functionalised with antibacterial agents.

Glass system	Antibacterial agent	Biological effect	Comments	Ref.
B_2O_3–SiO_2–Na_2O–ZnO	Zn	Antibacterial and antifungial	Zn^{2+} release 0.6–7.3 ppm	182
CaO–Na_2O–P_2O_5–AgO	Ag	Bactericidal	Ag release at 0.42–1.22 $\mu g\,mm^{-2}\,h^{-1}$)	183
15CaO–$5P_2O_5$–$80SiO_2$–CuO	Cu	N/A	Cu species release 1–50 ppm	165
P_2O_5–CaO–Na_2O–Ga_2O	Ga	Bactericidal	1 mol% Ga_2O_3 critical concentration; Ga ions of 10–40 ppm released	184

to be suitable for human osteoblast cell (HOC) culture while showing antibacterial effects against Gram-positive *Staph. aureus*. Hereby it was found that 4.0 mol% ZnO-containing glasses were most favourable for inducing HOC activity.[191] Also borosilicate glasses in the system B_2O_3–SiO_2–Na_2O–ZnO containing 15–40 wt% ZnO have been shown to exhibit antibacterial and antifungial activity against *E. coli* (Gram-negative), *S. aureus* (Gram-positive) and *C. krusei* (yeast).[182] The authors reported a Zn^{2+} ion concentration range between 0.6 and 7.3 ppm was released from the glass matrix depending on the glass composition.[182]

The reports mentioned above and studies summarised in Table 16.2 show successful cases of functionalisation of BGs towards antibacterial and antifouling biomaterials using different approaches, including inclusions of antibacterial nano-sized metal oxides, metallic nanoparticles or doping antibacterial ions into the glass network. Future research should determine which of these methods or their combinations are most suitable for the development of multi-functional BGs with antibacterial effects. Indeed, comparison of different BGs, for example incorporating different antibacterial ions, in similar infected bone defect models remains an important task for future studies.

16.7 BGs Beyond Orthopaedics and Bone Tissue Engineering (BTE)

In addition to orthopaedics, dentistry and BTE, new fields for BG use have been increasingly considered. These novel fields include soft tissue regeneration, cosmetics, wound healing, ophthalmology, peripheral nerve regeneration, and cancer treatment. A comprehensive description of BG use in these novel fields can be found in recently published literature[192,193] and the most recent developments are highlighted in Chapter 15 of this book. Within the scope of the present chapter, for completeness, a short overview of specific biological responses to ionic dissolution products relevant for some of these novel applications is presented in this section.

Wound healing is a new and emerging field where BGs, applied as fiber mesh or as particulate or fibrous inclusions in polymer-based graft materials, can stimulate angiogenesis and vascularisation of the wound hence accelerating the healing process while also providing antibacterial properties. One of the challenges for fabrication of functional skin grafts is achieving sufficient vascularisation.[194] Clearly, BGs containing angiogenic agents like Cu and Co represent an interesting group of materials for soft tissue repair.[87] In fact, the angiogenic potential of some TIIs can establish the key element for successful application of BGs in soft tissue engineering.

For nerve regeneration approaches (see also Chapter 18 of this book), for instance, polymer composite-based nerve guidance conducts (NGCs) including BGs containing ZnO and CeO_2 have been proposed,[195,196] whereby the release of Ce, Ca or Zn (which are known to be involved in peripheral nerve regeneration[197,198]) is believed to induce the nerve repair using these materials. PLGA (poly(D,L-lactide-*co*-glycolide)-based composites containing Zn-doped BG particles were shown to exhibit the desired cytocompatibility and tensile strength, being suitable for NGC development.[199] Moreover, fast resorbable phosphate glasses have been proposed as soluble temporary scaffolds which could be considered as a valuable alternative to the end-to-end suture for nerve regeneration therapies.[192] Considering the potential of phosphate glasses as carriers for TIIs, as discussed in Section 16.3, this particular application in nerve regeneration could open opportunities for further development of specific ion-doped phosphate glass compositions to make fibers.

In another innovative approach, BGs are being considered as materials for cancer treatment[200–204] utilising ferromagnetic (bioactive) glasses for hyperthermia treatment.[202–205] Fe-containing sol–gel-derived mesoporous glass scaffolds have been proposed for treatment of malignant bone disease using hyperthermia by inducing heat in the area of diseased bone and destroying tumor cells.[204] Since bioactive glass-based scaffolds show osteogenic properties they could be potentially used as templates for *in situ* bone tissue regeneration thus combining treatment of malignant bone and tissue regeneration in one procedure.[204] Even though in this application the ion release capability of the BG is not the primary function, it shows further potential application areas for BGs containing metallic ions.

16.8 Conclusion and Outlook

BGs of a broad range of compositions have been shown to be suitable carriers of therapeutic ions representing versatile vehicles for controlled ion release, whose kinetics can be tuned *via* the glass composition. There is large amount of experimental data confirming the statement that "ionic dissolution products released from bioactive glasses stimulate the genes of cells towards a path of regeneration and self-repair", as originally discussed by Hench.[6] Moreover, there is strong evidence, as summarised in this chapter, indicating that the biological activity of bioactive glasses can be enhanced by adding therapeutic ions, so widening the physiological functionality of BGs.

Figure 16.4 shows a schematic diagram of the biological response to BGs based on literature reports, as a graphical representation of the information provided in previous sections. Achieving control over the ion release profile is key for future developments aimed at creating advanced therapeutic ion-releasing BGs. Knowledge about ion release kinetics as a function of the BG composition will uncover the full potential of TIIs in biomedical applications, avoiding overdose-related cytotoxicity. Emerging evidence from experimental studies indicates that synergetic effects of single ions are critical in enhancing the biological performance of these materials. However, interactions, such as considering time-dependent variations of the concentration ratios of different ions, remain unknown. As indicated by Figure 16.4 it also becomes evident that combining multiple functionalities in one glass

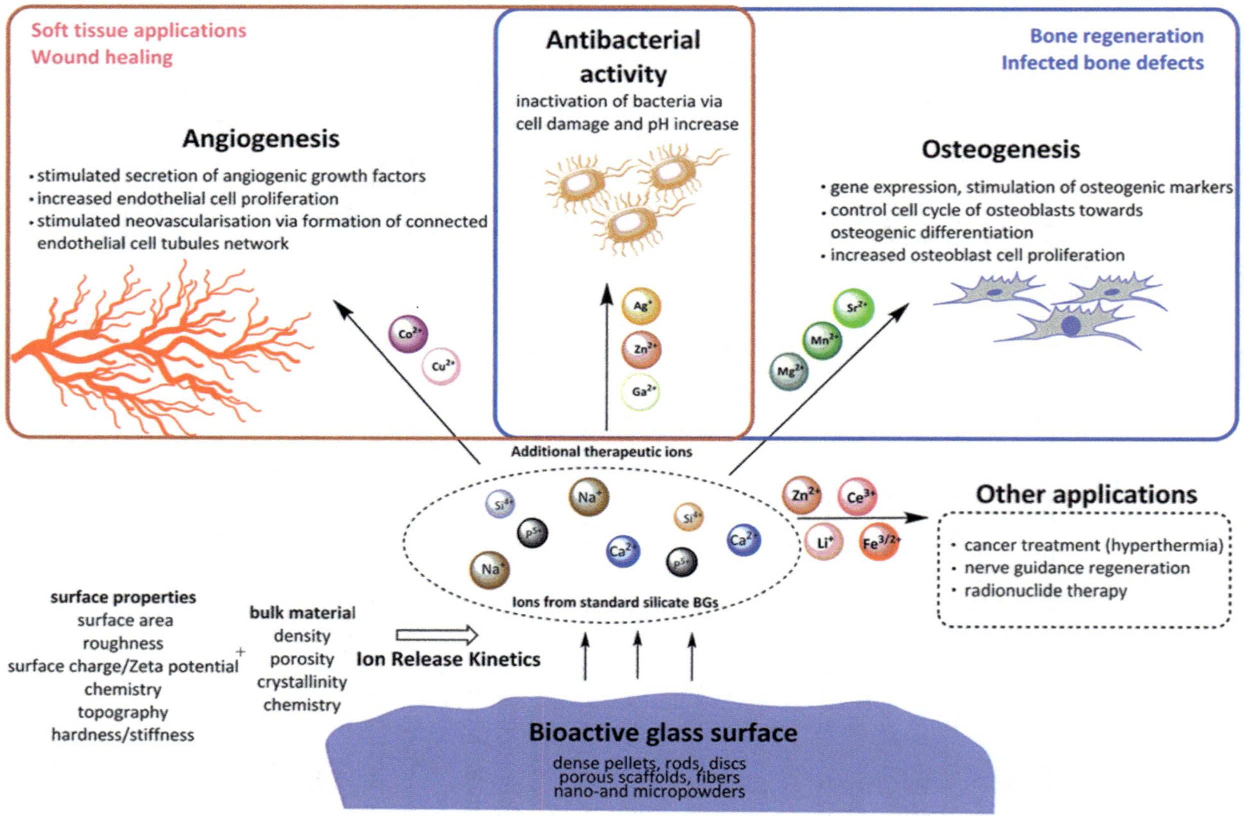

Figure 16.4 Schematic overview of biological responses to ionic dissolution products from bioactive glasses (BGs) and relevant fields of biomedical application. The release of therapeutic metal ions upon the degradation of the glass matrix is proposed to trigger several specific biological responses.
[Adapted from A. Hoppe *et al.*, A review of the biological response to ionic dissolution products from bioactive glasses and glass-ceramics, *Biomaterials*, **32**, 2757–2774, Copyright 2011 with permission from Elsevier.[1]]

composition using TIIs will open new application fields for BGs, for example wound healing and treatment of infected bone defects when combining, respectively, antibacterial properties with either angiogenic or osteogenic potential.

In order to fully exploit the effect of ions released from standard silicate glasses (mainly including Ca, P, Si, Na) and BGs with TIIs release, the following aspects are critical and should be considered for future research:

(i) Determination of critical concentration of single ions and their combinations,
(ii) Control of ion release kinetics and degradation of the glass and understanding of the mechanism involved,
(iii) Understanding the evolution and fate of the ions after being released from BG matrix,
(iv) Establishing of a model for the systematic assessment of the biological impact of ions, considering morphology, surface area, crystallinity and porosity of the BG materials, *e.g.* in the forms of coatings, granules, fibers, or scaffolds.

Future studies should focus on specific ions, considered individually and in combination, in both *in vitro* and *in vivo* models.

In particular, changing the glass chemical composition will result not only in the controlled release of specific ions but might also change the overall release kinetics of basic glass elements like Si, Ca, or P, which should be taken into account while investigating the overall biological performance of the material. These types of studies are also relevant for phosphate and borate glasses incorporating TIIs.

It is anticipated that the incorporation of TIIs with specific biological function will widen the application window of bioactive glasses towards novel fields in tissue regeneration, including soft tissue repair, wound healing and nerve guidance regeneration to name but a few (see also the related Chapters 15 and 18 of this book).

In order to further advance the field a multi-scale approach combining chemistry with novel manufacturing techniques (*e.g.* Additive Layer Manufacturing) for developing strong scaffolds, hierarchically designed materials and nano-scaled bioactive glasses could open up further application fields for bioactive glasses and their respective composites in combination with respective biopolymers. Indeed, the use of BGs in tissue regeneration using a multi-scale approach of mesoporosity, porous macrostructure and ion release, aimed at enhancing the biological impact of BGs has been discussed in the literature[186,206] (see also Chapter 17 of this book).

References

1. A. Hoppe, N. S. Güldal and A. R. Boccaccini, *Biomaterials*, 2011, **32**, 2757.

2. L. L. Hench and J. M. Polak, *Science*, 2002, **295**, 1014.
3. R. M. Day, *Tissue Eng.*, 2005, **11**, 768.
4. A. Gorustovich, J. Roether and A. R. Boccaccini, *Tissue Eng., Part B*, 2010, **16**, 199.
5. A. Leu and J. Leach, *Pharm. Res.*, 2008, **25**, 1222.
6. L. L. Hench, *J. Eur. Ceram. Soc.*, 2009, **29**, 1257.
7. I. D. Xynos, A. J. Edgar, L. D. K. Buttery, L. L. Hench and J. M. Polak, *J. Biomed. Mater. Res.*, 2001, **55**, 151.
8. A. Hoppe, V. Mourino and A. R. Boccaccini, *Biomater. Sci.*, 2013, **1**, 254.
9. V. Mouriño, J. P. Cattalini and A. R. Boccaccini, *J. R. Soc., Interface*, 2012, **9**, 401.
10. P. D. Saltman and L. G. Strause, *J. Am. Coll. Nutr.*, 1993, **12**, 384.
11. J. H. Beattie and A. Avenell, *Nutr. Res. Rev.*, 1992, **5**, 167.
12. F. Nielsen, *Biol. Trace Elem. Res.*, 1990, **26–27**, 599.
13. C. Ash and R. Stone, *Science*, 2003, **300**, 925.
14. P. Habibovic and J. E. Barralet, *Acta Biomater.*, 2011, **7**, 3013.
15. M. A. Saghiri, A. Asatourian, C. M. Sorenson and N. Sheibani, *J. Endod.*, 2015, **41**, 797.
16. M. A. Saghiri, A. Asatourian, J. Orangi, C. M. Sorenson and N. Sheibani, *Crit. Rev. Oncol. Hematol.*, 2015, **41**, 797.
17. K. H. Thompson and C. Orvig, *Science*, 2003, **300**, 936.
18. Z. L. Sun, J. C. Wataha and C. T. Hanks, *J. Biomed. Mater. Res.*, 1997, **34**, 29.
19. V. Mourino and A. R. Boccaccini, *J. R. Soc., Interface*, 2010, **7**, 209.
20. E. Hinoi, T. Takarada and Y. Yoneda, *J. Pharmacol. Sci.*, 2004, **94**, 215.
21. S. Maeno, Y. Niki, H. Matsumoto, H. Morioka, T. Yatabe, A. Funayama, Y. Toyama, T. Taguchi and J. Tanaka, *Biomaterials*, 2005, **26**, 4847.
22. P. J. Marie, *Bone*, 2010, **46**, 571.
23. P. Valerio, M. M. Pereira, A. M. Goes and M. F. Leite, *Biomed. Mater.*, 2009, 045011.
24. M. Julien, S. Khoshniat, A. Lacreusette, M. Gatius, A. Bozec, E. F. Wagner, Y. Wittrant, M. Masson, P. Weiss, L. Beck, D. Magne and J. Guicheux, *J. Bone Miner. Res.*, 2009, **24**, 1856.
25. E. Carlisle, *Calcif. Tissue Int.*, 1981, **33**, 27.
26. D. M. Reffitt, N. Ogston, R. Jugdaohsingh, H. F. J. Cheung, B. A. J. Evans, R. P. H. Thompson, J. J. Powell and G. N. Hampson, *Bone*, 2003, **32**, 127.
27. P. Delannoy, D. Bazot and P. J. Marie, *Metabolism*, 2002, **51**, 906.
28. M. D. Grynpas and P. J. Marie, *Bone*, 1990, **11**, 313.
29. P. J. Marie, P. Ammann, G. Boivin and C. Rey, *Calcif. Tissue Int.*, 2001, **69**, 121.
30. M. Shahnazari, N. A. Sharkey, G. J. Fosmire and R. M. Leach, *J. Bone Miner. Res.*, 2006, **21**, 1696.
31. P. J. Marie, *Bone*, 2006, **38**, 10.
32. J. Brandão-Neto, V. Stefan, B. B. Mendonça, W. Bloise and A. V. B. Castro, *Nutr. Res.*, 1995, **15**, 335.

33. M. Yamaguchi, *J. Trace Elem. Exp. Med.*, 1998, **11**, 119.
34. M. K. Choi, M. H. Kim and M. H. Kang, *Food Sci. Biotechnol.*, 2005, **14**, 242.
35. T. Uysal, A. Ustdal, M. F. Sonmez and F. Ozturk, *Angle Orthod.*, 2009, **79**, 984.
36. D. A. Barrio and S. B. Etcheverry, *Can. J. Physiol. Pharmacol.*, 2006, **84**, 677.
37. A. M. Cortizo, M. S. Molinuevo, D. A. Barrio and L. Bruzzone, *Int. J. Biochem. Cell Biol.*, 2006, **38**, 1171.
38. R. Buttyan, P. Chichester, B. Stisser, S. Matsumoto, M. A. Ghafar and R. M. Levin, *J. Urol.*, 2003, **169**, 2402.
39. S. Patntirapong, P. Habibovic and P. V. Hauschka, *Biomaterials*, 2009, **30**, 548.
40. K. Peters, H. Schmidt, E. R. Unger, M. Otto, Kamp, J. C. Kirkpatrick, *Software-Supported Image Quantification of Angiogenesis in an in Vitro Culture System: Application to Studies of Biocompatibility*, Elsevier, Oxford, ROYAUME-UNI, 2002.
41. T. Tanaka, I. Kojima, T. Ohse, J. R. Ingelfinger, S. Adler, T. Fujita and M. Nangaku, *Lab. Invest.*, 2005, **85**, 1292.
42. H. Xie and Y. J. Kang, *Curr. Med. Chem.*, 2009, **16**, 1304.
43. R. Uauy, M. Olivares and M. Gonzalez, *Am. J. Clin. Nutr.*, 1998, **67**, 952S.
44. J. Bartolomé, J. Moya, R. Couceiro, C. Gutiérrez-González, F. Guitián and A. Martinez-Insua, *Biomaterials*, 2015, **76**, 313.
45. A. Obata, Y. Takahashi, T. Miyajima, K. Ueda, T. Narushima and T. Kasuga, *ACS Appl. Mater. Interfaces*, 2012, **4**, 5684.
46. J. Makoukji, M. Belle, D. Meffre, R. Stassart, J. Grenier, G. G. Shackleford, R. Fledrich, C. Fonte, J. Branchu, M. Goulard, C. de Waele, F. Charbonnier, M. W. Sereda, E.-E. Baulieu, M. Schumacher, S. Bernard and C. Massaad, *Proc. Natl. Acad. Sci.*, 2012, **109**, 3973.
47. P. Clément-Lacroix, M. Ai, F. Morvan, S. Roman-Roman, B. Vayssière, C. Belleville, K. Estrera, M. L. Warman, R. Baron and G. Rawadi, *Proc. Natl. Acad. Sci. U. S. A.*, 2005, **102**, 17406.
48. S. Chigurupati, M. R. Mughal, E. Okun, S. Das, A. Kumar, M. McCaffery, S. Seal and M. P. Mattson, *Biomaterials*, 2013, **34**, 2194.
49. S. Das, S. Singh, J. M. Dowding, S. Oommen, A. Kumar, T. X. T. Sayle, S. Saraf, C. R. Patra, N. E. Vlahakis, D. C. Sayle, W. T. Self and S. Seal, *Biomaterials*, 2012, **33**, 7746.
50. A. Hartwig, *Mutat. Res., Fundam. Mol. Mech. Mutagen.*, 2001, **475**, 113.
51. R. K. Rude, H. E. Gruber, H. J. Norton, L. Y. Wei, A. Frausto and J. Kilburn, *Bone*, 2005, **37**, 211.
52. M. P. Staiger, A. M. Pietak, J. Huadmai and G. Dias, *Biomaterials*, 2006, **27**, 1728.
53. S. Tsuboi, H. Nakagaki, K. Ishiguro, K. Kondo, M. Mukai, C. Robinson and J. A. Weatherell, *Calcif. Tissue Int.*, 1994, **54**, 34.
54. H. Zreiqat, C. R. Howlett, A. Zannettino, P. Evans, G. Schulze-Tanzil, C. Knabe and M. Shakibaei, *J. Biomed. Mater. Res.*, 2002, **62**, 175.

55. M. A. Saghiri, A. Asatourian, J. Orangi, C. M. Sorenson and N. Sheibani, *Crit. Rev. Oncol. Hematol.*, 2015, **96**, 129.
56. R. Jugdaohsingh, K. L. Tucker, N. Qiao, L. A. Cupples, D. P. Kiel and J. J. Powell, *J. Bone Miner. Res.*, 2004, **19**, 297.
57. F. H. Nielsen and R. Poellot, *J. Trace Elem. Exp. Med.*, 2004, **17**, 137.
58. J. R. Henstock, L. T. Canham and S. I. Anderson, *Acta Biomater.*, 2015, **11**, 17.
59. Y. Yamasaki, Y. Yoshida, M. Okazaki, A. Shimazu, T. Uchida, T. Kubo, Y. Akagawa, Y. Hamada, J. Takahashi and N. Matsuura, *J. Biomed. Mater. Res.*, 2002, **62**, 99.
60. G. Boivin, P. Deloffre, B. Perrat, G. Panczer, M. Boudeulle, Y. Mauras, P. Allain, Y. Tsouderos and P. J. Meunier, *J. Bone Miner. Res.*, 1996, **11**, 1302.
61. P. J. Meunier, D. O. Slosman, P. D. Delmas, J. L. Sebert, M. L. Brandi, C. Albanese, R. Lorenc, S. Pors-Nielsen, M. C. De Vernejoul, A. Roces and J. Y. Reginster, *J. Clin. Endocrinol. Metab.*, 2002, **87**, 2060.
62. B. Habermann, K. Kafchitsas, G. Olender, P. Augat and A. Kurth, *Calcif. Tissue Int.*, 2010, **86**, 82.
63. C. Lang, C. Murgia, M. Leong, L.-W. Tan, G. Perozzi, D. Knight, R. Ruffin and P. Zalewski, *Am. J. Physiol.: Lung Cell. Mol. Physiol.*, 2007, **292**, L577.
64. I.-S. Kwun, Y.-E. Cho, R.-A. R. Lomeda, H.-I. Shin, J.-Y. Choi, Y.-H. Kang and J. H. Beattie, *Bone*, 2010, **46**, 732.
65. L. Finney, S. Vogt, T. Fukai and D. Glesne, *Clin. Exp. Pharmacol. Physiol.*, 2009, **36**, 88.
66. C. Gerard, L. J. Bordeleau, J. Barralet and C. J. Doillon, *Biomaterials*, 2010, **31**, 824.
67. G-f. Hu, *J. Cell. Biochem.*, 1998, **69**, 326.
68. J. P. Rodríguez, S. Ríos and M. González, *J. Cell. Biochem.*, 2002, **85**, 92.
69. J. C. Zhang, J. A. Huang, S. J. Xu, K. Wang and S. F. Yu, *Prog. Nat. Sci.*, 2003, **13**, 266.
70. K. Peters, H. Schmidt, R. Unger, G. Kamp, F. Pröls, B. Berger and C. J. Kirkpatrick, *Mol. Cell. Biochem.*, 2005, **270**, 157.
71. G. Chachami, G. Simos, A. Hatziefthimiou, S. Bonanou, P.-A. Molyvdas and E. Paraskeva, *Am. J. Respir. Cell Mol. Biol.*, 2004, **31**, 544.
72. G. L. Semenza, *Science*, 2007, **318**, 62.
73. C. Wan, S. R. Gilbert, Y. Wang, X. Cao, X. Shen, G. Ramaswamy, K. A. Jacobsen, Z. S. Alaql, A. W. Eberhardt, L. C. Gerstenfeld, T. A. Einhorn, L. Deng and T. L. Clemens, *Proc. Natl. Acad. Sci. U. S. A.*, 2008, **105**, 686.
74. P. J. Emans, F. Spaapen, D. A. M. Surtel, K. M. Reilly, A. Cremers, L. W. van Rhijn, S. K. Bulstra, J. W. Voncken and R. Kuijer, *Bone*, 2007, **40**, 409.
75. M. M. Azevedo, G. Jell, M. D. O'Donnell, R. V. Law, R. G. Hill and M. M. Stevens, *J. Mater. Chem.*, 2010, **20**, 8854.
76. W. Fan, R. Crawford and Y. Xiao, *Biomaterials*, 2010, **31**, 3580.

77. M. Dzondo-Gadet, R. Mayap-Nzietchueng, K. Hess, P. Nabet, F. Belleville and B. Dousset, *Biol. Trace Elem. Res.*, 2002, **85**, 23.
78. L. A. Haro Durand, G. E. Vargas, N. M. Romero, R. Vera-Mesones, J. M. Porto-López, A. R. Boccaccini, M. P. Zago, A. Baldi and A. Gorustovich, *J. Mater. Chem. B*, 2015, **3**, 1142.
79. E. M. Carlisle, *Science*, 1970, **167**, 279.
80. J. J. M. Damen and J. M. Ten Cate, *J. Dent. Res.*, 1992, **71**, 453.
81. B. R. McAuslan, W. G. Reilly, G. N. Hannan and G. A. Gole, *Microvasc. Res.*, 1983, **26**, 323.
82. F. H. Nielsen, *Nutr. Rev.*, 2008, **66**, 183.
83. I. Celardo, J. Z. Pedersen, E. Traversa and L. Ghibelli, *Nanoscale*, 2011, **3**, 1411.
84. R. E. Chapin, W. W. Ku, M. A. Kenney, H. McCoy, B. Gladen, R. N. Wine, R. Wilson and M. R. Elwell, *Fundam. Appl. Toxicol.*, 1997, **35**, 205.
85. L. L. Hench, *J. Mater. Sci.: Mater. Med.*, 2006, **17**, 967.
86. J. R. Jones, *Acta Biomater.*, 2013, **9**, 4457.
87. V. Miguez-Pacheco, L. L. Hench and A. R. Boccaccini, *Acta Biomater.*, 2015, **13**, 1.
88. C. Wu and J. Chang, *J. Controlled Release*, 2014, **193**, 282.
89. N. J. Lakhkar, I. H. Lee, H. W. Kim, V. Salih, I. B. Wall and J. C. Knowles, *Adv. Drug Delivery Rev.*, 2013, **65**, 405.
90. M. N. Rahaman, D. E. Day, B. S. Bal, Q. Fu, S. B. Jung, L. F. Bonewald and A. P. Tomsia, *Acta Biomater.*, 2011, **7**, 2355.
91. X. Liu, M. N. Rahaman and D. E. Day, *J. Am. Ceram. Soc.*, 2014, **97**, 3501.
92. R. J. Watters, R. F. Brown and D. E. Day, *Biomed. glass.*, 2015, **1**, 2299.
93. G. Kaur, O. P. Pandey, K. Singh, D. Homa, B. Scott and G. Pickrell, *J. Biomed. Mater. Res., Part A*, 2013, 1.
94. H. B. Pan, X. L. Zhao, X. Zhang, K. B. Zhang, L. C. Li, Z. Y. Li, W. M. Lam, W. W. Lu, D. P. Wang, W. H. Huang, K. L. Lin and J. Chang, *J. R. Soc., Interface*, 2010, **7**, 1025.
95. X. Liu, W. Huang, H. Fu, A. Yao, D. Wang, H. Pan, W. Lu, X. Jiang and X. Zhang, *J. Mater. Sci.: Mater. Med.*, 2009, **20**, 1237.
96. M. Erol-Taygun, K. Zheng and A. R. Boccaccini, *Int. J. Appl. Glass Sci.*, 2013, **4**, 136.
97. D. S. Brauer, *Angew. Chem., Int. Ed.*, 2015, **54**, 4160.
98. Q. A. Fu, M. N. Rahaman, H. L. Fu and X. Liu, *J. Biomed. Mater. Res., Part A*, 2010, **95A**, 164.
99. A. Kiani, N. J. Lakhkar, V. Salih, M. E. Smith, J. V. Hanna, R. J. Newport, D. M. Pickup and J. C. Knowles, *Philos. Trans. R. Soc., A*, 2012, **370**, 1352.
100. C. Stähli, M. Shah Mohammadi, K. E. Waters and S. N. Nazhat, *Acta Biomater.*, 2014, **10**, 3317.
101. A. Hoppe, B. Jokic, D. Janackovic, T. Fey, P. Greil, S. Romeis, J. Schmidt, W. Peukert, J. Lao, E. Jallot and A. R. Boccaccini, *ACS Appl. Mater. Interfaces*, 2014, **6**, 2865.

102. A. Hoppe, R. Meszaros, C. Stahli, S. Romeis, J. Schmidt, W. Peukert, B. Marelli, S. N. Nazhat, L. Wondraczek, J. Lao, E. Jallot and A. R. Boccaccini, *J. Mater. Chem. B*, 2013, **1**, 5659.
103. D. Zhang, H. Jain, M. Hupa and L. Hupa, *J. Am. Ceram. Soc.*, 2012, **95**, 2687.
104. S. M. Rabiee, N. Nazparvar, M. Azizian, D. Vashaee and L. Tayebi, *Ceram. Int.*, 2015, **41**, 7241.
105. A. Hoppe, B. Sarker, R. Detsch, N. Hild, D. Mohn, W. J. Stark and A. R. Boccaccini, *J. Non-Cryst. Solids*, 2014, **387**, 41.
106. L. L. Hench, I. D. Xynos and J. M. Polak, *J. Biomater. Sci., Polym. Ed.*, 2004, **15**, 543.
107. I. D. Xynos, A. J. Edgar, L. D. K. Buttery, L. L. Hench and J. M. Polak, *J. Biomed. Mater. Res.*, 2001, **55**, 151.
108. G. Jell, I. Notingher, O. Tsigkou, P. Notingher, J. M. Polak, L. L. Hench and M. M. Stevens, *J. Biomed. Mater. Res., Part A*, 2008, **86A**, 31.
109. Q. Fu, M. N. Rahaman, B. S. Bal and R. F. Brown, *J. Mater. Sci.: Mater. Med.*, 2009, **20**, 1159.
110. Q. Fu, M. N. Rahaman, B. S. Bal and R. F. Brown, *J. Biomed. Mater. Res., Part A*, 2010, **93A**, 1380.
111. Q. Fu, M. N. Ramahan, B. S. Bal, K. Kuroki and R. F. Brown, *J. Biomed. Mater. Res., Part A*, 2010, **95A**, 235.
112. V.-V. Välimäki, J. J. Yrjans, E. I. Vuorio and H. T. Aro, *Tissue Eng.*, 2005, **11**, 387.
113. M. Bosetti and M. Cannas, *Biomaterials*, 2005, **26**, 3873.
114. R. C. Bielby, I. S. Christodoulou, R. S. Pryce, W. J. P. Radford, L. L. Hench and J. M. Polak, *Tissue Eng.*, 2004, **10**, 1018.
115. I. Christodoulou, L. D. K. Buttery, G. Tai, L. L. Hench and J. M. Polak, *J. Biomed. Mater. Res., Part B*, 2006, **77B**, 431.
116. J. R. Jones, O. Tsigkou, E. E. Coates, M. M. Stevens, J. M. Polak and L. L. Hench, *Biomaterials*, 2007, **28**, 1653.
117. G. Jell and M. Stevens, *J. Mater. Sci.: Mater. Med.*, 2006, **17**, 997.
118. P. Valerio, M. M. Pereira, A. M. Goes and M. F. Leite, *Biomaterials*, 2004, **25**, 2941.
119. H. Li and Chang, *J. Acta Biomater.*, 2013, **9**, 5379.
120. M. Navarro, C. Aparicio, M. Charles-Harris, M. P. Ginebra, E. Engel, J. A. Planell Development of a biodegradable composite scaffold for bone tissue engineering: Physicochemical, topographical, mechanical, degradation, and biological properties. *Ordered Polymeric Nanostructures at Surfaces*, Springer-Verlag, Berlin, 2006, vol. 200, p. 209.
121. M. Alcaide, P. Portolés, A. López-Noriega, D. Arcos, M. Vallet-Regí and M. T. Portolés, *Acta Biomater.*, 2010, **6**, 892.
122. E. M. Brown, *Best Pract. Res., Clin. Endocrinol. Metab.*, 2013, **27**, 333.
123. V. G. Varanasi, K. K. Leong, L. M. Dominia, S. M. Jue, P. M. Loomer and G. W. Marshall, *J. Oral Implantol.*, 2012, **38**, 325.
124. J. H. Lopes, I. O. Mazali, R. Landers and C. A. Bertran, *J. Am. Ceram. Soc.*, 2013, **96**, 1464.

125. R. Detsch and A. R. Boccaccini, *J. Tissue Eng. Regener. Med.*, 2014, 29.
126. Ž. Mladenović, A. Johansson, B. Willman, K. Shahabi, E. Björn and M. Ransjö, *Acta Biomater.*, 2014, **10**, 406.
127. E. Gentleman, Y. C. Fredholm, G. Jell, N. Lotfibakhshaiesh, M. D. O'Donnell, R. G. Hill and M. M. Stevens, *Biomaterials*, 2010, **31**, 3949.
128. A. A. Gorustovich, T. Steimetz, R. L. Cabrini and J. M. P. López, *J. Biomed. Mater. Res., Part A*, 2010, **92A**, 232.
129. M. E. Santocildes-Romero, A. Crawford, P. V. Hatton, R. L. Goodchild, I. M. Reaney and C. A. Miller, *J. Tissue Eng. Regener. Med.*, 2015, **9**, 619.
130. C. Wu, Y. Zhou, C. Lin, J. Chang and Y. Xiao, *Acta Biomater.*, 2012, **8**, 3805.
131. K. O'Connell, M. Hanson, H. O'Shea and D. Boyd, *J. Non-Cryst. Solids*, 2015, **430**, 1.
132. M. Diba, F. Tapia, A. R. Boccaccini and L. A. Strobel, *Int. J. Appl. Glass Sci.*, 2012, **3**, 221.
133. V. G. Varanasi, E. Saiz, P. M. Loomer, B. Ancheta, N. Uritani, S. P. Ho, A. P. Tomsia, S. J. Marshall and G. W. Marshall, *Acta Biomater.*, 2009, **5**, 3536.
134. X. Chen, X. Liao, Z. Huang, P. You, C. Chen, Y. Kang and G. Yin, *J. Biomed. Mater. Res., Part B*, 2010, **93B**, 194.
135. A. Saboori, M. Rabiee, F. Moztarzadeh, M. Sheikhi, M. Tahriri and M. Karimi, *Mater. Sci. Eng., C*, 2009, **29**, 335.
136. R. L. Du, J. Chang, S. Y. Ni, W. Y. Zhai and J. Y. Wang, *J. Biomater. Appl.*, 2006, **20**, 341.
137. S. Haimi, G. Gorianc, L. Moimas, B. Lindroos, H. Huhtala, S. Räty, H. Kuokkanen, G. K. Sándor, C. Schmid, S. Miettinen and R. Suuronen, *Acta Biomater.*, 2009, **5**, 3122.
138. G. Lusvardi, G. Malavasi, L. Menabue, M. C. Menziani, A. Pedone, U. Segre, V. Aina, A. Perardi, C. Morterra, F. Boccafoschi, S. Gatti, M. Bosetti and M. Cannas, *J. Biomater. Appl.*, 2008, **22**, 505.
139. A. Oki, B. Parveen, S. Hossain, S. Adeniji and H. Donahue, *J. Biomed. Mater. Res., Part A*, 2004, **69A**, 216.
140. A. Balamurugan, G. Balossier, S. Kannan, J. Michel, A. H. S. Rebelo and J. M. F. Ferreira, *Acta Biomater.*, 2007, **3**, 255.
141. P. Balasubramanian, A. Strobel Leonie, U. Kneser and R. Boccaccini Aldo, Zinc-containing bioactive glasses for bone regeneration, dental and orthopedic applications, *Biomed. Glasses*, 2015, **1**, 2299.
142. M. Miola, C. V. Brovarone, G. Maina, F. Rossi, L. Bergandi, D. Ghigo, S. Saracino, M. Maggiora, R. A. Canuto, G. Muzio and E. Vernè, *Mater. Sci. Eng., C*, 2014, **38**, 107.
143. N. Zarkovic, Z. Ilic, M. Jurin, R. J. Schaur, H. Puhl and H. Esterbauer, *Cell Biochem. Funct.*, 1993, **11**, 279.
144. P. Han, C. Wu, J. Chang and Y. Xiao, *Biomaterials*, 2012, **33**, 6370.
145. M. Khorami, S. Hesaraki, A. Behnamghader, H. Nazarian and S. Shahrabi, *Mater. Sci. Eng., C*, 2011, **31**, 1584.

146. M. L. Brandi and P. Collin-Osdoby, *J. Bone Miner. Res.*, 2006, **21**, 183.
147. M. I. Santos and R. L. Reis, *Macromol. Biosci.*, 2010, **10**, 12.
148. E. Polykandriotis, A. Arkudas, S. Euler, J. P. Beier, R. E. Horch and U. Kneser, *Handchir. Mikrochir. Plast. Chir.*, 2006, **38**, 217.
149. A. Leu, S. M. Stieger, P. Dayton, K. W. Ferrara and J. K. Leach, *Tissue Eng., Part A*, 2009, **15**, 877.
150. S. Deb, R. Mandegaran and L. Di Silvio, *J. Mater. Sci.: Mater. Med.*, 2010, **21**, 893.
151. R. M. Day, A. R. Boccaccini, S. Shurey, J. A. Roether, A. Forbes, L. L. Hench and S. M. Gabe, *Biomaterials*, 2004, **25**, 5857.
152. A. L. Andrade, S. P. Andrade and R. Z. Domingues, *J. Biomed. Mater. Res., Part B*, 2006, **79B**, 122.
153. E. A. Ross, C. D. Batich, W. L. Clapp, J. E. Sallustio and N. C. Lee, *Kidney Int.*, 2003, **63**, 702.
154. J. Mahmood, H. Takita, Y. Ojima, M. Kobayashi, T. Kohgo and Y. Kuboki, *J. Biochem.*, 2001, **129**, 163.
155. S. K. Nandi, B. Kundu, S. Datta, D. K. De and D. Basu, *Res. Vet. Sci.*, 2009, **86**, 162.
156. H. Keshaw, G. Georgiou, J. J. Blaker, A. Forbes, J. C. Knowles and R. M. Day, *Tissue Eng., Part A*, 2009, **15**, 1451.
157. H. Y. Choi, J. E. Lee, H. J. Park and B. S. Oum, *Ophthal. Plast. Reconstr. Surg.*, 2006, **22**, 121.
158. R. M. Day, V. Maquet, A. R. Boccaccini, R. Jerome and A. Forbes, *J. Biomed. Mater. Res., Part A*, 2005, **75**, 778.
159. J. Barralet, U. Gbureck, P. Habibovic, E. Vorndran, C. Gerard and C. J. Doillon, *Tissue Eng., Part A*, 2009, **15**, 1601.
160. A. Ewald, C. Käppel, E. Vorndran, C. Moseke, M. Gelinsky and U. Gbureck, *J. Biomed. Mater. Res., Part A*, 2012, **100 A**, 2392.
161. M. M. Erol, V. Mouriño, P. Newby, X. Chatzistavrou, J. A. Roether, L. Hupa and A. R. Boccaccini, *Acta Biomater.*, 2012, **8**, 792.
162. C. Wu, Y. Zhou, M. Xu, P. Han, L. Chen, J. Chang and Y. Xiao, *Biomaterials*, 2013, **34**, 422.
163. C. Stähli, N. Muja and S. N. Nazhat, *Tissue Eng., Part A*, 2013, **19**, 548.
164. J. Bejarano, P. Caviedes and H. Palza, *Biomed. Mater.*, 2015, 10.
165. V. Aina, G. Cerrato, G. Martra, G. Malavasi, G. Lusvardi and L. Menabue, *Appl. Surf. Sci.*, 2013, **283**, 240.
166. H. Wang, S. Zhao, J. Zhou, Y. Shen, W. Huang, C. Zhang, M. N. Rahaman and D. Wang, *J. Mater. Chem. B*, 2014, **2**, 8547.
167. S. N. Rath, A. Brandl, D. Hiller, A. Hoppe, U. Gbureck, R. E. Horch, A. R. Boccaccini and U. Kneser, *PLoS One*, 2014, 9.
168. C. Stähli, M. James-Bhasin, A. Hoppe, A. R. Boccaccini and S. N. Nazhat, *Acta Biomater.*, 2015, **19**, 15.
169. C. Stähli, M. James-Bhasin and S. N. Nazhat, *J. Controlled Release*, 2015, **200**, 222.
170. N. Kong, K. Lin, H. Li and J. Chang, *J. Mater. Chem. B*, 2014, **2**, 1100.

171. R. K. Jain, P. Au, J. Tam, D. G. Duda and D. Fukumura, *Nat. Biotechnol.*, 2005, **23**, 821.
172. A. Hoppe, A. Brandl, O. Bleiziffer, A. Arkudas, R. E. Horch, B. Jokic, D. Janackovic and A. R. Boccaccini, *Mater. Sci. Eng., C*, 2015, **57**, 157.
173. C. Wu, Y. Zhou, W. Fan, P. Han, J. Chang, J. Yuen, M. Zhang and Y. Xiao, *Biomaterials*, 2012, **33**, 2076.
174. T. Zan, Z. J. Du, H. Li, Q. F. Li and B. Gu, *Front. Biosci., Landmark Ed.*, 2012, **17**, 2247.
175. A. Minchenko and J. Caro, *Mol. Cell. Biochem.*, 2000, **208**, 53.
176. C. Fleury, A. Petit, F. Mwale, J. Antoniou, D. J. Zukor, M. Tabrizian and O. L. Huk, *Biomaterials*, 2006, **27**, 3351.
177. C. Wu, R. Miron, A. Sculean, S. Kaskel, T. Doert, R. Schulze and Y. Zhang, *Biomaterials*, 2011, **32**, 7068.
178. Y. Lin, R. F. Brown, S. B. Jung and D. E. Day, *J. Biomed. Mater. Res., Part A*, 2014, **102**, 4491.
179. M. N. Rahaman, B. S. Bal and W. Huang, *Mater. Sci. Eng., C*, 2014, **41**, 224.
180. N. C. Lindfors, Bioactive glass S53P4 as a bone graft substitute in the treatment of osteomyelitis, in *Bioactive Glasses*, ed. H. O. Ylänen, Woodhead Publishing, 2011, ch. 9, p. 209.
181. M. Gubler, T. J. Brunner, M. Zehnder, T. Waltimo, B. Sener and W. J. Stark, *Int. Endod. J.*, 2008, **41**, 670.
182. L. Esteban-Tejeda, C. Prado, B. Cabal, J. Sanz, R. Torrecillas and J. S. Moya, *PLoS One*, 2015, **10**, e0132709, DOI: 10.1371/journal.pone.0132709.
183. S. P. Valappil, D. M. Pickup, D. L. Carroll, C. K. Hope, J. Pratten, R. J. Newport, M. E. Smith, M. Wilson and J. C. Knowles, *Antimicrob. Agents Chemother.*, 2007, **51**, 4453.
184. S. P. Valappil, D. Ready, E. A. A. Neel, D. M. Pickup, W. Chrzanowski, L. A. O'Dell, R. J. Newport, M. E. Smith, M. Wilson and J. C. Knowles, *Adv. Funct. Mater.*, 2008, **18**, 732.
185. S. P. Valappil, D. Ready, E. A. Abou Neel, D. M. Pickup, W. Chrzanowski, L. A. O'Dell, R. J. Newport, M. E. Smith, M. Wilson and J. C. Knowles, *Adv. Funct. Mater.*, 2008, **18**, 732.
186. A. J. Salinas and M. Vallet-Regí, *J. Non-Cryst. Solids*, 2016, **432**, 9.
187. M. Bellantone, H. D. Williams and L. L. Hench, *Antimicrob. Agents Chemother.*, 2002, **46**, 1940.
188. J. S. Moya, L. Esteban-Tejeda, C. Pecharromán, S. Mello-Castanho, A. C. dS and F. Malpartida, *Adv. Eng. Mater.*, 2011, **13**, B256.
189. A. W. Wren, A. Coughlan, P. Hassanzadeh and M. R. Towler, *J. Mater. Sci.: Mater. Med.*, 2012, **23**, 1331.
190. J. Ye, J. He, C. Wang, K. Yao and Z. Gou, *Biotechnol. Lett.*, 2014, **36**, 961.
191. S. Sánchez-Salcedo, S. Shruti, A. J. Salinas, G. Malavasi, L. Menabue and M. Vallet-Regí, *J. Mater. Chem. B*, 2014, **2**, 4836.
192. F. Baino, G. Novajra, V. Miguez-Pacheco, A. R. Boccaccini and C. Vitale-Brovarone, *J. Non-Cryst. Solids*, 2016, **432**, 15.

193. V. Miguez-Pacheco, L. L. Hench and A. R. Boccaccini, *Ceramic Bulletin*, 2015, **94**, 27.
194. L. Yildirimer, N. T. K. Thanh and A. M. Seifalian, *Trends Biotechnol.*, 2012, **30**, 638.
195. X. F. Zhang, S. Kehoe, S. K. Adhi, T. G. Ajithkumar, S. Moane, H. O'Shea and D. Boyd, *Mater. Sci. Eng., C*, 2011, **31**, 669.
196. X. F. Zhang, H. O'Shea, S. Kehoe and D. Boyd, *J. Mech. Behav. Biomed. Mater.*, 2011, **4**, 1266.
197. D. Williams, *Mater. Today*, 2004, **7**, 24.
198. V. Guarino, F. Causa and L. Ambrosio, *Expert Rev. Med. Devices*, 2007, **4**, 405.
199. S. Kehoe, X. F. Zhang and D. Boyd, *J. Mater. Sci.: Mater. Med.*, 2011, **22**, 945.
200. D. Cacaina, H. Ylanen, S. Simon and M. Hupa, *J. Mater. Sci.: Mater. Med.*, 2008, **19**, 1225.
201. S. A. Shah, M. U. Hashmi and S. Alam, *Mater. Sci. Eng., C*, 2011, **31**, 1010.
202. Y. M. J. Y. M. Jiang j. Ou, Z. H. Zhang and Q. H. Qin, *J. Mater. Sci.: Mater. Med.*, 2011, **22**, 721.
203. G. Li, S. Feng and D. Zhou, *J. Mater. Sci.: Mater. Med.*, 2011, **22**, 2197.
204. C. Wu, W. Fan, Y. Zhu, M. Gelinsky, J. Chang, G. Cuniberti, V. Albrecht, T. Friis and Y. Xiao, *Acta Biomater.*, 2011, **7**, 3563.
205. S. A. Shah, M. U. Hashmi, A. Shamim and S. Alam, *Appl. Phys. A: Mater. Sci. Process.*, 2010, **100**, 273.
206. I. Izquierdo-Barba, A. J. Salinas and M. Vallet-Regí, *Int. J. Appl. Glass Sci.*, 2013, **4**, 149.

CHAPTER 17

Mesoporous Bioactive Glasses in Tissue Engineering and Drug Delivery

MARÍA VALLET-REGÍ[a,b] AND ANTONIO J. SALINAS*[a,b]

[a] Dpto. Química Inorgánica y Bioinorgánica, Universidad Complutense, Instituto de Investigación Sanitaria, Hospital 12 de Octubre, i + 12, Madrid, Spain; [b] Centro Investigación Biomédica en Red Bioingeniería, Biomateriales y Nanomedicina (CIBER-BBN), Madrid, Spain
*Email: salinas@ucm.es

17.1 What are Mesoporous Bioactive Glasses

Bioactive glasses (BGs) bond with living tissues. For this reason they are a hot subject of research for use as biomaterials in several clinical applications, including as scaffolds in bone tissue engineering (BTE) and as matrixes in drug delivery systems (DDSs). During their historical evolution, these materials have been obtained by three methods of synthesis. Each method yields glasses with different textural properties which give them different capabilities as biomaterials. First, BGs were obtained by the traditional method of quenching a melt. During this chapter these initial BGs, discovered by Hench *et al.* in 1969,[1] will be denoted as melt-prepared glasses (MPGs). Investigations of bioactive MPGs using *in vitro* solutions established the essential role of Si–OH (silanol) groups in their bioactive response, formed when the glasses contact with physiological fluids or simulated body fluids (SBFs).[2] Hence, in 1991 Hench *et al.* proposed the synthesis of BGs by the sol–gel route, a wet chemistry method which yields to the glass network

RSC Smart Materials No. 23
Bioactive Glasses: Fundamentals, Technology and Applications
Edited by Aldo R. Boccaccini, Delia S. Brauer and Leena Hupa

Published by the Royal Society of Chemistry, www.rsc.org

at lower temperatures than used for MPG syntheses.[3] Glasses so obtained are denoted gel glasses (GGs) or sol–gel glasses (SGGs) and exhibit abundant silanol groups in their surface, although additional silanols are formed after soaking in simulated body fluids (SBFs). *In vitro* characterization of SGGs demonstrated that their quicker bioactive response, when compared with MPGs, was a consequence of both a surface rich in silanol groups and their excellent textural properties, including high surface area and pore volume.[4,5]

To obtain bioactive glasses with more controlled and reproducible textural properties than SGGs, glasses with ordered mesoporosity were synthesized and characterized for the first time in 2004 by Zhao *et al.* and then by Vallet-Regi *et al.*[6,7] These glasses are usually denoted template glasses (TGs) or mesoporous bioactive glasses (MBGs) and constitute the main subject of this chapter. MBG synthesis is based on the sol–gel method and supramolecular chemistry principles. Different from conventional SGGs, surfactants are added in this case during the glass synthesis. Surfactants are amphiphilic molecules able to self-assemble in aqueous solution producing mesophases that behave as structure-directing agents, *i.e.* templates, for the glass network. After removal of the surfactant by calcination or extraction, MBGs exhibiting amazing structural and textural features are obtained.[8–10] These glasses also contain abundant silanol groups on their surface that favor their bioactive response and confer on them the capability to be functionalized to tailor their properties on demand. Moreover, these glasses exhibit surface areas and pore volumes approximately twice those of SGGs with analogous composition. As a consequence, some MBGs have shown the quickest *in vitro* bioactive responses described for synthetic materials. Furthermore, MBGs exhibit ordered arrangements of mesopores within a narrow pore size distribution, both factors playing an essential role when they are intended to be used as scaffolds in tissue engineering of bone and as matrixes in drug delivery applications, as will be detailed in the following sections.

Figure 17.1 schematically depicts the synthesis procedures of the three families of BGs in a comparative way. The higher temperatures required for MPGs, the larger times and higher cost of the reactants required for SGG synthesis and the important control of the mesostructure with the use of surfactants in MBGs, should all be noted; the last opens new possibilities for use of this more recent family of BGs. As observed in Figure 17.1, the procedure to obtain MPGs is shorter, whereas the synthesis of SGGs and MBGs are time-consuming although in both cases the maximum temperature of synthesis required, *i.e.* around 700 °C, is almost the half that required for MPG synthesis.

The inspiration for the synthesis of MBGs were the SGGs, but also the pure silica mesoporous materials designed for the first time in the 1990's in an effort to design materials able to improve the performance of zeolites in catalysis.[11] Mesoporous materials were proposed as biomaterials for the first time by Vallet-Regi *et al.* in 2001 for use as matrixes in systems for the controlled release of drugs.[12] From then, such materials were profusely

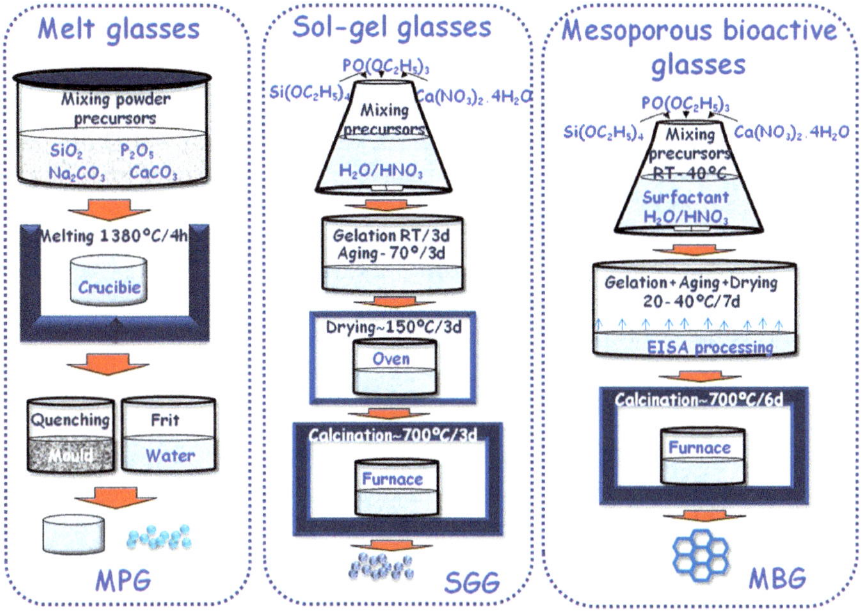

Figure 17.1 Schematic representation of the syntheses procedures for the three families of bioactive glasses: MPGs, SGGs and MBGs.

investigated to host and release an enormous variety of drugs with different biological actions.[13] Silica mesoporous materials exhibit two-dimensional (2D) or three-dimensional (3D) arrangements of mesoporous channels. Furthermore, they present excellent textural properties including surface areas larger than 1000 $m^2 g^{-1}$ and pore volumes over 1 $cm^3 g^{-1}$. In spite of these remarkable textural properties, and even taking into account that the surface of these materials is plentiful in silanol groups, silica mesoporous materials exhibit an *in vitro* bioactive response which is rather poor.[14] This behavior was explained by considering that pure silica mesoporous materials do not contain CaO or P_2O_5. Such components had been revealed as essential (CaO) or highly recommended (P_2O_5) for the bioactive responses observed in MPGs and SGGs.[15]

This was the reason why the Zhao and Vallet-Regi groups investigated the synthesis of mesoporous glasses in the SiO_2–CaO–P_2O_5 system. The objective was to obtain materials with textural properties close to silica mesoporous materials, but exhibiting enhanced *in vitro* bioactive response. The commonly used hydrothermal procedures of synthesis of silica mesoporous materials failed in the synthesis of mesoporous glasses containing CaO because this hinders the interactions between silica and the surfactant or precipitate in basic pH conditions. Thus, for the synthesis of SiO_2–CaO–P_2O_5 MBG powders, the evaporation induced self-assembly (EISA) method[16] is used, as observed in the right-hand part of Figure 17.1. The reactants used are generally the same as those for the synthesis of SGGs, that is, tetraethyl orthosilicate as the SiO_2 source, triethyl phosphate as the P_2O_5 source, and calcium nitrate tetrahydrate as the CaO source.

As will be described later, sometimes small amounts of oxides containing inorganic ions such as cerium(III), gallium(III), zinc(II) and many others are also included to improve the biological behavior of MBGs. These extra elements are generally added from the corresponding nitrate sources during the sol preparation.[17] The reactants are dissolved in a mixture of ethanol–water and adding nitric acid as a catalyst, and Pluronic® P123 as the surfactant. The initial surfactant concentration is lower than the critical micellar concentration (CMC), which is the concentration required for self-assembly of the surfactant. The sol obtained is poured in a Petri dish and left for several days at ambient, or slightly higher, temperature for solvent evaporation (the process sometimes denoted as 'aging'). With the solvent evaporation the surfactant concentration increases until it reaches the CMC, when the self-assembled surfactant behaves as a template during the hydrolysis and polycondensation of the silica precursor that produces the glass network. Resultant materials are thermally treated for surfactant and nitrate removal to obtain MBG powders. Other approaches for the MBG synthesis can be found in the literature.[18,19]

After the synthesis it is necessary to determine if the material obtained is actually a MBG. Figure 17.2 shows two experimental techniques commonly used for this purpose. On the left part of the figure, the low angle X-ray diffraction (LA-XRD) pattern obtained in the range between 0.6 and 6° in 2θ of a typical MBG is shown. This region is not usually investigated because the reflections of crystalline planes correspond to considerably smaller distances than those of families of mesoporous planes. Consequently, they are revealed by the presence of XRD maxima at larger 2θ values, generally in the 10 to 50° range. Because MBGs are glasses, they do not show diffraction

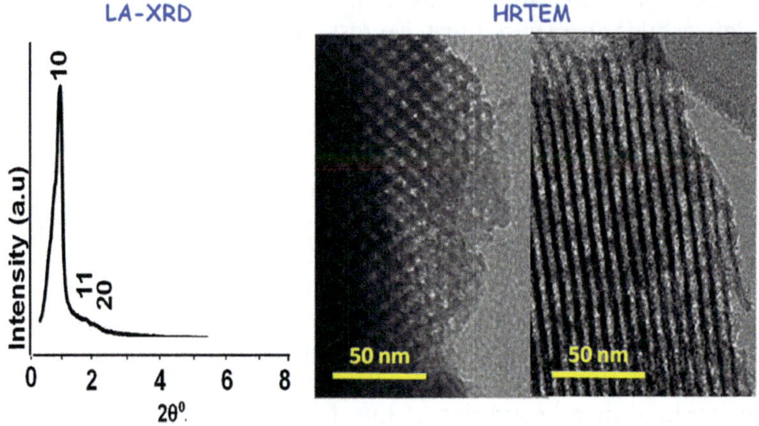

Figure 17.2 Experimental techniques used to demonstrate if a glass is really an MBG. Left: An LA-XRD pattern; right: HRTEM images along directions perpendicular and parallel, respectively, to the mesoporous channels.

maxima in the wide angle X-ray diffraction (WA-XRD) region. However, the maxima observed in the low angle region reveal the presence of a repetitive distance at the nanometric scale characteristic of the ordered mesoporous materials. It must be taken into account that following the International Union of Pure and Applied Chemistry (IUPAC) mesopores are those in the range 2 nm < pore diameter (D_P) < 50 nm. However, the majority of the mesoporous materials investigated have pores with D_P between 2 and 10 nm. The presence of a sharp diffraction maximum togetherwith other maxima of lower intensity in the LA-XRD pattern is indicative of a mesopore arrangement. Several types of mesoporous order can be obtained by the MBGs including cubic bicontinuous but, as in the mesoporous silica materials, the most common symmetry reached for MBGs is a 2D hexagonal plane group *p6mm* and diffraction maxima observed in the LA-XRD pattern in the left-hand part of Figure 17.2 are labeled with Miller indices according to this symmetry.

In addition, high resolution transmission electron microscopy (HRTEM) clearly allows for confirmation of the mesopore arrangement. For instance, in the HRTEM images in the right-hand part of Figure 17.2, mesopores order is observed in two directions, perpendicular and parallel to the mesoporous channels. Sometimes, only a partial order of mesopores is obtained. This mesostructure, commonly denoted worm-like, can be detected by HRTEM and inferred from the presence of a broad maximum in the LA-XRD pattern.

17.2 Textural Properties of the MBGs

MPGs are dense materials and consequently their surface area and porosity are negligible (below 1 $m^2 g^{-1}$). For this reason the textural characteristics of these glasses were not included in Figure 17.3. As will be explained in the next section, the bioactive response of MPGs comes from their special composition which makes them partially soluble, and not from their textural properties.

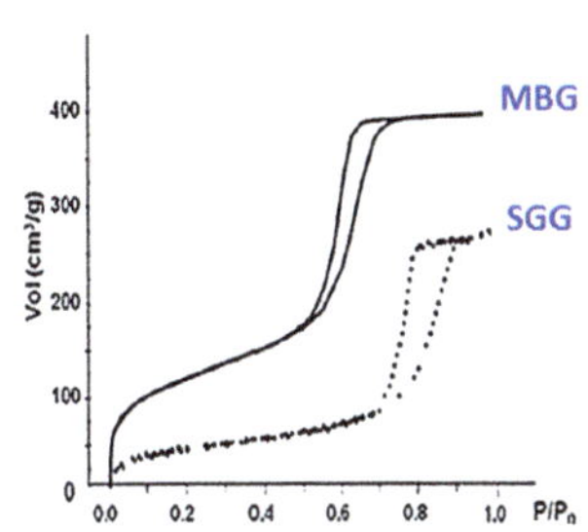

Family of BGs	Composition	SiO_2 content (mol%)	Surface area (m^2/g)	Pore volume (cm^3/g)
MBG	$CaO-P_2O_5-SiO_2$ $CaO-SiO_2$	58 – 95	200- 500	0.5 – 0.7
SGG	$CaO-P_2O_5-SiO_2$ $CaO-SiO_2$	50 – 95	100- 250	0.2 – 0.4

Figure 17.3 Textural properties of SGGs and MBGs exhibited in a comparative way. Left: N_2 adsorption isotherms of an SGG and an MBG. Right: Typical values of composition and textural properties for both families of glasses.

In the left part of Figure 17.3 typical nitrogen adsorption/desorption isotherms of a SGG and a MBG with analogous composition are included. Both curves are type IV isotherms, characteristic of mesoporous materials with a type H1 hysteresis loop characteristic of cylindrical pores open at both ends. These results demonstrate that both families of glasses exhibit the presence of mesopores. However, different to MBGs, in SGGs the D_P distribution is quite broad and, mainly, the pores are not ordered. However, although the isotherms of both types of glass show the same shape, higher gas volume was adsorbed by the MBG when compared to the SGG at the same relative pressure. In fact, as can be observed in the right-hand part of Figure 17.3, the textural parameters of MBGs are approximately twice that of SGGs. Thus, for the traditional SGGs the typical surface area values obtained by the Brunauer–Emmett–Teller method (S_{BET}) are between 100 and 250 $m^2\,g^{-1}$ and the pore volumes (V_P) lie in the range 0.2–0.4 $g\,cm^{-3}$. However for MBGs the characteristic S_{BET} values are from 200 to 500 $m^2\,g^{-1}$ and the pore volumes are usually in the range 0.5–0.7 $g\,cm^{-3}$.

These values show that MBGs, exhibit textural properties intermediate between SGGs and silica mesoporous materials which, as mentioned in the previous section, often reach and even surpass S_{BET} of 1000 $m^2\,g^{-1}$ and of V_P 1 $cm^3\,g^{-1}$. However, as will be explained in the next section, MBGs exhibit a very rapid *in vitro* bioactive response because, as distinct from mesoporous silica materials, they contain CaO in their composition.

SGGs and MBGs are different at the mesoscale, as can be seen in Figure 17.4 where several HRTEM images and the corresponding electron diffraction (ED) patterns are shown. As can be observed, the SGG and MBG HRTEM images at high magnification do not show any type of order. Accordingly, the ED patterns included as insets are typical of materials exhibiting an amorphous structure at the atomic scale. However, in the MBG

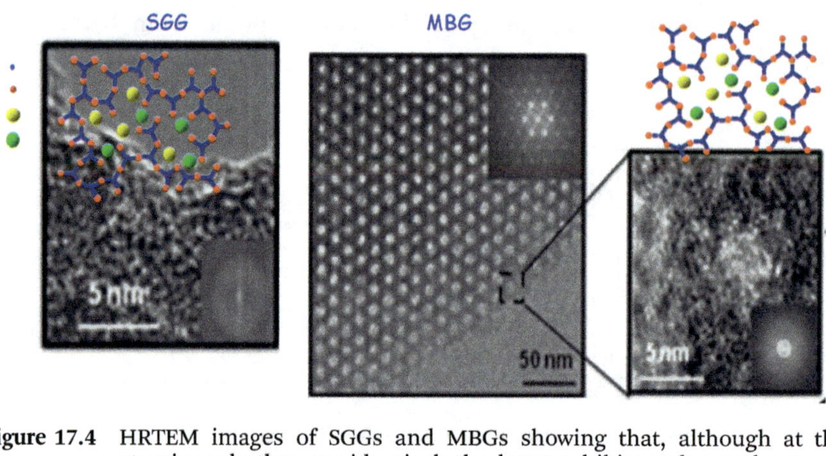

Figure 17.4 HRTEM images of SGGs and MBGs showing that, although at the atomic scale they are identical, the latter exhibits order at the mesostructural scale. The ED patterns corresponding to each image are included as an inset.

image at lower magnification a mesopores arrangement is clearly seen and the ED pattern confirms the hexagonal arrangement at the mesoscale. In conclusion, the three families of bioactive glasses, *i.e.* SGGs, MBGs, and MPGs, exhibit the amorphous structure of glasses at atomic scale, but MBGs also exhibit a mesopores arrangement which confers upon them outstanding properties, including very high pore volume and very rapid *in vitro* bioactive response, which make them optimal materials for the clinical applications that will be described in this chapter.

On the other hand, the influence of the synthesis parameters, including temperature and time of aging and drying and many others, on the textural properties of SGGs has been investigated. Particularly, it was found the CaO content has a great influence on the textural properties of SGGs. Thus, when this was increased, the surface area decreased and the pore volume increased.[20] However, for MBGs the most important effect of modifying the CaO proportion in the glass was the variation in the symmetry of the mesopores arrangement. For instance, when decreasing the CaO content, the mesostructure of an MBG changed from a 2D-hexagonal structure (*p6mm*) to a 3D-bicontinuous cubic structure ($Ia\bar{3}d$). In addition, the inverse variation occurred when the aging temperature used in the synthesis increased. As a result, with control of the CaO content and the aging temperature, it is possible to control the final mesostructure of a given MBG.[21]

17.3 Bioactivity of MBGs

What makes MPGs bioactive is their special composition—very different to that of non-bioactive glasses. The most important features for an MPG to exhibit bioactivity are network former content (SiO_2), as low as 45 mol%, and the presence of CaO and P_2O_5 in a proportion analogous to biological apatites. The composition of BG discovered by Hench, later denoted as Bioglass®, also included Na_2O as fourth component to decrease the melting temperature and making the glass manufacture cheaper.[1] This composition makes these glasses reactive enough in aqueous media to experience ion leaching to form a silica-rich layer on the surface essential to start the sequence of reactions that yield to the formation of nanocrystalline hydroxycarbonate apatite (HCA) indicative of a bioactive response.[22]

Because the sol–gel method uses relatively lower synthesis temperatures, it is not necessary to include Na_2O in the glasses as a melting agent. As a result, the majority of bioactive SGG compositions investigated belong to the SiO_2–CaO–P_2O_5 ternary system. Furthermore, it was demonstrated that the simultaneous presence of specific concentrations of SiO_2 and CaO in the glass was an essential requirement for bioactivity in MPGs and SGGs. However, the presence of amounts of P_2O_5 in the glass was just an alternative. Nevertheless, it was demonstrated that the presence of phosphorous in the glass modified the bioactivity mechanism, accelerating the HCA crystallization rate.[23]

Besides composition, other factors exert an important influence to obtain a rapid bioactive response in glasses. For instance, in SGGs the formation of further silanol groups after being soaked in physiological solutions was very important. Furthermore, the importance of the presence of mesopores was demonstrated to the rapid bioactive response of this family of glasses, although in SGGs they are not ordered. Moreover, SGGs exhibit a wide range of mesopore diameters which makes it difficult the control the reproducibility of their *in vitro* behavior. For instance, small variations in the glass composition, mostly belonging to the $CaO–P_2O_5–SiO_2$ system, for instance the CaO content, produced significant variations in the surface area and pore volume. The quickest bioactive response of SGGs, with respect MPGs, was demonstrated by the upper limits in the SiO_2 content to reach a bioactive response; this was 60 mol% for MPGs and over 90 mol% for SGGs.

In this scenario, unexpected results were obtained when the *in vitro* bioactivity of MBGs were tested in SBF. Because of the similarities between MBGs and SGGs, what was expected was an analogous reactivity of both families of glasses, perhaps somewhat increased in the MBGs by their more favorable textural properties. However, what was obtained was an appreciably faster bioactive response than the other two families of glasses as can be observed in Figure 17.5. The left-hand part of the figure shows two TEM micrographs and the corresponding ED patterns of the material formed on the surface of two glasses after soaking in SBF. At this stage it must be explained that although bioactivity is related to the *in vivo* behavior of a material after implantation, it has long been accepted that the *in vitro* bioactivity of a material can be related to the time required to form an HCA layer after being soaked in SBF. Such *in vitro* tests have been revealed as especially useful for glasses, mostly for comparative purposes allowing a first

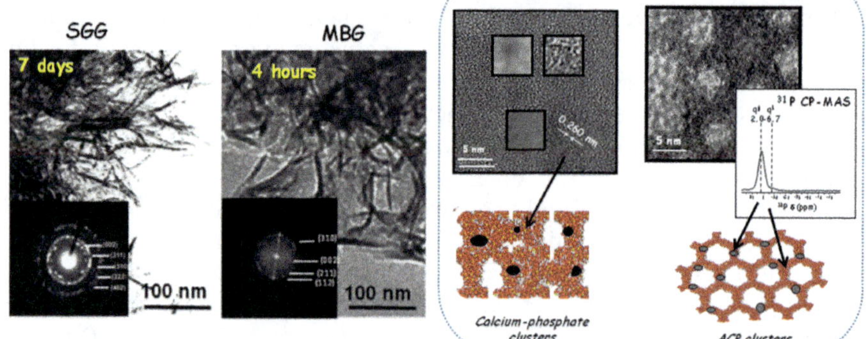

Figure 17.5 Left: TEM images and ED patterns of the apatite-like needle crystals formed after soaking an SGG and an MBG for different times in SBF. Right: The presence of calcium phosphate nuclei in both categories of glass is an important structural feature in terms of their *in vitro* behavior.

screening to select the more favorable compositions that will be subjected to ultimate *in vitro* and *in vivo* biocompatibility tests.

Therefore, the results included in the left part of Figure 17.5 demonstrate the considerably higher bioactive response of MBGs when compared with SGGs. In fact, as observed in the figure, the MBG required only 4 h in SBF to be coated by a material analogous to that formed on the SGG after 7 d of soaking. In both cases the presence of the needle-like crystals of HCA were detected at the times mentioned. In addition, in both cases the materials formed were in the nanometric range as it was demonstrated by the ED patterns which in both cases showed the presence of rings and some diffraction maxima.

The mesostructures of both families of glasses were investigated before soaking in SBF and the most relevant results obtained are those included in the right-hand part of Figure 17.5. As can also be seen, in SGGs as in MBGs the inclusions of P_2O_5 in the glass have similar effects due to binding to calcium-forming calcium phosphate (CaP) nuclei. Because calcium ions are withdrawn from the glass network, the four initial stages that are shown in Figure 17.6, and consequently the formation of amorphous calcium phosphate (ACP), are retarded. However, stage 5 that yields the formation of HCA is accelerated. Actually it was demonstrated that the material formed *in vitro* is a calcium deficient hydroxyapatite (CDHA) identical to biological apatites. Thus, these studies show the different *in vitro* bioactive responses of two SGGs, one with phosphorus and the other phosphorus-free, or of two MBGs with and without phosphorus in their composition, but no great difference in the kinetics of the bioactive responses of an MBG compared with an analogous SGG.

The *in vitro* studies in SBF allowed formulation of the Hench mechanism of bioactivity[24] for MPGs that was also extended to SGGs. The *in vitro* studies in SBF also allowed Izquierdo-Barba *et al.*[25] to formulate a mechanism of bioactivity for MBGs which exhibits the following differences compared with the Hench mechanism (see Figure 17.6):

(1) As a consequence of the higher surface area, the stages 1 to 3 are much greater in MBGs and the final result is a more acidic pH of 6.7 compared to 7.4 for MPGs/SGGs.
(2) The consequence of this lower surface pH was the formation of octacalcium phosphate (OCP), a phase described as intermediate between the ACP and the CDHA under *in vivo* conditions, but that was never detected before in *in vitro* assays of bioactivity.
(3) Finally a thicker layer of CDHA coating the MBG was formed in a few hours, a considerably shorter time than that required for MPGs or SGGs.

It must be taken into account that the increase in the kinetics of the bioactive response in going from MPGs to MBGs, through SGGs, is a fact but it is not a critical issue to selecting a glass for a specific application. The

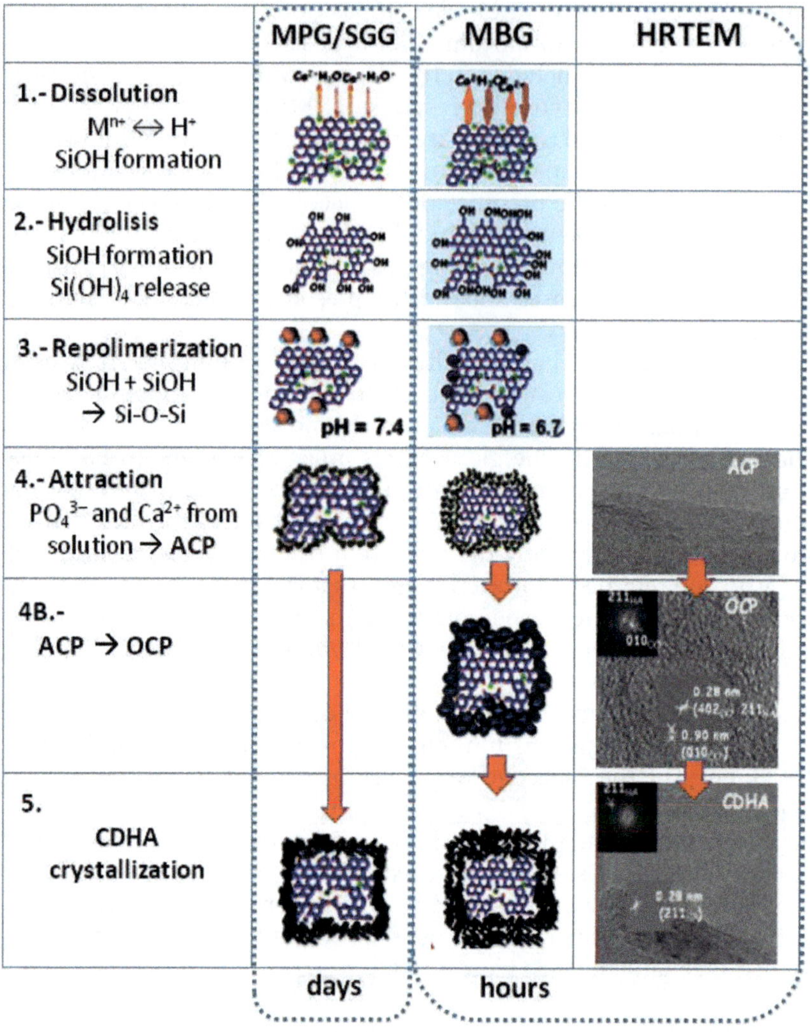

Figure 17.6 Schematic representation of the first five stages of the Hench mechanism (eleven stages in total) to describe the bioactive response of silica-based glasses and glass-ceramics.[24] It was established for MPGs, but it was extended to SGGs. The extremely rapid bioactive response of MBGs required a substantial modification of the mechanism proposed by Izquierdo-Barba *et al.*[25]

most important factors are the new physicochemical features coming from the new methods of synthesis which increase some parameters such as surface area and pore volume, and the new capabilities coming from the sol–gel process such as the possibility to obtain coatings for fibers which are not satisfactorily obtained by quenching of a melt. On the other hand, the great control of the pore morphology and size during MBGs synthesis brings the capability to decorate their surface with osteogenic molecules which, together with their remarkably quick bioactive response, converts to these materials into optimum candidates to construct scaffolds for BTE applications.[26,27]

17.4 Improvement of MBGs by Adding Inorganic Ions

Because MBGs are glasses their composition can be modified with the inclusion of extra oxides of cations (generally in very small proportions) featuring important biological activities, such as osteogenesis, antibacterial capacity, angiogenesis or cementogenesis.[28,29] For this reason, there is a general research trend aimed to upgrade MBG properties with some of the well-known beneficial ions because of their biological action.[30,31] Figure 17.7 depicts the approximate time required to exhibit *in vitro* bioactivity for the three families of BGs as well as the year of discovery of each family. The biological actions of the inorganic ions used to upgrade the glasses are also included. As can be seen in the figure, the increase in the kinetics of the bioactive response in going from MPGs to SGGs was not enough to justify a generalized use of SGGs. MBGs bring new capabilities, however. For instance, their mesoporous channels can be loaded with a variety of substances having therapeutic action.

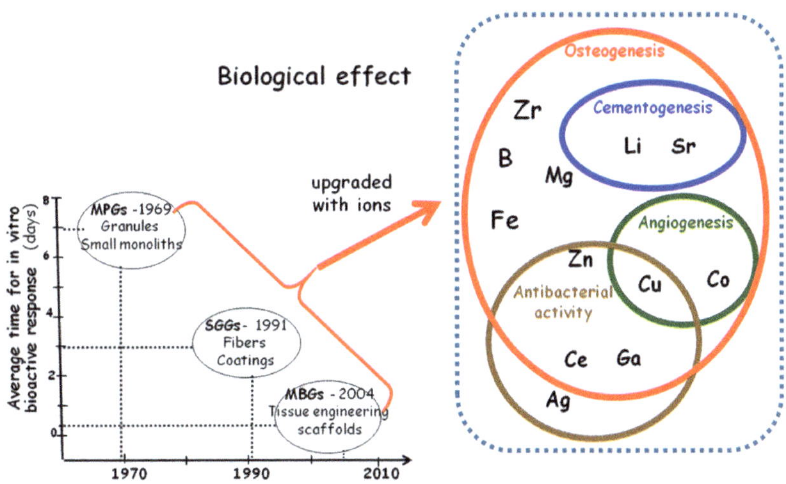

Figure 17.7 Left: The historical development of the three families of BGs *versus* the approximate time necessary for HCA formation after soaking the glass in SBF. The main clinical applications are also included. Right: The main biological activities attributed to the inorganic ions investigated to upgrade MBGs with therapeutic activity.
Modified from A. J. Salinas and M. Vallet-Regí, Glasses in bone regeneration: A multiscale issue, *J. Non-Cryst. Solids*, **432**, 9–14, Copyright (2016), with permission from Elsevier.[10]

Figure 17.7 shows a number of inorganic ions investigated in the synthesis of MBGs because they exhibit a positive biological action. Our group mainly focused in the effect of the inclusion of Ga, Ce and Zn.[17] However, other research groups, including those of Wu and Boccaccini,[32,33] have investigated the effect of including other inorganic ions in glasses.

However, after the inclusion of extra elements in the $CaO–P_2O_5–SiO_2$ system it is necessary to confirm that the important features which make MBGs useful for BTE and as DDSs are maintained. Evidently such features are high *in vitro* bioactivity and excellent textural properties. As an example, Figure 17.8 shows the results obtained after the characterization of an $80SiO_2–5P_2O_5–15CaO$ (mol%) MBG labeled with the letter B (for "Blank") before, and after adding different amounts of Ce_2O_3. In addition, for the glass containing 0.2% of Ce_2O_3, the scanning electron micrographs (SEM) and the corresponding energy dispersive X-ray spectra (EDS), before and after being soaked for one day in SBF, are also included. As can be deduced from Figure 17.7, in this specific investigation, the inclusion of cerium was in order to add osteogenic and antibacterial capabilities to the already excellent textural properties and bioactivity of glass B.

In the left-hand side of Figure 17.8 the LA-XRD patterns of B together with other glass compositions containing amounts of Ce_2O_3 from 0.2% to 3.5% as a fourth component are represented. MBGs lack long-range structural order

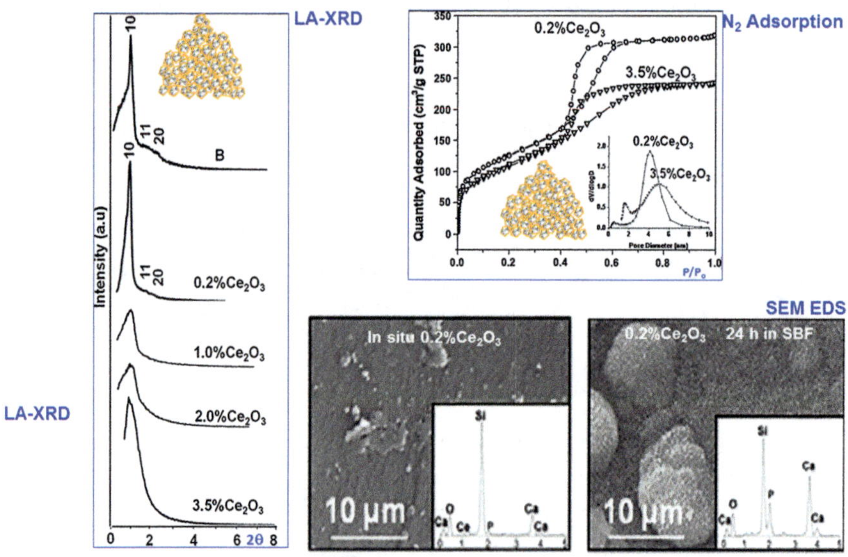

Figure 17.8 Characterization of MBG powders by LA-XRD and N_2 adsorption before and after inclusion of small amounts of Ce_2O_3. The SEM images and EDS spectra of the MBG containing a 0.2% of Ce_2O_3 before and after being soaked for one day in SBF are also included.
Modified from A. J. Salinas and M. Vallet-Regí, Glasses in bone regeneration: A multiscale issue, *J. Non-Cryst. Solids*, **432**, 9–14, Copyright (2016), with permission from Elsevier.[10]

because they are glasses and so do not exhibit diffraction maxima in the wide angle XRD (WA-XRD) patterns. As can be observed, all the glass compositions investigated present maxima in the LA-XRD patterns, indicative of mesoporous ordered arrangements. Moreover the LA-XRD pattern corresponding to unsubstituted glass (B) shows a sharp maximum at 1.4° in 2θ, that can be indexed to the (10) reflection of a 2D hexagonal phase. Also, the inclusion of Ce_2O_3 produces a decrease in the relative intensity of the diffraction maxima together with a widening of the (10) reflection, indicative of a decrease of the mesoporous order with the increase in the amount of Ce_2O_3. Observing this figure in detail, we can see that only for the glass containing 0.2% of Ce_2O_3 is the hexagonal ordered maintained whereas for Ce_2O_3 addition over 1% the LA-XRD patterns suggest that the hexagonal symmetry was lost, although a worm-like mesoporous order could be obtained since a broad maximum in this region was visible. HRTEM analysis (results not shown) confirmed results obtained by LA-XRD. Analogous results were obtained for similar substitutions of B with other oxides containing therapeutic ions, such as Ga_2O_3 and ZnO.[17]

Furthermore, for the clinical applications described in this chapter, these glasses must exhibit significant surface areas and pore volumes. As an example, in Figure 17.8, the N_2 adsorption isotherms of two Ce-doped (at 0.2% and 3.5%) MBGs and their pore size distributions are shown. From these measurements it was confirmed that the surface area and the pore volume for these MBG powders decreased from an S_{BET} of 515 $m^2\ g^{-1}$ and V_P of 0.58 $cm^3\ g^{-1}$ for the unsubstituted MBG, until reaching values over 300 $m^2\ g^{-1}$ and 0.27 $cm^3\ g^{-1}$, *i.e.* still higher to those of SGGs.

Finally, Figure 17.8 includes the SEM images and the corresponding EDS spectra of the MBG containing 0.2% Ce_2O_3 before and after being soaked for one day in SBF. The EDS spectrum of the pristine sample confirms the presence of cerium in the glass and yields values for the other glass components close to the nominal ones. As observed, after soaking for one day in SBF, the sample surface appears coated by a new material, that EDS determines is mainly composed of Ca and P. Other experimental techniques, including Fourier transform infrared (FTIR) spectroscopy, XRD, HRTEM, EDS, confirm that this material was a calcium apatite-like material indicative of the *in vitro* bioactive response. However, for all the substitutions investigated, a decrease of the *in vitro* bioactive response was observed with increase in the amount of substituent.

As a conclusion of this section it can be indicated that the approach to upgrade the biological behavior of a BG by adding therapeutic ions can modify the physicochemical properties of the initial glass. Specifically it can eliminate some properties that were the reason to select the particular glass as a biomaterial. For MBGs synthesized as powders, a slight decrease in the kinetics of the bioactive response together a decrease of around 40% in the textural properties are observed. However, these values are so high for pristine MBGs that the values exhibited by the substituted MBGs still remain useful for their intended use in BTE and as DDSs, as will be described.

17.5 Applications of the MBGs in Tissue Engineering of Bone

One of the most important challenges in BTE applications is to design materials to be used in critical bone defect regeneration. In small bone defects the solution is easier by using materials as granules. However, in the case of critical defects pieces are necessary which adjust in shape and size to the bone defect. Is must be considered that simultaneous to the development of glasses with characteristics more adapted to use as implants, there was a change in the paradigm of the role of bioactive materials used in applications such as orthopedics and dentistry. In the 1970's and 1980's there was a search for materials for the simple substitution and repair of osseous tissues. In 2002 Hench and Polak proposed the search of materials aimed to drive and favor the regeneration of bone that they denoted as third generation biomaterials.[34] In their studies they demonstrated that the silicon and calcium ions released from bioactive MPGs as granules stimulated the genes to direct the cells to form bone.

Third generation biomaterials can be directly included in the field of BTE. In the last few years there has existed a wide consensus that the biomaterials alone are not able to fulfill the body requirements when used for the treatment of bone defects. In these cases it is necessary to apply tissue engineering principles which are based on three basic pillars: cells, signal molecules and scaffolds. It is in this last pillar where the MBGs processed into 3D scaffolds are considered the optimal solution for most clinical necessities.[35] The scaffolds designed for this application must exhibit interconnected and hierarchical porosity, that is, to contain pores with D_P of several orders of magnitude. Thus, they must contain giant pores (channels) and macropores to allow internal angiogenesis and the interaction with cells. Nevertheless, the presence of nanometric pores like those exhibited by the MBGs is also required. Such pores will allow the inclusion of signal molecules that induce the formation of bone such as the bone morphogenetic proteins (BMPs), growth factors like vascular endothelial growth factor (VEGF), or different fractions of parathyroid hormone-related peptide (PTHrP).[36]

The first attempt to obtain pieces tailored for the clinical necessities was performed by Stucky *et al.*,[37] who prepared an injectable paste of a MBG able to set as a calcium phosphate bone cement. However, this material lacks the macroporous architecture required for a scaffold for BTE because it only exhibited random macroporosity and inferior pore interconnectivity. Yun *et al.* were pioneers in the incorporation of MBGs in scaffolds (MBG_Sc) to obtain materials with hierarchical, macro-nano porosity which could be optimum candidates for BTE applications.[18] With this purpose they use rapid prototyping equipment that allowed obtaining scaffolds with a computer-aided pre-designed macroporosity. The challenge of these investigations was that during the processing of the scaffolds, the materials should keep the mesoporosity and *in vitro* bioactivity of the MBGs previously obtained as powders. The ordered mesoporosity allows the adsorption and controlled release of the bone induction agents mentioned. Moreover, these mesoporous channels confer extremely high values of surface area and pore volume to MBGs which increase the already high levels of bioactivity of $CaO–P_2O_5–SiO_2$ MPGs or SGGs. Keeping in mind the essential properties of a 3D scaffold for BTE applications, several strategies have been proposed for the design of a macroporosity required for cell functions, such as bone cells ingrowths, nutrients supply and vascularization, as well as for adhesion and development of the bone cells. These strategies include foaming, freeze-drying, fiber bonding or rapid prototyping (RP) technologies.[38] Obviously, in all cases it is very important to confirm that the processing of MBG powders to obtain scaffolds does not eliminate the ordered mesoporosity or bioactivity.

Figure 17.9 shows some results of characterization of 3D scaffolds obtained by rapid prototyping. Raw materials were MBG powders previously obtained including up to 1% Ga_2O_3 in the $15CaO–5P_2O_5–80SiO_2$ (mol%) MBG. The synthesis conditions to obtain the scaffolds are detailed elsewhere.[39] The paste for the injection was obtained by adding the MBG powders to a solution of polycaprolactone (PCL) and dichloromethane. After processing the pieces with the external and channel sizes previously designed, MBG_Scs were obtained by drying for 2 h at 70 °C, to evaporate the dichloromethane, and heating at 500 °C for 3 h to remove the PCL. Evidently these manipulations and thermal treatments to obtain the scaffolds could destroy the mesoporosity arrangement and the bioactive response. For this reason a detailed characterization of the scaffolds is essential and Figure 17.9 includes the most significant results in this regard.

In the left of Figure 17.9 the LA-XRD patterns of B_Sc, 0.2%Ga_2O_3_Sc and 1%Ga_2O_3_Sc are shown. The first two show almost identical diffraction patterns to those observed in Figure 17.8 for Ce-substituted MBG powders. The sharp diffraction maxima at a 2θ of 1.4° with two poorly resolved peaks in the 2θ range 2.2–2.70°, can be assigned to (10), (11) and (20) reflections of an ordered 2-D hexagonal structure.[39] However, on increasing the amount of gallium no diffraction maxima were observed, suggesting a defective arrangement of mesopores. These results suggest that mesoporous order is not altered in B_Sc and 0.2%Ga_2O_3_Sc during the manufacturing process. However, in 1%Ga_2O_3_Sc the mesostructure do not withstand milling to obtain the powders, or the thermal treatment during scaffold fabrication. Therefore, MBG_Scs containing higher amounts of Ga_2O_3 were not prepared.

Figure 17.9 also includes the N_2 adsorption isotherms with pore size distributions and the textural properties S_{BET}, D_P and V_P of the two Ga-substituted scaffolds. Both curves are type IV isotherms characteristic of mesoporous materials with hysteresis loops of type H1, characteristic of cylindrical pores. The samples show a single-mode pore size distribution centered at 4.1 and 4.3 nm, respectively. S_{BET} and D_P of B_Sc and

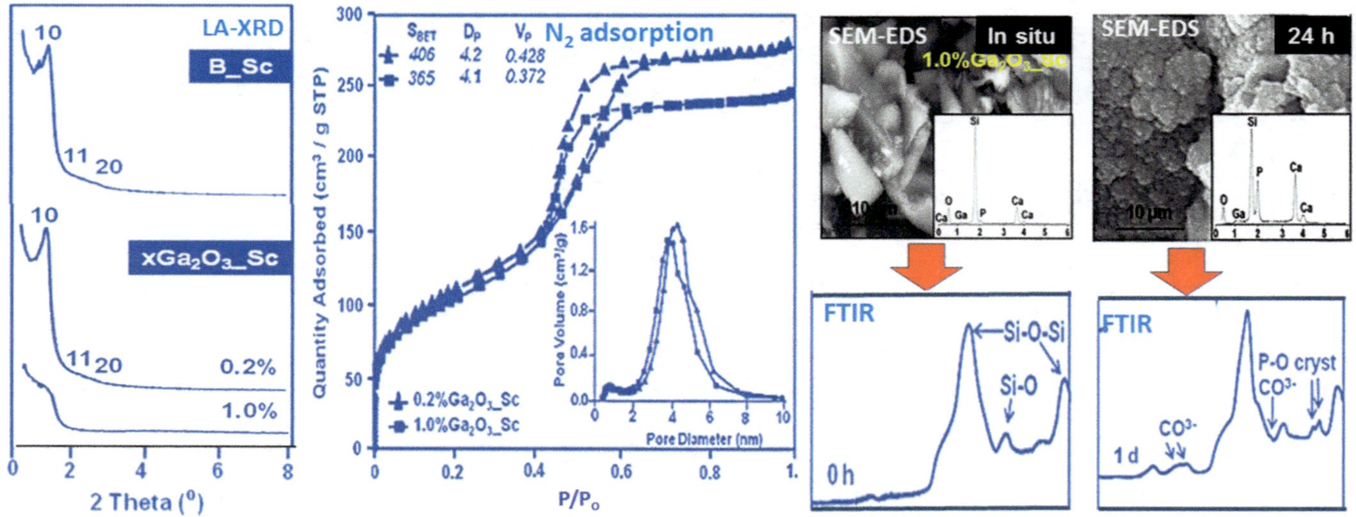

Figure 17.9 Characterization of the 3D scaffolds obtained from substituted MBGs. Left: LA-XRD patterns and N_2 adsorption isotherms together with the calculated textural parameters and pore size distribution. Right: SEM images together with EDS and FTIR spectra of the scaffold with 1% Ga_2O_3 before and after being soaked for 1 day in SBF.
Modified from A. J. Salinas and M. Vallet-Regí, Glasses in bone regeneration: A multiscale issue, *J. Non-Cryst. Solids*, **432**, 9–14, Copyright (2016), with permission from Elsevier.[10]

$0.2\%Ga_2O_3$_Sc are similar. In contrast, the disordered arrangement of mesopores observed by the XRD patterns for $1\%Ga_2O_3$_Sc produces a decrease in the textural properties.

Finally, at the right of Figure 17.9 are SEM micrographs together with the EDS and FTIR spectra of $1\%Ga_2O_3$_Sc before and after be soaked for 1 day in SBF. As can be observed by SEM, after the SBF treatment the scaffold is coated by a new material which EDS reveals is composed of Ca and P and new bands appear in the FTIR spectrum. These bands at 561 and 600 cm^{-1}, respectively, are indicative of phosphate in a crystalline environment.[40] Furthermore, bands around 1420 and 1460 cm^{-1} for carbonate indicate that a hydroxycarbonate apatite (HCA) was formed. These Ga-doped MBG scaffolds show a quick *in vitro* bioactive response. Thus, these results have demonstrated that even after doping with Ga_2O_3 and processing into scaffolds, both factors decrease the bioactive response, the final materials exhibit good enough bioactive response and textural properties to use in BTE applications.

As a summary of this section, Figure 17.10 depicts the role of each of the MBG_Sc elements essential to make them useful in BTE. These are: the hierarchical porosity including channels, macropores and mesopores, and a bioactive response. In addition, the biological role attributed to three

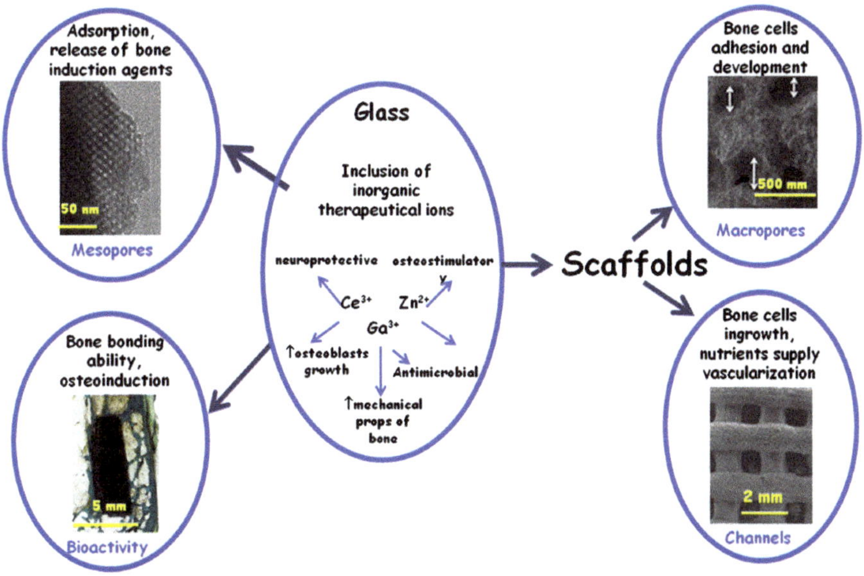

Figure 17.10 A schematic representation of the features that converts the MBG scaffolds into optimum candidates for BTE: mesopores to include biomolecules, macropores to favor the adhesion and proliferation of cells, bioactivity to exhibit osteoinduction, channels (or giant pores) for cell ingrowth and vascularization. In addition, as glasses they can easily incorporate extra ions with therapeutic actions.
Modified from A. J. Salinas and M. Vallet-Regí, Glasses in bone regeneration: A multiscale issue, *J. Non-Cryst. Solids*, **432**, 9–14, Copyright (2016), with permission from Elsevier.[10]

therapeutic ions: Ce^{3+}, Ga^{3+} and Zn^{2+}, from the twelve ions included in Figure 17.7, are highlighted.

17.6 Bactericide Capacity of MBG Scaffolds Enriched with Zn^{2+} or Ga^{3+}

From the previous sections it is clear that numerous inorganic ions have been added to the composition of the MBGs to improve their biological behavior, bringing new capabilities such as osteogenesis, angiogenesis or bactericide action. In this sense, post-operative bacterial infections are an important issue associated with the treatment of bone fractures and diseases with implants in BTE applications. This problem affects 2% of joint prostheses and 5% of fracture fixation devices implanted. Mostly prosthetic joint infections are produced by *Staphylococcus aureus* and *Staphylococcus epidermidis*. Furthermore, bacteria form a biofilm hindering antibiotics in reaching the infected area. This makes necessary a surgical intervention with further increase of the economic and human costs. For this reason new approaches must be attempted.

An important possible application of MBG scaffolds is to upgrade them by including in their composition ions such as Zn^{2+} or Ga^{3+} for which a bactericide effect was reported. From the studies described in the previous sections of this chapter it follows that, for relatively low concentrations of substituents, the obtained scaffold will maintain the mesoporosity that allows it to be loaded with osteogenic molecules and enough bioactivity to bond with living tissues without being encapsulated from the surrounding tissues by a fibrous capsule. However, it is necessary to check if the scaffolds actually exhibit a clear antibacterial capability against the bacteria more often found in the cases of prosthesis infections after surgery. Simultaneously, it would be required that the ions released to the medium do not convert into ions which are non-biocompatible to these scaffolds.

This was the case of the first studies (unpublished) of our group when we investigated the bactericide capacity of a MBG scaffold doped with Ga_2O_3. In this case in the results obtained, most of the Ga^{3+} ions remained in the glass network without exhibiting antibacterial capacity in the sample. To exert an effective bactericide action against *S. aureus*, the amount of Ga_2O_3 must be so high that it reaches levels not biocompatible with the osteoblasts. After this, we checked that only a low proportion of the Ga_2O_3 was released to the surrounding medium. The amount of Ga for the MBG to be bactericidal must be so high that it was toxic in the osteoblasts culture. At present we are investigating different MBG compositions including one which would enable Ga^{3+} to attain a bactericidal action within the biocompatible range.

On the other hand, successful results were obtained using ZnO as substituting component, as can be observed in Figure 17.11. This figure includes the results of the *in vitro* studies of biocompatibility and antibacterial capacity of a $15CaO–5P_2O_5–80SiO_2$ (mol%) MBG_Sc doped with 4% ZnO.[41]

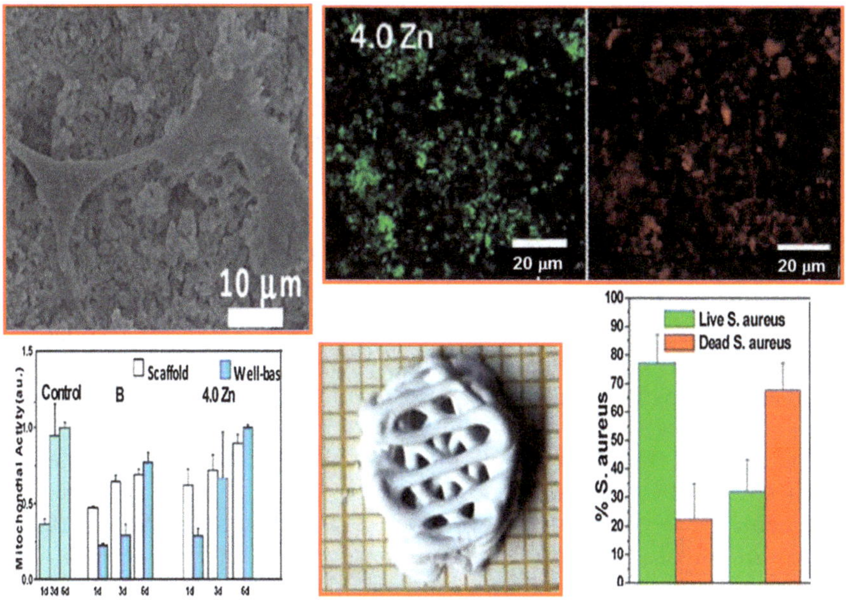

Figure 17.11 *In vitro* biocompatibiity in osteoblast cell cultures and bactericide action against *S. aureus* of a MBG scaffold containing 4.0% ZnO. The SEM image shows the cell morphology after 2 d. Cell proliferation was measured in terms of the mitochondrial activity at 1, 3 and 6 d. In addition confocal micrographs of *S. aureus* biofilms formed on MBG scaffold surfaces after 2 d are included. Viable (green) and non-viable (red) bacteria are shown.
Modified from A. J. Salinas and M. Vallet-Regí, Glasses in bone regeneration: A multiscale issue, *J. Non-Cryst. Solids*, **432**, 9–14, Copyright (2016), with permission from Elsevier.[10]

As can be observed in Figure 17.7, ZnO was included because Zn^{2+} ions exhibit positive biological effects, including their role as osteogenesis inductors and their antibacterial capability. As mentioned above, infection of the prosthesis is an important problem when a material is implanted into the body. This problem tends to increase because of the organization of the pathogen bacteria into biofilms resistant to many families of antibiotics.

As can be seen in the figure, unsubstituted MBG_Sc, *i.e.* B or blank exhibits a certain antibacterial capability against *S. aureus* because its silicon and calcium ions. However, the inclusion of 4% ZnO significantly increased the amount of dead bacteria compared with the control group. For the confocal microscopy images included in the figure, a fluorescence-based Live/Dead Backlight bacterial viability test was used because it provides information about the membrane integrity of *S. aureus.* Results observed clearly indicate that a 4.0% ZnO in the MBG hinders *S. aureus* biofilm formation.

Nevertheless, it is important to check if materials releasing such a high amount of Zn^{2+} ions are biocompatible. On the left of Figure 17.11 can be observed the osteoblast development investigated in both the well-base and in extracts of the culture medium after soaking the Zn-loaded scaffolds for 1, 3 and 6 days. For higher amounts of ZnO the resultant material was not biocompatible. These results indicate that when an extra element is included to induce a specific biological action, its concentration is a critical point that must be carefully investigated because it could induce negative effects preventing the use in implants.

17.7 Application of MBGs in Drug Delivery Systems

Biomaterials must play a multifunctional task in BTE applications, including the stimulation of osteogenesis and angiogenesis, as well as in anti-bacterial or anti-cancer treatment. This goal requires the additional participation of drugs with biological activity because the role of the inorganic therapeutic ions is not enough. The exceptional pore volume and surface area of MBGs compared with the other families of BGs allow considerably higher efficiencies when they are used to load drugs.

To achieve the multifunctional properties required of the MBGs, such as stimulators of osteogenesis or angiogenesis or exhibit antibacterial or anticancer activities, the inclusion of therapeutic ions is not enough. For this reason many drugs have also been investigated to enhance the required capabilities.[33] MBGs, even after being upgraded with therapeutic ions and being processed into 3D scaffolds, still shows surface areas and pore volumes higher than conventional SGGs. Furthermore, many drugs can be loaded with quite high efficiency into the MBGs. The release of drugs from MBGs follows a Fickian diffusion mechanism which can be influenced by the dissolution of the MBG, which would accelerate the process, and by apatite layer formation, which would cover the channels making the process difficult. In addition the numerous Si–OH groups inside the mesoporous channels can establish interactions with the drugs hosted.

In spite of this complexity the release of numerous drugs as well as growth factors and proteins from MBGs has been investigated. Among the drugs dexamethasone gentamicin, ampicillin, doxorubicin, tetracycline, triclosan can be mentioned, and many others. Moreover, bone morphogenetic proteins (BMPs) and vascular endothelial growth factor (VEGF) were also investigated. It must be indicated that in some cases like doxorubicin of VEGF, the loading efficiency reached 90%. A compressive review in this regard is in the ref. 33.

An accepted definition of a DDS is a formulation or a device that enables the introduction of a therapeutic substance in the body and improves its efficacy and safety by controlling the rate, time and place of the release of drugs in the body.[42] As said in the first section of this chapter, silica-based mesoporous materials are promising candidates wherein drugs can be confined because their excellent textural properties, stability, biocompatibility and the possibility to tailor their surface properties by the functionalisation of the silanol groups located at their surfaces to facilitate drug loading and release. Actually MBGs can be considered an improvement of these materials, where CaO and P_2O_5 were added to induce bioactivity and the oxide of a fourth element to upgrade the glass with new therapeutic actions.[43]

However, when MBGs are upgraded with a therapeutic ion and loaded with a drug, unexpected results can be obtained as a consequence of the ion–drug interactions. This happened when we investigated curcumin as model drug to be released from substituted MBGs. The role of curcumin was described in several clinical applications including the treatment of a variety of cancers, the suppression of myocardial infarction, the inhibition of HIV replication, wound healing and Alzheimer's disease.[44,45] However, pharmacokinetic performance of curcumin shows a poor bioavailability due to small absorption, rapid metabolism and fast systemic elimination, and the methods attempted to overcome this problem are not yet satisfactory.[46,47]

For this reason the release of curcumin from Ce-, Ga- and Zn-substituted MBGs was investigated. Previous studies have reported that the inclusion of extra elements can facilitate the drug uptake and release.[48,49] The results obtained confirmed that substituted MBGs maintained mesoporous order, high textural properties, high *in vitro* response and homogenous distribution of the extra element within the glass network.

The most important conclusions of these studies are outlined in Figure 17.12. The influence of the substituents on the drug uptake and release behavior, as well as on the *in vitro* bioactivity was tested. Oxide concentrations up to 2.0% for Ga and Ce and up to 4.0% for Zn were used as substituents to obtain systems designed to be used as DDSs. Because curcumin is a hard Lewis base it has high affinity towards hard Lewis acids like Ga^{3+} and Ce^{3+}. This fact facilitated drug loading as the highest gallium and

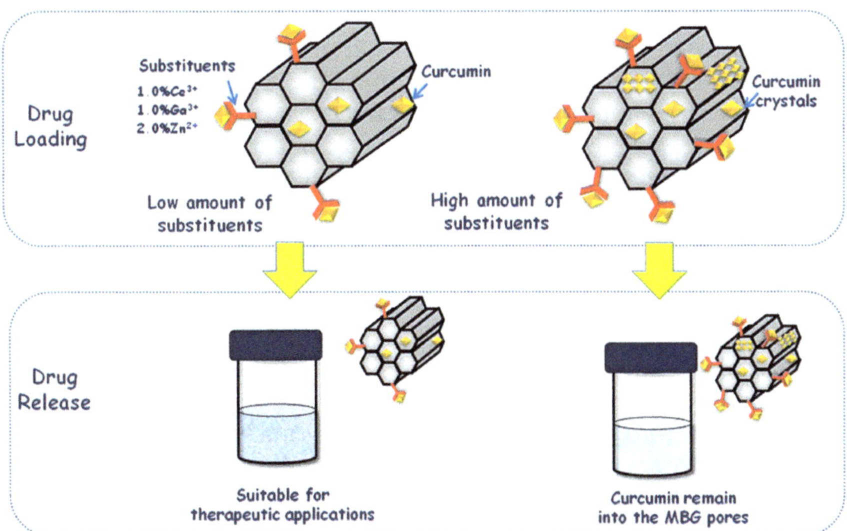

Figure 17.12 MBG-doped scaffolds investigated for the drug release of curcumin. High levels of interaction between extra ions and the drug enhances the loading process but decreases the drug delivery kinetics.

cerium substituted MBGs showed the maximum uptake. However, the interaction had negative impact on the *in vitro* release of drug from these matrixes. In addition, unsubstituted and the lowest substituted gallium, cerium and zinc MBGs showed optimum drug loading along with *in vitro* drug release inside the therapeutic limits, as well as rapid *in vitro* bioactive response proving them to be promising DDSs for curcumin.

In summary, before using substituted MBGs as DDSs it will be necessary for a thorough investigation of the possible interactions between the therapeutic ions and drugs, and how they can interact with each other influencing the kinetics of release of both.

17.8 Conclusions and Outlook

The development of the bioactive SGGs in the 1990's brought relatively small improvements with respect with traditional MPGs, most of them related to new ways of processing provided by the sol–gel method, such as obtaining of fibers or coatings. However, the third generation of BGs, *i.e.* MBGs, caused a huge advance in this field. These materials exhibit specific areas and pore volumes approximately twice those of SGGs. In addition, they exhibit mesopore arrangements within a very narrow distribution of pore diameters. All these textural features make them optimum candidates to host drugs or other molecules with biological activity because most of them can penetrate into the MBGs' mesopores which are close to 4 nm in size. Moreover, their exceptional textural properties surpass materials with the quickest *in vitro* bioactive responses reported.

For all these reasons MBGs are capable of supporting many actions and treatments. For instance, MBGs can be upgraded with therapeutic ions. They can be processed to obtain 3D scaffolds for BTE applications. They can be used as matrixes in DDSs. These actions provoke significant decreases in the textural properties and in the kinetics of the *in vitro* bioactive response, but these values are already so high in MBGs that in derived materials they are high enough to remain useful as implants. Sometimes the processing into scaffolds partially eliminates the mesoporous order but for some applications the maintenance of a worm-like order is sufficient.

However, much research work needs to be done before these still relatively new glasses can be clinically used in humans. Thus, it is clear that 3D scaffolds based on $CaO–P_2O_5–SiO_2$ MBGs upgraded with oxides of metals of biological significance is a promising subject for BTE. However, sometimes the extra elements bind so strongly with the glass network that they remain in the glass but without any biological action as they are not released. Other times the necessary concentrations of an element required to achieve the desired biological effect are so high that the MBG is no longer biocompatible. Other times, when osteogenic substances are loaded in the MBG scaffolds or when other drugs are loaded for use in DDSs, the interaction with the therapeutic ions can substantially modify the kinetics of release. These kinetics can be also modified for the partial solubility of the MBG. Another factor that must be considered comes from the high bioactivity of these materials. Thus, the quick formation of an HCA layer can hinder or slow down the kinetics of release of the biomolecules loaded into the mesopores or the release of the ions with antibacterial or other purposes included within the MBG. For all these reasons MBGs represent a good strategy to solve the problems caused by the loss of bone tissue.[50]

For all these reasons, many research efforts are focused in this field at the present. For instance, the ability of Ce^{3+}/Ce^{4+}-containing BGs to inhibit oxidative stress by mimicking the catalase enzyme activity was recently reported.[51] Another recent study traced the products of degradation of a MBG labeled with ^{45}Ca.[52] In any case, the inclusion of every specific inorganic ion into MBG scaffolds, including even the specific concentration of the cation, needs to be thoroughly investigated before being translated into clinical applications.

Acknowledgements

Financial support of Ministerio de Ciencia e Innovacion (MICINN) through the project MAT2012-35556, Ministerio de Economía y Competitividad, through the Spanish and European Network of Excellence CSO2010-11384-E and Instituto de Salud Carlos III through PI15/00978 project is acknowledged. The authors wish to thank also to the staff of the ICTS Centro Nacional de Microscopia Electrónica and Centro de Diffraction de Rayos X of Universidad Complutense de Madrid (Spain).

References

1. L. L. Hench, R. J. Splinter, W. C. Allen and T. K. Greenlee, Bonding mechanism at the interface of ceramics prosthetic materials, *J. Biomed. Mater. Res.*, 1971, **2**, 117–141, DOI: 10.1002/jbm.820050611.
2. L. L. Hench, Opening paper 2015 – some comments on Bioglass: four eras of discovery and development, *Biomed. Glasses*, 2015, **1**, 1–11, DOI: 10.1515/bglass-2015-0001.
3. R. Li, A. E. Clark and L. L. Hench, An investigation of bioactive glass powders by sol-gel processing, *J. Appl. Biomater.*, 1991, **2**, 231–239, DOI: 10.1002/jab.770020403.
4. M. Vallet-Regí, C. V. Ragel and A. J. Salinas, Glasses with medical applications, *Eur. J. Inorg. Chem.*, 2003, 1029–1042, DOI: 10.1002/ejic.200390134.
5. J. R. Jones, Review of bioactive glass: from Hench to hybrids, *Acta Biomater.*, 2013, **9**, 4457–4486, DOI: 10.1016/j.actbio.2012.08.023.
6. X. Yan, C. Yu, X. Zhou, J. Tang and D. Zhao, Highly ordered mesoporous bioactive glasses with superior *in vitro* bone-forming bioactivities, *Angew. Chem., Int. Ed.*, 2004, **43**, 5980–5984, DOI: 10.1002/anie.200460598.

7. A. Lopez-Noriega, D. Arcos, I. Izquierdo-Barba, Y. Sakamoto, O. Terasaki and M. Vallet-Regi, Ordered mesoporous bioactive glasses for bone tissue regeneration, *Chem. Mater.*, 2006, **18**, 3137–3144, DOI: 10.1021/cm060488o.
8. D. Arcos and M. Vallet-Regi, Sol–gel silica-based biomaterials and bone tissue regeneration, *Acta Biomater.*, 2010, **6**, 2874–2888, DOI: 10.1016/j.actbio.2010.02.012.
9. I. Izquierdo-Barba, A. J. Salinas and M. Vallet-Regi, Bioactive glasses: from macro to nano, *Int. J. Appl. Glass Sci.*, 2013, **4**, 149–161, DOI: 10.11111jag.12028.
10. A. J. Salinas and M. Vallet Regi, Glasses in bone regeneration: A multi-scale issue, *J. Non-Cryst. Solids*, 2016, **432**, 9–14, DOI: 10.1016/j.jnoncrysol.2015.03.025.
11. C. T. Kresge, M. E. Leonowitz, W. J. Roth, J. C. Vartuli and J. S. Beck, Ordered mesoporous molecular sieves synthesized by a liquid-crystal template mechanism, *Nature*, 1992, **359**, 710–712.
12. M. Vallet-Regi, A. Ramila, R. P. del Real and J. Perez-Pariente, A New Property of MCM-41: Drug Delivery System, *Chem. Mater.*, 2001, **13**, 308–311, DOI: 10.1021/cm0011559.
13. A. L. Doadrio, A. J. Salinas, J. M. Sanchez-Montero and M. Vallet-Regi, Drug Release from Ordered Mesoporous Silicas, *Curr. Pharm. Des.*, 2015, **21**, 6189–6213, DOI: 10.2174/1381612822666151106121419.
14. M. Vallet-Regi, L. Ruiz-Gonzalez, I. Izquierdo-Barba and J. M. Gonzalez-Calbet, Revisiting silica based ordered mesoporous materials: medical applications, *J. Mater. Chem.*, 2006, **16**, 26–31, DOI: 10.1039/b509744d.
15. A. J. Salinas, M. Vallet-Regı and I. Izquierdo-Barba, Biomimetic apatite deposition on calcium silicate gel glasses, *J. Sol-Gel Sci. Technol.*, 2001, **21**, 13–25, DOI: 10.1023/A:1011249330581.
16. C. J. Brinker, Y. Lu, A. Sellinger and H. Fan, Evaporation-induced self-assembly: nanostructures made easy, *Adv. Mater.*, 1999, **11**, 579–601, DOI: 10.1002/(SICI)1521-4095(199905)11:7.
17. A. J. Salinas, S. Shruti, G. Malavasi, L. Menabue and M. Vallet-Regi, Substitutions of cerium, gallium and zinc in ordered mesoporous bioactive glasses, *Acta Biomater.*, 2011, 7, 3452–3458, DOI: 10.1016/j.actbio.2011.05.033.
18. H. Yun, S. Kim and Y. Hyeon, Highly ordered mesoporous bioactive glasses with *Im3m* symmetry, *Mater. Lett.*, 2007, **61**, 4569–4572, DOI: 10.1016/j.matlet.2007.02.075.
19. H. Yun, S. Kim and Y. Hyeon, Preparation of 3D cubic ordered mesoporous bioactive glasses, *Solid State Sci.*, 2008, **10**, 1083–1092, DOI: 10.1016/j.solidstatesciences.2007.11.037.
20. M. Vallet-Regi, A. J. Salinas and D. Arcos, From the bioactive glasses to the star gels, *J. Mater. Sci.: Mater. Med.*, 2006, **17**, 1011–1017, DOI: 10.1007/s10856-006-0437-7.
21. A. Garcia, M. Cicuendez, I. Izquierdo-Barba, D. Arcos and M. Vallet-Regi, Essential role of calcium phosphate heterogeneities in 2D-Hexagonal

and 3D-Cubic SiO_2–CaO–P_2O_5 mesoporous bioactive glasses, *Chem. Mater.*, 2009, **21**, 5474–5484, DOI: 10.1021/cm9022776.
22. A. J. Salinas and M. Vallet-Regi, Evolution of ceramics with medical applications, *Z. Anorg. Allg. Chem.*, 2007, **633**, 1762–1773, DOI: 10.1002/zaac.200700278.
23. M. Vallet-Regi, A. J. Salinas, J. Ramirez-Castellanos and J. M. Gonzalez-Calbet, Nanostructure of bioactive sol–gel glasses and organic–inorganic hybrids, *Chem. Mater.*, 2005, **17**, 1874–1879, DOI: 10.1021/cm047956j.
24. L. L. Hench, Bioceramics: from concept to clinic, *J. Am. Ceram. Soc.*, 1991, **74**, 1487–1510, DOI: 10.1111/j.1151-2916.1991.tb07132.
25. I. Izquierdo-Barba, D. Arcos, Y. Sakamoto, O. Terasaki and A. Lopez-Noriega, High-performance mesoporous bioceramics mimicking bone Mineralization, *Chem. Mater.*, 2008, **20**, 3191–3198, DOI: 10.1021/cm800172x.
26. M. Vallet-Regi, I. Izquierdo-Barba and M. Colilla, Structure and functionalization of mesoporous bioceramics for bone tissue regeneration and local drug delivery, *Philos. Trans. R. Soc., A*, 2012, **370**, 1400–1421, DOI: 10.1098/rsta.2011.0258.
27. I. Izquierdo-Barba and M. Vallet-Regí, Mesoporous bioactive glasses: Relevance of their porous structure compared to that of classical bioglasses, *Biomed. Glasses*, 2015, **1**, 140–150, DOI: 10.1515/bglass-2015-0014.
28. C. Wu, Y. Luo, G. Cuniberti, Y. Xiao and M. Gelinsky, Three-dimensional printing of hierarchical and tough mesoporous bioactive glass scaffolds with a controllable pore architecture, excellent mechanical strength and mineralization ability, *Acta Biomater.*, 2011, **7**, 2644–2650, DOI: 10.1016/j.actbio.2011.03.009.
29. C. Wu and J. Chang, Mesoporous bioactive glasses: structure characteristics, drug/growth factor delivery and bone regeneration application, *Interface Focus*, 2012, **2**, 292–306, DOI: 10.1098/rsfs.2011.0121.
30. A. Hoppe, N. S. Güldal and A. R. Boccaccini, A review of the biological response to ionic dissolution products from bioactive glasses and glass-ceramics, *Biomaterials*, 2011, **32**, 2757–2774, DOI: 10.1016/j.biomaterials.2011.01.004.
31. V. Mouriño, J. P. Cattalini and A. R. Boccaccini, Metallic ions as therapeutic agents in tissue engineering scaffolds: an overview of their biological applications and strategies for new developments, *J. R. Soc., Interface*, 2012, **9**, 401–419, DOI: 10.1098/rsif.2011.0611.
32. A. Hoppe, V. Mouriño and A. R. Boccaccini, Therapeutic inorganic ions in bioactive glasses to enhance bone formation and beyond, *Biomater. Sci.*, 2013, **1**, 254–256, DOI: 10.1039/C2BM00116.
33. C. Wu and J. Chang, Multifunctional mesoporous bioactive glasses for effective delivery of therapeutic ions and drug/growth factors, *J. Controlled Release*, 2014, **193**, 282–295, DOI: 10.1016/j.jconrel.2014.04.026.
34. L. L. Hench and J. M. Polak, Third-generation biomedical materials, *Science*, 2002, **295**, 1014–1017, DOI: 10.1126/science.1067404.

35. A. J. Salinas and M. Vallet Regi, Bioactive ceramics: from bone grafts to tissue engineering, *RSC Adv.*, 2013, **3**, 11116–11131, DOI: 10.1039/c3ra00166k.
36. A. J. Salinas, P. Esbrit and M. Vallet Regi, A tissue engineering approach based on the use of bioceramics for bone repair, *Biomater. Sci.*, 2013, **1**, 40–51, DOI: 10.1039/c2bm00071g.
37. Q. H. Shi, J. F. Wang, J. P. Zhang, J. Fan and G. D. Stucky, Rapid-setting, mesoporous, bioactive glass cements that induce accelerated *in vitro* apatite formation, *Adv. Mater.*, 2006, **18**, 1038–1042, DOI: 10.1002/adma.200502292.
38. A. J. Salgado, O. P. Coutinho and R. L. Reis, Bone tissue engineering: state of the art and future trends, *Macromol. Biosci.*, 2004, **4**, 743–765, DOI: 10.1002/mabi.200400026.
39. S. Shruti, A. J. Salinas, G. Lusvardi, G. Malavasi, L. Menabue and M. Vallet-Regi, Mesoporous bioactive scaffolds prepared with cerium-, gallium- and zinc-containing glasses, *Acta Biomater.*, 2013, **9**, 4836–4844, DOI: 10.1016/j.actbio.2012.09.024.
40. J. C. Elliott, Structure and chemistry of the apatites and other calcium orthophosphates, *Studies in Inorganic Chemistry*, Elsevier, Amsterdam, 1994, vol. 18.
41. S. Sanchez-Salcedo, S. Shruti, A. J. Salinas, G. Malavasi, L. Menabue and M. Vallet-Regi, *In vitro* antibacterial capacity and cytocompatibility of SiO_2–CaO–P_2O_5 meso-macroporous glass scaffolds enriched with ZnO, *J. Mater. Chem. B*, 2014, **2**, 4836–4847, DOI: 10.1039/c4tb00403e.
42. K. K. Jain, *Methods in Molecular Biology: Drug Delivery Systems*, Humana Press, Totowa, NJ, USA, 2008, vol. 437.
43. A. Ito, H. Kawamura, M. Otsuka, M. Ikeuchi, H. Ohgushi, K. Ishikawa *et al.*, Zinc-releasing calcium phosphate for stimulating bone formation, *Mater. Sci. Eng., C*, 2002, **22**, 21–25, DOI: 10.1016/S0928-4931(02)00108-X.
44. P. Anand, C. Sundaram, S. Jhurani, A. B. Kunnumakkare and B. B. Aggarwal, Curcumin and cancer: an "old-age" disease with an "age-old solution, *Cancer Lett.*, 2008, **267**, 133–164, DOI: 10.1016/j.canlet.2008.03.025.
45. C. Nirmala and R. Puvanakrishnan, Protective role of curcumin against isoproterenol induced myocardial infarction in rats, *Mol. Cell. Biochem.*, 1996, **159**, 85–93, DOI: 10.1007/BF00420910.
46. A. B. Hegge, T. T. Nielsen, K. L. Larsen, E. Bruzell and H. H. Tonnesen, Impact of curcumin supersaturation in antibacterial photodynamic therapy-effect of cyclodextrin type and amount: studies on curcumin and curcuminoides XLV, *J. Pharm. Sci.*, 2011, **100**, 174–185, DOI: 10.1002/jps.23046.
47. X. Wei, Z. Du, X. Zheng, X. Cui, A. Conney and K. Zhang, Synthesis and evaluation of curcumin-related compounds for anticancer activity, *Eur. J. Med. Chem.*, 2012, **53**, 235–245, DOI: 10.1016/j.ejmech.2012.04.005.
48. G. Malavasi, E. Ferrari, G. Lusvardi, V. Aina, F. Fatini, C. Morterra *et al.*, The role of coordination chemistry in the development of innovative

gallium-based bioceramics: the case of curcumin, *J. Mater. Chem.*, 2011, **21**, 5027–5037, DOI: 10.1039/C0JM03421E.

49. S. Shruti, A. J. Salinas, E. Ferrari, G. Malavasi, G. Lusvardi, A. L. Doadrio *et al.*, Curcumin release from cerium, gallium and zinc containing mesoporous bioactive glasses, *Micropor. Mesopor. Mat.*, 2013, **180**, 92–101.
50. D. Arcos, A. R. Boccaccini, M. Bohner, A. Díez-Pérez, M. Epple, E. Gómez-Barrena *et al.*, The relevance of biomaterials to the prevention and treatment of osteoporosis, *Acta Biomater.*, 2014, **10**, 1793–1805, DOI: 10.1016/j.actbio.2014.01.004.
51. V. Nicolini, E. Gambuzzi, G. Malavasi, L. Menabue, M. C. Menziani, G. Lusvardi *et al.*, Evidence of Catalase Mimetic Activity in Ce^{3+}/Ce^{4+} Doped Bioactive Glasses, *J. Phys. Chem. B*, 2015, **119**, 4009–4019, DOI: 10.1021/jp511737b.
52. B. Sui, G. Shong and J. Sun, Evolution of a Mesoporous Bioactive Glass Scaffold Implanted in Rat Femur Evaluated by ^{45}Ca Labeling, Tracing, and Histological Analysis, *ACS Appl. Mater. Interfaces*, 2014, **6**, 3528–3535, DOI: 10.1021/am4056886.

CHAPTER 18

Bioactive Glasses for Nerve Regeneration

GIORGIA NOVAJRA,[a] FRANCESCO BAINO,[a] STEFANIA RAIMONDO,[b] JORIS LOUSTEAU,[a] DANIEL MILANESE[a] AND CHIARA VITALE-BROVARONE*[a]

[a] Institute of Materials Physics and Engineering, Department of Applied Science and Technology, Politecnico di Torino, Turin, Italy; [b] Department of Clinical and Biological Sciences, University of Turin, Italy
*Email: chiara.vitale@polito.it

18.1 Introduction

Peripheral nerve injuries, which can arise from mechanical stress, exposure to heat, cold, irradiation, burns, tumours or focal inflammation, can lead to a partial or total loss of motor, sensory and autonomic functions in the area of the body concerned.[1] Fortunately, peripheral nerve tissue has an intrinsic capability to regenerate, involving cell migration along the nerve gap favouring the axonal re-growth from the proximal to the distal nerve stump and the final target re-innervation. However, for severe tissue damage, the spontaneous regeneration process is often insufficient. In this case a surgical intervention is needed, such as end-to-end suture of the nerve stumps (*i.e.* gaps up to 5 mm) or the use of a nerve autograft (*i.e.* gaps up to 5 cm) or allograft (*i.e.* gaps longer than 5 cm).[1–3] Although considered as the gold standard treatment, the use of a nerve autograft shows the risk of random and non-specific re-innervation and in some cases the limited tissue availability leads to insufficient diameter and length for reconstruction.[1,3,4]

RSC Smart Materials No. 23
Bioactive Glasses: Fundamentals, Technology and Applications
Edited by Aldo R. Boccaccini, Delia S. Brauer and Leena Hupa

Published by the Royal Society of Chemistry, www.rsc.org

Moreover, it involves an additional surgical procedure for the tissue harvesting with associated patient morbidity and pain. The grafting with allogenic tissue, needed in the case of substantial nerve defects, requires immune suppression and can be the cause of opportunistic infections.[1,3]

The use of nerve guidance channels (or nerve guides) of natural (*i.e.* collagen, porcine small intestinal submucosa) or synthetic materials (*e.g.* polyglycolic acid (PGA), poly(DL-lactic acid) (PDLLA)–caprolactone, poly(vinyl alcohol) (PVA) hydrogel) represents an alternative to the grafting techniques for short nerve defects up to about 3 cm long.[2,5] Nerve guidance channels are tubular devices, which are implanted between the two nerve stumps, protecting the injured area from the scar tissue infiltration and favouring the healing process. In fact, plasma exudate from the nerve tissue, rich in neurotrophic factors and extracellular matrix precursor molecules, accumulates into the nerve guidance channel and then fibrin cables form between the two nerve stumps, allowing the migration of endothelial cells, fibroblasts and especially Schwann cells (SCs). SCs align to form the bands of Büngner to guide the axonal sprouts in moving from the proximal stump to the distal target and finally participate in the myelination of the regenerated axons.[3]

The fundamental technical requirement for a nerve guide are biocompatibility, a proper balance between biodegradability and mechanical stability, semi-permeability for nutrient passage and minimization of growth factor escape while preventing the infiltration of scar tissue and inflammatory cells.[2] Moreover the nerve guide has to be flexible, resistant to kinking and collapsing to allow joint movements without compromising the regeneration process and to meet the technical requirements for production, sterilization, long-term storage and surgical handling.[2] Several problems associated with the use of the current commercial nerve guides have been reported, like inadequate degradation time, severe swelling, adverse reactions caused by the degradation products, early collapse and low flexibility.[2] Moreover, the clinical effectiveness of the nerve guidance channels commercially available is now limited to nerve defects up to about 3 cm long and, as for autograft, the functional nerve recovery is not guaranteed and it is generally unsatisfactory.[1,5,6] During peripheral nerve regeneration over long distances inside a nerve guidance channel, inadequate formation of fibrin cables, insufficient neurotrophic support as well as insufficient SC number, migration and proliferation seem to be the main causes of unsatisfactory nerve healing.[3] For this reason, the investigation of new materials and designs for the development of novel nerve guidance channels is needed.

Alternative materials for the design of nerve guidance channels have been proposed, these include biological tissues (*e.g.* blood vessels, muscles, tendons), allogenic and xenogenic tissues (*e.g.* human amniotic membrane, porcine small intestinal submucosa), biologically-derived polymeric materials (*e.g.* collagen, fibrin, gelatin, chitin, chitosan, alginate, silk) and synthetic polymers.[7–10] Bioactive glasses have also been proposed for the development of devices for nerve fibre regeneration following peripheral nerve and also spinal cord injuries. As discussed later in this chapter, bioactive glasses have been used alone or in combination with polymeric materials for the development of tubular nerve guidance channels and synthetic grafts, as well as for the creation of additional features inside the nerve guides to improve the regeneration process. Bioactive glasses are also thought to have a beneficial effect on tissue healing since their ion dissolution products have been shown to stimulate angiogenesis and wound healing[11,12] while the role of specific ions in the nerve regeneration process is under investigation. A summary of the different applications of bioactive glasses for the treatment of peripheral nerve and spinal cord injuries is given in Figure 18.1.

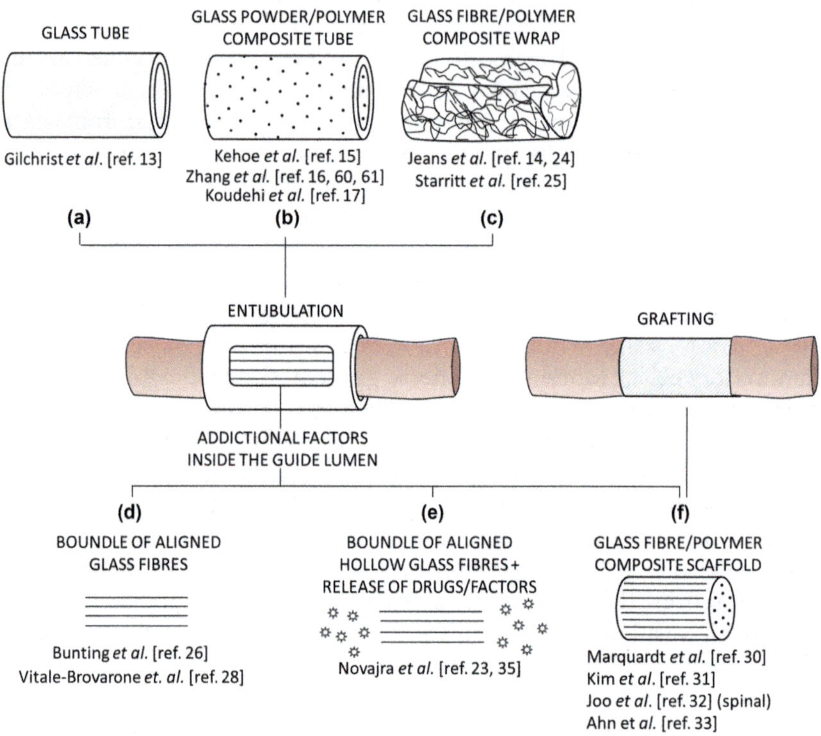

Figure 18.1 Summary of the different strategies proposed for the use of bioactive glasses for the treatment of peripheral nerve and spinal cord injuries. Bioactive glasses have been proposed as an alternative material for nerve entubulation alone (a: glass tube) or in the form of glass powder or fibres in combination with a polymer to obtain composite nerve wraps or guidance channels (b: glass powder–polymer composite tube and c: glass fibre–polymer composite wrap). Glass fibres have also been proposed as additional factors inside the lumen of a nerve guidance channel potentially carrying soluble factors to provide topographical and trophic cues for axonal regeneration for long nerve stumps (d: bundle of aligned glass fibres; e: bundle of aligned hollow glass fibres); moreover, glass fibres have been combined with polymers to obtain composite cylindrical scaffolds with anisotropic structure to be placed inside a nerve guidance channel or to be grafted between the two nerve stumps to guide axonal regeneration (f: glass fibre–polymer composite scaffold).

In this chapter, the possible strategies envisaged for the repair of peripheral nerve with the use of bioactive glasses are presented according to the material morphology in which the glassy material is used (bulk, powder or fibres). In Section 18.2, the development of a fully resorbable, glass-based nerve guidance channel is presented (Figure 18.1a). In Section 18.3, the use of micro- or nano-sized bioactive glass particles for the fabrication of polymer-based composite nerve guides with tailored physical, biological and mechanical properties is discussed (Figure 18.1b). Section 18.4 is focused on the use of bioactive glass fibres with micrometric diameters for nerve regeneration. In Section 18.4.1 a resorbable composite nerve wrap, containing bioactive glass fibres, is presented (Figure 18.1c). Moreover, as discussed in Sections 18.4.2 and 18.4.3, fibres of different bioactive glasses, which can both resorb during tissue healing and carry soluble factors, have been proposed for the creation of temporary topographical features inside the lumen of a tubular device or in a scaffold (graft) (Figure 18.1d–f) to support and improve the axonal regeneration outcome also for long nerve defects. Section 18.5 deals with the beneficial effect that the ion dissolution products of the bioactive glass can exert on the tissue healing process. Following these technical presentations of the state-of-the-art, the final section discusses the overall prospect of bioactive glasses for the release of nerve therapeutic ions. Since only one of the presented studies focused on spinal cord repair, some details on the traditional treatment of such injuries will be given when presenting this specific work (Section 18.4.2).

18.2 Glass Tube for Nerve Regeneration

Gilchrist *et al.*[13] first proposed the use of glass as a material for the treatment of nerve injuries in 1998. They developed a novel nerve guidance channel composed of a phosphate glass tube (length 4 cm, inner diameter 4 mm) with two holes drilled at the extremities for suturing purposes (Figure 18.2). The tube was tested *in vivo* for the repair of facial nerve of sheep. After the transection of the facial nerve, both nerve stumps were brought close to each other inside the lumen and held in place by suturing the tube to the epineurium. The treatment outcome, evaluated after 10 months, showed that the nerve reconnection occurred in all the animals with uniform diameter along the length with results comparable to those expected in direct end-to-end repair in a large animal model. The device was found to be fully dissolved after 10 months and no adverse reaction was provoked at the implantation site by the tube or by its degradation products. The authors concluded that this novel glass tube could be used as an alternative to the end-to-end suture when the equipment or the skills required for microsurgical repair are not available; moreover any intervention for the tube removal would not be needed thanks to the use of a soluble glass.

The glass composition was not reported by the authors in this work but Jeans *et al.*,[14] in a later work, commented on a previous study regarding a

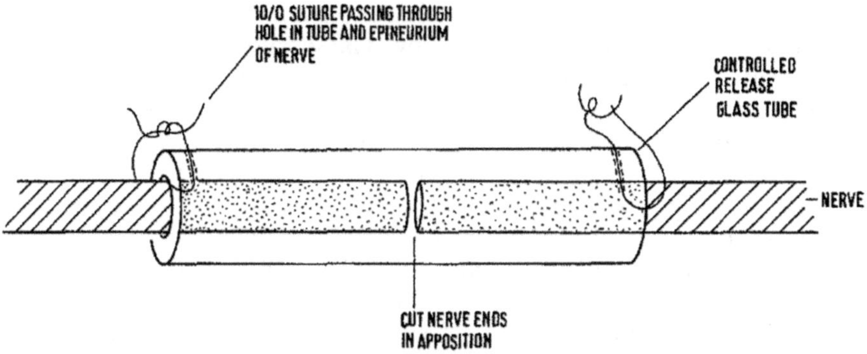

Figure 18.2 Diagram illustrating a biodegradable phosphate glass tube for the entubulation of transected nerve, showing two holes at the extremities for suturing the two nerve stumps inside the tube.
(Reprinted from T. Gilchrist *et al.*, *In vitro* nerve repair – *in vivo*. The reconstruction of peripheral nerves by entubulation with biodegradeable glass tubes – a preliminary report, *Brit. J. Plast. Surg.*, **51**, 231–237, Copyright 1998 with permission from Elsevier.)

glass tube used for nerve regeneration and referred to the material as Corglaes® (Giltech Ltd), a glass of the Na_2O–CaO–P_2O_5 system that can be designed with different compositions and dissolution rates.

The glass tube developed by Gilchrist *et al.*[13] showed its efficacy for the treatment of divided peripheral nerve without the presence of a nerve gap and, for this reason, the end-to-end suture was used as the reference surgical treatment in their study. However, the lack of semi-permeability and flexibility of the glass tube—two of the fundamental requirements for a nerve guidance channel—would probably represent a limitation for the treatment of injuries presenting a nerve defect. In fact, the absence of porosity on the tube walls would strongly limit the nutrient supply to the regenerating tissue across the nerve gap while the rigidity of the tube may lead to the escaping of the nerve stumps from the tube lumen during nerve regeneration due to inflexibility over the joints.[2]

18.3 Glass Powders for Nerve Regeneration

The use of bioactive glass powder has been proposed for the development of polymer–glass composite devices for peripheral nerve regeneration. The presence of the glass could be advantageous for better tailoring the mechanical properties as well as the degradation kinetics of the devices in order to guarantee both structural support and a full degradation that matches temporally with tissue regeneration. It can also affect the swelling of the device and influence some physical properties, such as the material wettability, which can favour a better cell–material interaction. In this context, micro- or nano-sized glass powder in combination with natural or synthetic polymers have been used for the fabrication of composite nerve guidance channels with tailored physical, biological and mechanical properties. Their use can be also advantageous due to the release of beneficial ions, as will be further discussed in Section 18.5.

In 2011 Kehoe *et al.*[15] and Zhang *et al.*[16] investigated the use of silicate glass micro-powder in the system SiO_2–CaO–ZnO–Na_2O–CeO_2 for the development of composite materials for nerve regeneration.

In a first study, Kehoe *et al.*[15] developed composite devices of different compositions using glass powder (<45 μm) in combination with poly(D,L-lactide-co-glycolide) (PLGA) and Pluronic F127. The aim was to obtain enhanced mechanical properties and material permeability, as well as to improve the control on the material resorption kinetics and the release of beneficial ions, such as calcium and zinc, compared to pure polymeric materials.[15] The guides were fabricated with a modified immersion–precipitation method, in which calcium alginate hydrogel rod of the desired diameter is immersed in a solution of PLGA and F127 with suspended glass powder causing the precipitation of the solution on the rod surface. The final composite PLGA–F127–glass tubes (diameter 1.5 mm, length 30 mm) were obtained after removal (sliding out) of the hydrogel rod and evaporation of the residual solvent. The presence of glass at high concentration exerted a negative effect on cell viability, probably due to the accumulation of the ions released into the solution. The optimum glass content in the composite material was found to be between 5.5 and 20 wt%, which allowed to both maximise the cell viability and to obtain an ultimate tensile strength comparable to that of the reference commercial device Neurolac® (poly(ε-caprolactone), 5–20 MPa).[15] Zhang *et al.*[16] further studied composite nerve guidance channels of PLGA–F127–glass in this selected compositional range (glass 0, 20 wt%, PLGA 12.5, 20 wt% and F127 0, 2.5, 5 wt%) and showed cytocompatibility and mechanical properties comparable to conventional nerve guidance channels (wet-state ultimate tensile strength 1–7 MPa and Young's modulus 5–203 MPa), with higher Young's modulus for the glass-containing conduits compared to the glass-free conduits.

In 2014 Koudehi *et al.*[17] fabricated composite nerve guidance channels of gelatin reinforced with glass nano-powders (less than 100 nm) in the SiO_2–CaO–P_2O_5–MgO system prepared with the sol–gel method. The tube was fabricated by dipping a special mandrel in a solution of gelatin with 30 wt% of suspended glass followed by a freeze-drying process and a final crosslink in a bath of glutaraldehyde solution to obtain porous walls (porosity about 70 vol%, pore size of 10–40 μm). Cytotoxicity *in vitro* with Chinese hamster ovary cells showed cell viability comparable to that of the control medium without the conduit. The guides (diameter 2 mm, length 12 mm) were sutured between the divided sciatic nerve of adult Wistar rats (10 mm gap) resulting in successful nerve regeneration for all the 15 animals operated upon with functional and histological properties comparable to a normal nerve after 3 postoperative months. Biocompatibility was good as only a thin layer of fibrous tissue around the tube and minimal inflammation were observed. Moreover the conduits were almost fully degraded 3 months after implantation.

18.4 Glass Fibres for Nerve Regeneration

Fibres of bioactive glass have been studied for the regeneration of both hard and soft tissues, such as bone,[18,19] tendon, ligament[20] and muscle.[21] Glass fibres with micrometric diameters are flexible and they can be produced as solid or hollow fibres[22,23] (Figure 18.3). They have been proposed for nerve regeneration for the fabrication of nerve wraps (to be folded around the divided nerve) as well as for the creation of topographical features inside the lumen of a nerve guidance channel or in a synthetic graft to guide

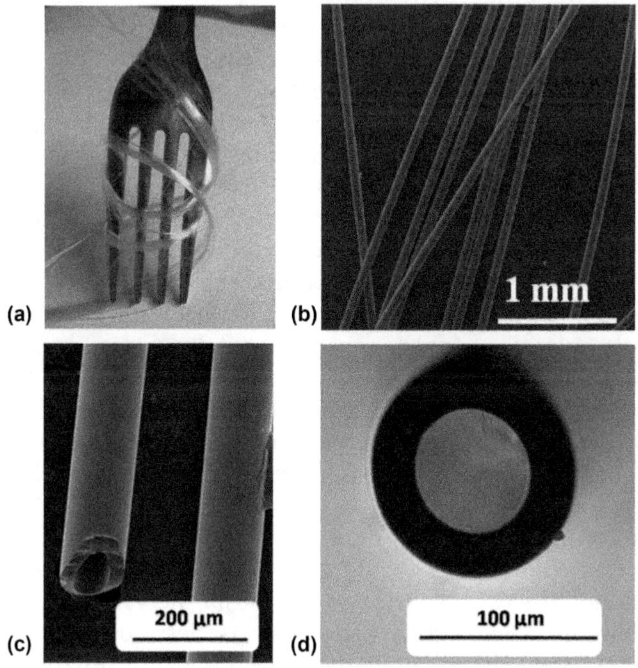

Figure 18.3 Examples of glass fibres with micrometric diameters showing (a) high flexibility (Reprinted from C. Vitale-Brovarone *et al.*, Novel phosphate glasses with different amounts of TiO_2 for biomedical applications Dissolution tests and proof concept of fibre drawing, *Mater. Sci. Eng.: C*, **31**, 434–442, Copyright 2011 with permission from Elsevier) and different morphology: (b) solid fibres (Reprinted from C. Vitale-Brovarone *et al.*, Novel phosphate glasses with different amounts of TiO_2 for biomedical applications Dissolution tests and proof concept of fibre drawing, *Materi. Sci. Eng.: C*, **31**, 434–442, Copyright 2011 with permission from Elsevier) and (c) and (d) hollow fibres (Reprinted from G. Novajra *et al.*, Resorbable hollow phosphate glass fibres as controlled release systems for biomedical applications, *Mater. Lett.*, **99**, 125–127, Copyright 2013 with permission from Elsevier).

axonal growth and improve the outcome of both nerve and spinal cord regeneration. Hollow glass fibres have also been proposed for the release of biologically active molecules like growth factors to enhance nerve regeneration.

18.4.1 Glass Fibre Wrap for the Treatment of Nerve Injuries

Some studies by Jeans *et al.*[14,24] in 2007 and Starritt *et al.*[25] in 2011 reported the use of fibres of the resorbable phosphate glass Corglaes® (Giltech Ltd) for the fabrication of a wrap to treat peripheral nerve injuries (Figure 18.4a). For fibre production, a melt spinning process was used in which liquid glass at appropriate viscosity passes through a metal bushings forming continuous fibre collected by an underlying rotating drum. To fabricate the wrap the fibres were assembled in sheets of non-woven meshes which were bonded using a biodegradable polymer in solution (the type of polymer was not

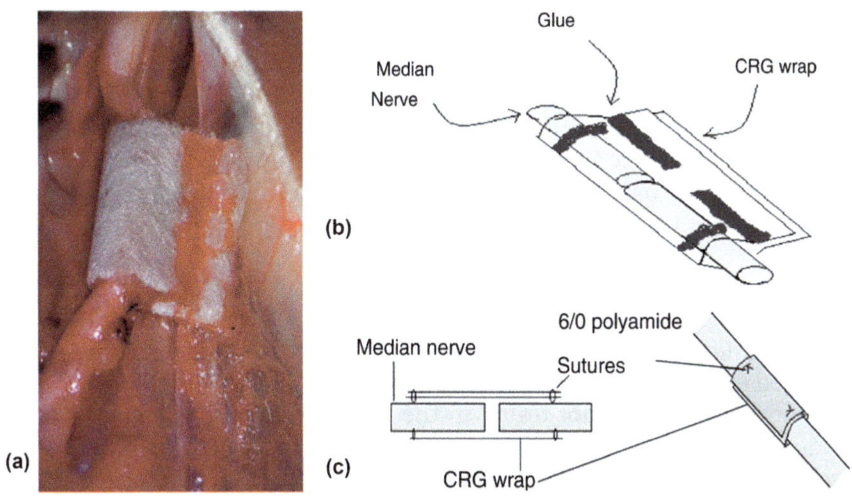

Figure 18.4 (a) Corglaes® fibre wrap used for the repair of the divided median nerve in the upper forelimb of sheep (Reprinted from L. A. Jeans *et al.*, Peripheral nerve repair by means of a flexible biodegradeable glass fibre wrap: A comparison with microsurgical epineurial repair, *J. Plast., Reconstr. Aesth. Surg.*, **60**, 1302–1308, Copyright 2007 with permission from Elsevier). (b) Diagram of wrap placed around divided median nerve and secured with glue (Reprinted from L. Jeans, An evaluation using techniques to assess muscle and nerve regeneration of a flexible glass wrap in the repair of peripheral nerves, *Acta Neurochirurg. Suppl.*, 2007, **100**, 25–28, with kind permission from Springer Science and Business Media); (c) Diagram of the technique used to secure the wrap with polyglactin sutures (Reprinted from L. Jeans, An evaluation using techniques to assess muscle and nerve regeneration of a flexible glass wrap in the repair of peripheral nerves, *Acta Neurochirurg. Suppl.*, 2007, **100**, 25–28, with kind permission from Springer Science and Business Media).

specified) obtaining a final texture like tissue paper. The wrap was flexible allowing the folding of the wrap around the divided nerve stumps without collapsing. The wrap also displayed some porosity allowing for the passage of molecules (*e.g.* nutrients) towards the site of repair.[14]

The wrap was studied *in vivo* for the treatment of divided median nerves in the upper forelimb[14,24] and facial nerves[25] of sheep. The divided nerve stumps were placed on the top of the wrap with no gap between them and the wrap was folded to form a tube around the nerve stumps. The wrap was secured in position by suturing it to the epineurium of both the nerve stumps[14,24] or by the use of fibrin[14,24,25] or poly(caprolactone) glue[24] (Figure 18.4b and c). The results of the different studies, evaluated at 6 months[25] or 7 months[14,24] after the treatment, showed that the use of the fibre wrap sustained a successful nerve regeneration and muscle re-innervation with results comparable to that of the standard microsurgical epineurial suturing. Moreover, no pull-out of the repair, foreign-body reaction or increased wound infection were observed and the wrap was not present macroscopically at the end of the test, thus not requiring any further intervention for its removal.[14,24,25] The authors concluded that the use of the glass fibre wrap can be advantageous compared to the microsurgical standard technique as it allows for shorter procedure (few minutes for the wrap suturing compared to about 25 minutes required by the microsurgical suturing). Besides it does not require any microsurgical equipment or skills, since only basic surgical knowledge and technique are required, and avoids the risk of damaging the nerve fascicles (*e.g.* fibrosis) caused by the suturing.

The nerve wrap fabricated with Corglaes® fibres overcame the technical limitations associated with the Corglaes® tube[13] (Section 18.2) since it was flexible and porous. However, as in the case of the glass tube, the nerve wrap was used for the treatment of divided nerve with stumps maintained close to each other and it was not tested in the presence of a nerve defect. Further studies would be necessary to evaluate the potential of the glass fibre wrap to support the tissue regeneration across a nerve gap.

18.4.2 Aligned Glass Fibres to Support Axonal Regeneration

One of the main problems related to nerve regeneration into a nerve guidance channel over a long distance seems to be insufficient formation of fibrin cables between the two nerve stumps, a key factor to guide cell migration and axonal growth.[3] For this reason, one objective of the research activities in this field consists of the introduction of additional features and functionalities to the nerve guides with the aim of enhancing the treatment outcome for long nerve defects also, limiting the need for autologous tissue and its associate drawbacks. In this context, the use of fibres of different materials, including bioactive glasses, showing micrometric or nanometric diameters is currently being investigated to support and guide the tissue regeneration, compensating for the lack of fibrin cable formation.

In 2005 Bunting *et al.*[26] proposed the use of a bundle of Bioglass® 45S5 ($45SiO_2$–$24.5Na_2O$–24.5 CaO–$6P_2O_5$; wt%) fibres aligned inside the lumen of a Silastic® conduit for nerve repair. They succeeded in drawing Bioglass® 45S5 by immersing a glass rod in molten glass of the same composition and by removing it at a certain speed to obtain fibres, whose diameter was controlled through the removal speed. The study of the *in vitro* interaction of SCs with Bioglass® 45S5 fibres (diameter 150–200 μm) showed that cells attached and spread to all the fibre types after 24 hours of culture and that they were sometimes aligned in longitudinal chains on the smallest calibre fibres. The authors inserted about 10 glass fibres (diameter 25 μm, length 0.5 cm) inside a Silastic® conduit, which was then tested *in vivo* for the repair of a 0.5 cm gap in the sciatic nerve of adult rats. Regenerating axons grew in minifascicles in well-vascularized connective tissue and the re-innervation of the distal stump after 2 and 4 weeks was comparable to that obtained with autograft repair and ten times greater than those observed for the repair with an empty Silastic® conduit and for untreated nerve gap. The hypothesis was that the fibres were able to facilitate nerve regeneration through a contact–guidance mechanism of outgrowing axons and their associated non-neuronal cells. The silicone nerve conduit used in this study has been the standard experimental model for many years. However, its use shows some limitations related to the lack of biodegradation and permeability that make it suitable for bridging short nerve defects.[27] It would be interesting to study the effect of 45S5 Bioglass® fibres inside a semi-permeable and degradable nerve guide, thus extending their potential application to the treatment of longer nerve defects.

In 2012 Vitale-Brovarone *et al.*[28] proposed the use of resorbable phosphate glass fibres ($50P_2O_5$–30CaO–$9Na_2O$–$3SiO_2$–3MgO–$2.5K_2O$–$2.5TiO_2$; mol%, coded as $TiPS_{2.5}$) to guide cell migration and axonal regeneration during nerve repair. Phosphate glasses, and especially metaphosphate glasses (50 mol% P_2O_5), have a structure composed of long chains of phosphate tetrahedra which enable fibre drawing.[22,29] These glasses can be designed to be soluble, with a resorption time that can be tailored on the basis of the specific application acting on the glass composition and the specific surface (*i.e.* fibre diameter) or by carrying out suitable thermal treatments able to stabilize the glass structure.[28] The fibres were produced following the preform drawing approach in which a preform (*i.e.* a glass rod) is heated at one of its extremities until the decrease in viscosity causes the formation of a neckdown region and then of a fibre that is wound on a rotating drum.[28] With a careful adjustment of the process parameters (*i.e.* preform diameter, preform feeding speed and drum rotation speed) a very precise control over the fibre diameters can be achieved. The authors investigated the cell–fibre interaction by *in vitro* culture of glial cells (neonatal olfactory bulb ensheathing cell line, NOBEC) and dorsal root ganglia (DRG) neurons in contact with a construct of aligned fibres of different diameters (25, 40 and 78 μm). Glial cells proliferated and spread onto all the fibre types, enveloping the fibres (Figure 18.5a and b) after 3 days. Also neurons adhered to the fibres with neuritis extending along the fibre axis

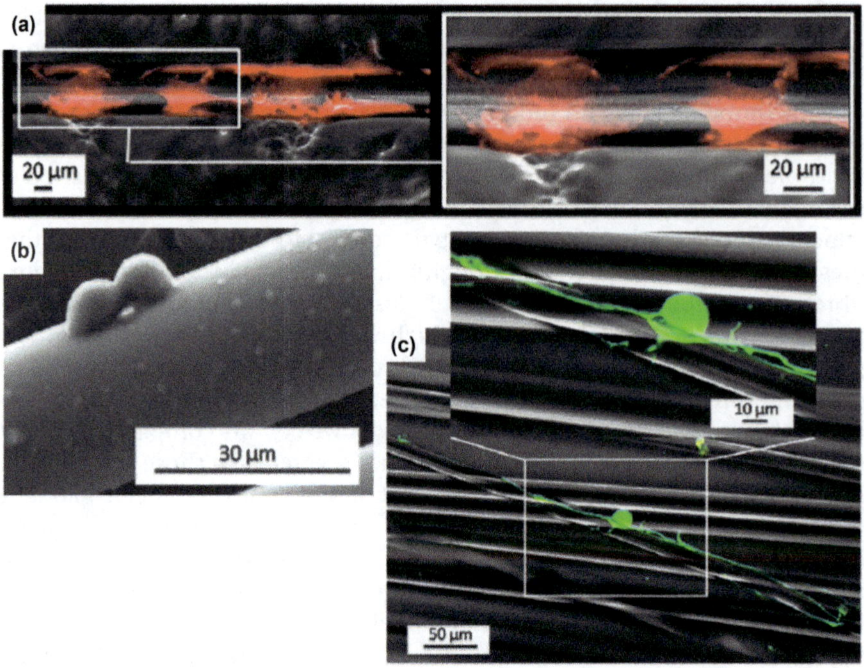

Figure 18.5 NOBEC glial cells after 3 days of culture (a) enveloping a glass fibres (confocal microscope image) and (b) showing active proliferation on phosphate glass fibres (scanning electron microscope images); (c) DRG neuron after 3 days of culture showing long neurites along the fibre axis direction (confocal microscope images).
(Reprinted from C. Vitale-Brovarone *et al.*, Phosphate glass fibres and their role in neuronal polarization and axonal growth direction, *Acta Biomater.*, **8**, 1125–1136, Copyright 2012 with permission from Elsevier.)

direction (Figure 18.5c), in contrast to the neurons cultured on control glass cover slips which showed shorter and non-oriented neurites. The authors concluded that this cell–fibre interaction can be exploited to promote and guide the tissue regeneration also for long nerve defects, where the lack of fibrin cables limits the cell migration and axonal regeneration. In the light of the good results obtained in the *in vitro* test, further studies are required to verify the efficacy of the proposed approach *in vivo.*

Another approach to guide axonal regeneration using glass fibres is to assemble the latter in an aligned configuration with the help of a polymer. In this case, the repair can be carried out with or without the use of a tubular nerve guidance channel, since the fibre–polymer construct can be either directly grafted between the two nerve stumps[30–32] or placed inside a nerve guidance channel.[33]

Marquardt *et al.*[30] in 2014 fabricated a composite scaffold of fibrin embedding fibres of resorbable 13-93 B3 borate glass ($53B_2O_3$–20CaO–$6Na_2O$–$12K_2O$–5MgO–$4P_2O_5$; wt%). Glass fibres of diameters ranging from 50 to 200 µm (pulled from the glass melt) and from 0.5 to 10 µm (unspecified drawing process) were used. Mixed cultures of neurons, glia and fibroblast resulted in decreased cell viability when in contact with glass fibres while the percentage of neurons increased compared with all cell types. Glass–fibrin composite scaffolds containing glass fibres in a random (fibre diameter 0.5–10 μm) or aligned configuration (fibre diameter 50–200 μm, length 10 mm) were investigated *in vitro* with DRG neurons. It was found that the neurite extension on the scaffolds containing random glass fibres was comparable to that observed on pure fibrin scaffold, while it proceeded in an oriented direction on the scaffold containing aligned glass fibres. Also in this case, further studies would be necessary to investigate the ability of this scaffold to support the nerve regeneration *in vivo*.

Another composite collagen–glass scaffold containing fibres of resorbable phosphate glass ($50P_2O_5$–$40CaO$–$5Na_2O$–$5Fe_2O_3$; mol%) was tested *in vivo* for both peripheral nerve and spinal cord repair.[31,32] Collagen gel was put on a filter paper and pressed to extrude water while the glass fibres were aligned on the sheet before extrusion was completed. The collagen–glass fibres sheet was then rolled to form a cylindrical scaffold, left to air-dry and cut at the desired length (Figure 18.6a and b).

In the case of peripheral nerve repair, Kim *et al.*[31] first studied the interaction of DRG neurons with these $50P_2O_5$–$40CaO$–$5Na_2O$–$5Fe_2O_3$ (mol%) glass fibres *in vitro*. They showed that the presence of aligned glass fibres (diameter 15 μm, length 20 mm) induced the growth of long neurites extending in both directions and showing linear-type axonal sprouting along the major axis; on the contrary, neurons cultured without glass fibres grew without any directionality. Then the authors studied the collagen–glass fibres composite scaffold (diameter 0.8 mm, length 3 mm) *in vivo* as a graft for the repair of rat sciatic nerve defect. After removing 3 mm of the nerve tissue, the preserved epineurium was used to wrap and hold in place the scaffold. Pure collagen scaffold was used as a control. Animals treated with the fibre-containing scaffolds showed faster nerve regeneration in the first weeks, while no differences were found at 12 weeks after implantation. This demonstrates that the presence of the glass fibres stimulated the early stage of repair (6 weeks) through the promotion of directional extension of axons along the fibre direction.

Joo *et al.*[32] studied the same type of collagen–glass fibres scaffold *in vivo* for the repair of divided rat spinal cord, demonstrating for the first time the efficacy of phosphate glass fibres for the treatment of thoracic spinal cord injury. Spinal cord injuries, which can be the consequence of vehicular and sports accidents, falls and different forms of violence (*e.g.* bullet wounds), can result in serious neurological deficits, compromise sensory and motor functions and involve other related problems such as bladder and kidney infections, cardiac and respiratory dysfunctions and bowel diseases.[34] Spinal cord does not possess the regenerating potential of the peripheral nerve tissue and no devices are actually available for the repair of spinal cord injuries. The surgical treatment of such injuries consists of stabilization and decompression of the spinal cord, drug therapy and rehabilitation to

Figure 18.6 Examples of phosphate glass fibres used in aligned configuration for the development of devices for nerve fibre regeneration. Collagen–glass fibres sheet (a) after dehydration (Reprinted from N.-Y. Joo *et al.*, Effects of phosphate glass fiber-collagen scaffolds on functional recovery of completely transected rat spinal cords, *Acta Biomater.*, **8**, 1802–1812, Copyright 2012 with permission from Elsevier) and (b) after cutting and rolling to obtain a three dimensional composite scaffold (Reprinted from *J. Tissue Eng. Regen. Med.*, **9**, 236–246 with permission from John Wiley and Sons. Copyright © 2012 John Wiley & Sons, Ltd); (c) phosphate glass fibres coated with carbon nanotubes; (d) aligned, wrapped around a poly(D,L-lactic acid) electrospun nanofiber mat and placed within a porous poly(D,L-lactic acid) cylindrical tube to obtain a device for the treatment of peripheral nerve injuries. (Reprinted from *Acta Biomater.*, **13**, 324–334, © (2015) with permission from Elsevier.)

stimulate the tissue plasticity. Truly satisfactory treatments or devices are still to be found. In the experimental study, the collagen–glass fibres scaffold (diameter 1.8 mm, length 3 mm) was implanted in the rat spinal cord gap (3 mm).[32] The presence of the fibres in the collagen scaffold resulted in an improved locomotor recovery compared to pure collagen scaffolds from 8 weeks after implantation. Moreover, while no axonal growth was observed into the pure collagen scaffolds at 12 weeks after implantation, some axons extended into the scaffold in two of the three rats treated with the collagen–glass fibres scaffolds and in one of them they crossed the scaffold reaching the lumbar enlargement. One of the rats treated with the fibre-containing scaffold showed a recovery of the normal bladder pattern. The fibres, which acted as physical directional guides for regenerating axons, were still present at the end of the 12-week test indicating a slow dissolution rate. This can be due to the presence of 5 mol% of Fe_2O_3 in the glass composition, which has a stabilization effect on the glass network. The authors believed that the high level of calcium and phosphate ions released from the fibres could play a role in the functional outcome following spinal cord injury repair and, for this reason, they aimed at further investigating the influence of the ionic dissolution products on the healing process as well as the use of faster dissolving glass compositions. Even if the presence of the fibres showed a beneficial effect on the functional recovery after complete spinal cord transection, they did not induce any reduction of the inflammatory responses nor increase in endogenous neurotrophic factors, which are essential in the subacute and chronic phase of spinal cord injury. Looking at forward applications, the authors concluded on the prospect of combining anti-inflammatory drugs and exogenous neurotrophic factors with phosphate glass fibres to further promote functional recovery.

Ahn *et al.*[33] modified the fibres used in the studies described above[31,32] ($50P_2O_5$–40CaO–$5Na_2O$–$5Fe_2O_3$; mol%) by chemically tethering carbon nanotubes (CNTs) onto their surface (Figure 18.6c) and studied their effect *in vitro*. PC12 cells, derived from a pheochromocytoma of the rat adrenal medulla, showed a good viability after being cultured in a medium containing the fibre dissolution products. Moreover, the presence of CNTs on the fibres' surfaces was responsible for a great improvement in the neurite extension from dorsal root ganglia. For nerve scaffold fabrication, CNT–glass fibres were aligned and fixed with poly(caprolactone) at the extremities, wrapped around a poly(D,L-lactic acid) electrospun nanofibre mat and placed within a porous poly(D,L-lactic acid) cylindrical tube (Figure 18.6d). The obtained device was tested *in vivo* for the repair of 10 mm transected sciatic nerve in rats showing that the presence of CNTs enhanced the clinical outcome in terms of the number of regenerating axons growing into the scaffold and the cross-sectional area of the re-innervated muscles. The authors hypothesized that the combination of glass fibres with CNTs can have a synergistic effect on nerve regeneration both providing a physical guidance (*i.e.* glass fibres) and stimulating cell adhesion and neurite extension (*i.e.* CNTs). The potential release of therapeutic molecules from CNTs could represent an additional scaffold functionality.

18.4.3 Hollow Glass Fibres for Growth Factor and Drug Release

Neurotrophic factor support is a necessary feature to ensure strong clinical and preclinical outcomes of nerve regeneration. After the entubulation of the stumps of a transected peripheral nerve, the presence of plasma exudate inside the conduit results in the accumulation of neurotrophic factors and molecules of the extracellular matrix, which contribute to the nerve regeneration process.[3] Also, it has been shown that the distal stump promotes the axonal regeneration by both cell migration and the release of neurotrophic factors.[4] For long nerve gaps, the reduction in the neurotrophic effects of the distal nerve stump can be one of the causes of poor nerve regeneration.[3] For this reason, the incorporation and controlled release of nerve growth factors in the nerve guidance channel can be a strategy to obviate to this problem.[3]

Novajra *et al.*[23,35] developed hollow fibres of resorbable $TiPS_{2.5}$ phosphate glass ($50P_2O_5$–30CaO–$9Na_2O$–$3SiO_2$–3MgO–$2.5K_2O$–$2.5TiO_2$; mol%) for nerve regeneration. The idea was to insert the hollow fibres inside a nerve guidance channel both to support and to direct the axonal growth, as previously showed *in vitro* in the form of solid fibres,[28] and to incorporate specific molecules or growth factors inside the fibre cavity to be released *in vivo*. In a first study, they demonstrated that the hollow glass fibres could be quickly filled with a testing liquid simply by dipping one of the fibre extremities in the liquid and exploiting the capillary action (Figure 18.7a and b). It was also observed that the testing solution was then released in about 24 hours *in vitro*.[23] In a further study, the authors investigated different strategies to tailor the release of a fluorescent model molecule from the fibres.[35] While the different fibre diameters (inner/outer diameters: 69/121 μm and 94/167 μm) and lengths (1.5, 2 and 2.5 mm) did not induce any significant differences in the molecule release kinetics, the use of different filling materials for the fibres was able to influence the molecule release rate (Figure 18.7c).

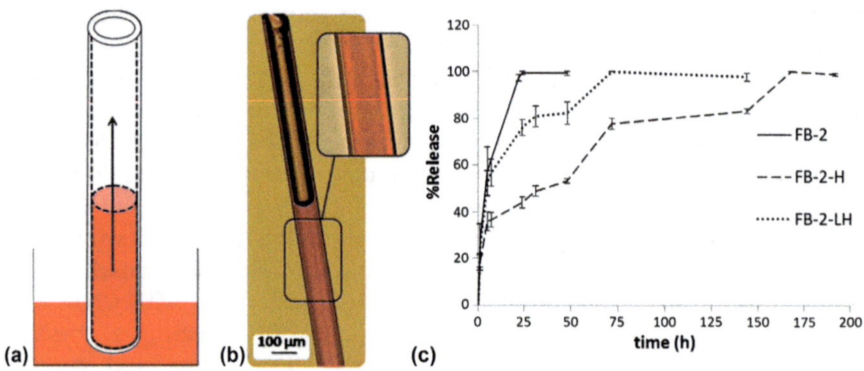

Figure 18.7 (a) Scheme of the filling process of hollow fibres exploiting the capillary action and (b) optical micrograph of a phosphate glass hollow fibre partially filled with a coloured testing solution (Reprinted from G. Novajra *et al.*, Resorbable hollow phosphate glass fibres as controlled release systems for biomedical applications, *Mater. Lett.*, **99**, 125–127, Copyright 2013 with permission from Elsevier); (c) release kinetic of a fluorescent model molecule (FD-20) from phosphate glass hollow fibres filled with different materials (FB-2: FD-20–phosphate buffered saline solution (PBS); FB-2-H: FD-20-containing hydrogel; FB-2-LH:lyophilized hydrogel with FD-20–PBS aqueous solution in the hydrogel porosity). (Reprinted from G. Novajra *et al.*, Novel systems for tailored neurotrophic factor release based on hydrogel and resorbable glass hollow fibers, *Mater. Sci. Eng.: C*, **36**, 25–32, Copyright 2014 with permission from Elsevier.)

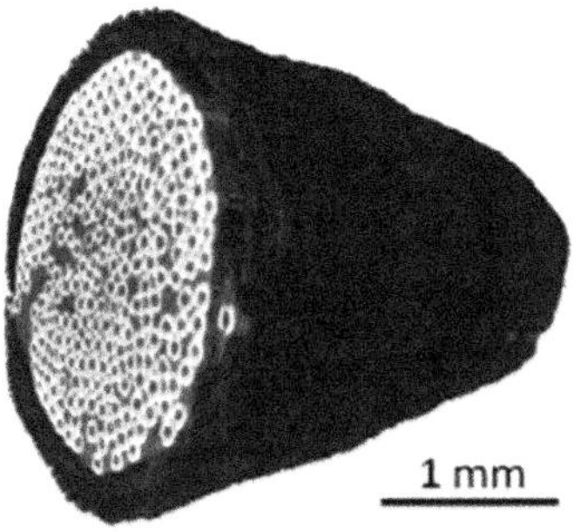

Figure 18.8 Poly(ε-caprolactone) tube containing a bundle of aligned hollow fibres filled with hydrogel able to incorporate and release drugs or growth factors.
(Reprinted from G. Novajra *et al.*, Novel systems for tailored neurotrophic factor release based on hydrogel and resorbable glass hollow fibers, *Mater. Sci. Eng.: C*, **36**, 25–32, Copyright 2014 with permission from Elsevier.)

If compared to liquid solution (complete release after 24 hours), the filling of the fibres with an agar–gelatin hydrogel, cross-linked with genipin, allowed a slower release. Moreover, different results were obtained according to the form of the hydrogel (*i.e.* hydrated or lyophylized hydrogel) and the model molecule (*i.e.* in liquid solution or in the hydrated polymer) inside the fibre lumen; 72 hours were needed for a complete release of the model molecule in liquid solution inside the porosity of the lyophilized hydrogel and 168 hours for the complete release of the model molecule in the hydrated hydrogel. An *in vitro* test showed that the fibre dissolution products (*viz.* 14% fibre weight loss after 28 days in culture medium) did not induce any negative effect on the growth of glial cells and on the expression of pro-/anti-apoptotic proteins.

The authors developed a nerve scaffold composed of a poly-(ε-caprolactone) tube containing a bundle of aligned hollow glass fibres filled with hydrogel with the aim of imparting to the scaffold both topographical and trophic cues (Figure 18.8). In fact, the aligned fibres create an anisotropic structure to orientate cell growth while the hollow cavity inside the fibres allows the incorporation of specific biomolecules or drugs for neurotrophic support, which are both key factors to improve the nerve regeneration in long nerve defects.

18.5 Potential of Bioactive Glass for Release of Nerve Therapeutic Ions

Bioactive glasses have been traditionally applied for the repair of mineralized tissues (bones and teeth) due to their ability to form a layer of surface apatite upon contact with biological fluids. This unique property, which allows a direct bond between bioactive glasses and hard tissues, is referred to as "bioactivity". As discussed in a couple of recent reviews,[36,37] 45S5 Bioglass® and other special glasses were found also to bond to soft tissues and their ion dissolution products can be exploited to induce a particular cell or tissue response in a broad range of applications, not restricted to bone-bonding. For these glasses the "bioactivity" is intended as the ability to stimulate a beneficial response from the host tissue.

Several studies demonstrated that ions released from bioactive glasses show an angiogenic effect, inducing an increased vascularization in the surrounding tissues which is important to promote the regeneration of all tissues, including nerves. An *in vivo* study by Day *et al.*[11] showed an increased neovascularization of 45S5 Bioglass®-containing scaffold implanted subcutaneously in rats. Lin *et al.*[12] tested some ointments containing micro- and nanopowders of sol–gel 58S glass ($58SiO_2$–33CaO–$9P_2O_5$; wt%) and particles of melt-derived 45S5 Bioglass® for the treatment of cutaneous wounds *in vivo.* They found that the wound healing process was promoted by the presence of the glass through the stimulation of the fibroblast proliferation, the growth of granulation tissue in rats, the production of the growth factors VEGF and FGF2 and the formation of new capillary microvessels by the seventh day. Rai *et al.*[38] found that the presence of nano-sized 45S5 Bioglass® particles in a composite scaffold for wound dressing was able to accelerate blood clotting time and to enhance the wettability, surface roughness and overall biocompatibility of the scaffold. Another strategy to promote angiogenesis is the development of cobalt-containing bioactive glasses able to mimic hypoxia conditions through the release of cobalt ions, thus activating pro- and anti-angiogenic genes.[39,40] Although the risks associated to cobalt toxicity *in vivo* still requires careful evaluation,[41] this approach is of interest.

In addition, the incorporation of specific ions showing antibacterial properties in the glass composition, such as silver, gallium, zinc and copper ions, can be useful for the prevention and treatment of bacterial infections that can occur after implantation of an external device. Valappil *et al.*[42,43] reported that 1 mol% of Ga_2O_3 in a phosphate glass was enough to exert an antibacterial effect due to the release of gallium ions. Zinc is known to exert a bacteriostatic action once released in ionic form from bioactive glasses,[44] although its antibacterial action is quite weak compared to that of gallium and silver, for instance. Released silver ions have shown a bactericidal action due to their role in the damage of bacterial RNA and DNA, hence inhibiting bacterial replication.[45] Silicate glass with 3 mol% Ag_2O_3 showed antimicrobial properties,[46,47] while silver-coated silicate glass particles exerted antibacterial and antifungal effects *in vitro.*[48] Ag-doped glasses were also incorporated in surgical sutures for the prevention of bacterial infections.[49,50] Ag-doped[51–54] and Cu-doped[55,56] phosphate glasses showed an antibacterial effect on the biofilm formation and on the growth of several types of bacteria. Some commercial products containing Ag-doped phosphate glass are already on the market for the control of infections. They are coupled to a polymeric adhesive for wound care film dressing (Antimicrobial

Arglaes® film, Antimicrobial Arglaes® Island, Medline Industries Inc.) and with alginate for topical powders (Arglaes® Powder, Medline Industries Inc.). Stoor *et al.*[57–59] demonstrated the antibacterial effect of a S53P4 glass ($53SiO_2$–$23Na_2O$–20CaO–$4P_2O_5$; wt%) implants (*i.e.* paste, granules, discs) on a wide range of oral pathogens in humans.

The angiogenic and antimicrobial effect of ionic release from bioactive glasses can be beneficial also in the repair of nerve tissue due to the importance of vascularization and prevention of infections for successful tissue regeneration. The influence of other specific ions in promoting the nerve healing process is currently being studied.

Zhang *et al.*[60,61] extensively discussed the specific action that calcium and zinc ions, released by bioactive glasses, can potentially exert on the nerve regeneration process. They studied the release profiles from glass in the SiO_2–Na_2O–CaO–ZnO–CeO_2 system and from composite nerve guidance channels containing glass micro-powder of the same composition, whose cytocompatibility and mechanical properties have been discussed in Section 18.3.[15,16] Adequate release values for calcium were found for glass powder soaked in simulated physiological fluid (Tris-HCl buffer) for 4 weeks, while for zinc they exceeded the concentration range that is considered to be beneficial for nerve regeneration, neuroprotection and antibacterial effect.[60] Ion release from a PGLA–F127–glass powder composite nerve guide was found appropriate to mediate therapeutic effects in peripheral nerve regeneration.[61] In order to obtain concentrations of the released ions that can be advantageous for nerve tissue healing, it is possible to vary several parameters, such as the pore size of the composite device in which the glass will be incorporated and the glass composition. In particular, the beneficial effect of the released ions is due to the role of calcium in the regulation of the nerve growth cone motility and neurite elongation. In addition, the authors reported the importance of zinc for effective wound healing and its positive influence on the immune system, its participation in neurotransmission (*i.e.* zinc ions are released to the synaptic space during neuronal activity), neuroprotection, Schwann cell proliferation and myelination process, as well as its antibacterial effect. Even if only a few studies have been carried out so far on cerium ions regarding their influence on the nerve healing process, they appear to play a role in the protection of nerves from oxidative stresses during regeneration.[60,61]

18.6 Conclusions and Outlook

Bioactive glasses have shown to be promising materials for the treatment of peripheral nerve and even spinal cord injuries. They have been proposed in bulk, powder and fibre form for the development of tubular devices or to impart additional functionalities inside the guide lumen and improve tissue regeneration even for long nerve gaps.

A nerve guidance channel and a glass fibre wrap of fully resorbable glass appeared to be a suitable alternative to the end-to-end suture (*i.e.* for the treatment of nerve injuries without the presence of a nerve defect) avoiding the need for the equipment and the skills required for microsurgical repair. The development of polymer-based composite tubes containing glass micro- or nanopowder showed good results in term of biocompatibility and regeneration outcome. The optimization of the glass/polymer ratio seems to be the key to tailor the physical, biological and mechanical properties of the nerve guidance channel.

The main purpose of undergoing scientific research activities in the field of peripheral nerve regeneration is to improve the performance of nerve guides, whose effectiveness is now limited to the treatment of nerve defects of about 3 cm, with the aim to replace the use of autografts through the introduction of additional features and functionalities to the devices. One of the strategies is the creation of anisotropic structures inside the tube lumen to properly support and direct the nerve regeneration. For this purpose, aligned glass fibres placed inside the lumen of a nerve guide or for the development of a nerve graft, allowed the support and the orientation of the axonal growth, both *in vitro* and *in vivo*, thus enhancing peripheral nerve and spinal cord regeneration. Another strategy to improve the outcome of the nerve guidance channels is the incorporation of drugs or exogenous neurotrophic factors to be released once the guide has been implanted. Resorbable hollow glass fibres have been developed, able to incorporate liquid solutions or hydrogels, potentially carrying growth factors or drugs for local release after implantation.

The promising results obtained with these glass-containing devices stimulate further investigation in this area and also the exploitation of bioglasses in other forms. For example, the functionalization of glass with specific molecules, as well as the use of mesoporous glasses, offering high specific surface area for ionic release and the possibility of drug incorporation and release, are research lines of important prospect. Until now, fibres with micrometric diameters (minimum diameter 0.5 μm) have been proposed for nerve regeneration, however glass nanofibers (*e.g.* fabricated by laser spinning or electrospinning), that can also be produced as mesoporous or hybrid materials, are also attractive for nerve healing applications since they can better mimic the fibrous structure of extracellular matrix. Finally, a more intensive investigation on the effect of calcium, zinc and cerium ions, as well as of other therapeutic ions, on the nerve regeneration process is crucial in order to develop novel glass compositions specifically designed to promote nerve healing.

References

1. R. Deumens, A. Bozkurt, M. F. Meek, M. A. E. Marcus, E. A. J. Joosten, J. Weis and G. Brook, *Prog. Neurobiol.*, 2010, **92**, 245.
2. S. Kehoe, X. F. Zhang and D. Boyd, *Injury*, 2012, **43**, 553.
3. W. Daly, L. Yao, D. Zeugolis, A. Windenbank and A. Pandit, *J. R. Soc., Interface*, 2012, **9**, 202.

4. J. S. Belkas, M. S. Shoichet and R. Midha, *Oper. Tech. Orthop*, 2004, **14**, 190.
5. A. Pabari, S. Y. Yang, A. M. Seifalian and A. Mosahebi, *J. Plast. Reconstr. Aesthetic Surg*, 2010, **63**, 1941.
6. M. F. Meek and J. H. Coert, *Ann. Plast. Surg.*, 2008, **60**, 466.
7. S. Geuna, P. Tos, P. Titolo, D. Ciclamini, T. Beningo and B. Battiston, *J. Brachial Plex Peripher Nerve Inj.*, 2014, **9**, 3.
8. X. Jiang, S. H. Lim, H. Q. Mao and S. Y. Chew, *Exp. Neurol.*, 2010, **223**, 86.
9. K. Haastert-Talini, S. Geuna, L. B. Dahlin, C. Meyer, L. Stenberg, T. Freier, C. Heimann, C. Barwig, L. F. V. Pinto, S. Raimondo, G. Gambarotta, S. R. Samy, N. Sousa, A. J. Salgado, A. Ratzka, S. Wrobel and C. Grothe, *Biomaterials*, 2013, **34**, 9886.
10. C. Meyer, S. Wrobel, S. Raimondo, S. Rochkind, C. Heimann, A. Shahar, O. Ziv-Polat, S. Geuna, C. Grothe and K. Haastert-Talini, *Cell Transplant*, 2016, **25**, 159.
11. R. M. Day, A. R. Boccaccini, S. Shurey, J. A. Roether, A. Forbes, L. L. Hench and S. M. Gabe, *Biomaterials*, 2004, **25**, 5857.
12. C. Lin, C. Mao, J. Zhang, Y. Li and X. Chen, *Biomed. Mater.*, 2012, **7**, 045017.
13. T. Gilchrist, M. A. Glasby, D. M. Healy, G. Kelly, D. V. Lenihan, K. L. McDowall and I. A. Miller, *Br. J. Plast. Surg.*, 1998, **51**, 231.
14. L. A. Jeans, T. Gilchrist and D. Healy, *J. Plast. Reconstr. Aesthetic Surg*, 2007, **60**, 1302.
15. S. Kehoe, X. F. Zhang and D. Boyd, *J. Mater. Sci.: Mater. Med.*, 2011, **22**, 945.
16. X. F. Zhang, H. O'Shea, S. Kehoe and D. Boyd, *J. Mech. Behav. Biomed. Mater.*, 2011, **4**, 1266.
17. M. F. Koudehi, A. A. I. Fooladi, K. Mansoori, Z. Jamalpoor, A. Amiri and M. R. Nourani, *J. Mater. Sci.: Mater. Med.*, 2014, **25**, 363.
18. Y. Gu, W. Huang, M. N. Rahaman and D. E. Day, *Acta Biomater.*, 2013, **9**, 9126.
19. R. M. Felfel, I. Ahmed, A. J. Parsons and C. D. Rudd, *J. Mech. Behav. Biomed. Mater.*, 2013, **17**, 76.
20. M. Bitar, V. Salih, V. Mudera, J. C. Knowles and M. P. Lewis, *Biomaterials*, 2004, **25**, 2283.
21. R. Shah, A. C. M. Sinanan, J. C. Knowles, N. P. Hunt and M. P. Lewis, *Biomaterials*, 2005, **26**, 1497.
22. C. Vitale-Brovarone, G. Novajra, D. Milanese, J. Lousteau and J. C. Knowles, *Mater. Sci. Eng., C*, 2011, **31**(2), 434.
23. G. Novajra, J. Lousteau, D. Milanese and C. Vitale-Brovarone, *Mater. Lett.*, 2013, **99**, 125.
24. L. Jeans, D. Healy and T. Gilchrist, *Acta Neurochir., Suppl.*, 2007, **100**, 25.
25. N. E. Starritt, S. A. J. Kettle and M. A. Glasby, *Laryngoscope*, 2011, **121**, 1614.
26. S. Bunting and L. Di Silvio, *J. Hand Surg.-Br. Eur.*, 2005, **30**, 242.
27. C. E. Schmidt and J. Baier Leach, *Annu. Rev. Biomed. Eng.*, 2003, **5**, 293.

28. C. Vitale-Brovarone, G. Novajra, J. Lousteau, D. Milanese, S. Raimondo and M. Fornaro, *Acta Biomater.*, 2012, **8**, 1125.
29. I. Ahmed, M. Lewis, I. Olsen and J. C. Knowles, *Biomaterials*, 2004, **25**, 501.
30. L. M. Marquardt, D. Day, S. E. Sakiyama-Elbert and A. B. Harkins, *J. Biomed. Mater. Res. A*, 2014, **102**, 2767.
31. Y. P. Kim, G. S. Lee, J. W. Kim, M. S. Kim, H. S. Ahn, J. Y. Lim, H. W. Kim, Y. J. Son, J. C. Knowles and J. K. Hyun, *J. Tissue Eng. Regener. Med.*, 2015, **9**, 236.
32. N. Y. Joo, J. C. Knowles, G. S. Lee, J. W. Kim, H. W. Kim, Y. J. Son and J. K. Hyun, *Acta Biomater.*, 2012, **8**, 1802.
33. H. S. Ahn, J. Y. Hwang, M. S. Kim, J. Y. Lee, J. W. Kim, H. S. Kim, U. S. Shin, J. C. Knowles, H. W. Kim and J. K. Hyun, *Acta Biomater.*, 2015, **13**, 324.
34. N. A. Silva, N. Sousa, R. L. Reis and A. J. Salgado, *Prog. Neurobiol.*, 2014, **114**, 25.
35. G. Novajra, C. Tonda-Turo, C. Vitale-Brovarone, G. Ciardelli, S. Geuna and S. Raimondo, *Mater. Sci. Eng., C*, 2014, **36**, 25.
36. V. Miguez-Pacheco, L. L. Hench and A. R. Boccaccini, *Acta Biomater.*, 2015, **13**, 1.
37. F. Baino, G. Novajra, V. Miguez-Pacheco, A. R. Boccaccini and C. Vitale-Brovarone, *J. Non-Cryst. Solids*, 2016, **432**, 15.
38. R. Rai, A. R. Boccaccini, J. C. Knowles, I. C. Locke, M. P. Gordge, A. McCormick, V. Salih, N. Mordon, T. Keshavarz and I. Roy, *AIP Conf. Proc.*, 2010, **1255**, 126.
39. C. Wu, Y. Zhou, W. Fan, P. Han, J. Chang, J. Yuen, M. Zhang and Y. Xiao, *Biomaterials*, 2012, **33**, 2076.
40. R. K. Jain, P. Au, J. Tam, D. G. Duda and D. Fukumura, *Nat. Biotechnol.*, 2005, **23**, 821.
41. X. Mao, A. A. Wong and R. W. Crawford, *Med. J. Aust.*, 2011, **194**, 649.
42. S. P. Valappil, D. Ready, E. A. Abou Neel, D. M. Pickup, W. Chrzanowski, L. A. O'Dell, R. J. Newport, M. E. Smith, M. Wilson and J. C. Knowles, *Adv. Funct. Mater.*, 2008, **18**, 732.
43. S. P. Valappil, D. Ready, E. A. Abou Neel, D. M. Pickup, L. A. O'Dell, W. Chrzanowski, J. Pratten, R. J. Newport, M. E. Smith, M. Wilson and J. C. Knowles, *Acta Biomater.*, 2009, **5**, 1198.
44. D. Boyd, H. Li, D. A. Tanner, M. R. Towler and J. G. Wall, *J. Mater. Sci.: Mater. Med.*, 2006, **17**, 489.
45. A. B. Lansdown, A. Williams, S. Chandler and S. Benfield, *J. Wound Care*, 2005, **14**, 155.
46. M. Bellantone, N. J. Coleman and L. L. Hench, *J. Biomed. Mater. Res.*, 2000, **51**, 484.
47. M. Bellantone, H. D. Williams and L. L. Hench, *Antimicrob. Agents Chemother.*, 2002, **46**, 1940.
48. A. W. Wren, A. Coughlan, P. Hassanzadeh and M. R. Towler, *J. Mater. Sci.: Mater. Med.*, 2012, **23**, 1331.

49. J. J. Blaker, S. N. Nazhat and A. R. Boccaccini, *Biomaterials*, 2004, **25**, 1319.
50. J. Pratten, S. N. Nazhat, J. J. Blaker and A. R. Boccaccini, *J. Biomater. Appl.*, 2004, **19**, 47.
51. I. Ahmed, D. Ready, M. Wilson and J. C. Knowles, *J. Biomed. Mater. Res. A*, 2006, **79**, 618.
52. S. P. Valappil, D. M. Pickup, D. L. Carroll, C. K. Hope, J. Pratten, R. J. Newport, M. E. Smith, M. Wilson and J. C. Knowles, *Antimicrob. Agents Chemother.*, 2007, **51**, 4453.
53. S. P. Valappil, J. C. Knowles and M. Wilson, *Appl. Environ. Microbiol.*, 2008, **74**, 5228.
54. M. M. Moss, D. M. Pickup, I. Ahmed, J. C. Knowles, M. E. Smith and R. J. Newport, *Adv. Funct. Mater.*, 2008, **18**, 634.
55. A. M. Mulligan, M. Wilson and J. C. Knowles, *Biomaterials*, 2003, **24**, 1797.
56. E. A. Abou Neel, I. Ahmed, J. Pratten, S. N. Nazhat and J. C. Knowles, *Biomaterials*, 2005, **26**, 2247.
57. P. Stoor, E. Soderling and J. I. Salonen, *Acta Odontol. Scand.*, 1998, **56**, 161.
58. P. Stoor, E. Soderling and R. Grenman, *J. Biomed. Mater. Res.*, 1999, **48**, 869.
59. P. Stoor, E. Soderling and R. Grenman, *J. Biomed. Mater. Res.*, 2001, **58**, 113.
60. X. F. Zhang, S. Kehoe, S. K. Adhi, T. G. Ajithkumar, S. Moane, H. O'Shea and D. Boyd, *Mater. Sci. Eng., C*, 2011, **31**, 669.
61. X. F. Zhang, A. Coughlan, H. O'Shea, M. R. Towler, S. Kehoe and D. Boyd, *Mater. Sci. Eng., C*, 2012, **32**, 1654.

CHAPTER 19

Bioactive Silicate Glass in Implantable Medical Devices: From Research to Clinical Applications

NIKO MORITZ[a] AND PEKKA K. VALLITTU*[a,b]

[a] Department of Biomaterials Science, Institute of Dentistry, University of Turku, Turku, Finland; [b] City of Turku, Welfare Division, Finland
*Email: pekka.vallittu@utu.fi

19.1 Background

The term "bioactive glasses" (BG) refers to certain compositions of melt-derived silicate glasses, originally discovered by Prof. Hench and colleagues in 1969.[1] The bioactive properties of these glasses include bone-bonding, soft-tissue bonding and antibacterial properties. When BGs are exposed to physiological environment, bone-bonding is achieved through a cascade of chemical reactions on the surface of the glass and concurrent cellular responses.[2–4] In a narrow compositional range, BGs are also capable of forming chemical bonds with soft-tissues.[5,6] Antibacterial properties of the glasses are attributed to the local rise of pH.[7,8] The US Food and Drug Administration (FDA) approved bioactive glasses Bioglass® 45S5 and BonAlive® S53P4 for certain clinical applications. These glasses are not suitable for hot working and can only be used to manufacture particulate and monoliths. BG particulate is primarily used as bone graft substitute or

RSC Smart Materials No. 23
Bioactive Glasses: Fundamentals, Technology and Applications
Edited by Aldo R. Boccaccini, Delia S. Brauer and Leena Hupa

Published by the Royal Society of Chemistry, www.rsc.org

filler[9–15] with an added value of antibacterial properties.[16,17] Reports on the clinical applications of BG implants in the form of monoliths are scarce.[18–20]

In principle, glasses possess adequate mechanical properties for many technical applications. In fact, the compressive strength of glass is comparable to that of steel and is higher than that of concrete. The tensile strength of glass is low but is still sufficient for many applications including structural elements in buildings. Brittleness is the major problem. Unlike steel, glass cannot accumulate high local stresses and cannot prevent the resultant formation and propagation of cracks.[21] In the case of BG exposed to the physiological environment, the formation of interfacial layers could lead to the development of surface defects. In turn, these defects could produce cracks which, upon the application of load, would propagate through the bulk of the glass causing catastrophic failure. Hence, BG as such is not suitable for the fabrication of load-bearing implants. Nevertheless, several ways to tackle this problem have been suggested. BG can be a part of a composite structure. For example, successful attempts to reinforce BG matrix with steel wires were reported.[22,23] It was also suggested, that a porous structure made of sintered BG particulate could be incorporated within the grooves of a load-bearing metallic implant.[24] Coating of metallic prosthetic devices with BG has been actively investigated. The theoretical advantage of a BG coating is similar to that of a hydroxyapatite (HA) coating. However, the higher bioactivity of BG compared with that of HA and other calcium phosphates could be beneficial for the incorporation of a BG-coated implant in bone.[4] Nevertheless, coating of metallic substrates with BG is challenging. Typical problems include the scaling of the substrates at the elevated temperatures, the loss of compositional control of the BG (*i.e.* due to partial crystallization of the glass), the transport of unwanted ions from the substrate into the glass and the difficulties in matching the thermal expansion coefficients of BG and metals.[25] Extensive research has been carried out to overcome these challenges.

BG particulate or fibers could be used as a dispersed phase within a biostable or a bioresorbable polymer matrix.[26–30] One of the anticipated applications of these materials is bone cements.[26,31]

Recently, an alternative concept was developed by Prof. Vallittu and colleagues. This concept involves a rigid and porous shell made of a biostable FRC with an internal reservoir for BG particulate.[32,33]

This chapter reviews two fundamentally different approaches to complementing medical devices with BG. We discuss metallic implants coated with BG and fiber-reinforced composite implants which rely on BG as an osteoconductive component.

19.2 Bioactive Glass Coatings on Metallic Implants

19.2.1 Spray Coating

Spray coating techniques have been commonly used to prepare hydroxyapatite (HA) coatings on medical implants. Spray coating techniques involve the injection of the stream of powder in a hot gas jet *e.g.* a DC plasma jet. The hot jet melts the powder and carries the molten particles towards the substrate. The particles collide with the substrate, solidify and form a coating.[34,35] Naturally, different spray coating techniques were tried for the fabrication of BG coatings on metal substrates.

The first attempt to coat 316L surgical stainless steel with BG using flame spraying was reported by Hench and colleagues in 1975.[36] Based on the original 45S4 composition they developed two new compositions, $45B_{15}S5$ and 45S5F, suitable for flame spraying. In the glass $45B_{15}S5$, the 15 wt% of SiO_2 was replaced by B_2O_3, while in the glass 45S5F half of the of CaO content by weight and 2 wt% of the SiO_2 were replaced by CaF_2.[37] The details are given in Table 19.1. These compositional modifications increased the coefficients of thermal expansion of the glasses to a value closer to that of 316L surgical stainless steel. Preclinical testing of the BG-coated femoral stem implant was performed in a monkey model. Results indicated good mechanical bond between the coating and the peri-implant bone, however uncontrolled microporosity of the coating was considered an issue which could possibly lead to the separation of the coating from the implant.[37] Later, different flame-spraying techniques were attempted to coat metallic implants with BG. Gabbi and colleagues used plasma spraying to coat Ti6Al4V substrates and implants with Biovetro® glass.[38] The compositional range of Biovetro® glass is shown in Table 19.1. X-ray diffractometric (XRD) analysis showed no crystallization of the glass. *In vitro* cell culture experiments and the implantation in the methaphysis of the rabbit tibia for up to 180 days, revealed the osteoconductive properties of the coated implants.[38] BGs of a similar type were applied to Ti6Al4V substrates and the cell viability and proliferation capacity on the surface of the coatings were assessed in *in vitro* cell-culture experiments.[39] The BG coatings were reported as stimulating the growth and differentiation of osteoblasts.[39] 45S5 BG was used by Lee and colleagues to coat Ti6Al4V substrates with an intermediate pure Ti layer by means of plasma spraying.[40] No crystallization of the glass was determined by XRD and the coating was able to precipitate calcium phosphate and form HA in Kokubo's SBF.[40] Schrooten and colleagues used a BG containing 1.5 wt% of CaF_2 to coat Ti6Al4V oral implants by reactive plasma spraying.[41,42] The glass composition is shown in Table 19.1. The adhesion of the coating to the substrate was considered adequate for load-bearing applications after immersion in Hank's SBF for two months.[42] Wheeler and colleagues studied the osteointegration of porous Ti6Al4V implants plasma-sprayed with HA or BG in the rabbit condyles.[43] Uncoated implants served as controls. The follow-up period was up to 16 weeks. The histomorphometric analysis demonstrated the superiority of BG-coated implants in terms of osteointegration while the biomechanical shear stress was similar in all three groups of implants.[43] Successful application of the high velocity suspension flame spraying (HVSFS) technique for the deposition of 45S5 BG coatings on Ti grade 2 substrates was reported by Altomare and colleagues.[44] Cattini and colleagues used BG in combination

Table 19.1 Compositions of selected bioactive glasses used for coating of metallic implants.

Glass	Composition (wt%)									Application
	Na_2O	K_2O	MgO	CaO	CaF_2	B_2O_3	Al_2O_3	P_2O_5	SiO_2	
45S5 Bioglass®	24.5	—	—	24.5	—	—	—	6.0	45.0	Basic composition Plasma spraying, HVSFS, immersion coating Laser cladding
S53P4 BonAlive®	23.0	—	—	20.0	—	—	—	4.0	53.0	Basic composition
45S5F	23.0	—	—	12.0	16.0	—	—	6.0	43.0	Flame spraying
$45B_{15}S5$	24.5	—	—	24.5	—	15.0	—	6.0	30.0	Flame spraying
52S4.6	21.0	—	—	21.0	—	—		6.0	52.0	Immersion coating Laser cladding
Biovetro® Gabbi *et al.*, 1996	7.0–24.0	2.0–8.0	0.1–2.0	9.0–20.0	—	—	0.1–2.0	4.0–8.0	49.0–53.0	Plasma spraying
Biovetro® Oliva *et al.*, 1998	7.0–24.0	0.5–6.0	1.0–3.0	8.0–42.0	—	—	0.1–2.0	0.0–7.0	46.0–72.0	Plasma spraying
Schrooten *et al.*, 2000	9.8	—	—	30.5	1.5	—	—	6.2	52.0	Plasma spraying
Cattini *et al.*, 2014	4.7	—	—	42.3	—	—	—	6.1	46.9	Plasma spraying (top layer)
HI-5	23.0	—	—	10.0	—	—	3.0	—	64.0	Enameling (ground layer)
HI-6	21.1	—	—	20.0	—	1.6	—	—	57.3	Enameling (top layer)
6P57	11.0	3.0	8.5	15.0	—	—	—	6.0	56.5	Enameling
6P61	10.3	2.8	7.2	12.6	—	—	—	6.0	61.1	Enameling, EPD
6P68	8.3	2.2	5.7	10.1	—	—	—	6.0	67.7	Enameling
1-98	6.0	11.0	5.0	22.0	—	1.0	—	2.0	53.0	Direct CO_2-laser sintering
S520	20.9	7.1	—	18.0	—	—	—	2.0	52.0	Laser cladding

with the ground coat of HA; the coatings were prepared by plasma spraying.[45]

In general, the variants of the spray coating technique seem to be promising for the coating of metallic implants with BG, especially, since the amorphous structure of the glass could be preserved. In this respect, the possibility of coating the implants with 45S5 BG is particularly attractive. However, over 40 years of research did not lead to a single clinical application of BG-coated implants.

19.2.2 Enameling

Enameling has been used for centuries to coat metals with glass. With this method, the suspension of glass powder is applied to the metallic substrate with subsequent heat-treatment. With respect to BG, it is the composition of the glass with 60% of silica by weight that facilitates bone-bonding and, at the same time, poses a challenge in obtaining a durable and mechanically stable coating. Indeed, due to its low silica content, the silica network of BG is open and ions, such as Al, Fe, Cr, Ni, Co, Mo, Ti and Ta, can pass through that network and retard or even inhibit the bioactivity of the glass.[37] Partial crystallization of the glass and mismatch of the thermal expansion coefficients of BG and metals are also characteristic of enameling techniques. In addition, the extensive oxidation of the titanium surface which occurs in the heat-treatment contributes to the poor attachment of coatings.[46] Typically, a ground coating has been used as a buffer layer between the metallic substrate and the BG coating.

Early unsuccessful attempts were undertaken by Hench and colleagues to coat metallic substrates using commercial enamel ground coats and top layers of 45S5.[37] Diffusion of the ions from the substrate into the coating was the reason for the failure.

Later, two early preclinical studies on the load-bearing BGC-coated hemiendoprosthesis in the rabbit hip were reported by Pajamäki and colleagues.[47,48] Custom-designed implants were prepared from Co–Cr–Mo alloy and Ti6Al4V alloy and coated with BG and bioactive glass-ceramics (BGC). The uncoated implants served as controls. A dual coating system was used to improve the adhesion of the coating to the Co–Cr–Mo alloy implants. The inactive ground coating was prepared from glass HI-5 (Table 19.1). This composition was optimized for better compatibility between the metal substrate and the coating in terms of thermal expansion. The methodology previously described by Andersson and colleagues[49] was applied for the compositional optimization. The top coating was prepared from glass HI-6 (Table 19.1). This coating was expected to promote bone-bonding. XRD analysis revealed that during the thermal processing, the glasses in both coating layers had partially crystallized to $Na_2CaSi_3O_9$. The exact composition of the glass used to coat the Ti6Al4V alloy implants was not disclosed apart from that the SiO_2–CaO–MgO–P_2O_5 system was used. The resulting BGC coating contained apatite, wollastonite and a glassy phase.

Both types of coated implants demonstrated undisputed bone-bonding. However, in the Co–Cr–Mo alloy implant group, the top coating in the dual coating system appeared to resorb at an undesirable rate. In addition, cracking of the BGC coatings on the Ti6Al4V alloy implants was noticed together with possible signs of delamination.

The same BGC was also used to coat keratoprosthesis made of commercially pure titanium (c.p.Ti).[50] The study was performed in a rabbit implantation model. The main challenge with keratoprostheses is the epithelial ingrowth between the corneal stroma and the implant which leads to implant failure. The BGC coating was expected to assist in resolving this issue. The results were encouraging; no significant epithelial ingrowth was detected in the follow-up period of up to 12 months, however, the quality of the coating was found to be inadequate for use in human applications.

Enameling techniques were further refined by Tomsia and colleagues in the SiO_2–Na_2O–K_2O–CaO–MgO–P_2O_5 glass system with the presence of a small amount of TiO_2 in some of the glasses.[51] The aim was to obtain BGs with thermal expansion coefficients close to that of Ti6Al4V without compromising bioactivity. With the strictly controlled processing parameters, coatings of 6P57, 6P61 and 6P68 BGs (Table 19.1) were successfully prepared on Ti6Al4V substrates.[52,53] The formation of a thin (~150 nm) nanostructured reaction layer composed of titanium silicide (Ti_5Si_3) nanoparticles was reported, facilitating a stable bond between the BG coating and the substrate.[52,54] Partial crystallization of the glasses (<5 vol%) was observed. The thermal stresses generated in the enameling process caused the formation of cracks in SBF, which could lead to delamination of the coatings.[53] It was suggested that the application of a ground layer prepared from inert glasses 6P68 or 6P61 could protect the substrate from the body fluids and increase the long-term stability of the coatings. Gradient coatings could then be prepared by subsequent application of more surface-reactive glass coatings,[55,56] with embedded hydroxyapatite and/or bioactive glass particles,[57] or a sol–gel-derived silica coating.[58] A new family of glasses in the same system was synthesized and the processing parameters were optimized for the coating of Ti-based and Co–Cr-based alloys. The glasses were used to coat commercial dental implants.[59] Using an optimized coating method, glass 6P61 was applied to Ti6Al4V substrates as a top coating in combination with a ground BGC coating in the system SiO_2–Al_2O_3–P_2O_5–K_2O–CaO–F^-, dental implants were coated.[60]

Low-alkaline BGs in the SiO_2–Na_2O–K_2O–CaO–P_2O_5 glass system were recently proposed as candidates for enameling of Ti6Al4V substrates.[61] These glasses contained 4.6 mol% of Na_2O or/and K_2O, 47.2 mol% SiO_2, 45.6 mol% CaO and 2.6 mol% P_2O_5. A crack-free interface between the coating and the substrate was achieved through a 5–6 μm thick interlayer of a titanium oxide. Partial crystallization of the glass did not prevent the precipitation of calcium phosphate in Kokubo's SBF.[61] Recently, Chakraborty and colleagues reported the coating of 316L surgical stainless steel with phosphate-free BG (PFBG); partial crystallization was detected.[62]

Enameling is a rather simple and inexpensive process for the creation of the glass coatings on metallic substrates. However the unresolved technical challenges apparently hamper the efforts to introduce the BG-coated implants into clinical practice.

19.2.3 Immersion Casting

In 1979, Hench and Buscemi patented a method to coat metallic implants with BG.[25] The method involves the immersion of a pre-oxidized and pre-heated metal substrate in the molten BG. The pre-heating of the substrates ensures that the total volume expansions of metal and BG are harmonized. The immersion takes place for 3–15 seconds to control the solubility of the oxide layer on the metal surface. The resultant composition gradient of metal oxide and BG results in the chemical bonding of the coating to the substrate.[37] The method was claimed to eliminate the problems typical of spray coating and enameling methods.[25,37] Immersion casting was used to coat substrates made of 316L surgical stainless steel,[25,63] Co–Cr alloy[64] and Ti6Al4V alloy.[65] The 52S4.6 BG composition (Table 19.1) was optimized for the coating of Co–Cr-alloys. In the study by West and colleagues, cylindrically-shaped Ti6Al4V alloy implants were coated by 45S5 BG. To improve the quality of the coatings, the substrates were pre-coated with boric acid in a reducing flame which resulted in the formation of a $TiBO_3$ glass layer.[65] Preclinical studies with the coated implants were performed in dogs. The implants were inserted transcortically in the mid-diaphyseal region of the femur. The animals were followed-up for to 24 weeks. The implants in the retrieved bone samples were subjected to push-out tests. In the push-out tests, the failure predominantly occurred at the BG–metal interface. Insufficient quality of the coating, *i.e.* the presence of bubbles at the BG–metal interface, indicated the need for further refinement of the coating method.[65] Undue variability of the $TiBO_3$ layer rendered the coatings unsuitable for clinical use.[37]

19.2.4 Laser Cladding

Advantages of laser cladding were demonstrated in the metallurgical field.[66] The heat-treatment occurs locally, thus the size of heat-affected zone in reduced thermal distortion is minimized; good fusion bonding of the coating is anticipated. In laser cladding, a carrier gas is used to inject the powder of the precursor material into the interaction zone, where the powder is affected by the laser beam. The laser beam creates a molten pool of the material on the surface of the metallic substrate. The position of the laser beam is fixed while the substrate is moving at a constant speed. The procedure is carried out in shielding gas to avoid the oxidation of the substrate. As the substrate moves away from the interaction zone, the molten pool is rapid quenched.[67–69] The laser cladding technique was applied to the production of CaP coatings on Ti6Al4V alloy substrates and promising *in vitro* results in terms of MG-63 osteoblast-like cellular activity on the surface of the coatings was reported.[67,68]

Comesaña and co-workers used laser cladding to produce BG coatings on Ti6Al4V alloy substrates.[69] Their selection of glasses included 45S5, 52S4.6 and S520 (Table 19.1). A Nd:YAG laser emitting at the wavelength of 1064 nm with a maximum power of 500 W was employed in the experiment. The wetting characteristics of BG S520 were discovered to be more suitable for the laser cladding process. Nevertheless, crystallization of the glass was detected. Apatite formation was discovered upon the immersion of the coated specimens in SBF.[69]

19.2.5 Enameling by Direct Laser Treatment

Carbon dioxide (CO_2) lasers are widely used in industrial applications. The typical emission wavelength of CO_2 lasers is 10.6 μm in the infra-red region of the spectra. High power output CO_2 lasers are typically used for cutting and welding of metals such as stainless steel, aluminum or copper. Lasers with moderate power output are used for processing of plastics and wood. Laser marking of different materials is another common application of CO_2 lasers.

Direct laser treatment with a CO_2 laser was reported as an alternative method to coat metallic implants with BG.[70–72] In this method, the laser beam is applied directly to the specimen to create the coating. This method is different from indirect laser treatment methods, *i.e.* pulsed laser deposition, when a target is affected by laser irradiation in order to achieve a thin film on a substrate placed separately. The benefits of the method are similar to those in laser cladding; heat treatment occurs locally and rapidly and primarily affects the coating.[72] The laser radiation is absorbed within the first atomic layers of the metallic substrates.[66] Therefore, especially if shielding gas is used in the interaction zone, the oxidation of substrate is expected to be lower than in conventional heat treatment. In addition, structural changes occur in Ti or Ti-alloys upon heat-treatment above 500–600 °C in a conventional furnace.[73] These structural changes may alter the mechanical strength and fatigue resistance of the implant making it unusable for load-bearing applications. In direct laser treatment, these structural changes can be avoided.

During the direct laser treatment, the position of the laser is fixed and the specimens are scanned through the focused beam using a computer-controlled motorized stage (Figure 19.1). The size of the interaction zone is limited to around 200 μm and the pattern of stage movement can be controlled with high precision.[72] If needed, the power output of the laser could be synchronized with the scanning pattern.[72] Consequently, the coating may be applied within the grooves,[74,75] recesses or thoroughgoing holes[24] made in the implantable devices. Consequently, the implant could be integrated into the bone by both bioactive and biological fixation. In this respect, a BG coating could stimulate rapid bone ingrowth into the implant texture and

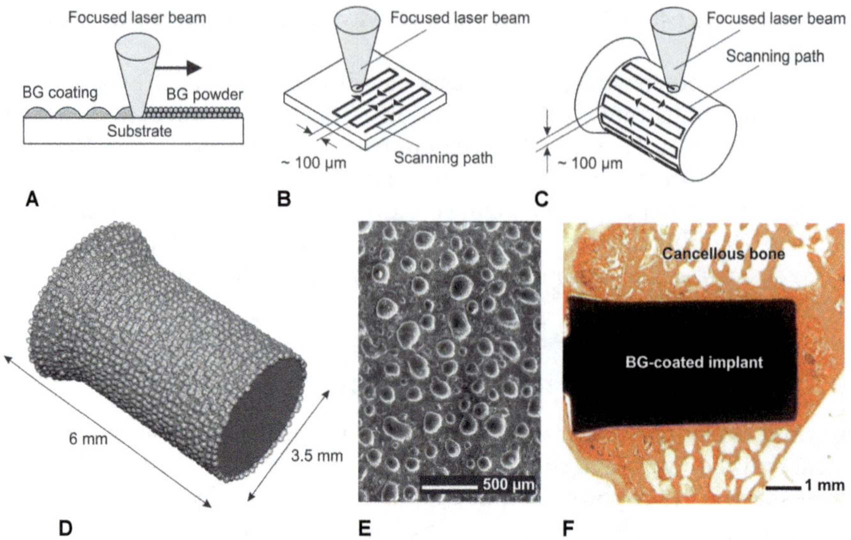

Figure 19.1 Enameling by direct laser treatment: (A) The technique for the preparation of BG coatings by direct laser treatment; (B) Schematic illustration of the set-up for coating flat specimens; (C) Schematic illustration of the set-up for coating cylindrical implants; (D) Computer rendering of the cylindrical implants coated with BG; (E) SEM micrograph showing the typical appearance of the surface of the BG-coated implants; (F) Histological section of a BG-coated implant incorporated within cancellous bone after 8 weeks of implantation.

provide the bioactive fixation of the implant within the first weeks after implantation. As the glass resorbs, the long-term stability of the implant would be maintained by the bone ingrown into the grooves, recesses or thoroughgoing holes.

As an additional benefit, the BG coating produced by the direct laser treatment is not uniform, but consists of separate glass droplets of around 100–200 μm in diameter (Figure 19.1E), thus the coating could be less prone to delamination than the conventional uniform coating. At the same time, the coating itself provides a suitable texture for the biological fixation.[70–72]

In the studies involving direct laser treatment with a CO_2 laser, the substrates and implants made of c.p.Ti[70,71] and Ti6Al4V[74,75] were coated with the bioactive (1-98) or inert glass (flat) from the suspension of glass powder (<45 μm) in ethanol by a dip-coating technique. Glass 1-98 has a broad viscosity range which allows hot working and repeated heat-treatment without the risk of uncontrolled crystallization of the glass.[76] The composition of the glass is presented in Table 19.1. After dip-coating and drying, the deposited layer of glass powder was irradiated with the focused beam from a 25 W CO_2 laser. The spot size of the focused laser beam was around 200 μm. Under computer control, the laser beam was selectively applied to the desired areas on the surface of the implants. The laser processing was carried out in an N_2 atmosphere. The coating set-up is shown in Figure 19.1A–C.

Compositional and structural changes in the glass-coated specimens were studied by scanning electron microscopy (SEM), energy dispersive spectroscopy (EDS) and X-ray diffraction (XRD). The compositions BG and inert glass coatings were equivalent to those of the respective precursor glasses. No crystallization was detected by XRD.[70] *In vitro* reactivity experiments were performed. These experiments, aimed at the assessment of the apatite-forming capacity of the BG-coated specimens, involved the immersion of the specimens in Kokubo's SBF for up to 14 days. Control materials included flat glass-coated specimens as well as granules and plates prepared from the original 1-98 and flat glasses. As expected, the rate of apatite formation on the surface of the BG-coatings was comparable to that on the original 1-98. No apatite formation was observed on flat glass coatings and original flat glass.[70]

In the consequent *in vivo* study, BG-coated c.p.Ti cylinders were implanted into femoral condyles of the rabbit with a follow-up period of 8 weeks. The implants are shown in Figure 19.1D. This experimental setup assumed no load-bearing. In histological analysis, the BG-coated implants developed significantly more interfacial bone contact than the control, uncoated grit-blasted Ti cylinders.[71] Histological section is shown in Figure 19.1F. The implants were also subjected to biomechanical testing in torsion. Torsional testing was chosen based on the studies on transcortical HA-coated cylindrical implants performed by Cook and colleagues.[77,78] No statistically significant differences in the ultimate torque were detected between the BG-coated implants and the control, uncoated grit-blasted Ti. However, the primary mode of failure of BG-coated implants was at the coating–substrate interface while the BG–bone interface was intact.[71]

Further preclinical studies were performed.[75] The BG-coating was applied by direct laser treatment within the longitudinal grooves of a custom-designed intercalary femoral endoprosthesis made of Ti6Al4V alloy (Figure 19.2). Uncoated grit-blasted implants served as control. The segmental bone replacement model of the canine femur was used to study the performance of the implants. The load-bearing conditions affected the performance of the BG coatings. After a 3 month follow-up period, the typical reaction layers formed on the surface of the coating but, unexpectedly, also at the interface between the metal substrate and the coating. At some locations, the coating delaminated. In the biomechanical testing of the implants in torsion, the recorded torque of the coated implants was inferior to that of the controls.[75] Hence, contrary to expectation, no benefit of BG coating for osteointegration of the implants could be confirmed.

19.2.6 Electrophoretic Deposition

Electrophoretic deposition (EPD) is an electrochemical method which involves the application of an electric field between the electrodes in an EPD cell containing particles suspended in a liquid medium. The transport of charged particles towards an oppositely-charged electrode results in

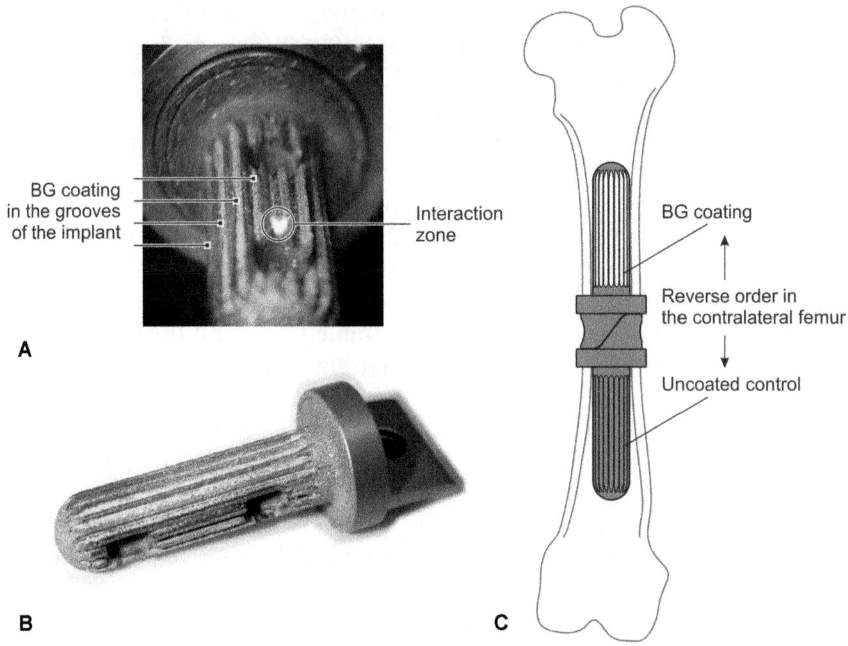

Figure 19.2 BG coating of a custom-designed intercalary femoral endoprosthesis. (A) Preparation of the coating; (B) BG-coated implant; (C) schematic illustration of the implant placement in the canine femur.

deposition of the coatings. The deposited coatings are subsequently sintered at elevated temperatures. The technique was applied to create HA coatings on substrates made of titanium[79,80] Ti6Al4V alloy and 316L stainless steel.[80] Later, the EPD technique was further optimized for the creation of uniform and reproducible BG coatings on metallic substrates. These substrates included stainless steel (type 304)[81] and Nitinol.[81,82] Glass 6P61 was used in combination with nanostructured HA to create functionally graded coatings on Ti6Al4V alloy substrates.[83] In more recent developments, Pishbin and colleagues employed the Taguchi experimental design method for the optimization of EPD process parameters for 45S5 BG coatings.[84] The popularity of the EPD as an alternative to enameling and spraying methods is based on a higher degree of control over the nano- and microstructure, stoichiometry, microscopic and macroscopic dimensions in the process of creating bioactive coatings and biomedical nanostructures.[85]

19.2.7 Clinical Relevance of Bioactive Coatings on Metallic Implants

The goal of BG-coating development was to introduce a coating which is superior to HA in terms of implant osteointegration.[43] However, efforts were hampered by technical challenges and, to the best of our knowledge, metallic implants coated with BG have not been introduced in the clinic. With HA, the situation is different; HA-coated implants are being used in patients on a routine basis.

The methods of depositing HA on metal implants include plasma spraying, sputter deposition, sol–gel coating, electrophoretic deposition and biomimetic precipitation.[86,87] For orthopaedic cementless arthroplasty implants, plasma spraying is the most common method.[88] The anticipated clinical benefits of HA-coating are: rapid osteointegration and improved fixation,[77,78,89] bridging of the gaps between the implant and the surrounding bone[90] and decrease in the initial micromotion[91,92] and replacement of the resorbed HA by newly formed bone.[93]

Yet, the clinical benefits of HA-coating for arthroplasty implants are being debated.[94,95] There are vast data that support the use of HA-coated implants. Preclinical studies demonstrated that HA-coated implants were strongly bonded to the peri-implant bone.[96] The ability of HA coatings to promote implant incorporation with bone was established.[93] In particular, it was shown that under continuous loading, HA-coated implants were even capable of converting the peri-implant, micromotion-induced fibrous tissues to bone.[97] Moreover, the osteopenic condition of the host bone did not affect the fixation of the HA-coated implants.[98]

Favorable clinical results with the use of HA-coated, cementless arthroplasty implants were reported.[99–103] In a study involving radio-stereometric analysis (RSA), the low frequency of subsidence was noted in connection with the HA-coated femoral stems, which was seen as a sign of possible favorable long-term results.[104] High degree of bone apposition was seen around retrieved HA-coated implants indicating the stability of the bone–implant interface.[105–107] In addition, HA-coating can act as a system for the delivery of biological agents, *e.g.* for the prevention of infections.[108]

In contrast to these encouraging results, the absence of a clinical or radiological advantage of HA-coated femoral stems in comparison with the identical non-coated sandblasted or roughened stems was reported in several randomized trials or smaller observational studies.[109–114] Enhanced bone remodeling observed for the porous HA-coated stems was not complemented with a clinical advantage when compared with the similar uncoated stems.[110] Concerns were expressed over the delamination of the coatings resulting in third-body polyethylene wear with subsequent peri-implant bone loss and implant loosening.[115–117] A meta-analysis of clinical studies of the HA-coated femoral stems in the primary total hip arthroplasty was performed by Gandhi and colleagues.[118] They selected 9 papers for systematic review out of 630 papers identified. The main finding was the absence of clinical benefits in the use of HA-coated implants over uncoated ones.[118]

Paulsen and colleagues assessed the Danish arthroplasty registry data on almost 8000 stems and 12 000 cups to establish the effect of HA-coatings on the risk of implant revision.[119] Their findings suggest that, in the middle-term, HA-coatings are not associated with a reduced risk of implant revision compared with the non-coated counterparts.[119] Lazarinis and colleagues performed a risk analysis of the data on over 8000 acetabular cups with and without HA-coatings.[117] The data was obtained from the Swedish hip arthroplasty register. They concluded that HA-coating was a risk factor for cup revision due to aseptic loosening.[117] Later, the same research group performed a study on the effects of HA-coating on the survival of uncemented femoral stems.[120] The data on 4772 femoral stems was obtained from the Swedish hip arthroplasty register. Results of the study indicated that in the 10 year follow-up period, the performance of HA-coated uncemented implants was equal to that of non-coated ones. Therefore, the positive effect of HA coating could not be confirmed.[120] Recently, a large-scale review of uncemented total hip arthroplasty procedures registered in Scandinavian arthroplasty registers was carried out.[121] The data on 116 000 procedures in Denmark, Sweden and Norway were analyzed. The authors concluded that within the 15 year follow-up period there was no clinically relevant effect of HA-coating on the risk of revision.[121] These findings cast a reasonable doubt on the use of HA-coated orthopaedic implants. Furthermore, it was suggested that the additional economic burden related to the creation of HA coatings may be unjustified.[117]

Taking into consideration the data available on HA-coated implants, one could speculate that, even if the technical issues related to BG-coated metallic implants are resolved one day, there is still a possibility that the benefit of the enhanced peri-implant bone formation stimulated by the BG may be of negligible clinical relevance, especially in the long-term. However, one should keep in mind that the data derived from the implant registers should be considered with caution as these data do not provide the comprehensive information needed to judge the performance of the implants.

19.3 Fiber-reinforced Composite Implants

19.3.1 Load-bearing Fiber-reinforced Composite Implants

In skeletal reconstructions, load-bearing fiber-reinforced composite (FRC) implants are considered potential alternatives to metallic implants.[122] In contrast to metals, FRC implants allow physiological load-sharing between implant and host bone, which could reduce the risk of adverse peri-implant bone resorption due to the "stress-shielding" phenomenon.[122,123] In addition, FRC-based implants create fewer artifacts with modern diagnostic imaging, such as computed tomography (CT) and magnetic resonance imaging (MRI), and do not interfere with radiotherapy.

Early FRC implants were predominantly prepared from epoxy matrix reinforced with carbon fibers,[124–126] however, the use of other polymer matrices was also reported.[127] Carbon fibers have higher stiffness : mass ratio than glass fibers, therefore, carbon fibers were used as a reinforcement phase in the early FRC implants.[124] However, the generation of carbon debris observed in preclinical studies[128–131] and cases of failed clinical performance of the early FRC implants[132,133] led to the decline of interest in the FRC technology.

Recently, a new concept of load-bearing FRC implants was introduced. The implant structure is based on bisphenol-*A*-glycidyldimethacrylate (BisGMA), a typical polymer matrix used in dental restorations, reinforced with E-glass fibers.[134,135] The use of FRC as dental implants has been studied extensively.[136–138] In preclinical studies, various types of FRC implants were investigated for non-load-bearing cranial reconstructions[139,140] and load-bearing implant applications.[141,142] In particular, FRC intramedullary nails reinforced with continuous unidirectional glass fibers were surgically implanted in rabbit femurs.[141] In the selected animal model, surgical stabilization of bone metastasis in the subtrochanteric region of the femur was simulated. Performance of the FRC intramedullary nails evaluated by biomechanical testing and histomorfometry was equal to that of the control titanium implants.[141] In a further development, the structure of the FRC intramedullary nails was refined by the addition of biaxial braiding. The implants were reported sustaining the femurs with surgically created defects.[142] The experimental set-up is shown in Figure 19.3. Further optimization of the FRC implants for load-bearing applications was performed within a large-scale EU-funded project NEWBONE.[143] In craniofacial reconstructions, FRC implants have been successfully used in clinical practice.[32,33]

Finite element modelling of total hip replacement stems demonstrated that "stress-shielding" and the stresses at the bone–implant interface are interconnected.[144] Therefore, upon the application of physiological loads, the reduction in the implant rigidity aimed at matching the flexural properties of implant and bone will create significant interfacial stresses. Consequently, in order to perform without failure, more flexible FRC intramedullary implants need to be firmly attached to the host bone either by bone ingrowth or by bioactive fixation. BG particulate embedded within the surface layer of an implant made of a fiber-reinforced composite (FRC) could enhance the osteointegration of the implant.[137,138,141,142] Typically, the BG particulate is mixed with the matrix polymer and then mechanically exposed to the surface of the implant. Alternatively, the BG particulate is glued to the surface of the implant. However, the absence of a chemical bond between the BG particulate and polymer matrix was reported resulting in disintegration of the granules.[141,142] In the case of an intramedullary implant, when micromotion occurs, the loose BG granules adjacent to the implant surface may create a rolling surface and cause implant failure. Silanation of the BG particulate could be an attractive option to achieve the chemical attachment of the granules to the polymer matrix.[31,145,146] Another possibility is etching of the polymer layer on top of the BG granules by Excimer laser.[147] FRC plugs (truncated cone, with an 8-mm length and base diameters of 3.6 and 5 mm) with BG particulate (S53P4) exposed by laser-etching were implanted in the femoral shaft of the minipig. As seen in Figure 19.3, after the follow-up period of 8 weeks, the implants were firmly attached to the cortical bone.[143]

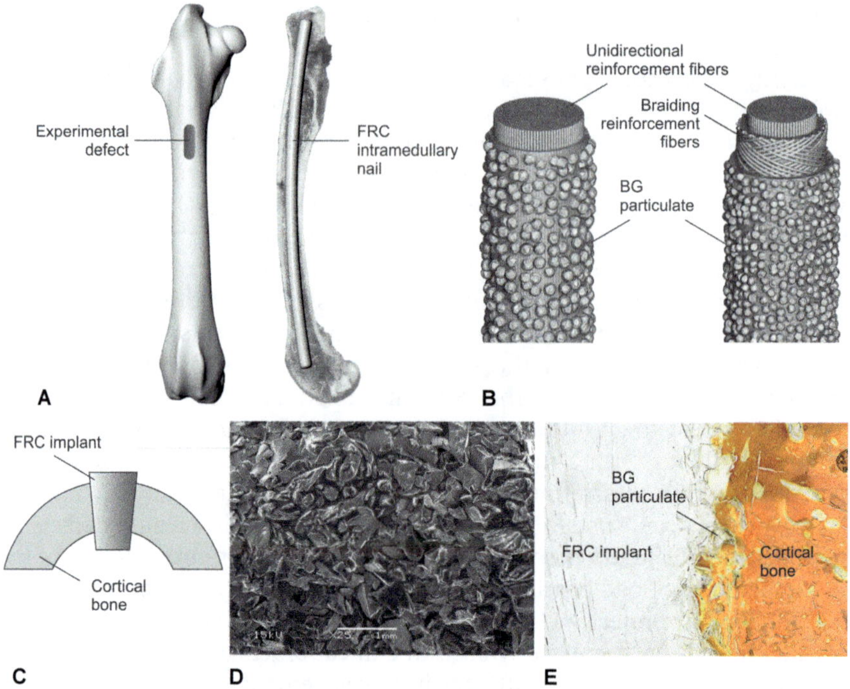

Figure 19.3 Experimental FRC implants with BG surface. (A) Placement of the FRC intramedullary nails in the rabbit femur; (B) Computer rendering of the FRC intramedullary nails; (C) Placement of FRC plugs in the cortical bone; (D) SEM micrograph showing the appearance of the surface of the FRC plugs with BG particulate exposed by laser etching; (F) Histological section of a BG-coated plug incorporated within cortical bone after 8 weeks of implantation.

FRC technology offers a vast capacity in the optimization of load-bearing implants. However, extensive research is needed before the load-bearing FRC implants could be introduced in clinical practice.

19.3.2 Fiber-reinforced Calvarial Implants

Clinical use of FRC materials in implantology started with cranial implants because of the urgent need of antimicrobial implants. Non-metallic FRC implants also enable increasing utilization of magnetic resonance imaging in identifying large numbers of infections related to autologous bone flaps and implants of other synthetic materials.[148,149] In implantology, radiopaque materials like titanium and zirconia cause severe artifacts in cone beam computer tomography images.[150]

There are structural differences in the implants made of FRC and bioactive glass for the lamellar cranial bone applications, and those made for oral and orthopaedic applications in which the implants have to withstand higher mechanical loads. Cranial implants have sandwich structure, where

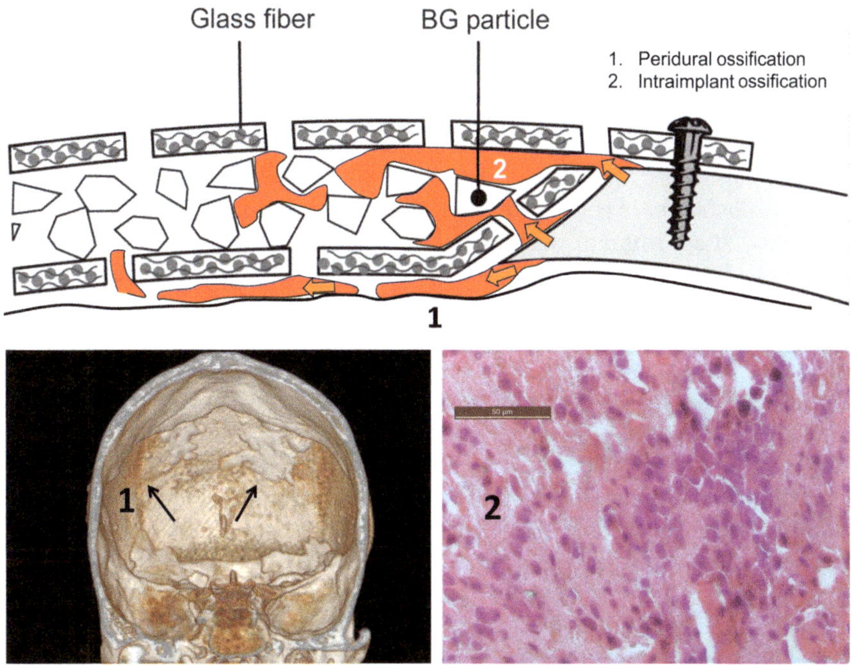

Figure 19.4 Schematic drawing (upper) of the lamellar structure of FRC–BG implant and two paths of osseointegration (intraimplant ossification, peridural ossification). Computer CT scan demonstrates peridural ossification (lower left) and histological image (HE-staining) new forming bone inside of the implant (lower right).

there is a space for free particles of bioactive glass between the outer and inner laminates (Figure 19.4). Load-bearing FRC implants which have been studied so far have had surfaces of polymer matrix and reinforcing glass fibers with exposed particles of BG. The idea to imitate the lamellar structure of the cranial bone has influenced the structure the FRC implant. Like the cortical cranial bone, the outer and inner surface laminate of the implant are responsible for its durability. In turn, the site of cancellous bone in the natural cranial bone is loaded with particles of BG in the implant. The mechanism of biological function of this kind of implant is based on absorption of blood and cells with other autologous bone-forming components, such as growth factors, into the interlaminate space where they are able to interact with particles of bioactive glass. For this, laminates are mesh-like perforated FRC laminates with perforation size smaller than the particles inside the implant. Fixation of the implant is made by titanium or polymer screws from the overlaying marginal area of the implant.

The FRC–BG calvarial composite implant with bacteriostatic properties from the bioactive glass has been developed in two stages: early development stage implants with first implantation operations in 2007, and authority-approved types of implant launched in 2014. Both types of FRC implants contain bioactive glass as one component and biostable FRC as a framework of the implant. In the implants of the early development stage, underneath a dense FRC outer surface laminate there was a porous dimethacrylate bioactive glass-containing scaffold-like FRC structure acting as a reservoir for BG and providing room for new bone to be formed. The scaffold structure of the implant was filled with particles of BG of S53P4 type. Perfusion of blood with osteogenic cells from the refreshed bone margins to the implant in the early stage implant occurs predominantly from the margins towards the center of the implant by capillary forces. Leaching of BG begins after becoming in contact with the extracellular liquid, *i.e.* plasma of blood, and therefore blood penetration to the implant is expected to be crucial for promoting bone growth and the antimicrobial properties of the implant. However, it was found that perfusion of blood from the margins only to center of implant was not effective enough, and the center part of the implant with particles of BG did not become wetted with blood. As a result, there was not any bone growth in the center of the implant. This changed the structure of the implant to one being perforated throughout the inner and outer laminates as described earlier. It has recently been demonstrated that when blood from the refreshed bone margins penetrates to the implant, osteoblasts, collageneous fibers and bone could be found in the inner parts of the implant.[151] In addition, CT has demonstrated bone growth on the inner surface of the implant. Therefore, bone formation with FRC–BG cranioplasty implants has been divided to *intraimplant ossification* and *peridural ossification* (Figure 19.4). The clinical outcome of presently a few hundred cases, and other research findings, suggest that FRC–BG calvarial implants are useful in treatments of decompressive craniectomies and traumas of cranial bones of paediatric and adult patients.

19.4 Conclusions and Outlook

The chapter reviewed various methods of applying BG coatings to metallic substrates for potential use in implantable medical devices. BG coatings are challenging to apply and their *in vivo* performance is unpredictable. So far, despite over 40 years of research and development efforts, BG-coated metallic implants have not been applied in the clinic. Moreover, there is emerging evidence of the lack of mid- and long-term clinical benefits of HA coatings on cementless arthroplasty implants. Based on this data, one could speculate that BG-coated metallic implants could be of marginal clinical significance. However, FRC implants have the potential to overtake metallic implants in non-load-bearing and load-bearing applications. This development could downplay the need for BG-coated metallic implants. FRC technology offers a vast capacity in the optimization of load-bearing implants.

Indeed, compared with metallic implants, FRC implants may provide more physiological load-sharing between bone and implant and reduce the risk of adverse bone remodeling due to "stress-shielding". FRC implants create fewer artifacts with modern diagnostic imaging. In addition, the polymer matrix can contain bioactive components, such as BG, as well as active molecules, such as antibiotics, to promote implant incorporation and prevent infections. However, extensive research is needed before load-bearing FRC implants could be introduced in clinical practice. A major breakthrough in the clinical use of non-load-bearing FRC implants occurred in the cranial bone reconstructions of large skull bone defects. These defects result from trauma, tumors, infections and craniotomies, *e.g.* in the treatment of cerebral insults. For the patients, the defects cause functional and aesthetic problems. The cranial FRC implants mimic the lamellar structure of the cranial bone with the shell of the implant made of FRC sheets while the inner space is filled with BG granules. With this arrangement, BG plays an essential role for the osteointegration of the implants. Indeed, excellent clinical outcomes were demonstrated for FRC–BG cranial implants applied in hundreds of paediatric and adult patients. We believe that, the knowledge gained in the clinical use of non-load-bearing cranial FRC implants will create a paradigm shift in respect to the clinical use of load-bearing FRC implants leading to the widespread clinical applications of the FRC implants.

References

1. L. L. Hench, R. J. Splinter, W. C. Allen and T. K. Greenlee, Bonding mechanisms at the interface of ceramic prosthetic materials, *J. Biomed. Mater. Res. Symp.*, 1971, **2**, 117–141.
2. L. L. Hench and J. K. West, Biological applications of bioactive glasses, *Life Chem. Rep.*, 1996, **13**, 187–241.
3. L. L. Hench, I. D. Xynos and J. M. Polak, Bioactive glasses for *in situ* tissue regeneration, *J. Biomater. Sci., Polym. Ed.*, 2004, **15**(4), 543–562.
4. V. V. Välimäki and H. T. Aro, Molecular basis for action of bioactive glasses as bone graft substitute, *Scand. J. Surg.*, 2006, **95**(2), 95–102.
5. J. Wilson, F. J. Schoen, G. H. Pigott and L. L. Hench, Toxicology and bioactivity of Bioglasses, *J. Biomed. Mater. Res.*, 1981, **15**(6), 805–817.
6. J. Wilson and D. Noletti, Bonding of soft tissues to Bioglass, in *Handbook of Bioactive Ceramics*, ed. T. Yamamuro *et al.*, CRC Press, Boca Raton, 1990, vol. I, pp. 235–244.
7. P. Stoor, V. Kirstilä, E. Söderling, I. Kangasniemi, K. Herbst and A. Yli-Urpo, Interactions between bioactive glass and periodontal pathogens, *Microb. Ecol. Health Dis.*, 1996, **9**(3), 109–4.
8. P. Stoor, E. Söderling and J. I. Salonen, Antibacterial effects of a bioactive glass paste on oral microorganisms, *Acta Odontol. Scand.*, 1998, **56**(3), 161–165.
9. M. Peltola, K. Aitasalo, J. Suonpää, M. Varpula and A. Yli-Urpo, Bioactive glass S53P4 in frontal sinus obliteration: A long-term clinical experience, *Head Neck*, 2006, **28**(9), 834–841.

10. N. C. Lindfors, J. T. Heikkilä, I. Koski, K. Mattila and A. J. Aho, Bioactive glass and autogenous bone as bone graft substitutes in benign bone tumors, *J. Biomed. Mater. Res., Part B*, 2009, **90**(1), 131–136.
11. N. C. Lindfors, I. Koski, J. T. Heikkilä, K. Mattila and A. J. Aho, A prospective randomized 14-year follow-up study of bioactive glass and autogenous bone as bone graft substitutes in benign bone tumors, *J. Biomed. Mater. Res., Part B*, 2010, **94**(1), 157–164.
12. N. C. Lindfors, P. Hyvönen, M. Nyyssönen, M. Kirjavainen, J. Kankare, E. Gullichsen and J. Salo, Bioactive glass S53P4 as bone graft substitute in treatment of osteomyelitis, *Bone*, 2010, **47**(2), 212–218.
13. J. Frantzen, J. Rantakokko, H. T. Aro, J. Heinanen, S. Kajander, E. Gullichsen, E. Kotilainen and N. C. Lindfors, Instrumented spondylodesis in degenerative spondylolisthesis with bioactive glass and autologous bone: A prospective 11-year follow-up, *J. Spinal Disord. Tech.*, 2011, **24**(7), 455–461.
14. J. Heikkilä, J. Kukkonen, A. Aho, S. Moisander, T. Kyyrönen and K. Mattila, Bioactive glass granules: A suitable bone substitute material in the operative treatment of depressed lateral tibial plateau fractures: A prospective, randomized 1 year follow-up study, *J. Mater. Sci.: Mater. Med.*, 2011, **22**(4), 1073–1080.
15. P. Stoor, J. Pulkkinen and R. Grénman, Bioactive glass S53P4 in the filling of cavities in the mastoid cell area in surgery for chronic otitis media, *Ann. Otol., Rhinol., Laryngol.*, 2010, **119**(6), 377–382.
16. J. McAndrew, C. Efrimescu, E. Sheehan and D. Niall, Through the looking glass; bioactive glass S53P4 (BonAlive®) in the treatment of chronic osteomyelitis, *Ir. J. Med. Sci.*, 2013, **182**(3), 509–511.
17. C. L. Romanò, N. Logoluso, E. Meani, D. Romanò, E. De Vecchi, C. Vassena and L. Drago, A comparative study of the use of bioactive glass S53P4 and antibiotic-loaded calcium-based bone substitutes in the treatment of chronic osteomyelitis: a retrospective comparative study, *Bone Joint J.*, 2014, **96-B**(6), 845–850.
18. K. R. Rust, G. T. Singleton, J. Wilson and P. J. Antonelli, Bioglass middle ear prosthesis: long-term results, *Am. J. Otol.*, 1996, **17**(3), 371–374.
19. P. Stoor, E. Söderling and R. Grénman, Bioactive glass S53P4 in repair of septal perforations and its interactions with the respiratory infection-associated microorganisms *Haemophilus influenzae* and *Streptococcus pneumoniae*, *J. Biomed. Mater. Res.*, 2001, **58**(1), 113–120.
20. M. Peltola, I. Kinnunen and K. Aitasalo, Reconstruction of orbital wall defects with bioactive glass plates, *J. Oral Maxillofac. Surg.*, 2008, **66**(4), 639–646.
21. E. Bon, Structural opportunities for glass, Thesis of Master of Engineering in high performance structures, Massachusetts Institute of Technology, 2003.
22. G. Gheysen, P. Ducheyne, L. L. Hench and P. de Meester, Bioglass composites: a potential material for dental application, *Biomaterials*, 1983, **4**(2), 81–84.

23. E. Schepers, M. De Clercq and P. Ducheyne, Histological and histomorphometrical analysis of bioactive glass and fibre reinforced bioactive glass dental root implants, *J. Oral. Rehabil.*, 1988, **15**(5), 473–487.
24. H. Ylänen, H. Aro, K. Karlsson, A. Yli-Urpo, M. Hupa and E. Nordström, Novel composite and its use, Patent Application US 2001/0041942 A1, 2001.
25. L. L. Hench and P. Buscemi, Method of bonding a bioglass to metal, Patent US 4159358 A, 1979.
26. J. T. Heikkilä, A. J. Aho, I. Kangasniemi and A. Yli-Urpo, Polymethylmethacrylate composites: disturbed bone formation at the surface of bioactive glass and hydroxyapatite, *Biomaterials*, 1996, **17**(18), 1755–1760.
27. H. Niiranen, T. Pyhältö, P. Rokkanen, M. Kellomäki and P. Törmälä, In vitro and in vivo behavior of self-reinforced bioabsorbable polymer and self-reinforced bioabsorbable polymer/bioactive glass composites, *J. Biomed. Mater. Res., Part A*, 2004, **69**(4), 699–708.
28. T. Niemelä, H. Niiranen, M. Kellomäki and P. Törmälä, Self-reinforced composites of bioabsorbable polymer and bioactive glass with different bioactive glass contents. Part I: Initial mechanical properties and bioactivity, *Acta Biomater.*, 2005, **1**(2), 235–242.
29. M. Huttunen, P. Törmälä, P. Godinho and M. Kellomäki, Fiber-reinforced bioactive and bioabsorbable hybrid composites, *Biomed. Mater.*, 2008, **3**(3), 034106.
30. J. Kulkova, A. A. Abdulmajeed, E. Könönen and T. O. Närhi, Biofilm medium leads to apatite formation on bioactive surfaces, *J. Appl. Biomater. Funct. Mater.*, 2013, **11**(2), 95–98.
31. S. Shinzato, T. Nakamura, T. Kokubo and Y. Kitamura, Bioactive bone cement: Effect of silane treatment on mechanical properties and osteoconductivity, *J. Biomed. Mater. Res.*, 2001, **55**(3), 277–284.
32. K. M. Aitasalo, J. M. Piitulainen, J. Rekola and P. K. Vallittu, Craniofacial bone reconstruction with bioactive fiber-reinforced composite implant, *Head Neck*, 2014, **36**(5), 722–728.
33. J. M. Piitulainen, J. P. Posti, K. M. Aitasalo, V. Vuorinen, P. K. Vallittu and W. Serlo, Paediatric cranial defect reconstruction using bioactive fibre-reinforced composite implant: early outcomes, *Acta Neurochir.*, 2015, **157**(4), 681–687.
34. R. W. Lacefield, Hydroxylapatite coatings, in *An Itroduction to Bioceramics*, ed. L. L. Hench and J. Wilson, World Scientific, Singapore, New Jersey, London, Hong-Kong, 1993, pp. 223–238.
35. R. B. Heimann, Thermal spraying of biomaterials, *Surf. Coat. Technol.*, 2006, **201**(5), 2012–2019.
36. L. L. Hench, H. A. Paschall, W. C. Allen and G. Piotrowski, Interfacial behavior of ceramics implants, *NBS Spec. Publ.*, 1975, **415**, 19–35.
37. L. L. Hench and Ö. H. Andersson, Bioactive glass coatings, in *An Itroduction to Bioceramics*, ed. L. L. Hench and J. Wilson, World Scientific, Singapore, New Jersey, London, Hong-Kong, 1993, pp. 239–259.

38. C. Gabbi, A. Cacchioli, B. Locardi and E. Guadagnino, Bioactive glass coating: physicochemical aspects and biological findings, *Biomaterials*, 1995, **16**(7), 515–520.
39. A. Oliva, A. Salerno, B. Locardi, V. Riccio, F. Della Ragione, P. Iardino and V. Zappia, Behaviour of human osteoblasts cultured on bioactive glass coatings, *Biomaterials*, 1998, **19**(11–12), 1019–1025.
40. T. M. Lee, E. Chang, B. C. Wang and C. Y. Yang, Characteristics of plasma-sprayed bioactive glass coatings on Ti-6A1-4V alloy: an in vitro study, *Surf. Coat. Technol.*, 1996, **79**(1–3), 170–177.
41. J. Schrooten, H. Van Oosterwyck, J. Vander Sloten and J. A. Helsen, Adhesion of new bioactive glass coating, *J. Biomed. Mater. Res.*, 1999, **44**(3), 243–252.
42. J. Schrooten and J. A. Helsen, Adhesion of bioactive glass coating to Ti6A14V oral implant, *Biomaterials*, 2000, **21**(14), 1461–1469.
43. D. L. Wheeler, M. J. Montfort and S. W. McLoughlin, Differential healing response of bone adjacent to porous implants coated with hydroxyapatite and 45S5 bioactive glass, *J. Biomed. Mater. Res.*, 2001, **55**(4), 603–612.
44. L. Altomare, D. Bellucci, G. Bolelli, B. Bonferroni, V. Cannillo, L. De Nardo, R. Gadow, A. Killinger, L. Lusvarghi, A. Sola and N. Stiegler, Microstructure and in vitro behaviour of 45S5 bioglass coatings deposited by high velocity suspension flame spraying (HVSFS), *J. Mater. Sci.: Mater. Med.*, 2011, **22**(5), 1303–1319.
45. A. Cattini, D. Bellucci, A. Sola, L. Pawłowski and V. Cannillo, Functional bioactive glass topcoats on hydroxyapatite coatings: Analysis of microstructure and in-vitro bioactivity, *Surf. Coat. Technol.*, 2014, **240**, 110–117.
46. J. Hautaniemi, Oxidation and porcelain veneering of Ti and Pd alloys, Academic Dissertation, University of Turku, Finland, 1993.
47. J. Pajamäki, Bioactive glass and glass-ceramic interfacial reactions to bone, Academic Dissertation, University of Tampere, 1994.
48. J. Pajamäki, S. Lindholm, O. Andersson, K. Karlsson, A. Yli-Urpo and R. P. Happonen, Glass-ceramic-coated titanium hip endoprosthesis. Experimental study in rabbits, *Arch. Orthop. Trauma. Surg.*, 1995, **114**(2), 119–122.
49. Ö. H. Andersson, K. H. Karlsson, K. Kangasniemi and A. Yli-Urpo, Models for physical properties and bioactivity of phosphate opal glasses, *Glastech. Ber.*, 1988, **61**(10), 300–305.
50. R. J. Linnola, R. P. Happonen, Ö. H. Andersson, E. Vedel, A. U. Yli-Urpo, U. Krause and L. Laatikainen, Titanium and bioactive glass-ceramic coated titanium as materials for keratoprosthesis, *Exp. Eye Res.*, 1996, **63**(4), 471–478.
51. A. Pazo, E. Saiz and A. P. Tomsia, Silicate glass coatings on Ti-based implants, *Acta Mater.*, 1998, **46**(7), 2551–2558.
52. D. R. Bloyer, J. M. Gomez-Vega, E. Saiz, J. M. Mcnaney, R. M. Cannon and A. P. Tomsia, Fabrication and characterization of a bioactive glass coating on titanium implant aloys, *Acta Mater.*, 1999, **47**(15), 4221–4224.

53. E. Saiz, M. Goldman, J. M. Gomez-Vega, A. P. Tomsia, G. W. Marshall and S. J. Marshall, *In vitro* behaviour of silicate glass coatings on Ti6Al4V, *Biomaterials*, 2002, **23**(17), 3749–3756.
54. T. Oku, K. Suganuma, L. R. Wallenberg, A. P. Tomsia, J. M. Gomez-Vega and E. Saiz, Structural characterization of the metal/glass interface in bioactive glass coatings on Ti-6Al-4V, *J. Mater. Sci.: Mater. Med.*, 2001, **12**(5), 413–417.
55. J. M. Gomez-Vega, E. Saiz, A. P. Tomsia, T. Oku, K. Suganuma, G. W. Marshall and S. J. Marshall, Novel bioactive glass functionally graded coatings on Ti6Al4V, *Adv. Mater.*, 2000, **12**(12), 894–898.
56. S. Foppiano, S. J. Marshall, G. W. Marshall, E. Saiz and A. P. Tomsia, Bioactive glass coatings affect the behavior of osteoblast-like cells, *Acta Biomater.*, 2007, **3**(5), 765–771.
57. J. M. Gomez-Vega, E. Saiz, A. P. Tomsia, G. W. Marshall and S. J. Marshall, Bioactive glass coatings with hydroxyapatite and Bioglass particles on Ti-based implants. I. Processing, *Biomaterials*, 2000, **21**(2), 105–111.
58. J. M. Gomez-Vega, A. Hozumi, E. Saiz, A. P. Tomsia, H. Sugimura and O. Takai, Bioactive glass-mesoporous silica coatings on Ti6Al4V throgh enameling and triblock-copolymer-templated sol-gel processing, *J. Biomed. Mater. Res.*, 2001, **56**(3), 382–389.
59. S. Lopez-Esteban, E. Saiz, S. Fujino, T. Oku, K. Suganuma and A. P. Tomsia, Bioactive glass coatings for orthopedic metallic implants, *J. Eur. Ceram. Soc.*, 2003, **23**, 2921–2930.
60. E. Verné, C. Fernández Vallés, C. Vitale Brovarone, S. Spriano and C. Moisescu, Double-layer glass-ceramic coatings on Ti6Al14V for dental implants, *J. Eur. Ceram. Soc.*, 2004, **24**(9), 2699–2705.
61. A. Sola, D. Bellucci and V. Cannillo, Enameled coatings produced with low-alkaline bioactive glasses, *Surf. Coat. Technol.*, 2014, **248**, 1–8.
62. J. Chakraborty, S. Sengupta, S. Ray, S. Ghosh, R. Kapoor, S. P. Gouri, G. Gopal Pande and S. Datta, Multifunctional gradient coatings of phosphate-free bioactive glass on SS316L biomedical implant materials for improved fixation, *Surf. Coat. Technol.*, 2014, **240**, 437–443.
63. J. K. West, G. Gonzales, LaTorre and L. L. Hench, Optimization of rapid immersion coating of bioactive glass on 316L stainless steel, in *Bioceramics*, ed. P. Ducheyne and D. Christiansen, Butterworth-Heinemann, Oxford, 1993, vol. 6, pp. 405–410.
64. W. R. Lacefield and L. L. Hench, The bonding of Bioglass to a cobalt-chromium surgical implant alloy, *Biomaterials*, 1986, 7(2), 104–108.
65. J. K. West, A. E. Clark, M. B. Hall and G. E. Turner, *In vivo* bone-bonding study on Bioglass-coated titanium alloy, in *Handbook of Bioactive ceramics*, ed. T. Yamamuro, L. L. Hench and J. Wilson, CRC Press, Boca-Raton, 1990, pp. 161–166.
66. W. M. Steen, *Laser Materials Processing*, Springer-Verlag, London, 1991.
67. F. Lusquiños, A. De Carlos, J. Pou, J. L. Arias, M. Boutinguiza, B. León, M. Pérez-Amor, F. C. Driessens, K. Hing, I. Gibson, S. Best and

W. Bonfield, Calcium phosphate coatings obtained by Nd:YAG laser cladding: physicochemical and biologic properties, *J. Biomed. Mater. Res., Part A*, 2003, **64**(4), 630–637.
68. A. De Carlos, F. Lusquiños, J. Pou, B. León, M. Pérez-Amor, F. C. Driessens, K. Hing, S. Best and W. Bonfield, In vitro testing of Nd:YAG laser processed calcium phosphate coatings, *J. Mater. Sci.: Mater. Med.*, 2006, **17**(11), 1153–1160.
69. R. Comesaña, F. Quintero, F. Lusquiños, M. J. Pascual, M. Boutinguiza, A. Durán and J. Pou, Laser cladding of bioactive glass coatings, *Acta Biomater.*, 2010, **6**(3), 953–961.
70. N. Moritz, E. Vedel, H. Ylänen, M. Jokinen, M. Hupa and A. Yli-Urpo, Characterisation of bioactive glass coatings on titanium substrates produced using a CO_2 laser, *J. Mater. Sci.: Mater. Med.*, 2004, **15**(7), 787–794.
71. N. Moritz, S. Rossi, E. Vedel, T. Tirri, H. Ylänen, H. Aro and T. Närhi, Implants coated with bioactive glass by CO_2-laser, an in vivo study, *J. Mater. Sci.: Mater. Med.*, 2004, **15**(7), 795–802.
72. N. Moritz, Bioactive ceramic coatings on medical implants: with special reference to CO_2-laser treatment, Academic Dissertation, University of Turku, 2005.
73. M. Textor, C. Sittig, V. Frauchiger, S. Tosatti and D. M. Brunette, Properties and biological significance of natural oxide films on titanium and its alloys, in *Titanium in Medicine*, ed. D. M. Brunette *et al.*, Springer-Verlag, Berlin, 2001, pp. 171–230.
74. P. Keränen, Use of bioactive glasses as porous implant surface inlays, a laser-deposited coating and peri-implant bone graft extender, Academic Dissertation, University of Helsinki, 2012.
75. H. T. Aro, P. Keränen, N. Moritz, T. J. Mäkinen, J. J. Alm and B. Kommonen, Bioactive glasses as laser-deposited implant coating and as extender of allogenic bone grafting in cementless intercalary reconstruction of the canine femur (manuscript), in Use of bioactive glasses as porous implant surface inlays, a laser-deposited coating and peri-implant bone graft extender, ed. P. Keränen, Academic Dissertation, University of Helsinki, 2012.
76. A. Itälä, E. G. Nordström, H. Ylänen, H. T. Aro and M. Hupa, Creation of microrough surface on sintered bioactive glass microspheres, *J. Biomed. Mater. Res.*, 2001, **56**(2), 282–288.
77. S. D. Cook, G. C. Baffes, A. J. Palafox, M. W. Wolfe and A. Burgess, Torsional stability of HA-coated and grit-blasted titanium dental implants, *J. Oral. Implantol.*, 1992, **18**(4), 354–358.
78. S. D. Cook, G. C. Baffes, M. W. Wolfe and A. J. Palafox, A comparison of femoral and mandibular animal models for the evaluation of HA-coated implants, *J. Oral. Implantol.*, 1992, **18**(4), 359–365.
79. A. Stoch, A. Brozek, G. Kmita, J. Stoch, W. Jastrzebski and A. Rakowska, Electrophoretic coating of hydroxyapatite on titanium implants, *J. Mol. Struct.*, 2001, **596**(1–3), 191–200.

80. M. Wei, A. J. Ruys, B. K. Milthorpe, C. C. Sorrell and J. H. Evans, Electrophoretic deposition of hydroxyapatite coatings on metal substrates: a nanoparticulate dual-coating approach, *J. Sol-Gel Sci. Technol.*, 2001, **21**(1), 39–48.
81. D. Krause, B. Thomas, C. Leinenbach, D. Eifler, E. J. Minaya and A. R. Boccaccini, The electrophoretic deposition of Bioglass particles on stainless steel and Nitinol substrates, *Surf. Coat. Technol.*, 2006, **200**(16–17), 4835–4845.
82. A. R. Boccaccini, E. J. Minay and D. Krause, Bioglass coatings on superelastic NiTi wires by electrophoretic deposition (EPD), *Electrophor. Depos. Fundam. Appl. II Key Eng. Mater.*, 2006, **314**, 219–224.
83. D. Stojanovic, B. Jokic, D. j. Veljovic, R. Petrovic, P. S. Uskokovic and D. j. Janackovic, Bioactive glass-apatite composite coating for titanium implant synthesized by electrophoretic deposition, *J. Eur. Ceram. Soc.*, 2007, **27**(2–3), 1595–1599.
84. F. Pishbin, A. Simchi, M. P. Ryana and A. R. Boccaccini, A study of the electrophoretic deposition of Bioglass® suspensions using the Taguchi experimental design approach, *J. Eur. Ceram. Soc.*, 2010, **30**(14), 2963–2970.
85. A. R. Boccaccini, S. Keim, R. Ma, Y. Li and I. Zhitomirsky, Electrophoretic deposition of biomaterials, *J. R. Soc., Interface*, 2010, **7**, S581–S613.
86. J. E. Ellingsen, P. Thomsen and S. P. Lyngstadaas, Advances in dental implant materials and tissue regeneration, *Periodontology*, 2000–2006, **41**, 136–156.
87. R. Junker, A. Dimakis, M. Thoneick and J. A. Jansen, Effects of implant surface coatings and composition on bone integration: a systematic review, *Clin. Oral. Implants. Res.*, 2009, **20**(Suppl 4), 185–206.
88. A. A. Campbell, Bioceramics for implant coatings, *Mater. Today*, 2003, **6**(11), 26–30.
89. K. A. Thomas, J. F. Kay, S. D. Cook and M. Jarcho, The effect of surface macrotexture and hydroxylapatite coating on the mechanical strengths and histologic profiles of titanium implant materials, *J. Biomed. Mater. Res.*, 1987, **21**(12), 1395–1414.
90. K. Søballe, E. S. Hansen, H. Brockstedt-Rasmussen, C. M. Pedersen and C. Bünger, Hydroxyapatite coating enhances fixation of porous coated implants. A comparison in dogs between press fit and noninterference fit, *Acta Orthop. Scand.*, 1990, **61**(4), 299–306.
91. P. K. Stephenson, M. A. Freeman, P. A. Revell, J. Germain, M. Tuke and C. J. Pirie, The effect of hydroxyapatite coating on ingrowth of bone into cavities in an implant, *J. Arthroplasty*, 1991, **6**(1), 51–58.
92. K. Søballe, S. Overgaard, E. S. Hansen, H. Brokstedt-Rasmussen, M. Lind and C. Bünger, A review of ceramic coatings for implant fixation, *J. Long-Term Eff. Med. Implants*, 1999, **9**(1–2), 131–151.
93. K. Søballe, Hydroxyapatite ceramic coating for bone implant fixation. Mechanical and histological studies in dogs, *Acta Orthop. Scand., Suppl.*, 1993, **255**, 1–58.

94. A. Eskelinen, V. Remes, I. Helenius, P. Pulkkinen, J. Nevalainen and P. Paavolainen, Uncemented total hip arthroplasty for primary osteoarthritis in young patients: a mid-to long-term follow-up study from the Finnish Arthroplasty Register, *Acta Orthop.*, 2006, 77(1), 57–70.
95. G. J. Hooper, A. G. Rothwell, M. Stringer and C. Frampton, Revision following cemented and uncemented primary total hip replacement: a seven-year analysis from the New Zealand Joint Registry, *J. Bone Jt. Surg., Br. Vol.*, 2009, **91**(4), 451–458.
96. R. G. Geesink, K. de Groot and C. P. Klein, Bonding of bone to apatite-coated implants, *J. Bone Jt. Surg., Br. Vol.*, 1988, **70**(1), 17–22.
97. K. Søballe, E. S. Hansen, H. Brockstedt-Rasmussen and C. Bunger, Hydroxyapatite coating converts fibrous tissue to bone around loaded implants, *J. Bone Jt. Surg., Br. Vol.*, 1993, **75**(2), 270–278.
98. K. Søballe, E. S. Hansen, H. Brockstedt-Rasmussen, V. E. Hjortdal, G. I. Juhl, C. M. Pedersen, I. Hvid and C. Bünger, Fixation of titanium and hydroxyapatite-coated implants in arthritic osteopenic bone, *J. Arthroplasty*, 1991, **6**(4), 307–316.
99. R. G. Geesink, Hydroxyapatite-coated total hip prostheses. Two-year clinical and roentgenographic results of 100 cases, *Clin. Orthop. Relat. Res.*, 1990, **261**, 39–58.
100. W. J. Donnelly, A. Kobayashi, M. A. Freeman, T. W. Chin, H. Yeo, M. West and G. Scott, Radiological and survival comparison of four methods of fixation of a proximal femoral stem, *J. Bone Jt. Surg., Br. Vol.*, 1997, **79**(3), 351–360.
101. L. D. Dorr, Z. Wan, M. Song and A. Ranawat, Bilateral total hip arthroplasty comparing hydroxyapatite coating to porous-coated fixation, *J. Arthroplasty*, 1998, **13**(7), 729–736.
102. J. C. Theis and C. Ball, Medium-term results of cementless hydroxyapatite-coated primary total hip arthroplasty: a clinical and radiological review, *J. Orthop. Surg.*, 2003, **11**(2), 159–165.
103. S. Singh, S. P. Trikha and A. J. Edge, Hydroxyapatite ceramiccoated femoral stems in young patients. A prospective ten-year study, *J. Bone Jt. Surg., Br. Vol.*, 2004, **86**(8), 1118–1123.
104. J. Kärrholm, H. Malchau, F. Snorrason and P. Herberts, Micromotion of femoral stems in total hip arthroplasty. A randomized study of cemented, hydroxyapatite-coated, and porous-coated stems with roentgen stereophotogrammetric analysis, *J. Bone Jt. Surg., Am. Vol.*, 1994, **76**(11), 1692–1705.
105. T. W. Bauer, R. C. Geesink, R. Zimmerman and J. T. McMahon, Hydroxyapatite-coated femoral stems. Histological analysis of components retrieved at autopsy, *J. Bone Jt. Surg., Am. Vol.*, 1991, **73**(10), 1439–1452.
106. R. J. Furlong and J. F. Osborn, Fixation of hip prostheses by hydroxyapatite ceramic coatings, *J. Bone Jt. Surg., Br. Vol.*, 1991, **73**(5), 741–745.

107. D. C. Hardy, P. Frayssinet, A. Guilhem, M. A. Lafontaine and P. E. Delince, Bonding of hydroxyapatite-coated femoral prostheses. Histopathology of specimens from four cases, *J. Bone Jt. Surg., Br. Vol.*, 1991, **73**(5), 732–740.
108. Y. Ghani, M. J. Coathup, K. A. Hing and G. W. Blunn, Development of a hydroxyapatite coating containing silver for the prevention of peri-prosthetic infection, *J. Orthop. Res.*, 2012, **30**(3), 356–363.
109. R. H. Rothman, W. J. Hozack, A. Ranawat and L. Moriarty, Hydroxyapatite-coated femoral stems. A matched-pair analysis of coated and uncoated implants, *J. Bone Jt. Surg., Am. Vol.*, 1996, **78**(3), 319–324.
110. E. J. McPherson, L. D. Dorr, T. A. Gruen and M. T. Saberi, Hydroxyapatite-coated proximal ingrowth femoral stems. A matched pair control study, *Clin. Orthop.*, 1995, **315**, 223–230.
111. M. Tanzer, S. Kantor, L. Rosenthall and J. D. Bobyn, Femoral remodeling after porous-coated total hip arthroplasty with and without hydroxyapatite-tricalcium phosphate coating: a prospective randomized trial, *J. Arthroplasty*, 2001, **16**(5), 552–558.
112. Y. H. Kim, J. S. Kim, S. H. Oh and J. M. Kim, Comparison of porous-coated titanium femoral stems with and without hydroxyapatite coating, *J. Bone Jt. Surg., Am. Vol.*, 2003, **85**(9), 1682–1688.
113. J. Parvizi, P. F. Sharkey, W. J. Hozack, F. Orzoco, G. A. Bissett and R. H. Rothman, Prospective matched-pair analysis of hydroxyapatite-coated and uncoated femoral stems in total hip arthroplasty. A concise follow-up of a previous report, *J. Bone Jt. Surg., Am. Vol.*, 2004, **86**(4), 783–786.
114. J. Sanchez-Sotelo, D. G. Lewallen, W. S. Harmsen, J. Harrington and M. E. Cabanela, Comparison of wear and osteolysis in hip replacement using two different coatings of the femoral stem, *Int. Orthop.*, 2004, **28**(4), 206–210.
115. R. D. Bloebaum, D. Beeks, L. D. Dorr, C. G. Savory, J. A. DuPont and A. A. Hofmann, Complications with hydroxyapatite particulate separation in total hip arthroplasty, *Clin. Orthop. Relat. Res.*, 1994, **298**, 19–26.
116. E. W. Morscher, A. Hefti and U. Aebi, Severe osteolysis after third-body wear due to hydroxyapatite particles from acetabular cup coating, *J. Bone Jt. Surg., Br. Vol.*, 1998, **80**(2), 267–272.
117. S. Lazarinis, J. Kärrholm and N. P. Hailer, Increased risk of revision of acetabular cups coated with hydroxyapatite, *Acta Orthop.*, 2010, **81**(1), 53–59.
118. R. Gandhi, J. R. Davey and N. N. Mahomed, Hydroxyapatite coated femoral stems in primary total hip arthroplasty: a meta-analysis, *J. Arthroplasty*, 2009, **24**(1), 38–42.
119. A. Paulsen, A. B. Pedersen, S. P. Johnsen, A. Riis, U. Lucht and S. Overgaard, Effect of hydroxyapatite coating on risk of revision after primary total hip arthroplasty in younger patients. Findings from the Danish Hip Arthroplasty Registry, *Acta Orthop.*, 2007, **7**(5), 622–628.

120. S. Lazarinis, J. Kärrholm and N. P. Hailer, Effects of hydroxyapatite coating on survival of an uncemented femoral stem. A Swedish Hip Arthroplasty Register study on 4,772 hips, *Acta Orthop.*, 2011, **82**(4), 399–404.
121. N. P. Hailer, S. Lazarinis, K. T. Mäkelä, A. Eskelinen, A. M. Fenstad, G. Hallan, L. Havelin, S. Overgaard, A. B. Pedersen, F. Mehnert and J. Kärrholm, Hydroxyapatite coating does not improve uncemented stem survival after total hip arthroplasty!, *Acta Orthop.*, 2015, **86**(1), 18–25.
122. S. L. Evans and P. J. Gregson, Composite technology in load-bearing orthopaedic implants, *Biomaterials*, 1998, **19**(15), 1329–1342.
123. R. Huiskes, H. Weinans and B. van Rietbergen, The relationship between stress shielding and bone resorption around total hip stems and the effects of flexible materials, *Clin. Orthop. Relat. Res.*, 1992, **274**, 124–134.
124. G. W. Hastings, Carbon fibre composites for orthopaedic implants, *Composites*, 1978, **9**(3), 193–197.
125. K. Tayton, C. Johnson-Nurse, B. McKibbin, J. Bradley and G. Hastings, The use of semirigid carbon-fibre-reinforced plastic plates for fixation of human fractures. Results of preliminary trials, *J. Bone Jt. Surg., Br. Vol.*, 1982, **64**(1), 105–111.
126. M. S. Ali, T. A. French, G. W. Hastings, T. Rae, N. Rushton, E. R. Ross and C. H. Wynn-Jones, Carbon fibre composite bone plates. Development, evaluation and early clinical experience, *J. Bone Jt. Surg., Br. Vol.*, 1990, **72**(4), 586–591.
127. S. Ramakrishna, J. Mayer, E. Wintermantel and K. W. Leong, Biomedical applications of polymer-composite materials: a review, *Compos. Sci. Technol.*, 2001, **61**(9), 1189–1224.
128. N. Gillett, S. A. Brown, J. H. Dumbleton and R. P. Pool, The use of short carbon fibre reinforced thermoplastic plates for fracture fixation, *Biomaterials*, 1985, **6**(2), 113–121.
129. J. Kettunen, A. Mäkelä, H. Miettinen, T. Nevalainen, M. Heikkilä, P. Törmälä and P. Rokkanen, Fixation of femoral shaft osteotomy with an intramedullary composite rod: an experimental study on dogs with a two-year follow-up, *J. Biomater. Sci., Polym. Ed.*, 1999, **10**(1), 33–45.
130. K. A. Jockisch, S. A. Brown, T. W. Bauer and K. Merritt, Biological response to chopped-carbon-fiber-reinforced peek, *J. Biomed. Mater. Res.*, 1992, **26**(2), 133–146.
131. A. Minovic, I. Milowsev, V. Pisot, A. Cor and V. Antolic, Isolation of polyacetal wear particles from periprosthetic tissue of isoelastic femoral stems, *J. Bone Jt. Surg., Br. Vol.*, 2001, **83**(8), 1182–1190.
132. S. Allcock and M. A. Ali, Early failure of a carbon-fiber composite femoral component, *J. Arthroplasty*, 1997, **12**(3), 356–358.
133. F. Adam, D. S. Hammer, S. Pfautsch and K. Westermann, Early failure of a press-fit carbon fiber hip prosthesis with a smooth surface, *J. Arthroplasty*, 2002, **17**(2), 217–223.

134. P. K. Vallittu, Flexural properties of acrylic resin polymers reinforced with unidirectional and woven glass fibers, *J. Prosthet. Dent.*, 1999, **81**(3), 318–326.
135. P. K. Vallittu and C. Sevelius, Resin-bonded, glass fiber-reinforced composite fixed partial dentures: a clinical study, *J. Prosthet. Dent.*, 2000, **84**(4), 413–418.
136. A. M. Ballo, L. V. Lassila, P. K. Vallittu and T. O. Närhi, Load bearing capacity of bone anchored fiber-reinforced composite device, *J. Mater. Sci.: Mater. Med.*, 2007, **18**(10), 2025–2031.
137. A. M. Ballo, E. A. Akca, T. Ozen, L. Lassila, P. K. Vallittu and T. O. Närhi, Bone tissue responses to glass fiber-reinforced composite implants - a histomorphometric study, *Clin. Oral Implants Res.*, 2009, **20**(6), 608–615.
138. A. M. Ballo, I. Cekic-Nagas, G. Ergun, L. Lassila, A. Palmquist, P. Borchardt, J. Lausmaa, P. Thomsen, P. K. Vallittu and T. O. Närhi, Osseointegration of fiber-reinforced composite implants: histological and ultrastructural observations, *Dent. Mater.*, 2014, **30**(12), e384–e395.
139. S. M. Tuusa, M. J. Peltola, T. Tirri, L. V. Lassila and P. K. Vallittu, Frontal bone defect repair with experimental glass-fiber-reinforced composite with bioactive glass granule coating, *J. Biomed. Mater. Res., Part B*, 2007, **82**(1), 149–155.
140. S. M. Tuusa, M. J. Peltola, T. Tirri, M. A. Puska, M. Röyttä, H. Aho, J. Sandholm, L. V. Lassila and P. K. Vallittu, Reconstruction of critical size calvarial bone defects in rabbits with glass-fiber-reinforced composite with bioactive glass granule coating, *J. Biomed. Mater. Res., Part B*, 2008, **84**(2), 510–519.
141. D. S. Zhao, N. Moritz, P. Laurila, R. Mattila, L. V. Lassila, N. Strandberg, T. Mäntylä, P. K. Vallittu and H. T. Aro, Development of a multi-component fiber-reinforced composite implant for load-sharing conditions, *Med. Eng. Phys.*, 2009, **31**(4), 461–469.
142. N. Moritz, N. Strandberg, D. S. Zhao, R. Mattila, L. Paracchini, P. K. Vallittu and H. T. Aro, Mechanical properties and in vivo performance of load-bearing fiber-reinforced composite intramedullary nails with improved torsional strength, *J. Mech. Behav. Biomed. Mater.*, 2014, **40**, 127–139.
143. NEWBONE: Development of load-bearing fibre reinforced composite based non-metallic biomimetic bone implants EU 6th Framework Programme, NMP, Integrated Project for SMEs, Contract no. 026279, Final Report; 2010.
144. R. Huiskes, Failed innovation in total hip replacement. Diagnosis and proposals for a cure, *Acta Orthop. Scand.*, 1993, **64**(6), 699–716.
145. V. A. Koleganova, S. M. Bernier, S. J. Dixon and A. S. Rizkalla, Bioactive glass/polymer composite materials with mechanical properties matching those of cortical bone, *J. Biomed. Mater. Res., Part A*, 2006, **77**(3), 572–579.

146. O. Oral, L. V. J. Lassila, O. G. Kumbuloglu and P. K. Vallittu, Bioactive glass particulate filler composite: Effect of coupling of fillers and filler loading on some physical properties, *Dent. Mater.*, 2014, **30**(5), 570–577.
147. J. Kulkova, N. Moritz, H. Huhtinen, R. Mattila, I. Donati, E. Marsich, S. Paoletti, P. K. Vallittu, Bioactive glass surface for fiber reinforced composite implants via surface etching by Excimer laser 2016, (manuscript in submission).
148. S. Honeybul and K. M. Ho, Decompressive craniectomy for severe traumatic brain injury: the relationship between surgical complications and the prediction of an unfavourable outcome, *Injury*, 2014, **45**(9), 1332–1339.
149. K. D. Martin, B. Franz, M. Kirsch, W. Polanski, M. von der Hagen, G. Schackert and S. B. Sobottka, Autologous bone flap cranioplasty following decompressive craniectomy is combined with a high complication rate in pediatric traumatic brain injury patients, *Acta Neurochir.*, 2014, **156**(4), 813–824.
150. N. Kuusisto, P. K. Vallittu, L. V. Lassila and S. Huumonen, Evaluation of intensity of artefacts in CBCT by radio-opacity of composite simulation models of implants in vitro, *Dentomaxillofac. Radiol.*, 2015, **44**(2), 20140157.
151. J. P. Posti, J. M. Piitulainen, L. Hupa, S. Fagerlund, J. Frantzén, K. M. Aitasalo, V. Vuorinen, W. Serlo, S. Syrjänen and P. K. Vallittu, A glass fiber-reinforced composite - bioactive glass cranioplasty implant: A case study of an early development stage implant removed due to a late infection, *J. Mech. Behav. Biomed. Mater.*, 2015, **55**, 191–200.

CHAPTER 20

Glass Materials in Interventional Radiology and Interventional Oncology

SHARON KEHOE,[a] NANCY KILCUP,[b] ROBERT ABRAHAM[a,c] AND DANIEL BOYD*[a,b,c,d]

[a] ABK Biomedical Inc. 1344 Summer Street, Suite 212, Halifax, Nova Scotia B3H 0A8, Canada; [b] School of Biomedical Engineering, Dalhousie University, PO BOX 15000, Halifax, Nova Scotia B3H 4R2, Canada; [c] Department of Diagnostic Imaging and Interventional Radiology, QE II Health Sciences Centre, Dalhousie University, Halifax, Nova Scotia B3H 3A7, Canada; [d] Department of Applied Oral Sciences, Dalhousie University, PO BOX 15000, Halifax, Nova Scotia B3H 4R2, Canada
*Email: d.boyd@dal.ca

20.1 Transarterial Embolization with Embolic Particles

Transarterial embolization (TAE) is a minimally invasive procedure performed by interventional radiologists; the procedure is designed to selectively deliver materials into targeted vessels, in order to occlude blood flow and/or deliver therapeutic agents directly into a target tissue (*e.g. a tumor*). TAE is effective for the targeted treatment of parenchymal tumors in addition to numerous other conditions, including gastrointestinal hemorrhage, hemoptysis, epistaxis, post-partum hemorrhage and tumor hemorrhage. TAE may also be deployed prior to surgical resection of hypervascular lesions

RSC Smart Materials No. 23
Bioactive Glasses: Fundamentals, Technology and Applications
Edited by Aldo R. Boccaccini, Delia S. Brauer and Leena Hupa

Published by the Royal Society of Chemistry, www.rsc.org

in order to reduce peri-operative bleeding, and has also become a lifesaving intervention in cases of acute hemorrhage.[1,2]

Generally, TAE involves the percutaneous introduction of a small caliber catheter into an accessible artery (*e.g. femoral or radial arteries*). Subsequent to vessel access, the catheter is directed to the target tissue using steerable guide-wires. In certain instances, parent catheters (*i.e. 4–5 Fr. in caliber*) are used to access larger arteries and smaller diameter catheters (*i.e. 2–3 Fr. micro-catheters*) may be placed, co-axially, through the larger diameter catheters to gain access to much smaller arteries. Once the catheter is verified at the target site, an embolic material is injected through the catheter and is either: (i) deposited just beyond the catheter tip (*i.e. proximal occlusion*) or (ii) carried by blood flow deeper into the arterial bed causing distal penetration and occlusion.[1] In respect of the latter, the level of arterial occlusion is dependent on the particle size distribution of the particles selected, where smaller particles are associated with deeper penetration into the target tissue, leading to a higher likelihood of tissue infarction. In the case of the former, particles are deposited more proximally resulting in occlusion at the proximal level but this allows for distal tissue perfusion through collateral channels. A specific example of where TAE has provided excellent clinical performance is in the treatment of hypervascular tumors and arteriovenous malformations, or more particularly in the embolization of uterine leiomyoma (*i.e. uterine fibroids*) which are benign monoclonal tumors of the uterus composed of smooth muscle cells, and an extracellular matrix comprising collagen, fibronectin and proteoglycans,[3] and are a primary driver of hysterectomy.

20.2 'Bland' Embolization of Hypervascular Tumors/ Uterine Fibroids

Uterine fibroids (UFs) are the most common tumor of the pelvis for women, and it is claimed that >60% of women over the age of 45 are effected by UFs.[4] These benign tumors cause severe symptoms, including (*but not limited to*) heavy menstrual bleeding, pain, and an increase in nocturia.[5] Following from these data, it is estimated that 30–70% of the 600 000 hysterectomies performed annually in the US are performed to eliminate UFs.[6–8] However, TAE is increasingly utilized as an intervention to successfully treat parenchymal-based processes and hypervascular tumors, such as uterine fibroids, and is associated with significantly reduced recovery times and healthcare costs.[9,10] Uterine fibroid embolization (UFE) is a minimally invasive alternative to hysterectomy with 98–100% success rate, and efficacy rates in the order of 85–94%.[11–13] Being a TAE procedure, catheters are placed into both uterine arteries, either sequentially or concurrently. Subsequent to vessel access, particles in the size distribution range of *ca.* 300–500 μm or *ca.* 500–700 μm, are delivered through the catheters to occlude the uterine arteries (*i.e. those arteries which provide direct blood flow to the fibroids).* The uterus itself is unaffected by the embolization event, as it receives collateral blood flow from numerous pelvic collateral pathways. Embolization of the uterine arteries results in targeted and specific fibroid infarction and resolution of patient symptoms.[9] The embolic agents utilized for UFE are clinically described as 'bland' agents since they are simply designed to occlude blood flow. Typically, these materials are provided as either: (i) irregularly shaped particles synthesized from polyvinyl alcohol (PVA) or (ii) microspheres, typically synthesized from PVA, trisacryl gelatin (TAG), or hydrogel-based materials.[14] However, it is important to note that microspherical particles are increasingly common, and are preferred since they eliminate issues of particle aggregation, which can lead to false endpoints or suboptimal outcomes.

20.3 Clinical Challenges in TAE and Potential Areas for Glass Technology Innovation

As stated, TAE with microspheres is an effective technique in the treatment of hypervascular and arteriovenous malformations. However, new material technologies are required to enhance clinical utility, and to enable the personalization and optimization of TAE outcomes. One of the principal challenges relating to embolization with micro-spherical particles is the issue of radiolucency. Conventional polymeric materials lack inherent radiopacity, and as such they limit the ability to define both the temporal and spatial distribution of particles during an embolization procedure. Being radiolucent, conventional materials require blending with contrast media prior to delivery such that the contrast media can act as a surrogate measure of the temporal and spatial distribution of microspheres within the target tissue. Nonetheless, the inability to directly monitor embolic microspheres "preclude[s] ideal spatial delivery" and does not permit accurate assessment of their terminal locations.[15] Conversely, the provision of an intrinsically radiopaque embolic microsphere for TAE offers the potential to:[15–17]

i. Define the true spatial distribution of embolic materials in a target tissue,
ii. Provide real-time intraprocedural feedback,
iii. Optimize, personalize and improve TAE techniques and, as a consequence,
iv. Standardize material choice and procedural technique with a high degree of confidence.

It is in this context that glass materials may provide a significant advantage.

20.4 Glass Materials as Image-able Beads for TAE

Recently, multicomponent zinc–silicate glasses have been designed and prepared by flame spheronization (*in multiple particle size distributions ranging from 40 to 540 μm*) for embolization therapies.[18–21] The approach of using glass materials in TAE provides particles which exhibit non-compressible behavior (*under embolization conditions*). Contrastingly, much of the embolic literature speaks to the need for microspheres to exhibit compressibility; it does so, however, in the context of aiding particle delivery through a catheter (*where internal diameters are of the order of ca. 700 μm*) rather than in direct terms of patient safety. However, it is known that some of the root causes relating to technical failures in TAE, namely deformation and lack of durability of occlusion, are directly related to the compressibility of conventional materials.[22,23] To substantiate this point, conventional embolic biomaterials have relatively high compressibility of up to 30% (*without undergoing any fragmentation*), however many conventional materials can also swell upon mixing with contrast media to result in a 10–40% increase in diameter.[24] The consequence of these changes leads to clinical uncertainty with respect to appropriate size distributions for clinical use. Furthermore, since excessive compressibility can also lead to deformation of the embolic agent *in situ*, there is enhanced potential for embolic instability leading to subsequent redistribution of microspheres and revascularization of targeted tissue.[22,23] Accordingly, it is accepted that non-compressible particles may ameliorate such challenges and provide guaranteed particle size distributions to optimize clinical decision making.

Notwithstanding the benefits of predictable morphology and definite particle size distributions, the principal advantage relating to the use of glass microspheres for TAE relates to image-ability. Glass materials have a deep history of being utilized as clinical materials, especially in dentistry and orthopedics where they have conferred radiopacity into a variety of medical materials (*e.g. restorative dental composites and bone cements*). Despite this fact, glass materials have not been heavily considered as imageable beads for TAE until recently. Kehoe *et al.* have recently published a series of articles relating to glass microspheres based on the multi-component SiO_2–CaO–ZnO–La_3O_2–TiO_2–MgO–SrO–Na_2O system for TAE applications. This system provides glasses that melt below 1500 °C, and contain sufficient La_3O_2 (*up to 0.068 mole fraction*) to offer multi-modal imageability (*e.g. X-ray fluoroscopy and CT*) for TAE applications.[18,20,21,25,26] Structural analysis of these glasses, using ^{29}Si MAS NMR, has indicated that the peak maxima for these glasses are between −82.2 ppm to −86.2 ppm; chemical shifts associated with silicate glass networks predominantly comprise Q^2 and Q^3 structural units.[21] Composition–structure–property analysis has shown that lanthanum is a significant compositional factor in respect of stabilizing the structure of the silicate network whilst concurrently having a positive influence on radiopacity. However, it is also noted in the literature that La can have deleterious effects on melt characteristics and as such an upper concentration threshold of 0.068 has been identified for inclusion in the glass formulation.[18,21] This observation (*i.e. network stabilization as a result of La inclusion*) has also been corroborated by glass transition and density analysis, which shows that:

i. The glass transition temperature (T_g) within this system is in the range of 597 to 690 °C and is heavily dependent on the La_2O_3 : ZnO ratio; specifically the replacement of La_2O_3 by ZnO decreases T_g and T_p[1]

ii. Density variations from 3.17 to 3.78 g cm^{-3} have also been recorded and are highly sensitive to La_2O_3 and ZnO content.[18,21]

It has also been established that the integrity of the glass network structure (*based on predominant Q speciation*) and its contiguous properties remain unchanged as a result of either (i) exposure to 30–60 kGy gamma irradiation or (ii) exacerbated environmental storage conditions (*up to 24 months real-time equivalent aging*).[27] These features substantiate the suitability of these materials as medical devices from a manufacturing perspective.

From a biocompatibility standpoint, the Zn–Si glass system developed by Kehoe *et al.*, provides for cell viabilities (*i.e. elution assay with L929 cells*) in the range 61–92% as compared with the tissue culture water controls.[21] Interestingly, regression analysis shows that La_2O_3 is more cytotoxic than ZnO and as such careful consideration of composition is required in order to design a medical product that provides for cytocompatibility yet also provides for ease of melt, spheronization capabilities and multimodal imageability in a clinical setting. Given the propensity for glass biomaterials to release constituent metal cations under physiological conditions, the ion release characteristics and genotoxic potential for these materials has also been clarified in the literature.[25,31] As one would expect, the ions released from the Zn–Si microspheres are dependent upon the surface area to volume ratio. However, it has been ascertained that the cumulative release of all constituent ions from this glass system does not cause a genotoxic response *in vitro*.[19]

Examinations of the imaging characteristics of Zn–Si glass microspheres were originally reported based on CT bench-top experimentation. This research identified a particular formulation of glass,[†] as having a radiopacity of *ca.* 8000 HU, which is over three times the contrast provided by conventional dilutions of contrast media (*ca.* 2400 HU).[18] It has also been shown, using advanced MRI imaging studies, that Zn–Si glass microspheres exhibit diamagnetic susceptibility with bulk magnetic susceptibility (BMS) contrast, resulting in very low *T*1 and *T*2 contrast (*to confound most clinical scans*) and permit visualization and quantification of these glass microspheres *in situ* using R2* mapping protocols.[26] However, developing the data set relating to (i) the imageability and (ii) the host response of Zn–Si microspheres required significant pre-clinical evaluations to evaluate clinical utility.

Correspondingly, conventional imaging (*intraprocedural fluoroscopy and computed tomography, CT*) and histopathological techniques, have been utilized in the literature to examine these characteristics *in vivo* using a

---

[†]$0.562SiO_2$–$0.035CaO$–$0.188ZnO$–$0.068La_3O_2$–$0.042TiO_2$–$0.035MgO$–$0.035SrO$–$0.035Na_2O$ (*mole fraction*).

variety of pre-clinical models, including the swine renal artery model.[20] The published studies in this regard have evaluated, in real-time, the spatial and temporal distribution of Zn–Si microspheres, as well as their ease of deliverability, embolization effectiveness and the local host response (*versus TAG microspheres as the control article*). Multiple animal models have been utilized to examine an array of particle size distributions of 40–540 μm. However, more recent publications have emphasized a particle size distribution of 40–150 μm for use in *in vivo* studies since this particle size distribution represents the worst-case scenario in respect of it being:

i. The highest surface area to volume ratio used for embolization procedures and, as such, is an important consideration in relation to the generation of degradation by-products from glasses under physiological conditions (*which may lead to adverse host responses*),
ii. The most challenging to image using conventional imaging modalities, both from a spatial and temporal distribution standpoint,
iii. The most likely to migrate to distal locations.[28,29]

From the published data, it can be seen that Zn–Si microspheres provide direct intraprocedural feedback using standard diagnostic imaging techniques. The specific locoregional distribution of the Zn–Si microspheres is clearly visible on high-resolution radiographs and CT (Figures 20.1 and 20.2) without contiguous imaging artifacts, the latter point being important as imaging artifacts from high radiopacity materials can complicate imaging.

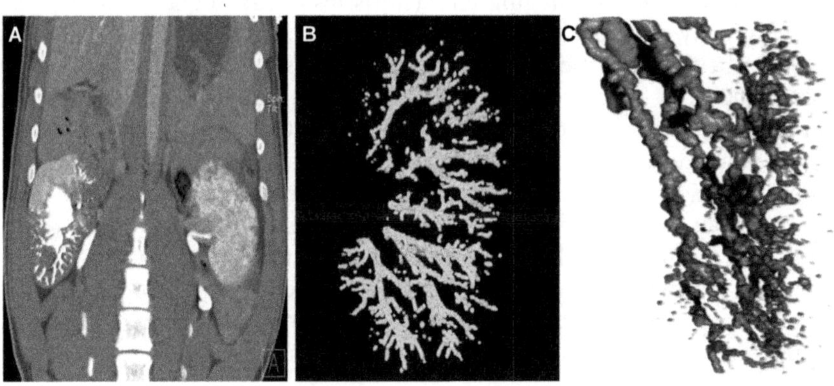

Figure 20.1 (A) Whole animal CT (right kidney coronal plane) and (B) micro-CT scans (whole kidney embolization, *acute assessment*) and (C) selective renal artery embolization (*30-day assessment*) of swine kidneys embolized with 40–150 μm Zn–Si glass microspheres.
Reprinted with permission from Kehoe *et al.*, "Imageable" Zinc-Silicate Glass Microspheres For Transarterial Embolization: A Renal Artery Embolization Study, *J. Biomed. Glasses*, **1**, 70–79.

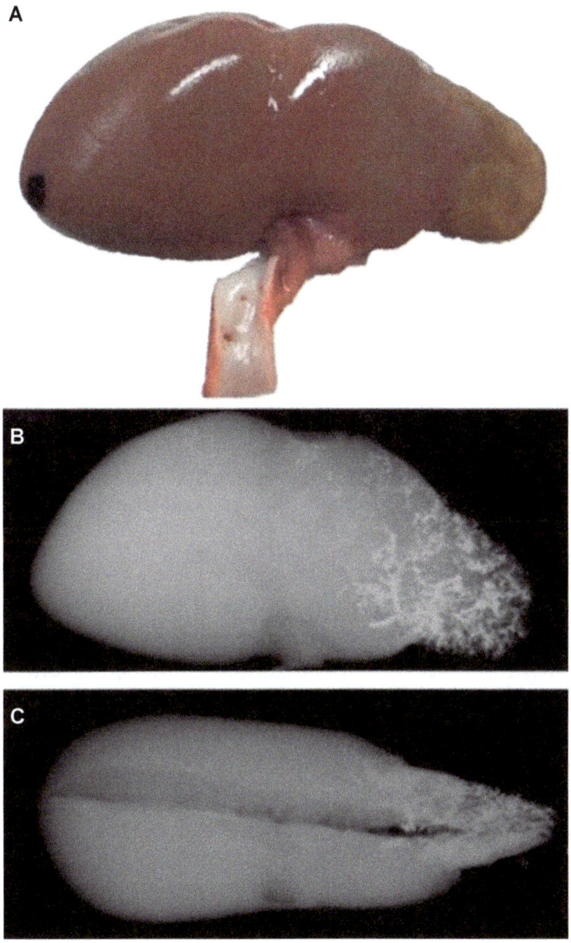

Figure 20.2 Representative images of gross pathology and high-resolution kidney radiographs. (A) Harvested kidney 29 days post-implantation and images of the (B) anterior and (C) lateral views of a left swine kidney embolized with 40–150 µm Zn–Si glass microspheres.
Reprinted with permission from Kehoe *et al.*, "Imageable" Zinc-Silicate Glass Microspheres For Transarterial Embolization: A Renal Artery Embolization Study, *J. Biomed. Glasses*, **1**, 70–79.

In the clinical setting, Zn–Si microspheres enable the definition of spatial distribution within a target tissue (Figures 20.1 and 20.3), consequently permitting the optimization, personalization and improvement of TAE techniques. It has also been noted from the pre-clinical studies that Zn–Si microspheres offered more predictable occlusion than the conventional microspheres (*control articles*), a feature which may be attributable to the non-compressible nature of the glass microspheres *versus* the high compressibility of control articles.

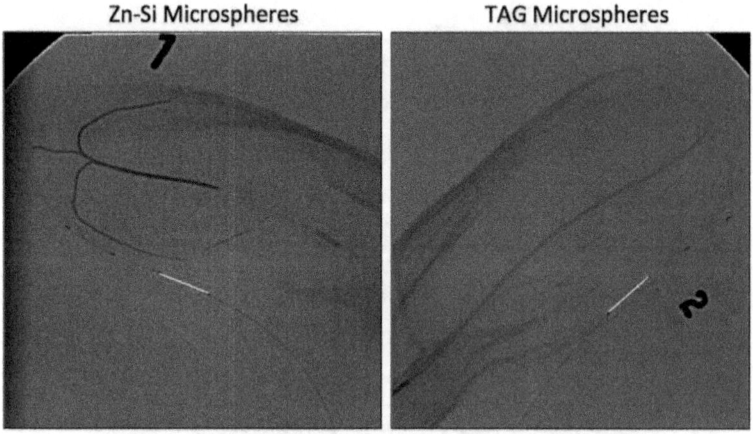

Figure 20.3 Fluoroscopic visibility of the Zn–Si glass microspheres in a rabbit auricular artery *versus* conventional TAG microspheres, demonstrating spatial distribution for 300–500 μm Zn–Si microspheres.

In examining the host responses associated with Zn–Si glasses *in vivo*, it is instructive to first identify the mechanism of occlusion and host response characteristics of existing embolization materials. PVA and TAG materials have been shown to provide a combination of biological (*including thrombus formation, cell adhesion, and inflammation*) and mechanical mechanisms of occlusion.[22,30,31] From a host response standpoint, no necrosis or other abnormalities are reported in the literature arising from the multiple preclinical studies of Zn–Si microspheres. Histopathological examination from swine renal artery models reveals that there is substantial equivalence between the mechanism of occlusion for both Zn–Si glass microspheres and conventional materials, with direct contact between the vessel wall and the glass illustrating obvious mechanical occlusion, and thrombus formation filling the gaps between adjacent microspheres and the vascular walls (black arrows, Figure 20.4A).

Further histopathological evaluations, focused on inflammatory response, show that comparable levels of mononuclear cell infiltration are associated with both conventional microspheres and Zn–Si microspheres.[16] In addition, the published histopathological staining of tissues containing Zn–Si microspheres (Figure 20.4C) show only minor disruptions to the elastic laminae, in a manner consistent with other embolization products.[32–34]

In summary, the Zn–Si glass microspheres examined in the literature in recent years demonstrate the potential utility of glass materials in applications such as TAE where the intrinsic properties of a glass can: (i) enable unique material characteristics and, as a consequence, (ii) drive additional research endeavors in other areas of interventional radiology. While the provision of an imageable bead for bland embolization is an important development from the clinical standpoint, significant contributions have been made by glass technologies in the treatment of malignant tumors, and

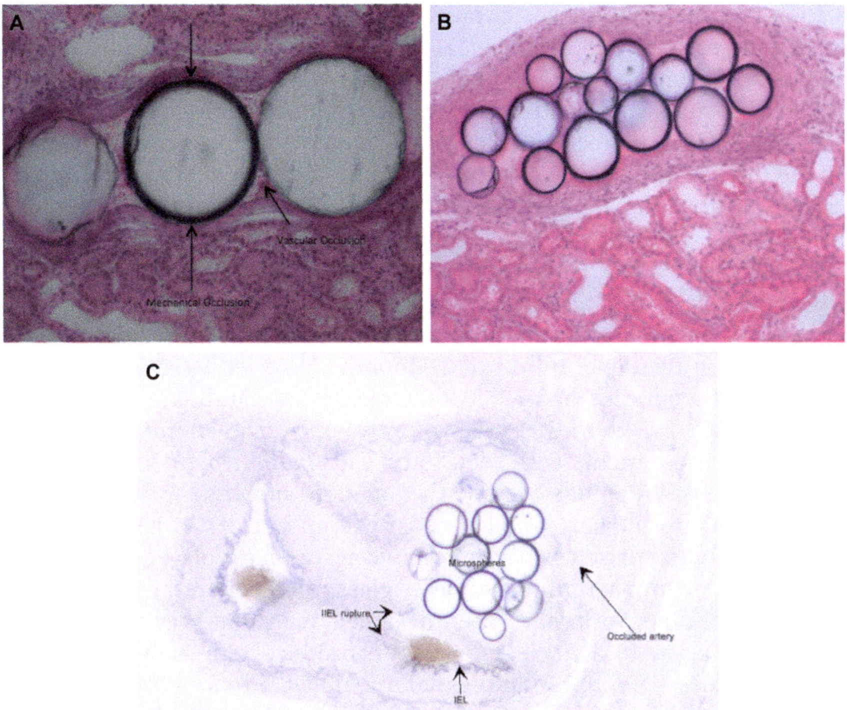

Figure 20.4 (A) Acute (*magnification 20×*) and (B and C) subchronic (*magnification 20×*) histopathologic evaluation of swine RAE with 40–150 and 300–500 µm Zn–Si glass microspheres.
Reprinted with permission from Kehoe *et al.*, "Imageable" Zinc-Silicate Glass Microspheres For Transarterial Embolization: A Renal Artery Embolization Study, *J. Biomed. Glasses*, **1**, 70–79.

many opportunities exist to expand the design and utility of glasses in such indications.

20.5 Embolization of Malignant Tumors

In recent years, microspheres that deliver therapeutic payloads of either: (i) selective internal radiation, or (ii) selected chemotherapeutic agents, have played a much larger role in the treatment of malignant tumors, especially those associated with the liver.[35] The clinical context for use of TAE in the treatment of malignant tumors arises due to the significant global concerns associated with hepatocellular carcinoma (HCC), and metastatic disease (*in particular metastatic colorectal cancer, or mCRC*) of the liver.[36]

HCC is the sixth most common organ cancer, and has a very poor prognosis; the ratio of mortality to incidence being 0.95.[37] HCC has a rapidly rising incidence in the United States and Europe, largely driven by the burden of advanced hepatitis C virus (HCV) and nonalcoholic steatohepatitis (NASH) cases.[36,38] Surgical resection, liver transplantation or focal tumor ablation are the only known curative options, and may be effectively deployed at an early stage.[39] However, HCC is insidious in nature, and >60% of patients are diagnosed at a late stage, where the median survival rate is less than 1 yr; as such, these patients are not eligible for curative therapies.[40] In addition to HCC, colorectal cancer (CRC) is the third most common cancer and the third leading cause of cancer death in the United States.[41] Critically however, 20% of all CRC patients will have liver involvement at the time of CRC diagnosis, while *ca.* 60% of remaining patients will go on to develop liver metastases.[42,43] Arising from this situation, the literature notes that: (i) unresectable liver metastases account for much of the morbidity and mortality for patients with CRC and, (ii) prognosis for 'unresectable' patients is poor, with a median survival of 22 months.[42] As with HCC, surgical resection can be curative but <25% of patients are candidates for curative interventions.[43] Based on these data, control of liver involvement is key for survival in these patients with metastatic CRC (*with hepatic involvement*);[44] for these patients, TAE has become the standard of care.

It is in the intermediate to advanced stage, in both HCC and mCRC patient populations, that glass based embolic agents have played a significant clinical role. Yttrium–aluminum–silicate glass microspheres, containing ^{89}Y within the glass network, have been designed such that they may undergo neutron activation in a nuclear reactor to produce radioactive ^{90}Y microspheres for locoregional radiation therapy, delivered *via* a TAE approach. Once the therapeutic dose for the patient is determined, the beads undergo quality assurance steps and are then packaged in shielded containers to be transported to the treatment site for a pre-determined treatment date (*based on a decay period, to provide the necessary radiation activity for a specific dose*). Briefly, at the time of administration, radioactive ^{90}Y glass beads are delivered *via* a transarterial catheter-based approach, targeting the hepatic arterial supply of the tumor (*Detailed literature on the technical and methodological aspects of this procedure have been comprehensively covered elsewhere.*[45]). The microspheres are carried by blood flow and, as a result, preferentially target tumor tissue (*since tumors are more hypervascular than surrounding normal liver tissue*). Importantly, the dual blood supply of the liver benefits arterial therapy as normal liver is predominately supplied by portal venous supply whereas liver cancers are predominately supplied by the hepatic artery.[1] Thus, selective arterial infusion targets the tumor rather than normal hepatic parenchyma.[46] This treatment modality is a subcategory of TAE and is referred to as transarterial radioembolization (TARE) or selective internal radiation therapy (SIRT).

TheraSphere® ^{90}Y glass microspheres (BTG International), are a clinically available form of the Y–Al–Si glass for TARE and have a mean diameter of 20–30 μm. The primary mode of therapy derived from activated ^{90}Y glass microspheres is based on beta radiation emission—an ionizing radiation related to high speed, high energy electrons. The energy emitted is determined per sphere, and is typically on the order of 2500 Bq. Usually, 1–2 million beads are present in each treatment dose; accordingly the administered dose to the tumor is typically on the order of *ca.* 120 Gy, but can be modulated (*or reduced*) *via* a number of factors including natural decay processes. One particular benefit of beta radiation arising from these glass beads is the small diameter of the associated therapeutic effect; specifically, the radiation is highly effective for an average diameter of 2.5 mm around each microsphere (*but may be effective for up to ca. 10 mm*) after which there is no further damage to tissue.[45,47] This provides a safety advantage and minimizes injury to adjacent organs. Following from these unique attributes, ^{90}Y glass materials allow for targeted and localized treatment of hepatic tumors with doses as high as 150 Gy, even though healthy liver cannot tolerate doses greater than 50 Gy.[48] Numerous publications have discussed the technical aspects of TARE.[45,47,49] and an emerging literature comprising large retrospective and prospective studies continues to substantiate the safety and therapeutic benefit of ^{90}Y radioembolization for patients with both HCC[50–52] and mCRC.[53–56] More recently, ^{90}Y microspheres have been deployed to downsize tumors with new interventions such as radiation segmentectomy; such procedures are also making it possible for some patients to become candidates for the curative treatments of surgical resection or transplant.[57–60]

20.6 Glass Materials for Selective Therpauetic Efficacy in Malignant Tumors

The administration and delivery of *in situ* radiation, with modification of dose being based on tumor and target volume, defines TARE (*in part*) as a brachytherapy-based procedure. Following from its therapeutic effect, a glass microsphere indicated for the *in situ* irradiation of HCC or mCRC must satisfy certain fundamental criteria, and provide for:[61]

i. Neutron activation and conversion of ^{89}Y to ^{90}Y. As such, the glass network must be composed of high purity reagents *and* free of elements/impurities that form unwanted activation by-products during neutron activation.
ii. Once activated, the concentration of therapeutic element (*i.e.* ^{90}Y) within the glass network must provide the required level of specific activity.
iii. Chemical resistance to body fluids, to ensure that no activation by-products (*i.e.* radioactive isotopes) are released as a result of network dissolution within the physiological milieu.

In addition to these fundamental requirements, ideally, glass materials for TARE procedures should also satisfy a further set of characteristics, and exhibit:

i. Intrinsic radiopacity for real-time intraprocedural feedback regarding spatial and temporal distribution of microspheres within the target tissue (*e.g. X-ray fluoroscopy, CT/Cone-beam CT*).

ii. Spherical morphology for enhanced clinical utility, safety and efficacy.
iii. No toxicity against surrounding healthy liver tissue and should explicitly have no carcinogenic potential.

In the USA, ^{90}Y glass microspheres (*TheraSpheres®*) are regulated by a humanitarian device exception from the Food and Drug Administration (FDA) for the exclusive treatment of unresectable HCC. These microspheres are also approved for the treatment of unresectable liver tumors within the EU and in a selection of other jurisdictions. ^{90}Y is a near-pure beta emitter and has a half-life of 64.2 hours, with 94% of the energy being emitted in the first 11 days after activation, rendering it an ideal *in situ* radiation agent.[62] Arising from these characteristics, and the limited depth of penetration of the radiation energy, shielding and isolation are not required for the patient post-implantation. Each milligram dose of TheraSphere® contains between 22 000 and 73 000 microspheres, with the upper size limit (*of particle size distribution*) of the microspheres permitting delivery to the tumors *via* the hepatic artery, and the lower size limit (*of particle size distribution*) preventing the microspheres from passing from the arterial to the venous circulation.[62] This particle size distribution ensures that microspheres remain trapped within the vasculature of the tumors and deliver a selective radiation dose directly to the targeted tissue.[63]

20.7 Composition–Structure–Property Relationships for Yttrium Aluminosilicate Glasses

To better understand the composition–structure–property relationships of the ^{90}Y glass, multiple studies have been performed on the clinically relevant yttrium aluminosilicate glass systems (*i.e. those where* Y_2O_3 *is present in 25–55 wt%,* Al_2O_3 *13–35 wt% and* SiO_2 *30–55 wt%*). In summary, this literature[64–66] has confirmed some fundamental composition–structure–property relationships. Infrared spectroscopy of glasses within this system confirms the structure of the glasses as a continuous closed network. Using the Archimedean method to determine density, the density values for these glasses are reported to range from 2.86 to 3.81 $g\,cm^{-3}$ with marked increases in density associated with substitutions of SiO_2 and Al_2O_3 for Y_2O_3, likely attributable to the large difference in molecular weights between Y_2O_3, SiO_2, and Al_2O_3. The softening and glass transition temperatures (T_s *and* T_g, *respectively*) are reported to range from 933 to 947 °C and 884 to 895 °C, respectively, with T_s only varying slightly with composition. The thermal expansion co-efficient, however, was observed to increase with increasing Y_2O_3 content and decreases with increasing SiO_2 content. The refractive index (*ranging from 1.717 to 1.572*) showed a similar trend to both density and thermal expansion co-efficient in that the corresponding refractive index for the glass tended to increase with Y_2O_3 content. Similarly, the Vickers Hardness Number (VHN) for all YAS glasses were shown to vary from 6.36 to 8.28 GPa, with glasses of high Y_2O_3 content yielding harder glasses.

Measurements of the chemical durability for yttrium aluminosilicate glass networks (*with multiple morphological characteristics e.g. sphere, monoliths, and irregular particles*) have been undertaken in the literature.[64,65] These experiments are necessary to determine that microspheres have sufficient durability so as not to cause errant release of active isotopes *in situ*. Experimental procedures are varied in the literature, and include, placing weighed amounts of glass powder in distilled water and gently agitating the samples for an immersion period of up to 15 days.[64] Dissolution rates are also frequently determined using glass rectangular samples (*e.g.* $0.5 \times 0.75 \times 1$ *cm*) in 1N HCl at 30 °C and 50 °C, again for periods of up to 15 days.[64] Overall, the chemical durability of ^{90}Y glasses has been shown to be very high, and is regarded as comparable to that of fused silica as validated by weight loss, atomic absorption (AA) spectroscopy, energy dispersive spectroscopy (EDS) and scanning electron microscopy (SEM).[64] However, dissolution characteristics are observed to increase with increasing the Al_2O_3 content while, contrastingly, in 1N HCl, an increase in dissolution rates are reported with increasing Y_2O_3 content (and increasing temperature).[65] The wt% Y released from ^{90}Y glasses is observed to be only 0.02–0.13% of total Y_2O_3 content,[65] with a negligible dissolution rate in HCl of between 2.1×10^{-7} and 54.6×10^{-7} g cm^{-2} min^{-1} being reported in the literature.[67] Table 20.1 summarizes the wt% Y_2O_3 and Al_2O_3 content on the surface of 'as-prepared' glass microspheres (and *irregular powders*). The results suggest

Table 20.1 Wt% Y_2O_3 and Al_2O_3 on the surface of $40Y_2O_3$–$20Al_2O_3$–$40SiO_2$ (wt%) glass in the form of YAS glass microspheres, powders and sol–gel derived powders in comparison to the parent wt% values in the parent (*initial*) YAS glass.

Extraction conditions	Y_2O_3 (wt%)	Al_2O_3 (wt%)
Initial YAS glass	40	20
CM microspheres[a]		
DI water, 37 °C	40.5	18.9
Saline, 37 °C	41.4	19.1
Saline, 50 °C	41.8	18.7
12N HCl, 37 °C	18.4	10.6
CM powder[a]		
DI water, 37 °C	39.4	19.7
12N HCl, 37 °C	23.4	12.8
SG powder[b]		
DI water, 37 °C	41.2	19.5
Saline, 37 °C	41.4	18.9
12N HCl, 37 °C	7.2	0.7

[a]CM denotes conventional melt-quenched.
[b]SG denotes sol–gel-derived.

that the microspheres and/or powders, whether conventional melt-quenched or sol–gel-derived, impart no statistically significant difference on the chemical durability of the glass under either simulated physiological or accelerated conditions.[65] Finally, extremely low levels of released radiation (nCi g^{-1}) were calculated on the basis of the dissolution rates for the $40Y_2O_3$–$20Al_2O_3$–$40SiO_2$ (wt%) glass, with the maximum amount of radiation released being 1000 times lower than that considered clinically acceptable (*i.e. 5 rads*).[65]

With respect to the processing of irregular glass particles prepared *via* the conventional melt-quench technique to form microspheres, it is useful to note that two different high temperature processes are typically used; (a) flame spheronization and (b) spheronization by gravitational fall in a tubular furnace. In the first process, glass microspheres of varying particle size distributions may be obtained by re-melting (irregularly shaped) glass powders within a hot flame, utilizing torch systems with plasma, oxy-acetylene, and oxy-propane fuelled flames for this purpose.[64,66] This process is typically referred to as a flame spheronization process, whereby the process variables (*including, but not limited to, powder feed rate, flame temperature and length, etc.*) must be adjusted for each type of glass and particle size range in order to obtain a high degree of spherical output. The second process introduces glass powders at the top of a vertical tubular furnace, and allows them to fall through the inside of the furnace.[68] This process is termed spheronization by gravitational fall in a tubular furnace, though is commonly associated with extremely poor product yields and is consequently not feasible for the purposes of large-scale manufacturing.

20.8 Clinical Efficacy of ^{90}Y Glass Microspheres Therapy in Oncology

The use of ^{90}Y glass microspheres in TARE has enabled the selective locoregional delivery of radiation therapy for patients suffering from both primary *and* metastatic liver tumors. It is important to note that while the basic glass technology has existed for several decades, the delayed acceptance of these biomaterials for TARE has been related to safety concerns arising from the deposition of radioactive glass spheres within the pulmonary and gastrointestinal anatomy causing adverse side effects. However, recent technical and methodological improvements to the interventional procedure have ameliorated these issues.[45,47] Today, based on an evidence-based approach, TARE is considered appropriate for patients with (i) unresectable primary or metastatic liver cancer, (ii) a liver dominant tumor burden and (iii) a life expectancy of >3 months.[69,70] On examination of the efficacy of radioactive glass beads to provide a therapeutic effect, Edeline *et al.* have noted that the main knowledge limitation, from a clinical standpoint, is the lack of phase three randomized controlled trials comparing ^{90}Y treatments against other transarterial interventions.[71] Nevertheless, much clinical research relating to safety and efficacy of ^{90}Y radioembolization exists and is based on excellent retrospective, non-randomized, or pilot prospective studies (*An excellent summary of such data is provided elsewhere*[56]). The data arising from these studies demonstrate that the ^{90}Y microsphere therapy is safe and efficacious. Indeed a recent consensus report from the 'Radioembolization Brachytherapy Oncology Consortium' identified that, while ^{90}Y microsphere therapy is complex (requiring exceptional diligence and meticulous angiographic technique), the consortium stated that 'sufficient evidence exists to support the safety and effectiveness of ^{90}Y microsphere therapy''.[69] Despite its developing successes, there remain areas of controversy and opportunities for future developments, both from a clinical and a materials standpoint. In respect of the former, response evaluation criteria, liver volume changes as a function of treatment, and the efficacy of radiation segmentectomy are examples of critical areas under investigation. However, from a materials standpoint significant research and development opportunities arise from an analysis of existing challenges associated with ^{90}Y microsphere therapy and more generally in utility of glass-based materials to treat malignant tumors. Crucially, these materials challenges also intersect with key clinical questions. If the materials community can overcome such challenges, significant room for enhanced outcomes may be enabled.

20.9 Emerging Opportunities for Therapeutic Glasses in Interventional Oncology

A procedure known as 'drug-eluting bead transarterial chemoembolization' or DEB-TACE, was also developed to treat liver tumors, with the specific objective of providing targeted locoregional drug delivery while simultaneously increasing tumor–drug exposure time and minimizing systemic chemotherapeutic concentrations. Again the materials on offer for this procedure are largely PVA- and hydrogel-based materials and, as such, carry all of the associated limitations relating to compressibility and imageability in a clinical setting. Nonetheless, based on substantiating clinical evidence for lower complication rates and improved survivability, DEB-TACE has emerged as the preferred treatment option for patients with intermediate HCC.[72–74] Currently none of the clinically available embolic agents are provided with the therapeutic drugs pre-loaded which means that existing microspheres must be drug-loaded immediately prior to a procedure, requiring significant logistical coordination between the pharmacy and the interventional oncology team to ensure proper drug loading.[75–77] This process is a time consuming and technically demanding process that must be properly coordinated with the hospital pharmacy in order to provide the optimum therapy.[78,79] If done incorrectly the desired therapeutic levels may not be provided to the patient. Once loaded, the drug-infused particles have a limited shelf life of up to 24 hours, which provides only a small window in which to utilize these particles and unforeseen delays may result in wasted material.[80] More importantly, drug loading causes changes to the particle's physical properties[81] and it remains unclear if subsequent drug release results in further changes to physical properties, such changes being potential contributing factors to suboptimal therapy.[82] Designing a biomaterial for DEB-TACE with the therapeutic agent preloaded in a microsphere is critical to overcoming the limitations of the current DEB-TACE materials.

In a review of tissue engineering and biomaterial design principles, Place *et al.* described a shift in the design paradigm of biomaterials from one of complexity to one of simplicity. They argue that research should focus on designing biomaterials that elicit desired biological responses by triggering cellular functions that unlock the native healing properties of the body through simple structural and/or chemical composition.[83] Complementary to this perspective and specifically for hard tissue applications, biologically active glasses have been designed to deliver therapeutic inorganic ions (TII). However, the biological role of TIIs released from biologically active glasses can extend far beyond hard tissue mineralization. A number of reviews have summarized the different TIIs incorporated into bioactive glasses for improved angiogenesis as well as osteogenesis.[84–87] Evidence from *in vitro* studies show that addition of copper (Cu) and cobalt (Co) ions to cell culture media can modulate the production of factors associated with angiogenesis: HIF-α and VEGF. Wu *et al.* cultured human bone marrow stromal cells on a mesoporous bioactive glass containing 5 mol% Cu demonstrated increased production of VEGF and of HIF-α, biological markers for angiogenesis. In fact, when these cells were exposed to extracts of the same glass they demonstrated a concentration-dependent increase in VEGF and HIF-α.[88] Lee *et al.* cultured a human osteosarcoma cell line (MG63A) on titanium phosphate glass disks doped with 0, 5, 10 and 15 mol% of CoO, substituted for $CaCO_3$. VEGF concentration (normalized by number of cells) was significantly increased for all glass samples, with the highest observed VEGF concentration occurring in the 5 and 10% Co-doped glasses, which suggests that there is an optimum Co concentration for stimulating VEGF expression.[89]

In oncology applications the therapeutic paradigm for biomedical glasses shifts from tissue repair to targeted tumor destruction. Building on the philosophy of simplicity and considering the versatility of glasses to provide for predictable delivery of TIIs, these materials offer a vast number of compositional possibilities capable of meeting the requirements for treating malignant tumors whilst concurrently providing imageable beads for personalized and optimized clinical outcomes. Critically, the vectored delivery of TIIs from glass networks for oncology represents an evolution in the use of TIIs. The question is: what TIIs should be included in glasses for oncology applications?

To determine what TIIs should be included in a glass delivery system, it is important to first evaluate metal-based chemotherapeutic agents. These drugs may give insight into possible TIIs for inclusion into bioactive glasses. Historically, metal-containing drugs have been used to treat cancer as far back as the 16th century.[90] Modern use of metals in cancer therapy stems from the discovery of the cytotoxic effects of *cis*-diamminedichloroplatinum(II), known today as cisplatin.[91] From this discovery, other platinum-based chemotherapeutics have been developed that have significant clinical use worldwide and include carboplatin and oxaliplatin.[92,93] Since the discovery of cisplatin and its analogues many metals have been used as targets for chemotherapy drugs.[94,95] Much research has been done investigating the role of metals and their complexes in the treatment of disease. In general, the therapeutic effect of a metal is due to its ability to modulate the activity of enzymes and/or to alter the molecular structure of biologically important molecules.[95,96] Another important effect of metals is their ability to generate reactive oxygen species (ROS), which has been connected to the cytotoxic effect of metal compounds on cancer cells.[94–96]

Many different metal-containing compounds, other than cisplatin and its derivatives have been studied as chemotherapy agents.[97] However, the toxic effect of metal-based drugs is typically non-selective, which contributes to a general reluctance in developing metal-based therapies.[95] In a review article in Science, Thompson and Orvig address the need to develop new metal-based therapies with the understanding that it is the dose that differentiates a therapy from a toxin.[98] Some examples of toxic compounds that have therapeutic benefits when used in the appropriate doses are arsenic trioxide (As_2O_3), used to treat acute promyelocytic leukemia,[99] and antimony trioxide (Sb_2O_3), which is active against human leukemia cell lines.[100] In these two examples, dose determines if the compound is toxic or therapeutic. Due to the *in vitro* and *in vivo* success of experimental therapies based on transition metal therapies there has been a drive to develop new cancer therapies based on these systems.[101]

In a glass designed for DEB-TACE, the ideal TII should be homogenously incorporated into the glass network and must also have the ability to be released from the glass at therapeutic levels over a preferred treatment window. Keeping in mind the desire to provide a radiopaque material for DEB-TACE, it is also desirable that the TII might contribute positively to the radiopacity of the glass system during delivery. However, the most critical and important feature of the ideal TII is that it will have demonstrated effectiveness treating cancer. In this regard, Kilcup *et al.* have explored one approach where V_2O_5 has been utilized in a glass formulation to provide for potential therapeutic effects against cancer cell lines.[102] The rationale for this approach lies in the fact that vanadium-based salts and organometallic compounds have been shown to possess excellent properties against particular cancers and cancer cell lines, including reduced tumor growth and prevention of metastatic growth.[103] The mechanisms underpinning such responses are largely unknown, however the following has been identified. Soluble ammonium monovanadate (NH_4VO_3) causes apoptosis in MCF7 breast cancer cell lines[104] whilst elsewhere sodium metavanadate ($NaVO_3$(aq)) can prevent cell proliferation in the human prostate cancer cell line (PC-3) by production of reactive oxygen species (ROS).[105] Kilcup *et al.*, determined that the addition of vanadium led to silicate network polymerization, while simultaneously and surprisingly increasing V release capabilities. Solid-state ^{51}V NMR demonstrated that vanadium was present in the network as sodium metavanadate-like structures. Since sodium metavanadate is highly soluble, it is probable that these structures enabled hydrolytic degradation and therefore causes the increased V release. HepG2 cells exposed to dissolution by-products derived from these glasses demonstrate concentration-dependent cytotoxic responses, where increased V concentration results in significantly greater cell death. Using flow cytometry, it was determined that the mechanism of cell death was an apoptosis pathway activated by caspase 8, indicating that the selective release of targeted concentrations of V may provide a means to cause the self-destruction of cancer cells in HCC patients. Of course, biomedical glasses offer a unique ability to mediate host–material responses to support the delivery of TIIs, which can be controlled through compositional changes. From this perspective, it is interesting to note that the periodic table is full of elements previously disregarded for inclusion in bioactive glasses for skeletal applications due to their toxic nature. However, from an interventional oncology perspective, it is the potentially toxic nature of these elements and their ability to cause tissue damage that makes these elements potential candidates for inclusion in bioactive glasses. However, the metabolism of potentially toxic elements must be considered to avoid unwanted systemic accumulation. The subsequent testing of associated hypotheses may ultimately advance the utility of bioactive glasses in non-traditional fields such as interventional oncology.

20.10 Conclusions and Outlook

Composition–structure–property relationships for a series of imageable glasses have proved instrumental in the development of a new range of materials for embolization therapies. These imageable glass microspheres permit the definition of temporal and spatial distribution in a target tissue, consequently permitting optimization, personalization and improvement of TAE techniques. Imageable yttrium 90-loaded glass microspheres for TARE are now warranted in order to significantly enhance patient safety. These new glasses must comprise non-toxic radiopaque elements that do not form unwanted impurities from neutron activation by-products. Glass systems also offer the ability to overcome limitations with current embolization microspheres for TACE since they can be provided "drug-preloaded" with TIIs, eliminating the loading step and the associated alterations in physical properties. The vectored delivery of TIIs from glass networks for oncology represents the next evolution of this concept. The clinical utility of biomedical glasses beyond hard tissue repair, as such, is rich in product commercialization opportunities.

References

1. *Transcatheter Embolization and Therapy*, ed. M. Lee and A. Watkinson, 2010, Springer, London.

2. K. Osuga *et al.*, Principles and techniques of transcatheter embolotherapy for peripheral vascular lesions, *Radiat. Med.*, 2006, **24**(4), 309–314.
3. W. H. Parker, Etiology, symptomatology, and diagnosis of uterine myomas, *Fertil. Steril.*, 2007, **87**(4), 725–736.
4. S. Okolo, Incidence, aetiology and epidemiology of uterine fibroids, *Best Pract. Res. Clin. Obstet. Gynaecol.*, 2008, **22**(4), 571–588.
5. J. B. Spies *et al.*, Uterine artery embolization for leiomyomata, *Obstet. Gynecol.*, 2001, **98**(1), 29–34.
6. J. K. Gupta *et al.*, Uterine artery embolization for symptomatic uterine fibroids, *Cochrane Database Syst. Rev.*, 2012, (5), CD005073.
7. J. K. Gupta *et al.*, Uterine artery embolization for symptomatic uterine fibroids, *Cochrane Database Syst. Rev.*, 2014, (12), CD005073.
8. J. K. Gupta *et al.*, Uterine artery embolization for symptomatic uterine fibroids, *Cochrane Database Syst. Rev.*, 2006, (1), CD005073.
9. S. M. van der Kooij *et al.*, Uterine artery embolization vs hysterectomy in the treatment of symptomatic uterine fibroids: 5-year outcome from the randomized EMMY trial, *Am. J. Obstet. Gynecol.*, 2010, **203**(2), 105e1–105e13.
10. S. M. van der Kooij *et al.*, The effect of treatment preference and treatment allocation on patients' health-related quality of life in the randomized EMMY trial, *Eur. J. Obstet. Gynecol. Reprod. Biol.*, 2013, **169**(1), 69–74.
11. S. C. Goodwin and J. B. Spies, Uterine Fibroid Embolization, *N. Engl. J. Med.*, 2009, **361**(7), 690–697.
12. S. C. Goodwin *et al.*, Uterine artery embolization for treatment of leiomyomata – Long-term outcomes from the FIBROID registry, *Obstet. Gynecol*, 2008, **111**(1), 22–33.
13. S. C. Goodwin and G. C. Wong, Uterine artery embolization for uterine fibroids: a radiologist's perspective, *Clin. Obstet. Gynecol.*, 2001, **44**(2), 412–424.
14. G. P. Siskin *et al.*, Embolic agents used for uterine fibroid embolization, *Am. J. Roentgenol.*, 2000, **175**(3), 767–773.
15. K. V. Sharma *et al.*, Development of "Imageable" Beads for Transcatheter Embolotherapy, *J. Vasc. Interv. Radiol.*, 2010, **21**(6), 865–876.
16. M. R. Dreher *et al.*, Radiopaque Drug-Eluting Beads for Transcatheter Embolotherapy: Experimental Study of Drug Penetration and Coverage in Swine, *J. Vasc. Interv. Radiol.*, 2012, **23**(2), 257–264.
17. X. J. Lu *et al.*, Research of novel biocompatible radiopaque microcapsules for arterial embolization, *Int. J. Pharm.*, 2013, **452**(1–2), 211–219.
18. S. Kehoe *et al.*, Predicting the thermal responses and radiopacity of multicomponent zinc-silicate bioglasses: A focus on ZnO, La2O3, SiO2 and TiO2, *J. Non-Cryst. Solids*, 2012, **358**(23), 3388–3395.
19. M. S. Hasan, S. Kehoe and D. Boyd, Temporal analysis of dissolution by-products and genotoxic potential of spherical zinc-silicate

bioglass: "imageable beads" for transarterial embolization, *J. Biomater. Appl.*, 2014, **29**(4), 566–581.
20. S. Kehoe *et al.*, "Imageable" Zinc-Silicate Glass Microspheres For Transarterial Embolization: A Renal Artery Embolization Study, *Biomed. Glasses*, 2015, **1**(1), 70–79.
21. S. Kehoe *et al.*, Mixture designs to assess composition-structure-property relationships in SiO(2)-CaO-ZnO-La(2)O(3)-TiO(2)-MgO-SrO-Na(2)O glasses: potential materials for embolization, *J. Biomater. Appl.*, 2013, **28**(3), 416–433.
22. J. B. Spies, Uterine artery embolization for fibroids: Understanding the technical causes of failure, *J. Vasc. Interv. Radiol.*, 2003, **14**(1), 11–14.
23. J. B. Spies, What evidence should we demand before accepting a new embolic material for uterine artery embolization?, *J. Vasc. Interv. Radiol.*, 2009, **20**(5), 567–570.
24. R. K. Ryu, Uterine artery embolization: Current implications of embolic agent choice, *J. Vasc. Interv. Radiol.*, 2005, **16**(11), 1419–1422.
25. S. Kehoe *et al.*, Novel radiopaque embolic agent for uterine fibroid embolization: determination of radiopacity and biological evaluation; cytocompatibility, intracutaneous reactivity, and lecal effects after implantation, *J. Vasc. Interv. Radiol.*, 2011, **23**(3), S33.
26. R. J. Abraham *et al.*, Magnetic resonance imaging characteristics of imageable embolic microspheres, *J. Vasc. Interv. Radiol.*, 2015, **26**(2), S79–S80.
27. S. Kehoe *et al.*, Effects of gamma-irradiation and accelerated aging on composition-structure-property relationships for radiopaque embolic microspheres, *J. Non-Cryst. Solids*, 2014, **402**, 84–90.
28. S. Stampfl *et al.*, Arterial distribution characteristics of Embozene particles and comparison with other spherical embolic agents in the porcine acute embolization model, *J. Vasc. Interv. Radiol.*, 2009, **20**(12), 1597–607.
29. S. Stampfl *et al.*, Inflammation and recanalization of four different spherical embolization agents in the porcine kidney model, *J. Vasc. Interv. Radiol.*, 2008, **19**(4), 577–586.
30. G. P. Siskin *et al.*, Pathologic evaluation of a spherical polyvinyl alcohol embolic agent in a porcine renal model, *J. Vasc. Interv. Radiol.*, 2003, **14**(1), 89–98.
31. S. Stampfl *et al.*, Biocompatibility and recanalization characteristics of hydrogel microspheres with polyzene-F as polymer coating, *Cardiovasc. Intervent. Radiol.*, 2008, **31**(4), 799–806.
32. I. M. Germano *et al.*, Histopathological follow-up study of 66 cerebral arteriovenous malformations after therapeutic embolization with polyvinyl alcohol, *J. Neurosurg.*, 1992, **76**(4), 607–614.
33. G. S. Davidson and K. G. Terbrugge, Histologic long-term follow-up after embolization with polyvinyl alcohol particles, *AJNR Am. J. Neuroradiol.*, 1995, **16**(4 suppl), 843–846.

34. H. Arakawa *et al.*, Endovascular embolization of the swine rete mirabile with Eudragit-E 100 polymer, *AJNR Am. J. Neuroradiol.*, 2007, **28**(6), 1191–1196.
35. J. Kritzinger *et al.*, Hepatic embolotherapy in interventional oncology: Technology, techniques, and applications, *Clin. Radiol.*, 2013, **68**(1), 1–15.
36. H. B. El-Serag, CURRENT CONCEPTS Hepatocellular Carcinoma, *N. Engl. J. Med.*, 2011, **365**(12), 1118–1127.
37. J. Ferlay *et al.*, Cancer incidence and mortality worldwide: sources, methods and major patterns in GLOBOCAN 2012, *Int. J. Cancer*, 2015, **136**(5), E359–E386.
38. R. Salgia and A. G. Singal, Hepatocellular Carcinoma and Other Liver Lesions, *Med. Clin. North Am.*, 2014, **98**(1), 103–118.
39. J. M. Llovet, A. Burroughs and J. Bruix, Hepatocellular carcinoma, *Lancet*, 2003, **362**(9399), 1907–1917.
40. S. F. Altekruse, K. A. McGlynn and M. E. Reichman, Hepatocellular Carcinoma Incidence, Mortality, and Survival Trends in the United States From 1975 to 2005, *J. Clin. Oncol.*, 2009, **27**(9), 1485–1491.
41. R. Siegel, C. DeSantis and A. Jemal, Colorectal cancer statistics, 2014, *Ca-Cancer J. Clin.*, 2014, **64**(2), 104–117.
42. M. F. Mulcahy *et al.*, Radioembolization of Colorectal Hepatic Metastases Using Yttrium-90 Microspheres, *Cancer*, 2009, **115**(9), 1849–1858.
43. A. R. Sasson and E. R. Sigurdson, Surgical treatment of liver metastases, *Semin. Oncol.*, 2002, **29**(2), 107–118.
44. D. J. Bentrem, R. P. DeMatteo and L. H. Blumgart, Surgical therapy for metastatic disease to the liver, *Annu. Rev. Med.*, 2005, **56**, 139–156.
45. R. Salem and K. G. Thurston, Radioembolization with (90)Yttrium microspheres: A state-of-the-art brachytherapy treatment for primary and secondary liver malignancies part 1: Technical and methodologic considerations, *J. Vasc. Interv. Radiol.*, 2006, **17**(8), 1251–1278.
46. K. T. Sato *et al.*, The Role of Tumor Vascularity in Predicting Survival after Yttrium-90 Radio embolization for Liver Metastases, *J. Vasc. Interv. Radiol.*, 2009, **20**(12), 1564–1569.
47. R. Salem and K. G. Thurston, Radioembolization with (90)Yttrium microspheres: A state-of-the-art brachytherapy treatment for primary and secondary liver malignancies - Part 2: Special topics, *J. Vasc. Interv. Radiol.*, 2006, **17**(9), 1425–1439.
48. A. S. Kennedy *et al.*, Pathologic response and microdosimetry of Y-90 microspheres in man: Review of four explanted whole livers, *Int. J. Radiat. Oncol., Biol., Phys.*, 2004, **60**(5), 1552–1563.
49. D. M. Liu *et al.*, Angiographic considerations in patients undergoing liver-directed therapy, *J. Vasc. Interv. Radiol.*, 2005, **16**(7), 911–935.
50. R. Salem *et al.*, Radioembolization for Hepatocellular Carcinoma Using Yttrium-90 Microspheres: A Comprehensive Report of Long-term Outcomes, *Gastroenterology*, 2010, **138**(1), 52–64.

51. V. Mazzaferro *et al.*, Yttrium-90 Radioembolization for Intermediate-Advanced Hepatocellular Carcinoma: A Phase 2 Study, *Hepatology*, 2013, **57**(5), 1826–1837.
52. B. Sangro *et al.*, Survival After Yttrium-90 Resin Microsphere Radioembolization of Hepatocellular Carcinoma Across Barcelona Clinic Liver Cancer Stages: A European Evaluation, *Hepatology*, 2011, **54**(3), 868–878.
53. R. A. Sharma *et al.*, Radioembolization of liver metastases from colorectal cancer using yttrium-90 microspheres with concomitant systemic oxaliplatin, fluorouracil, and leucovorin chemotherapy, *J. Clin. Oncol.*, 2007, **25**(9), 1099–1106.
54. M. Cosimelli *et al.*, Multi-centre phase II clinical trial of yttrium-90 resin microspheres alone in unresectable, chemotherapy refractory colorectal liver metastases, *Br. J. Cancer*, 2010, **103**(3), 324–331.
55. A. Hendlisz *et al.*, Phase III Trial Comparing Protracted Intravenous Fluorouracil Infusion Alone or With Yttrium-90 Resin Microspheres Radioembolization for Liver-Limited Metastatic Colorectal Cancer Refractory to Standard Chemotherapy, *J. Clin. Oncol.*, 2010, **28**(23), 3687–3694.
56. A. Saxena *et al.*, A systematic review on the safety and efficacy of yttrium-90 radioembolization for unresectable, chemorefractory colorectal cancer liver metastases, *J. Cancer Res. Clin. Oncol.*, 2014, **140**(4), 537–547.
57. L. Kulik *et al.*, Yttrium-90 microspheres (TheraSphere (R)) treatment of unresectable hepatocellular carcinoma: Downstaging to resection, RFA and bridge to transplantation, *Hepatology*, 2006, **44**(4), 396a.
58. L. M. Kulik *et al.*, Yttrium-90 microspheres (Therasphere (R)) treatment of unresectable hepatocellular carcinoma: Downstaging and bridge to transplantation, *Hepatology*, 2005, **42**(4), 389a.
59. M. Inarrairaegui *et al.*, Response to radioembolization with yttrium-90 resin microspheres may allow surgical treatment with curative intent and prolonged survival in previously unresectable hepatocellular carcinoma, *Eur. J. Surg. Oncol.*, 2012, **38**(7), 594–601.
60. M. Vouche *et al.*, Radiation lobectomy: Time-dependent analysis of future liver remnant volume in unresectable liver cancer as a bridge to resection, *J. Hepatol.*, 2013, **59**(5), 1029–1036.
61. L. L. Hench *et al.*, Glass and Medicine, *Int. J. Appl. Glass Sci.*, 2010, **1**(1), 104–117.
62. A. Kennedy, Radioembolization of hepatic tumors, *J. Gastrointest. Oncol.*, 2014, **5**(3), 178–189.
63. R. J. Lewandowski and R. Salem, Yttrium-90 radioembolization of hepatocellular carcinoma and metastatic disease to the liver, *Semin. Intervent. Radiol.*, 2006, **23**(1), 64–72.
64. M. J. Hyatt and D. E. Day, Glass Properties in the Yttria-Alumina-Silica System, *J. Am. Ceram. Soc.*, 1987, **70**(10), C283–C287.

65. E. M. Erbe and D. E. Day, Chemical durability of Y2O3-Al2O3-SiO2 glasses for the in vivo delivery of beta radiation, *J. Biomed. Mater. Res.*, 1993, **27**(10), 1301–1308.
66. G. J. Ehrhardt and D. E. Day, Therapeutic Use of Y-90 Microspheres, *Nucl. Med. Biol.*, 1987, **14**(3), 233–242.
67. D. E. Day and G. J. Ehrhardt, *Glass Microspheres*, 1988, USA.
68. M. R. Ghahramani, A. A. Garibov and T. N. Agayev, Production and quality control of radioactive yttrium microspheres for medical applications, *Appl. Radiat. Isot.*, 2014, **85**, 87–91.
69. A. Kennedy *et al.*, Recommendations for radioembolization of hepatic malignancies using yttrium-90 microsphere brachytherapy: a consensus panel report from the radioembolization brachytherapy oncology consortium, *Int. J. Radiat. Oncol., Biol., Phys.*, 2007, **68**(1), 13–23.
70. B. Sangro *et al.*, Radioembolization for hepatocellular carcinoma: a review of the evidence and treatment recommendations, *Am. J. Clin. Oncol.*, 2011, **34**(4), 422–431.
71. J. Edeline *et al.*, Yttrium-90 microsphere radioembolization for hepatocellular carcinoma, *Liver Cancer*, 2015, **4**(1), 16–25.
72. J. F. Dufour *et al.*, Intermediate hepatocellular carcinoma: current treatments and future perspectives, *Ann. Oncol.*, 2013, **24**(suppl 2), ii24–ii29.
73. J. Lammer *et al.*, Prospective randomized study of doxorubicin-eluting-bead embolization in the treatment of hepatocellular carcinoma: results of the PRECISION V study, *Cardiovasc. Intervent. Radiol.*, 2010, **33**(1), 41–52.
74. K. Huang *et al.*, Doxorubicin-eluting Bead versus Conventional Transarterial Chemoembolization for the Treatment of HCC: a Meta-Analysis, *J. Gastroenterol. Hepatol.*, 2013, **29**, 920–925.
75. P. Huppert, T. Wenzel and H. Wietholtz, Transcatheter arterial chemoembolization (TACE) of colorectal cancer liver metastases by irinotecan-eluting microspheres in a salvage patient population, *Cardiovasc. Intervent. Radiol.*, 2014, **37**(1), 154–164.
76. K. H. Lee *et al.*, Doxorubicin-loaded QuadraSphere microspheres: plasma pharmacokinetics and intratumoral drug concentration in an animal model of liver cancer, *Cardiovasc. Intervent. Radiol.*, 2010, **33**(3), 576–582.
77. A. L. Lewis *et al.*, DC bead: In vitro characterization of a drug-delivery device for transarterial chemoembolization, *J. Vasc. Interv. Radiol.*, 2006, **17**(2), 335–342.
78. INSTRUCTIONS FOR USE: DC Bead Drug Delivery Embolisation System. 2012 [cited 2012 November 22]; Available from: http://assets.biocompatibles.com/products/uploads/Files/instructions/dcbead/DC-Bead-IFU-English.pdf.
79. HepaSphere™ Microspheres (With Doxorubicin) IFU- Canada (English/French). 2011 [cited 2015 May 5]; Available from: http://www.merit.com/products/media.aspx?type=ifu&id=327961.

80. K. Kato and E. Takayama, Das Entwasserungsverhalten des Natriummetavanadatdihydrats und die Kristallstruktur des [beta]-Natriummetavanadats, *Acta Crystallogr., Sect. B: Struct. Sci., Cryst. Eng. Mater.*, 1984, **40**(2), 102–105.
81. A. L. Lewis *et al.*, Doxorubicin eluting beads - 1: effects of drug loading on bead characteristics and drug distribution, *J. Mater. Sci.: Mater. Med.*, 2007, **18**(9), 1691–1699.
82. J. B. Spies *et al.*, Spherical polyvinyl alcohol versus tris-acryl gelatin microspheres for uterine artery embolization for leiomyomas: Results of a limited randomized comparative study, *J. Vasc. Interv. Radiol.*, 2005, **16**(11), 1431–1437.
83. E. S. Place, N. D. Evans and M. M. Stevens, Complexity in biomaterials for tissue engineering, *Nat. Mater.*, 2009, **8**(6), 457–470.
84. V. Mourino, J. P. Cattalini and A. R. Boccaccini, Metallic ions as therapeutic agents in tissue engineering scaffolds: an overview of their biological applications and strategies for new developments, *J. R. Soc., Interface*, 2012, **9**(68), 401–419.
85. S. Bose *et al.*, Understanding of dopant-induced osteogenesis and angiogenesis in calcium phosphate ceramics, *Trends Biotechnol.*, 2013, **31**(10), 594–605.
86. A. Hoppe, N. S. Guldal and A. R. Boccaccini, A review of the biological response to ionic dissolution products from bioactive glasses and glass-ceramics, *Biomaterials*, 2011, **32**(11), 2757–2774.
87. N. J. Lakhkar *et al.*, Bone formation controlled by biologically relevant inorganic ions: role and controlled delivery from phosphate-based glasses, *Adv. Drug. Deliv. Rev.*, 2013, **65**(4), 405–420.
88. C. Wu *et al.*, Copper-containing mesoporous bioactive glass scaffolds with multifunctional properties of angiogenesis capacity, osteostimulation and antibacterial activity, *Biomaterials*, 2013, **34**(2), 422–433.
89. I. H. Lee *et al.*, Development, characterisation and biocompatibility testing of a cobalt-containing titanium phosphate-based glass for engineering of vascularized hard tissues, *Mater. Sci. Eng., C*, 2013, **33**(4), 2104–2112.
90. P. Koepf-Maier, Complexes of metals other than platinum as antitumour agents, *Eur. J. Clin. Pharmacol.*, 1994, **47**(1), 1–16.
91. B. Rosenber *et al.*, Platinum Compounds - A New Class Of Potent Antitumour Agents, *Nature*, 1969, **222**(5191), 385.
92. S. M. Cohen and S. J. Lippard, Cisplatin: From DNA damage to cancer chemotherapy, *Prog. Nucleic Acid Res. Mol. Biol.*, 2001, **67**, 93–130.
93. A. Stein and D. Arnold, Oxaliplatin: a review of approved uses, *Expert Opin. Pharmacother.*, 2012, **13**(1), 125–137.
94. B. Desoize, Metals and metal compounds in cancer treatment, *Anticancer Res.*, 2004, **24**(3a), 1529–1544.
95. R. Huang, A. Wallqvist and D. G. Covell, Anticancer metal compounds in NCI's tumor-screening database: putative mode of action, *Biochem. Pharmacol.*, 2005, **69**(7), 1009–1039.

96. B. Desoize, Cancer and metals and metal compounds: part II–cancer treatment, *CRC Cr. Rev. Oncol.-Hem.*, 2002, **42**(3), 213–215.
97. M. Frezza *et al.*, Novel Metals and Metal Complexes as Platforms for Cancer Therapy, *Curr. Pharm. Des.*, 2010, **16**(16), 1813–1825.
98. K. H. Thompson and C. Orvig, Boon and bane of metal ions in medicine, *Science*, 2003, **300**(5621), 936–939.
99. B. T. Huang *et al.*, The early addition of arsenic trioxide versus high-dose arabinoside is more effective and safe as consolidation chemotherapy for risk-tailored patients with acute promyelocytic leukemia: multicenter experience, *Med. Oncol.*, 2012, **29**(3), 2088–2094.
100. P. Sharma *et al.*, Perspectives of antimony compounds in oncology, *Acta Pharmacol. Sin.*, 2008, **29**(8), 881–890.
101. K. B. Garbutcheon-Singh *et al.*, Transition Metal Based Anticancer Drugs, *Curr. Top. Med. Chem.*, 2011, **11**(5), 521–542.
102. N. Kilcup, Zinc-Silicate Glasses for Transarterial Chemoembolization of Hepatocellular Carcinoma, in *School of Biomedical Engineering*, Dalhousie University, Nova Scotia, Canada, 2015.
103. A. M. Evangelou, Vanadium in cancer treatment, *Crit. Rev. Oncol. Hemat.*, 2002, **42**(3), 249–265.
104. R. S. Ray *et al.*, Vanadium mediated apoptosis and cell cycle arrest in MCF7 cell line, *Chem.-Biol. Interact.*, 2006, **163**(3), 239–247.
105. T.-T. Liu *et al.*, Reactive-oxygen-species-mediated Cdc25C degradation results in differential antiproliferative activities of vanadate, tungstate, and molybdate in the PC-3 human prostate cancer cell line, *JBIC, J. Biol. Inorg. Chem.*, 2012, **17**(2), 311–320.

CHAPTER 21

Sol–Gel-Derived Glasses for Bone Tissue Engineering

WILLIAM C. LEPRY AND SHOWAN N. NAZHAT*

Department of Mining and Materials Engineering, McGill University, Montreal, QC, Canada
*Email: showan.nazhat@mcgill.ca

21.1 Introduction

Bone is a complex, hierarchically organized tissue made of both inorganic and organic phases. It is the main component of the skeletal system and its health relies on various elements, of which many come from the diet. Bone degeneration, fractures and osteoporosis affect millions of people worldwide and since bone is a living tissue with vasculature, it can often heal itself, but in the case of severe injuries and diseases that lead to large defects, intervention with bone grafts, or "scaffolds", is needed. Indeed, in the USA alone, 500 000 bone graft procedures are performed annually and worldwide it is estimated that 2.2 million bone graft procedures are performed each year.[1]

After blood, bone is the most widely transplanted tissue in the human body.[2] The clinical 'gold standards' for repairing bone tissue are traditionally donor-based, which can be autografts from the patient (preferred), allografts from human donors, or xenografts from animal sources.[1,3,4] These grafts are advantageous because the material matches the complex hierarchical tissue structure of native bone. Although there is little chance of rejection with autografts, donor site morbidity and limited supply present some issues of concern. On the other hand, the factors that impact allografts and xenografts include limited donor supply, particularly in the former, as well as the

RSC Smart Materials No. 23
Bioactive Glasses: Fundamentals, Technology and Applications
Edited by Aldo R. Boccaccini, Delia S. Brauer and Leena Hupa

Published by the Royal Society of Chemistry, www.rsc.org

potential for disease transmission and rejection in both cases, which causes these grafts to undergo thorough processing before implantation.[4–6] These disadvantages have created a great push to develop tissue-engineered grafts and are some of the main drivers for the recent growth of the bone tissue engineering field.

Many different engineered materials have been developed in order to help repair bone, such as metals, ceramics, glasses, polymers and their combinations. Although many of these materials have had success, there are still many opportunities to improve the rate of bone healing by using new compositions and forms. Bioactive glasses have emerged as a leader in their field attributable to their ability to release ions which allows them to both chemically and physically bond to native bone.[7,8]

As discussed in Chapter 5, the sol–gel process is a robust, solution-based process that is the basis for creating many nanostructured and nanoporous forms. The process relies on hydrolysis of precursors, typically metal alkoxides, followed by condensation reactions to form the initial sol, which is a stable dispersion of colloidal particles, or polymers, in a solvent. Polycondensation reactions continue to form over time, creating a three-dimensionally connected network known as a gel, which is essentially the glass network. This gel can then be dried either in the atmosphere creating a xerogel, in which some of the pore structure collapses, or through critical point drying using supercritical CO_2 where the fine porosity and surface area of the gel is preserved.

The sol–gel method offers advantages when compared to traditional high temperature glass processing as it can be carried out at room temperature making it easier to vary the composition, although a final calcination step is still needed to obtain the glass. Furthermore, since the raw materials are typically in aqueous form, uniform mixing and homogeneity are easier to achieve. Moreover, near end-shape forming is achievable with sol–gel-produced materials.[9] It is also possible to add and encapsulate various elements such as drugs, or incorporate inorganic ions that can further assist in bone healing, making sol–gel-derived glasses suitable for bone tissue repair. This chapter will give a brief overview of bone tissue and discuss the various types of sol–gel-derived glasses used to help repair that tissue.

21.2 Bone

In an adult human, the skeletal system is composed of 206 bones of various shapes and sizes which can generally be classified into three main categories: long (*e.g.* femur or tibia), short (*e.g.* vertebra) and flat (*e.g.* skull).[10,11] The function, composition, and size of bones can vary greatly and, working together, they provide structural support, transmit mechanical forces, protect vital organs, and store minerals within the body.[11,12] An overview of the structure of bone can be seen in Figure 21.1. A single bone is composed of two components: cortical and cancellous bone. Cortical bone, also called compact bone, is located on the outside of all bones and possesses a dense

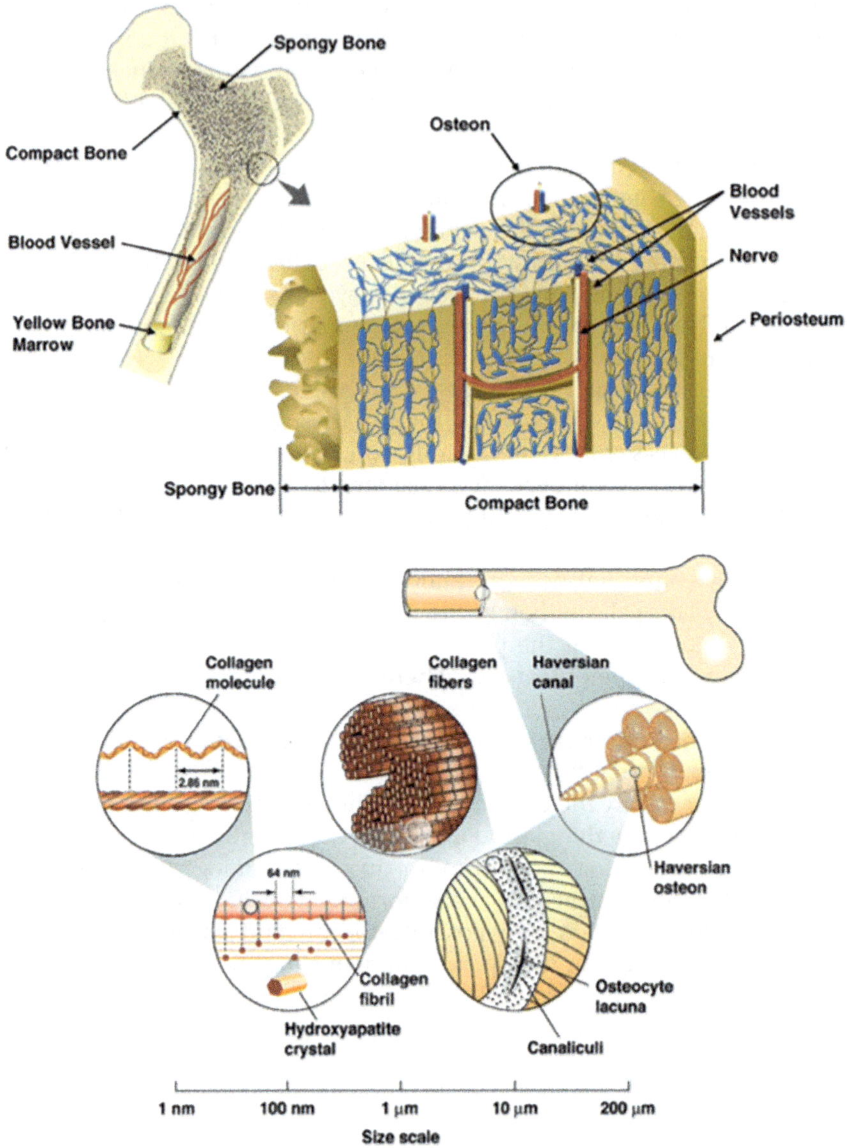

Figure 21.1 The hierarchical structural organization of bone including cortical ('compact') and cancellous ('spongy') bone, collagen fibre assemblies of collagen fibrils, and bone mineral crystals.
(Reproduced with permission from R. K. Nalla *et al.*, Role of microstructure in the aging-related deterioration of the toughness of human cortical bone, *Mater. Sci. Eng., C*, **26**, 1251–1260, Copyright 2006 with permission from Elsevier).

structure with low porosity making up approximately 80% of the total bone mass. Alternatively, internal bone structure is highly porous and called cancellous bone but is also known as trabecular or spongy bone.[10,13]

Cancellous bone accounts for about 20% of the total bone mass in an adult skeleton and possesses significantly more surface area.[14–18] The trabeculae that make up cancellous bone appear sponge-like, and consist of the basic cellular structures: rod–rod, rod–plate, or plate–plate with a typical trabecular rod being about 50–300 μm in diameter.[19]

Approximately 65 wt% of bone is an inorganic, mineral phase called hydroxycarbonate apatite (HCA, $(Ca_5(PO_4,CO_3)_3(OH))$.[20,21] The high inorganic content makes bone one of the most mineralized tissues in the body along with enamel and dentine.[22,23] The remaining 35 wt% of bone is composed of the organic phase which mainly consists of type I collagen (>90 wt%) along with minor quantities of non-collagenous proteins, proteoglycans, and water.[11,20,24] These highly organized phases play a critical role in the properties of bone as the mineral phase provides compressive strength and stiffness while the organic phase provides toughness and ductility.[12,20]

At the cellular level, bone is made up of osteoblasts (bone-forming cells) osteocytes (mature bone cells), and osteoclasts (bone resorbing cells). Bone is constantly being remodelled through the functions of osteoblasts and osteoclasts,[11,12,19] which can be controlled through the action of chemical factors.[25–27] Osteoblasts can be activated by increased secretion of the osteoid, the unmineralized portion of organic bone extracellular matrix (ECM), which is stimulated by the emission of growth hormones by various glands.[28] There are also many bone matrix non-collagenous proteins, such as osteonectin and bone sialoprotein, that play a significant role in bone mineralization and have broad functions that include control of cell proliferation and differentiation, cell–matrix interactions, and mediation of HCA deposition.[26,27] An imbalance of cellular action can lead to osteoporosis, a skeletal disorder characterized by low bone density and deterioration of the microarchitecture of bony tissue that results in an increase in fracture risk and affects millions of people worldwide.[29–31] Bone health relies on many different factors and since bone is a highly mineralized tissue, inorganic elements from the diet or ones released from synthetic bone scaffold materials, like glass, play a crucial role in its health and will be discussed below.

21.2.1 Calcium

Calcium (Ca) is synonymous with bone health. Over 99% of all calcium in the body is found in bones in the form of HCA making it a crucial structural component.[32] This element also mediates many effects at cellular and tissue levels, such as acting as a signal transduction molecule for producing cellular responses which, if not maintained, can cause bone disease.[33] Hypercalcemia occurs when osteoclast activity increases causing the release of ECM-stored calcium and leading to a decrease in bone density.[34] The opposite response can occur when bone production exceeds bone resorption, requiring additional calcium to form new bone mineral which is supplied by the free calcium, thus decreasing HCA formation.[33] The amount of calcium and its effect on skeletal production is thought to have a threshold zone, meaning that if calcium levels are below the threshold zone then increasing its levels will increase bone production but if calcium levels are already above the threshold zone then increasing Ca will likely not have an effect on bone production (for adults the suggested dietary intake is about 1500 mg day^{-1}).[32,33] For these reasons, calcium is often a major component in bioactive glasses and is one of the most important elements in bone health that has been extensively studied and reviewed.[35–39]

21.2.2 Phosphorous

Most of the phosphorous in the body is in the form of phosphates, which are generally classified into two categories: orthophosphates (PO_4^{3-}) and polyphosphates which can take many forms. Inorganic phosphate (P_i), also known as orthophosphate, is the second most abundant mineral element in the human body and plays an essential role in many bodily processes including bone mineralization.[33,40,41] In fact, most of the phosphates in the body (≈85%) are present in bone and teeth as orthophosphate[32,33] and constitute an important regulator of skeletal development, mineral metabolism, and intracellular signaling.[40,42] During osteoblastic differentiation, phosphate provides a cell signaling function[43,44] and can also control mineralization processes by moderating gene expression.[45] Orthophosphate couples with calcium ions to form HCA in bone and other mineralized tissues such as teeth and enamel, as well as extracellular fluid and soft tissues.[40,41] Contrary to singular orthophosphates, polyphosphates, and in particular pyrophosphate ($P_2O_7^{4-}$), are well known to inhibit mineralization, and thus bone growth.[33]

21.2.3 Silica

Although silicon (Si) does not occur within the body, it has been shown to play an important role in bone health.[46–48] Dietary silica can come from many sources including plant-based foods, beer, and some mineral water which is broken down by the gastrointestinal tract in the form of orthosilicic acid.[48] Since silica (SiO_2) is the main network former of most glasses, Si release is very important when understanding mineralization. Si is essential to skeletal development[49–51] by promoting matrix synthesis and the differentiation of osteoblastic cells,[52,53] where its ionic release from glass has been shown to stimulate osteoblast proliferation and differentiation *in vitro*.[54] Advanced mapping techniques have recently shown that Si is located in only the mineralized tissue areas and thus may play a biological role in bone formation due to its distribution near these mineralized sites.[55]

21.3 Bioactive Glasses

Bioactive glasses are a category of glasses that bond to living tissues by forming a layer of HCA on their surface which allows them to both chemically and physically bond to native bone tissue through well-defined mechanisms.[7,8,56,57] A glass is considered to be bioactive if a layer of HCA forms on its surface after immersion in a physiological environment. Typically, *in vitro*, this environment is simulated body fluid (SBF) which is a solution of ions close to that of human blood plasma. This method of *in vitro* testing was developed by Kokubo and colleagues[58] and has since been considered the standard for testing the mineral-forming ability of bioactive glasses.[59] However, due to mismatches between *in vitro* and *in vivo* mineralization results, the validity of this technique has been brought into question[60] and there is a push to create a uniform testing technique to better compare HCA formation between different materials.[61] Despite this, SBF is still frequently used to test the bioactivity of glasses. However, this definition is slowly changing as these glasses are beginning to be considered for applications outside the skeletal tissues and for soft tissue repair.[62]

Bioactive glasses are traditionally made by the melt-quench method, which involves the mixing of powdered precursor materials, such as silica, sodium oxide, and calcium oxide. The mixture is then heated until melting at high temperatures (>1000 °C depending on the glass) followed by the rapid cooling ("quenching") to maintain the amorphous form.[63] Once cooled, the glasses are typically ground to the desired particle size. However, these particles typically have low surface areas and porosities. Furthermore, it is often difficult to incorporate metal ions into these glasses since it further increases the melting temperature.

The chemical (composition) and physical (textural) properties of glasses are the two main determinants of bioactivity in a glass. The composition of the glass will determine the network connectivity (N_C), which is a measure of the bridging oxygens per network former (typically Si). N_C has been calculated to predict the bioactivity of glasses[64,65] where an increase in the amount of bridging oxygens has been correlated with more chemically durable and less bioactive glasses. On the other hand, the textural properties of a glass also influence its bioactivity. It is known that an increase in porosity and surface area will improve the bioactivity of glasses since these allow for greater interaction with the physiological environment prompting quicker dissolution rates and thus HCA conversion. Using the sol–gel process allows for the optimization of both these parameters and the tailoring of glass bioactivity which helps overcome some of the drawbacks of typical melt-derived glasses.

21.3.1 Sol–Gel-Derived Bioactive Glasses

As discussed in Chapter 5, the sol–gel method is a flexible, solution-based approach to making glass. After hydrolysis, the precursors undergo condensation reactions to form the initial sol which is then followed by additional polycondensation reactions which forms the gel (*i.e.* the glass network). While melt-derived glasses require a SiO_2 content of ≤60 mol% in order to be bioactive, HCA formation is possible up to 90 mol% SiO_2 content when produced using the sol–gel process.[66] This is due to the excess –OH groups that are incorporated into the glass network during processing, which lowers N_C to a greater extent.[67] Furthermore, inherently, sol–gel-processed glasses have nanoscale porosity due to the elimination of the organic precursor species during drying. This greatly increases their specific surface area allowing for their rapid degradation and HCA conversion when placed in physiological environments.[67]

The first sol–gel-derived bioactive glasses were created in the early 1990's based on the ternary system $[(100-x)SiO_2-(x-4)CaO-(4)P_2O_5$, $x=40$, 30, 20 (mol%)].[66,68] Fourier transform infrared (FTIR) spectroscopy of these glasses post-immersion in Tris buffer solution and SBF, identified HCA-associated phosphate peak formation. Attributable to the enhanced textural properties of this sol–gel-derived composition, these glasses more rapidly converted to bone-like mineral when compared to melt-derived glasses.[69] One of the compositions from the previous study became a 'standard' sol–gel composition "58S" $[(60)SiO_2-(36)CaO-(4)P_2O_5$, (mol%)] and has since been well studied.[69,70]

Christodoulou *et al.* showed that the ionic release products from a sol–gel-derived 58S glass exposed to primary human osteoblasts promoted the upregulation and gene expression of common osteogenic markers such as alkaline phosphatase, osteocalcin, and osteopontin. However, the results were not statistically significant when compared to non-58S-treated well plates.[71] This study also identified that a narrow range of 58S glass dissolution products (Si >50 ppm) were able to enhance the occurrence of mature, terminally-differentiated osteoblast with prolonged periods in culture.

Another commonly studied composition is "70S30C" $[(70)SiO_2-(30)CaO$, (mol%)]. Although this composition does not contain phosphorous, it has been shown to be bioactive.[72] Radev and coworkers showed this composition to be bioactive by the formation of phosphate peaks according to FTIR after soaking in 1.5× SBF.[73] Lin and coworkers were able to tailor the nanoporosity of this composition by adding specific amounts of trimethylethoxysilane (TMES) during the sol–gel process.[74] Further work studied the effect of protein interactions with these glasses and found that modal pore diameters 6 nm were accessible to fibrinogen while <2 nm diameters inhibited the penetration.[75] Along with the addition of TMES, the glass surface could also be controlled by its composition, gelling pH, and final calcination temperature, all of which can affect its bioactivity. Zhang *et al.* found that sol–gel-derived 70S glass particles could be encapsulated in alginate to help with murine embryonic stem cells' expansion and differentiation using a bioreactor system.[72] This therapy showed promising results for bone transplantation where the glass particles could improve the bone bonding rate by rapidly forming HCA within the implanted matrix.

One of the main advantages of the sol–gel process is that it negates the use of sodium (Na), which is often added as flux to help lower the melting temperature in melt-derived systems. However, this is not needed in sol–gel processing since the glass network is essentially formed at room temperature.[9] Notwithstanding this, Na-containing sol–gel-derived glasses have been made which are thought to help increase the dissolution rate of glasses. The 'gold standard' bioactive glass composition Bioglass® ["45S5"; (46.1)SiO_2–(26.9)CaO–(24.4)Na_2O–(2.6)P_2O_5, (mol%)], is a melt-derived composition that was invented by Larry Hench in 1969 and has since been extensively well studied and used for many clinical applications.[9,56,57,76] To improve the textural properties of this glass, the sol–gel process has been implemented. For example, Pirayesh *et al.* created 45S5 glass-ceramic particles through this process[77] where a calcination temperature of 700 °C resulted in 80% crystallinity. However, upon immersion in SBF, an HCA layer was detected after 21 days, similar to that of the melt-derived glass. Also, the HCA layer for the sol–gel glass appeared to be finer than that of the melt-derived version.[77] Cacciotti *et al.* also created a gel-derived 45S5 glass and demonstrated that the calcination temperature and dwell time influenced the crystalline phase evolution and microstructure, thus influencing the dissolution rate which is important for bone repair.[78]

Faurea and researchers recently developed a new method to create a sol–gel-derived 45S5 glass by replacing the commonly-used nitric acid catalyst with citric acid.[79] These glasses showed increased Ca and P amounts using energy dispersive spectroscopy (EDXA) and scanning electron microscopy (SEM) within 4 h, thus demonstrating a significantly more rapid HCA formation compared with the melt-derived equivalent. Furthermore, the more mild processing condition of the citric acid route is attractive as it allows for the addition and encapsulation of biomolecules (*e.g.* drugs, growth factors, *etc.*) that can help accelerate bone growth.[79] Similarly, by using a multi-step, nitrate-free method, Rezabeigi *et al.* created a 45S5 sol–gel-derived glass with 10 times greater surface area than that of melt-derived glasses.[80]

While sol–gel-derived glasses show rapid conversion to HCA when compared to their melt-derived counterparts, they often have low strength (2–3 MPa),[81] and thus are mainly used only in low-load sites. Furthermore, they are often limited to particulate form after calcination since it is difficult to maintain a crack free monolith during drying. To overcome these limitations, the conversion of these glasses to three-dimensional (3D) scaffolds may tailor the mechanical and dissolution properties, and potentially widen their potential range of applications.

21.4 Scaffold Production

Three-dimensional scaffolds offer a number of advantages for bone tissue engineering applications as these can more closely match the morphology of native bone, acting as a template for new bone growth. The ideal scaffold should allow for cell migration, bone growth, and vascularization by possessing an interconnected macroporous network with pore diameters >100 μm.[82,83] It should also stimulate osteogenesis so that the scaffold can be grown *in vitro* then implanted into the patient, have mechanical properties similar to that of the native bone tissue and help with new tissue growth.[84] More importantly, post-implantation, the ideal engineered scaffold should be able to maintain its strength during degradation at the rate at which the new bone forms.[85,86] Typically, scaffolds produced using the sol–gel route are made either by creating foams or fibres.

21.4.1 Foam Scaffolds

Foam scaffolds are created by adding a surfactant during the sol–gel process and vigorously agitating the sol (usually using a high velocity blade) until it begins to bubble, followed by rapid gelation with the help of an acidic catalyst. This technique is not to be confused with scaffolds created using the polymer foam replication method, which uses a sacrificial polymer foam as the template for the scaffold creation.[87,88] Depending on the amount and type of surfactant, the hierarchal structure can be controlled with interconnected macropores generally ranging from 10 to 500 μm which allows for tissue ingrowth, and mesopores from 2 to 50 nm that allow for the release of ion products and increased bioactivity (Figure 21.2).[69,89,90] These 3D foams closely mimic human trabecular bone and, depending on the nature of interconnected pores, degradation rates can be controlled, thus enhancing cellular differentiation and proliferation.

Three-dimensional foam scaffolds were first created by Sepulveda *et al.* using the two common sol–gel compositions, 58S and 70S30C.[69,70] These foam scaffolds possessed high surface areas with pore structures similar to that of trabecular bone and as a consequence of their increased surface area and porosity, these glasses demonstrated significantly more rapid HCA conversion rates when compared to a melt-derived 45S5 glass.[69] Further research was performed by Jones *et al.* to optimize the processing parameters in order to control the hierarchical structure of the foams.[90] Using the 58S composition, it was found that the structure can be simply and consistently controlled by adjusting the amount of water added during processing. Other factors, such as calcination temperature, also affected the final structure. Furthermore, for large-scale commercial production, air injection was found to help create larger foam scaffolds.[91]

Since mechanical cues can also help dictate bone healing, Jones *et al.* found that increasing the sintering temperature of the foams increased the mechanical properties, allowing the creation of a foam with a similar compressive strength (2.26 MPa) to that of trabecular bone (2–12 MPa).[81] Even with this increase the interconnected pore diameters remained near 100 μm which is thought to be large enough to allow for cell migration.[81] It has also been shown that the pore sizes of 70S30C foam scaffolds under flowing SBF remained fairly constant while the interconnect pores decreased in size slightly as observed with micro-computed tomography (μCT).[92] However, the glasses became more crystallised with increasing calcination temperature, highlighting the need to balance increased mechanical strength and maintenance of the amorphous nature. It is generally recognised that

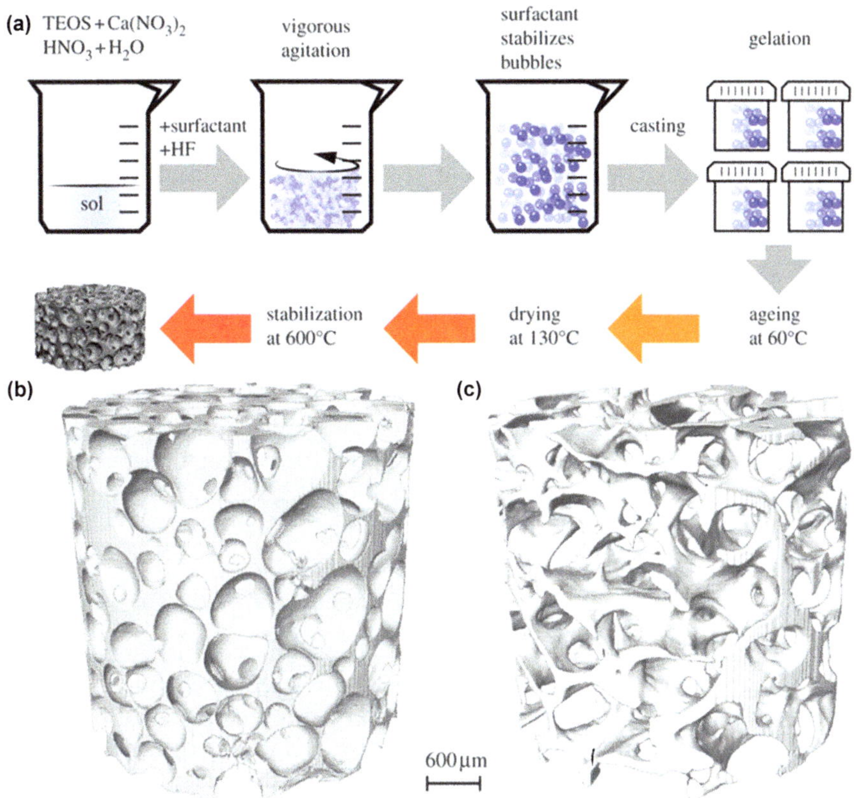

Figure 21.2 An overview of (a) the sol–gel foam processing route with μCT images of (b) the final foam structure and (c) natural trabecular bone found near human femur proximal to the knee.
(Adapted with permission from Martin *et al.*, Characterizing the hierarchical structures of bioactive sol–gel silicate glass and hybrid scaffolds for bone regeneration, *Philos. Trans. R. Soc., A*, 2012, **370**, 1422–1443 by permission of the Royal Society.)

amorphous structures are more bioactive than crystalline forms, attributable to a more even dissolution mechanism.[9]

Valerio *et al.* examined the response of seeded primary human osteoblasts to a 58S foam.[93] Compared to the cell cultures without foam samples, it was found that viable osteoblasts were observed within the scaffolds and their proliferation increased suggesting that these foams offered promise as a bone tissue material.[93] 45S5 foam scaffolds were created by Chen *et al.* and sintered at 1000 °C for 2 h producing a glass-ceramic with significantly increased strength.[94] It was also found that ion release was more rapid when compared to 3D scaffolds produced through the polymer foam replication technique. The scaffolds also supported the proliferation of seeded MG63 osteoblast-like cells, but migration into the foams was limited due to the mainly closed pore structure.[94]

Sol–gel-derived foam scaffolds have also been tested *in vivo*. Midha *et al.* implanted 70S30C foams in a rat tibia defect model using three different pre-treatment methods: dry, pre-wetted in serum-free α-MEM (minimum essential medium) for 5 min or preconditioned in α-MEM for 3 days.[95] The purpose of the preconditioning was to help alleviate the effects of the rapid pH increase during scaffold implantation due to the increased surface area. Bone ingrowth was found to be poor for the dry and wetted foams and only the preconditioned sample showed greater than 60% material–bone contact, similar to that of two commercially available bone substitutes: a melt-derived bioactive glass NovaBone® and a porous silicon-substituted hydroxyapatite Actifuse®. It was found that the preconditioned 70S30C scaffolds underwent degradation while being replaced with new bone, unlike the commercial products.[95] However, while these scaffolds show promise for bone repair, their lack of strength retention properties and brittleness limit their applications to non-weight-bearing applications. For this reason, the production of polymer-based composites or hybrid scaffolds is being considered.

21.4.2 Fibrous Scaffolds

A common route to produce fibrous scaffolds or "meshes" is through electrospinning which uses an electrical charge to draw very fine (typically on the micro- or nano-scale) fibres *via* a spinning jet from a liquid or sol.[96] When creating fibres from the sol–gel process, polymer binding agents are often required to produce the viscoelastic solution needed in order to prevent the spinning jet from breaking-up. The first electrospun silica-based bioactive glass nanofibres derived from the sol–gel process were made using polyvinylbutyral (PVB) as the organic binder.[97] Here the PVB was added to the sol–gel mixture [(70)SiO_2–(25)CaO–(5)P_2O_5; (mol%)] to obtain the optimum viscosity for electrospinning. These fibres possessed around 2–3 orders of magnitude higher surface areas when compared to melt-derived fibres and formed a layer of HCA in one day. Furthermore, the sol–gel fibres significantly enhanced alkaline phosphatase response with bone marrow-derived stem cells (BMSCs) when compared to the normal PCL fibre.[97] A similar method was also implemented using the 58S glass with collagen creating hybrid fibres that showed excellent bioactivity in SBF and the ability for MG63 cells to grow on their surface.[98] Lu *et al.* demonstrated that hybrid polyvinyl alcohol (PVA)–70S30C sub-micron fibres had increased mechanical hardness and elastic modulus, as tested by nanoindentation, making them good candidates for bone tissue engineering scaffolds.[99]

While polymer addition is often needed to correctly adjust the viscosity for the solution to be electrospun, it is not always required when working with the sol–gel process. Recently researchers have created 3D cotton-wool-like bioactive glasses for bone tissue repair.[100] The 70S30C sol was heated, causing evaporation, which increased the viscosity and created a solution suitable for electrospinning while the Ca^{2+} and NO_3^- ions made the solution

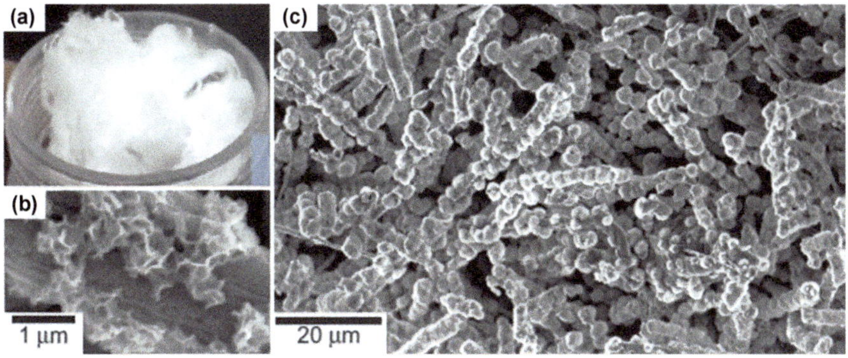

Figure 21.3 (a) Cotton-wool-like fibres (70S30C), (b) growth of HCA after 12 h in SBF, (c) growth of HCA after 72 h in SBF.
(Adapted from G. Poologasundarampillai *et al.*, Cotton-wool-like bioactive glasses for bone regeneration, *Acta Biomater.*, 2014, **10**(8), 3733–3746.)

conductive allowing for this process to occur. After immersion in SBF, apatite layer formation initiated at around 12 h, according to X-ray diffraction (XRD) (Figure 21.3). Cell compatibility studies using MC3T3-E1 pre-osteoblasts demonstrated good attachment and an MTT assay showed no toxicity. This type of flexible, 3D scaffold has good workability with the potential of filling complex defects to promote new bone growth.

21.5 Mesoporous Bioactive Glasses

Drug loading is an important concept for bone repair since the main complication of bone reconstruction surgeries is osteomyelitis, which is the inflammation of bone or bone marrow, usually due to infection.[101] While this is typically treated through antibiotics, wound drainage, surgical debridement or even implant removal, there is no guarantee of their effectiveness and these treatments often require follow up surgery.[102] Another approach is to introduce a drug into the glass to be released at the implant site (Figure 21.4).[103] This has many advantages including higher delivery efficiency, reduced toxicity, and does not require any further medical procedures.[102]

While sol–gel bioactive glasses have a wider compositional range and greater bioactivity than melt-derived bioactive glasses, their non-uniform micropore distribution is inadequate for consistent drug loading and release.[104] In order to have controlled drug release, an organized microstructure is required which has led to the creation of mesoporous bioactive glasses (MBGs) that possess highly ordered mesoporous channel structure along with highly specific surface areas.

During the sol–gel process non-ionic block copolymers, such as $EO_{20}PO_{70}EO_{20}$ (P123), $EO_{106}PO_{70}EO_{106}$ (F127), and cetyltrimethyl ammonium bromide (CTAB) are added to act as structure-directing agents. These

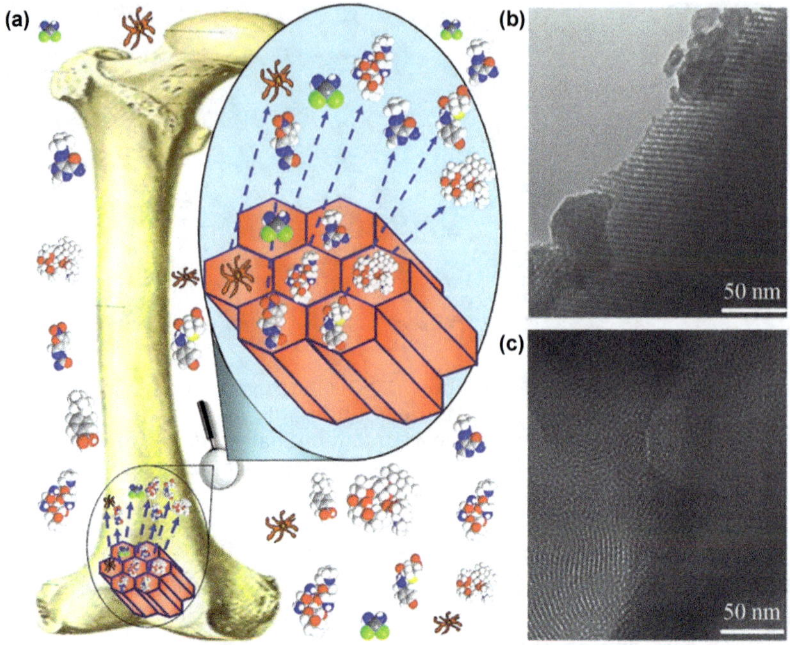

Figure 21.4 (a) A schematic of using MBGs as drug delivery devices for bone tissue engineering, (b) & (c) transmission electron microscopy images of P123 and CTAB-induced MBGs in (a). The P123 produced a more ordered microstructure.
((a) Reproduced from M. Vallet-Regí, Ordered Mesoporous Materials in the Context of Drug Delivery Systems and Bone Tissue Engineering, *Chem. - Eur. J.*, 2006, **12**,5934–5943 with permission from John Wiley and Sons. Copyright © WILEY-VCH Verlag GmbH & Co. KGaA, Weinheim and (b) and (c) adapted with permission from C. Wu and J. Chang, Mesoporous bioactive glasses: structure characteristics, drug/growth factor delivery and bone regeneration application, *Interface Focus*, 2012, **2**,292–306 by permission of the Royal Society.)

agents self-organize into micelles which link with the hydrolysed silica precursors to self-assemble and form the well-ordered mesophase (Figure 21.4b and c).[101] This system undergoes an evaporation-induced self-assembly process which is loosely defined as the spontaneous organization of materials through non-covalent interactions without external mediation resulting in the creation of patterned porous or nanocomposite materials.[105] In the case of bioactive glass the resultant dried material is then typically sintered, resulting in the final well-ordered, mesoporous glass structure containing high surface area and porosity.[106]

The first MBGs were invented by Yu *et al.*[107,108] and demonstrated a structure that contained a highly ordered hexagonal arrangement of one-dimensional channels as confirmed by transmission electron microscopy (TEM) and XRD. MGBs formed a layer of HCA in SBF in as little as 8 h, compared to 3 days in glasses made without surfactants. It has also been possible to control the mesopore structure and size, as well as surface area and pore volume, by varying the structure-directing agent.[107] The various morphologies have been well summarized by Wu *et al.*[101]

It is not only the ability to load drugs into sol–gel glasses but also the ability for the drug to be released in a controlled, sustained manner over time to prevent infection while the bone defect heals. Typically, the drug is loaded into the mesopores of the MBG through a soaking process which allows for the molecules to be adsorbed.[102,109,110] The efficiency of the drug loading is also influenced by the composition of the glass. For example, it was found that increasing the Ca content allowed for higher drug loading but resulted in a slower release profile.[102,111] This may be due to the chelating of the drugs with the calcium ion on the pore wall which hinders release.[102] MBGs have been shown to allow for a 3-fold increase of gentamicin uptake when compared to sol–gel glasses, a decrease in the initial burst release of the drug, and to induce rapid HCA conversion.[111]

MBGs have been tested *in vivo* in a rat femur model.[112] MBGs created with $^{45}CaCl_2$ were implanted in critical-sized rat femur defects (3×3 mm) for 1 day and their degradation and bone conversion behaviour were observed for 1, 4, 8, and 12 weeks. Compared to the control, an empty bone defect, significantly more bone was formed at the 4, 8, and 12 week time points. Furthermore, the researchers were able to trace the Ca release profile which showed the very small amount of Ca ions (<9.63%) released during degradation could be directly integrated into new bone matrix while the other was detected in blood in various organs.[112] This is one of the initial studies showing the possible risks and benefits of using MBGs *in vivo* for bone tissue repair.

In addition to particulate form, MBGs can also be made into fibres,[113] spheres,[114] and 3D scaffolds[115] allowing for tailored morphologies for bone tissue engineering applications. Fibres based on $(70)SiO_2$–$(25)CaO$–$(5)P_2O_5$ (mol%) made using the electrospinning technique were able to maintain their hierarchical characteristics.[113] Incorporated drugs also demonstrated controlled release based on the pore size inclusions. In order to develop a drug/protein delivery system, MBG spheres were created with $CaCl_2$ cross-linked alginate.[114] These spheres formed an apatite layer on their surface after immersion in SBF and the thickness of this layer determined the amount of bovine serum albumin released, allowing for controlled drug delivery. These spheres also supported the attachment and growth of human bone mesenchymal stem cells.[114] 3D scaffolds are possible through the foam replication technique and this method allows for the creation of large micropores with the textured mesoporous glass surface.[116–118] It is also possible to mix the MBGs with methylcellulose and then use rapid prototyping to print 3D structures.[119] Recently, PVA–MBG scaffolds have demonstrated improved mineralization and mechanical properties when compared to the methylcellulose composites.[120]

21.6 Sol–Gel-Derived Coatings for Biomedical Applications

Glasses are often not suitable for orthopaedic applications since their low strength prohibits them from being used in weight-bearing locations. Therefore, metals and high density ceramics, such as alumina, are commonly used in orthopaedic procedures. Some standard surgical metals include stainless steel 316 L, cobalt-based alloys, and titanium alloys such as Ti–6Al–4V.[121,122] However, the wear of these materials must be negligible since debris particles can cause loosening of the implant, also known as aseptic loosening, which is a major cause of implant failure. This can be caused by insufficient initial fixation, mechanical loss of fixation, or biologic loss of fixation caused by particle-induced osteolysis around the implant.[123]

The coating of these inert materials with bioactive ones has been carried out to improve the fixation of these devices, for instance the alumina hip socket cup or the metal femoral stem. One commonly used method is plasma spraying of hydroxyapatite which works relatively well for metals, but can also damage the implant surface and often provides a heterogeneous coating, especially with complex implant shapes. Thus, an alternative lies in coating these materials with sol–gel-derived glasses which may help improve their bioactive properties and prevent corrosion without compromising their bulk mechanical properties. Due to the aqueous nature of the sol–gel process, it is possible to coat many different types of materials with a variety of compositions, in order to improve their long-term attachment. The most common technique is dip-coating, which involves submersing the substrate or material in the sol–gel then removing it at a controlled rate to ensure uniform thickness. This process can be repeated to increase the layer thickness or to coat with different materials. It is thought that the HCA formation induced from these coatings can help block the micro-cracks and pore channels that form in the coatings over time, thus extending their protective effect.[124]

Titanium disks have been coated with a silver-doped 70S30C sol–gel glass composition for potential dental applications.[125] These coatings improved the apatite-forming ability in SBF, however increasing amounts of Ag limited this response. On the other hand, improved antibacterial affects were also observed with the silver doping. Dou and colleagues created sol–gel-derived glass-ceramic 45S5 coating for a magnesium AZ31 alloy.[126] The coating was able to decrease the mass loss of the magnesium alloy substrate from 78.04 to 2.31% in a 7 day test in SBF and was also shown to form an HCA layer. However, cracking of the layer is thought to limit the potential of this particular coating and the coating thickness also plays an important role.

Zhang and researchers were able to coat a Mg alloy with mesoporous 45S5 bioactive glass-ceramic using the dip-coating technique.[127] When loads greater than 25 MPa were applied, the coating failed to protect the substrate and no apatite formation was found. However, if the applied force was

<25 MPa the coated sample was able to deposit an apatite layer and thus provided protective effects for the substrate. Huang and colleagues coated an AZ31 magnesium alloy with a sol–gel-derived mesoporous 58S bioactive glass.[124] This coating not only improved the wetting ability of the substrate but also improved the apatite-forming ability after submersion in SBF, which is thought to further help prevent corrosion.

Hamadouche *et al.* showed the *in vitro* and *in vivo* effects of coating 58S and 77S ((80)SiO_2–(16)CaO–(4)P_2O_5; (mol%)) sol–gel glasses on an alumina substrate.[128] SEM analysis after submersion in SBF showed HCA formation within one week. The 58S-coated sample showed a significant increase in the percentage of bone tissue in contact with the substrate after 3 weeks (45.1%) and 24 weeks (87.8%) when implanted into the femurs of New Zealand rabbits. The 77S composition also demonstrated new bone growth but not at the same rate as the 58S composition. In addition to metals and ceramics, sol–gel bioactive glass coatings can also be used for polymers. MBG glass was coated on top of polylactide scaffold in order to create a sustained antibiotic drug (Gentamicin sulfate) release system for bone repair.[129] While the neat polylactide scaffold released the entire drug within 3 h, the MBG coated scaffold exhibited a sustained, two-stage release profile along with increasing the bioactivity of the scaffold.

21.7 Other Sol–Gel Glasses

As an alternative to silicate-based glass-forming systems, both phosphate- and borate-based glasses have been developed for bone tissue engineering applications.[76,130] These glasses typically have lower chemical durability when compared to silicate-based glasses and are thus more readily soluble. The compositions of the glasses can also incorporate a greater range of metal ions which, when released, can act as bone stimulating agents. While silicate-based sol–gel glasses have been well studied, sol–gel glasses based on borate and phosphate have not been investigated to the same extent. Phosphate-based glasses have similar compositions to those of the inorganic component of bone, and their degradations are highly controllable from hours to weeks by adjusting the compositional oxides.[131] However, their poor HCA formation characteristics make it difficult for them to be used for hard tissue repair.

Similar to phosphate-based glasses, borate-based glasses have lower chemical durability due to their lower network connectivity and have shown great promise in bone tissue applications, both *in vitro* and *in vivo*,[132,133] yet the processing of these glasses using the sol–gel route has not been explored. Recently, Lepry and Nazhat created the first sol–gel-derived borate glasses (SGBGs) that were shown to be highly bioactive.[134] The conversion of these SGBGs to HCA initiated in as little as 30 minutes after immersion in SBF which was at least 25 times more rapid than the melt-derived equivalent composition (Figure 21.5). This rapid HCA conversion was due to the high

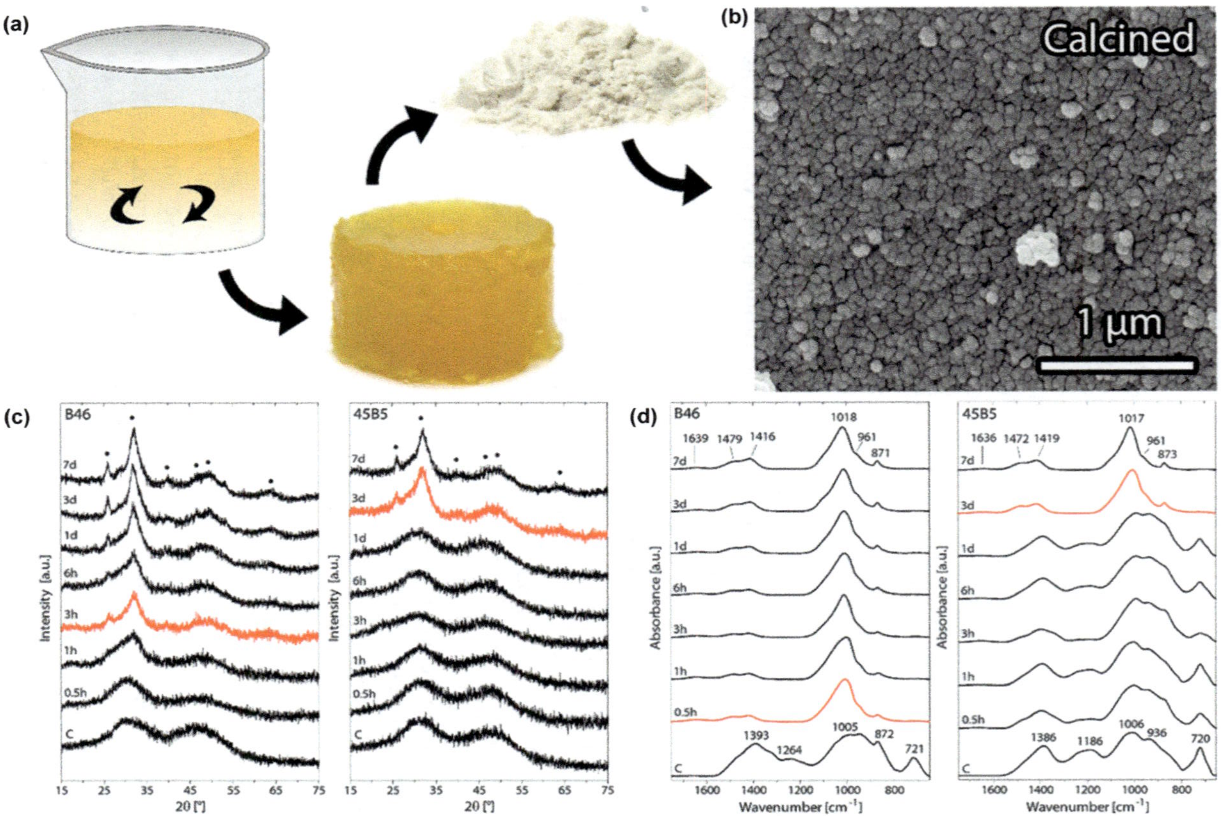

Figure 21.5 (a) A schematic showing the processing of SGBGs (b) resultant nanoporous surface after calcination, (c) XRD diffractographs showing typical HCA peaks ("•" JCPDS 9-0432) in B46 SGBG after 3 h compared to day 3 in 45B5, (d) FTIR showing typical phosphate peaks indicative of HCA formation within 30 min in SBF.
(Adapted with permission from W. C. Lepry and S. N. Nazhat, Highly Bioactive Sol-Gel-Derived Glasses, *Chem. Mater.*, 2015, **27**, 4821–4831 Copyright (2015) American Chemical Society).

surface areas and pore volumes of these glasses coupled with the lower chemical durability of the borate glass, allowing for rapid ion release. Further optimization studies are currently underway to explore the potential of these glasses for hard tissue repair.

21.8 Conclusions and Outlook

Bone repair and replacement is a major health concern affecting many people across the world. While traditional melt-derived bioactive glasses have shown promise in treating some of these defects, the sol–gel process allows for a more tailored and rapid approach to bone tissue repair. Since the precursor materials are aqueous-based, incorporation of metallic ions is often easier, allowing for a wide range of compositions and targeted therapeutic ion effects. The resultant high specific surface areas and porosities, which can be controlled through processing, increase the conversion rates to HCA, thus potentially translating to more rapid healing times. These enhanced textural properties also allow for the incorporation of drugs for specific healing effects. Furthermore, the aqueous nature of the process serves as a flexible starting point to produce various 3D forms such as foams, fibres, and ordered mesoporous glasses. This also allows for the coating of commonly used implants to induce bioactivity and increase biocompatibility.

While most sol–gel glasses are silica-based, new materials, such as those based on borate, have recently been examined and have shown remarkable HCA conversion rates. However, much more work is needed to understand the gel formation mechanisms and how the different processing parameters affect the final textural and bioactive properties of these glasses. Future trends will continue to optimize the sol–gel processing parameters and expand the compositions in order to create tailored bone tissue engineering glasses.

References

1. C. Laurencin, Y. Khan and S. F. El-Amin, Bone graft substitutes, *Expert Rev. Med. Devices*, 2006, **3**(1), 49–57.
2. A. Van Heest and M. Swiontkowski, Bone-graft substitutes, *Lancet.*, 1999, **353**, S28–S29.
3. J. S. Silber, D. G. Anderson, S. D. Daffner, B. T. Brislin, J. M. Leland, A. S. Hilibrand *et al.*, Donor site morbidity after anterior iliac crest bone harvest for single-level anterior cervical discectomy and fusion, *Spine*, 2003, **28**(2), 134–139.
4. C. G. Finkemeier, Bone-Grafting and Bone-Graft Substitutes, *J. Bone. Joint. Surg.*, 2002, **84**(3), 454–464.
5. M. A. Greenwald, M. J. Kuehnert and J. A. Fishman, Infectious Disease Transmission during Organ and Tissue Transplantation, *Emerging Infect. Dis.*, 2012, **18**(8), e1.

6. T. Boyce, J. Edwards and N. Scarborough, Allograft bone: the influence of processing on safety and performance, *Orthop. Clin. North Am.*, 1999, **30**(4), 571–581.
7. L. L. Hench, The story of Bioglass®, *J. Mater. Sci.: Mater. Med.*, 2006, **17**(11), 967–978.
8. L. L. Hench, R. J. Splinter, W. Allen and T. Greenlee, Bonding mechanisms at the interface of ceramic prosthetic materials, *J. Biomed. Mater. Res.*, 1971, **5**(6), 117–141.
9. J. R. Jones, Review of bioactive glass–from Hench to hybrids, *Acta Biomater.*, 2013, **9**(1), 4457–4486.
10. D. Mitton, C. Roux and P. Laugier, Bone overview, *Bone Quant. Ultrasound.*, 2011, 1–28.
11. B. Clarke, Normal bone anatomy and physiology, *Clin. J. Am. Soc. Nephrol.*, 2008, **3**(Supplement 3), S131–S139.
12. M. Doblaré, J. Garcıa and M. Gómez, Modelling bone tissue fracture and healing: a review, *Eng. Fract. Mech.*, 2004, **71**(13), 1809–1840.
13. G. A. Rodan, Introduction to bone biology, *Bone*, 1992, **13**(Supplement 1), S3–S6.
14. D. Burr, Targeted and nontargeted remodeling, *Bone*, 2002, **30**(1), 2–4.
15. R. Eastell, L. Mosekilde, S. F. Hodgson and B. L. Riggs, Proportion of human vertebral body bone that is cancellous, *J. Bone Miner. Res.*, 2009, **5**(12), 1237–1241.
16. E. F. Eriksen, D. W. Axelrod, F. Melsen *Bone Histomorphometry*, Raven Press, New York, 1994.
17. T. A. Guise and G. R. Mundy, Cancer and bone, *Endocr. Rev.*, 1998, **19**(1), 18–54.
18. R. K. Nalla, J. J. Kruzic, J. H. Kinney, M. Balooch, J. W. Ager Iii and R. O. Ritchie, Role of microstructure in the aging-related deterioration of the toughness of human cortical bone, *Mater. Sci. Eng., Proc. Conf.*, 2006, **26**(8), 1251–1260.
19. J.-Y. Rho, L. Kuhn-Spearing and P. Zioupos, Mechanical properties and the hierarchical structure of bone, *Med. Eng. Phys.*, 1998, **20**(2), 92–102.
20. S. Weiner and H. D. Wagner, The material bone: structure-mechanical function relations, *Annu. Rev. Mater. Sci.*, 1998, **28**(1), 271–298.
21. A. S. Posner, Crystal chemistry of bone mineral, *Physiol. Rev.*, 1969, **49**(4), 760–792.
22. L. L. Hench, *New Materials and Technologies for Healthcare*, World Scientific, 2011.
23. J. Jones and A. Clare, *Bio-Glasses: An Introduction*, John Wiley & Sons, 2012.
24. M. Young, Bone matrix proteins: their function, regulation, and relationship to osteoporosis, *Osteoporosis Int.*, 2003, **14**(3), 35–42.
25. P. G. Robey and A. L. Boskey, Asbmr, *The Composition of Bone. Primer on the Metabolic Bone Diseases and Disorders of Mineral Metabolism*, John Wiley & Sons, Inc., 2009, ch. 6, pp. 32–38.
26. J. P. Bilezikian, L. G. Raisz and T. J. Martin, *Principles of Bone Biology, Two-volume Set*, Academic Press, 2008.

27. J. O. Hollinger, A. Srinivasan, P. Alvarez, E. Hsu, S. McBride and S. Eppell *et al.*, 5.522 – Bone Tissue Engineering: Growth Factors and Cytokines. in *Comprehensive Biomaterials*, Editor-in-Chief D. Paul, Elsevier, Oxford, 2011, pp. 281–301.
28. B. Datta, *Textbook of Pathology*, Jaypee Brothers Medical, 2008.
29. C. Christodoulou and C. Cooper, What is osteoporosis? *Postgrad. Med. J.*, 2003, **79**(929), 133–138.
30. N. Harvey, E. Dennison and C. Cooper, Osteoporosis: impact on health and economics, *Nat. Rev. Rheumatol.*, 2010, **6**(2), 99–105.
31. Consensus development conference: diagnosis, prophylaxis and treatment of osteoporosis. *Am. J. Med.*, 1993, **94**, 646–650.
32. J. Z. Ilich and J. E. Kerstetter, Nutrition in bone health revisited: a story beyond calcium, *J. Am. Coll. Nutr.*, 2000, **19**(6), 715–737.
33. N. J. Lakhkar, I.-H. Lee, H.-W. Kim, V. Salih, I. B. Wall and J. C. Knowles, Bone formation controlled by biologically relevant inorganic ions: Role and controlled delivery from phosphate-based glasses, *Adv. Drug Delivery Rev.*, 2013, **65**(4), 405–420.
34. J. P. Bilezikian, Management of Acute Hypercalcemia, *N. Engl. J. Med.*, 1992, **326**(18), 1196–1203.
35. R. P. Heaney, J. Gallagher, C. Johnston, R. Neer, A. Parfitt and G. Whedon, Calcium nutrition and bone health in the elderly, *Am. J. Clin. Nutr.*, 1982, **36**(5), 986–1013.
36. J. Z. Ilich, R. Brownbill and L. Tamborini, Bone and nutrition in elderly women: protein, energy, and calcium as main determinants of bone mineral density, *Eur. J. Clin. Nutr.*, 2003, **57**(4), 554–565.
37. J. P. Bonjour, Calcium and phosphate: a duet of ions playing for bone health, *J. Am. Coll. Nutr.*, 2011, **30**(5 Supplement 1), 438S–448SS.
38. G. M. Chan, Dietary calcium and bone mineral status of children and adolescents, *Arch. Pediatr. Adolesc. Med.*, 1991, **145**(6), 631.
39. C. C. Johnston Jr, J. Z. Miller, C. W. Slemenda, T. K. Reister, S. Hui, J. C. Christian *et al.*, Calcium supplementation and increases in bone mineral density in children, *N. Engl. J. Med.*, 1992, **327**(2), 82–87.
40. B. L. Foster, K. A. Tompkins, R. B. Rutherford, H. Zhang, E. Y. Chu, H. Fong *et al.*, Phosphate: Known and potential roles during development and regeneration of teeth and supporting structures, *Birth Defects Res., Part C*, 2008, **84**(4), 281–314.
41. B. Hoac, T. Kiffer-Moreira, J. L. Millán and M. D. McKee, Polyphosphates inhibit extracellular matrix mineralization in MC3T3-E1 osteoblast cultures, *Bone*, 2013, **53**(2), 478–486.
42. E. Takeda, H. Yamamoto, K. Nashiki, T. Sato, H. Arai and Y. Taketani, Inorganic phosphate homeostasis and the role of dietary phosphorus, *J. Cell. Mol. Med.*, 2004, **8**(2), 191–200.
43. H. Tenenbaum, Role of organic phosphate in mineralization of bone in vitro, *J. Dent. Res.*, 1981, **60**(3 suppl), 1586–1589.
44. G. R. Beck, Inorganic phosphate as a signaling molecule in osteoblast differentiation, *J. Cell. Biochem.*, 2003, **90**(2), 234–243.

45. G. R. Beck, B. Zerler and E. Moran, Phosphate is a specific signal for induction of osteopontin gene expression, *Proc. Natl. Acad. Sci.*, 2000, **97**(15), 8352–8357.
46. E. M. Carlisle, Silicon: a requirement in bone formation independent of vitamin D1, *Calcif. Tissue Int.*, 1981, **33**(1), 27–34.
47. R. Jugdaohsingh, Silicon and bone health, *J. Nutr. Health. Aging*, 2007, **11**(2), 99.
48. S. Sripanyakorn, R. Jugdaohsingh, R. P. Thompson and J. J. Powell, Dietary silicon and bone health, *Nutr. Bull.*, 2005, **30**(3), 222–230.
49. W. Landis, D. Lee, J. Brenna, S. Chandra and G. Morrison, Detection and localization of silicon and associated elements in vertebrate bone tissue by imaging ion microscopy, *Calcif. Tissue Int.*, 1986, **38**(1), 52–59.
50. K. Schwarz and D. B. Milne *Growth-Promoting Effects of Silicon in Rats*, 1972.
51. E. M. Carlisle, Silicon: an essential element for the chick, *Science*, 1972, **178**(4061), 619–621.
52. P. E. Keeting, M. J. Oursler, K. E. Wiegand, S. K. Bonde, T. C. Spelsberg and B. L. Riggs, Zeolite a increases proliferation, differentiation, and transforming growth factor β production in normal adult human osteoblast-like cells in vitro, *J. Bone Miner. Res.*, 1992, **7**(11), 1281–1289.
53. C. M. Botelho, R. A. Brooks, S. M. Best, M. A. Lopes, J. D. Santos, N. Rushton *et al.*, Human osteoblast response to silicon-substituted hydroxyapatite, *J. Biomed. Mater. Res., Part A*, 2006, **79A**(3), 723–730.
54. I. D. Xynos, A. J. Edgar, L. D. Buttery, L. L. Hench and J. M. Polak, Ionic products of bioactive glass dissolution increase proliferation of human osteoblasts and induce insulin-like growth factor II mRNA expression and protein synthesis, *Biochem. Biophys. Res. Commun.*, 2000, **276**(2), 461–465.
55. S. Boonrungsiman, S. Fearn, E. Gentleman, L. Spillane, R. Carzaniga, D. W. McComb *et al.*, Correlative spectroscopy of silicates in mineralised nodules formed from osteoblasts, *Nanoscale*, 2013, **5**(16), 7544–7551.
56. L. L. Hench, Bioceramics: from concept to clinic, *J. Am. Ceram. Soc.*, 1991, **74**(7), 1487–1510.
57. L. L. Hench, Biomaterials: a forecast for the future, *Biomaterials*, 1998, **19**(16), 1419–1423.
58. T. Kokubo, H. Kushitani, S. Sakka, T. Kitsugi and T. Yamamuro, Solutions able to reproduce in vivo surface-structure changes in bioactive glass-ceramic A-W3, *J. Biomed. Mater. Res.*, 1990, **24**(6), 721–734.
59. T. Kokubo and H. Takadama, How useful is SBF in predicting in vivo bone bioactivity? *Biomaterials*, 2006, **27**(15), 2907–2915.
60. M. Bohner and J. Lemaitre, Can bioactivity be tested in vitro with SBF solution? *Biomaterials*, 2009, **30**(12), 2175–2179.
61. A. L. B. Maçon, T. B. Kim, E. M. Valliant, K. Goetschius, R. K. Brow, D. E. Day *et al.*, A unified in vitro evaluation for apatite-forming ability of bioactive glasses and their variants, *J. Mater. Sci.: Mater. Med.*, 2015, **26**(2), 1–10.

62. V. Miguez-Pacheco, L. L. Hench and A. R. Boccaccini, Bioactive glasses beyond bone and teeth: Emerging applications in contact with soft tissues, *Acta Biomater.*, 2015, **13**, 1–15.
63. J. E. Shelby, *Introduction to Glass Science and Technology*, Royal Society of Chemistry, 2005.
64. R. G. Hill and D. S. Brauer, Predicting the bioactivity of glasses using the network connectivity or split network models, *J. Non-Cryst. Solids*, 2011, **357**(24), 3884–3887.
65. R. Hill, An alternative view of the degradation of bioglass, *J. Mater. Sci. Lett.*, 1996, **15**(13), 1122–1125.
66. R. Li, A. Clark and L. Hench, An investigation of bioactive glass powders by sol-gel processing, *J. Appl. Biomater.*, 1991, **2**(4), 231–239.
67. S. Lin, C. Ionescu, K. J. Pike, M. E. Smith and J. R. Jones, Nanostructure evolution and calcium distribution in sol–gel derived bioactive glass, *J. Mater. Chem.*, 2009, **19**(9), 1276–1282.
68. M. Pereira, A. Clark and L. Hench, Calcium phosphate formation on sol-gel-derived bioactive glasses in vitro, *J. Biomed. Mater. Res.*, 1994, **28**(6), 693–698.
69. P. Sepulveda, J. R. Jones and L. L. Hench, Bioactive sol-gel foams for tissue repair, *J. Biomed. Mater. Res.*, 2002, **59**(2), 340–348.
70. P. Sepulveda, J. R. Jones and L. L. Hench, Characterization of melt-derived 45S5 and sol-gel–derived 58S bioactive glasses, *J. Biomed. Mater. Res.*, 2001, **58**(6), 734–740.
71. I. Christodoulou, L. D. Buttery, P. Saravanapavan, G. Tai, L. L. Hench and J. M. Polak, Dose-and time-dependent effect of bioactive gel-glass ionic-dissolution products on human fetal osteoblast-specific gene expression, *J. Biomed. Mater. Res., Part B*, 2005, **74**(1), 529–537.
72. J. Zhang, M. Wang, J. M. Cha and A. Mantalaris, The incorporation of 70s bioactive glass to the osteogenic differentiation of murine embryonic stem cells in 3D bioreactors, *J. Tissue Eng. Regener. Med.*, 2009, **3**(1), 63–71.
73. L. Radev, K. Hristova, V. Jordanov, M. Fernandes and I. Salvado, In vitro bioactivity of 70 wt.% SiO2—30 wt.% CaO sol-gel glass, doped with 1, 3 and 5 wt.% NbF5, *Open Chem.*, 2012, **10**(1), 137–145.
74. S. Lin, C. Ionescu, E. M. Valliant, J. V. Hanna, M. E. Smith and J. R. Jones, Tailoring the nanoporosity of sol-gel derived bioactive glass using trimethylethoxysilane, *J. Mater. Chem.*, 2010, **20**(8), 1489–1496.
75. S. Lin, W. Van den Bergh, S. Baker and J. R. Jones, Protein interactions with nanoporous sol–gel derived bioactive glasses, *Acta Biomater.*, 2011, **7**(10), 3606–3615.
76. M. N. Rahaman, D. E. Day, B. Sonny Bal, Q. Fu, S. B. Jung, L. F. Bonewald *et al.*, Bioactive glass in tissue engineering, *Acta Biomater.*, 2011, **7**(6), 2355–2373.
77. H. Pirayesh and J. A. Nychka, Sol–Gel Synthesis of Bioactive Glass-Ceramic 45S5 and its in vitro Dissolution and Mineralization Behavior, *J. Am. Ceram. Soc.*, 2013, **96**(5), 1643–1650.

78. I. Cacciotti, M. Lombardi, A. Bianco, A. Ravaglioli and L. Montanaro, Sol–gel derived 45S5 bioglass: synthesis, microstructural evolution and thermal behaviour, *J. Mater. Sci.: Mater. Med.*, 2012, **23**(8), 1849–1866.
79. J. Faure, R. Drevet, A. Lemelle, N. B. Jaber, A. Tara, H. El Btaouri *et al.*, A new sol–gel synthesis of 45S5 bioactive glass using an organic acid as catalyst, *Mater. Sci. Eng., Proc. Conf.*, 2015, **47**, 407–412.
80. E. Rezabeigi, P. M. Wood-Adams and R. A. Drew, Synthesis of 45S5 Bioglass® via a straightforward organic, nitrate-free sol–gel process, *Mater. Sci. Eng., Proc. Conf.*, 2014, **40**, 248–252.
81. J. R. Jones, L. M. Ehrenfried and L. L. Hench, Optimising bioactive glass scaffolds for bone tissue engineering, *Biomaterials*, 2006, **27**(7), 964–973.
82. T. Freyman, I. Yannas and L. Gibson, Cellular materials as porous scaffolds for tissue engineering, *Prog. Mater. Sci.*, 2001, **46**(3), 273–282.
83. N. Okii, S. Nishimura, K. Kurisu, Y. Takeshima and T. Uozumi, In Vivo Histological Changes Occurring in Hydroxyapatite Cranial Reconstruction. Case Report, *Neurol. Med. Chir.*, 2001, **41**(2), 100–104.
84. J. R. Jones, O. Tsigkou, E. E. Coates, M. M. Stevens, J. M. Polak and L. L. Hench, Extracellular matrix formation and mineralization on a phosphate-free porous bioactive glass scaffold using primary human osteoblast (HOB) cells, *Biomaterials*, 2007, **28**(9), 1653–1663.
85. H. Ohgushi and A. I. Caplan, Stem cell technology and bioceramics: from cell to gene engineering, *J. Biomed. Mater. Res.*, 1999, **48**(6), 913–927.
86. T. Takezawa, A strategy for the development of tissue engineering scaffolds that regulate cell behavior, *Biomaterials*, 2003, **24**(13), 2267–2275.
87. Q. Fu, M. N. Rahaman, B. S. Bal, R. F. Brown and D. E. Day, Mechanical and in vitro performance of 13–93 bioactive glass scaffolds prepared by a polymer foam replication technique, *Acta Biomater.*, 2008, **4**(6), 1854–1864.
88. H. R. Ramay and M. Zhang, Preparation of porous hydroxyapatite scaffolds by combination of the gel-casting and polymer sponge methods, *Biomaterials*, 2003, **24**(19), 3293–3302.
89. M. de Barros Coelho and M. Magalhães Pereira, Sol–gel synthesis of bioactive glass scaffolds for tissue engineering: effect of surfactant type and concentration, *J. Biomed. Mater. Res., Part B*, 2005, **75**(2), 451–456.
90. J. R. Jones and L. L. Hench, Factors affecting the structure and properties of bioactive foam scaffolds for tissue engineering, *J. Biomed. Mater. Res., Part B*, 2004, **68**(1), 36–44.
91. J. R. Jones, S. Ahir and L. L. Hench, Large-scale production of 3D bioactive glass macroporous scaffolds for tissue engineering, *J. Sol-Gel Sci. Technol.*, 2004, **29**(3), 179–188.
92. S. Yue, P. D. Lee, G. Poologasundarampillai and J. R. Jones, Evaluation of 3-D bioactive glass scaffolds dissolution in a perfusion flow system with X-ray microtomography, *Acta Biomater.*, 2011, **7**(6), 2637–2643.

93. P. Valerio, M. Guimaraes, M. Pereira, M. Leite and A. Goes, Primary osteoblast cell response to sol-gel derived bioactive glass foams, *J. Mater. Sci.: Mater. Med.*, 2005, **16**(9), 851–856.
94. Q.-Z. Chen and G. A. Thouas, Fabrication and characterization of sol–gel derived 45S5 Bioglass®–ceramic scaffolds, *Acta Biomater.*, 2011, 7(10), 3616–3626.
95. S. Midha, T. B. Kim, W. van den Bergh, P. D. Lee, J. R. Jones and C. A. Mitchell, Preconditioned 70S30C bioactive glass foams promote osteogenesis in vivo, *Acta Biomater.*, 2013, **9**(11), 9169–9182.
96. Z.-M. Huang, Y.-Z. Zhang, M. Kotaki and S. Ramakrishna, A review on polymer nanofibers by electrospinning and their applications in nanocomposites, *Compos. Sci. Technol.*, 2003, **63**(15), 2223–2253.
97. H. W. Kim, H. E. Kim and J. C. Knowles, Production and Potential of Bioactive Glass Nanofibers as a Next-Generation Biomaterial, *Adv. Funct. Mater.*, 2006, **16**(12), 1529–1535.
98. H. W. Kim, J. H. Song and H. E. Kim, Bioactive glass nanofiber–collagen nanocomposite as a novel bone regeneration matrix, *J. Biomed. Mater. Res., Part A*, 2006, **79**(3), 698–705.
99. H. Lu, T. Zhang, X. P. Wang and Q. F. Fang, Electrospun submicron bioactive glass fibers for bone tissue scaffold, *J. Mater. Sci.: Mater. Med.*, 2009, **20**(3), 793–798.
100. G. Poologasundarampillai, D. Wang, S. Li, J. Nakamura, R. Bradley, P. Lee *et al.*, Cotton-wool-like bioactive glasses for bone regeneration, *Acta Biomater.*, 2014, **10**(8), 3733–3746.
101. C. Wu and J. Chang, Mesoporous bioactive glasses: structure characteristics, drug/growth factor delivery and bone regeneration application, *Interface Focus*, 2012, **2**(3), 292–306.
102. L. Zhao, X. Yan, X. Zhou, L. Zhou, H. Wang, J. Tang *et al.*, Mesoporous bioactive glasses for controlled drug release, *Microporous Mesoporous Mater.*, 2008, **109**(1–3), 210–215.
103. V. Mouriño and A. R. Boccaccini, Bone tissue engineering therapeutics: controlled drug delivery in three-dimensional scaffolds, *J. R. Soc. Interface*, 2010, **7**(43), 209–227.
104. D. Arcos, A. López-Noriega, E. Ruiz-Hernández, O. Terasaki and M. Vallet-Regí, Ordered Mesoporous Microspheres for Bone Grafting and Drug Delivery, *Chem. Mater.*, 2009, **21**(6), 1000–1009.
105. C. J. Brinker, Y. Lu, A. Sellinger and H. Fan, Evaporation-induced self-assembly: nanostructures made easy, *Adv. Mater.*, 1999, **11**(7), 579–585.
106. D. Arcos and M. Vallet-Regí, Sol–gel silica-based biomaterials and bone tissue regeneration, *Acta Biomater.*, 2010, **6**(8), 2874–2888.
107. X. Yan, C. Yu, X. Zhou, J. Tang and D. Zhao, Highly Ordered Mesoporous Bioactive Glasses with Superior In Vitro Bone-Forming Bioactivities, *Angew. Chem., Int. Ed.*, 2004, **43**(44), 5980–5984.
108. X. Yan, X. Huang, C. Yu, H. Deng, Y. Wang, Z. Zhang *et al.*, The in-vitro bioactivity of mesoporous bioactive glasses, *Biomaterials*, 2006, **27**(18), 3396–3403.

109. A. Doadrio, E. Sousa, J. Doadrio, J. P. Pariente, I. Izquierdo-Barba and M. Vallet-Regı, Mesoporous SBA-15 HPLC evaluation for controlled gentamicin drug delivery, *J. Controlled Release*, 2004, **97**(1), 125–132.
110. M. Vallet-Regí, Ordered mesoporous materials in the context of drug delivery systems and bone tissue engineering, *Chem. – Eur. J.*, 2006, **12**(23), 5934–5943.
111. W. Xia and J. Chang, Well-ordered mesoporous bioactive glasses (MBG): a promising bioactive drug delivery system, *J. Controlled Release*, 2006, **110**(3), 522–530.
112. B. Sui, G. Zhong and J. Sun, Evolution of a Mesoporous Bioactive Glass Scaffold Implanted in Rat Femur Evaluated by 45Ca Labeling, Tracing, and Histological Analysis, *ACS Appl. Mater. Interfaces*, 2014, **6**(5), 3528–3535.
113. Y. Hong, X. Chen, X. Jing, H. Fan, B. Guo, Z. Gu *et al.*, Preparation, bioactivity, and drug release of hierarchical nanoporous bioactive glass ultrathin fibers, *Adv. Mater.*, 2010, **22**(6), 754–758.
114. C. Wu, Y. Zhang, X. Ke, Y. Xie, H. Zhu, R. Crawford *et al.*, Bioactive mesopore-glass microspheres with controllable protein-delivery properties by biomimetic surface modification, *J. Biomed. Mater. Res., Part A*, 2010, **95**(2), 476–485.
115. C. Wu, Y. Zhou, W. Fan, P. Han, J. Chang, J. Yuen *et al.*, Hypoxia-mimicking mesoporous bioactive glass scaffolds with controllable cobalt ion release for bone tissue engineering, *Biomaterials*, 2012, **33**(7), 2076–2085.
116. Y. Zhu, C. Wu, Y. Ramaswamy, E. Kockrick, P. Simon, S. Kaskel *et al.*, Preparation, characterization and in vitro bioactivity of mesoporous bioactive glasses (MBGs) scaffolds for bone tissue engineering, *Microporous Mesoporous Mater.*, 2008, **112**(1), 494–503.
117. H. S. Yun, S. E. Kim, Y. T. Hyun, S. J. Heo and J. W. Shin, Hierarchically mesoporous–macroporous bioactive glasses scaffolds for bone tissue regeneration, *J. Biomed. Mater. Res., Part B*, 2008, **87**(2), 374–380.
118. X. Li, X. Wang, H. Chen, P. Jiang, X. Dong and J. Shi, Hierarchically porous bioactive glass scaffolds synthesized with a PUF and P123 co-templated approach, *Chem. Mater.*, 2007, **19**(17), 4322–4326.
119. A. García, I. Izquierdo-Barba, M. Colilla, C. L. De Laorden and M. Vallet-Regí, Preparation of 3-D scaffolds in the SiO 2–P 2 O 5 system with tailored hierarchical meso-macroporosity, *Acta Biomater.*, 2011, 7(3), 1265–1273.
120. C. Wu, Y. Luo, G. Cuniberti, Y. Xiao and M. Gelinsky, Three-dimensional printing of hierarchical and tough mesoporous bioactive glass scaffolds with a controllable pore architecture, excellent mechanical strength and mineralization ability, *Acta Biomater.*, 2011, 7(6), 2644–2650.
121. A. Durán, A. Conde, A. G. Coedo, T. Dorado, C. Garcia and S. Ceré, Sol–gel coatings for protection and bioactivation of metals used in orthopaedic devices, *J. Mater. Chem.*, 2004, **14**(14), 2282–2290.

122. C. Garcia, S. Cere and A. Duran, Bioactive coatings prepared by sol–gel on stainless steel 316L, *J. Non-Cryst. Solids*, 2004, **348**, 218–224.
123. Y. Abu-Amer, I. Darwech and J. C. Clohisy, Aseptic loosening of total joint replacements: mechanisms underlying osteolysis and potential therapies, *Arthritis Res. Ther.*, 2007, **9**(1), S6.
124. K. Huang, S. Cai, G. Xu, M. Ren, X. Wang, R. Zhang *et al.*, Sol–gel derived mesoporous 58S bioactive glass coatings on AZ31 magnesium alloy and in vitro degradation behavior, *Surf. Coat. Technol.*, 2014, **240**, 137–144.
125. M. Catauro, F. Bollino, F. Papale and S. V. Ciprioti, Investigation on bioactivity, biocompatibility, thermal behavior and antibacterial properties of calcium silicate glass coatings containing Ag, *J. Non-Cryst. Solids*, 2015, **422**, 16–22.
126. Y. Dou, S. Cai, X. Ye, G. Xu, K. Huang, X. Wang *et al.*, 45S5 bioactive glass–ceramic coated AZ31 magnesium alloy with improved corrosion resistance, *Surf. Coat. Technol.*, 2013, **228**, 154–161.
127. F. Zhang, S. Cai, G. Xu, S. Shen, Y. Li, M. Zhang *et al.*, Corrosion behavior of mesoporous bioglass-ceramic coated magnesium alloy under applied forces, *J. Mech. Behav. Biomed. Mater.*, 2016, **56**, 146–155.
128. M. Hamadouche, A. Meunier, D. C. Greenspan, C. Blanchat, J. P. Zhong, G. P. La Torre *et al.*, Bioactivity of sol-gel bioactive glass coated alumina implants, *J. Biomed. Mater. Res.*, 2000, **52**(2), 422–429.
129. M. Zhu, L. Zhang, Q. He, J. Zhao, G. Limin and J. Shi, Mesoporous bioactive glass-coated poly (l-lactic acid) scaffolds: a sustained antibiotic drug release system for bone repairing, *J. Mater. Chem.*, 2011, **21**(4), 1064–1072.
130. N. J. Lakhkar, I.-H. Lee, H.-W. Kim, V. Salih, I. B. Wall and J. C. Knowles, Bone formation controlled by biologically relevant inorganic ions: role and controlled delivery from phosphate-based glasses, *Adv. Drug Delivery Rev.*, 2013, **65**(4), 405–420.
131. E. A. A. Neel, D. M. Pickup, S. P. Valappil, R. J. Newport and J. C. Knowles, Bioactive functional materials: a perspective on phosphate-based glasses, *J. Mater. Chem.*, 2009, **19**(6), 690–701.
132. X. Liu, M. N. Rahaman and D. E. Day, In Vitro Degradation and Conversion of Melt-Derived Microfibrous Borate (13-93B3) Bioactive Glass Doped with Metal Ions, *J. Am. Ceram. Soc.*, 2014, **97**(11), 3501–3509.
133. A. M. Deliormanlı, X. Liu and M. N. Rahaman, Evaluation of borate bioactive glass scaffolds with different pore sizes in a rat subcutaneous implantation model, *J. Biomater. Appl.*, 2014, **28**(5), 643–653.
134. W. C. Lepry and S. N. Nazhat, Highly Bioactive Sol-Gel-Derived Borate Glasses, *Chem. Mater.*, 2015, **27**(13), 4821–4831.

Subject Index